(in order of decreasing priority)

Example	*IUPAC Name*	*Common Name*	*Text Reference (page)*
$CH_3CH_2CH_2CO_2H$	Butanoic acid	*n*-Butyric acid	278, 317
$CH_3CO_2CH_2CH_3$	Ethyl ethanoate	Ethyl acetate	293, 306
$CH_3CH_2CONH_2$	Propanamide	Propionamide	312
$CH_3C{\equiv}N$	Ethanenitrile	Acetonitrile	288
CH_3CHO	Ethanal	Acetaldehyde	241
$CH_3\overset{\underset{\Vert}{O}}{C}CH_2CH_3$	2-Butanone	Methyl ethyl ketone	241
CH_3CH_2OH	Ethanol	Ethyl alcohol	196, 197
C_6H_5OH	Phenol	Phenol	199, 208
$CH_3CH_2CH_2NH_2$	1-Propanamine[a]	*n*-Propylamine	353
$CH_3CH{=}CH_2$	1-Propene	Propylene	78
$CH_3C{\equiv}CCH_3$	2-Butyne	Dimethylacetylene	107
$CH_3CH_2CH_2CH_2CH_2CH_3$	Hexane	*n*-Hexane	39, 52
$CH_3CH_2{-}O{-}CH_2CH_3$	Ethoxyethane	Diethyl ether	222
CH_3CH_2I	Iodoethane	Ethyl iodide	174

[a] Chemical Abstracts name.

Organic Chemistry

Organic Chemistry

A Brief Course

Sixth Edition

Walter W. Linstromberg
University of Nebraska at Omaha

Henry E. Baumgarten
University of Nebraska—Lincoln

D. C. HEATH AND COMPANY • Lexington, Massachusetts Toronto

Cover photograph: Marc Bernheim/Woodfin Camp & Associates

Published simultaneously in Canada.

Printed in the United States of America.

International Standard Book Number: 0-669-12660-8

Library of Congress Catalog Card Number: 86-81385

Preface

The revision of a textbook for a sixth edition would appear to present to its authors somewhat of a challenge. One might suppose that after five editions little remains for improvement in matters of fact, judgment, and presentation. Not true. Organic chemistry is a living, dynamic science and in the brief time between editions many changes occur. A revision thus presents an opportunity to bring the text up to date, to increase its value as a teaching tool, and to make the new edition better than the previous one. To this end some changes in the organization have been made, certain sections have been rewritten to include new material, and others have been relocated to make for better continuity.

As in the preparation of the previous editions, we reviewed the entire text for conformity with our current understanding of both fact and theory. We have revised the material at a level consistent with a brief course. Although some new chemistry and new examples of older chemistry have been introduced, most of the changes in the sixth edition were made with the objective of clarifying or simplifying the presentation and were concentrated in the first thirteen chapters of the text.

We, and numerous users of previous editions, believe that elementary organic chemistry can be taught most effectively to students in the brief course by means of the chemistry of the various functional groups. Therefore, the core portion of the sixth edition of this text has again been organized around organic functional groups. In addition to this core, there are separate chapters on lipids, carbohydrates, amino acids and proteins, nucleic acids, natural products, and organic spectroscopy—all of considerable importance to students in the agricultural, biological, and medical sciences.

Spectroscopy, as in the previous edition, is the final chapter in this text. It is a complete and up-to-date chapter on a subject with an intrinsic application to organic chemistry. An instructor may choose to introduce the material in Chapter 18 as early as Chapter 6, "The Halogen Compounds," or at any point thereafter. We believe that the purpose of any textbook, even a short one, is to aid rather than restrict the creative efforts of the instructor, and should be of sufficient length and depth to offer a choice of material best suited to his or her objectives, schedule, and students.

Changes in this Edition

As well as completely reviewing and revising the text, we have increased by more than twenty-five percent the number of exercises and problems. These appear both within and at the ends of chapters. Those within the chapters are intended to test comprehension of specific points made in that section of the text. Exercises at the ends of chapters are more comprehensive and include both drill and thought questions. A hint or clue to the solution of a problem is given when considered desirable.

A feature new to this edition is the inclusion of a number of carefully selected worked-out examples that illustrate each important new concept as it is met. A well-chosen worked-out example can instruct better than a lengthy explanation. Each illustrative example is followed by one or more of a similar type for the student to solve. As before, exercises requiring the knowledge and use of spectroscopic information are grouped at the end of the exercises in each chapter beginning with Chapter 6 and are identified with an asterisk (*).

The sections dealing with the thiols and thio ethers were removed from Chapter 13, "Amines and Nitrogen Compounds," and included in Chapter 7, "Alcohols, Phenols, and Ethers."

The order of presentation of former Chapters 11 and 12 has been reversed so that the chapter on bifunctional acids is continuous with the chapters on carboxylic acids and their derivatives. The chapter on lipids is now adjacent to the other chapters dealing with our principal nutrients.

Supplements

- *Study Guide with Problems and Solutions for Organic Chemistry, A Brief Course,* prepared by Professor Robert L. Zey, contains solutions to all of the text exercises and numerous additional exercises, including some of a very simple review nature.
- *Organic Experiments,* Sixth Edition, by Walter W. Linstromberg and Henry E. Baumgarten, contains thirty-nine experiments complete with report pages. This laboratory guide has a strong emphasis on safety and many applications of natural products and biologically important compounds. An instructor's guide accompanying *Organic Experiments* contains answers to questions posed in the material and also gives chemical and equipment needs for each experiment.

Acknowledgments

The revision of the sixth edition, as in previous ones, was made easier by the helpful comments and suggestions of many people. We are especially grateful to Daniel H. O'Brien of Texas A & M University, who read a major portion of the manuscript and whose suggestions were incorporated into this edition. We also acknowledge with thanks the thoughtful comments of David M. Olsen of Merced College, Louis A. Jones of North Carolina State University, and James Brown of Middle Tennessee State University. We are indebted to G. L. Lange of the University of Guelph, whose assistance in this, as in previous editions, was sought and graciously given. Other helpful suggestions were given by Lynn G. Wiedermann, Black Hawk State College; Jim Schammerhorn, Murray State College; George J. Gradel, Philadelphia College of Textiles and Science; Alexander G. Bednekoff, Pittsburgh State University; Anne Loeb, College of Lake County; Charles D. Yates, State Technical Institute at Memphis; K. M. Elsen, Mount Mary College; James K. Wood, University of Nebraska at Omaha; Robert Boxer, Georgia Southern College; Francis Scalzi, Hiram College; Dale Beck, Cape Fear Technical Institute; George Wittenberg, Arizona State University; John McLeod, South Carolina State University; Richard Sasim, Millersville University; and Kenneth Kolb, Bradley University. We wish to thank our colleagues at the University of Nebraska for their assistance and support. Finally, we wish to thank the editorial staff of D. C. Heath and Company for its assistance and for providing the catalyst from time to time in the form of a gentle prodding to complete the manuscript on schedule.

As always, we continue to think that this text is written to serve the needs of instructors in brief courses in organic chemistry, and we invite all users and potential users to share with us their ideas for improvement of the text. Specific suggestions for change are always welcome.

Walter W. Linstromberg

Henry E. Baumgarten

Introduction

What is organic chemistry—this course of study you are about to begin? In what respects is organic chemistry different from the inorganic chemistry you have just recently studied? Why is organic chemistry taught as a separate branch of chemistry? The term *organic* and the answers to these questions have an interesting historical connotation.

Organic chemistry, according to the definition given it by chemists in the eighteenth and early nineteenth centuries, dealt only with those substances of natural *organic* origin—that is, products of plants or animals. It was believed by these early chemists that the production of an organic compound required the "vital force" of a living organism. The science of chemistry was thus separated into two broad divisions: **inorganic** and **organic.** However, in 1828 the German chemist Friedrich Wöhler[1] heated ammonium cyanate, NH_4OCN, and obtained urea, $(H_2N)_2CO$, an end product of protein metabolism that is excreted in the urine of mammals. This first laboratory synthesis of an organic compound from an inorganic salt disproved the vital force theory and the interrelationship of the two branches of chemistry was clearly recognized. As is often the case, names persist even though they are no longer descriptive. Organic chemistry is best defined today as *the chemistry of the compounds of carbon.*

Progress in the development of organic chemistry as a separate branch of science was slow. Although a number of substances of organic origin such as acetic acid and ethyl alcohol were known even to the ancients, many of the organic compounds isolated by early chemists were much more complex than the simple inorganic salts with which they were familiar. Moreover, analytical procedures for determining the composition and structure of an organic compound were long and laborious and the laboratory equipment available at the time was very crude. Consider, for example, the common (and notorious) steroid cholesterol, $C_{27}H_{46}O$, which was first isolated from gallstones in 1775. Elucidation of the cholesterol structure was not accomplished until 1932—a time lapse of one hundred fifty-seven years! Today, with the aid of the electron microscope, with infrared, ultraviolet, nuclear magnetic resonance, and mass spectrometers, the analysis of organic compounds and the determination of their structures take place in a matter of days or even hours. The impetus given to the advancement of organic chemistry by the development of electronic equipment has been tremendous. To date, over two million compounds of carbon have been synthesized or studied, and the list of known organic compounds continues to grow daily. In many instances the organic chemist not only has recreated structures found in nature, but has improved upon them as well. Chemists have also synthesized many compounds that have no counterparts in nature. Indeed, the synthesis of new and useful compounds is one of the most important aspects of organic chemistry. It was

[1]Friedrich Wöhler (1800–1882), Professor of Chemistry, University of Göttingen.

thus that we obtained the synthetic fibers nylon, Orlon, and Dacron, from which much of our clothing, carpeting, and upholstery materials are made; the local and general anesthetics used in the operating room; the elastomers we use as substitutes for rubber; the films and molded articles of plastic; many medicines and drugs; latex paints and surface coatings; and photographic film. These are but a few of the more familiar products. Other synthetics that you will read about in this text would make a long and impressive list.

The credits we accord the research chemist for the many products that make our lives more healthful, more comfortable, and more enjoyable certainly are well deserved. However, in fairness it must be said that occasionally a synthetic product proves to be a mixed blessing. DDT is a case in point. While the use of this powerful insecticide has eradicated legions of insect pests once costly and dangerous to humans, it has at the same time found its way into the ecosystem with deleterious results. Each new synthetic product, therefore, may have its price, and scientists, in their technological and economic zeal, must be ever vigilant to preserve a natural balance in the environment. To do otherwise would threaten the existence of all living things—ourselves included.

It is highly desirable that we have a knowledge of organic chemistry, if for no other reason than to enjoy a greater appreciation of the world around us. From the list of products cited above it is obvious that hardly a phase of our daily lives is not related to or dependent upon this tremendously important and fascinating science.

The cultural aspect of a knowledge of organic chemistry is an added bonus when a working knowledge of organic chemistry is a requirement for professions in which it plays an integral part. The study of medicine, nursing, pharmacy, home economics, agriculture, and all related fields includes organic chemistry as a prerequisite.

As was pointed out in earlier editions of this text, learning organic chemistry is similar to learning a foreign language. The alphabet we use is a simple one and consists only of C, H, O, N, I, S, P, Cl, Br, and a few additional inorganic "letters." The "vocabulary," on the other hand, contains over two million "words"! To learn these words and to be able to use them in simple phrases (one-step reactions), then later in sentences (multi-step reactions), and finally in paragraphs (bio-organic sequences), requires that you learn a great deal of new chemistry. Do not despair. You will not be asked to learn the names of two million organic compounds. Indeed, many of the compounds you will learn to name and to use are members of the same family and are quite easily recognized. In this regard, the study of organic chemistry is unlike the study of a foreign language because organic chemistry is beautifully organized and has none of the illogical idiomatic peculiarities of spoken languages.

Most students who enroll in the organic chemistry course find the subject easy to understand, interesting and even exciting to study. However, nearly all find it a difficult subject to remember. The need for getting off to a good start in your organic chemistry course cannot be overemphasized. Therefore, it is in order to offer you a few suggestions that will make the study of organic chemistry orderly, meaningful, and interesting. Read your assignments carefully. Learn each new concept and term as you come to it. Use the reviews, exercises, and problems that you will find within the body of the text and at the end of each chapter. Most important, by the extensive use of scratch pad and pencil, apply through practice what you have learned. *This is the only way you will remember the material you study*. Again, as in the study of a foreign language we need to remember what we have learned in the previous lesson in order to use and build upon it in the next.

In the beginning it may appear confusing that the laboratory preparation of one class of compounds frequently requires the use of another class not yet studied. It will soon become obvious that in this manner the synthesis of one substance automatically employs a reaction of the other. Since the study of organic chemistry involves principally learning the **names, structures, preparation,** and **reactions** of many compounds, the whole scheme of presentation is one of integrating these constituents. As in a jigsaw puzzle, the pieces will fit into place as you proceed through the course.

Contents

16 Nucleic Acids 436

17 Natural Products 455

18 Determination of Molecular Structure; Spectroscopy 474

1 General Principles

The analysis of any plant or animal product reveals that it always contains carbon and hydrogen and usually also oxygen and/or nitrogen. In many organic compounds sulfur, phosphorus, and the halogens are also found. In some cases even a metallic element makes up part of a naturally occurring organic structure.[1] It would be correct to say that a very small number of elements (fewer than a dozen) comprise most organic compounds. This is small indeed when compared to the large number of different elements found in inorganic substances. How is it possible for such a small number of elements to account for over two million organic compounds? To obtain the answer to this question we need to examine the manner in which the atoms in compounds are held together. It is not enough to know the number and kind of atoms an organic compound contains. We must be able to translate each molecular formula into a structure. When we have done this, we will see that carbon is nearly unique among the elements. Not only does the carbon atom bond to each of the elements cited above, but is also shows a predilection for bonding to other carbon atoms. The unique role of the carbon atom in the formation of organic compounds will be understood better after we have reviewed certain principles that are basic to the study of organic chemistry. Some of the concepts that we will discuss in this first chapter, though not entirely new to students who have already taken general chemistry, are worthy of review. Others, unique to organic chemistry, will be introduced in order to help the student understand the chemistry in later chapters. Success in learning organic chemistry is almost assured if one has a working knowledge of the principles upon which the subject is based.

[1]Iron is present in the hemoglobin of blood to the extent of 0.34%, magnesium comprises about 2.7% of the chlorophyll found in green plants, and cobalt is present in Vitamin B_{12} to the extent of 4.34%.

1.1 Atomic Structure

In your introductory course in chemistry you learned that an atom consists of a positively charged nucleus surrounded by negative electrons equal in number to the charge on the nucleus. You also learned that the electrons are distributed around the nucleus in energy levels called **principal quantum shells** designated by numbers 1, 2, 3, etc., in each case starting with the shell nearest the nucleus. If n is the number of the principal shell, the number of electrons needed to completely fill that shell with electrons is $2n^2$. Thus the first principal shell can hold a maximum of $2(1)^2$ or two electrons, the second principal shell a maximum of $2(2)^2$ or eight electrons, etc. Electrons in the first shell have the lowest energy, the energy increasing with shell number.

Principal shells are divided into **subshells** designated, in order of filling, by the letters *s, p, d,* and *f*. The first principal shell has only one subshell, designated as the 1*s*. The second principal shell has two subshells designated 2*s* and 2*p* subshells, and the third principal shell consists of three subshells, namely, 3*s*, 3*p*, and 3*d*. Inasmuch as our present study is limited mainly to compounds of carbon formed with elements of low atomic numbers, we need not concern ourselves at this time with higher subshells. The subshells are further divided into **atomic orbitals.** An orbital may be described as the region in the space surrounding an atomic nucleus that is most likely to be occupied by an electron. Imagine an electron in the *s* orbital as a point of light capable of being photographed. A time exposure on a photographic film then would reveal a circular cloud, dense toward the center and diffuse at its outer boundary. Next imagine a *surface* within the electron cloud. The probability of finding the electron at any point on this surface is everywhere the same. Furthermore, this surface will enclose a volume in which there is a high probability of finding the electron most of the time. For an *s* orbital the surface we have described will be a sphere. Therefore, we will use the sphere as a geometrical representation for the *s* orbital. All *s* orbitals, or orbitals in *s* subshells, are spheres, with their centers at the nucleus of the atom (Fig. 1.1). However, the shapes of orbitals vary. The first principal shell has only the 1*s* orbital. The second principal shell has, in addition to the 2*s* orbital, three *p* orbitals. The *p* orbitals are of equal energy and are dumbbell-shaped with a lobe on either side of the atomic nucleus (Fig. 1.2). The axes of the *p* orbitals are perpendicular to each other and lie along the coordinate axes with the nucleus at the origin. The orbitals are differentiated as $2p_x$, $2p_y$, or $2p_z$, where the small subscripts refer to the *x, y,* and *z* axes. Each 2*p* orbital has a **nodal plane** defined by the two coordinate axes perpendicular to the orbital axis. Thus the $2p_x$ orbital has the *yz* plane as a nodal plane (Fig. 1.2). The probability of finding a p_x electron in this nodal plane is zero. Each orbital, regardless of its designation, has a maximum capacity of two electrons. When the orbital is filled, the two electrons must be oriented in opposite directions or must have **opposite spin.** Further, in any principal shell electrons occupy orbitals of lower energy first—that is, they occupy *s* orbitals before *p* orbitals, *p* before *d*, etc. Moreover, an orbital in any subshell is not occupied by a pair of electrons until each orbital of the subshell is occupied by one electron. Table 1.1 shows graphically how electrons are distributed around the nuclei of the first ten elements.

FIGURE 1.1 The s Orbital

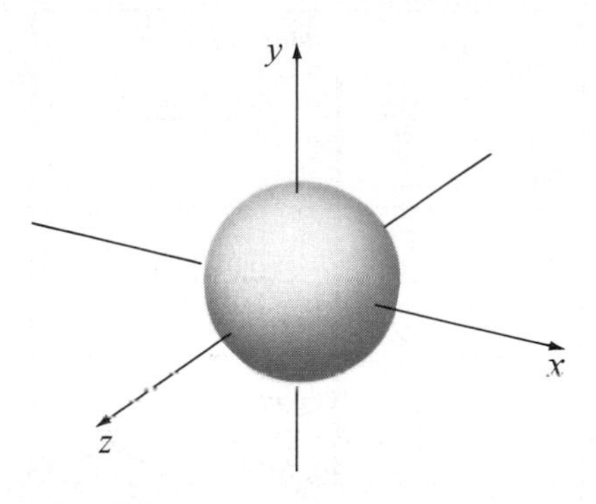

The spins of electrons are represented in Table 1.1 by small arrows pointing in opposite directions. The empty (broken line) circles in the diagram represent orbitals potentially available to electrons of higher energy. The normal state of any atom is that in which the electrons are as close to the nucleus as possible. This is called the **ground state** and is the lowest-energy state; thus all electrons are in the lowest-energy orbitals. An excited state is a higher-energy state in which one or more electrons have been promoted, by the absorption of energy, to orbitals further removed from the atomic nucleus.

FIGURE 1.2 Atomic $2p_x$, $2p_y$, and $2p_z$ Orbitals with Axes Mutually Perpendicular

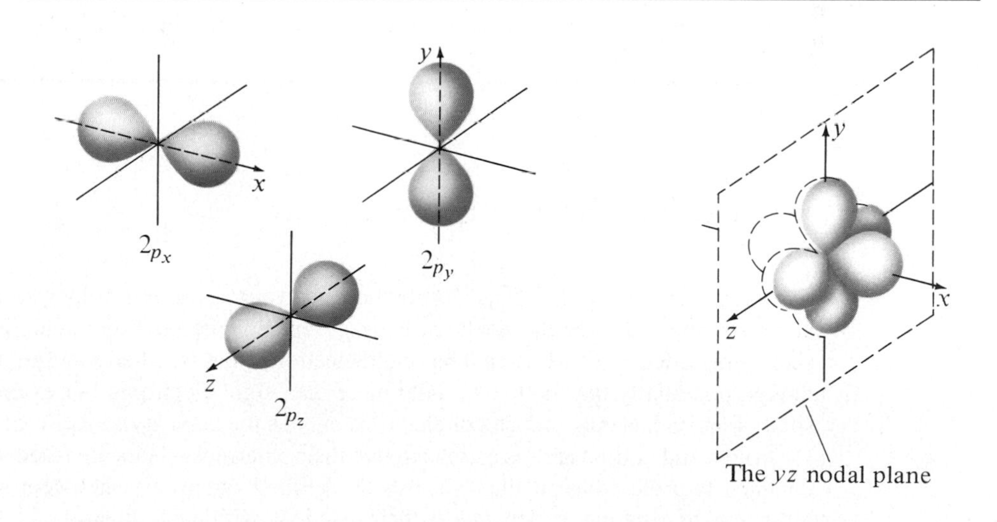

TABLE 1.1 Schematic Representation of Electron Orbitals for the First Ten Elements of the Periodic Table

Element	At. No.	Normal					Excited					Valence
		1st shell	2nd shell				1st shell	2nd shell				
		$1s$	$2s$	$2p_x$	$2p_y$	$2p_z$	$1s$	$2s$	$2p_x$	$2p_y$	$2p_z$	
H	1	↓										1
He	2	↓↑										0
Li	3	↓↑	↓									1
Be	4	↓↑	↓↑	○	○	○	↓↑	↓	↓	○	○	2
B	5	↓↑	↓↑	↓	○	○	↓↑	↓	↓	↓	○	3
C	6	↓↑	↓↑	↓	↓	○	↓↑	↓	↓	↓	↓	4
N	7	↓↑	↓↑	↓	↓	↓						3
O	8	↓↑	↓↑	↓↑	↓	↓						2
F	9	↓↑	↓↑	↓↑	↓↑	↓						1
Ne	10	↓↑	↓↑	↓↑	↓↑	↓↑						0

Exercise 1.1 Show the electron distribution of the three elements in the third principal quantum shell that appear in the same groups of the periodic table as beryllium, boron, and carbon (see Table 1.2).

Covalent Bonding

The only elements with "filled" outermost, or valence, shells belong to the family of elements referred to as the noble gases in group 0 of the periodic table. The outermost shell is regarded as filled when it has eight electrons, that is, filled s and p subshells, even though potentially the shell could hold more than eight electrons. For example, the third shell of argon has only an octet of electrons but has the capacity for $2(3)^2$ or 18 electrons. Hydrogen and helium are exceptions in that their outermost shells are filled when the shell contains two electrons in the $1s$ subshell. A filled outermost shell represents a stable arrangement, and many elements in their reactions with each other exhibit a tendency to fill their outermost shells through a transfer or sharing of electrons. Electrons outside the shell of the next lower noble gas are called **valence electrons** and are the electrons

TABLE 1.2 An Abbreviated Periodic Table

	1	2	3	4	3	2	1	0
1	^{1}H							^{2}He
2	^{3}Li	^{4}Be	^{5}B	^{6}C	^{7}N	^{8}O	^{9}F	^{10}Ne
3	^{11}Na	^{12}Mg	^{13}Al	^{14}Si	^{15}P	^{16}S	^{17}Cl	^{18}Ar
4	^{19}K	^{20}Ca					^{35}Br	^{36}Kr
5	^{37}Rb	^{38}Sr					^{53}I	^{54}Xe
6	^{55}Cs	^{56}Ba					^{85}At	^{86}Rn

Note: The numbers across the top are the usual valences of the elements in that column.

involved in transfers or sharing. Either transfers or sharing results in the formation of chemical bonds. Since bond formation by electron sharing occurs only when electron transfer is impossible, we will first consider the formation of bonds by electron transfer.

1.2 Electrovalent, or Ionic, Bonds

When sodium and chlorine enter into chemical combination, the product, sodium chloride, is a solid compound in which both sodium and chlorine appear as ions, or charged particles. By contributing its one outermost (valence) electron to the chlorine atom, sodium is left deficient by one electron and becomes a sodium ion with a charge of plus one. Chlorine, by acquiring the extra electron from sodium, becomes a chloride ion with a charge of minus one.

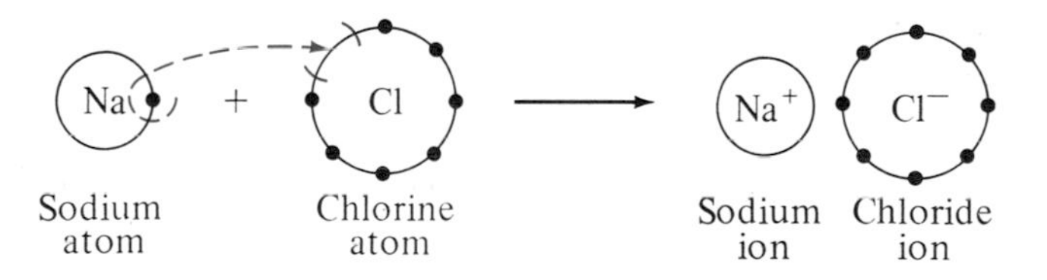

After the transfer of the electron is completed, both sodium and chloride ions have filled outer shells of electrons. The sodium ion has an electronic **configuration** equivalent to that of the noble gas (Ne) preceding it in the periodic table, and the chloride ion has the configuration of the noble gas (Ar) immediately following (Table 1.2).

In the sodium chloride crystal each sodium ion is surrounded by six chloride ions, and each chloride ion is surrounded by six sodium ions in a cubic structure (Fig. 1.3). These ions of opposite charge are held together by strong, attractive electrostatic forces, or **electrovalent bonds.**

Thus many inorganic compounds are formed by transfers of electrons between the elements to produce, not molecules, but aggregates of charged particles. Crystals of inorganic salts usually are hard substances with high melting points, because the electrostatic forces that hold the ions together in the crystal are great and not easily broken. When

FIGURE 1.3 The Spatial Arrangement of Sodium and Chloride Ions in a Crystal of Sodium Chloride (Common Salt)

The small spheres represent sodium ions, and the large spheres chloride ions.

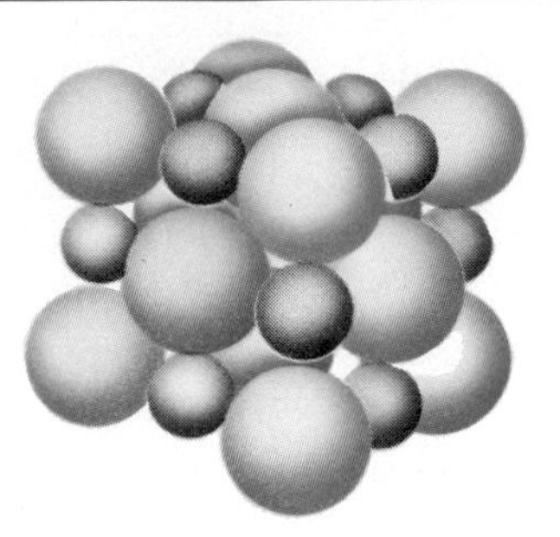

inorganic salts dissolve in water, the ions present in the crystals combine with water molecules to produce solutions that conduct an electric current. These properties are not generally characteristic of organic compounds.

Ionic bonding occurs only when one element is able to lose one or two (rarely three) electrons and the other element is able to accept one or two (very rarely three) electrons. Such electron transfers occur only between elements on opposite sides of the periodic table—that is, between the most active metals in Groups IA, IIA, and some in IIIA and the most active nonmetals in Groups VIA and VIIA and nitrogen. In order for carbon to form ionic bonds with another element, carbon would have to gain or lose four electrons. Because of the great energy requirements for such gain or loss to occur this is not possible, and carbon forms bonds by electron-sharing, as discussed in the next section.

Exercise 1.2 In Table 1.1 we see that beryllium, like helium and neon, has no unpaired electrons in the ground state, yet forms a chloride whereas helium and neon do not. Explain. What is the molecular formula for beryllium chloride?

Exercise 1.3 Supply a number in place of the *x* in the following combinations of elements to indicate the formulas of actual compounds.

(a) PH_x (b) CCl_x (c) $NaNH_x$ (d) SiO_x (e) $HCCl_x$

1.3 Covalent Bonds

A covalent single bond forms between two atoms through the *sharing* of a pair of electrons. Two atomic orbitals, one from each atom, overlap and form *two* new orbitals called **molecular orbitals,** which encompass the nuclei of both atoms. The two molecular orbitals are quite different in energy, and only the lower-energy orbital, or **bonding orbital,** is used in the formation of covalent bonds. We will not need to use the higher-energy orbital, or **antibonding orbital,** in most of our study of organic chemistry; however, it is useful to remember that, whenever we mix two or more orbitals to form molecular orbitals or hybridized orbitals (Sec. 1.7), we always get back the same number of new orbitals as the number of old orbitals we start with. In the most common case, one electron from each bonded atom occupies the bonding molecular orbital to form the two-electron covalent bond. As in the case of a filled atomic orbital, the electrons in the pair occupying the molecular orbital also must have opposite spins. Now each electron, which before bonding was attracted only to its own nucleus, is also subject to the attraction of a second nucleus. This additional attractive force gives the bond strength and makes for a stable arrangement. Ions are not formed in this type of union because there is no transfer of electrons.

The simplest example of the covalent single bond is found in the hydrogen molecule. Each hydrogen atom with a single electron in its 1*s* orbital can supply one of the shared electron pair that fills the molecular orbital. The shape of the resulting molecular orbital is no longer spherical but, as might be imagined, more sausage-shaped. The bond formed between the two hydrogen atoms is cylindrically symmetrical about its axis and is called a ***σ* (sigma) bond.** It is a very strong bond and results in a very stable molecule. In Fig. 1.4 the approximate shapes of both the atomic and molecular orbitals and the energy relationships between them are shown.

FIGURE 1.4 Atomic and Molecular Orbitals of Hydrogen and Their Energy Relationships

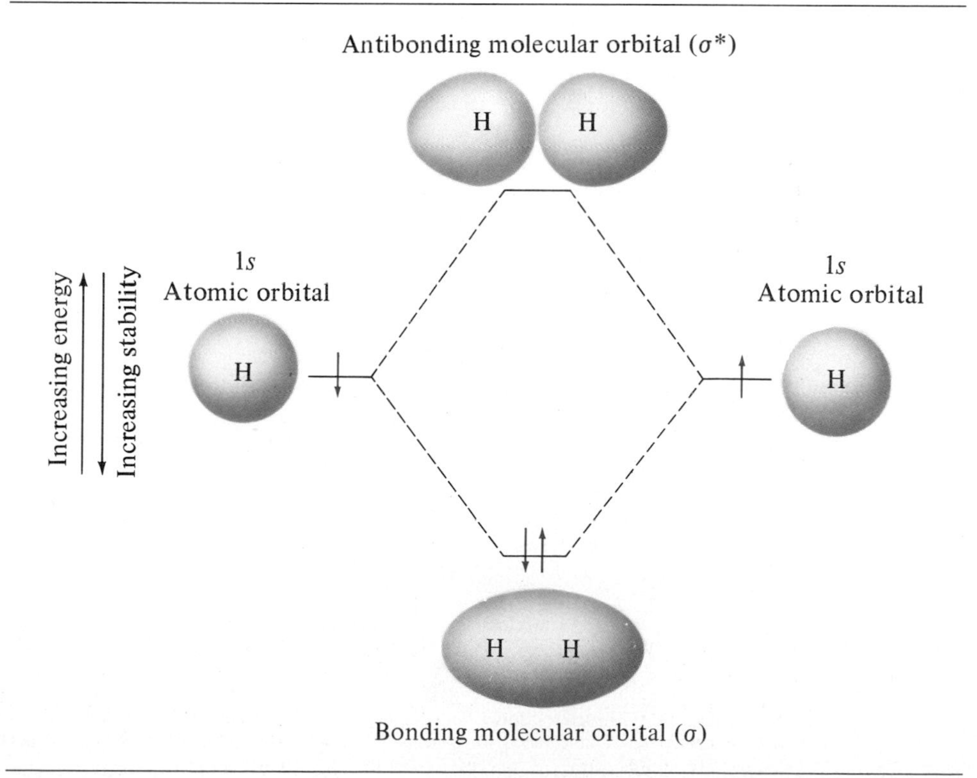

1.4 Multiple Bonds and Lewis Structures

A number of atoms share more than one pair of electrons. Nitrogen, for example, is capable of sharing three pairs of electrons to form three covalent bonds. In the nitrogen molecule, N_2, each nitrogen atom supplies three electrons to a common linkage. In ammonia, NH_3, nitrogen shares three electrons with those of three hydrogen atoms. Oxygen is capable of providing two electrons for two shared electron pairs, as in H_2O and CO_2.

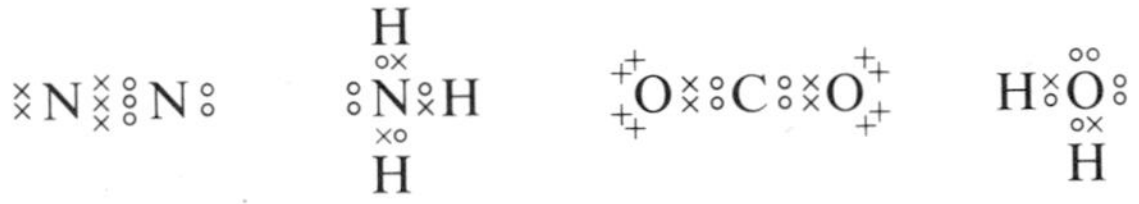

Nitrogen	Ammonia	Carbon dioxide	Water
×N⁑N∘ (electron-dot)	H, ∘N∘H, H (electron-dot)	O∷C∷O (electron-dot)	H∘O∘, H (electron-dot)

Structures drawn in the above manner are called **Lewis structures.** Each shared pair of electrons is in a **bonding** molecular orbital. Each unshared pair (one pair on each nitrogen atom, two pairs on each oxygen atom) is in a **nonbonding** molecular orbital. The nature of the orbitals involved in double and triple bonds will be discussed in later chapters (Secs. 3.2 and 3.15).

Since electrons are the same regardless of their origin, we can use a short line (—) to represent the shared electron pair that forms the covalent single bond, double lines for the two shared pairs that form the covalent double bond, and triple lines for the three shared pairs that represent the covalent triple bond.[2] Figure 1.5 shows electron-pair bonding for elements that comprise most organic compounds.

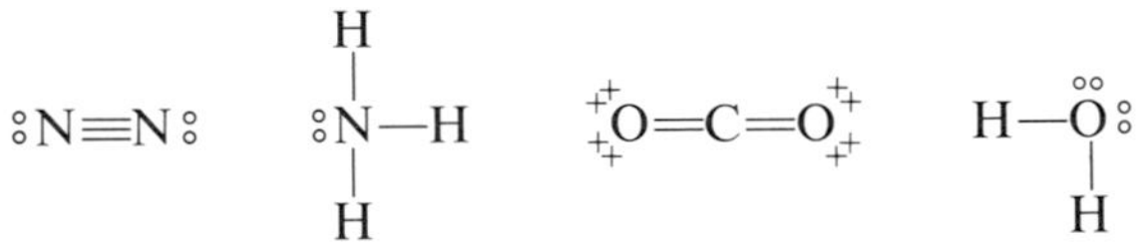

Nitrogen	Ammonia	Carbon dioxide	Water
:N≡N:	H–N(–H)–H with lone pair	O=C=O with lone pairs	H–O–H with lone pairs

EXAMPLE

Redraw the following as Lewis structures showing all valence electrons.

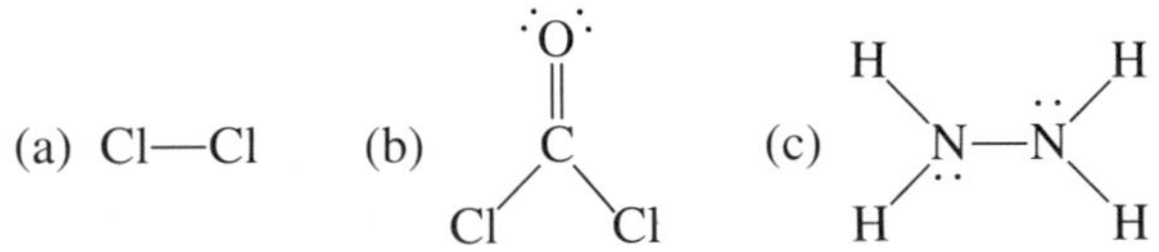

(a) Cl—Cl (b) $Cl_2C{=}O$ (c) $H_2N{-}NH_2$

Chlorine Phosgene Hydrazine

[2]In some texts Lewis structures drawn with combinations of lines and dots are called Kekulé structures. To avoid confusion with the Kekulé forms for benzene (Sec. 4.1), we prefer to use the term Lewis structures for all structures written according to the rules given in this section.

Solution Although electrons are the same regardless of which atom contributes them to the formation of a bond, it will help our accounting for all electrons to use a different set of symbols to represent the electrons of each atom in the molecule.

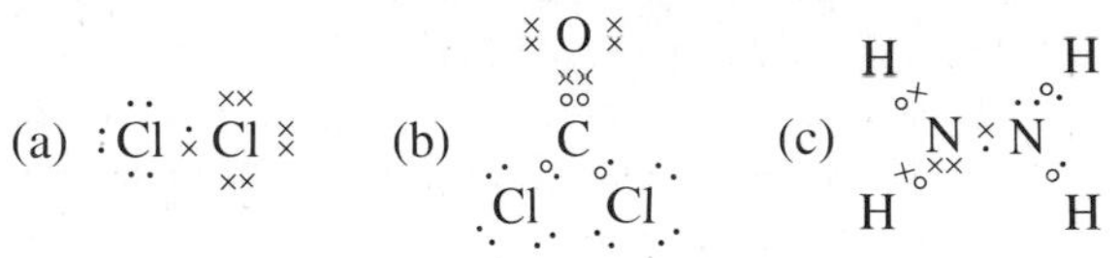

FIGURE 1.5 Common Valences of Elements in Organic Compounds

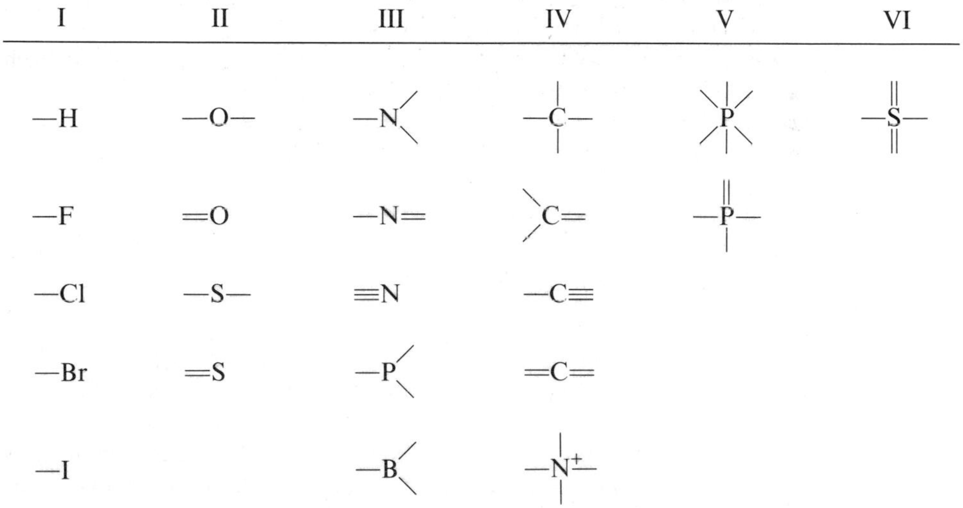

1.5 Coordinate Covalence. Formal Charges

In our discussion of the covalent bond we have emphasized the idea of electron-pair sharing as a give-and-take proposition in which each atom sharing in the bond makes an equivalent contribution. Frequently structures are formed in which both electrons comprising the shared pair of a covalent bond come from only one of the atoms. Such an arrangement is possible when one atom has an empty valence orbital and the orbital of the other contains an unshared pair of electrons. This type of bond is called a **coordinate covalent,** or **semipolar, bond.** An excellent illustration of the coordinate covalent bond is found in the boron fluoride–ammonia complex. Using dots to represent the valence electrons of boron and hydrogen, small circles for the five outermost electrons of nitrogen, and small x's for the seven electrons of fluorine, we can show how B, N, F, and H are bonded to each other in this structure.

BF_3 (Lewis dot structure) + NH_3 (Lewis dot structure) ⟶ $F_3B{:}NH_3$ (Lewis dot structure)

Boron fluoride   Ammonia   Boron fluoride–ammonia complex

$F_3B \leftarrow NH_3$ or $F_3\overset{-}{B}—\overset{+}{N}H_3$

The bond with an arrowhead indicates a coordinate covalent, or semipolar, bond between nitrogen and boron, with nitrogen as the donor of the electron pair and boron as the acceptor. Since nitrogen has bestowed upon boron a charge equivalent to one electron and lost a corresponding amount, we say that nitrogen has a **formal charge** of +1 and boron a formal charge of −1 as shown. All other bonds shown in the structure are of the ordinary covalent type. Coordinate covalent bonding is most important in inorganic complexes.

Learning to draw and use Lewis structures is an essential element in the study of organic chemistry. A Lewis structure for almost any molecule or ion can be drawn by following these general rules:

1. Let the atomic symbol of the atom represent its nucleus plus all electrons *except* those in the valence shell. Determine and show the total number of valence electrons for the entire molecule or ion by counting the valence electrons that can be contributed by each neutral atom. Subtract one electron if the species is a univalent positive ion, two if divalent, and so on. Add one electron if the species is a univalent negative ion, two if divalent, and so on. The following illustration shows how the valence electrons of some common species may be determined.

Molecule or ion	Electrons from atoms	*minus*	Cationic charge	*plus*	Anionic charge	*equals*	Total valence electrons
CH_4	$\cdot\dot{\underset{\cdot}{C}}\cdot$ + 4 H· = 4 + 4(1)	−	0	+	0	=	8
N_2	2 · $\ddot{\underset{\cdot}{N}}$· = 2(5)	−	0	+	0	=	10
NH_4^+	·$\ddot{\underset{\cdot}{N}}$· + 4 H· = 5 + 4(1)	−	1	+	0	=	8
CO_3^{2-}	·$\dot{\underset{\cdot}{C}}$· + 3 :$\dot{\underset{\cdot}{O}}$: = 4 + 3(6)	−	0	+	2	=	24
NO_3^{1-}	·$\ddot{\underset{\cdot}{N}}$· + 3 :$\dot{\underset{\cdot}{O}}$: = 5 + 3(6)	−	0	+	1	=	24

2. Draw the atomic symbols in approximately the correct geometry (Sec. 1.7). Fill in single, double, and triple bonds using a short line (—) for *each pair* of electrons in the bonds. Add the unshared electrons (nonbonding pairs and/or single electrons) using one dot for each electron. The total number of valence electrons must be assigned. If a proper accounting and distribution is made, then each atom should, to the greatest extent possible, have a filled valence shell: eight electrons for first-row atoms and eight, ten, or twelve electrons for second-row atoms. (*Note:* Remember that a covalent bond counts as two electrons *for each atom* in organic chemistry.) The following examples illustrate Rule 2.

The structure for nitrogen is:

$:N{\equiv}N:$ (8 + 8 valence electrons) rather than $:\ddot{N}{-}N:$ (8 + 6) or $:\dot{\ddot{N}}{-}\ddot{N}:$ (6 + 6)

The structure for carbon dioxide is:

$:\ddot{O}{=}C{=}\ddot{O}:$ (8 + 8 + 8) rather than $:\ddot{\underset{..}{O}}{-}C{=}\ddot{O}:$ (8 + 6 + 8)

or $:\ddot{\underset{..}{O}}{-}C{-}\ddot{\underset{..}{O}}:$ (8 + 4 + 8)

3. Find the formal charge on each atom. First, assign to each atom *all* of its unshared *valence* electrons and *half* of the electrons in its covalent bond pairs. Then subtract this total number of electrons from the number of *valence* electrons present in the *isolated unbonded atom*. For example, the formal charge on each atom in the ammonium ion NH_4^+,

```
     H
     |+
  H—N—H
     |
     H
```

is determined as follows:

(a) Each hydrogen is assigned one e^- from the covalent bond. Hydrogen has only one electron in its valence shell. Thus the formal charge on each H = 1 − 1 = 0
(b) Nitrogen is assigned one e^- from each covalent bond for a total of four electrons. Nitrogen has five electrons in its valence shell. Thus the formal charge on N = 5 − 4 = +1

EXAMPLE

Which atom in the cyanide ion bears the formal charge?

Solution The cyanide ion is written as: $[:C{\equiv}N:]^-$

Formal charge = valence electrons − (unshared electrons + $\frac{1}{2}$ shared electrons)

For C,	4	minus	$(2 + \frac{6}{2}) = -1$
For N,	5	minus	$(2 + \frac{6}{2}) = 0$

1.6 Condensed Structural Formulas

Structural formulas or graphic representations of organic compounds are necessary to show the geometric arrangement of the atoms and to communicate necessary structural information. However, if we were to use only full Lewis structures for every organic molecule in this text, then learning organic chemistry would be made much more difficult than it needs to be. Therefore, we will use Lewis structures only in those situations where the full structure is necessary or desirable. In most instances we will abbreviate structural formulas by: (1) not showing all the bonds, (2) omitting unshared electron pairs, (3) omitting atomic symbols and showing only a graphic representation of the carbon framework. Such abbreviated structural formulas are called **condensed formulas.** The following examples illustrate the various ways in which a Lewis structure may be condensed.

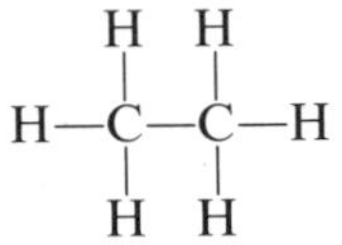

Ethane: $CH_3{-}CH_3$ CH_3CH_3 C_2H_6

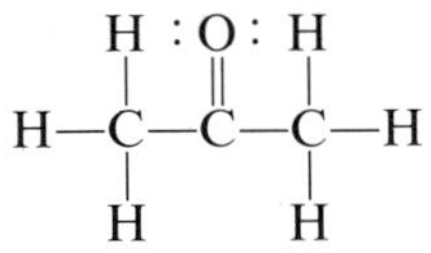

Acetone: 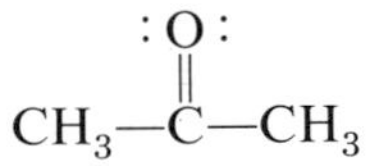 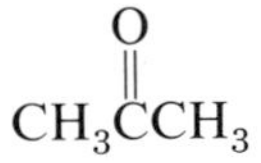 $CH_3{-}\overset{\displaystyle :\!O\!:}{\overset{\|}{C}}{-}CH_3$ $CH_3\overset{\displaystyle O}{\overset{\|}{C}}CH_3$ CH_3COCH_3  C_3H_6O

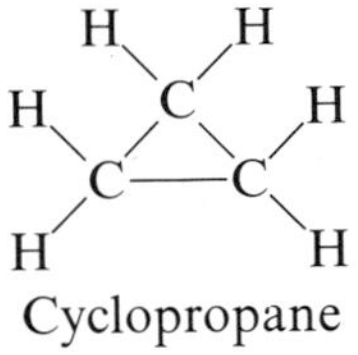

Cyclopropane: 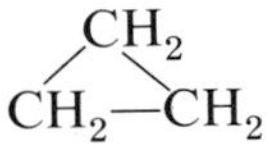 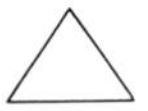 C_3H_6

Although framework representations such as

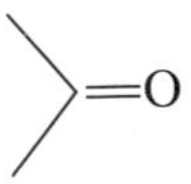

for acetone and

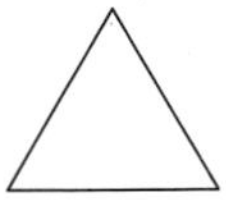

for cyclopropane will be described in more detail in Secs. 3.13 and 5.5, a few simple rules will be given here to allow you to use these convenient structures in your studies. A *single* line projecting from a structure represents a carbon atom with three attached hydrogens. A *double* line projecting from a structure is understood to be directed to a carbon with two attached hydrogens. All lines in the skeletal framework are understood to be directed to other carbon atoms unless they are clearly shown to be directed toward oxygen, nitrogen, or other non-carbon atoms as in the acetone structure above. An example will help to illustrate these guidelines. The two structures below represent isoprene, the important structural component of natural rubber and a host of other natural products.

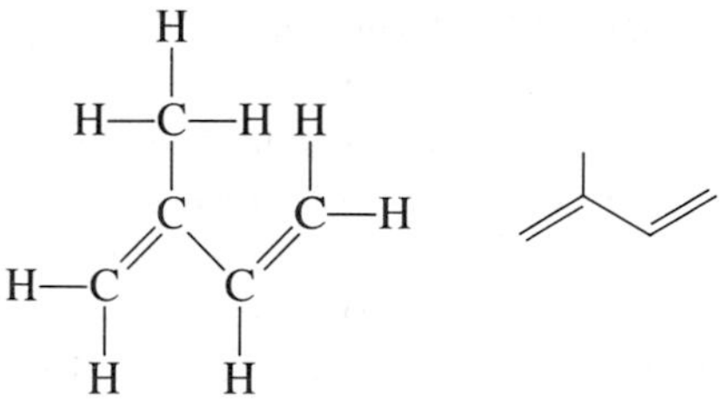

Isoprene

Exercise 1.4 Expand the framework structures below into complete structural formulas showing all carbon and hydrogen atoms.

(a)  *(1,3-butadiene)*

(b) 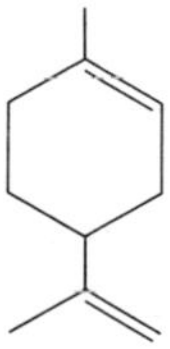 *(limonene)*

The Covalence of Carbon

Nearly all bonds that bind atoms together to form organic compounds are of the covalent type. The unit particles in most organic compounds are molecules, not ions of opposite charge as in inorganic salts. Since the attractive forces between molecules are much weaker than those between ions in salts, many organic substances are liquids with low boiling points. They ordinarily have low melting points if they are crystalline solids. Organic substances, for the most part, are insoluble or only slightly soluble in water. If slightly soluble, their aqueous solutions seldom conduct an electric current, unless they are the salts of organic acids or salts of organic bases. Reactions between organic molecules are usually slow and take place only when molecules collide with sufficient kinetic energy to break, rearrange, or form new bonds. For this reason organic reactions frequently require long periods of heating and the presence of a catalyst before any appreciable changes occur. For example, a covalently bound halogen atom may show no tendency to precipitate as silver halide when merely shaken with a silver nitrate solution. On the other hand, the reaction of an ionic halide with silver nitrate is immediate.

1.7 Hybridized Orbitals, Bond Angles, and the Geometry of Molecules

Since the covalence that an element generally exhibits is usually determined by the number of unpaired electrons in half-filled orbitals in its valence shell, the case of carbon deserves special mention. In nearly all organic compounds, carbon exhibits a covalence of four. How can this be when Table 1.1 shows carbon has but two unpaired electrons in its outer shell? The answer is that the first excited state of carbon is attained by promoting one of the 2*s* electrons to the higher-energy, vacant $2p_z$ orbital as shown below. Now there are four singly occupied orbitals, one 2*s* and three 2*p*. If these four orbitals are mixed (mathematically), an excited atom with four equivalent orbitals results. The energies of the four equivalent orbitals are equal and intermediate in value between the energy of a 2*s* orbital and that of a 2*p* orbital. The total energy (for all four) is the same as the sum of the energies of the 2*s* and three 2*p* orbitals. These blended orbitals are said to be *hybridized* and are called sp^3 (pronounced s-p-three) orbitals. Chemical theory predicts that the four sp^3 hybrid orbitals should be directed from the carbon atom toward the corners of a tetrahedron (Fig. 1.6).

The formation of four sp^3 hybrid orbitals for carbon may be illustrated graphically by the following simple diagram.

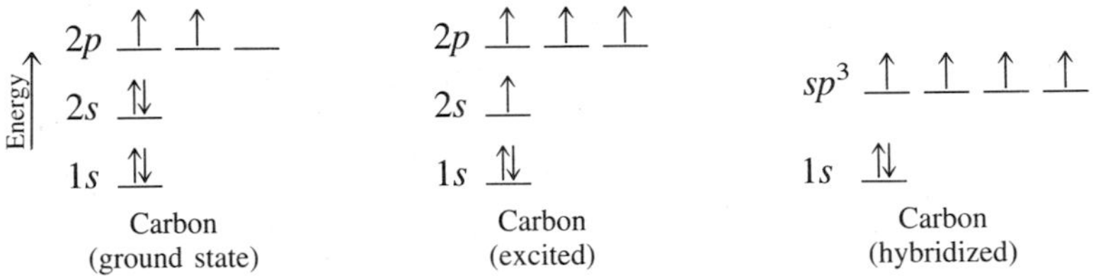

Energy ↑	Carbon (ground state)	Carbon (excited)	Carbon (hybridized)
	2*p* ↑ ↑ _	2*p* ↑ ↑ ↑	sp^3 ↑ ↑ ↑ ↑
	2*s* ↑↓	2*s* ↑	
	1*s* ↑↓	1*s* ↑↓	1*s* ↑↓

If the four hybrid sp^3 orbitals are allowed to overlap with the 1*s* orbitals of four hydrogen atoms, four bonding molecular orbitals of the sigma type will be formed. The resultant molecule, CH_4, should have the carbon atom at the center and four hydrogen atoms at the vertices of a regular tetrahedron. Experimental data are in agreement with this prediction. Thus, in methane, CH_4, the principal constituent of natural gas, the four covalent C—H bonds are directed toward the vertices of a regular tetrahedron (Fig. 1.7).[3]

[3] A regular tetrahedron is easily constructed by drawing a cube and connecting all nonadjacent corners.

With carbon in the center, these four sigma bonds make angles of 109.5° with each other. They are of the same length (1.09 Å).[4] The energy required to break any one of them is the same (102 kcal).[5] The bonds, therefore, *must* be equivalent.

FIGURE 1.6 sp^3-Hybridized Orbitals: (a) An sp^3-Hybridized Orbital, (b) Conventional Symbol Used to Represent an sp^3 Orbital, (c) Four Tetrahedral sp^3 Orbitals of the Carbon Atom

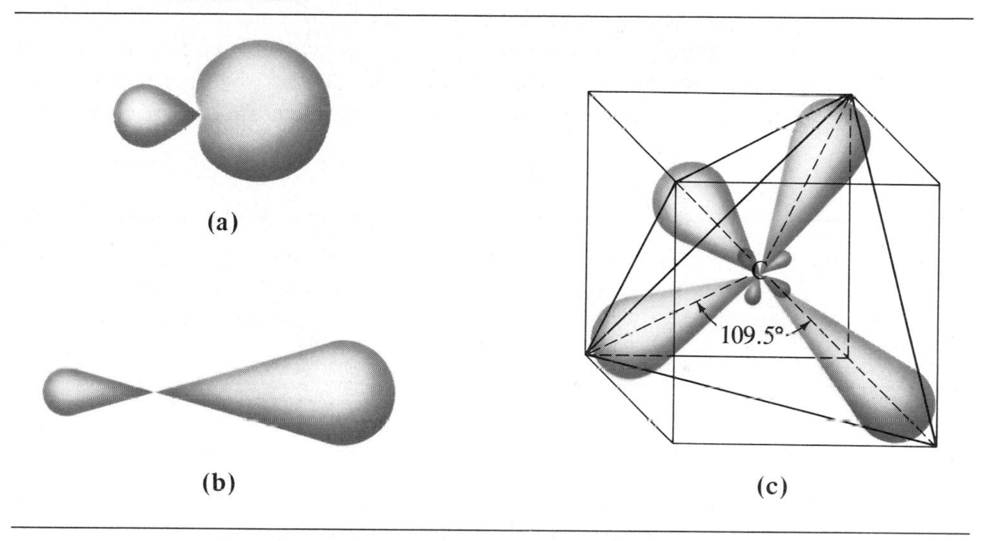

FIGURE 1.7 Structural Representations of Methane: (a) Ball and Stick Model in Regular Tetrahedron, (b) Tetrahedral Structure Showing Electrons Involved in Bonding and Bond Angles, (c) Stuart Model

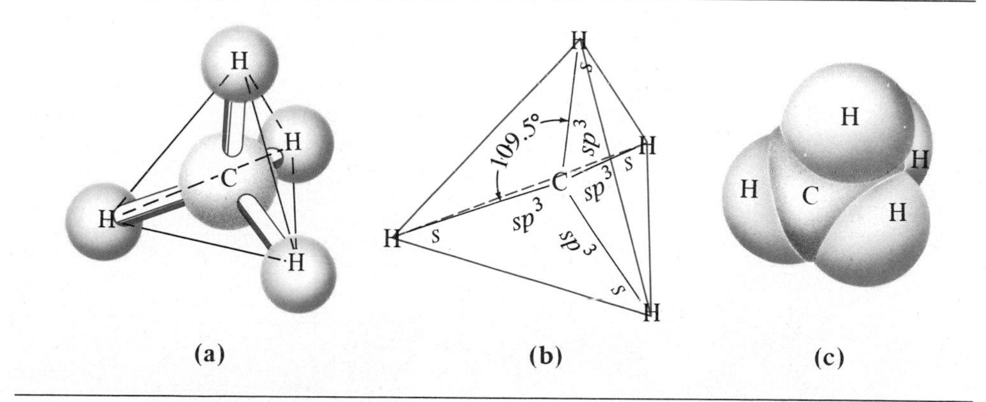

[4]The angstrom (Å) is equal to 10^{-8} centimeter.

[5]The kilocalorie (kcal) is equal to 1,000 calories.

What have we gained by the process of hybridization, which required the use of the higher-energy excited-state orbitals? First, we were able to form four *equivalent* bonds from carbon to hydrogen, rather than two. Thus we lowered the energy of the molecule more by forming two extra bonds than we lost by using the higher-energy orbitals. This alone would not justify hybridization, for we could have used the 1*s* and three 2*p* excited-state orbitals without hybridization to form four nonequivalent covalent bonds. However, the use of hybridized orbitals permitted the formation of *stronger* bonds because of the better overlapping properties of sp^3 orbitals. Second, the tetrahedral geometry of the sp^3 orbitals allows the maximum *separation in space* of the four bonds. Since the electron pairs in one bond repel those in the other bonds, when the bonds are farther apart, the energy of the system will be lower and the molecule will be more stable.

Exercise 1.5 Using circles for an *s* orbital, a figure-eight shape for a *p* orbital, and a tear drop shape for a hybrid orbital, draw a sketch that illustrates the overlapping of the following pairs of orbitals *lying along the x-axis*. (a) Hydrogen 1*s* and fluorine $2p_x$, (b) carbon sp^3 and carbon sp^3.

In the case of boron, hybridization gives three sp^2 hybrid orbitals, which are directed away from the boron atom toward the corners of an equilateral triangle (bond angles = 120°). Thus boron trifluoride is a planar molecule with the geometry shown below.

Energy ↑

Boron (ground state)	Boron (excited state)	Boron (hybridized)
2*p* ↑ _ _	2*p* ↑ ↑ _	2*p* _
2*s* ⇅	2*s* ↑	sp^2 ↑ ↑ ↑
1*s* ⇅	1*s* ⇅	1*s* ⇅

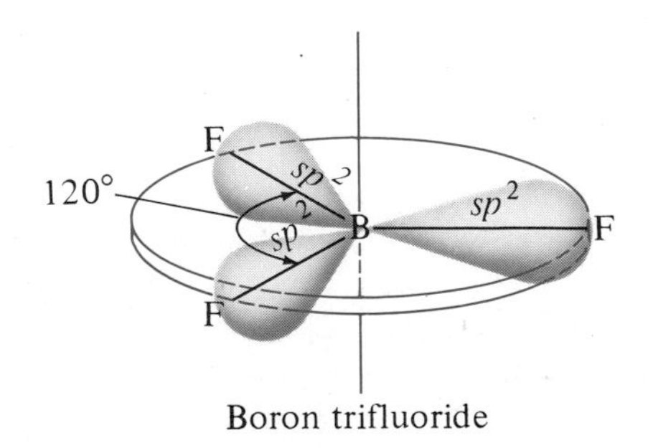

Boron trifluoride

The hybridization of beryllium gives two *sp* hybrid orbitals that are directed away from the beryllium atom with a bond angle of 180°. Thus beryllium hydride is a linear molecule as shown.

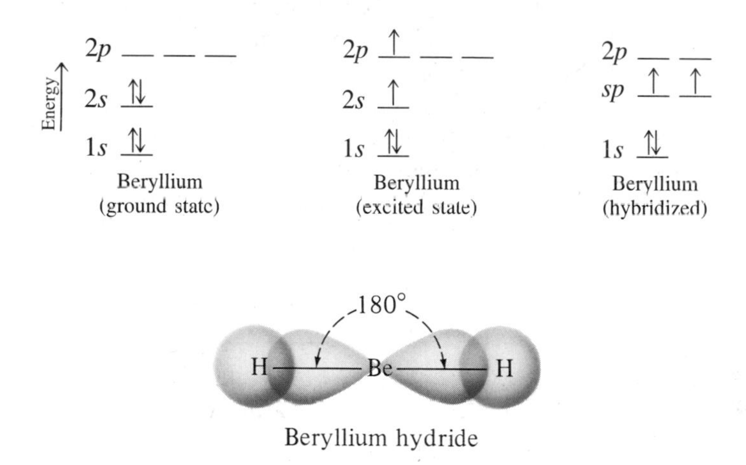

Beryllium hydride

Were it not for hybridization of orbitals as described in the preceding sections, one would expect the covalence of carbon to be two and that of boron to be one; beryllium would have no electrons available for chemical combination. This we know to be contrary to fact.

Although our discussion of hybridization and its effect on bond angles might lead to the conclusion that there is no easy, nonmathematical way to predict the geometry of molecules, there is a very simple procedure for making such predictions. This involves a model of molecular structure called the **valence-shell electron-pair repulsion** (VSEPR) model. This awesome-sounding but simple model assumes that electrostatic repulsions between pairs of valence-shell electrons cause these pairs to spread out in space in such a way as to minimize the repulsion. Of course, these pairs of valence-shell electrons are those that are involved in the single, double, or triple bonds of the molecule and the unshared pairs of electrons (nonbonding electrons) on any of the bonded atoms. To use this model, first draw the two-dimensional Lewis structure for the molecule; then draw a picture or build a simple three-dimensional model of the molecule in which the valence electrons (bond pairs *and* unshared electron pairs) are as far apart in space as possible. Consider, for example, the water molecule, H—Ö—H (with two unshared pairs on O). According to the VSEPR model, there are four pairs of valence-shell electrons to distribute—exactly the same number as in methane. Therefore, the water molecule should have roughly the same geometry as methane (Fig. 1.7), with the two unshared pairs of oxygen replacing two of the bond pairs of methane. The H—O—H bond angle would be predicted to be near 109.5°. However, because electron repulsion is greater between pairs of unshared electrons than between shared pairs we might expect the angle between the unshared pairs to be somewhat larger than 109.5° and the H—O—H angle to be somewhat smaller. This is indeed the case, as the H—O—H angle is about 104.5°. Similarly, the geometry of the ammonia molecule, $:NH_3$, is also a distorted tetrahedron with H—N—H angles of about 106.8°.

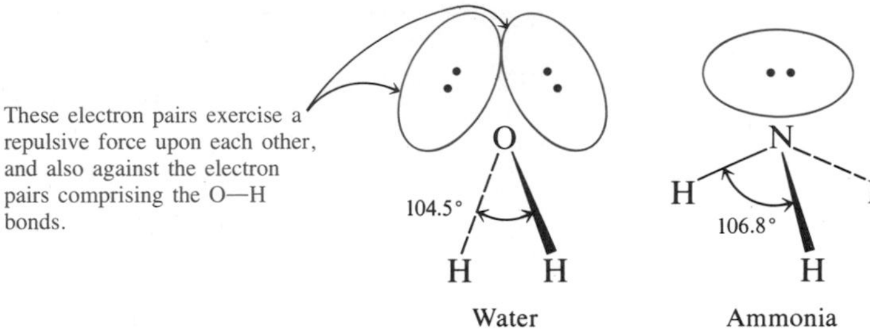

These electron pairs exercise a repulsive force upon each other, and also against the electron pairs comprising the O—H bonds.

Water Ammonia

Exercise 1.6 Using the concept of VSEPR discussed above, predict the geometry of the following: (a) NH_4^+ (b) BF_4^- (c) PH_3 (d) CaF_2 (e) $AlCl_3$

1.8 Electronegativity and Polar Covalent Bonds

Electronegativity has been defined by Pauling as the power of an atom in a molecule to attract electrons to itself. Thus when two atoms in a molecule have the same electronegativity, the bonding electrons are equally shared. Simple examples are the hydrogen (H—H) and chlorine (Cl—Cl) molecules. However, when two atoms of a molecule have different electronegativities, the bonding pair will be drawn somewhat closer to the more electronegative atom. Such displacement, however small, results in a polar bond and confers a partial negative charge (delta minus) on the more electronegative element and a partial positive charge (delta plus) on the less electronegative (more electropositive) element. An example is the hydrogen chloride (H—Cl) molecule.

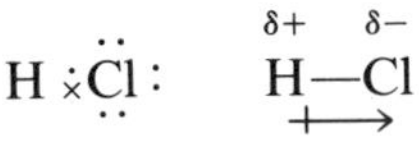

$$\mathrm{H} \overset{\cdot}{\times} \ddot{\underset{\cdot\cdot}{\mathrm{Cl}}}: \qquad \overset{\delta+}{\mathrm{H}}—\overset{\delta-}{\mathrm{Cl}}$$

This slight shift in electron density is indicated by an arrow directed toward the more electronegative element. Opposite the arrowhead the shaft is crossed by a short vertical bar as in a plus sign.

Since chemistry largely involves the interaction of polar molecules and ions with each other, electronegativity is one of the most important concepts in chemistry. Unfortunately, there is, as yet, no general agreement as to the best method of determining electronegativities nor as to the best values of the electronegativities themselves. The values

TABLE 1.3 Electronegativities of Selected Elements

H 2.2						
Li 0.98	Be 1.57	B 2.04	C 2.55	N 3.04	O 3.44	F 3.98
Na 0.93	Mg 1.31	Al 1.61	Si 1.90	P 2.19	S 2.38	Cl 3.16
K 0.82	Ca 1.00					Br 2.96
						I 2.66

Increasing electronegativity ⟶ (left to right)

Increasing electronegativity ⟶ (bottom to top)

given in Table 1.3 are based on the original method of Pauling. Electronegativities probably are not constant but depend somewhat on the nature of the hybridization of the atoms involved and any charges on these atoms. Note that electronegativities increase as we proceed from left to right or from bottom to top through the periodic arrangement of the elements. Thus, fluorine is the most electronegative element and the alkali metals the most electropositive.

Exercise 1.7 Indicate with a small arrow (+→) the polarity of the bond in each of the following molecules: (a) PH_3 (b) BF_3 (c) ICl (d) CH_3Li

Exercise 1.8 From the data in Table 1.3 tell which bond in each of the following pairs is more polar. Show the direction of polarization by placing $\delta+$ and $\delta-$ signs on the appropriate atoms. (a) The H—Br or the H—I bond? (b) The P—H or the N—H bond?

1.9 Structural Formulas and Isomerism

As was stated in the introductory section, it is not enough simply to determine the number and kinds of elements comprising an organic compound because this would allow us to write only a molecular formula. The reason there are over two million organic compounds is that many with the same molecular formula have different atomic arrangements or structures. We have already encountered methane, whose molecular formula is CH_4, and have learned that in methane the four hydrogens are bonded to a central carbon atom. There is only one structure possible for methane; only one for ethane, C_2H_6; and only one for propane, C_3H_8—the first three members of a family of compounds we shall come to know as the alkanes (Chap. 2). The next member of this family has the molecular formula C_4H_{10}. Now how can the four carbon and ten hydrogen atoms be arranged so that each carbon has four bonds and each hydrogen only one? We soon discover that there are two possibilities. In one structure, no carbon would be bonded to more than *two* others in a continuous chain structure, whereas in the second structure, one carbon could be attached to *three* others in a branched-chain structure.

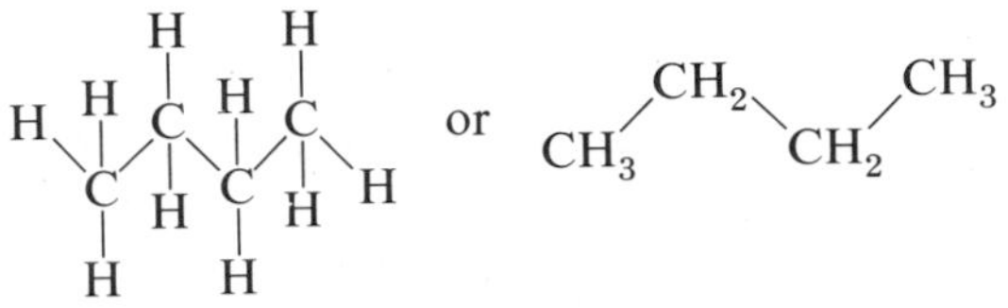

Normal butane, bp, −0.5°C
(a continuous chain structure)

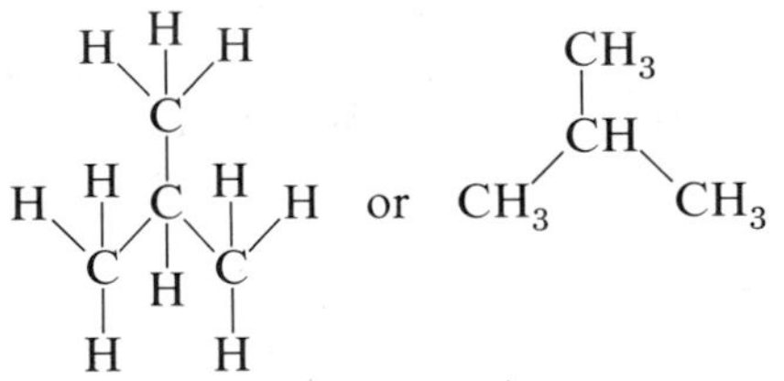

Isobutane (Methylpropane) bp, −11.7°C
(a branched-chain structure)

In order to draw these two structures, we simply have allowed the sp^3 hybrid orbitals of carbon atoms to overlap those of other carbons, as well as with *s* orbitals of hydrogen atoms, to form the four sigma bonds that each carbon is entitled to.

Although both compounds have the *same molecular formula,* they have *different structures* and are called **isomers** (Gr. *isos,* same; *meros,* part). Because they have different structures, they are different compounds with different physical and chemical properties.

Exercise 1.9 Is there more than one structure possible for dichloromethane, CH_2Cl_2, if its geometry is tetrahedral? Would more than one structure be possible if its structure were pyramidal with carbon at the apex and hydrogen and chlorine atoms at the corners of the base?

Another pair of gaseous hydrocarbons has the molecular formula C_4H_8. By following the rules of valence for these two elements and bonding the carbon atoms in a continuous **cyclic** arrangement, we are able to draw two different C_4H_8 structures.

```
    H  H                         H  H
    |  |                         | /
H—C—C—H               H       C—H
    |  |                    \   /
H—C—C—H                  C
    |  |                    /   \
    H  H              H—C———C—H
                          |       |
                          H       H
```

Cyclobutane, bp, 13.0°C   Methylcyclopropane, bp, 4.0°C

In Chapter 3 you will discover that this same molecular formula will permit four additional structures. Although the bonds involved in these structures will not all be of the sigma type as in the previous examples, the six compounds represented by the molecular formula will all be isomers, nevertheless.

You can see from the preceding examples that carbon is capable of sharing an electron pair not only with hydrogen and other elements, but also with another carbon atom. You also have seen that it is capable not only of forming continuous and branched-chain structures, but cyclic ones as well. The tetrahedral nature of the carbon atom and its ability to bond to other carbons makes possible a great number of organic compounds by the combination of a relatively few different atomic species. As the number of carbon atoms is increased, the number of possible isomers multiplies rapidly.

There are three different structures possible for C_5H_{12}, five for C_6H_{14}, and nine for C_7H_{16}. Think what a chore it would be to write all *seventy-five* structures possible for $C_{10}H_{22}$! Being told that condensed forms (Sec. 1.6) usually are used in writing organic formulas is welcome news indeed.

EXAMPLE

Write five structures for C_6H_{14}.

Solution We begin by writing a chain of *six* carbon atoms, each carbon with four bonds. For the moment, we will set the hydrogens aside.

```
  |  |  |  |  |  |
—C—C—C—C—C—C—
  |  |  |  |  |  |
```

Place hydrogens on each free bond to give us isomer number 1.

```
         H   H   H   H   H   H
         |   |   |   |   |   |
(1)  H—C—C—C—C—C—C—H
         |   |   |   |   |   |
         H   H   H   H   H   H
```

Hexane

For isomers 2 and 3 we will need to write two chains of *five* carbons each.

```
             |                          |
           —C—                        —C—
   |   |   |   |   |          |   |   |   |   |
 —C—C—C—C—C—              —C—C—C—C—C—
   |   |   |   |   |          |   |   |   |   |
```

The sixth carbon may be placed either on the second carbon (counting from either end) or on the middle carbon. Again fill all unoccupied bonds with hydrogen.

```
            H                                   H
            |                                   |
          H—C—H                               H—C—H
            |                                   |
        H   |   H   H   H                H   H  |   H   H
        |   |   |   |   |                |   |  |   |   |
(2)  H—C—C—C—C—C—H           (3)  H—C—C—C—C—C—H
        |   |   |   |   |                |   |  |   |   |
        H   H   H   H   H                H   H  H   H   H
```

2-Methylpentane 3-Methylpentane

For isomers 4 and 5 we will write two continuous chains of *four* carbons each. Now we can either place the two remaining carbons both on the second carbon (again counting from either end), or we can place one carbon each on the second and third ''links'' of the chain.

```
           |                       |
         —C—                     —C—
     |   |   |   |           |   |   |   |
   —C—C—C—C—           —C—C—C—C—
     |   |   |   |           |   |   |   |
         —C—                         —C—
           |                           |
```

Again fill all unoccupied bonds with hydrogen.

```
                 H                                    H
                 |                                    |
              H—C—H                                H—C—H
                 |                                    |
           H     |   H  H                      H      |   H  H
           |     |   |  |                      |      |   |  |
(4)     H—C—C—C—C—H                (5)     H—C—C—C—C—H
           |     |   |  |                      |      |   |  |
           H     |   H  H                      H      H   |  H
                 |                                        |
              H—C—H                                    H—C—H
                 |                                        |
                 H                                        H
        2,2-Dimethylbutane                       2,3-Dimethylbutane
```

Had we placed a two-carbon (ethyl) group on the second carbon of the four-carbon chain, we would simply have drawn isomer 3 again. For an organic molecule with many isomers, there may be several combinations of carbon chains that give the same isomer, and we must be careful to eliminate duplicate structures. If we always begin with the longest possible chain and work down through shorter chains, we will usually find it easy to spot and eliminate duplicates.

Exercise 1.10 Rewrite the five structural isomers of C_6H_{14} in an acceptable condensed form. (*Hint:* See butane examples in this section.)

The determination of a structure for an unknown organic compound may, at first thought, appear to be a formidable task. Quite often it is. However, in recent years the availability of sophisticated instruments for measuring certain physical properties has made it much easier for the organic chemist to classify a compound and to determine its structure. In addition, these physical measurements give some idea of its chemical properties, and a clue to its synthesis as well.

Among the most useful instrumental methods employed by present-day organic chemists in the determination of structure are several methods based on absorption spectroscopy. These will be described in Chapter 18.

1.10 Classification of Organic Compounds by Functional Groups

Chemical reactions involving organic compounds center, for the most part, about some unique structural feature known as a **functional group.** This group not only bestows a characteristic behavior upon the molecule, but also identifies it as belonging to a certain family of compounds or **homologous series.** The simplest example of functional groups may be found in the four families of hydrocarbons: the alkanes, alkenes, alkynes, and arenes (or aromatic hydrocarbons). For example, there are three hydrocarbons having only two carbon atoms in the molecule: ethane, ethylene (ethene), and acetylene (ethyne). As may be seen from their Lewis structures, these molecules differ principally in their functional groups. Ethane has no functional group. Ethylene has a carbon–carbon double bond as its functional group, and acetylene has a carbon–carbon triple bond.

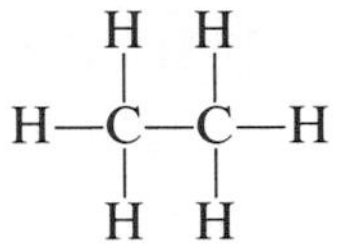

H—C—C—H (with H above and below each C)

Ethane
(an alkane)

H₂C=CH₂ drawn as H, H \ C=C / H, H

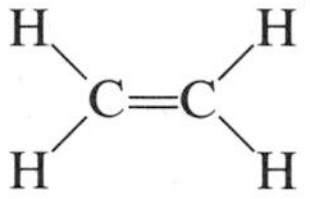

Ethylene
(an alkene)

H—C≡C—H

Acetylene
(an alkyne)

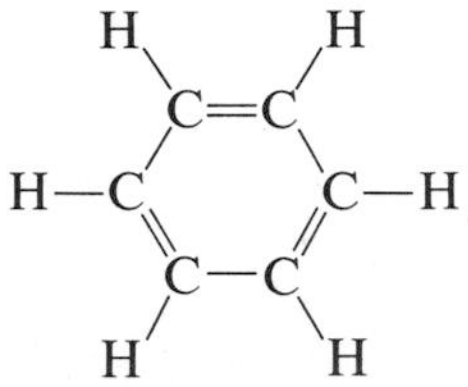

H—C ring with C=C, C—C (benzene, each C bearing H)

Benzene
(an arene)

The fourth class of hydrocarbons, the arenes or aromatic hydrocarbons, may appear to have three double bonds and thereby belong to the alkene family, but we will learn later that this structural unit, alternating single and double bonds in a six-membered ring, has unique properties and is the functional group of the arenes. It is called the **benzene ring.** In the next several chapters we will find that each of these four classes of hydrocarbons has its own unique chemistry, and that this chemistry is largely dependent upon the chemical behavior of the functional groups present.

To illustrate the principle of homology, let us look at the first two members of the alcohol family. The alcohols may be considered to be derived from the familiar inorganic compound, water. The first two members, or **homologs,** of this family have the structures shown.

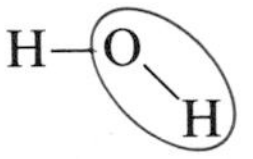

H—O—H

Water

H—C(H)(H)—O—H or CH_3—O—H

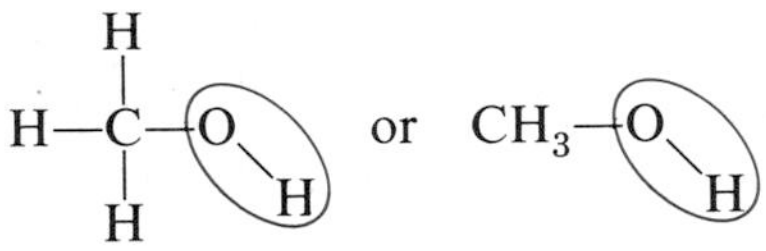

Methanol (methyl alcohol)

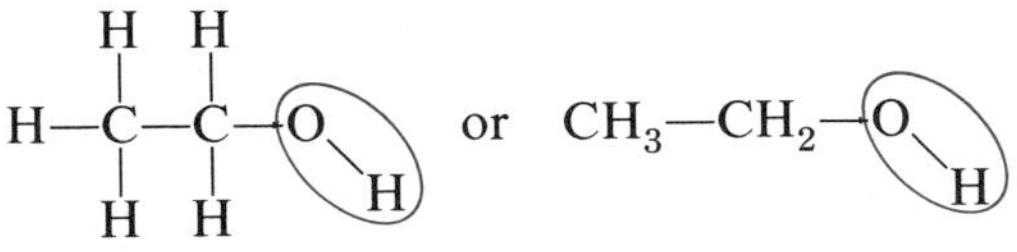

H—C(H)(H)—C(H)(H)—O—H or CH_3—CH_2—O—H

Ethanol (ethyl alcohol)

It is clear from their structures that a feature common to both alcohols is the **hydroxyl groups,** —OH, which is the functional group that characterizes the alcohol family. By simply increasing the length of the hydrocarbon chain each time by an increment of —CH_2— we can write formulas for other members of the series. The size and geometry of the unreactive portion of the structure may modify, but usually does not alter, the reactions characteristic of the functional group. This greatly simplifies the study of organic chemistry. Functional groups not only allow us to recognize a compound as belonging to a certain class or family, and to name it, but also to define its chemical properties. In the following chapters you will learn how to build into a structure the desired functional group, how to replace it by another, and in some cases how to eliminate it entirely. Table 1.4 lists the principal families, or classes, of compounds encountered in organic chemistry, the corresponding functional groups, and a common example within each class.

TABLE 1.4 Classes of Compounds and Their Functional Groups

Class Name	Functional Group	General Formula*	Common Example	Example Name
Alkanes	None	R—H	CH_3—CH_3	Ethane
Alkenes	>C=C<	R_2C=CR_2	CH_2=CH_2	Ethylene
Alkynes	—C≡C—	R—C≡C—R	H—C≡C—H	Acetylene
Arenes	C_6 ring (—C=C— alternating)	C_6R_6	C_6H_6	Benzene
Alkyl Halides	—F, —Cl, —Br, —I	R—X	CH_3—I	Methyl iodide
Alcohols	—O—H	R—O—H	CH_3—CH_2—O—H	Ethanol
Ethers	—O—	R—O—R	CH_3—CH_2—O—CH_2—CH_3	Diethyl ether
Aldehydes	—C(=O)—H	R—C(=O)—H	CH_3—C(=O)—H	Acetaldehyde
Ketones	>C=O	R_2C=O	$(CH_3)_2$C=O	Acetone
Carboxylic Acids	—C(=O)—O—H	R—C(=O)—O—H	CH_3—C(=O)—O—H	Acetic acid
Primary Amines	—N(H)H	R—N(H)H	CH_3—N(H)H	Methylamine
Secondary Amines	—N(R)H	R—N(R)H	CH_3—CH_2—N(CH_3)H	Methylethylamine
Tertiary Amines	—N(R)R	R—N(R)R	CH_3—N(CH_3)CH_3	Trimethylamine

*The R used in a general formula refers to that portion of the molecule other than the functional group. Often R = C_nH_{2n+1}. R is called the *alkyl* group. Example:

CH_3— (methyl) CH_3—CH_2— (ethyl)

Exercise 1.11 There are two isomers of the next higher homolog after ethyl alcohol. Write structural formulas for each.

Exercise 1.12 There are six known organic compounds for the molecular formula C_3H_6O, representing four different families. Draw Lewis structures for as many as you can. (*Remember:* Oxygen may have one double bond or two single bonds.)

1.11 Classification of Organic Compounds by Framework

Another frequently used classification of organic compounds is that based on the molecular framework of the molecules, i.e., that part of the molecule other than the functional group (Fig. 1.8). We have seen that carbon can, by bonding to other carbon atoms, form a continuous chain of carbon atoms in which each carbon is bonded to hydrogen and/or to other carbons. Such open-chain compounds, whether ''straight'' or branched, are called **acyclic.** Acyclic compounds may again be divided into **saturated** (all carbons singly bonded to four other atoms), and **unsaturated** (some carbons bearing double or triple bonds).

Cyclic, as the name implies, means ring compounds. If the ring is comprised of only carbon atoms, it is called **carbocyclic.** Here again the carbons in the ring may bear four single bonds or be multiply bonded. Cyclic compounds having the same properties as acyclic compounds are called **alicyclic** compounds to differentiate them from the **aromatic** hydrocarbon benzene, C_6H_6, and its derivatives.

FIGURE 1.8 Classification of Organic Compounds Based on Molecular Framework

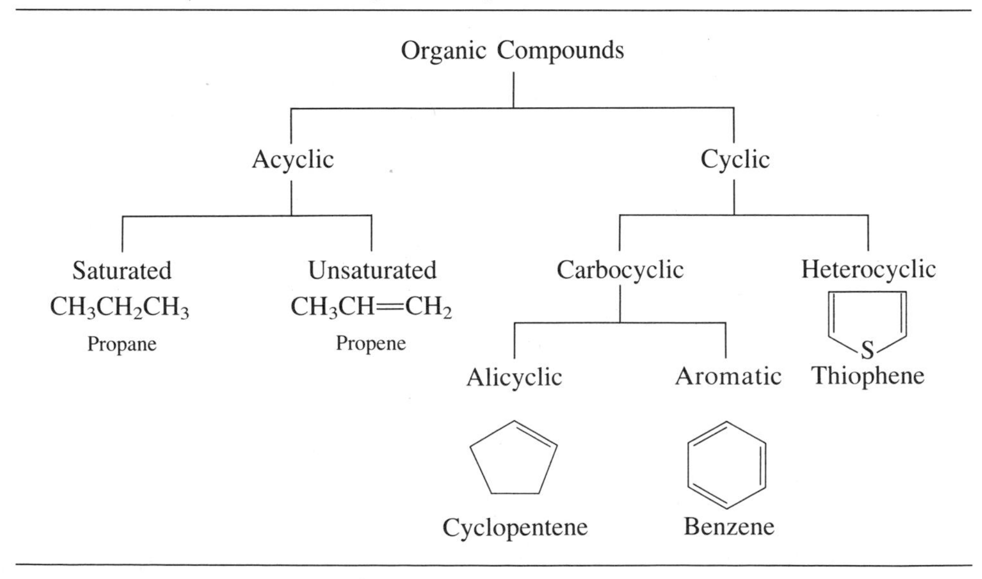

Heterocyclic rings contain some element other than carbon in the cyclic framework, usually nitrogen, oxygen, or sulfur.

A simplified diagram of this classification is given in Figure 1.8, along with examples of each class. The various classes we have just defined will be described in Chapters 2 through 4.

Exercise 1.13 Draw (a) an acyclic unsaturated structure and (b) a heterocyclic structure for C_4H_8O.

1.12 Chemical Reactivity. Reaction Mechanisms

Although ionic compounds may undergo reactions by dissociation into component ions and recombination to yield products, reactions of organic molecules generally take place when the molecules collide with sufficient kinetic energy to break, rearrange, or form bonds. When such reactions are carried out, there is usually a net heat change in going from reactants to products. If heat is evolved, the reaction is said to be **exothermic;** if heat is absorbed, the reaction is **endothermic.** The net heat change is called the **heat of reaction** and is symbolized by ΔH, the capital Greek delta signifying *change*. The breaking of a chemical bond requires energy. Conversely, the formation of a new bond yields energy. It would seem, then, that if the formation of bonds in the products were to yield a greater amount of energy than that expended in the breaking of bonds in the reactants, that is, if the reaction were **exothermic,** the reaction should proceed spontaneously. This is not the case, however. Consider the following analogy: Your automobile is parked at the top of a steep hill. Its energy is higher than it would be if the automobile were at the bottom of the hill, for if you were to release the brake, the automobile would roll down the hill to a position of lower energy. Thus we say that the *potential energy* of the automobile is greater at the top of the hill than at the bottom.

Now if your automobile were on a plateau just beyond the western side of the continental divide, it would be at a much higher elevation than Kansas City, Missouri; therefore, it would also be in a position of greater potential energy. However, the ride to Kansas City would not be simply one of coasting downhill. Some energy must be expended to climb over the divide as well as to overcome the intervening hills. Thus it is with chemical reactions; for although we will often write a single, simple equation for an organic reaction, in reality the reaction may be quite complex, requiring several steps as the various bonds are broken and new bonds are formed. For example, the reaction we might write as

$$A + B \longrightarrow C + D$$

could involve the following typical two-step sequence:

(1) $A + B \longrightarrow$ Transition State I $\longrightarrow$ Reaction Intermediate
(2) Reaction Intermediate $\longrightarrow$ Transition State II $\longrightarrow C + D$

For such a sequence we can draw a **reaction profile** or **reaction coordinate diagram** of the type shown in Fig. 1.9. The change in the potential energy of the system is plotted (along the ordinate) against the progress of the reaction (along the abscissa) or the **reaction coordinate.** As the reaction begins and the reactants, A and B, come together, bonds *begin* to break or form; and the energy increases to a first maximum at **transition state** I. The energy then falls as the initial breakage and formation of bonds is completed to give a **reaction intermediate.** The energy then rises again, as further bond breakage and formation begins, to a new maximum at transition state II. Finally, the energy falls to the level of that of the products, C and D, as all bond breakage and formation is completed.

In these equations and drawings the transition states will be found at energy *maxima*. Their structures are not well defined because they have some partially formed or partially broken bonds. They exist for only a very short time, and we cannot make physical measurements on them. Reaction intermediates are found at energy *minima* between two transition states. They have definite structures with fully formed bonds. Sometimes they can be isolated, but even if not, they can often be observed by physical methods, especially by spectroscopic techniques (Chap. 18). Not all reaction coordinate diagrams look exactly like Fig. 1.9; some will be simpler and others more complex. Although these diagrams are only qualitative approximations of the courses of chemical reactions, their use makes it easier to understand why many chemical reactions behave as they do. In these diagrams the height of the energy barrier (the ''hump'') represents the **activation energy** or the minimum energy needed to get the reaction to ''go.'' If the activation energy for any step in the reaction is high, at any instant only a few of the many molecules

FIGURE 1.9 Reaction Coordinate Diagram

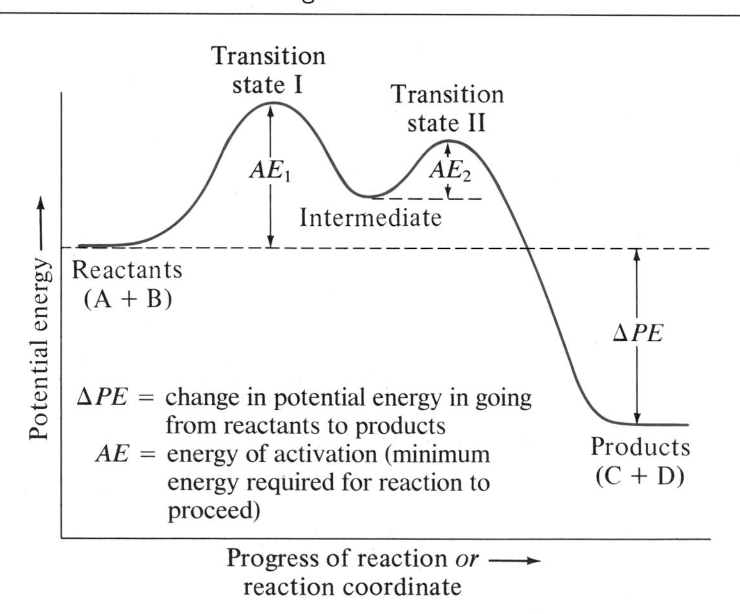

present will have the energy needed to get over the barrier, and the reaction will be slow. Thus the activation energy of a reaction affects its *rate*. The principal reason we heat organic reaction mixtures is to increase the number of molecules with energies greater than the energy barrier, and thereby speed up the reaction. If the activation energy for one step is markedly higher than that for any of the other steps in the overall reaction, then that step not only will be the slowest step but also will tend to determine the rate for the overall reaction. As in a crowded traffic lane, no one moves faster than the slowest car. Thus the step with the highest activation energy is appropriately called the *rate-determining step*.

The detailed description of a reaction in which all transition states and reaction intermediates are specified is called a **reaction mechanism.** In other words, the equation for a reaction tells us *what* takes place, and the reaction mechanism tells us *how* it takes place.

Among the various *reaction intermediates* involved in organic reactions, there are three that are especially important for the chemistry covered in this text: **carbocations, carbanions,** and **free radicals.** Originally, carbocations were called carbonium ions, and this term is still occasionally used today. All three species are very reactive and difficult, if not impossible, to isolate; therefore, they are sometimes referred to as *reactive intermediates*. The relationship between these species is illustrated in the following schematic equation:

$$\underset{\text{Methane}}{CH_4} \xrightarrow{-H^+} \underset{\substack{\text{Methyl}\\ \text{anion}\\ \text{(Carbanion)}}}{H_3C:^-} \xrightarrow{-e} \underset{\substack{\text{Methyl}\\ \text{radical}}}{H_3C\cdot} \xrightarrow{-e} \underset{\substack{\text{Methyl}\\ \text{cation}}}{H_3C^+}$$

Although each of these reactive intermediates has only a transitory existence, some related species of higher molecular weight are of greater stability and have been studied by physical methods.

Both methyl radical and methyl cation are electron deficient in that they lack a full octet of valence electrons, although only the cation carries a positive charge. The state of hybridization in methyl cation is sp^2 (see Sec. 1.7 and Fig. 3.1), precisely what is needed to permit maximum separation of the electrons in the three bonds. Therefore, like most other carbocations, methyl cation is planar. The state of hybridization in methyl radical is less certain, but we can assume that it is approximately sp^2 with the *odd,* unpaired electron in a $2p$ orbital, and that the radical is approximately planar. The negatively charged methyl anion is approximately sp^3 hybridized and has a pyramidal geometry similar to that of ammonia, NH_3:, with the unshared pair of electrons in an sp^3-like orbital. Reactive intermediates of these types are very important in understanding organic reactions and will be encountered frequently in your study of organic chemistry.

In the following drawings we are trying to represent three dimensions in two. For this purpose these conventions are frequently employed in this text: a solid line represents a bond in the plane of the paper; a dotted line represents a bond to an atom behind the plane of the paper; and a wedge represents a bond to an atom in front of the plane of the paper.

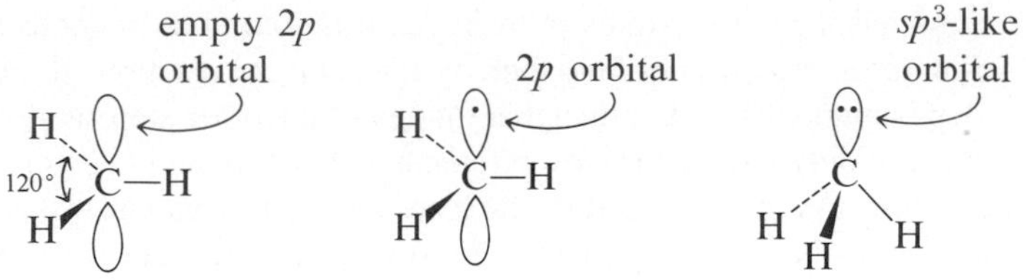

empty $2p$ orbital — $2p$ orbital — sp^3-like orbital

120°

Methyl cation — Methyl radical — Methyl anion

1.13 Acids and Bases. Electrophiles and Nucleophiles

Examination of the presently accepted mechanisms of organic reactions shows that many of them are best described as acid–base reactions. Since acids and bases can be defined in several ways, we will review the definitions of interest to us.

As defined by Brønsted and Lowry, an acid is a substance that supplies protons to a base, and a base is a substance that accepts protons from an acid. Examples of Brønsted–Lowry acids that are familiar to you include the strong mineral acids hydrochloric acid (HCl) and sulfuric acid (H_2SO_4), as well as the weak organic acid, acetic acid ($CH_3—CO_2H$). Typical Brønsted–Lowry bases are the strong base, hydroxide ion (OH^-), and the weak base, ammonia (NH_3). The strengths of Brønsted–Lowry acids and bases are measured by the position of the equilibrium:

$$\underset{\text{acid}}{HA} + \underset{\text{base}}{B:} \overset{K}{\rightleftharpoons} \underset{\text{conjugate base of HA}}{A:^-} + \underset{\text{conjugate acid of B:}}{BH^+}$$

An acid–base pair related through the loss or gain of a proton is called a conjugate pair (L. *conjugare,* to unite). Thus the conjugate acid and conjugate base in the above equilibrium are the products resulting from the proton exchange, and K is the equilibrium constant. When we compare the strengths of organic acids and bases, we will use the Brønsted–Lowry definition. However, the Brønsted–Lowry concept of acid–base behavior as one of proton transfer is not as useful in understanding organic chemical behavior as is the more general definition of acids and bases proposed by G. N. Lewis. The Lewis definition of acids and bases is based on the *sharing of an electron pair* between acid and base. A Lewis acid is a species that accepts an electron pair from a Lewis base, and a Lewis base is a species that donates an electron pair to a Lewis acid. Typical Lewis acids are the proton (H^+), silver ion (Ag^+), boron trifluoride (BF_3), aluminum chloride ($AlCl_3$), and carbocations (such as $CH_3{}^+$)—that is, positive ions and species containing atoms with unfilled octets or electron-deficient atoms. Lewis bases are essentially the same as Brønsted–Lowry bases. Typical Lewis bases are hydroxide ion (OH^-), ammonia

(:NH_3), chloride ion (Cl^-), water (H_2O), and carbanions (such as $CH_3{:}^-$)—that is, negative ions, species having unshared electron pairs, or in some cases, species with loosely bound shared pairs. A typical Lewis acid–base reaction is the formation of the boron trifluoride–ammonia complex (Sec. 1.5).

In organic chemistry reagents that are electron-seeking (Lewis acids) and can form a covalent bond by accepting an electron pair *from a carbon atom* are called **electrophiles** (electron-loving). Reagents that are electron-donating (Lewis bases) and can form a covalent bond to carbon by donating a pair of electrons *to a carbon atom* are called **nucleophiles** (nucleus-loving). It is important not to confuse the terms basicity and nucleophilicity. As stated above, basicity is measured by the position of an *equilibrium* and is the affinity of a base for a proton. **Nucleophilicity** is the affinity of a base for a carbon atom and is measured by the *rate* of a reaction in which the nucleophile forms a bond to a carbon atom in a standard substrate.[6]

Exercise 1.14 Classify the following species as Lewis acids or Lewis bases:

(a) $:\ddot{\underset{..}{F}}:^-$

(b) $CH_3:\ddot{\underset{..}{O}}:CH_3$

(c) $:\ddot{\underset{..}{Br}}^+$

(d) $H:\underset{\underset{H}{..}}{C}::\underset{\underset{H}{..}}{C}:H$

(e) $AlCl_3$

(f) HS^-

(g) CN^-

(h) $C_2H_5\ddot{\underset{..}{O}}H$

1.14 Resonance Structures

There are many molecules, ions, and radicals whose physical and chemical properties are not satisfactorily represented by a single Lewis structure. A familiar example is the nitrate ion, NO_3^-.

$$\left[{}^-:\ddot{\underset{..}{O}}-\overset{+}{N}(=\ddot{O}:)(-\ddot{\underset{..}{O}}:^-) \longleftrightarrow {}^-:\ddot{\underset{..}{O}}-\overset{+}{N}(-\ddot{\underset{..}{O}}:^-)(=\ddot{O}:) \longleftrightarrow \ddot{O}=\overset{+}{N}(-\ddot{\underset{..}{O}}:^-)(-\ddot{\underset{..}{O}}:^-) \right] \equiv \overset{2/3-}{O}\!=\!=\!\overset{+}{N}(\overset{2/3-}{O})(\overset{2/3-}{O})$$

I II III IV

[6] The term *substrate* refers to the organic compound that is undergoing a reaction with a given reagent.

As shown, we can draw three equivalent Lewis structures with two negatively charged, singly bonded oxygen atoms and a third doubly bonded oxygen. Since all oxygen atoms are alike, one is no more deserving of a negative charge than any other. Therefore, in the actual ion the negative charges are distributed equally over *all* oxygen atoms—that is, there is a unit positive charge on nitrogen and two-thirds of a negative charge on each oxygen. This fractional charge distribution cannot be shown by a single Lewis structure; therefore, we draw structure IV as an approximation of the actual structure of the nitrate ion. The broken lines represent a fractional bond. However, structures such as IV are not as easy to use as Lewis structures in most chemical applications, and for convenience we would represent the nitrate ion as I, II, or III. In general, whenever two or more Lewis structures can be written for a molecule, ion, or radical that differ only in the distribution of electrons, the structure of that species *may* not be satisfactorily represented by any one of the Lewis structures, but may be better described as a **resonance hybrid** of them all. In such situations all of the reasonable Lewis structures (called **contributing forms**) are written as shown, connected with a *double-headed* arrow and enclosed in a bracket. The meaning of this symbolism [$\leftrightarrow$] is that the actual structure is a hybrid, or some blend or mix, of all the structures shown. It is *not* to be confused with an equilibrium ($\rightleftarrows$) between several independently existing species. None of the species I, II, or III exists—what exists is a species intermediate among them all.

The reason that the resonance hybrid exists rather than an equilibrium mixture of I, II, and III is that the hybrid is of lower energy and more stable than any of the contributing forms. A factor contributing to the increased stability of many hybrids is the dispersal or distribution of electron surpluses or deficiencies over as many atoms as are able to accept them. In the hybrid nitrate ion, the net negative charge is distributed over all three electronegative oxygen atoms, rather than only two.

Resonance structures will be used frequently in this text; therefore, a few rules will be given to guide you in drawing such structures.

1. The positions of the *nuclei* of the atoms must be the same in all contributing forms. The only changes are in electron assignments. Bond lengths and bond angles must be the same.

2. All resonance structures must have the same number of *paired* electrons.

3. In deciding which of two or more resonance structures is the most important—that is, the most stable—the following guidelines should be applied *in the order given*.
   (a) Structures with filled octets in second-row atoms are more important than those with unfilled octets. This is true even if it causes a positive charge to appear on an electronegative atom.
   (b) Structures of neutral molecules with no separation or minimum separation of charge are more important than those with considerable charge separation. Charge separation is more favorable when negative charges are on electronegative elements, and positive charges are on electropositive elements.
   (c) A structure with a greater number of bonds is usually more important than one with a lesser number.

4. Resonance stabilization of a species is greatest when there are two or more *equivalent* structures that can be said to be of greatest importance—that is, of lowest energy. Conversely, a single Lewis structure is least likely to be satisfactory when two or more equivalent structures of lowest energy can be written.

The application of these rules may be illustrated with the following examples of species you will meet later in the text: (1) ethene, (2) benzene, (3) allyl cation.

	More important		Less (or equally) important	Comment
(1)	$H_2C{=}CH_2$	⟷	$H_2\overset{+}{C}{-}\ddot{C}^{-}H_2$	Charge separation on atoms of same electronegativity. Resonance is not important.
(2)	C_6H_6 (benzene, Kekulé structure)	⟷	C_6H_6 (benzene, Kekulé structure)	No charge separation. Two forms have same structure, are equivalent. Resonance is very important.
(3)	$H_2C{=}CH{-}\overset{+}{C}H_2$	⟷	$H_2\overset{+}{C}{-}CH{=}CH_2$	No charge separation. Two forms have same structure, are equivalent. Resonance is very important.

Note the use of the small curved arrows. These curved arrows are employed in two ways: as sort of a bookkeeping device to keep track of electrons as we construct resonance structures, and as a means of showing the general flow of electrons in some reaction mechanisms.

To save time and conserve space we will not write all the resonance forms of a species unless it is necessary to do so. Instead we will use the most important resonance structure to represent the structure of the species.

Summary

1. Bonding in molecules will be such that the molecule has the lowest overall energy; i.e., the maximum number of bonds will form, the strongest possible bonds will form, and the arrangement of atoms will be such as to minimize repulsions among bond pairs and unshared pairs of electrons.
2. The rules for covalent bonding are: (a) ionic bonding is not possible, (b) the atomic orbitals on the bonded atoms must overlap, (c) two electrons are available to form the pair bond (at least in organic chemistry), (d) for molecules using only *s* and *p* orbitals in bonding there will be no more than eight electrons in the valence shell (exceptions are sulfur and phosphorus, which may use *d* orbitals).
3. Carbon has four electrons available for covalent bonding and has a great tendency to bond to itself in straight-chain, branched-chain, and cyclic structures.
4. When carbon is bonded to four other atoms or groups, it usually does so through four sp^3-hybridized orbitals. The bonds so formed are directed to the vertices of a tetrahedron.
5. Compounds having the same molecular formula, but different structures, are isomers.
6. Organic reactions nearly always involve some structural feature known as a functional group.
7. A reaction mechanism is a detailed description of a reaction that includes all transition states and intermediates.
8. Carbocations, carbanions, and free radicals are important intermediates in organic reactions.
9. Electrophiles are reagents that can form a covalent bond by accepting an electron pair from a carbon atom. Nucleophiles are reagents that can form a covalent bond by donating a pair of electrons to a carbon atom.
10. Whenever two or more Lewis structures may be written for a molecule, ion, or radical, the actual structure may be a resonance hybrid of all contributing structures.

Supplementary Exercises

1.15 Using dots, small circles, or asterisks as a bookkeeping device simply to show the source of electrons, write Lewis structures for the following. Each atom (except hydrogen) should have a complete octet of electrons.

(a) H_2S
(b) N_2O_5
(c) ethane, C_2H_6
(d) methyl alcohol
(e) formaldehyde, CH_2O
(f) chloroform, $CHCl_3$
(g) formic acid, CH_2O_2
(h) dimethyl ether, C_2H_6O
(i) benzene, C_6H_6
(j) methylamine, CH_5N
(k) hydroxylamine, $HONH_2$
(l) hydrogen peroxide, H_2O_2
(m) dichloromethane, CH_2Cl_2
(n) HI
(o) CO
(p) Cl_2

1.16 Draw structures for the following substances in their approximate geometry and classify them as: (1) ionic (2) covalent and polar (3) covalent and nonpolar.

(a) C_2H_6
(b) CH_3Br
(c) CCl_4
(d) IBr
(e) CO_2
(f) C_2H_2
(g) $NaNO_3$
(h) acetone, C_3H_6O
(i) CCl_2F_2
(j) BF_3
(k) CH_3OH
(l) $C_2H_5OC_2H_5$
(m) CH_3Li
(n) $NaBH_4$
(o) KBr

1.17 The number in parentheses accompanying the following is the number of isomeric structures permitted for that molecular formula. Can you draw them?

(a) C_3H_8 (1)
(b) C_3H_7Cl (2)
(c) $C_2H_4Cl_2$ (2)
(d) C_5H_{12} (3)
(e) C_2H_3N (2)
(f) C_3H_7N (6)

1.18 Draw carbon chain skeletons for the nine isomers of C_7H_{16}. Draw four bonds for each carbon, but do not add the hydrogen atoms.

1.19 Predict the molecular geometry of the following molecules.

(a) formaldehyde, CH_2O
(b) acetylene, C_2H_2
(c) nitrate ion, NO_3^-
(d) perchlorate ion, ClO_4^-

1.20 Using the symbol (R—) to represent the hydrocarbon segment(s) attached to the functional group of each example, draw Lewis structures for the following classes of compounds. Place electrons on all atoms comprising the functional group until you have accounted for all valence electrons. (a) an organic acid (b) an aldehyde (c) a ketone (d) an alcohol (e) an ether (f) a primary amine (g) a secondary amine

1.21 Calculate the formal charge on each atom in the following substances.

(a) $CH_3{-}N{\equiv}C{-}\ddot{\underset{..}{O}}:$

(b)
$$\begin{array}{ccccccc} & & H & & :\ddot{O}: & & H & & \\ & & | & & | & & | & & \\ H & - & C & - & \underset{..}{S} & - & C & - & H \\ & & | & & & & | & & \\ & & H & & & & H & & \end{array}$$

(c) $CH_3{-}N{\equiv}C:$

1.22 Draw structures for the contributing forms to the following resonance hybrids. (a) NO_2 (b) SO_2 (c) bicarbonate ion, HCO_3^- (d) acrolein, $H_2C{=}CH{-}CHO$ (e) hydrazoic acid, HN_3 (f) ozone, O_3

1.23 Halothane,

$$\begin{array}{ccccccc} & & F & & H & & \\ & & | & & | & & \\ F & - & C & - & C & - & Br \\ & & | & & | & & \\ & & F & & Cl & & \end{array}$$

has largely replaced diethyl ether as a general anesthetic. Redraw the structure of halothane and, using the crossed arrow convention ($\leftrightarrow$), show the polarity of each bond. Now sever the molecule with a vertical broken line between the carbon atoms and indicate on which "half" the $\delta+$ and $\delta-$ belong.

1.24 The molecular formula, $C_4H_{10}O$, is that for *four* alcohols and *three* ethers. Using C—OH and C—O—C as starting points, draw all seven isomers. Use dashes for shared pairs of electrons, dots for unshared pairs.

1.25 Label the following pairs of structures as isomers or resonance forms.

(a) 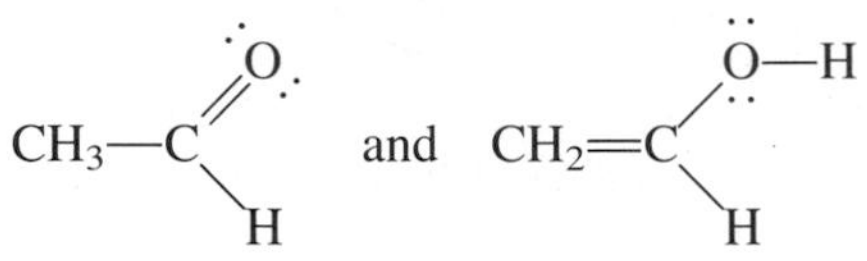 $CH_3—C(=\ddot{O}:)—H$ and $CH_2=C(—\ddot{O}—H)—H$

(b) $CH_3—C(=\!:\!O\!:)—CH_3$ and $CH_3—\overset{+}{C}(—:\ddot{O}:^-)—CH_3$

(c) $H_2C=C(H)—\ddot{Cl}:$ and $H_2\overset{-}{\ddot{C}}—C(H)=\ddot{Cl}^+$

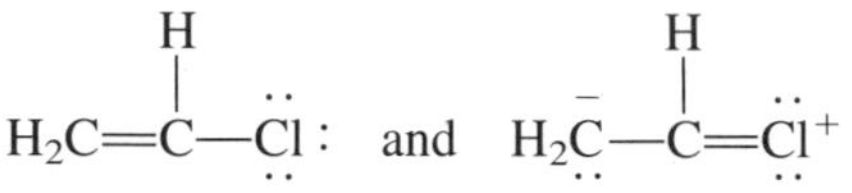

(d) 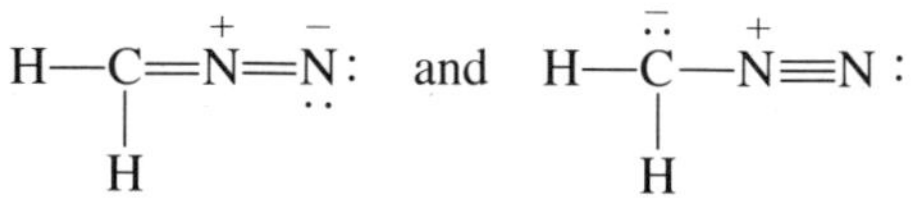 $H—C(H)=\overset{+}{N}=\overset{-}{\ddot{N}}:$ and $H—\overset{-}{\ddot{C}}(H)—\overset{+}{N}\equiv N:$

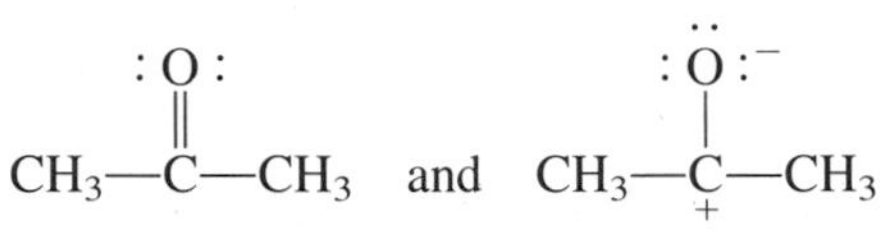

(e) $CH_2=C=CH_2$ and $CH_3—C\equiv C—H$

(f) $CH_3—C\equiv N:$ and $CH_3—\overset{+}{N}\equiv\overset{-}{C}:$

(g) 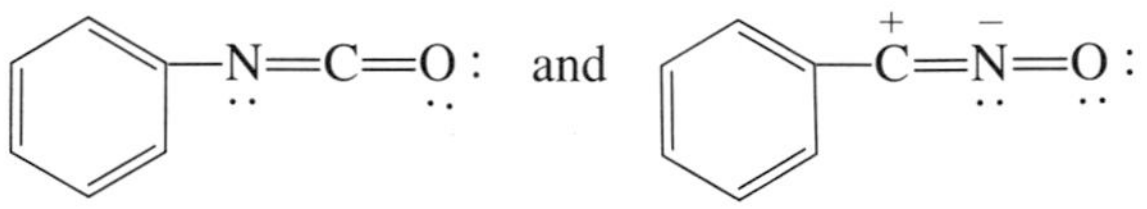 $C_6H_5—\ddot{N}=C=\ddot{O}:$ and $C_6H_5—\overset{+}{C}=\overset{-}{\ddot{N}}=\ddot{O}:$

1.26 Circle each functional group in the following important molecules. Copy the structural formulas; do not mark up your book. These formulas will be used again in later chapters.

(a) 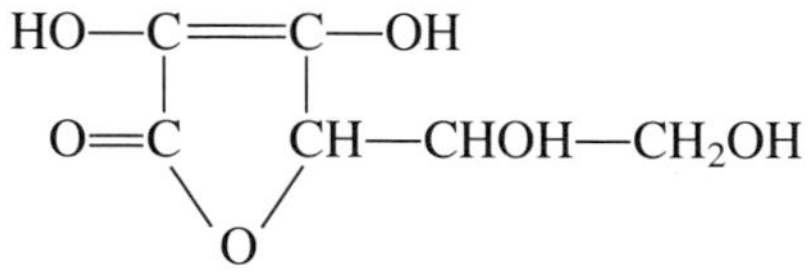 HO—C═C—OH; O═C; CH—CHOH—CH$_2$OH; O

Ascorbic acid
(Vitamin C)

(b) 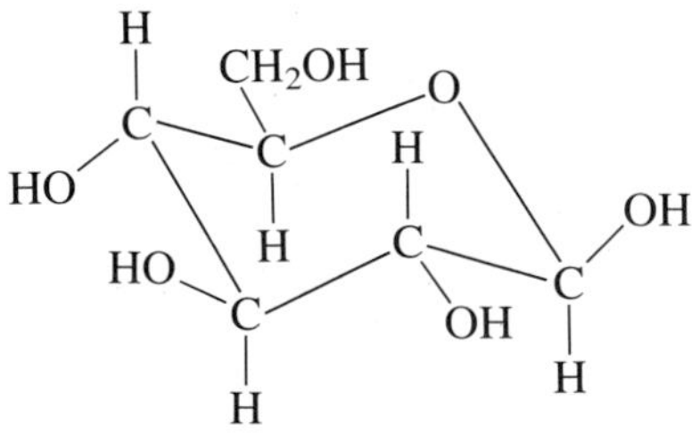 H; CH$_2$OH; O; C; C; H; HO; C; OH; H; HO; C; C; OH; H; H; H

β-D-Glucose
(sugar)

(c)

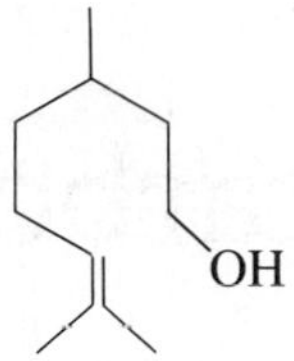

OH

Citronellol
(oil of lemon grass)

(d)

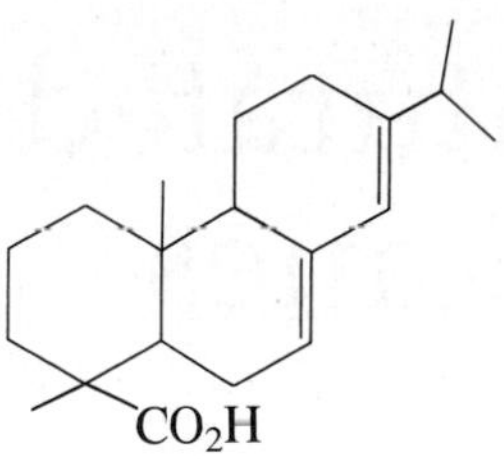

CO_2H

Abietic acid
(pine rosin)

(e)

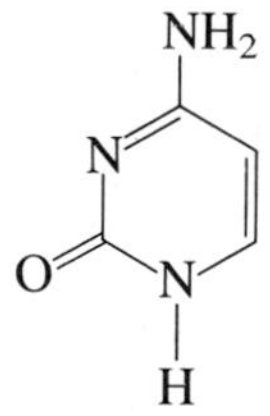

NH_2

N

O

N

H

Cytosine
(found in DNA)

1.27 Expand structures (c) and (d) in Exercise 1.26 into complete structures showing all carbon and hydrogen atoms.

2 Saturated Hydrocarbons: Alkanes

The hydrocarbons, as the name reveals, are compounds containing only carbon and hydrogen. The various structural forms, which combinations of these two elements permit, provide us with the simplest introduction possible to the study of organic chemistry. These useful compounds are isolated from petroleum, our principal fossil fuel, and therefore comprise a great energy source. However, through reactions whereby the hydrogen atoms are substituted by other groups, a large number of other useful compounds result. Thus the hydrocarbons, by name and structure, have a fundamental relationship to other organic substances.

2.1 Structure and Formulas of the Alkanes

The **alkanes,** or **saturated hydrocarbons,** are compounds composed of carbon and hydrogen in which each carbon atom is covalently linked to four other atoms by single electron-pair bonds. The alkanes also are frequently referred to as the **paraffins** (L. *parum,* little; *affinis,* affinity) because of their relative inertness.

The molecular formulas for the alkanes are easily obtained from the general formula C_nH_{2n+2}, where n is the number of carbon atoms present. Each member of a family, or homologous series, differs from its immediate relatives by a methylene group, $—CH_2$. However, the structures for isomers that have the same molecular formula, as we have already noted (Sec. 1.9), are written in different ways. Table 2.1 lists the formulas and structures of the alkanes containing one to five carbon atoms. Higher homologs are listed in Table 2.3.

TABLE 2.1 Formulas of the Alkanes

n	No. of Isomers	Molecular Formula	Projection Formula and Name	Condensed Formula
1	None	CH_4	H | H—C—H | H Methane	CH_4
2	None	C_2H_6	H H | | H—C—C—H | | H H Ethane	CH_3CH_3
3	None	C_3H_8	H H H | | | H—C—C—C—H | | | H H H Propane	$CH_3CH_2CH_3$
4	2	C_4H_{10}	H H H H | | | | H—C—C—C—C—H | | | | H H H H Normal butane	$CH_3CH_2CH_2CH_3$
4	2	C_4H_{10}	H H H | | | H—C——C——C—H | | | H H—C—H H | H Isobutane	CH_3 \\ CH—CH_3 CH_3 /
5	3	C_5H_{12}	H H H H H | | | | | H—C—C—C—C—C—H | | | | | H H H H H Normal pentane	$CH_3(CH_2)_3CH_3$

TABLE 2.1 *Continued*

n	No. of Isomers	Molecular Formula	Projection Formula and Name	Condensed Formula
5	3	C_5H_{12}	$\begin{array}{ccccccccc} & & & \text{H} & & & & & \\ & \text{H} & & \text{H—C—H} & & \text{H} & & \text{H} & \\ \text{H—} & \text{C} & \text{——} & \text{C} & \text{——} & \text{C} & \text{—} & \text{C} & \text{—H} \\ & \text{H} & & \text{H} & & \text{H} & & \text{H} & \end{array}$ Isopentane	$(CH_3)_2CHCH_2CH_3$
5	3	C_5H_{12}	$\begin{array}{ccccccc} & & & \text{H} & & & \\ & \text{H} & & \text{H—C—H} & & \text{H} & \\ \text{H—} & \text{C} & \text{——} & \text{C} & \text{——} & \text{C} & \text{—H} \\ & \text{H} & & \text{H—C—H} & & \text{H} & \\ & & & \text{H} & & & \end{array}$ Neopentane	$C(CH_3)_4$

2.2 Nomenclature of the Alkanes

Names of the first four members of the alkane family have no simple derivation and must be memorized. However, naming the higher members of the series does follow a system, and learning them is relatively easy. The number of carbon atoms present in a continuous chain is indicated by a Greek prefix that is followed by the suffix *-ane:*

Penta = 5; C_5H_{12} is **pentane**

Hexa = 6; C_6H_{14} is **hexane,** etc., as on p. 59, Table 2.3

The unbranched, or continuous, chain is called **normal** and must be indicated in the name as a prefix, *n-,* to differentiate it from its branched-chain isomers:

$$CH_3—CH_2—CH_2—CH_2—CH_3$$

n-Pentane

$$\begin{array}{l} CH_3 \\ \quad \diagdown \\ \quad CH—CH_2—CH_3 \\ \quad \diagup \\ CH_3 \end{array}$$

Isopentane, or
2-Methylbutane

$$\begin{array}{c} CH_3 \\ | \\ CH_3—C—CH_3 \\ | \\ CH_3 \end{array}$$

Neopentane, or
2,2-Dimethylpropane

In the sections that follow we will learn how to name highly branched structures, but before we can do this we must first learn the names of some of the "branches."

2.3 Alkyl Groups

Alkyl groups, as such, have no independent existence but are simply structural derivatives useful in the naming of compounds (Table 2.2).

Removal of one of the hydrogen atoms of an alkane produces an alkyl group. The residual group is then named by replacing the *-ane* suffix of the parent hydrocarbon by *-yl:*

$$\begin{array}{c}\text{H}\\ |\\ \text{H—C—H}\\ |\\ \text{H}\\ \text{Methane}\end{array} \text{ becomes } \begin{array}{c}\text{H}\\ |\\ \text{H—C—}\\ |\\ \text{H}\\ \text{Methyl}\end{array} \text{ or more simply written as } CH_3\text{—}$$

$$\begin{array}{c}\text{H H}\\ |\ \ |\\ \text{H—C—C—H}\\ |\ \ |\\ \text{H H}\\ \text{Ethane}\end{array} \text{ becomes } \begin{array}{c}\text{H H}\\ |\ \ |\\ \text{H—C—C—}\\ |\ \ |\\ \text{H H}\\ \text{Ethyl}\end{array} \text{ or simply } C_2H_5\text{—}$$

For the third member of the series, propane, there is only one structure as a saturated hydrocarbon, but there are two different **propyl groups.** This is possible because the two end carbons are attached to only *one* other carbon whereas the middle carbon is attached to *two* others.

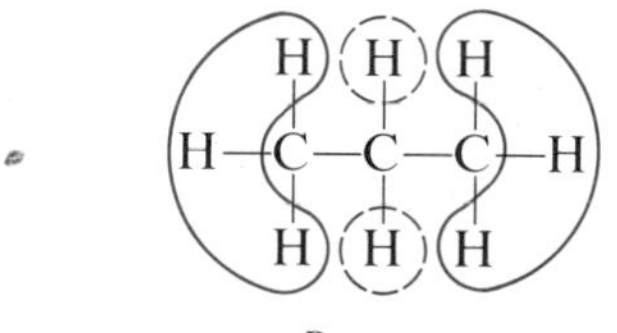

Propane

A carbon bonded to only one other carbon is called **primary,** one bonded to two carbons is called **secondary** (*sec*) and to three carbons, **tertiary** (*tert*). The terms "primary," "secondary," and "tertiary" are sometimes simply designated by the symbols 1°, 2°, and 3°. Hydrogens bonded to these carbons are designated according to the types of carbons to which they are attached. Thus, a primary carbon has bonded to it *three* primary hydrogens and *one* alkyl group. A secondary carbon has *two* secondary hydrogens and *two* alkyl groups. A tertiary carbon is bonded to only *one* tertiary hydrogen, but to *three* alkyl groups.

The six primary hydrogens shown bonded by solid lines in the above formula for propane are all equivalent. Removal of any one of these six hydrogen atoms gives the **normal propyl,** or *n*-propyl, group:

$$\begin{array}{c}\text{H H H}\\ |\ \ |\ \ |\\ \text{H—C—C—C—}\\ |\ \ |\ \ |\\ \text{H H H}\end{array} \text{ or } CH_3\text{—}CH_2\text{—}CH_2\text{— or } n\text{-}C_3H_7\text{—}$$

TABLE 2.2 Commonly Used Alkyl Groups

Hydrocarbon	Group	Example of Common Usage
H | H—C—H | H Methane	H | H—C— | H Methyl	CH_3I Methyl iodide (Iodomethane)
H H | | H—C—C—H | | H H Ethane	CH_3—CH_2— Ethyl	C_2H_5OH Ethyl alcohol (Ethanol)
H H H | | | H—C—C—C—H | | | H H H Propane	CH_3—CH_2—CH_2— *n*-Propyl	$CH_3CH_2CH_2Br$ *n*-Propyl bromide (1-Bromopropane)
	H | CH_3—C—CH_3 | Isopropyl	H | CH_3—C—CH_3 | OH Isopropyl alcohol (2-Propanol)
H H H H | | | | H—C—C—C—C—H | | | | H H H H *n*-Butane	$CH_3CH_2CH_2CH_2$— *n*-Butyl	$CH_3CH_2CH_2CH_2$—Cl *n*-Butyl chloride (1-Chlorobutane)
	CH_3—CH_2—CH—CH_3 | *sec*-Butyl	CH_3—CH_2—CH—Br | CH_3 *sec*-Butyl bromide (2-Bromobutane)
H H H | | | H—C—C—C—H | | | H | H CH_3 Isobutane	CH_3—CH—CH_2— | CH_3 Isobutyl	CH_3—CH—CH_2—Br | CH_3 Isobutyl bromide (1-Bromo-2-methylpropane)
	CH_3 | CH_3—C— | CH_3 *tert*-Butyl	CH_3 | CH_3—C—Br | CH_3 *tert*-Butyl bromide (2-Bromo-2-methylpropane)

Note that for brevity the normal propyl group can be shown simply as n-C_3H_7—. Removal of one of the secondary hydrogens shown within the broken circles in the formula for propane gives the **isopropyl** group:

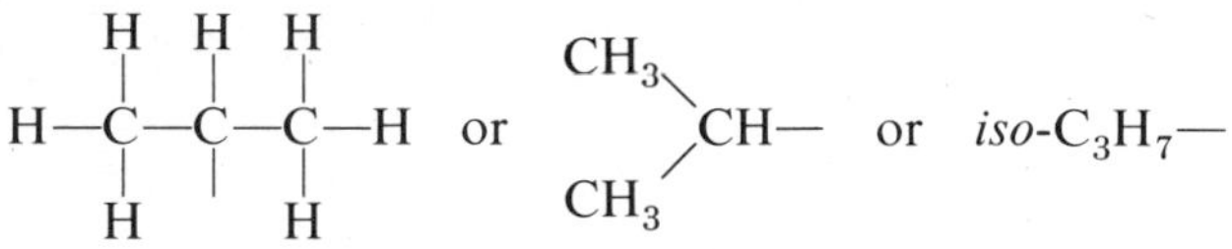

$$H-\underset{H}{\overset{H}{C}}-\overset{H}{C}-\underset{H}{\overset{H}{C}}-H \quad \text{or} \quad (CH_3)_2CH- \quad \text{or} \quad iso\text{-}C_3H_7-$$

Obviously the general formula for an alkyl group is C_nH_{2n+1}. The number of different alkyl groups that can be produced from any alkane depends upon how many different primary, secondary, or tertiary hydrogen atoms can be replaced. We can illustrate this very easily using the two isomeric butanes.

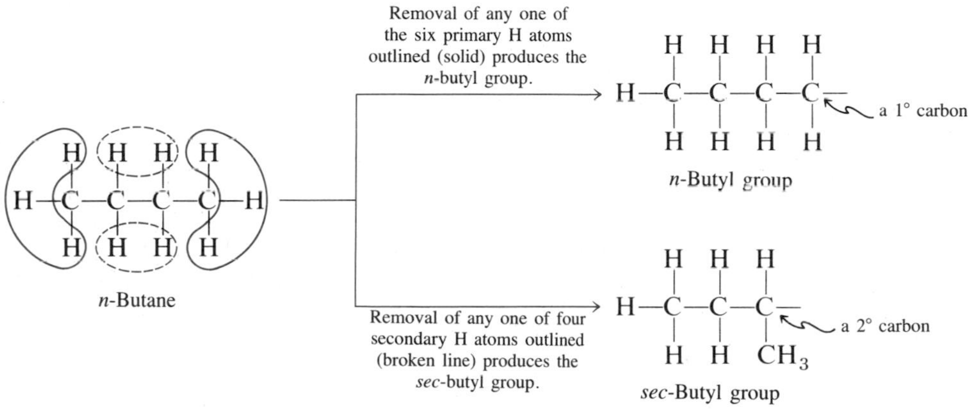

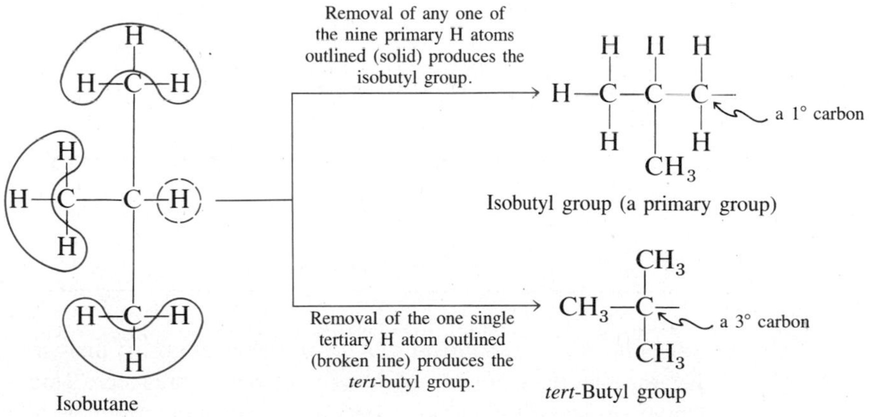

n-Butane

Removal of any one of the six primary H atoms outlined (solid) produces the *n*-butyl group.

n-Butyl group (a 1° carbon)

Removal of any one of four secondary H atoms outlined (broken line) produces the *sec*-butyl group.

sec-Butyl group (a 2° carbon)

Isobutane

Removal of any one of the nine primary H atoms outlined (solid) produces the isobutyl group.

Isobutyl group (a primary group) (a 1° carbon)

Removal of the one single tertiary H atom outlined (broken line) produces the *tert*-butyl group.

tert-Butyl group (a 3° carbon)

An *iso* isomer is one that has a methyl substituent on the next to last carbon atom. The isopropyl group is the only *iso* group that is also a secondary group.

The prefix *neo* is used only for the hydrocarbons neopentane and neohexane and for the **neopentyl** group.

$$\underset{\text{Neopentane}}{CH_3-\overset{\overset{CH_3}{|}}{\underset{\underset{CH_3}{|}}{C}}-CH_3} \qquad \underset{\text{Neohexane}}{CH_3-\overset{\overset{CH_3}{|}}{\underset{\underset{CH_3}{|}}{C}}-CH_2-CH_3} \qquad \underset{\text{Neopentyl group}}{CH_3-\overset{\overset{CH_3}{|}}{\underset{\underset{CH_3}{|}}{C}}-CH_2-}$$

Branched-chain hydrocarbons containing six or more carbon atoms have so many isomers that to attempt to characterize all of them by simple prefixes such as iso, *sec, tert,* and the like is hopeless. The introduction of functional groups into the molecule further complicates any attempts to use such prefixes for any but simple molecules. What is required is a systematic, unambiguous set of nomenclature rules of such power that when any organic molecule, no matter how complex, is named by using the rules, *one and only one* name will result. This ideal has not yet been reached largely because chemists and others either cannot or will not follow the rules in their entirety.

Foibles of human nature aside, the International Union of Pure and Applied Chemistry (IUPAC) has provided us with a system of nomenclature that is accepted by almost all chemists and will give each organic molecule a unique name if the system is properly applied. This system is called the **IUPAC system** (or occasionally the **Geneva system** or **IUC system,** after earlier versions). With the increasing use of computers in preparing indices and searching the chemical literature, some time-honored rules have been changed recently. The use of common names is being increasingly discouraged, although they are still used for relatively simple structures and *must be learned.*

Exercise 2.1 Identify each carbon atom in the structure below as: (a) primary (b) secondary (c) tertiary.

$$CH_3-\overset{\overset{CH_3}{|}}{\underset{\underset{CH_3}{|}}{C}}-\overset{\overset{H}{|}}{\underset{\underset{H}{|}}{C}}-\overset{\overset{H}{|}}{\underset{\underset{CH_3}{|}}{C}}-CH_3$$

2.4 Rules for Naming the Alkanes

1. Select the *longest continuous chain* of carbon atoms. If there are two chains of equal length, take the one with the *larger* number of branches. This is the **parent chain.**
2. Count the number of carbon atoms in the parent chain and give the chain the name of the normal alkane (called the **parent hydrocarbon**) having that number of carbon atoms. The compound will be named as a derivative of the parent hydrocarbon. (*Note:* The prefix *n-* is not used in this system.)

3. Number the parent chain from one end to the other so that the **substituents** (the groups or atoms that have replaced hydrogen in the chain) can be located by number. (See Rule 5.)
4. Arrange all the substituent group names in alphabetical order and place them in front of the parent hydrocarbon name, indicating the *number* of each type of group present by prefixes and the location (position) of each group by an Arabic numeral.

Prefixes:	di = 2	hexa = 6	deca = 10
	tri = 3	hepta = 7	undeca = 11
	tetra = 4	octa = 8	dodeca = 12
	penta = 5	nona = 9	

Ignore the above prefixes, and the prefixes *sec*- and *tert*-, in establishing the alphabetical order (i.e., **e**thyl before di**m**ethyl). The unitalicized prefixes, cyclo-, iso-, and neo-, are considered part of the substituent name and are included in establishing alphabetical order. If two groups are attached to the same carbon atom of the parent chain, repeat the Arabic numeral.
5. The substituents must be given the lowest numbers possible. In comparing two numbering schemes, use the scheme that has the lower *first* number for a substituent. If the first numbers are the same in both schemes, use the lower second number, etc. If the numbers are the same in both schemes, use the scheme that gives the lower first number to the substituent that will appear first in the name (lower in the alphabet).
6. Each homologous series of compounds has a characteristic *suffix*. The suffix for the alkanes is *-ane*.
7. Write the name as one word, separating Arabic numerals by commas and numerals from alkyl or other group names by hyphens.

EXAMPLE

We can illustrate each of the above rules of nomenclature by naming the hydrocarbon whose structure is shown below.

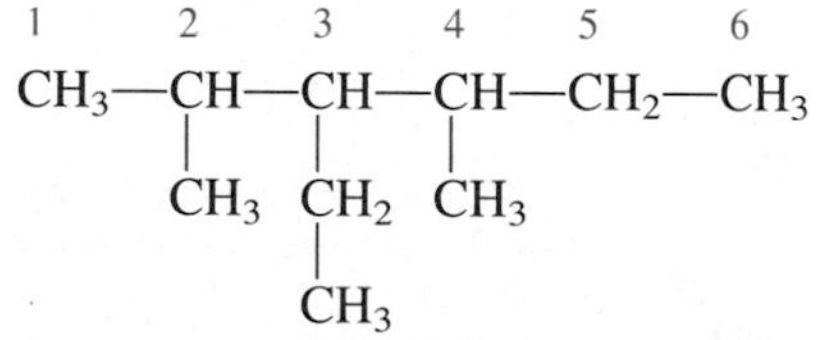

```
 1     2     3     4     5      6
CH3—CH—CH—CH—CH2—CH3
      |     |     |
     CH3  CH2  CH3
            |
           CH3
```

Solution

Rule 1, 2 and 6

On inspection we find that the longest chain is six carbons in length whether we count from left to right, right to left, or begin with the bottom-most carbon, go up and turn a corner. Therefore, the name of the parent chain is **hexane.**

Rule 1, 3 and 5

We number the chain from left to right in order to include the greatest number of substituents and to assign to them the lowest numbers.

Rule 4 and 7

Finally, we name the compound as a hexane with three substituents, two of which are alike. We name and locate the substituents, separating numbers from names with hyphens, and numbers from numbers with commas. According to IUPAC rules then, the structure above is **3-ethyl-2,4-dimethylhexane.**

Exercise 2.2 Are the following alternative numbering schemes acceptable for the illustration example just given? Why?

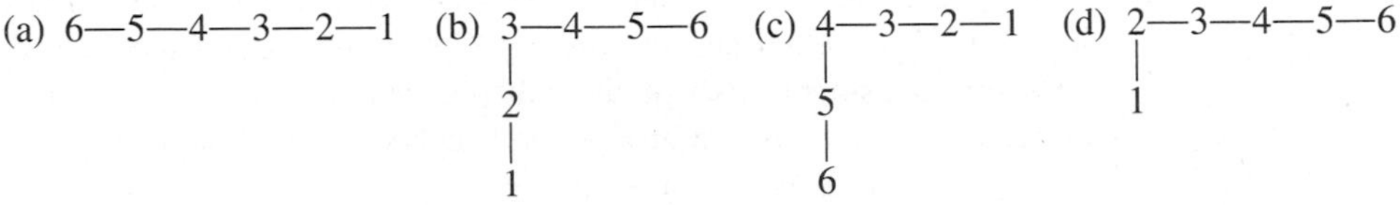

```
(a) 6—5—4—3—2—1   (b) 3—4—5—6   (c) 4—3—2—1   (d) 2—3—4—5—6
                      |             |             |
                      2             5             1
                      |             |
                      1             6
```

Let us apply these rules of nomenclature to a few examples:

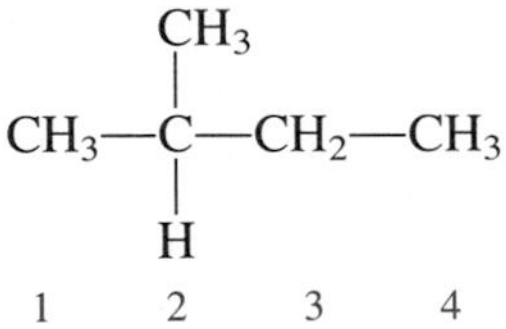

```
       CH3                         CH3       CH3
       |                           |         |
 CH3—C—CH2—CH3              CH3—C—CH2—C—CH3
       |                           |         |
       H                           H         CH3
  1    2    3    4              5    4    3    2   1
```

2-Methylbutane
(Isopentane)

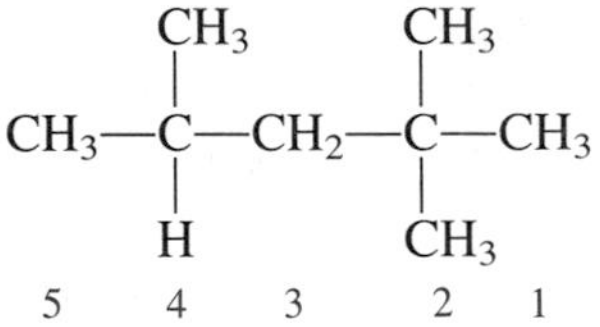

2,2,4-Trimethylpentane
(*Not* 2,4,4-Trimethylpentane)

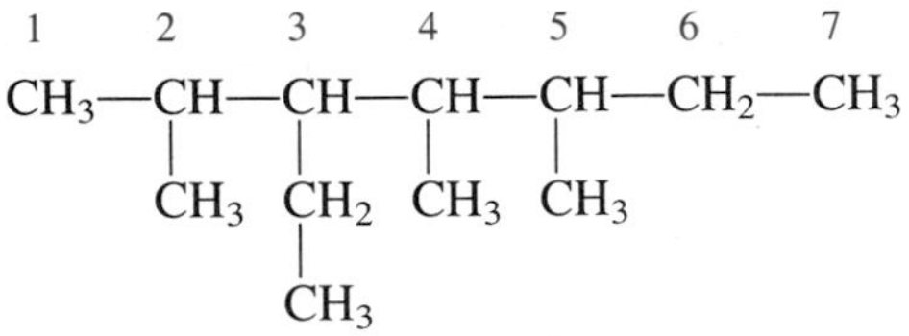

```
  1     2     3     4     5     6     7
 CH3—CH—CH—CH—CH—CH2—CH3
       |     |     |     |
      CH3   CH2   CH3   CH3
             |
            CH3
```

3-Ethyl-2,4,5-trimethylheptane
(*Not* 2,4,5-Trimethyl-3-ethylheptane)

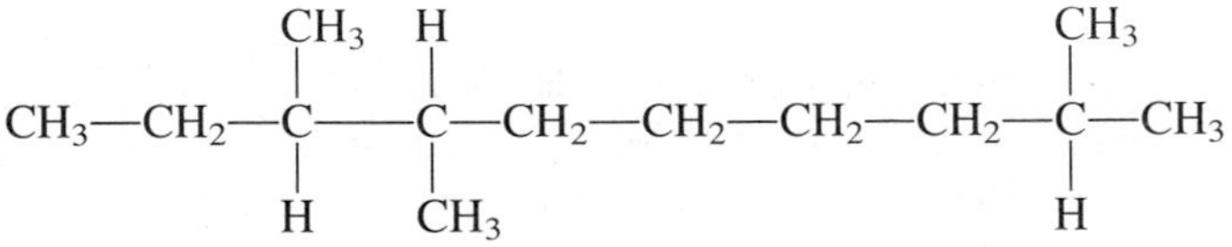

```
              CH3   H                                CH3
              |     |                                |
 CH3—CH2—C——C—CH2—CH2—CH2—CH2—C—CH3
              |     |                                |
              H     CH3                              H
```

2,7,8-Trimethyldecane
(*Not* 3,4,9-Trimethyldecane)

Exercise 2.3 Name the following compound by the IUPAC system.

```
        H
        |
CH3—C—CH2—CH2—CH2—CH3
        |
CH3—C—H
        |
        CH3
```

Exercise 2.4 Expand the following condensed formula into a more workable form and name the structure according to IUPAC rules.

$$(CH_3)_2CH—CH(CH_3)—CH_2—C(C_2H_5)_2CH_3$$

2.5 The Ubiquitous R Group

Many of the compounds we will be discussing in this text will consist of some alkyl group attached to some functional group. In Sec. 1.10 we pointed out that the chemistry of organic compounds is largely the chemistry of functional groups. Thus, much of the time our attention will be focused on the functional group and not on the particular hydrocarbon framework to which the functional group is attached. We need to be able to describe the chemistry of functional groups in general terms that apply to almost any compound containing that functional group. To do this, we will use the standard abbreviation, R—, to represent any monovalent alkyl or cycloalkyl group. Later we will expand the use of R to mean any hydrocarbon group. Of course, the nature of the alkyl group often does affect the chemistry of the attached functional group. Thus, when necessary or desirable, we may put limits on R by simply stating what R may or may not be. Another common generic abbreviation is X—, which we will use to represent any of the halogens (Cl, Br, I, and often, but not always, F). These abbreviations and the names we associate with them are illustrated with the following formulas.

R—	R—H	R—X	R—Cl (R = *tert*)
an alkyl group	an alkane	an alkyl halide	a *tert*-alkyl chloride

Examples of substituents other than (or in addition to) alkyl groups are amino, $—NH_2$; nitro, $—NO_2$; chloro, —Cl; bromo, —Br:

$H_2N—CH_2CH_2—OH$	$CH_3CH_2—NO_2$	$CH_3CH_2CH_2CH_2—Br$
2-Aminoethanol	Nitroethane	1-Bromobutane

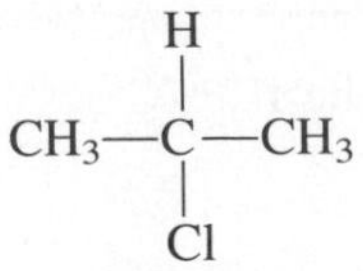

$$CH_3-\underset{\underset{Cl}{|}}{\overset{\overset{H}{|}}{C}}-CH_3$$

2-Chloropropane
(Isopropyl chloride)

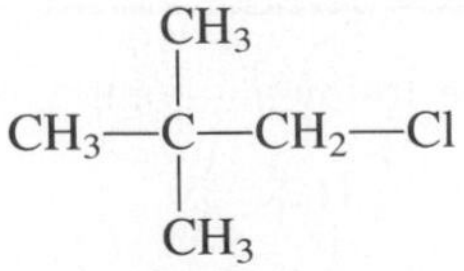

$$CH_3-\underset{\underset{CH_3}{|}}{\overset{\overset{CH_3}{|}}{C}}-CH_2-Cl$$

1-Chloro-2,2-
dimethylpropane
(Neopentyl chloride)

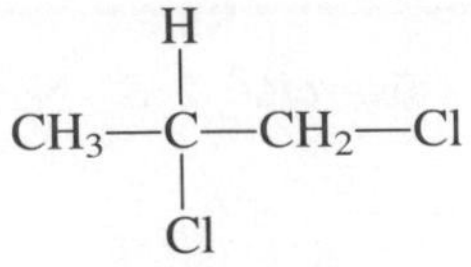

$$CH_3-\underset{\underset{Cl}{|}}{\overset{\overset{H}{|}}{C}}-CH_2-Cl$$

1,2-Dichloropropane
(Propylene chloride)

Exercise 2.5 Abbreviations of many types are used in organic chemistry to simplify the drawing of structural formulas. Abbreviations for all the simple alkyl groups through C_4 are among those most frequently used, and the following abbreviations are almost universally accepted. Draw structural formulas for these groups, using simple logic to deduce what they are.

(a) *t*-Bu	(c) *i*-Pr	(e) Me-	(g) *n*-Pr
(b) Et-	(d) *s*-Bu	(f) *i*-Bu	(h) *n*-Bu

Exercise 2.6 A student named the structure below incorrectly as 1-chloro-2-ethyl-2-methylpropane.

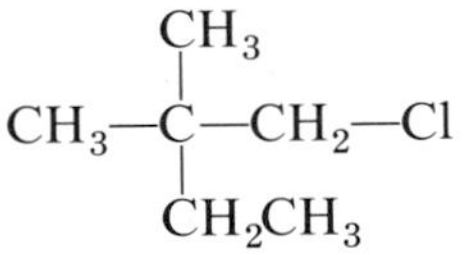

$$CH_3-\underset{\underset{CH_2CH_3}{|}}{\overset{\overset{CH_3}{|}}{C}}-CH_2-Cl$$

Tell why this name is objectionable and assign a correct name to the compound.

2.6 Conformational Isomers

The single bond between carbon atoms in a molecule permits rotation of these atoms as if the shared pair of electrons were a pivot point between two tetrahedra (Fig. 2.1).

Frequently this ability to revolve around the line joining two carbons is called **free rotation.** How free or easy it is depends upon the nature of the atoms or groups occupying the other three corners of each tetrahedron. If these groups are large or bulky, free rotation may be not only inhibited, but prevented entirely. Again, if groups on adjacent carbons are of high electron density such as chlorine or bromine, they will tend to repel each other and will therefore occupy positions as far from each other as possible. The different spatial arrangements made possible by rotation about a single bond are called **conformations,** and a small energy barrier must be overcome in order for adjacent carbon atoms to rotate from one conformation to another. The conformational isomers possible for ethane, and the potential energy changes required when carbon atom 2 is rotated through succes-

FIGURE 2.1 Ball and Stick Model of Ethane Showing Arrangement of Atoms in (a) Staggered and (b) Eclipsed Conformations

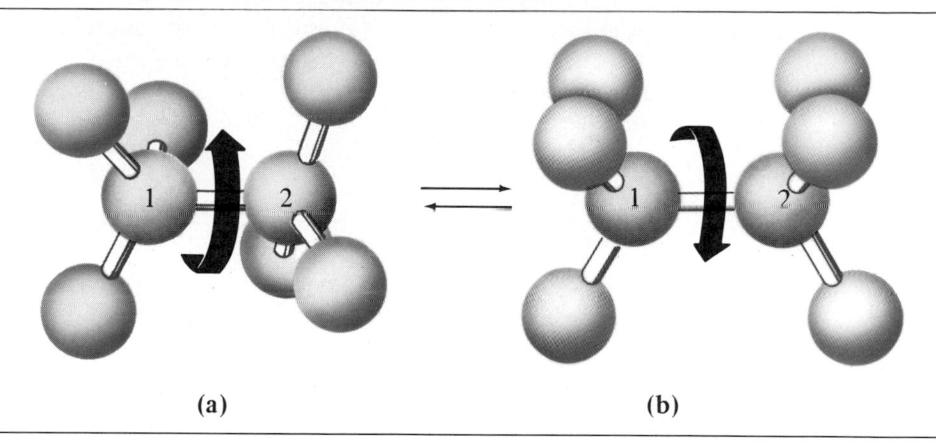

sive 60° angles, may be shown more clearly by using the representations first devised by Professor M. S. Newman of Ohio State University. In the **Newman projections** of the ethane molecule (Fig. 2.2), the viewer is looking along the axis connecting the two carbon

FIGURE 2.2 Potential Energy Changes Required to Rotate Carbon Atoms of Ethane Through Successive 60° Angles

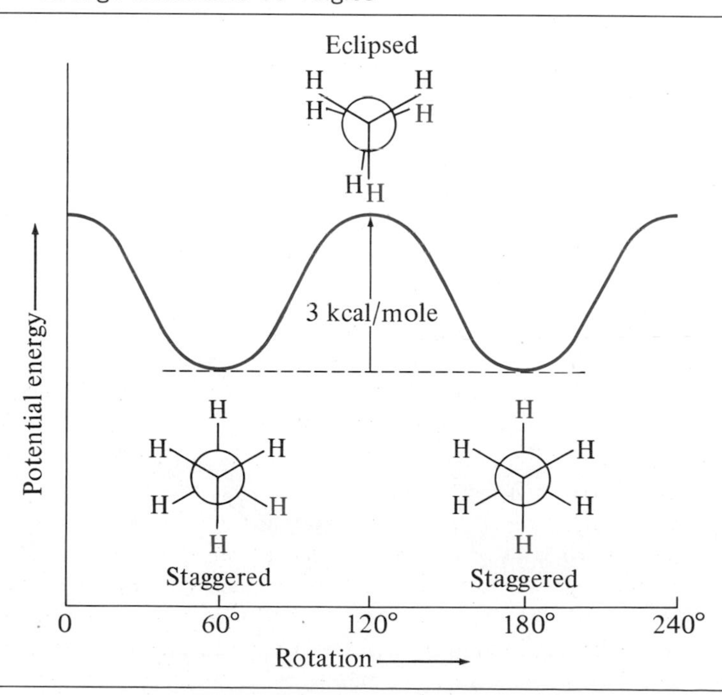

atoms. The conformational isomers or **conformers** of 1,2-dichloroethane are shown in Figure 2.3. Note the three conformations shown and the **dihedral angles** between the chlorine atoms in each conformation: (a) **anti** (180°), (b) **gauche,** or skew (60°), and (c) **eclipsed** (0°). The Newman projections for the three conformations of 1,2-dichloroethane may be drawn as follows:

θ = dihedral angle

anti　　gauche　　eclipsed

The repulsive forces between the chlorine atoms would clearly be the greatest for the eclipsed and least for the anti conformer. Thus the order of stability of the conformations shown is: anti > gauche > eclipsed.

It is important that we distinguish between such structural arrangements made possible by a rotation about bonds and those made possible by breaking and rearranging bonds. Regardless of the number of different conformations a molecule may assume, it still retains the same identity and all its physical and chemical properties. This is not always the case when bonds are broken and rearranged. For example, were we to break the bonds connecting the chlorine and the hydrogen atoms to the number 2 carbon of 2-chlorobutane and exchange the positions of the chlorine and the hydrogen, the result would be a 2-chlorobutane but, as we shall see in Chapter 5, a different 2-chlorobutane from the one we altered.

FIGURE 2.3 Scale Models Showing (a) Anti, (b) Gauche, and (c) Eclipsed Conformations of 1,2-Dichloroethane

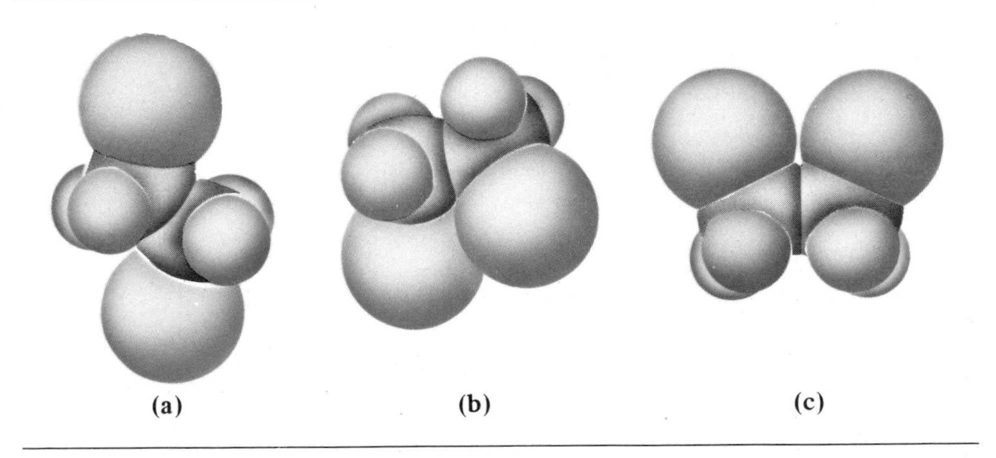

FIGURE 2.4 Ball and Stick Models of *n*-Pentane*

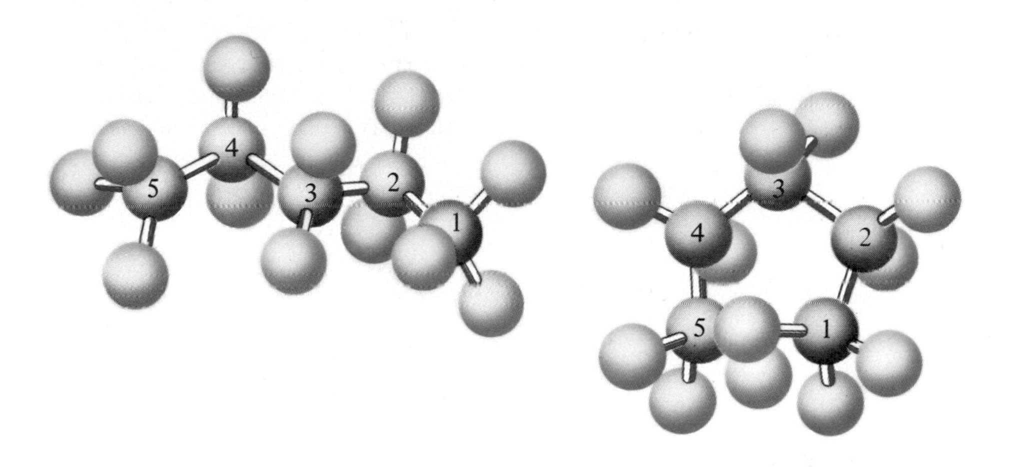

*The model at right shows the "head-to-tail" arrangement possible because of singly-bonded carbons.

Free rotation about singly bonded carbons permits hydrocarbons in a continuous chain to be anything but "straight." They may zigzag, turn corners, spiral, and even cyclize in a head-to-tail arrangement. Note the near cyclic form normal pentane (Fig. 2.4) can assume when carbon atoms 2 and 3 are each rotated through an angle of 109.5°. For this reason we must inspect a carbon chain carefully before numbering it. The longest chain is not necessarily the horizontal one.

Exercise 2.7 As was stated in Section 1.12, conformational isomers are often shown by pseudo-three-dimensional or perspective formulas in which the dotted line is the bond below, the wedge the bond above, and the solid line the bond within the plane of the paper. Consider the structure below a representation of *n*-butane.

CH_3 H
CH_3
H----C—C
H H

Draw both the pseudo-three-dimensional and Newman projections of (a) the anti conformation and (b) the gauche conformation. Which conformation is more likely for *n*-butane? Why?

2.7 Cycloalkanes

The cycloalkanes are saturated hydrocarbons that are ring structures. Much of the chemistry of the cycloalkanes is essentially the same as that of the open-chain alkanes. However, the cyclic nature of these ring compounds confers upon them special properties, isomerism, and nomenclature. Many cyclic structures containing up to thirty or more carbon atoms in a ring have been isolated or synthesized, but we will confine our discussion to those shown below.

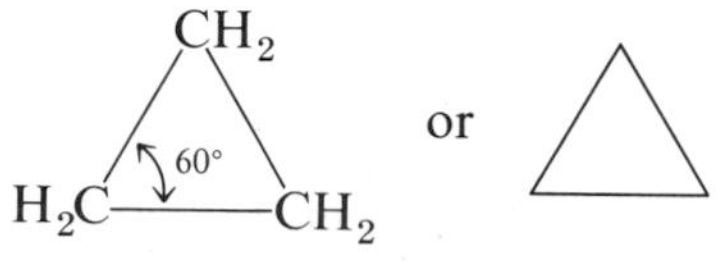

CH_2 60° H_2C CH_2 or

Cyclopropane

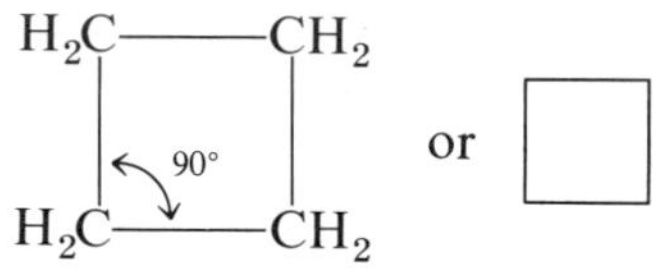

H_2C CH_2 90° H_2C CH_2 or

Cyclobutane

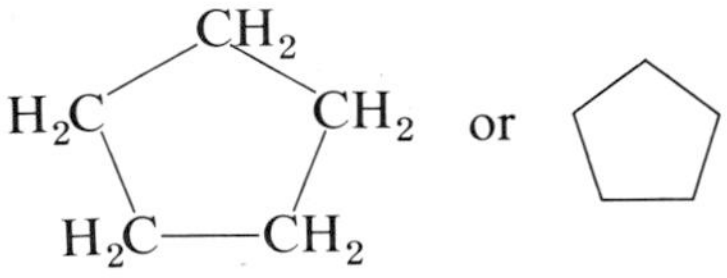

CH_2 H_2C CH_2 H_2C CH_2 or

Cyclopentane

All bond angles = 109.5° or

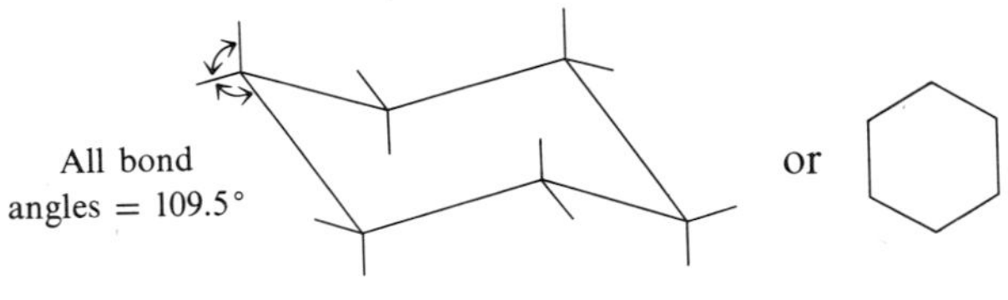

Cyclohexane

Note that in the examples shown the prefix ''cyclo'' becomes one word with the parent hydrocarbon. The parent hydrocarbon in the cycloalkanes is the ring structure; therefore, only the ring carbons are counted to determine the name of the parent cycloalkane. The ring carbon atoms are numbered in such a way as to give the smallest possible numbers to any substituents, following the general procedure used with the alkanes. The names of alkyl substituents are joined to that of the ring to become one word. A few examples will help make this clear.

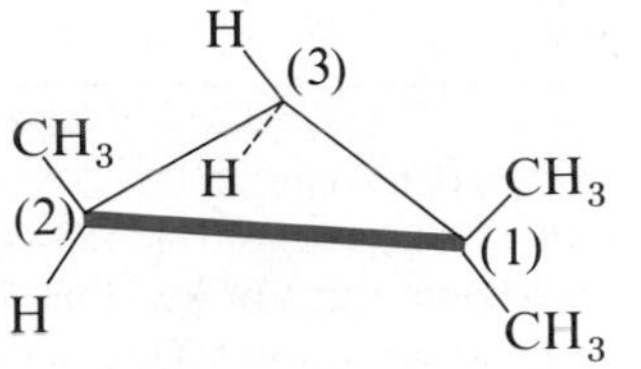

1,1,2-Trimethylcyclopropane 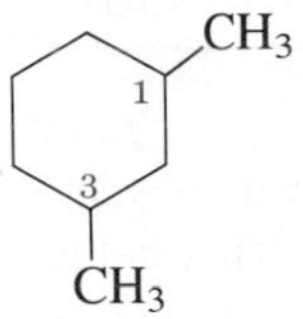 1,3-Dimethylcyclohexane 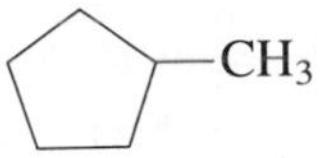 Methylcyclopentane

Although the name 1,3-dimethylcyclohexane, given for the second example above, is correct as it stands, it is nevertheless ambiguous. Whenever there are two or more substituents on different carbons of a cycloalkane ring, a new type of isomerism becomes *possible*, as described in the next section.

Exercise 2.8 The general formula for cycloalkanes is C_nH_{2n} where $n \geq 3$. Draw structures for six different cycloalkanes with the molecular formula C_5H_{10}. If you have difficulty drawing more than five, read the next section and try again.

2.8 *cis,trans*-Isomerism in Cycloalkanes

Cycloalkanes having two or more substituents on the ring may show a form of **stereoisomerism** (Chap. 5) called *cis,trans*-**isomerism.** This form of isomerism in the cycloalkane series is caused by restricted rotation about the single bonds of the ring. The *cis*-isomer has both substituents on the same side of the ring (top or bottom), whereas the *trans*-isomer has one substituent on each side of the ring. This type of isomerism is illustrated with the *cis*- and *trans*-isomers of 1,2-dimethylcyclopropane. In naming such isomers the prefix, *cis* or *trans,* is italicized and placed in front of the cycloalkane name. A hyphen separates *cis* and *trans* from the rest of the name.

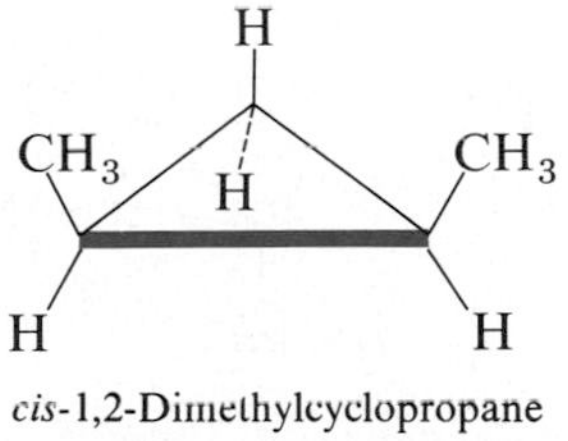

cis-1,2-Dimethylcyclopropane 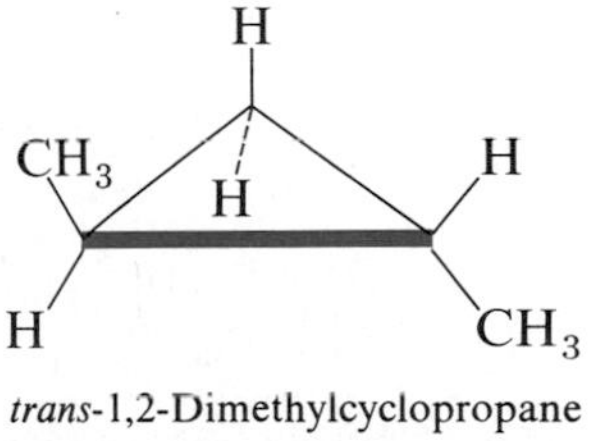 *trans*-1,2-Dimethylcyclopropane

Other examples of this important type of isomerism are given in the following section and in Sec. 5.8.

Exercise 2.9 Draw and name six cyclobutanes of formula C_6H_{12}.

2.9 Ring Structure and Strain Theory

Many of the natural substances with which early chemists worked contained 5- and 6-membered rings. They believed that cycloalkane rings of less than five or more than six carbon atoms would possess too much strain to be stable. This belief was based on the assumption that all carbons within the ring lie in the same plane and that the bond angle deviation from the tetrahedral value of 109.5° would be too great for stability. Although 3-membered rings must be planar, we will see that rings of four or more carbons need not be planar.

In order to bring the terminal carbons of propane together and form a bond between them, the tetrahedral angle of 109.5° between carbon–carbon bonds must be changed to 60°, which is the value for the angles in an equilateral triangle. Each valence bond in cyclopropane is thus distorted from the normal tetrahedral angle by $\frac{1}{2}(109.5°–60°)$ or 24.75°. A bond angle deviation of this magnitude does indeed place the ring structure of cyclopropane under considerable strain. However, cycloalkanes of six or more carbons do not need to have all carbons in the same plane, but can "pucker" or fold slightly to retain the tetrahedral angle. Any **angle strain** within the ring is thus relieved. Even 4- or 5-membered rings show a tendency to pucker in order to provide a greater ring stability, their structures being those affording the greatest relief from angle strain. Therefore, the planar structures that we commonly use to represent the cycloalkanes may not show the true geometry of these cyclic molecules.

The actual geometry and bond angles of cyclobutane, cyclopentane, and cyclohexane are shown below. Each nonplanar form exists in equilibrium with an exactly equivalent conformer.

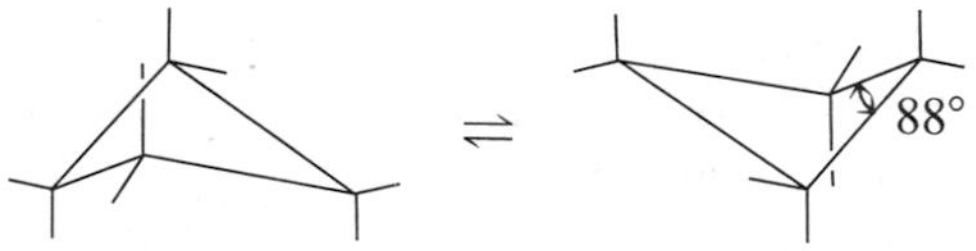

⇌ 88°

Cyclobutane

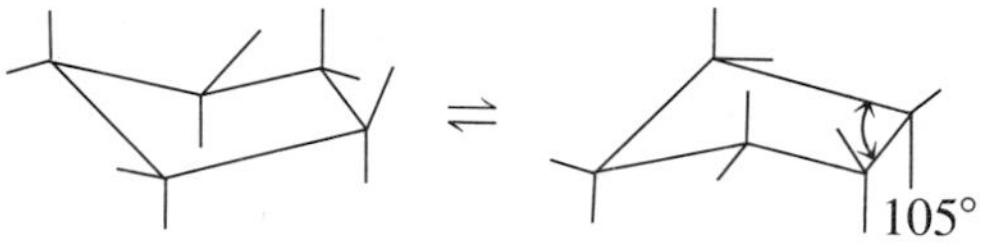

⇌ 105°

Cyclopentane

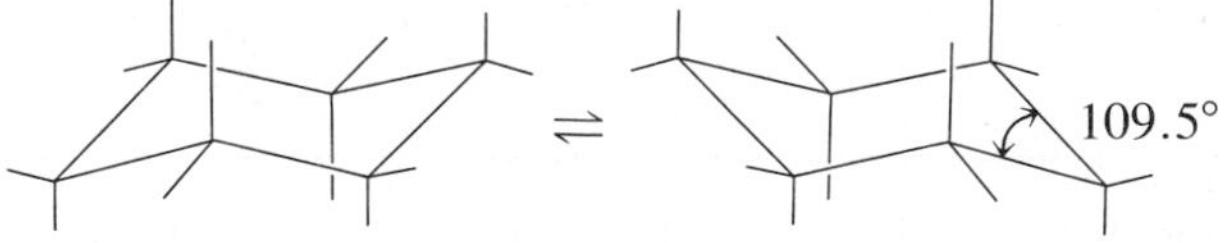

⇌ 109.5°

Cyclohexane

FIGURE 2.5 Boat and Chair Conformations of Cyclohexane.

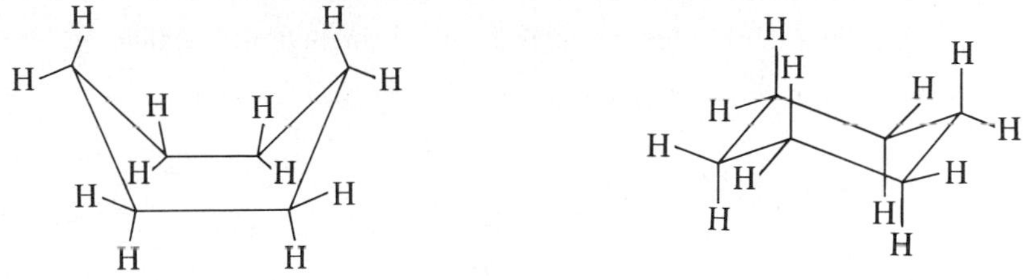

You may question how the puckered rings of cyclobutane and cyclopentane could be more stable than their planar forms when their bond angles show a greater deviation from the tetrahedral value than when their rings have a planar geometry. The reason for this anomaly lies in another type of strain. In the planar forms of both cyclobutane and cyclopentane, there is a total hydrogen eclipsing between each pair of adjacent carbons. This "eclipsing" strain between hydrogens is relieved at the expense of a little angle strain in order to keep the total strain at a minimum.

Inasmuch as 6-membered rings are so prevalent in natural substances, cyclohexane deserves special consideration. Although both **boat** and **chair** conformations are possible for cyclohexane and are free of angle strain, the **chair** form is an essentially strainfree conformation in which all angles are close to tetrahedral, and all hydrogen atoms on adjacent carbons are staggered (Fig. 2.5). In the chair conformation the hydrogen atoms or other substituents will be in either of two distinct positions. In one position, six bonds lie roughly in the "plane" of the ring or around the equator of the molecule, and are called **equatorial bonds.** The other six bonds point in directions parallel to an axis drawn perpendicular to the plane of the ring and are called **axial bonds** (Fig. 2.6). Each carbon atom of cyclohexane has one equatorial and one axial bond. Substituents are called axial or equatorial substituents depending on the positions they occupy in a given conformer. These positions are shown by the letter subscripts on the hydrogen atoms in Fig. 2.6.

FIGURE 2.6 Equatorial and Axial Bonds in Cyclohexane

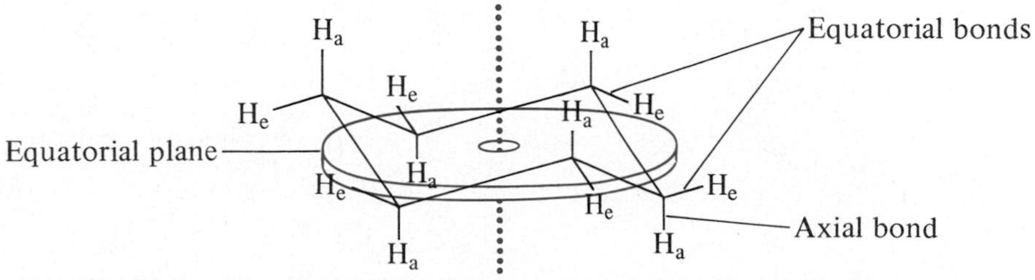

Equatorial bonds

Equatorial plane

Axial bond

EXAMPLE

Draw two structures for 1,*cis*-2,*cis*-4-trimethylcyclohexane.

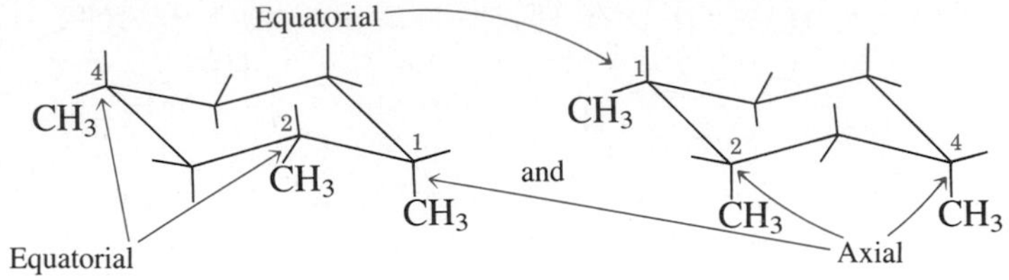

Solution The prefix *cis* in the name requires that all three methyl groups be on the same side of the equatorial plane.

As may be seen from Fig. 2.7, there are two chair conformations of cyclohexane that are exactly equivalent. These two conformations exist in a mobile equilibrium. One conformer becomes the other by the simple expedient of flipping carbon number 1 up and carbon number 4 down. However, you will note that the solid spheres representing the

FIGURE 2.7 The Equilibrium Between Axial and Equatorial Bonds in Cyclohexane.

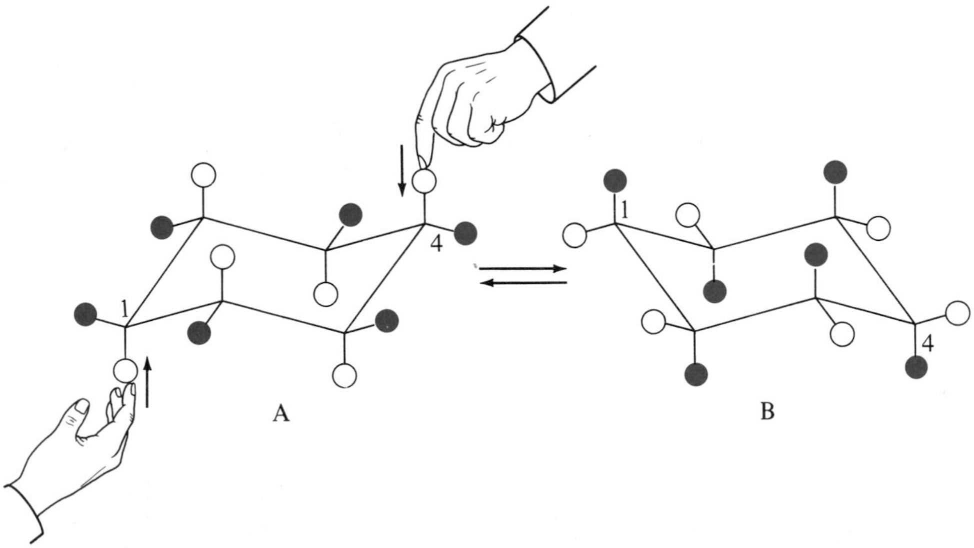

equatorial hydrogens and the open circles representing the axial hydrogens in conformer A will occupy the opposite positions in conformer B.

Substituents on alternate carbons in 1,3-diaxial positions of the cyclohexane ring are relatively close to each other. The 1,3- designation as used here refers to relative positions on the ring, *not* to the loci of substituents. The more stable chair conformation of a cyclohexane derivative, therefore, is usually that with the largest substituents in equatorial positions. Let us illustrate this point by drawing the two chair forms of *trans*-1,2-dichlorocyclohexane. Note that in I there are four unfavorable 1,3-diaxial interactions between H and Cl when the chlorines are on axial bonds, whereas in II the chlorine atoms are farther removed on equatorial bonds and the only 1,3-diaxial interactions are between the relatively smaller hydrogen atoms.

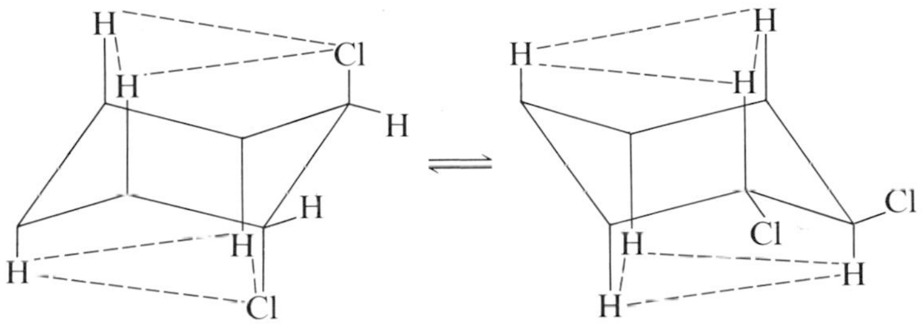

I (less stable)

II (more stable)

1,3-diaxial interactions shown by dotted lines

1,3-Diaxial interactions are perhaps even more clearly evident when structures I and II are represented by Newman projections.

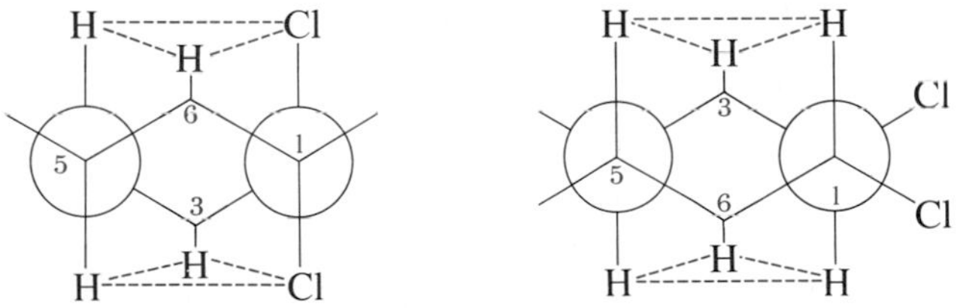

I

II

Exercise 2.10 Explain why the *trans*-isomer of 1,2-dichlorocyclohexane is more stable than the *cis*, whereas the *cis*-isomer of 1,3-dichlorocyclohexane is more stable than the *trans*. (*Hint:* Draw chair forms and try both axial and equatorial bonds for placing substituents.)

2.10 Physical Properties of the Alkanes

The first four members of the alkane family are gases. Those containing five to seventeen carbon atoms are liquids at room temperature, while those members having eighteen or more carbon atoms are solids. In general terms, the boiling points and melting points of organic compounds should be directly related to the interactions among the molecules: the stronger the interactions, the higher the boiling or melting points. For neutral molecules these interactions (the forces holding the molecules together in the liquid or solid state) include van der Waals forces (dispersion forces), dipole-dipole interactions, and hydrogen bonding. For the relatively low-boiling and low-melting alkanes, only the van der Waals forces are important. These forces, although weak and diminishing very rapidly with distance, result from the constant circulation of the electron cloud about the nuclei that creates temporary molecular dipoles (regions of positive and negative charge). The electron motion is correlated in such a way as to keep the electrons as far apart as possible, but there is always a net attraction among the molecules. The strength of this attraction depends largely on the area available for contact with another molecule: small or compact ball-shaped molecules will have little contact; long chains can have considerable contact; and broad, flat molecules will have the greatest contact. Thus boiling points tend to rise with increasing molecular weight (molecular size) and to be lower among branched-chain isomers that offer a lower surface area for contact. The effect of different geometric configurations on boiling points can be illustrated with the following three examples.

$C(CH_3)_4$
Neopentane
(bp = 9.45°, a gas)

$H_3C{-}CH_2{-}CH_2{-}CH_2{-}CH_3$
n-Pentane
(bp = 36°)

$(CH_2)_5$ (ring)
Cyclopentane
(bp = 49°)

All hydrocarbons are insoluble in water but dissolve easily in most organic solvents. They are colorless and tasteless. When pure, they are odorless. The odor of natural gas is not that of methane, its principal constituent, but that of methyl sulfide, $(CH_3)_2S$, a contaminant purposely added to allow for the detection of leaks. Table 2.3 summarizes the properties of some representative members of the alkane series.

Exercise 2.11 Boiling points of 99°, 118°, and 107° are those for the following isomers: 2,2,4-trimethylpentane, 4-methylheptane, and 2,2-dimethylhexane. Which boiling point is the property of each?

TABLE 2.3 Physical Properties of Some Normal Saturated Hydrocarbons

Name	Formula	mp, °C	bp, °C	Specific Gravity	Normal State
Methane	CH_4	−182.6	−161.4	—	gas
Ethane	C_2H_6	−183.3	−88.6	—	gas
Propane	C_3H_8	−189.7	−42.1	0.5005	gas
n-Butane	C_4H_{10}	−135.4	−0.5	0.5788	gas
n-Pentane	C_5H_{12}	−129.7	36.1	0.6262	liquid
n-Hexane	C_6H_{14}	−95	69.0	0.6603	liquid
n-Heptane	C_7H_{16}	−90.6	98.4	0.6837	liquid
n-Octane	C_8H_{18}	−56.8	125.6	0.7025	liquid
n-Decane	$C_{10}H_{22}$	−29.7	174.1	0.7300	liquid
n-Pentadecane	$C_{15}H_{32}$	10	270.6	0.7685	liquid
n-Octadecane	$C_{18}H_{38}$	28.2	316.1	0.7768	solid
Eicosane	$C_{20}H_{42}$	36.8	343	0.7886	solid

2.11 Chemical Properties of the Alkanes

The only reactions the saturated hydrocarbons are capable of undergoing without a rupture of carbon–carbon bonds are reactions in which hydrogen of the parent hydrocarbon is replaced by some other atom or group. Such replacements are called **substitution reactions.** Generally, the hydrocarbons are unreactive toward most reagents but are attacked by halogens and oxygen under certain conditions. Reactions with halogens are substitution reactions; attack by oxygen results in the breaking of carbon–carbon and carbon–hydrogen bonds. When alkanes are heated to extreme temperatures in the absence of air (a treatment referred to as *pyrolysis,* or *cracking*), the carbon–carbon and carbon–hydrogen bonds are also ruptured to give a variety of fragmentary products. Catalytic cracking of long-chain hydrocarbons as carried out in the petroleum industry makes possible the production of great quantities of motor fuel.

A. Oxidation (Combustion). Alkanes are resistant to ordinary chemical oxidizing reagents such as potassium permanganate, $KMnO_4$, and potassium dichromate, $K_2Cr_2O_7$, but may be oxidized to carbon dioxide and water when ignited in the presence of an excess of oxygen. Great quantities of heat energy called *heats of combustion* are released per mole of hydrocarbon burned, and it is this reaction that makes hydrocarbons so useful as fuels. As may be seen from the following three reaction equations, each CH_2 increment in the hydrocarbon chain adds approximately 156 kcal to the heat of combustion.

$$CH_4 + 2\ O_2 \longrightarrow CO_2 + 2\ H_2O + 211\ \text{kcal}$$

$$C_2H_6 + 3\tfrac{1}{2}\ O_2 \longrightarrow 2\ CO_2 + 3\ H_2O + 368\ \text{kcal}$$

$$C_3H_8 + 5\ O_2 \longrightarrow 3\ CO_2 + 4\ H_2O + 526\ \text{kcal}$$

Exercise 2.12 The boiling point of *n*-pentane is 36.2°, and the heat of combustion *for the liquid* is given in the literature as 833 kcal/mole. Why is this value *less* than that obtained by the simple addition of CH_2 heat equivalents to any of the values given in the preceding three examples?

Incomplete combustion of alkanes results in the formation of carbon (soot) and the very dangerous carbon monoxide.

$$2\ CH_4 + 3\ O_2 \longrightarrow 2\ CO + 4\ H_2O$$

$$CH_4 + O_2 \longrightarrow C + 2\ H_2O$$

Although both of the preceding reactions should be avoided if we are to heat our homes safely and economically, the reaction that produces soot (carbon black) is of vital importance to our tire and rubber industry.

B. Substitution (Halogenation). At ordinary temperatures and in the absence of light, chlorine does not react with saturated hydrocarbons. At elevated temperatures and in the presence of sunlight or ultraviolet light, the hydrogen atoms are replaced by one or more chlorine atoms. A substitution of hydrogen by chlorine is know as a **chlorination** reaction. The reaction pathway by which chlorine substitutes for hydrogen in a hydrocarbon has been well established as one involving the formation of **free radicals.** Recall that a free radical is an atom or group having an odd unpaired electron and is represented by the presence of a single dot on the symbol for the atom or group, for example, R· and Cl·. Often when free radicals react with organic compounds, one product of the reaction is a new free radical. Thus, the reaction may be a self-propagating **chain reaction** that continues until all materials are consumed or until two free radicals meet and unite. A reaction pathway followed in this manner is referred to as a **free-radical chain mechanism.** The first steps in the chlorination of methane may be shown as follows:

Chain initiation step

$$Cl:Cl \xrightarrow{\text{sunlight}} 2\ Cl\cdot$$

Chain propagation steps (repeated over and over)

STEP 1

$$H{-}\overset{\displaystyle H}{\underset{\displaystyle H}{\overset{|}{\underset{|}{C}}}}{-}H + Cl\cdot \longrightarrow HCl + H{-}\overset{\displaystyle H}{\underset{\displaystyle H}{\overset{|}{\underset{|}{C}}}}\cdot$$

Methyl radical

STEP 2

$$H{-}\overset{\displaystyle H}{\underset{\displaystyle H}{\overset{|}{\underset{|}{C}}}}\cdot + Cl_2 \longrightarrow H{-}\overset{\displaystyle H}{\underset{\displaystyle H}{\overset{|}{\underset{|}{C}}}}{-}Cl + Cl\cdot$$

Chloromethane (Methyl chloride)

Note that one of the products in Step 2 (the chlorine radical $Cl\cdot$) is also one of the reactants in Step 1. Therefore, the two steps are repeated over and over (1, 2, 1, 2, 1, 2, 1, 2, . . .) until something happens to break the chain. Usually, this *chain termination step* involves the combining of two free radicals without forming the new free radical needed to continue the chain.

Chain termination

$$CH_3\cdot + CH_3\cdot \longrightarrow CH_3 \quad CH_3$$

We will now consider the reaction coordinate diagrams (Sec. 1.11) for the chlorination of methane (Fig. 2.8) to show how these may be used to visualize a reaction sequence. Recall that the rate of reaction depends on the activation energy (Sec. 1.1). From the diagrams we see clearly that the first step is endothermic and that both the rate and ΔH of the reverse reaction ($CH_3\cdot + HCl \rightarrow CH_4 + Cl\cdot$) are more favorable than those of the forward reaction. However, in the highly exothermic second step the reverse reaction

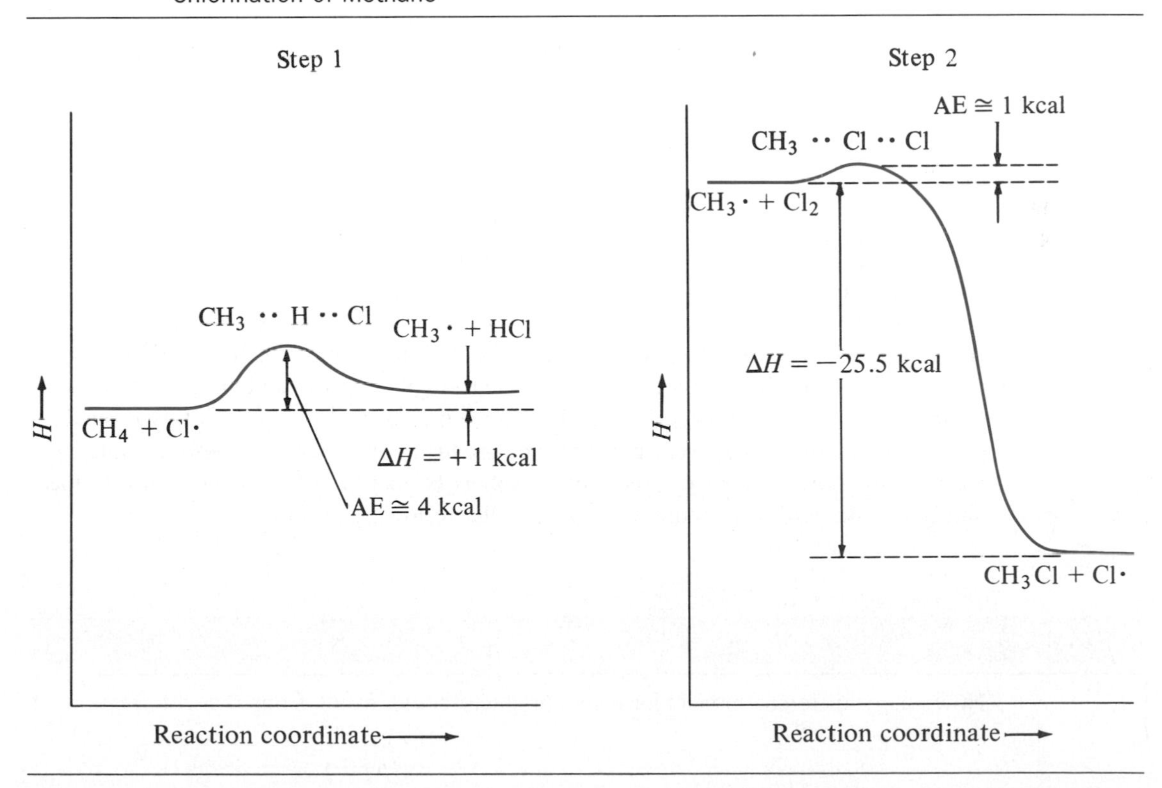

FIGURE 2.8 Reaction Profiles for the Chain Propagation Steps in the Chlorination of Methane

($CH_3Cl + Cl\cdot \rightarrow CH_3\cdot + Cl_2$) would be very slow, highly endothermic, and improbable. Thus the success of free radical chlorination must be due to the nearly irreversible second step.

The chlorination of methane is not restricted to the exclusive conversion of all methane to methyl chloride, but proceeds until two, three, or four hydrogen atoms of the methane molecules have been substituted to yield the useful products shown. The chlorination of a hydrocarbon may result in a mixture of products.

$$CH_4 \xrightarrow{Cl_2} CH_3Cl \xrightarrow{Cl_2} CH_2Cl_2 \xrightarrow{Cl_2} CHCl_3 \xrightarrow{Cl_2} CCl_4$$

Methane; Chloromethane (Methyl chloride); Dichloromethane (Methylene chloride); Trichloromethane (Chloroform); Tetrachloromethane (Carbon tetrachloride)

The chlorination of either methane or ethane in which only one hydrogen atom has been replaced can result in only one monochlorinated product because all hydrogen atoms in these two alkanes are of the primary type. Propane, on the other hand, can yield two different monochlorinated products because it has secondary hydrogen as well as primary hydrogen atoms that may be replaced (Sec. 2.3). This being the case, a question arises as to which propyl chloride we obtain if propane is chlorinated. Actually, both *n*-propyl chloride and isopropyl chloride are produced but not in equal amounts.

$$CH_3CH_2CH_3 + Cl_2 \xrightarrow{uv} CH_3CH_2CH_2Cl + CH_3CHClCH_3$$

Propane; 1-Chloropropane (*n*-Propyl chloride) 45%; 2-Chloropropane (Isopropyl chloride) 55%

Generally speaking, a tertiary hydrogen is more readily replaced than a secondary one, and a secondary more readily than a primary hydrogen.

Bromine also reacts with the alkanes when the reaction is catalyzed by high-frequency radiation, but does so much more slowly than chlorine. Iodine fails to react when catalyzed in this manner, but fluorine reacts with explosive violence. The reasons for this difference in the reaction behavior among members of the same family is a matter of energy relationships between reactants and products. To understand these relationships, we need to know what energy requirements must be met in order for molecules to meet and react. We will learn more about this in the following section.

EXAMPLE

How many different monochlorinated products could result from the chlorination of 2-methylpentane?

Solution

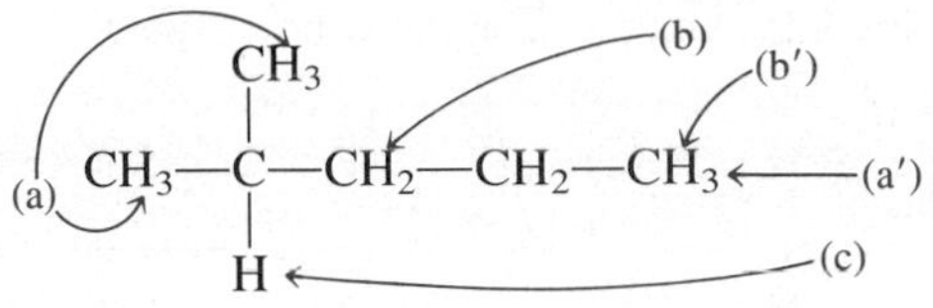

$$CH_3-\overset{CH_3}{\underset{H}{C}}-CH_2-CH_2-CH_3$$

(a) (b) (b′) (a′) (c)

There are six equivalent primary hydrogens designated by (a). Replacement of any one by chlorine gives the same product. Replacement of any one of three (a′) primary hydrogens gives another product. Either of two secondary (b) hydrogens can be replaced to give a third product. Either of two secondary (b′) hydrogens can be replaced to give a fourth product. Only one tertiary (c) hydrogen can be replaced. Total products = 5.

2.12 Bond Dissociation Energies and Heats of Reaction

The heats of reaction for many chemical reactions can be calculated from a knowledge of bond dissociation energies, the energies required to break specific bonds in a molecule. The dissociation energies for a number of different bonds are given in Table 2.4.

TABLE 2.4 Bond Dissociation Energies (kcal/mole at 25°)

H—H	104	F—F	38	O═O	119
H—F	135	Cl—Cl	58	O—H	111
H—Cl	103	Br—Br	46	C═O[a]	175
H—Br	87	I—I	36	H_3C—H	104
H—I	71				
H_3C—F	108	$CH_3\overset{H}{\underset{H}{C}}-Cl$	81.5	$CH_3\overset{H}{\underset{H}{C}}-H$	98
H_3C—Cl	83.5	$CH_3CH_2\overset{H}{\underset{H}{C}}-Cl$	81	$CH_3CH_2\overset{H}{\underset{H}{C}}-H$	98
H_3C—Br	70	$(CH_3)_2\overset{H}{C}-Cl$	81	$(CH_3)_2\overset{H}{C}-H$	95
H_3C—I	56	$(CH_3)_3C$—Cl	80	$(CH_3)_3C$—H	92

[a] CH_2═O

The following example illustrates how Table 2.4 may be used.

EXAMPLE

Calculate ΔH for the monochlorination of methane.

Solution (1) Write the equation to show exactly which bonds are to be broken and which new bonds are to be formed.

$$\begin{matrix} & \text{H} & & & & & \text{H} & & \\ \text{H}- & \text{C} & -\text{H} & + & \text{Cl}-\text{Cl} & \longrightarrow \text{H}- & \text{C} & -\text{Cl} & + \text{H}-\text{Cl} \\ & \text{H} & & & & & \text{H} & & \end{matrix}$$

104 + 58 83.5 + 103 (kcal)

(2) Add bond energies on both sides of the equation. Breaking the C—H bond and the Cl—Cl bond requires 162 kcal; formation of the C—Cl and the H—Cl bonds yields a total of 186.5 kcal.

(3) Subtract the heat liberated through bond formation from the heat expended in bond dissociation. The difference is ΔH for the reaction.

$$\Delta H = 162 - 186.5 = -24.5 \text{ kcal}$$

The sign is negative when the net result is *exothermic*.
The sign is positive when the net result is *endothermic*.

2.13 Synthesis of Alkanes and Cycloalkanes

Most simple alkanes and cycloalkanes used in the laboratory are obtained from petroleum (Sec. 2.15); therefore, it is seldom necessary to synthesize these substances. Nevertheless, there are numerous methods for the synthesis of saturated hydrocarbons, most of which involve elimination of a functional group. Some of these will be described in later chapters. In this section we will describe some selected procedures for converting alkyl halides to hydrocarbons, as this gives us the opportunity to introduce you to a very important class of compounds, the **organometallic compounds,** R—M, in which R has its usual meaning and M is any metal (or metal complex) capable of forming ionic or covalent bonds to carbon. Because metals are much less electronegative than carbon (Table 1.3), carbon–metal bonds may be ionic or covalent or somewhere in between, depending on the metal; thus, we say that these bonds have considerable ionic character.

$$[\text{R—M} \longleftrightarrow \text{R}:^{-}\text{M}^{+}] \quad \text{or} \quad \overset{\delta-}{\text{R}}\text{—}\overset{\delta+}{\text{M}}$$

When M = Li, Na, or K, the bond is highly ionic. When M = Mg, Zn, Cu, or Hg (metals with electronegativities of 1.3, 1.65, 1.9, and 2.0, respectively), the bond is polar but partially covalent. In all of these organometallic species the organic part of the molecule will tend to behave as if it were a carbanion (Sec. 1.12). Carbanions are strong bases and will react with even weak Brønsted–Lowry acids such as water in an acid–base reaction. This is the basis of the first synthesis described in this section. In these syntheses (and in any syntheses described later) be sure to observe whether or not the number of carbon atoms in the product is the *same* or *different* from that in the starting material(s). This will help you to decide which synthesis, among several, to use in solving a given exercise.

A. Reduction of Alkyl Halides (via the Hydrolysis of a Grignard Reagent). In 1901 Victor Grignard discovered one of the most useful of all chemical reactions. He prepared the organometallic compound methylmagnesium iodide by treating methyl iodide with magnesium turnings in anhydrous ether (diethyl ether).

$$\underset{\text{Methyl iodide}}{CH_3—I} + Mg \xrightarrow{\text{ether}} \underset{\text{Methylmagnesium iodide}}{CH_3—MgI}$$

The overall equation for the preparation of alkylmagnesium halides, or Grignard reagents, as they came to be known, can be written as

$$R—X + Mg \xrightarrow{\text{ether}} RMgX$$

where X— is a common generic abbreviation for any halogen. The reaction also proceeds satisfactorily with higher molecular weight alkyl iodides, bromides, or chlorides. Alkyl fluorides are not used to prepare Grignard reagents. Grignard reagents are strong bases and react readily with water to give alkanes. The **hydrolysis** of methylmagnesium iodide yields methane; that of ethylmagnesium bromide yields ethane. Note that the number of carbon atoms in the product is the same as that in the starting material.

$$CH_3MgI + H_2O \longrightarrow CH_4 + HO—Mg—I$$

$$C_2H_5MgBr + H_2O \longrightarrow C_2H_6 + HO—Mg—Br$$

In practice, hydrolysis is usually accomplished by the use of dilute hydrochloric acid, which converts the basic magnesium halide, HO—Mg—X, to the water-soluble magnesium halide, MgX_2.

B. Coupling of Alkyl Halides. Alkyllithium compounds can be prepared by a procedure similar to that used for Grignard reagents.

$$R—X + 2\,Li \xrightarrow{\text{ether}} R—Li + LiX$$

$$\underset{n\text{-Butyl bromide}}{CH_3—CH_2—CH_2—CH_2—Br} + 2\,Li \xrightarrow{\text{ether}} \underset{n\text{-Butyllithium}}{CH_3—CH_2—CH_2—CH_2—Li} + LiBr$$

Both Grignard reagents and organolithium reagents are widely used in the laboratory and in industry.

Organolithium compounds react with cuprous iodide to yield a different class of organometallic compounds known as **cuprates** (or lithium dialkylcopper compounds).

$$2\ CH_3\text{—}Li + CuI \longrightarrow Li(CH_3)_2Cu + LiI$$

Methyllithium; Lithium dimethylcuprate

These relatively new organometallic compounds have been found to react with alkyl halides to give alkanes in which the total number of carbon atoms is the sum of those in the alkyl halide plus those in *one* of the two alkyl groups in the cuprate, as shown in the following example.

$$CH_3\text{—}CH_2\text{—}CH_2\text{—}CH_2\text{—}Br + Li(CH_3)_2Cu \longrightarrow$$

n-Butyl bromide (4 carbons); Lithium dimethylcuprate (1 carbon)

$$CH_3\text{—}CH_2\text{—}CH_2\text{—}CH_2\text{—}CH_3 + CH_3Cu + LiBr$$

n-Pentane (5 carbons)

Exercise 2.13 If 1-bromopropane were the only organic reagent available, what would be the longest hydrocarbon chain obtainable via a lithium–cuprate coupling? Show the reactions involved.

Cyclopropane, which has been used as an inhalation anesthetic, may be prepared by a head-to-tail coupling reaction of 1,3-dibromopropane—a reaction that probably proceeds through a organozinc compound as an intermediate. The reaction is limited to the preparation of cyclopropanes only.

$$BrCH_2\text{—}CH_2\text{—}CH_2Br + Zn \longrightarrow \text{cyclo-}(CH_2)_3 + ZnBr_2$$

1,3-Dibromopropane; Cyclopropane

2.14 Hydrocarbons from Natural Gas and Petroleum

Our chapter on alkanes would not be complete without a brief discourse on the importance of natural gas and petroleum. Most of us think of petroleum in terms of energy, and this indeed is what petroleum is—stored "solar energy" from a remote geological era. From petroleum we obtain the fuels to heat our homes, to till our fields, to run our factories, and to power our automobiles and nearly every other means of transport. However, petroleum provides us with much more than the hydrocarbons we use as fuels. From this blackish, viscous, crude material come thousands of different chemicals that range from extremely light gases to semisolid carbonaceous materials such as asphalt and paraffin wax. From these raw materials are synthesized the chemicals needed to manufacture such diverse articles as automobile and bicycle tires, computer tapes and diskettes, phonograph records, "miracle drugs," food supplements, detergents, dyestuffs, insecticides and herbicides, fiberglass insulation, photographic films and developers, synthetic fabrics, paints and adhesives, and numerous other products. The manufacture of the various everyday commodities listed above, along with the great demand for liquid fuels, has resulted in the consumption of petroleum products that far exceeds our domestic production. Although it is unlikely that we will run out of petroleum in the near future as is sometimes predicted, nevertheless it is clear that our petroleum reserves are not infinite and we cannot afford to spend this rich legacy in a prodigal manner. To expect new discoveries to continue to match consumption indefinitely is unrealistic, and other sources of both energy and carbon-based chemicals must be developed and utilized, and other ways found to recycle and reclaim many of the materials that we now throw away. Table 2.5 shows the different petroleum fractions and the products obtainable from each.

TABLE 2.5 Petroleum Fractions

Fraction	Number of Carbon Atoms	Approx. bp Range, °C	Principal Uses
Gas	1–4	20	fuel
Light naphtha	4–10	20–150	motor fuel, solvents, chemical feedstock
Heavy naphtha		150–200	
Kerosene	9–16	175–275	jet and diesel fuel, fuel oil
Gas oil	15–25	200–400	diesel fuel, fuel oil, cracking stock
Lubricating oil	>26	>400	lubricants, mineral oil, cracking stock
Heavy fuel oil	>26	>400	boiler fuel, paraffin wax, cracking stock
Asphalt			paving, roofing, waterproofing

2.15 Natural Gas

Natural gas varies in composition according to the source from which it is taken, but consists largely of methane with lesser amounts of ethane and propane (along with nitrogen, carbon dioxide, hydrogen sulfide, and occasionally helium). Methane is also found in coal mines, where, if mixed with air, it leads to the deadly explosive mixture called "fire damp." Methane also results from the anaerobic bacterial decomposition of plant and animal materials, for example, under water in swamps and marshes ("marsh gas") or in sewage disposal plants (where it may be collected and burned as fuel). Natural gas sold as a fuel is purified by chemical means or by fractional distillation at low temperature under pressure. The final product is largely methane (about 90%). Although ethane is probably too valuable to burn as a fuel, by law there must be some ethane in natural gas to maintain its BTU[1] rating. The other components are nitrogen and small amounts of propane and the other low molecular weight alkanes.

Refinery gases contain high percentages of propane and butane. These are very stable and cannot be polymerized as can their corresponding olefins (Chap. 3). Liquefied petroleum gas (LPG), sold in large cylinders for use as a fuel in remote places or in small cylinders for soldering torches, is predominantly propane. Very little butane is used as a fuel because it commands a much better price as a base stock for a wide variety of petrochemicals.

2.16 Octane Number

The supply of light naphtha (straight-run gasoline, the C_6–C_9 fraction obtained by simple distillation of petroleum) has long been inadequate to meet the demand. Moreover, straight-run gasoline is a low-quality motor fuel unsuitable for use in high-compression automobile engines. In the internal combustion engine the fuel–air mixture is compressed by the piston, then ignited by the "firing" of the spark plug. Proper operation requires that the fuel burn smoothly outward from the plug, causing the steady formation and expansion of the combustion products that drive the piston. If the fuel–air mixture spontaneously detonates prematurely under high compression, the engine is said to "knock." To provide a measure of the performance of automotive gasolines, the **octane number** scale was devised. Two fuels, one with excellent and one with poor antiknock properties, were chosen as standards. The good fuel, 2,2,4-trimethylpentane ("isooctane"), was assigned an octane number of 100, and the poor fuel, heptane, was assigned on octane number of 0. The octane number of any fuel was defined as the percentage of "isooctane" in a blend of the two standards that gives the same antiknock performance as the fuel under test in

[1] British Thermal Unit. A unit of heat used in the engineering system, defined as the heat necessary to raise the temperature of a pound of water from 63° to 64° Fahrenheit.

a test engine. Thus a typical gasoline mixture with an octane number of 88 is comparable in antiknock properties to a blend of 88% of 2,2,4-trimethylpentane and 12% heptane. The very-low molecular weight alkanes, highly branched alkanes, and aromatic hydrocarbons have high octane numbers, while the straight-chain alkanes have low octane ratings: methane > 120; ethane, 118; propane, 112.5; cumene (isopropylbenzene), 113.0; toluene (methylbenzene), 103.2; pentane, 61.7; hexane, 24.8; octane, −19.0.

Octane ratings of motor fuels may be improved by the use of small amounts of certain additives, principal among which has been tetraethyllead, $(C_2H_5)_4Pb$. To prevent the fouling of the engine by lead deposits, 1,2-dibromoethane is also added to convert the lead to lead dibromide, which is volatile at engine and exhaust temperatures. However, lead salts "poison" the catalytic converters used on many contemporary automobiles to reduce the levels of unburned fuels and other smog-producing substances, and the emission of lead compounds into the atmosphere has caused some concern. Unfortunately, "unleaded" gasoline with reasonable antiknock properties is both more expensive and more energy-consuming to make, because of the extra steps that must be taken in the refining process. Thus the search for new, safe fuels and fuel additives continues. Two such additives that have shown some promise are methanol (methyl alcohol, CH_3—OH, octane number > 100) and methyl *tert*-butyl ether (CH_3—O—$C(CH_3)_3$, octane number 115–135).

Exercise 2.14 Reform *n*-heptane into an isomeric branched structure that would perform as satisfactorily in an automobile engine as does isooctane. Name your product.

2.17 The Manufacture of Gasoline

The manufacture of a modern gasoline is not a simple matter. Although crude oil is the raw material, it contains a very limited amount of the hydrocarbons that are suitable for immediate use as gasoline components. In order to increase the supply of high-octane gasoline and to provide the low molecular weight alkanes, alkenes, and arenes needed as starting materials by the chemical industry, a number of procedures have been developed by petroleum chemists to convert straight-chain or high molecular weight alkanes and cycloalkanes into more useful hydrocarbons. High molecular weight hydrocarbons are "cracked" into smaller molecules by heating the hydrocarbons to high temperatures, usually in the presence of a catalyst. Three processes are used at present: (1) thermal cracking or pyrolysis (Gr. *pyros,* fire; *lysis,* a loosening) carried out without a catalyst at high temperatures (850–900°), most useful for making small alkenes; (2) catalytic cracking carried out in the presence of a catalyst in a reactor called a "cat cracker" and favoring formation of branched alkanes, cycloalkanes, and branched-chain alkenes; and (3) hydrocracking carried out in the presence of a catalyst and hydrogen and leading largely to branched-chain alkanes. Steam cracking is pyrolysis in the presence of water, largely for safety reasons. Because the presence of branched-chain alkanes, alkenes, and aromatic hydrocarbons in gasoline greatly increases its octane number, other processes have been developed to increase the content of these species. One of these processes, known as *isomerization,* causes the rearrangement of straight-chain molecules into branched-chain structures. Small branched-chain alkanes and alkenes may then be combined by a process called **alkylation** to produce larger molecules that fall in the gasoline range. Catalytic **reforming** brings about not only isomerization of small hydrocarbons but also the cyclization of straight-chain alkanes and alkenes of sufficient length. The **dehydrogenation** of (or loss of hydrogen atoms from) the cyclized products, and any cycloalkanes already present, into aromatic compounds is called **aromatization.** Platinum catalysts have proven to be especially useful in reforming, leading to the name "platforming." Examples of these various processes are given in the following equations; however, these examples have been made simple for clarity. The actual products may be numerous and will be dependent on the presence or absence of catalysts, the time and temperature of reaction, the pressure of reactants, and other factors.

Cracking

$$C_{15}H_{32} \xrightarrow[\text{heat}]{\text{catalyst}} C_8H_{18} + C_7H_{14} \text{ (an alkane + an alkene)}$$

$$C_8H_{18} \xrightarrow[\text{heat}]{\text{catalyst}} \underset{n\text{-Butane}}{CH_3CH_2CH_2CH_3} + \underset{\text{2-Butene}}{CH_3{-}CH{=}CH{-}CH_3}$$

$$\text{or } \underset{\text{1-Butene}}{CH_3{-}CH_2{-}CH{=}CH_2}$$

Isomerization

$$n\text{-}C_4H_{10} \xrightarrow[\text{heat}]{\text{catalyst}} \underset{\text{Isobutane}}{CH_3{-}\overset{\overset{\displaystyle H}{|}}{\underset{\underset{\displaystyle CH_3}{|}}{C}}{-}CH_3}$$

$$\underset{\text{1-Butene}}{CH_3{-}CH_2{-}CH{=}CH_2} \xrightarrow[\text{heat}]{\text{catalyst}} \underset{\text{Isobutene}}{CH_3{-}\overset{\overset{\displaystyle CH_3}{|}}{C}{=}CH_2}$$

Alkylation

$$\underset{\text{Isobutane}}{CH_3{-}\overset{\overset{\displaystyle CH_3}{|}}{\underset{\underset{\displaystyle CH_3}{|}}{C}}{-}H} + \underset{\text{Isobutene}}{H_2C{=}\underset{\underset{\displaystyle CH_3}{|}}{C}{-}CH_3} \xrightarrow[\text{heat}]{\text{catalyst}} \underset{\text{2,2,4-Trimethylpentane}}{CH_3{-}\overset{\overset{\displaystyle CH_3}{|}}{\underset{\underset{\displaystyle CH_3}{|}}{C}}{-}CH_2\overset{\overset{\displaystyle H}{|}}{\underset{\underset{\displaystyle CH_3}{|}}{C}}{-}CH_3}$$

Reforming (Aromatization)

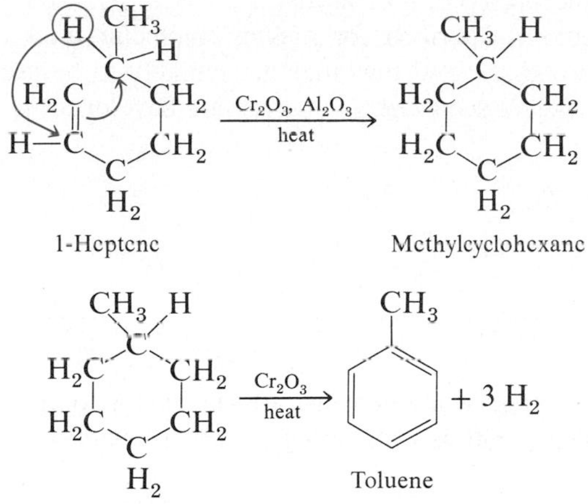

1-Heptene $\xrightarrow[\text{heat}]{Cr_2O_3,\ Al_2O_3}$ Methylcyclohexane

Methylcyclohexane $\xrightarrow[\text{heat}]{Cr_2O_3}$ Toluene $+ 3\ H_2$

2.18 Synthetic Liquid Fuels

Since optimistic estimates of coal reserves have suggested that by the year 2000 only 2% of the coal actually in the ground will have been consumed, there has been considerable discussion of the feasibility and economics of converting coal into a petroleumlike product, one of the so-called ''synfuels.'' Coal contains high molecular weight aromatic compounds that can be cracked and rearranged into coal tar in the process of making coke for the manufacture of steel. However, at present coal tar is a minor source of aromatic hydrocarbons. Nevertheless, it is still the only good source of two important chemicals, naphthalene and pyridine (Chap. 4). The major interest in ''coal liquefaction'' is in the conversion of the carbon in coal (whose empirical formula is approximately CH) into fuels and organic raw materials. In an *indirect liquefaction* process known as the Fischer–Tropsch process, synthesis gas (carbon monoxide and hydrogen) is made from coke and steam. Addition of more hydrogen and reaction in the presence of a catalyst at high temperatures and pressures yield hydrocarbons suitable for fuels.

$$C + H_2O \xrightarrow{300°C} CO + H_2$$

$$n\ CO + (2n + 1)\ H_2 \xrightarrow[250°C]{ThO_2} C_nH_{2n+2} + n\ H_2O$$

An attractive alternative to the Fischer–Tropsch process is the conversion of synthesis gas to methanol (Sec. 7.4). Methods for conversion of methanol, not only to gasoline but also to other useful materials such as ethylene, propylene, acetic acid, and ethylene glycol (Sec. 7.13), are already known.

Direct liquefaction processes have been proposed, and even employed during wartime emergencies. These avoid the preliminary conversion of carbon to carbon monoxide and consist primarily of the direct reaction of carbon with hydrogen under heat and pressure in the presence of catalysts.

The methods described for making synthetic fuels are themselves high-energy-consuming processes, and thus they are unlikely to be used on a large scale until safe nuclear or efficient solar energy sources are developed.

Summary

1. The alkanes, or saturated hydrocarbons, contain only carbon and hydrogen. They are also called paraffins. All carbon atoms in alkanes are singly bonded to four other atoms.
2. The **general formula for an alkane** is $\mathbf{C_nH_{2n+2}}$.
3. The names for the alkanes end in "*ane*." Alkanes may be named unambiguously by the IUPAC system.
4. The **general formula for an alkyl group** is $\mathbf{C_nH_{2n+1}}$. Groups are named after the parent hydrocarbon by changing the suffix *ane* to *yl*. The prefixes iso, *sec,* and *tert* are used, depending upon which hydrogens have been removed from the parent compound.
5. Saturated hydrocarbons may have straight (that is, in a continuous chain), branched, or cyclic structures.
6. Chemically, the alkanes are relatively inert. The only reactions that occur without rupture of the carbon–carbon bond are those of substitution. Typical reactions include:

   (a) Oxidation (combustion)

$$C_nH_{2n+2} + \frac{(3n + 1)}{2}O_2 \longrightarrow n\ CO_2 + (n + 1)\ H_2O$$

   (b) Halogenation

$$R{-}H + X_2 \xrightarrow[\text{light}]{\text{uv}} R{-}X + HX$$

7. The energy required to break a chemical bond is called its dissociation energy. The same amount of energy is released when the same bond is formed.
8. The heat of reaction, ΔH, is the difference in the heat content of reactants and products. ΔH is negative when the reaction is exothermic and positive when endothermic.
9. Cycloparaffins are ring compounds of the general formula, C_nH_{2n}, which can exist as *cis,trans*-isomeric pairs. Small rings show the effects of angle and eclipsing strain; large rings are free of angle strain but may show eclipsing strain. The chair form of cyclohexane is strainfree.

10. Alkanes may be prepared by:
    (a) Coupling of alkyl groups.

$$\text{R—Br} + \text{Li(R')}_2\text{Cu} \longrightarrow \text{R—R'}$$

    (b) Reduction of an alkyl halide via a Grignard reagent.

$$\text{R—X} + \text{Mg} \xrightarrow[\text{ether}]{\text{anhydrous}} \text{RMgX} \xrightarrow[\text{(HX)}]{\text{hydrolysis}} \text{R—H} + \text{MgX}_2$$

11. The ability of a gasoline to perform well in an internal combustion engine is measured by its "octane number."
12. The octane numbers of hydrocarbons increase with chain branching, unsaturation, and cyclization. Octane numbers of motor fuels may also be increased by the use of "additives."
13. High-molecular weight hydrocarbons (above C_9) not suitable for use as gasoline are catalytically "cracked" or broken into smaller molecules. The latter are rearranged (isomerized), cyclized (aromatized), or combined (alkylated) to form compounds of high "octane number."

Supplementary Exercises

2.15 Assign acceptable names (common or IUPAC) to the following structures.

(a) 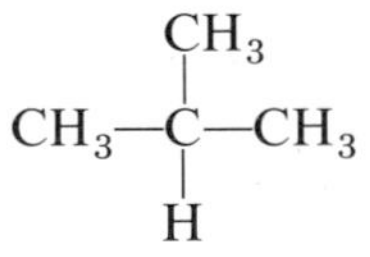 CH_3—C(CH_3)(H)—CH_3

(b) [cyclohexane ring with H and CH_3 on one carbon (CH_3 down), and CH_3 and H on the adjacent carbon (H down)]

(c) CH_3—C(CH_3)(H)—CH_2—CH_2—CH_3

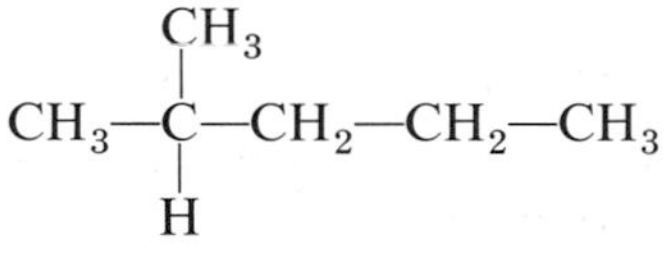

(d) 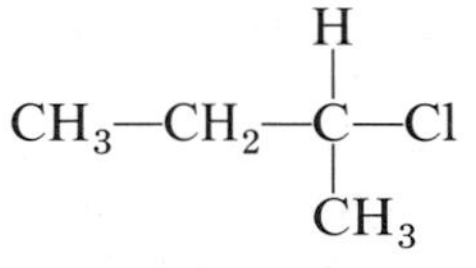 CH_3—CH_2—C(H)(CH_3)—Cl

(e) 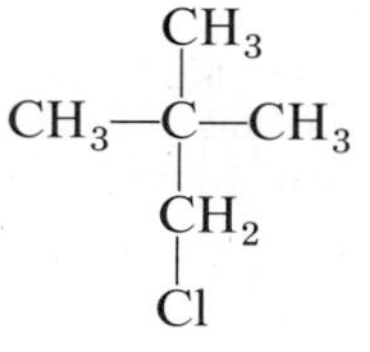 CH_3—C(CH_3)(CH_2Cl)—CH_3

(f) 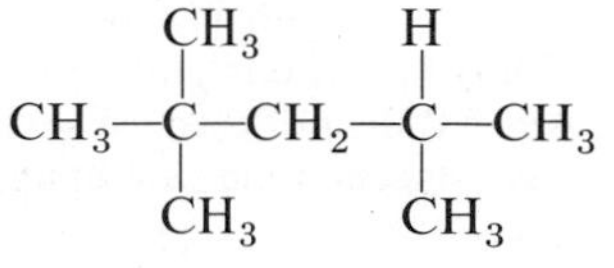 CH_3—C(CH_3)(CH_3)—CH_2—C(H)(CH_3)—CH_3

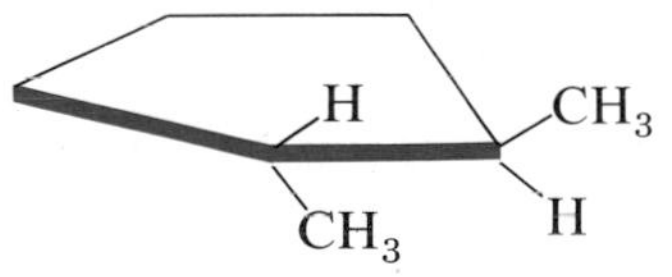

2.16 Which of the following structures are the same? What are their IUPAC names?

(a) 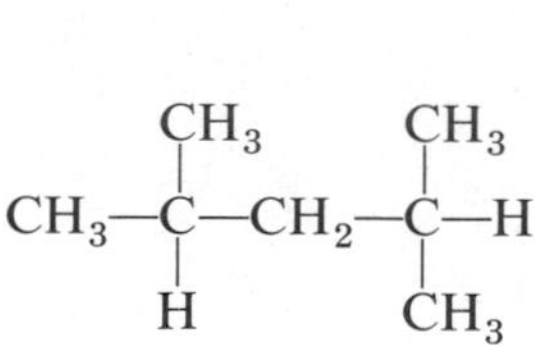
$$\begin{array}{ccccc} & CH_3 & & CH_3 & \\ & | & & | & \\ CH_3- & C & -CH_2- & C & -H \\ & | & & | & \\ & H & & CH_3 & \end{array}$$

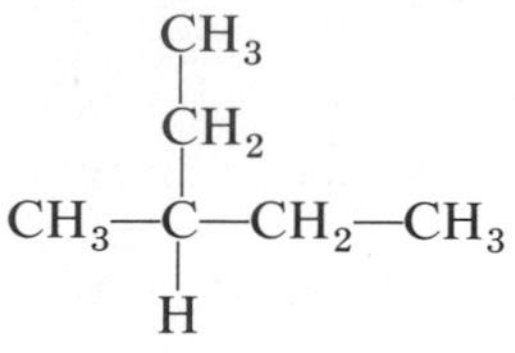

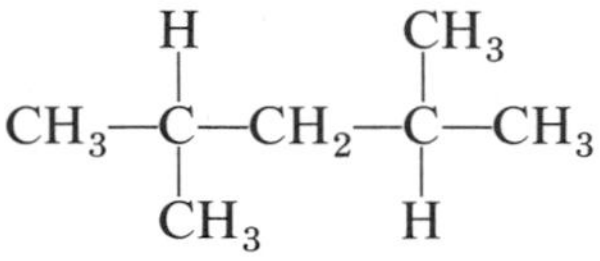

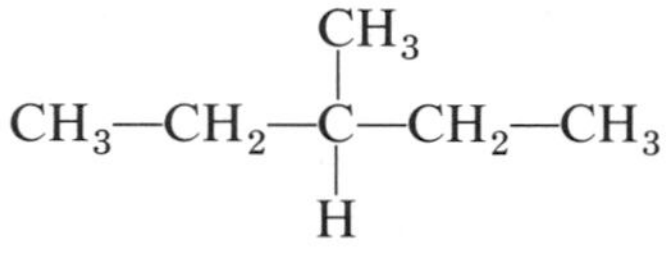

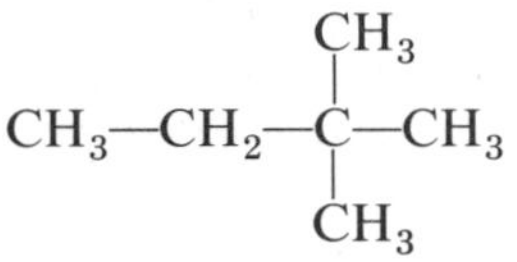

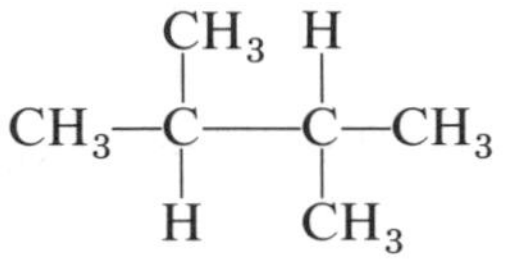

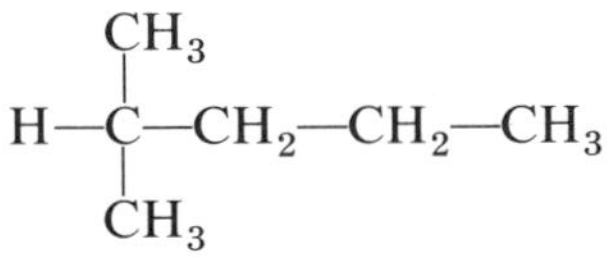

(b)
$$\begin{array}{ccc} & CH_3 & \\ & | & \\ & CH_2 & \\ & | & \\ CH_3- & C & -CH_2-CH_3 \\ & | & \\ & H & \end{array}$$

(c)
$$\begin{array}{ccccc} & H & & CH_3 & \\ & | & & | & \\ CH_3- & C & -CH_2- & C & -CH_3 \\ & | & & | & \\ & CH_3 & & H & \end{array}$$

(d)
$$\begin{array}{ccc} & CH_3 & \\ & | & \\ CH_3-CH_2- & C & -CH_2-CH_3 \\ & | & \\ & H & \end{array}$$

(e)
$$\begin{array}{ccc} & CH_3 & \\ & | & \\ CH_3-CH_2- & C & -CH_3 \\ & | & \\ & CH_3 & \end{array}$$

(f)
$$\begin{array}{ccccc} & CH_3 & & H & \\ & | & & | & \\ CH_3- & C & - & C & -CH_3 \\ & | & & | & \\ & H & & CH_3 & \end{array}$$

(g)
$$\begin{array}{ccc} & CH_3 & \\ & | & \\ H- & C & -CH_2-CH_2-CH_3 \\ & | & \\ & CH_3 & \end{array}$$

(h)
$$\begin{array}{ccc} & CH_3 & \\ & | & \\ CH_3-CH_2- & C & -CH_2-CH_3 \\ & | & \\ & CH_3 & \end{array}$$

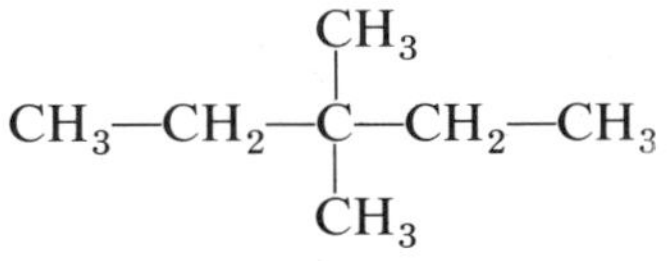

2.17 The following names are incorrect. In each case tell why the name is wrong and give the correct form.

(a) 2-isopropylhexane
(b) 2-methyl-3-ethylbutane
(c) 4-methylpentane
(d) 3,5,5-trimethylhexane
(e) 3,4-methylhexane
(f) 4-methyl-5-ethylheptane
(g) dimethylcyclohexane
(h) 3-methyl-4-chloropentane
(i) 1,3-dimethylcyclopropane
(j) 2-ethylpropane
(k) chloropropane
(l) 2,3,5,5-tetramethylhexane

2.18 Write structural formulas for the following. Also assign systematic names to compounds (c)–(f).

(a) 2-methylbutane
(b) 2,2,4-trimethylhexane
(c) neopentyl chloride
(d) *tert*-butyl chloride
(e) isobutyl chloride
(f) chloroform
(g) 1,3-dibromopropane
(h) 3-ethyl-2-methylpentane

2.19 Draw the structures for all the monochlorinated isomers that could result from the free radical chlorination of 2-methylbutane (isopentane). If the approximate ratios of rates at attack of chlorine on hydrogens located in primary, secondary, and tertiary positions are 1.0:3.8:5.8, approximately what percent yield of each isomeric monosubstituted product might be expected from isopentane? *Hint:* The percent yield of 1-chloropropane (p. 62) was derived from

$$\frac{6 \times 1}{(6 \times 1) + (2 \times 3.8)} \times 100 = 45\%$$

2.20 Give all structures possible for $C_3H_6Cl_2$. Name each according to the IUPAC system.

2.21 Write structures for each of the following condensed formulas. Name each according to IUPAC rules.

(a) *sec*-C_4H_9Cl
(b) $CH_3CHBrCH_2Br$
(c) $(CH_3)_2CHCH_2C(CH_3)_3$
(d) $CH_3CH_2C(CH_3)_2CH_2CH_3$
(e) $(CH_3)_2CHBr$
(f) $(CH_3)_3CH$
(g) $CH_3CH(CH_3)(CH_2)_4CH_3$
(h) $CH_3C(CH_3)_2C(CH_3)_2CH_2CH_3$
(i) *tert*-C_4H_9Br
(j) $(CH_3)_4C$

2.22 Draw structures for all the monochlorinated products possible when each of the following is reacted with chlorine in the presence of sunlight.

(a) 2-methylpentane
(b) 2,2-dimethylpropane
(c) 2,2,4-trimethylpentane
(d) 2,2,3,3-tetramethylbutane
(e) 2,2,3-trimethylpentane

2.23 Which compounds in question 2.22 can yield only one product?

2.24 Show how the percentage of each product was obtained in the following equation. (*Hint:* See Exercise 2.19)

$$CH_3-\underset{\displaystyle H}{\overset{\displaystyle CH_3}{\overset{|}{\underset{|}{C}}}}-CH_3 + Cl_2 \xrightarrow[25^\circ]{\text{light}} CH_3-\underset{\displaystyle H}{\overset{\displaystyle CH_3}{\overset{|}{\underset{|}{C}}}}-CH_2-Cl + CH_3-\underset{\displaystyle Cl}{\overset{\displaystyle CH_3}{\overset{|}{\underset{|}{C}}}}-CH_3$$

1-Chloro-2-methylpropane (61%)   2-Chloro-2-methylpropane (39%)

2.25 Complete the following reactions (the equations should be balanced):

(a) $C_2H_5Br + Mg \xrightarrow{\text{ether}}$

(b) Product of (a) + $H_2O \longrightarrow$

(c) Cyclohexane + $Cl_2 \xrightarrow{\text{light}}$

(d) $CH_3Li + H_2O \xrightarrow{H^+}$

(e) C_3H_8 (bottled gas) + $O_2 \longrightarrow$

(f) Neopentane + $Cl_2 \xrightarrow{\text{light}}$

2.26 Calculate ΔH for the following reaction.

$$CH_3-H + Br-Br \longrightarrow CH_3-Br + H-Br$$

Draw and properly identify each part of a reaction coordinate diagram similar to that shown in Fig. 2.5 for this reaction. The activation energy for the reaction of CH_4 and $Br\cdot$ is approximately 18 kcal/mole and for the reaction of $CH_3\cdot$ and Br_2 is approximately 2 kcal/mole. Which step in this reaction would you expect to be the rate-determining step?

2.27 Calculate ΔH for the following reactions: (Consult Table 2.4)

(a) $2\ H_2 + O_2 \longrightarrow 2\ H_2O$

(b) $C_2H_6 + Cl_2 \longrightarrow C_2H_5Cl + HCl$

(c) $C_3H_8 + Cl_2 \longrightarrow (CH_3)_2CHCl + HCl$

2.28 You have available the following alkyl halides: methyl bromide, ethyl bromide, *n*-butyl bromide, and any inorganic reagents of your choosing. Write equations showing how longer-chain alkanes could be synthesized using one or more of these alkyl halides.

2.29 An alkyl halide (A) forms a Grignard reagent, which on treatment with water yields isopentane. When A is allowed to react with lithiumdimethyl cuprate, 2,3-dimethylbutane is formed. What is the structure of A? Write the equations for the reactions described.

2.30 Draw both chair conformations of the *cis* and *trans* forms of the following compounds and, in each case, indicate which of the two conformations is the more stable, or state if the two conformations are equivalent.

(a) 1,2-dimethylcyclohexane
(b) 1,3-dimethylcyclohexane
(c) 1,4-dimethylcyclohexane
(d) 1-*tert*-butyl-4-methylcyclohexane

2.31 The following thermochemical data are for cyclohexane (strainfree, as you will recall):

$$C_6H_{12} + 9\ O_2 \longrightarrow 6\ CO_2 + 6\ H_2O \qquad \Delta H = -944.80 \text{ kcal/mole}$$

When measured under the same conditions, ΔH for the combustion of cyclobutane is -655.86 kcal/mole. Write a balanced equation for the combustion of cyclobutane. Calculate the total strain energy (in kcal/mole) of cyclobutane. (*Hint:* What is the increment of ΔH per —CH_2— in an *unstrained* cycloalkane?)

2.32 Draw the structures of the following:

(a) the most stable conformer of 1-bromopropane
(b) the most stable conformer of *tert*-butylcyclohexane
(c) the product obtained in greatest yield from the bromination of 2-methylbutane
(d) the pentane isomer most likely to have the lowest boiling point

2.33 Although methanol (methyl alcohol, CH_3OH) and 2,2,4-trimethylpentane have about the same octane number, octane number is not the only factor in the choice of a fuel. For which of these fuels would you have to have the larger gas tank if you were to drive your automobile using solely that fuel? You will need the following data. Heats of combustion (at 25°C for combustion of the liquid fuel to gaseous products; i.e., any corrections have already been made): methanol, 143 kcal/mole; 2,2,4-trimethylpentane, 1192 kcal/mole. Density: methanol, 0.791 g/ml; 2,2,4-trimethylpentane, 0.692 g/ml. Show your reasoning.

2.34 Write balanced equations for the reactions that you might expect from the following processes.

(a) the thermal cracking of pentane (all alkanes and alkenes)
(b) the isomerization of hexane (no cracking)
(c) the reforming of octane (cyclized, aromatized products only)

2.35 List the following compounds in an order of increasing octane number.

(a) *n*-pentane
(b) *n*-hexane
(c) 2-methylbutane
(d) *n*-octane
(e) methylcyclohexane
(f) 2,2,4-trimethylpentane

2.36 According to the chemistry handbooks, the boiling points of the C_3 through C_6 cycloalkanes are about 12° higher than their straight-chain analogs, and their specific gravities are about 0.1 unit higher. Give an explanation for these differences.

2.37 The highly strained molecule, bicyclobutane, can be made by reacting the dihalide, $C_4H_6Cl_2$, with sodium. What would the structure of the dihalide have to be?

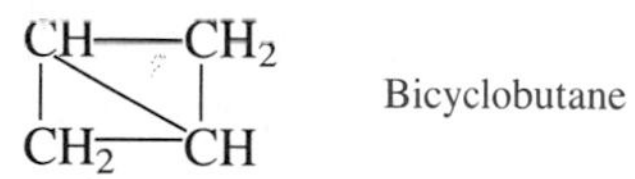

2.38 The Newman projections of cyclohexane derivatives are usually drawn as shown on page 57. However, at first glance it appears that there are only three bonds drawn to carbon atoms 3 and 6. In structure I, in what directions relative to the plane of the page are the ''missing'' bonds on atoms 3 and 6?

3 Unsaturated Hydrocarbons: Alkenes and Alkynes

The unsaturated hydrocarbons include two classes of compounds. In one class are hydrocarbons that contain carbon–carbon double bonds,

$$-\overset{|}{C}=\overset{|}{C}-$$

and are known as **alkenes,** or **olefins.** The second class of unsaturated hydrocarbons includes those that contain carbon–carbon triple bonds, $-C \equiv C-$, and are known as **alkynes,** or **acetylenes.** Both classes are described as unsaturated because they do not contain the maximum possible number of hydrogen atoms. They are not as "filled up" or saturated with hydrogen as they could be and are potentially capable of bonding to additional atoms. Unlike the saturated hydrocarbons, these unsaturated compounds have more than one pair of electrons shared between two carbon atoms. As a result of such unsaturation, both the olefins and the acetylenes are much more reactive than the alkanes. They combine readily with other compounds or with themselves and yield many useful products.

Part I: The Alkenes (Olefins)

As was just pointed out, the alkenes or olefins are characterized by the presence of double bonds between adjacent carbon atoms. The general formula C_nH_{2n} corresponds to an open-chain alkene only if *one* double bond is present in the molecule, for the same general formula also represents the monocyclic cycloalkanes. As a class these compounds are still commonly referred to as the olefins (L. *oleum,* oil; *ficare,* to make); however, the name alkene is used in systematic nomenclature. Both common and systematic names will be used in this chapter.

3.1 Formulas and Nomenclature of the Alkenes

Systematic names for members of the alkene family are formed by replacing the suffix *ane* of the corresponding alkane with *ene*. Common names are often employed to name the simplest members of the alkenes. Such common names are formed by replacing the suffix *ane* of the corresponding alkane with *ylene*. The following are examples of both systems of nomenclature.

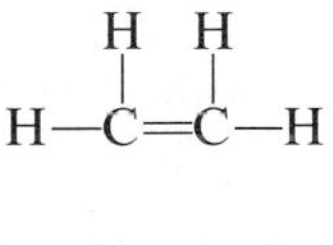

```
   H  H              H  H           CH3   H
   |  |              |  |              \  |
 H—C=C—H       CH3—C=C—H              C=C—H
                                      /
                                   CH3

   Ethene          Propene         2-Methylpropene
 (Ethylene)      (Propylene)       (Isobutylene)
```

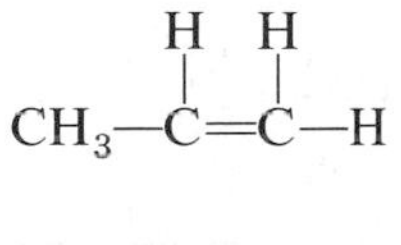

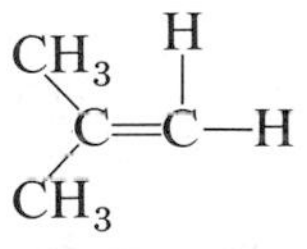

The IUPAC rules for naming the alkenes follow those for the alkanes, except for the following modifications:

(1) Instead of simply numbering the longest carbon chain, the longest chain *that includes the functional group* is numbered. In the case of the alkenes, the functional group is the carbon–carbon double bond,

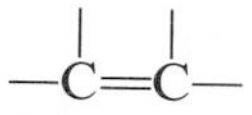

```
  |  |
 —C=C—
```

Change the *ane* of the corresponding alkane to *ene*.

(2) Numbering begins from the end of the chain that will give the smaller number to the carbon atom holding the functional group. Since the double bond in an alkene appears between two carbons, only the carbon atom with the lower number need be designated. This number precedes the name and is separated from it by a hyphen.

(3) The functional group is located by a number each time it appears. The prefixes *di, tri,* and so on appear before *ene* in alkenes with more than one double bond and are incorporated within the name.

(4) In cycloalkenes the double bond is numbered 1,2. Since this is always the case, no number for the double bond is given in the name. If more than one double bond is present in the ring, they are given the lowest numbers possible, and the numbers are given in the name. The ring is numbered in the direction that gives the lower number to the first substituent encountered.

A few examples here and on the next page will help make these points clear.

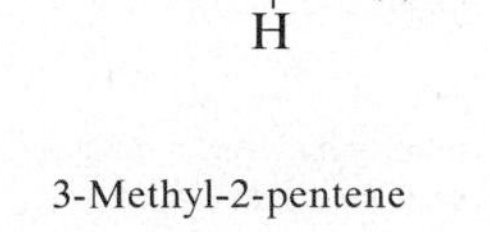

```
      (5)CH3                      CH3               H—C————C—H
         |                         |                 ‖(2) (3)‖
      (4)CH2                    (1)C  (6)            ‖(1) (4)‖
      (3)|  (2)                 //  \               H—C  (5) C—H
  CH3—C=C—CH3             (2)CH     CH—CH3             \   /
          |  (1)              |      |  (5)              C
          H               (3)CH2   CH2                   H2
                               \  /
                             (4)CH2

 3-Methyl-2-pentene    1,6-Dimethylcyclohexene    1,3-Cyclopentadiene
```

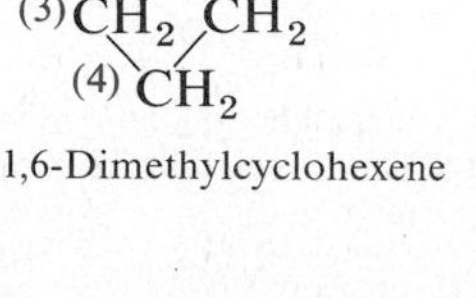

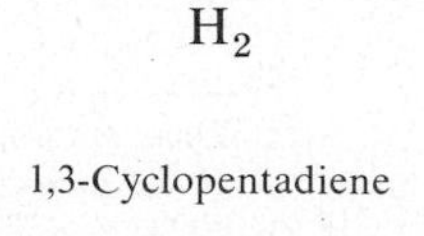

```
  H  H
  |  |
CH3—C═C—CH3
```
2-Butene

```
             H
             |
CH3—CH2—C═CH2
(4)  (3)  (2) (1)
```
1-Butene

```
 H  H  H  H
 |  |  |  |
H—C═C—C═C—H
 (1)(2)(3)(4)
```
1,3-Butadiene

Exercise 3.1 Name each of the following compounds according to IUPAC rules:

(a) $CH_2{=}C(CH_3){-}CH{=}CH_2$ (b) $CH_2{=}CH{-}CH{=}C(CH_3)_2$ (c) 1-methylcyclopentene ring structure with $-CH_3$

3.2 The Double Bond

A double bond represents two pairs, or four electrons, shared between two carbon atoms. Our simple structural formulas show these as two equivalent covalent bonds, and this formulation can be justified mathematically. However, much of the chemical behavior of alkenes is more conveniently described in terms of a double bond in which the two bonds are not equivalent. You learned in Section 1.7 that, when a hybrid is formed from one $1s$ and three $2p$ orbitals, four sp^3-hybridized orbitals results. However, if only *two* of the three $2p$ orbitals are combined with the $1s$ orbital, then three sp^2 hybrid orbitals are formed and one $2p$ orbital is left (Sec. 1.7). This is the situation for each carbon in ethene. The axes of the three sp^2 orbitals formed lie in the same plane and form angles of 120° with each other. The axis of the remaining $2p$ orbital is perpendicular to the plane (Fig. 3.1). When two carbon atoms with bonding orbitals of this type unite, an sp^2 orbital from each carbon overlaps to form a carbon–carbon sigma orbital. The two $2p$ orbitals overlap in regions above and below the plane to form two molecular orbitals of a new type called **π (pi) orbitals.** The *bonding* π orbital resembles a cloud lying above and below the plane of the molecule (Fig. 3.2)[1]. Now, of the four shared electrons that form the carbon–carbon double bond, two are assigned to the sigma orbital and two to the π orbital. A bond formed in this manner is called a **pi** (π) bond to distinguish it from the ordinary covalent or **sigma** (σ) bond. Thus, a double bond can be regarded as made up of one sigma C—C bond and one pi C—C bond.

The effect of such double bonding is threefold. First, two carbon atoms are nearer each other when joined by a double bond than when joined by a single bond; the distance between two doubly bonded carbon atoms is 1.35 Å compared to 1.54 Å for the C—C

[1]There is formed also a higher energy *antibonding* π orbital, which has a nodal plane perpendicular to the plane of the molecule and which bisects the molecule between the carbon atoms.

FIGURE 3.1 The Three sp^2 Orbitals of Carbon (in the Horizontal Plane) with the Remaining p Orbital Perpendicular to the Plane*

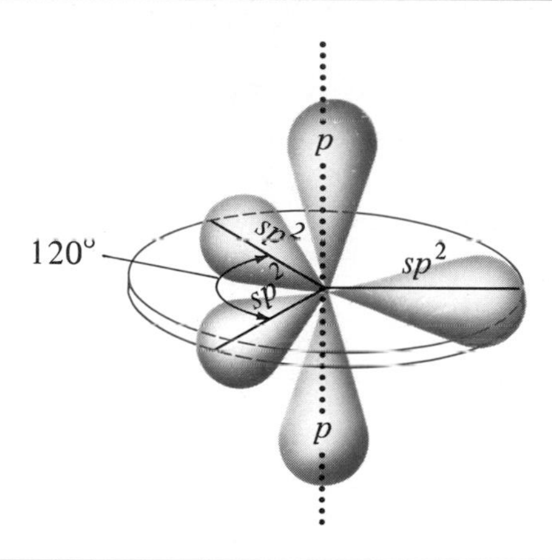

*The shape of an sp^2 orbital is similar to that of an sp^3 orbital (Fig. 1.5) except that the front lobe (the larger lobe) is somewhat larger in the sp^2 orbital and the back lobe is somewhat smaller. The orbital shapes used in this figure are those conventionally employed by chemists to represent the orbitals and were chosen to simplify the drawings.

single bond. Second, carbon atoms doubly bonded and the atoms attached to them are now held *in the same plane*. Free rotation around the bond joining the two carbon atoms is no longer possible, because considerable energy is required to break the π bond. This restriction in a plane makes possible a variation in the space arrangements of groups attached to the carbon atoms and a type of **stereoisomerism** called ***cis,trans*-isomerism** (Sec. 5.8). The *cis*-isomer has like groups on the same side of the molecule, and the

FIGURE 3.2 Pi Orbitals: (a) Pi Orbitals, (b) Schematic Representation of the Carbon–Carbon Double Bond of Ethene in Perspective Showing Five σ Bonds and One π Bond

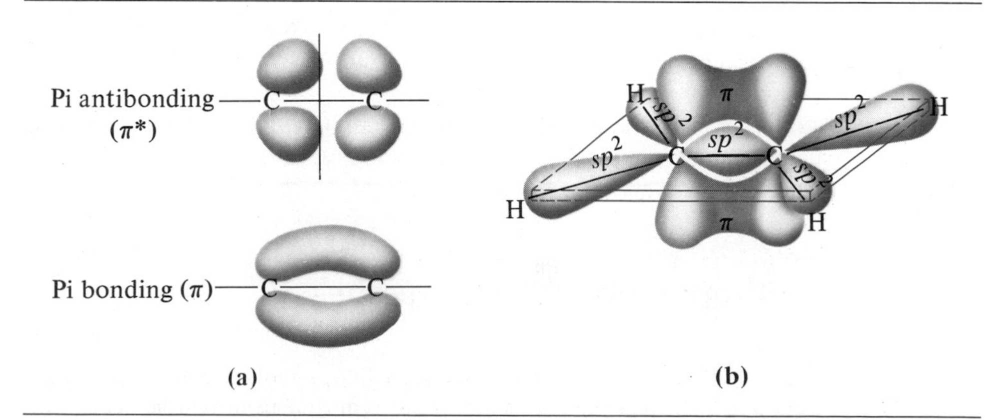

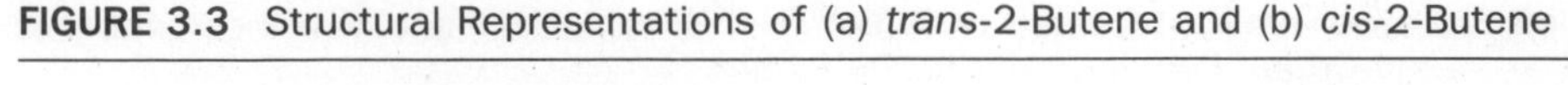

FIGURE 3.3 Structural Representations of (a) *trans*-2-Butene and (b) *cis*-2-Butene

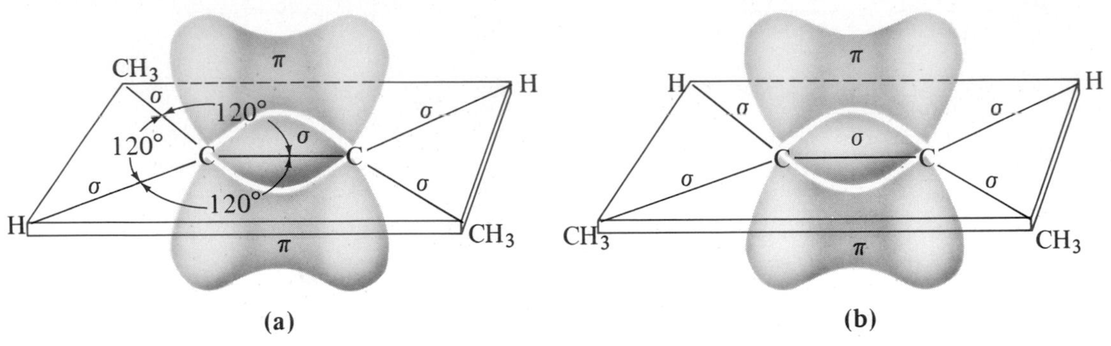

trans-isomer has like groups on opposite sides of the molecule (Fig. 3.3). Thus, beginning with C_4H_8, a greater number of isomers is possible for any member of the alkene family than was possible for the corresponding alkane, C_4H_{10}, not simply because the position of the double bond in the carbon chain may vary, but also because the groups attached to it may be *cis* or *trans*. When the two groups (or atoms) attached to either of the doubly bonded carbon atoms are identical (Fig. 3.4), then, of course, *cis,trans*-isomerism is not possible.

A third effect manifest in doubly bonded carbon compounds is that the π electrons are not as firmly held as those comprising the σ bond, but are more readily available to an electrophilic or electron-seeking reagent. Thus, the region above and below the double bond is electron rich, and this availability of π electrons accounts for the greater reactivity of the alkenes compared to the alkanes. Alkenes tend to behave as Lewis bases toward electrophilic reagents.

Exercise 3.2 Which of the following compounds may be either *cis* or *trans*? Which can not show *cis-trans* isomerism?

(a) $CH_3—CH=CH_2$
(b) $CH_3—CH=CH—CH=CH_2$
(c) 2-Methyl-2-butene
(d) 2-Hexene

Exercise 3.3 Draw three different structures for C_5H_{10} in which the continuous chain is five carbons long. Name each according to IUPAC rules.

Preparation of the Alkenes

The double bond is usually introduced into a molecule by some type of **elimination reaction** in which groups or atoms are removed from two adjacent carbons.

FIGURE 3.4 Structural Representation of 1-Butene (*cis,trans*-Isomerism Not Possible)

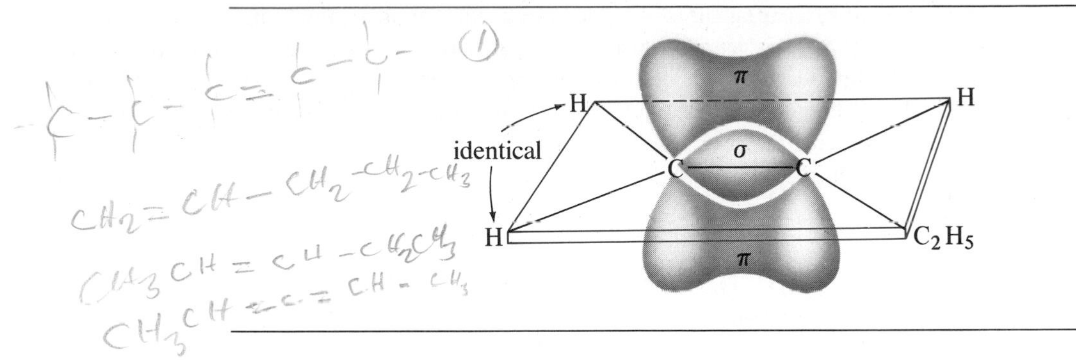

$$\underset{\displaystyle A\quad B}{-\overset{|}{\underset{|}{C}}-\overset{|}{\underset{|}{C}}-} \longrightarrow AB + -\overset{|}{C}=\overset{|}{C}-$$

In the petroleum industry the elimination of hydrogen atoms from alkanes is called dehydrogenation and is accomplished by thermal or catalytic cracking (Sec. 2.17). Both continuous and branched alkenes are produced in this manner. However, the two olefins made in greatest amounts are ethylene and propylene. Ethylene is the starting material for about 30% of all petrochemicals. Almost any alkane can be thermally cracked to ethylene and propylene (plus other products). In the United States ethane and propane from natural gas are the preferred feedstocks, while in Europe the gasoline (naphtha) fractions from petroleum are favored. In a laboratory preparation of an alkene, the double bond is usually introduced by eliminating a small molecule such as water, hydrogen halide, or ammonia. Such elimination reactions are generally catalyzed by ions such as H^+, for the removal of water, and OH^-, for the elimination of hydrogen halide, but the double bond may also result from the pyrolytic decomposition of certain ammonium salts (Sec. 13.4B). Many reactions of both types are known, but we shall consider here only two of the most general procedures for the preparation of alkenes.

3.3 Dehydration of Alcohols

Elimination of the elements of water from adjacent carbon atoms of an alcohol will lead to the formation of an alkene. Industrially, water is eliminated from an alcohol by passing the alcohol vapor over heated alumina, Al_2O_3.

$$H-\overset{\displaystyle H}{\overset{|}{\underset{\displaystyle H}{\underset{|}{C}}}}-\overset{\displaystyle H}{\overset{|}{\underset{\displaystyle OH}{\underset{|}{C}}}}-H \xrightarrow[350\text{–}360^\circ C]{Al_2O_3} H-\overset{\displaystyle H}{\overset{|}{C}}=\overset{\displaystyle H}{\overset{|}{C}}-H + H_2O$$

Ethyl alcohol Ethene

In the laboratory, **dehydration** of an alcohol is usually accomplished by heating it with either concentrated sulfuric or phosphoric acid. The acid functions primarily as the catalyst, as may be seen from an inspection of the mechanism for this elimination. The reaction is thought to take place in the following manner. In the first stage of the reaction, the acid (shown here as the hydrogen ion) protonates the electron-rich hydroxyl group of the alcohol molecule to form an alkyl-substituted oxonium ion.

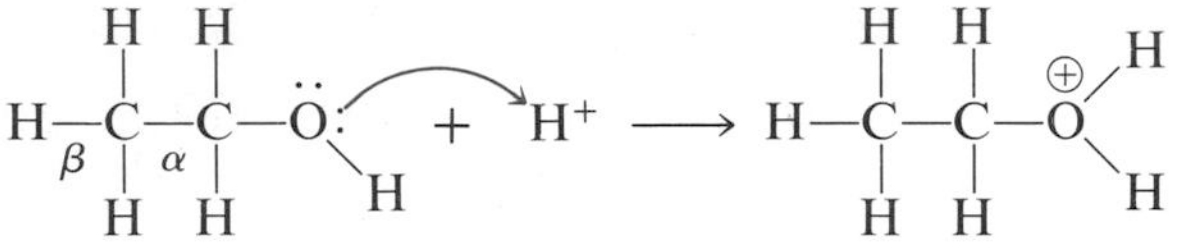

$$\underset{\text{Ethanol}}{H-\overset{H}{\underset{H}{\overset{|}{\underset{|}{C}}}}_{\beta}-\overset{H}{\underset{H}{\overset{|}{\underset{|}{C}}}}_{\alpha}-\ddot{O}:-H} + H^{+} \longrightarrow \underset{\text{An alkyl oxonium ion}}{H-\overset{H}{\underset{H}{\overset{|}{\underset{|}{C}}}}-\overset{H}{\underset{H}{\overset{|}{\underset{|}{C}}}}-\overset{\oplus}{O}H_2}$$

Next, some base present (B: = bisulfate, HSO_4^-, or another molecule of alcohol) abstracts a proton from the β-carbon[2] atom, and a molecule of water departs to form the alkene.

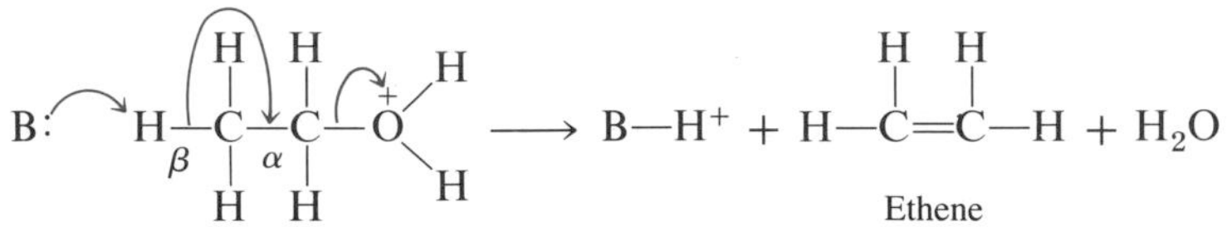

$$B: \; H-\overset{H}{\underset{H}{\overset{|}{\underset{|}{C}}}}_{\beta}-\overset{H}{\underset{H}{\overset{|}{\underset{|}{C}}}}_{\alpha}-\overset{+}{O}H_2 \longrightarrow B-H^{+} + \underset{\text{Ethene}}{H-\overset{H}{\overset{|}{C}}=\overset{H}{\overset{|}{C}}-H} + H_2O$$

The reaction is reversible, and the reaction conditions are chosen to favor the forward reaction. Generally, strong acids and high temperatures favor alkene formation. Removal of the usually volatile alkene will force the reaction to the right. Unless removed, some alkenes can react with the sulfuric acid and water to reform the alcohol as illustrated in Sec. 3.8C. The net effect of heating an alcohol with sulfuric acid may be shown by the following equation for the overall reaction.

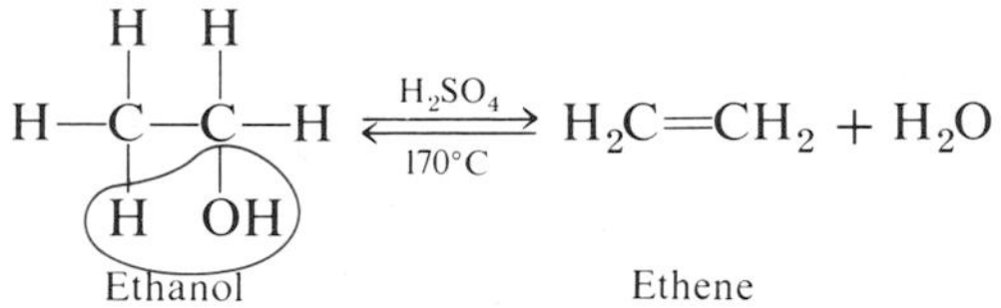

$$\underset{\text{Ethanol}}{H-\overset{H}{\underset{H}{\overset{|}{\underset{|}{C}}}}-\overset{H}{\underset{OH}{\overset{|}{\underset{|}{C}}}}-H} \underset{170°C}{\overset{H_2SO_4}{\rightleftharpoons}} \underset{\text{Ethene}}{H_2C=CH_2} + H_2O$$

The elimination of water from a tertiary alcohol proceeds via a carbocation intermediate.

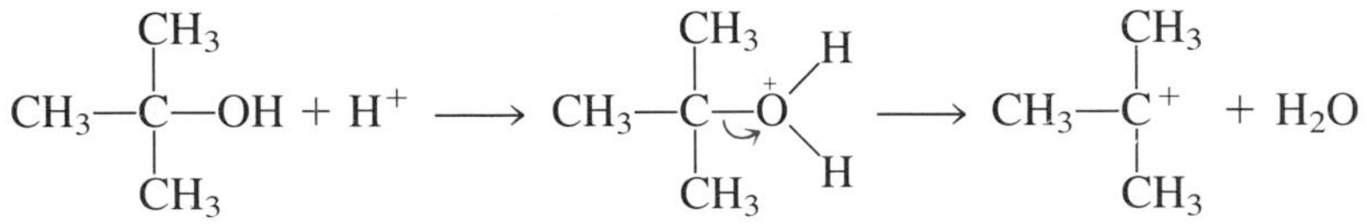

$$\underset{\text{2-Methyl-2-propanol}}{CH_3-\overset{CH_3}{\underset{CH_3}{\overset{|}{\underset{|}{C}}}}-OH} + H^{+} \longrightarrow CH_3-\overset{CH_3}{\underset{CH_3}{\overset{|}{\underset{|}{C}}}}-\overset{+}{O}H_2 \longrightarrow \underset{\textit{tert}\text{-Butyl carbocation}}{CH_3-\overset{CH_3}{\underset{CH_3}{\overset{|}{\underset{|}{C^{+}}}}}} + H_2O$$

[2]Here we are again trying to describe a structural relationship in words, using Greek letters rather than numbers. Unfortunately, there are no rules telling us when to use Greek letters rather than numbers; however, Greek letters are generally used to describe structural positions relative to some functional group. The carbon atom bonded to the functional group (hydroxyl in this example) is designated α, the next adjacent carbon β, the next γ, and so on.

$$CH_3-\overset{\overset{CH_3}{|}}{\underset{\underset{CH_2-H}{|}}{C^+}} + B: \longrightarrow CH_3-\overset{\overset{CH_3}{|}}{\underset{\underset{CH_2}{\|}}{C}} + B-H^+$$

2-Methyl-2-propene

Exercise 3.4 Draw the structures for the two alkenes possible when 2-butanol

$(CH_3-CH_2CH(OH)-CH_3)$

is heated with a strong acid. Which product will likely be formed in greater amount?

3.4 Dehydrohalogenation of Alkyl Halides

The elimination of the elements of hydrogen halide (HX) from adjacent carbons is called **dehydrohalogenation.** This reaction is another common method that leads to the preparation of an alkene.

$$H-\overset{\overset{H}{|}}{\underset{\underset{H}{|}}{C}}-\overset{\overset{H}{|}}{\underset{\underset{Br}{|}}{C}}-\overset{\overset{H}{|}}{\underset{\underset{H}{|}}{C}}-H + KOH \xrightarrow[\text{solution}]{\text{alcoholic}} H-\overset{\overset{H}{|}}{C}=\overset{\overset{H}{|}}{C}-\overset{\overset{H}{|}}{\underset{\underset{H}{|}}{C}}-H + KBr + H_2O$$

2-Bromopropane Propene

Since this reaction requires a very strong base, the dehydrohalogenation is usually carried out using a solution of potassium hydroxide in alcohol. The elimination of hydrogen halide appears to take place in the following manner.

Hydrogen as a proton is removed from the β-carbon atom by a strong base, shown in the equation as the hydroxide ion. At the same time the halogen on the α-carbon atom moves off as a halide ion, assisted by its attraction to the solvent molecules. This migration is accompanied by a simultaneous shifting of an electron pair.

$$OH^- \quad H-\overset{\overset{H}{|}}{\underset{\underset{H}{\beta|}}{C}}-\overset{\overset{H}{|}}{\underset{\underset{Br}{\alpha|}}{C}}-CH_3 \xrightarrow{C_2H_5OH} H-\overset{\overset{H}{|}}{C}=\overset{\overset{H}{|}}{C}-CH_3 + Br^- + H_2O$$

Dehydrohalogenation is thus a concerted reaction in which the carbon–hydrogen and the carbon–halogen σ bonds are broken at the same time the new carbon–carbon π bond is formed. Later (Sec. 6.5), a more detailed description will be given of the changes that can occur during the course of a reaction of the foregoing type, which is called an E2 elimination reaction.

When either of two β-carbon atoms can supply the hydrogen to be eliminated with the halogen atom, two isomeric alkenes are possible. If such is the case, the alkene most likely to result will be the more stable one. This appears to be the one with *the greater number of alkyl groups attached to the doubly bonded carbons*. Thus, for example, two alkenes are possible when 2-bromobutane is heated with alcoholic potassium hydroxide, but the amount of one isomer produced exceeds that of the other in a ratio of 4:1.

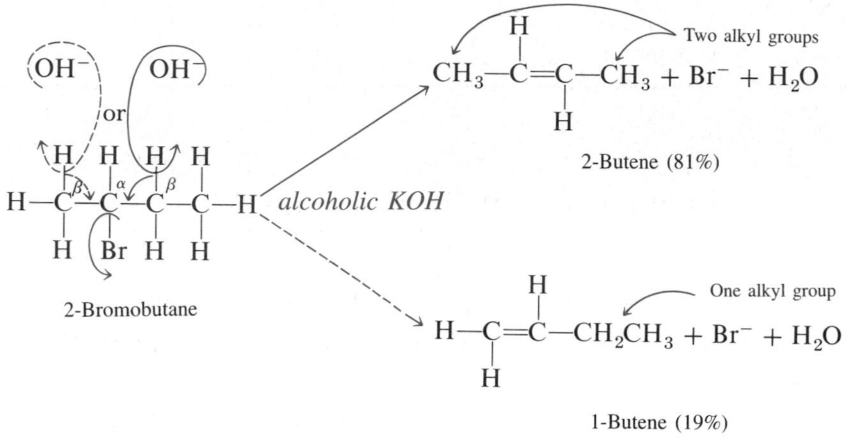

$$H{-}CH_2{-}CHBr{-}CH_2{-}CH_3 \xrightarrow{\text{alcoholic KOH}} CH_3{-}CH{=}CH{-}CH_3 + Br^- + H_2O$$

2-Bromobutane (OH⁻ or OH⁻; α, β)

2-Butene (81%) — Two alkyl groups

$$\dashrightarrow H_2C{=}CH{-}CH_2CH_3 + Br^- + H_2O$$

1-Butene (19%) — One alkyl group

EXAMPLE

Two isomeric alkenes are produced when 2-chloro-2-methylbutane is heated with alcoholic KOH. The yield of one is 86%, the other 14%. Which of the isomers is produced in greater amount?

Solution An inspection of the 2-chloro-2-methylbutane structure shows that hydrogen can be removed from either β-carbon. Removal of hydrogen from C-3 along with the halogen will produce the more stable alkene.

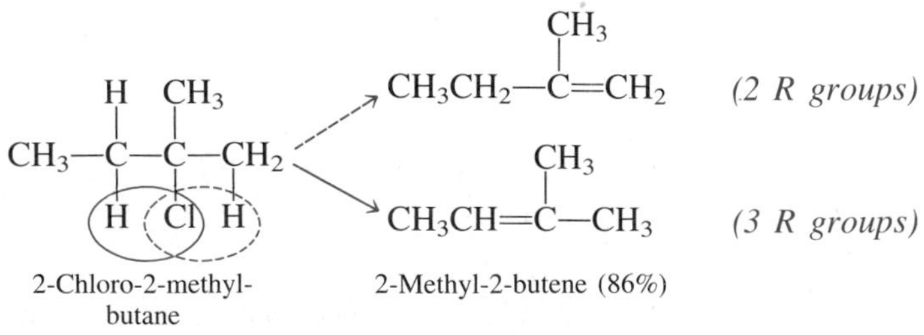

$$CH_3{-}CH_2{-}CCl(CH_3){-}CH_3 \dashrightarrow CH_3CH_2{-}C(CH_3){=}CH_2 \quad (2\ R\ groups)$$

$$CH_3{-}CH_2{-}CCl(CH_3){-}CH_3 \rightarrow CH_3CH{=}C(CH_3){-}CH_3 \quad (3\ R\ groups)$$

2-Chloro-2-methyl-butane

2-Methyl-2-butene (86%)

Exercise 3.5 Two isomeric alkenes are produced when 2-bromo-2-methylpentane is heated with alcoholic KOH. Give the name and structure of the isomer produced in greater amount.

3.5 Physical Properties of the Alkenes

The physical properties of the alkenes are not too unlike those of the corresponding alkanes. The lower members of the alkene series are gases, while the higher members are liquids and solids. The alkenes have slightly higher densities than the alkanes (Table 3.1), and are more soluble in water. The higher densities of the alkenes may be attributed to their geometry. The increase in water solubility may be attributed to the attraction of the π electrons for the polar water molecule.

TABLE 3.1 Physical Constants of the Alkenes

Name	Formula	mp, °C	bp, °C	Density as Liquid
Ethene	$CH_2{=}CH_2$	−169.2	−103.7	—
Propene	$CH_3CH{=}CH_2$	−182.3	−47.4	0.5193
1-Butene	$CH_3CH_2CH{=}CH_2$	−185.4	−6.3	0.5951
2-Butene	$CH_3CH{=}CHCH_3$	*cis* −138.91	3.7	0.6213
		trans −105.55	0.9	0.6042
2-Methylpropene (Isobutylene)	$CH_3{-}C(CH_3){=}CH_2$	−140.35	−6.9	0.5942
1,3-Butadiene	$CH_2{=}CH{-}CH{=}CH_2$	−108.91	−4.4	0.6211
2-Methyl-1,3-butadiene (Isoprene)	$CH_2{=}C(CH_3){-}CH{=}CH_2$	−146	35	0.6810
1-Pentene	$CH_2{=}CH(CH_2)_2CH_3$	−138	30.0	0.6405
1-Hexene	$CH_2{=}CH(CH_2)_3CH_3$	−139.82	63.4	0.6731

Reactions of the Alkenes

The chemical behavior most characteristic of the alkenes is the ability to form derivatives by addition to the double bond.

$$\overset{|}{\underset{}{-C}}=\overset{|}{C}- + AB \longrightarrow -\overset{|}{\underset{A}{\underset{|}{C}}}-\overset{|}{\underset{B}{\underset{|}{C}}}-$$

In such reactions the addition of hydrogen, the halogens (Cl_2, Br_2, I_2), the halogen acids (HCl, HBr), the hypohalous acids (HOCl, HOBr), sulfuric acid, and water all result in the formation of saturated compounds. Here the π bond plays the role of an electron pair donor. The principal reactions of the alkenes are discussed in the following sections.

3.6 Addition of Hydrogen

Hydrogenation. Alkenes add hydrogen under pressure and in the presence of a catalyst to produce saturated hydrocarbons. A hydrogenation carried out in this manner is called **catalytic hydrogenation.**

$$H-\overset{H}{\overset{|}{C}}=\overset{H}{\overset{|}{C}}-H + H_2 \xrightarrow[\text{pressure, heat}]{\text{Ni}} H-\overset{H}{\overset{|}{\underset{H}{\underset{|}{C}}}}-\overset{H}{\overset{|}{\underset{H}{\underset{|}{C}}}}-H$$

Ethene Ethane

Without a catalyst the rate of hydrogenation is negligible even though the addition of hydrogen to the double bond is an exothermic reaction, that is, even though the products are more stable and have lower energy than the reactants. The function of the catalyst appears to be to lower the very high energy of activation (Sec. 1.12) of the overall reaction by changing the mechanism of the reaction to one with a lower overall energy of activation. Such a change would increase the rate of hydrogenation. The exact mechanism of hydrogenation is complex and apparently involves metal–hydrogen and metal–carbon bonds. The schematic picture shown in Fig. 3.5 is sufficient for our purposes and shows the tendency for both hydrogens to add to the double bond from the same side—that is, from the side complexed with the catalyst surface by interaction with the π electrons. This mode of addition is called *syn* addition (Gr. *syn,* together) and *usually* results in the formation of only the *cis*-isomer if isomeric products are possible. For example, the hydrogenation of 1,2-dimethylcyclopentene yields only *cis*-1,2-dimethylcyclopentane, as shown in Fig. 3.5.

The best *heterogeneous* (insoluble) catalysts for the hydrogenation of unsaturated compounds are finely divided transition metals such as platinum and palladium, which are known to absorb large quantities of gaseous hydrogen. Nickel catalysts are also good and

FIGURE 3.5 Schematic Representation of the Hydrogenation of 1,2-Dimethylcyclopentene

1,2-Dimethylcyclopentene

cis-1,2-Dimethylcyclopentane

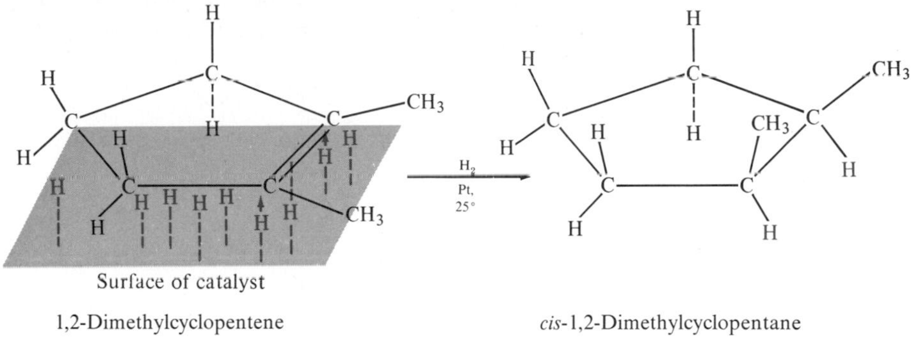

less expensive, but usually require temperatures higher than those needed with platinum or palladium. In recent years *homogeneous* (soluble) catalysts have been used increasingly for hydrogenation. These catalysts are organometallic complexes such as $RhCl(P(C_6H_5)_3)_3$ (Wilkinson's catalyst). Hydrogenation of unsaturated compounds is a useful and common reaction both in the laboratory and in industry. A hydrogenation reaction of great commercial importance is that used to convert edible vegetable oils into semisolid shortenings and butterlike margarines (Sec. 12.4). This process is referred to as "hardening."

Exercise 3.6 Draw structures for three alkenes that on hydrogenation would yield 2-methylbutane. Name each according to IUPAC rules.

3.7 Addition of Halogens

Bromination. An alkene reacts with bromine to add the halogen across the double bond to place one bromine atom on each carbon. The alkene is usually added to a carbon tetrachloride solution of bromine. The reddish color of bromine in carbon tetrachloride disappears as the bromine adds to the alkene. The reaction is thus a test reaction if the presence of the double bond in an unknown compound is suspected. The mechanism by which bromine adds to ethene may be explained as follows.

Ethene, although a nonpolar molecule, may, because of the mobility of the π electrons, cause a polarization of the bromine molecule—that is, a displacement of the pair of electrons joining the bromine atoms toward one of the atoms. One bromine with six electrons combines with the π electrons of ethene to produce a positively charged intermediate, leaving the other bromine as a negative bromide ion. The positively charged intermediate formed with the bromine may be either a carbocation, (a) or (c), or a bromonium ion, (b).

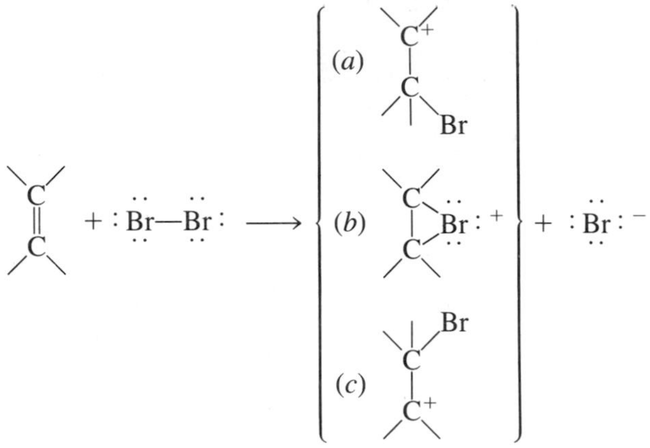

$$\text{C}{=}\text{C} + :\ddot{\text{Br}}{-}\ddot{\text{Br}}: \longrightarrow \left\{ \begin{array}{ll} (a) & \text{C}^{+}{-}\text{C}{-}\text{Br} \\ (b) & \text{C}{-}\text{C}\ \text{bridged by}\ \text{Br}:^{+} \\ (c) & \text{Br}{-}\text{C}{-}\text{C}^{+} \end{array} \right\} + :\ddot{\text{Br}}:^{-}$$

The negative bromide ion then combines with the positively charged intermediate. Experimental studies show that the bromide ion approaches from a direction opposite that of the bromine already bonded.

This mode of addition is called *anti*-addition.

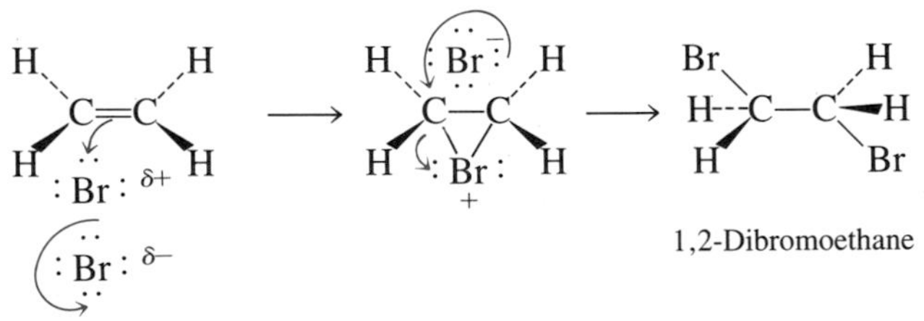

$$\text{H}_2\text{C}{=}\text{CH}_2 + \text{Br}^{\delta+}{-}\text{Br}^{\delta-} \longrightarrow \text{H}_2\text{C}{-}\text{CH}_2\ (\text{bridged}\ \text{Br}^{+}) + \text{Br}^{-} \longrightarrow \text{BrH}_2\text{C}{-}\text{CH}_2\text{Br}$$

1,2-Dibromoethane

The existence of the intermediate bromonium ion or the carbocation has been established by the addition of bromine to an alkene in the presence of other negative ions. In such additions one bromine remains attached to one of the carbons, but a negative ion other than bromide can make the *anti* attack. The following reaction illustrates a "mixed" addition of this type.

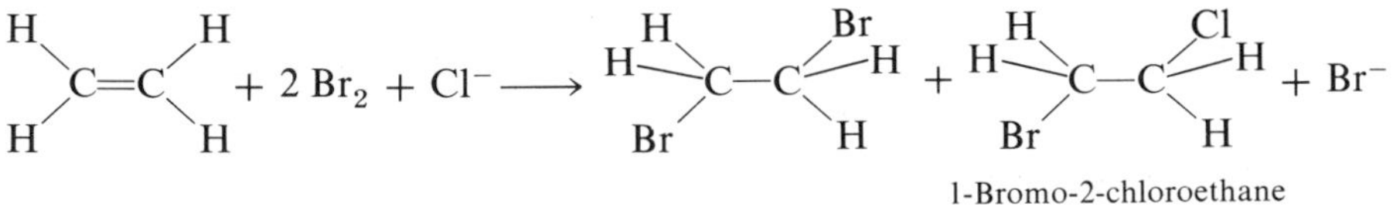

$$\text{H}_2\text{C}{=}\text{CH}_2 + 2\,\text{Br}_2 + \text{Cl}^{-} \longrightarrow \text{BrH}_2\text{C}{-}\text{CH}_2\text{Br} + \text{BrH}_2\text{C}{-}\text{CH}_2\text{Cl} + \text{Br}^{-}$$

1-Bromo-2-chloroethane

The *anti*-addition of bromine to the double bond of cyclopentene is illustrated schematically in Fig. 3.6. Whenever isomeric products are possible, *anti*-addition will give the *trans*-isomer.

FIGURE 3.6 Schematic Representation of the Bromination of Cyclopentene

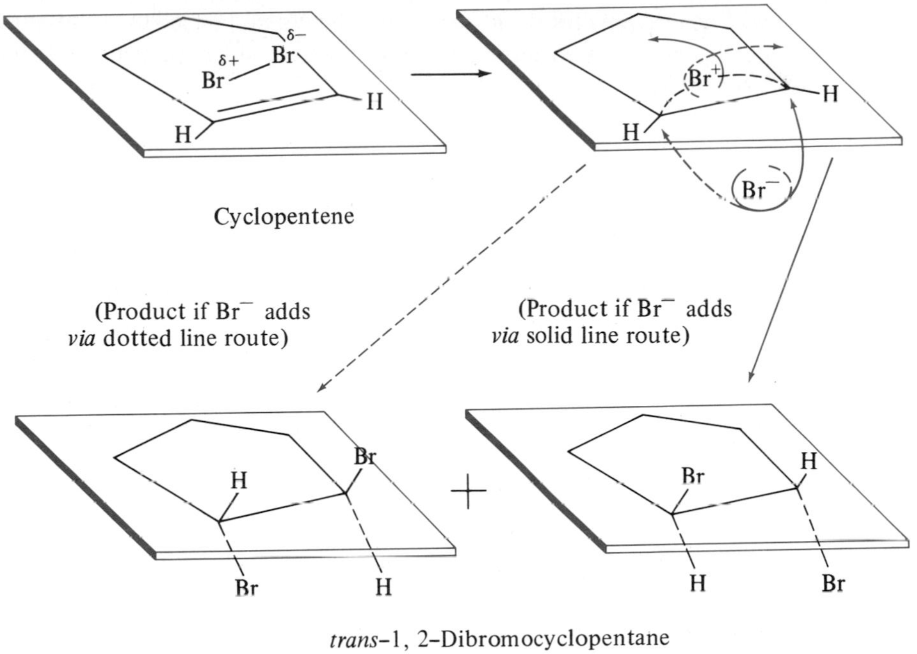

Cyclopentene

(Product if Br^- adds *via* dotted line route)

(Product if Br^- adds *via* solid line route)

trans-1, 2-Dibromocyclopentane

Exercise 3.7 Draw the chair conformation of the product that results when bromine is added to cyclohexene.

3.8 Addition of Acids to Unsymmetrical Alkenes (Markovnikov's Rule)

A. Addition of Hydrogen Halide. Hydrogen halides add readily to alkenes to produce alkyl halides, R—X. Thus hydrogen bromide adds to ethene to produce ethyl bromide according to the following equation.

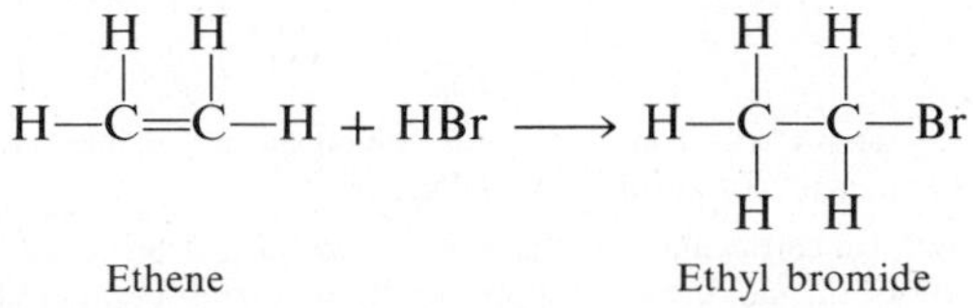

$$\mathrm{H_2C{=}CH_2} + \mathrm{HBr} \longrightarrow \mathrm{H{-}CH_2{-}CH_2{-}Br}$$

Ethene — Ethyl bromide

In this particular reaction there is only one way in which the halogen acid can add across the double bond, and therefore only one product is obtainable. This is not the case with an unsymmetrical alkene in which the groups bonded to the unsaturated carbon atoms are different. For example, hydrogen bromide adds to propene to produce *mainly* isopropyl bromide.

$$\underset{\text{Propene}}{CH_3-\overset{H}{\underset{\delta+}{C}}=\overset{H}{\underset{\delta-}{C}}-H} + H^+ \; :\ddot{Br}:^- \longrightarrow \underset{\substack{\text{2-Bromopropane}\\ \text{(Isopropyl bromide)}\\ \text{bp } 59.4^\circ}}{CH_3-\overset{H}{\underset{Br}{C}}-CH_3}$$

Note that in this reaction the bromide ion seeks the central rather than the terminal carbon atom. The Russian chemist Markovnikov in 1871 formulated an empirical rule regarding this mode of addition. The rule simply states: ''*In the addition of acids to unsymmetrical alkenes, the negative portion of the species added will seek the carbon atom holding the fewer hydrogen atoms.*'' This empirical rule has few exceptions.

The reason for the success of the Markovnikov rule is easily explained. As was pointed out in Sec. 3.2, the double bond is an electron-rich bond. As such, it tends to behave like a Lewis base and reacts with electrophilic (electron-seeking) reagents, such as hydrogen bromide.

$$H-\ddot{Br}: + CH_3-\underset{H}{C}=\underset{H}{C}-H \nearrow CH_3-\underset{H}{\overset{+}{C}}-\underset{H}{\overset{H}{C}}-H + :\ddot{Br}:^- \longrightarrow \underset{\text{Major product}}{CH_3-\overset{Br}{CH}-CH_3}$$

$$\searrow\!\!\!\!\times \; CH_3-\underset{H}{\overset{H}{C}}-\underset{H}{\overset{H}{C^+}} + :\ddot{Br}:^- \longrightarrow \underset{\text{Minor product}}{CH_3-CH_2-CH_2-Br}$$

In the first step a carbocation is formed, which then reacts with bromide ion to form the product.[3] The rate at which a carbocation forms in a reaction has been found to depend on its stability, and the stability of a carbocation increases with the number of alkyl substituents on the positive carbon atoms, or in the order: tertiary $>$ secondary $>$ primary.[4]

[3]The dotted arrow with a cross through it indicates a possible but unfavorable reaction path that is observed to a small extent or not at all in the laboratory.

[4]The *class* of a carbocation, carbanion, or radical is determined by the number of substituents on the positive carbon atom, negative carbon atom, or atom bearing the odd electron, respectively: *tertiary,* 3 substituents; *secondary,* 2 substituents; *primary,* 1 substituent.

Therefore, the secondary carbocation with two methyl groups is formed as an intermediate more readily than the primary carbocation bearing only one ethyl group.

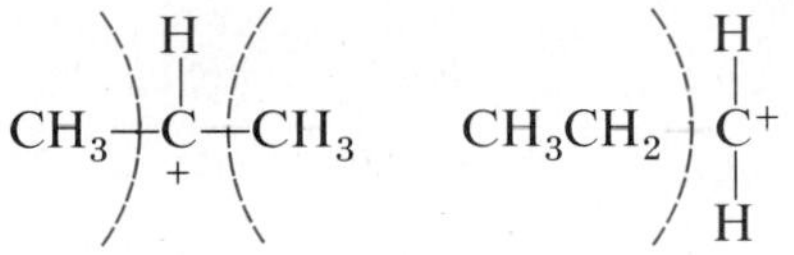

$$CH_3-\overset{H}{\underset{+}{C}}-CH_3 \qquad CH_3CH_2-\overset{H}{\underset{H}{C^+}}$$

A secondary carbocation — A primary carbocation

Exercise 3.8 What product would be expected if 2-methyl-2-butene were reacted with HBr?

Although hydrogen bromide adds to unsymmetrical alkenes in accordance with the Markovnikov rule in the dark and in the absence of peroxides, in the presence of light or peroxides hydrogen bromide adds to alkenes in an *anti*-Markovnikov manner. The mechanism of the addition under these conditions cannot proceed through a carbocation intermediate. The *anti*-Markovnikov mode of addition of HBr can be explained in terms of a free radical chain mechanism. Indicating the peroxide involved simply as R—O:O—R, we may show the different steps in the "peroxide effect" as follows:

$$R-O\!:\!O-R \longrightarrow 2\,R-O\cdot$$

$$R-O\cdot + HBr \longrightarrow R-OH + Br\cdot$$

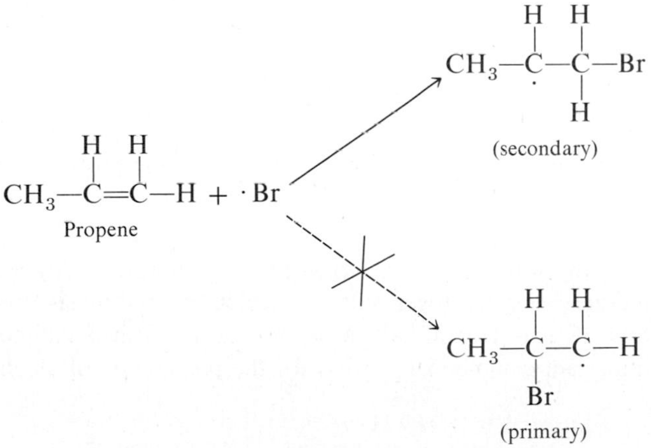

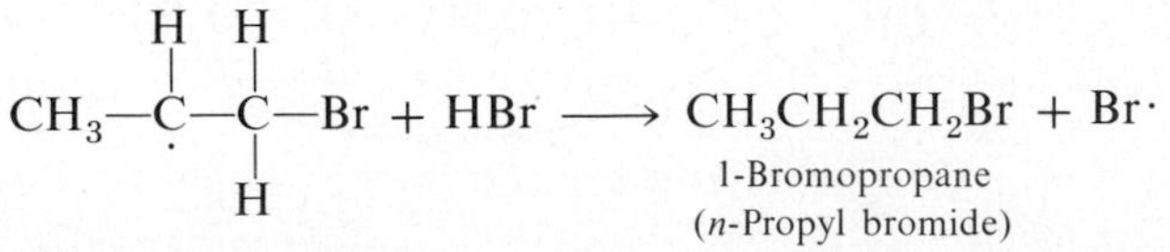

$$CH_3-\overset{H}{C}=\overset{H}{C}-H + \cdot Br \longrightarrow CH_3-\overset{H}{\underset{\cdot}{C}}-\overset{H}{\underset{H}{C}}-Br \text{ (secondary)}$$

Propene

$$CH_3-\overset{H}{C}=\overset{H}{C}-H + \cdot Br \not\longrightarrow CH_3-\overset{H}{\underset{Br}{C}}-\overset{H}{\underset{\cdot}{C}}-H \text{ (primary)}$$

$$CH_3-\overset{H}{\underset{\cdot}{C}}-\overset{H}{\underset{H}{C}}-Br + HBr \longrightarrow CH_3CH_2CH_2Br + Br\cdot$$

1-Bromopropane (*n*-Propyl bromide) bp 71°

Since the relative stabilities of free radicals parallel those of carbocations, namely *tertiary > secondary > primary,* the secondary radical will form almost to the exclusion of the primary radical.

Stability: $R_3C\cdot > R_2CH\cdot > RCH_2\cdot > H_3C\cdot$

While the peroxide-catalyzed addition of hydrogen halide to an alkene is a useful reaction, unfortunately it works only with hydrogen bromide.

Exercise 3.9 Two alkyl halides are possible when isobutylene, $(CH_3)_2C{=}CH_2$, is treated with HBr. Under what conditions would each predominate?

B. Addition of Hypohalous Acids. Markovnikov's rule may be extended to cover the addition of other acids. The addition of chlorine or bromine to an alkene, when carried out in the presence of water, has the net result of adding a hypohalous acid, HOX. The reaction results in the formation of *halohydrins*. Halohydrins have a hydroxyl group and a halogen atom on adjacent carbons. The addition of an aqueous solution of chlorine to propene produces propylene chlorohydrin.

$$CH_3{-}CH{=}CH_2 + H{:}\ddot{O}{:}Cl \longrightarrow CH_3{-}CH(OH){-}CH_2Cl$$

Propene → 1-Chloro-2-propanol (Propylene chlorohydrin)

In this addition, you will note, chlorine is the positive part of the species added.

C. Addition of Sulfuric Acid (Hydration of Alkenes). Alkenes react with sulfuric acid to produce alkyl hydrogen sulfates, which, on hydrolysis, yield alcohols. Hydration (addition of water) of an alkene is thus accomplished indirectly in this manner and is an important commercial process for the preparation of alcohols from petroleum products.

$$CH_3{-}CH{=}CH_2 + H^+ \; {:}\ddot{O}{-}SO_3H \longrightarrow CH_3{-}CH(OSO_3H){-}CH_3$$

Isopropyl hydrogen sulfate

$$CH_3-\underset{\displaystyle OSO_3H}{\overset{\displaystyle H}{\underset{|}{\overset{|}{C}}}}-CH_3 + \underset{\displaystyle H}{\underset{|}{O}}-H \longrightarrow CH_3-\underset{\displaystyle OH}{\overset{\displaystyle H}{\underset{|}{\overset{|}{C}}}}-CH_3 + H_2SO_4$$

2-Propanol
(Isopropyl alcohol)

Exercise 3.10 Which alkene would be required to produce 2-methyl-2-propanol via the acid-catalyzed hydration reaction?

D. "Anti-Markovnikov" Hydration of Alkenes. Because the addition of ionic reagents to isolated carbon–carbon double bonds follows Markovnikov's rule (Sec. 3.8), the alcohol that results from the hydration of an alkene is always the more highly branched of the two possibilities. As a consequence, the only primary alcohol that can be prepared in this way is ethanol from ethylene. However, in many instances the "anti-Markovnikov" hydration product can be made by an indirect hydration of alkenes called **hydroboration,** a reaction discovered in 1956 by H. C. Brown.[5]

Hydroboration consists of the addition of the B—H bond to an alkene, a rapid and quantitative reaction. The usual reagent is diborane, B_2H_6, which is formed from sodium borohydride and boron trifluoride according to the following equation.

$$3\,NaBH_4 + 4\,BF_3 \longrightarrow 2\,B_2H_6 + 3\,NaBF_4$$

Sodium borohydride — Diborane

Diborane is an unusual molecule with *three-center* bonds and a *bridging* hydrogen atom. However, we need not be concerned with the bonding in this molecule, for it usually behaves as if it were BH_3. In its reaction with an alkene, the electron pair–deficient boron of BH_3 becomes attached to the less substituted carbon of the double bond, apparently largely because of steric factors. At the same time, one of the hydrogen atoms of borane is donated to the other doubly bonded carbon. Borane has three hydrogens to confer in this manner, and thus the end product becomes a trialkyl-substituted boron. The latter is not isolated but is subsequently treated with an alkaline solution of hydrogen peroxide to yield the alcohol and boric acid. The net result is the *syn*-addition of the elements of water

[5]Herbert C. Brown (1912–), Purdue University, shared Nobel Prize in Chemistry for 1979 for his research on organoboron compounds.

across the carbon–carbon double bond to produce an alcohol in which the addition is "anti" Markovnikov. The reaction is illustrated using propylene as our starting material.

$$CH_3{-}CH{=}CH_2 + H{-}BH_2 \longrightarrow CH_3CH_2CH_2{-}BH_2$$

n-Propyl boron hydride

$$CH_3CH_2CH_2{-}BH_2 + 2\ CH_3{-}CH{=}CH_2 \longrightarrow (CH_3CH_2CH_2)_3B$$

Tri-*n*-propylborane

$$(CH_3CH_2CH_2)_3B + 3\ H_2O_2 \xrightarrow{NaOH} 3\ CH_3CH_2CH_2OH + B(OH)_3$$

n-Propyl alcohol

The stereochemistry of this reaction is shown by the following example.

$$\text{1-Methylcyclopentene} \xrightarrow{B_2H_6} \xrightarrow[(NaOH)]{H_2O_2} \textit{trans}\text{-2-Methylcyclopentanol}$$

1-Methylcyclopentene — *trans*-2-Methylcyclopentanol

Exercise 3.11 Which alkene must be your starting material in the hydroboration reaction to produce 3-methyl-2-butanol?

3.9 Oxidation with Permanganate. The Baeyer Test for Unsaturation

A. The Baeyer Test. Alkenes are oxidized with cold dilute potassium permanganate to form glycols. In the absence of other easily oxidizable groups that also react with $KMnO_4$, this reaction serves as an easily recognizable test for alkenes because as the oxidant is consumed, its bright purple color disappears, and a brown precipitate of manganese dioxide forms. The reaction is the basis for the Baeyer Test for unsaturation. Ethylene reacts with potassium permanganate to yield ethylene glycol, the widely used radiator antifreeze. The name *glycol* is used to describe any compound with hydroxyl groups on adjacent carbon atoms.

Cyclic alkenes are oxidized to *cis*-glycols, which indicates that the permanganate ion adds in a *syn* fashion to be subsequently hydrolyzed to the glycol. Cyclopentene reacts with potassium permanganate to give *cis*-1,2-cyclopentanediol.

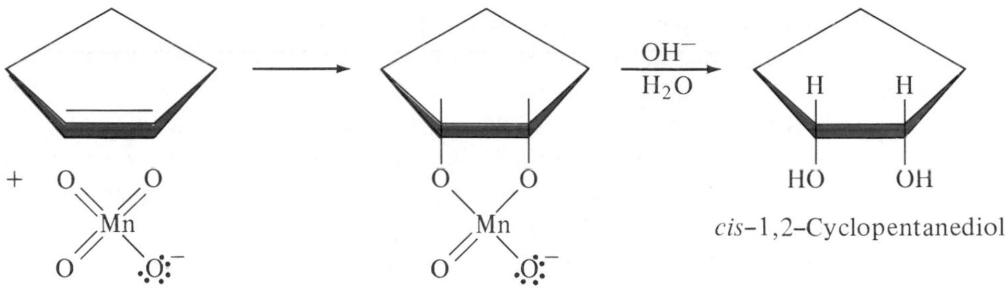

$$\text{cyclopentene} + MnO_4^- \longrightarrow \text{cyclic manganate ester} \xrightarrow[H_2O]{OH^-} \textit{cis}\text{-1,2-Cyclopentanediol}$$

cis-1,2-Cyclopentanediol

B. Oxidation with Ozone (O_3). Most alkenes react with ozone to form cyclic peroxide intermediates called **ozonides.** Ozonides may be explosive compounds and thus are not isolated but are directly decomposed with zinc metal and aqueous acetic acid or, preferably, by treatment with dimethyl sulfide ($CH_3—S—CH_3$). The ozonolysis of an alkene yields two products, each with a doubly bonded oxygen where the doubly bonded carbon atoms had originally been. Examination of the reaction products thus establishes the location of the double bond. The recent introduction of safer laboratory procedures has made ozonolysis, long used principally for diagnostic purposes, a useful preparative method.

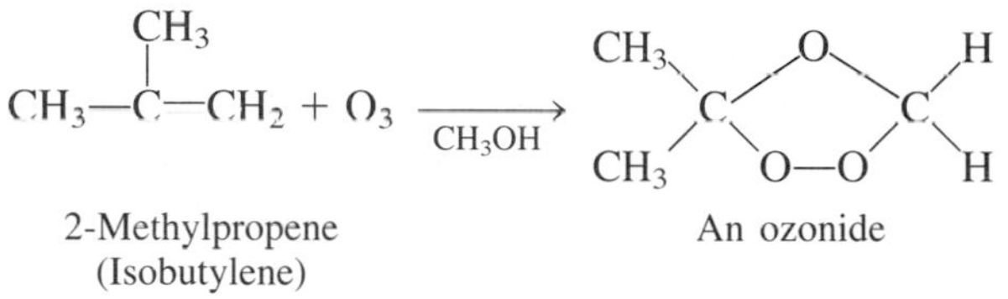

$$CH_3—\underset{}{\overset{CH_3}{\overset{|}{C}}}{=}CH_2 + O_3 \xrightarrow[CH_3OH]{} \text{An ozonide}$$

2-Methylpropene (Isobutylene)

An ozonide

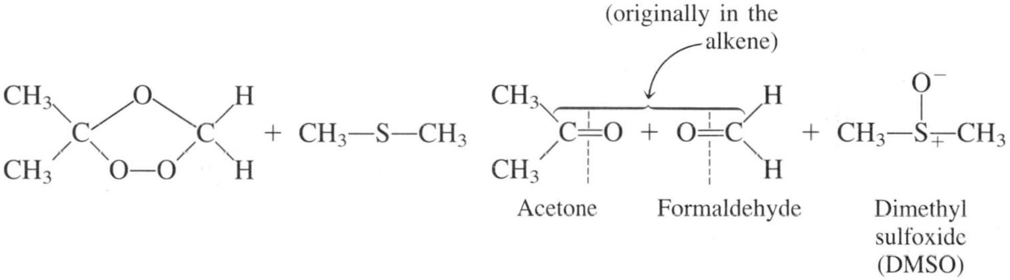

$$\text{Ozonide} + CH_3—S—CH_3 \longrightarrow (CH_3)_2C{=}O + O{=}CH_2 + CH_3—\overset{O^-}{\overset{|}{S^+}}—CH_3$$

(originally in the alkene)

Acetone — Formaldehyde — Dimethyl sulfoxide (DMSO)

Exercise 3.12 In each case deduce the structure of the alkene that gave on ozonolysis (a) only acetone, $(CH_3)_2{=}O$; (b) $(C_2H_5)CH_3C{=}O$ and formaldehyde; (c) only acetaldehyde, $CH_3—CHO$.

Ozone is a highly reactive gas that will oxidize many substances at temperatures not high enough for them to be oxidized by atmospheric oxygen. The deterioration of rubber with a resultant loss of elasticity is due largely to ozone oxidation. This is especially apparent when rubber compounds are in close proximity to electrical discharges since ozone can be produced in this manner.

3.10 Polymerization or Self-Addition

Polymerization is a process in which molecules of a low-molecular-weight compound **(monomer)** react with themselves over and over again to form **polymers** or large, long-chain molecules with many *recurring structural units*. A number of simple derivatives of ethylene may be polymerized to yield useful polymeric products, such as Teflon, Orlon, polypropylene, or polyethylene itself, as shown in the following equation.

$$\underset{\text{(Ethylene)}}{\underset{\text{Monomer}}{(n+2)\ CH_2{=}CH_2}} \xrightarrow[\text{or initiator}]{\text{catalyst}} \underset{\text{(Polyethylene)}}{\underset{\text{Polymer}}{X{-}CH_2{-}CH_2{\leftarrow}CH_2{-}CH_2{\rightarrow_n}CH_2{-}CH_2{-}Y}}$$

Note that the polymer consists of a long chain (n = 100–1,000) of *ethylene* units, each representing one molecule of monomer. This type of polymerization is called **chain** or **addition** polymerization because it proceeds by a chain reaction involving free-radical, anionic, or cationic intermediates. Polymerization proceeding by a free-radical chain mechanism is both very common and relatively simple and will be used to illustrate the chain polymerization process.

Free-radical polymerization is **initiated** by generating a few free radicals from some good source of radicals, such as an organic peroxide, R—O—O—R, which has a relatively weak —O—O— bond that will break on mild heating to form two alkoxy radicals. Addition of the alkoxy radical to the monomer, ethylene, starts the growth of the chain, which proceeds to **propagate** itself by repeated addition to further monomer units until the growth of the chain is **terminated** by some competitive process. The following series of equations illustrates the *initiation, propagation,* and *termination* steps.

Initiation: Free radicals are produced by an initiator.

$$R{-}O{:}O{-}R \longrightarrow 2\ R{-}O\cdot \qquad \text{(alkoxy radical)}$$

$$R{-}O\cdot + \underset{\text{Ethylene}}{CH_2{=}CH_2} \longrightarrow R{-}O{-}CH_2{-}CH_2\cdot$$

Propagation: Continued addition of alkene produces a larger free radical.

$$R{-}O{-}CH_2{-}CH_2\cdot + n\ CH_2{=}CH_2 \longrightarrow R{-}O{\leftarrow}CH_2{-}CH_2{\rightarrow_n}CH_2{-}CH_2\cdot$$

Termination: Two free radicals finally couple or disproportionate.

$$R{-}O{\leftarrow}CH_2{-}CH_2{\rightarrow_n}CH_2{-}CH_2\cdot + \cdot CH_2{-}CH_2{\leftarrow}CH_2{-}CH_2{\rightarrow_m}O{-}R$$

Couple:

$$R{-}O{\leftarrow}CH_2{-}CH_2{\rightarrow_{n+m+2}}O{-}R$$

Disproportionate:

$$R{-}O{\leftarrow}CH_2{-}CH_2{\rightarrow_m}CH{=}CH_2 + CH_3{-}CH_2{\leftarrow}CH_2{-}CH_2{\rightarrow_n}O{-}R$$

Polyethylene

Table 3.2 lists some of the ethylene derivatives that have been used to prepare useful polymeric materials. Some are simple alkenes, but others contain other functional groups that modify the properties of the polymer, making possible all the applications shown in the table. In some instances more than one monomer will be used in a polymerization reaction to make a **copolymer;** for example, the **copolymerization** of styrene (Table 3.2) and butadiene (Sec. 3.11) is used to make most of the synthetic rubber for automobile tires. The impact of the synthetic polymers on contemporary life has been tremendous, virtually revolutionizing whole industries such as the packaging, textile, and paint industries. New products such as "no-stick" cooking utensils, thermal clothing rivaling goose down, soft contact lenses, artificial joints, can be added to an ever-growing list of synthetic polymeric products.

TABLE 3.2 Common Addition Polymers from $R{-}CH{=}CH_2 \rightarrow {+}CH(R){-}CH_2{+}$

Structure of R	Name of Monomer	Name of Polymer	Use
$-H$	Ethylene	Polyethylene	Film, conduit, rubberlike articles
$-CH_3$	Propylene	Polypropylene	Molded and extruded plastics, film, fibers for garments and carpeting
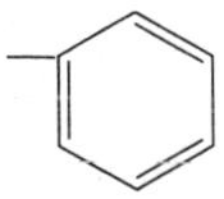 $-C_6H_5$ (phenyl)	Styrene	Polystyrene	Insulation, breakage-proof packaging, molded articles
$-Cl$	Vinyl chloride	Polyvinyl chloride	Electrical insulation, films, rubberlike articles
$-F$	Vinyl fluoride	Polyvinyl fluoride	Surface coatings
$-CN$	Vinyl cyanide (Acrylonitrile)	Polyacrylonitrile (Orlon)	Fibers for garments, carpeting
$-O-C(=O)-CH_3$	Vinyl acetate	Polyvinyl acetate	Films, fibers, molded articles
$-O-CH(C_3H_7)OCH-CH_2-$	Vinyl butyral	Polyvinyl butyral	Laminate in safety glass

Exercise 3.13 Polyisobutylene (which is used in inner tubes) is prepared by the cationic chain polymerization of isobutylene, $(CH_3)_2C{=}CH_2$. Assume that you have a good source of a carbocation R^+ to use as an initiator and write a mechanism for this polymerization.

3.11 Dienes

Dienes in which the two double bonds are separated by one or more carbon atoms behave very much like the simple alkenes, and such double bonds are called **isolated** double bonds. Dienes, or polyenes, which have a system of alternating double and single bonds, may show reactions not characteristic of isolated double bonds, and such systems are called **conjugated** double bonds. The simplest conjugated system is 1,3-butadiene. It is also probably the most important diene, for it is used as a copolymer in the preparation of much of our synthetic rubber such as styrene–butadiene rubber and polybutadiene. 1,3-Butadiene is also used in the manufacture of such high impact strength plastics as the ABS (*A*crylonitrile–*B*utadiene–*S*tyrene) polymers.

One important characteristic of conjugated dienes is their greater stability than that of isolated dienes with the same carbon skeleton. We can explain this greater stability by a closer examination of 1,3-butadiene. For example, when we show butadiene as

$$CH_2{=}CH{-}CH{=}CH_2$$

or

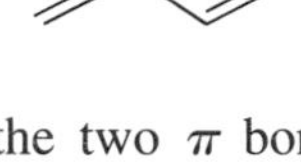

we are showing its structure as if the two π bonds were independent of each other. Actually, they are not entirely independent. If we redraw 1,3-butadiene as shown below, we can see that the $2p$ orbitals comprising the π system can overlap between carbons 2 and 3 as well as between 1 and 2 or 3 and 4. Thus the π electrons can move to a limited extent over the whole system and are said to be delocalized. The delocalization of the π electrons not only adds somewhat to the stability of the butadiene molecule but also alters its chemical behavior. However, the extent of π bond delocalization in butadiene is sufficiently small that the simple Lewis structure is a satisfactory representation (as shown at the top of the following page).

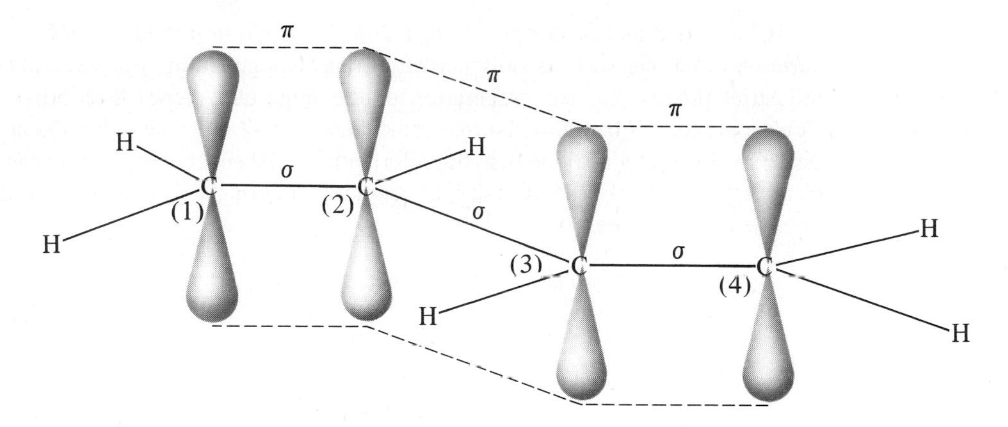

To illustrate the reactivity of butadiene, we will examine possible reactions of the compound with bromine.

$$\overset{(4)}{CH_2}{=}\overset{(3)}{CH}{-}\overset{(2)}{CH}{=}\overset{(1)}{CH_2} + :\ddot{Br}{-}\ddot{Br}: \longrightarrow$$

$$\left[CH_2{=}CH{-}\overset{+}{C}H{-}CH_2Br \longleftrightarrow \overset{+}{C}H_2{-}CH{=}CH{-}CH_2Br \right]$$

I ⟷ II

I $\xrightarrow{Br^-}$ $CH_2{=}CH{-}CHBr{-}CH_2Br$

3,4-Dibromo-1-butene
III

II $\xrightarrow{Br^-}$ $BrCH_2{-}CH{=}CH{-}CH_2Br$ (H and H trans)

trans-1,4-Dibromo-2-butene
IV

The reaction of butadiene with bromine should yield cation I. However, cation I differs from cation II only in the arrangement of electrons (the shift of a pair of electrons converts I into II, as shown). Thus I and II are a pair of resonance structures contributing to a hybrid structure for the cationic intermediate (Sec. 1.13). The true structure of the cation is that of neither I nor II, but a hybrid of the two. Resonance hybrids may react as though they had the structure of only one of the contributing forms, or they may react as if they were a mixture of several of the forms. In the present case the positive charge is almost equally distributed between the two carbon atoms shown with positive charges in I and II; therefore, bromide ion adds to the cationic intermediate to give products derived from both forms I and II to yield 3,4-dibromo-1-butene (III) and *trans*-1,4-dibromo-2-butene (IV). The mode of addition leading to structure III is called **1,2-addition;** that leading to structure IV is called **1,4-addition.**

Conjugated dienes also give both 1,2- and 1,4-addition products in free-radical chain *addition* reactions such as polymerization and halogenation. For example, the current industrial process for the preparation of the important diene, 2-chloro-1,3-butadiene, "chloroprene," is based on the free-radical addition of chlorine to butadiene in the vapor phase, followed by alkaline dehydrochlorination. The success of the process is due to the discovery of a catalyst (CuCl) for the isomerization of the 1,4- to the 3,4-isomer of the intermediate dihalide.

$$CH_2{=}CH{-}CH{=}CH_2 \xrightarrow[300°]{Cl_2} CH_2{=}CH{-}CH(Cl){-}CH_2Cl \ \text{(3,4-dichloro-1-butene)} \xrightarrow[\text{alkali } 85°]{\text{dilute}} CH_2{=}CH{-}C(Cl){=}CH_2 \ \text{(2-chloro-1,3-butadiene (chloroprene))}$$

$$CH_2{=}CH{-}CH{=}CH_2 \xrightarrow[300°]{Cl_2} Cl{-}CH_2{-}CH{=}CH{-}CH_2{-}Cl \ \text{(cis- and trans-1,4-dichloro-2-butene)} \xrightarrow{CuCl} \text{3,4-dichloro-1-butene}$$

Chloroprene can be polymerized through a 1,4-addition of diene units into a synthetic rubber called "neoprene." Neoprene, unlike natural rubber, is resistant to attack by greases and oils and finds numerous uses in the automotive industry. It is used primarily as insulation and in sheathing for wires and cables, in hose and hydraulic lines, and in many other rubber components that are likely to come into contact with lubricants.

Exercise 3.14 Write a mechanism for the *free-radical chain addition* of chlorine to 1,3-butadiene that will explain the products obtained.

Exercise 3.15 The polymerization of butadiene with free-radical catalysts usually produces a type of polymer in which 80% of the recurring units are 1,4-units, $-CH_2-CH{=}CH-CH_2-$, and 20% are 1,2-units:

$$\left(CH_2{-}\underset{\displaystyle CH{=}CH_2}{\underset{|}{CH}}{-} \right)$$

Write a mechanism that will explain the formation of both types of recurring unit.

3.12 The Diels–Alder Reaction

One of the most important reactions of conjugated dienes is the Diels–Alder[6] reaction:

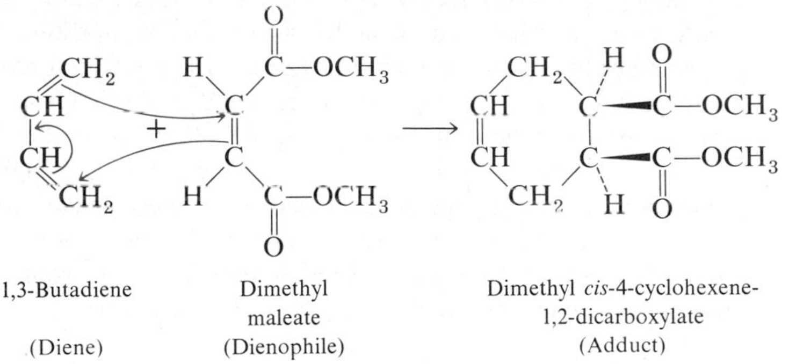

1,3-Butadiene (Diene) + Dimethyl maleate (Dienophile) ⟶ Dimethyl *cis*-4-cyclohexene-1,2-dicarboxylate (Adduct)

The reaction involves a **diene** and a **dienophile** to give an **adduct** by a concerted [4 + 2]-**cycloaddition** reaction that requires no catalyst. The designation [4 + 2] indicates that a 4-π-electron system combines with a 2-π-electron system during the course of the reaction. Although the dienophile can be a simple alkene (at the expense of higher reaction temperatures and elevated pressure), electron-attracting groups, such as —COR (aldehyde or ketone), —CO_2R (carboxylic acid or its ester), and —C≡N, facilitate the reaction. The reaction is **stereospecific** in that the *cis* or *trans* configuration of the diene and of the dienophile is preserved in the product. Thus, in the example shown, the two ester groups are *cis* relative to the double bond in the dienophile and *cis* relative to the cyclohexene ring in the adduct. This reaction is widely used in the laboratory and in industry for the synthesis of cyclohexene derivatives.

Exercise 3.16 Draw the structures of the dienes and dienophiles that would react to give the following products.

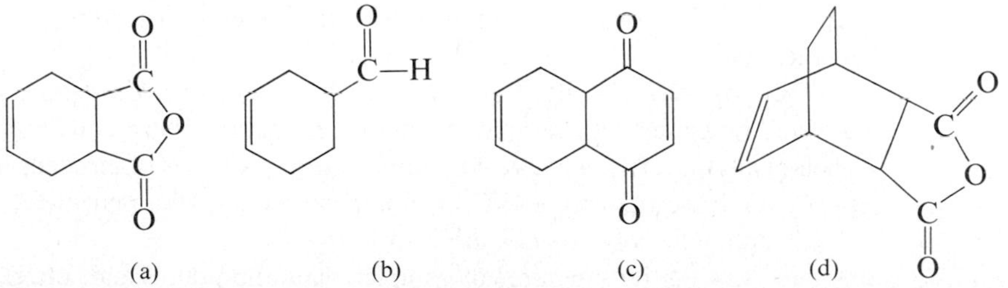

(a) (b) (c) (d)

[6]Otto Diels (1876–1954), University of Kiel, Nobel Prize in 1950. Kurt Alder (1902–1958), a student of Diels before going to the University of Cologne, Nobel Prize in 1950.

3.13 Polyenes and Terpenes

Although the lower alkenes do not occur as such in nature, other alkenes occur widely in a variety of complex forms. For example, although the diene, isoprene (2-methyl-1,3-butadiene), does not occur naturally, a large number of natural products appear to be built from **isoprene units.** One important class of these natural products is the **terpenes** (Gr. *terebinthos,* turpentine tree), which are sometimes called **isoprenoid** compounds because their carbon skeletons are multiples of the C_5 isoprene unit. This unit appears in most natural substances in a regular head-to-tail sequence, although in some cases a head-to-head or tail-to-tail arrangement is found. The widespread occurrence of this structural unit has led to the **isoprene rule,** which states that the most probable structure of a terpene is that which allows its carbon skeleton to be divisible into isoprene units. This rule has been useful in deriving the structures not only of terpenes, but also of a number of other natural substances.

$$\text{head (h)} \longrightarrow \underset{\displaystyle CH_3}{CH_2{=}\underset{|}{C}{-}CH{=}CH_2} \longleftarrow \text{tail (t)}$$

Isoprene

$$-C-\overset{\displaystyle C}{\overset{|}{C}}-C-C- \qquad\qquad -C-\overset{\displaystyle C}{\overset{|}{C}}-C-C\,\vdots\,C-\overset{\displaystyle C}{\overset{|}{C}}-C-C-$$

Isoprene unit Terpene unit

Many of the terpenes are hydrocarbons, while others are alcohols, ethers, aldehydes, ketones, or carboxylic acids. Terpenes, for the most part, are fragrant compounds and can usually be separated from other plant materials by gentle heating or steam distillation. A number are classified as "essential oils"[7] and are used in perfumes, flavoring agents, and medicinals.

Terpenes are classified on the basis of the number of **terpene units** present in the molecule, where a terpene unit is *two* isoprene units. These units may appear in either an open-chain (acyclic) or cyclic structure. Common classes of terpenes include the monoterpenes (C_{10}), sesquiterpenes (C_{15}), diterpenes (C_{20}), triterpenes (C_{30}), and tetraterpenes (C_{40}), although polyterpenes are also found.

In drawing the structures of complex natural products and, often, simple compounds as well, the use of Lewis structures becomes tedious and may hide structural relationships. Therefore, abbreviations are employed in which individual carbon and hydrogen atoms are seldom shown. This is done most frequently for cyclic compounds, but is being done

[7]Essential oils are "oils of essence"—that is, volatile and pleasantly scented.

with increasing frequency for acyclic compounds as well. In these drawings each line represents *two bonded carbon atoms* unless it is explicitly shown to be otherwise. Hydrogen atoms are not shown unless required to show the geometry of the molecule, in which instances they *must* be shown with the usual H symbol and never just a line, which would be interpreted to mean a methyl group! Often the geometry of an acyclic molecule is shown in a specific, perhaps distorted, conformation to show the relationship of this molecule to some cyclic analog. These abbreviations are illustrated in the following examples of two terpenes, *limonene* and *citronellol*. The junction of separate isoprene units in each structure in this section is indicated by a dashed line.

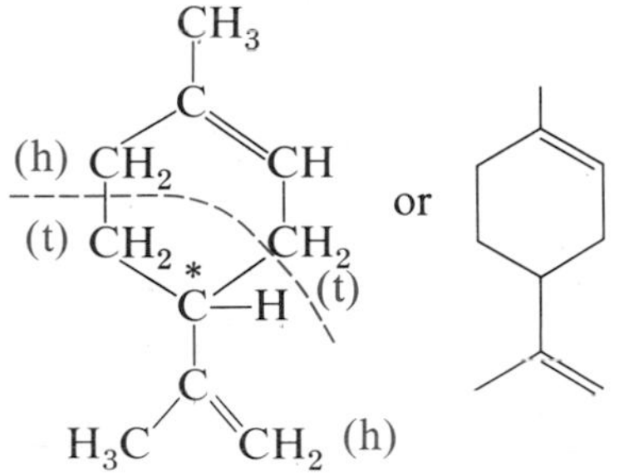

CH_3 (h) CH_2 CH (t) CH_2 * CH_2 (t) C—H C H_3C CH_2 (h) or

Limonene

CH_3 C—H H_2C CH_2 H_2C CH_2OH C—H C H_3C CH_3 or OH

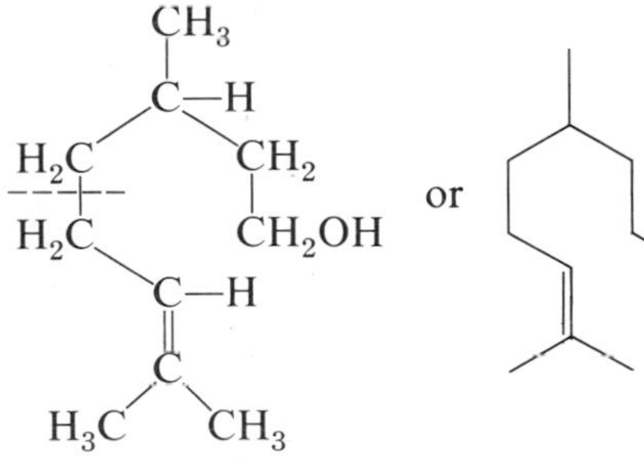

Citronellol

Fused ring systems may be drawn in several ways, the best of which try to picture the actual geometry of the molecule (see α-pinene and camphor in Figure 3.7).

The structures of a number of terpenes together with their names and sources are given in Figure 3.7. Some of these are articles of commerce, and are obtained from natural sources. For example, the diterpene, α-pinene, is a chief constituent of turpentine and is obtained from pine tree sap. Camphor used in medicine and industry is obtained from the bark of the Formosan camphor tree. Other terpenes are very important in biosynthetic processes. Thus, the sesquiterpene *farnesol* is the precursor of the triterpene *squalene,* which is itself the precursor in the biosynthesis of the important steroidal alcohol *cholesterol*. The diterpene Vitamin A is apparently formed in the body by the oxidation of the central double bond of the tetraterpene β-carotene, the yellow coloring matter of carrots.

FIGURE 3.7 Representative Terpenes

Terpenes (one terpene unit)

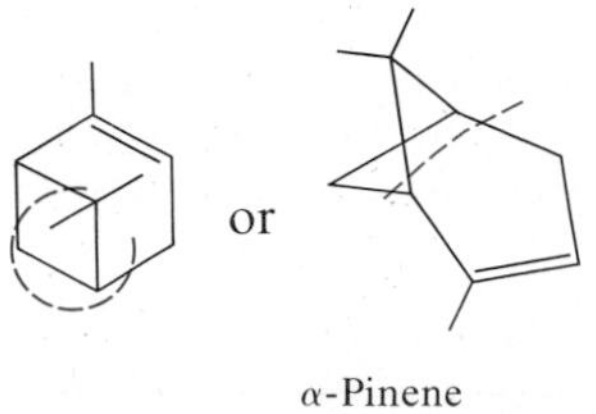

or

α-Pinene
(oil of turpentine)

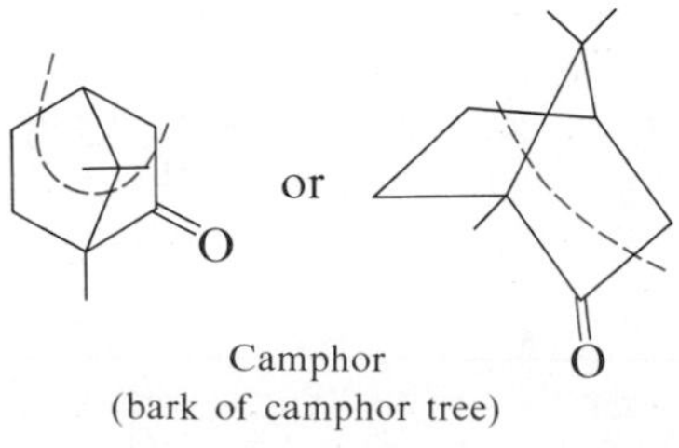

or O O

Camphor
(bark of camphor tree)

FIGURE 3.7 *Continued*

Sesquiterpenes (one and a half terpene units)

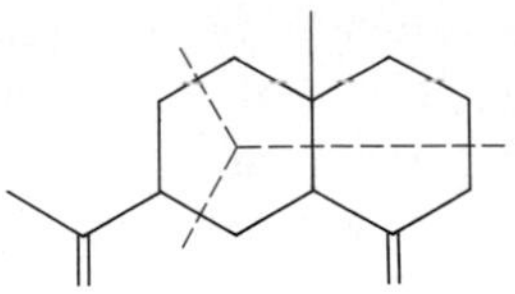

β-Selinene
(oil of celery)

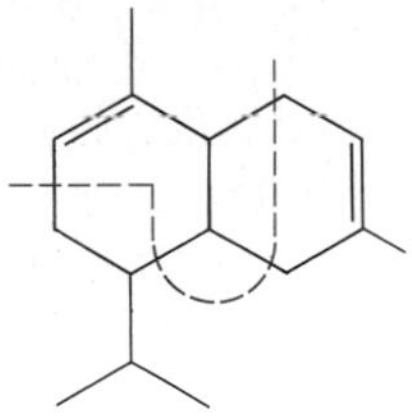

Cadinene
(oil of cade in juniper
and cedar oils)

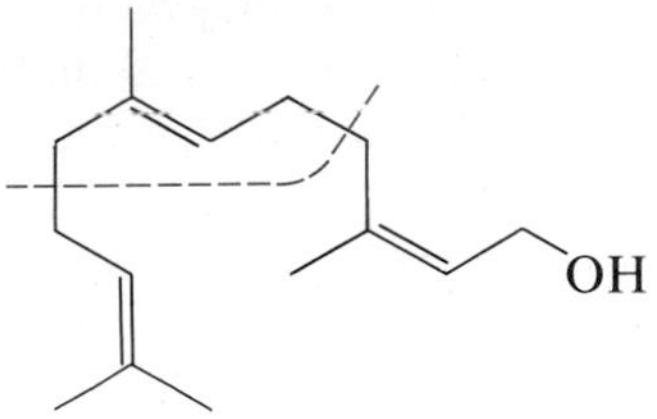

Farnesol
(lily of the valley)

Diterpenes (two terpene units)

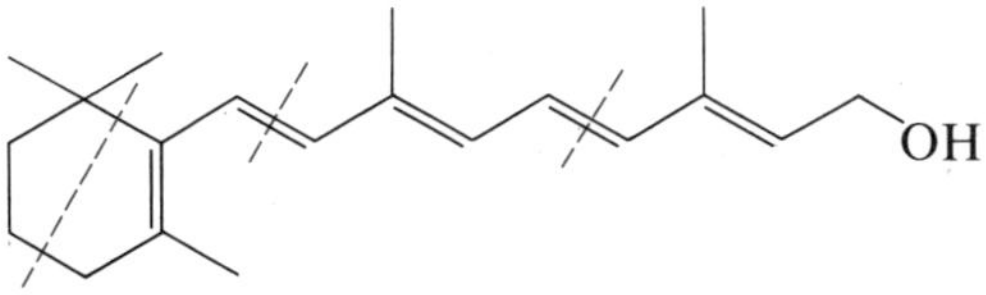

Vitamin A
(retinol)

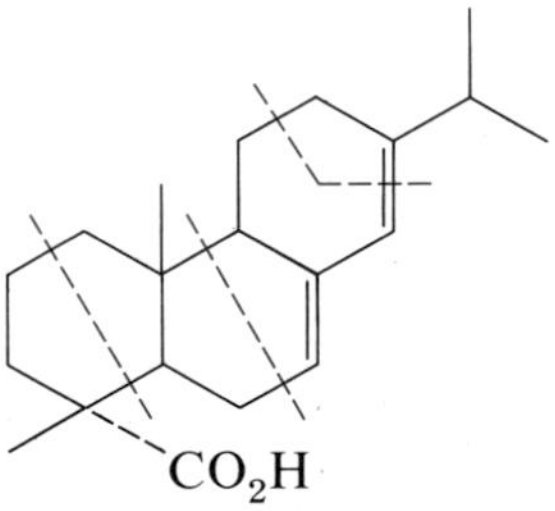

Abietic acid
(pine rosin)

Triterpenes (three terpene units)

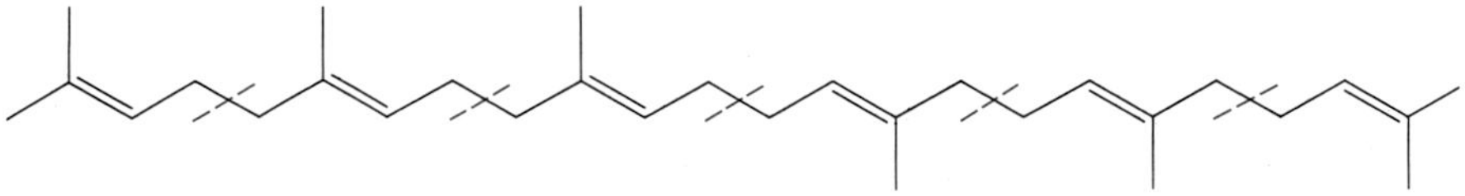

Squalene (shark liver oil)

Tetraterpenes (four terpene units)

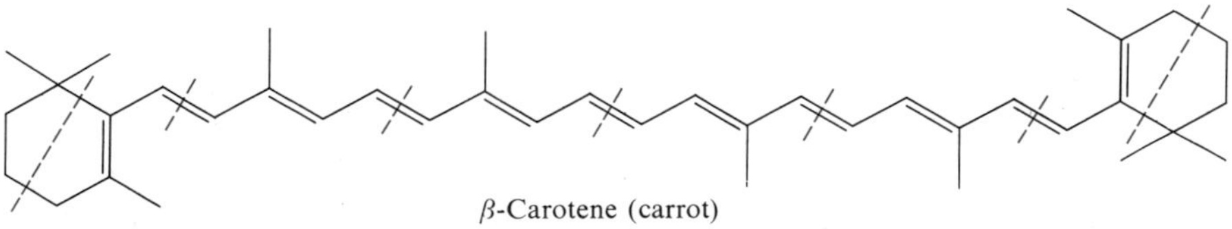

β-Carotene (carrot)

Exercise 3.17 Locate the isoprene units in the terpene caryophyllene (oil of cloves), using broken lines as in the text examples. To which class of terpenes does caryophyllene belong?

Part II: The Alkynes (Acetylenes)

The alkynes, or acetylenes, are unsaturated hydrocarbons that contain a triple bond or three pairs of shared electrons between adjacent carbon atoms. The general formula for an alkyne is C_nH_{2n-2} if only one triple bond is present in the structure. The same general formula fits that of an open-chain diene. The first member of the alkyne family is acetylene, C_2H_2, whose structure may be written as

$$H{-}C{\equiv}C{-}H \quad \text{or} \quad H\overset{\times}{\cdot}C\vdots\vdots C\overset{\cdot}{\times}H$$

3.14 Formulas and Nomenclature of the Alkynes

The alkynes are named systematically by the same rules of nomenclature that were used to name the alkenes. The triple bond is indicated by the suffix *yne* and is located by number in the longest chain that contains it. The alkyl derivatives of acetylene, in which one or both of the singly bonded hydrogens of acetylene have been replaced by alkyl groups, are named simply as alkyl acetylenes. A few examples will help make this clear.

$H{-}C{\equiv}C{-}H$
Ethyne
(Acetylene)

$CH_3{-}C{\equiv}C{-}H$
Propyne
(Methylacetylene)

$CH_3{-}C{\equiv}C{-}CH_3$
2-Butyne
(Dimethylacetylene)

$\overset{(1)}{CH_3}{-}\overset{(2)}{C}{\equiv}\overset{(3)}{C}{-}\overset{(4)}{CH_2}{-}\overset{(5)}{CH_3}$
2-Pentyne
(Methylethylacetylene)

$\overset{(5)}{CH_3}{-}\overset{(4)}{CH_2}{-}\overset{(3)}{CH}(CH_3){-}\overset{(2)}{C}{\equiv}\overset{(1)}{C}{-}H$
3-Methyl-1-pentyne
(*sec*-Butylacetylene)

$\overset{(1)}{CH_2}{=}\overset{(2)}{CH}{-}\overset{(3)}{CH}{\equiv}\overset{(4)}{CH}$
1-Butene-3-yne
(Vinylacetylene)

3.15 The Carbon–Carbon Triple Bond

When carbon is attached to only two other atoms, hybridization of one 1*s* and one 2*p* orbital will give two *sp* orbitals. Overlap of two hydrogen 1*s* and four carbon *sp* orbitals will form three sigma bonds and leave two 2*p* orbitals at right angles to each other on each carbon (Sec. 1.7). Overlap of the 2*p* orbitals will form two π bonds. Now, instead of a planar structure with a cloud of electrons above and below the plane, as in the case of the alkenes, we have a linear structure with a cloud not only above and below, but also in front and behind the axis. Here the electrons of both π bonds overlap and blend into a cylindrical sleeve about the axis of the molecule (Fig. 3.8).

Exercise 3.18 In the following formula for vinyl acetylene, name the hybrid orbitals representing the labeled sigma bonds (a), (b), (c), and (d). Also, indicate the size of bond angles θ and ϕ.

$$H_2C\underset{(a)}{=}CH\underset{(b)}{-}C\underset{(c)}{\equiv}C\underset{(d)}{-}H \quad (\theta \text{ between (a) and (b)}; \ \phi \text{ between (c) and (d)})$$

FIGURE 3.8 Structural Representation of the Carbon–Carbon Triple Bond of Ethyne in Perspective (a) Showing Three σ Bonds and Two π Bonds and (b) As Seen if Viewed along the Carbon–Carbon Axis

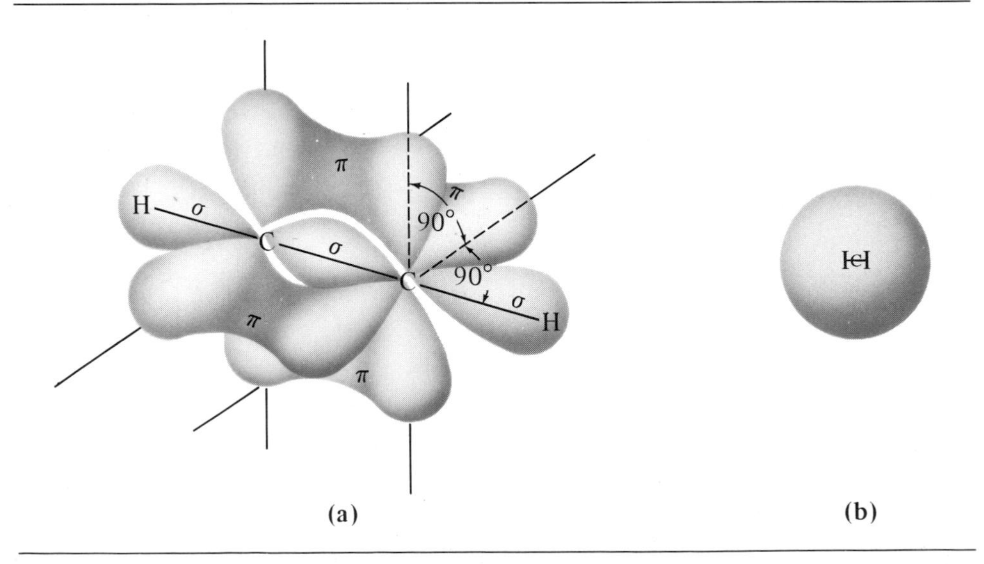

3.16 Occurrence and Properties of the Alkynes

The alkynes are somewhat more reactive than the alkenes, and the lower molecular weight members are not found in nature. However, the triple bond does occur as a structural feature in some natural substances. An example is the antibiotic *mycomycin,* which has the remarkable structure

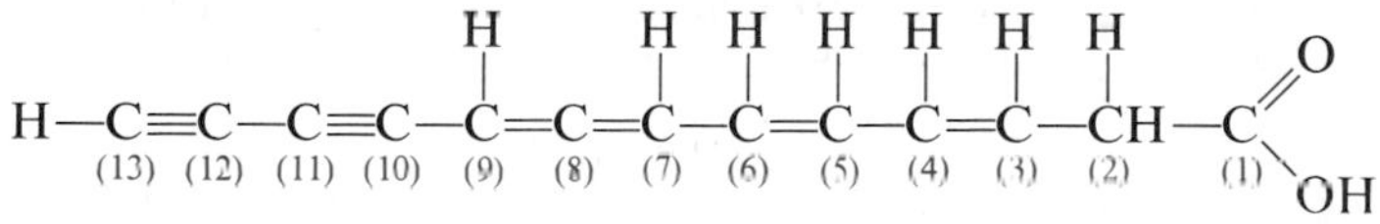

$$\underset{(13)}{\mathrm{H{-}C}}{\equiv}\underset{(12)}{\mathrm{C}}{-}\underset{(11)}{\mathrm{C}}{\equiv}\underset{(10)}{\mathrm{C}}{-}\underset{(9)}{\mathrm{CH}}{=}\underset{(8)}{\mathrm{C}}{=}\underset{(7)}{\mathrm{CH}}{-}\underset{(6)}{\mathrm{CH}}{=}\underset{(5)}{\mathrm{CH}}{-}\underset{(4)}{\mathrm{CH}}{=}\underset{(3)}{\mathrm{CH}}{-}\underset{(2)}{\mathrm{CH_2}}{-}\underset{(1)}{\mathrm{C}}(=\mathrm{O})\mathrm{OH}$$

It has the rather impressive name **3,5,7,8-tridecatetraene-10,12-diynoic acid.** An unusual group of similar highly unsaturated, linear compounds are secreted by various species of fungi. These compounds are straight-chain polyacetylenes of eight to fourteen carbon atoms in length with one or more oxygen-containing functional groups.

The chemical properties of the alkynes, with some exceptions, are very similar to those of the corresponding alkenes. The reactions are largely those of addition. One exception is notable, however. *The hydrogen atoms in terminal acetylenes (1-alkynes) are relatively acidic*. Thus, acetylene will react with the strong base, sodamide ($NaNH_2$), to form the acetylide carbanion, but ethylene and ethane do not form carbanions with this base. It is usually easiest to explain the relative acidities of organic acids in terms of the relative stabilities of the corresponding anions: the more stable the anion, the more acidic the organic acids. The three carbanions of interest are the following:

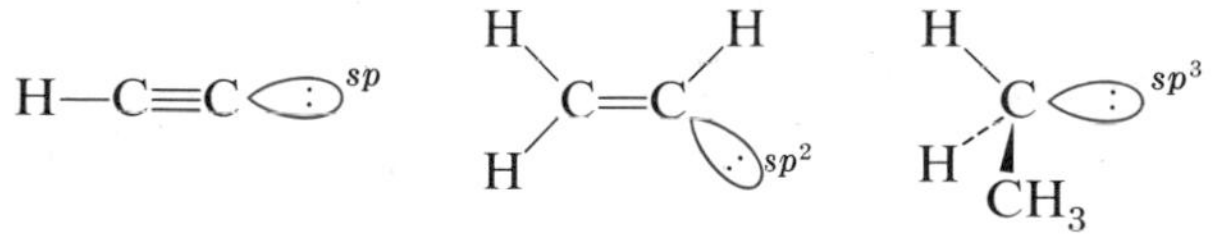

$\mathrm{H{-}C{\equiv}C}^{:}$ (sp) — Acetylide anion

$\mathrm{H_2C{=}CH}^{:}$ (sp^2) — Vinyl anion

$\mathrm{CH_3{-}CH_2}^{:}$ (sp^3) — Ethyl anion

We can see that the lone-pair electrons in the acetylide anion are farther from the bond pairs than those in ethyl or vinyl anions; thus the inter-electron repulsions should be smaller, and the ion more stable. An alternative, and more rigorous, explanation is that electrons in sp orbitals will be slightly closer to the nucleus than those in sp^2 or sp^3 orbitals. Thus, the electrostatic attraction between these electrons and the nucleus will be greatest in the sp case. This attraction lowers the energy of the anion, causing the acetylide anion to be the most stable of the three anions.

Preparation of the Alkynes

The alkynes are prepared in the laboratory by reactions very much like those employed for the preparation of alkenes. *Two simple molecules* must be eliminated from adjacent carbons in a saturated compound in order to introduce the triple bond into the molecule.

3.17 Preparation of Acetylene

Acetylene (ethyne) once was one of the most important of industrial raw materials. During the last decade or so it has been replaced by cheaper and more easily handled substances such as ethylene, propene, and butadiene. In countries with abundant coal supplies and cheap electricity, it may be prepared by reaction between calcium carbide and water. Calcium carbide is prepared by heating a mixture of coke and limestone in an electric furnace.

$$CaCO_3 \xrightarrow{\text{heat}} CaO + CO_2$$

$$3\,C + CaO \xrightarrow{2000°C} \underset{\text{Calcium carbide}}{CaC_2} + CO$$

$$CaC_2 + 2\,H_2O \longrightarrow \underset{\text{Acetylene}}{H{-}C{\equiv}C{-}H} + \underset{\text{Calcium hydroxide}}{Ca(OH)_2}$$

Acetylene may also be obtained from the thermal cracking of alkanes or the dehydrodimerization of methane at high temperatures.

$$6\,CH_4 + O_2 \xrightarrow{1500°} 2\,HC{\equiv}CH + 2\,CO + 10\,H_2$$

3.18 Preparation of Higher Alkynes

Higher members in the series may be prepared by one of two general methods: (1) elimination of two molecules of hydrogen halide from adjacent carbon atoms in a saturated compound; or (2) replacement of acetylenic hydrogen of acetylene by an alkyl group.

A. Elimination of Halogen Acid from Vicinal Dihalides.[8] Notice that the product of the first HBr elimination is a substituted vinyl bromide. The vinyl halides are unreactive compounds and a stronger base (NH_2^-) is used to remove the second HBr molecule.

$$\underset{\text{1,2-Dibromopropane}}{CH_3{-}\overset{H}{\underset{Br}{C}}{-}\overset{Br}{\underset{H}{C}}{-}H} + KOH \xrightarrow[\text{solution}]{\text{alcoholic}} \underset{\text{1-Bromo-1-propene (a vinyl halide)}}{CH_3{-}\overset{H}{C}{=}\overset{Br}{C}{-}H} + KBr + H_2O$$

[8] Vicinal (L. *vicinalis*, neighboring). Vicinal dihalides have halogen atoms on adjacent carbon atoms.

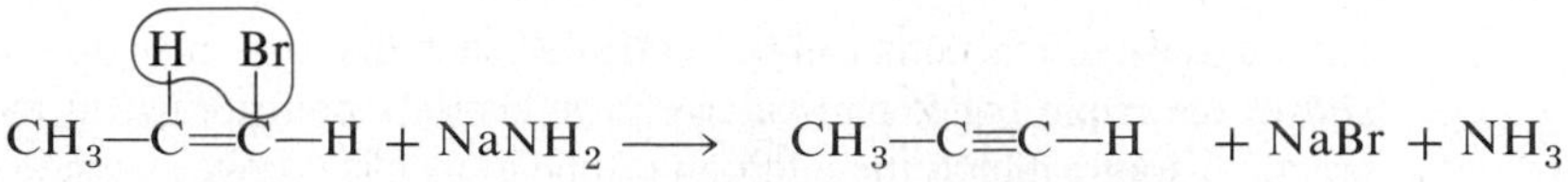

$$CH_3{-}\overset{H\ \ Br}{C{=}C}{-}H + NaNH_2 \longrightarrow CH_3{-}C{\equiv}C{-}H + NaBr + NH_3$$

Propyne (Methylacetylene)

B. Replacement of Acetylenic Hydrogen by an Alkyl Group. The first step in this synthesis is the preparation of the sodium salt of acetylene by reaction with sodamide. The sodium acetylide then is coupled with an alkyl halide.

$$H{-}C{\equiv}C{-}H + NaNH_2 \xrightarrow{\text{liquid } NH_3} H{-}C{\equiv}C\overset{-}{:}Na^+$$

Sodium acetylide

$$H{-}C{\equiv}C\overset{-}{:}Na^+ + C_2H_5I \longrightarrow H{-}C{\equiv}C{-}CH_2CH_3 + Na^+I^-$$

Ethyl iodide — 1-Butyne

The second acidic hydrogen, if available, may also be replaced by repeating the process. An alkyne of known structure can thus be synthesized.

$$H{-}C{\equiv}C{-}CH_2CH_3 + NaNH_2 \xrightarrow[NH_3]{\text{liquid}} Na^+:\bar{C}{\equiv}C{-}CH_2CH_3$$

$$C_2H_5Br + \overset{+}{Na}:\bar{C}{\equiv}C{-}CH_2CH_3 \longrightarrow CH_3CH_2{-}C{\equiv}C{-}CH_2CH_3 + NaBr$$

3-Hexyne

3.19 Reactions of the Alkynes

The chemical reactions of the alkynes, like those of the alkenes, are largely reactions of addition. The mode of addition of hydrogen halides to unsymmetrical alkynes, you will note, also follows Markovnikov's rule.

A. Addition of Br_2, HBr, and H_2 to Acetylenes.

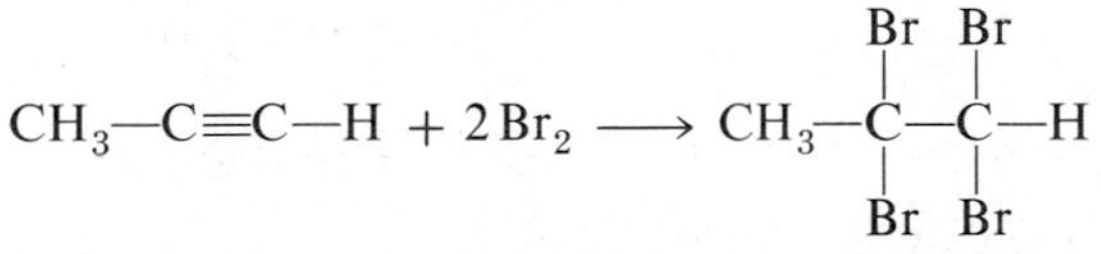

$$CH_3{-}C{\equiv}C{-}H + 2\,Br_2 \longrightarrow CH_3{-}CBr_2{-}CBr_2{-}H$$

Propyne — 1,1,2,2-Tetrabromopropane

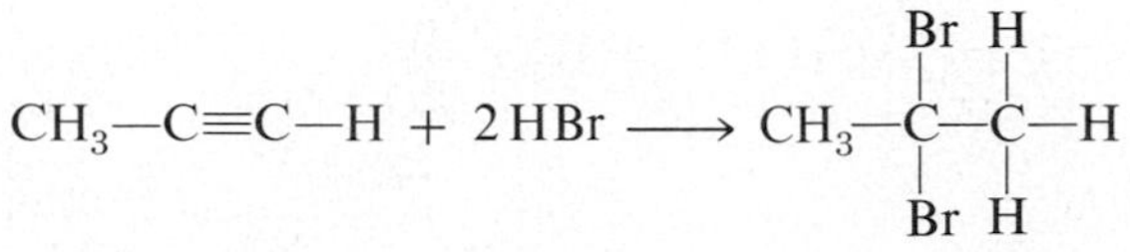

$$CH_3{-}C{\equiv}C{-}H + 2\,HBr \longrightarrow CH_3{-}CBr_2{-}CH_2{-}H$$

2,2-Dibromopropane

$$CH_3{-}C{\equiv}C{-}H + 2\,H_2 \xrightarrow{\text{catalyst}} CH_3CH_2CH_3$$

Propane

Although these reactions can be controlled such that one or two moles of the reagent will add to the triple bond, particularly in industrial applications, in the laboratory the only practical application is the addition of one mole of hydrogen in the *syn* fashion to give the *cis*-alkene using a special, prepared, slightly ''poisoned'' catalyst called Lindlar's catalyst.

$$\underset{\text{3-Hexyne}}{CH_3CH_2{-}C{\equiv}C{-}CH_2CH_3} \xrightarrow[\text{Pd–Pb/BaSO}_4\text{, quinoline}]{H_2} \underset{\textit{cis}\text{-3-Hexene}}{(CH_3CH_2)(H)C{=}C(H)(CH_2CH_3)}$$

B. Addition Reactions NOT Given by Olefins. Water and carboxylic acids will add across carbon–carbon triple bonds—but not across carbon–carbon double bonds. These useful reactions are catalyzed by mercury salts, such as mercuric sulfate, and are usually carried out in an acidic medium. Although we may write these reactions as proceeding through a vinyl alcohol or **enol** (*-ene* for the carbon–carbon double bond, *-ol* for alcohol) as an intermediate, vinyl alcohols, if formed, usually rearrange rapidly to aldehydes or ketones (Sec. 8.4).

$$H{-}C{\equiv}C{-}H + H_2O \xrightarrow[H_2SO_4]{HgSO_4} \left[\underset{\text{An enol}}{H{-}CH{=}CH{-}OH}\right] \longrightarrow \underset{\text{Acetaldehyde}}{CH_3{-}CH{=}O}$$

$$H{-}C{\equiv}C{-}H + CH_3{-}C({=}O){-}OH \xrightarrow[75^\circ C]{Hg^{2+}} \underset{\text{Vinyl acetate}}{CH_2{=}CH{-}O{-}C({=}O){-}CH_3}$$

Since this addition follows Markovnikov's rule, the hydration of 1-alkynes yields methyl ketones,

$$CH_3{-}C({=}O){-}R$$

The hydration of nonterminal alkynes is useful only if the alkyne is symmetrical—that is, has two identical groups on the triply bonded carbon atoms.

$$H{-}C{\equiv}C{-}CH_2CH_3 + H_2O \xrightarrow[H_2SO_4]{Hg^{2+}}$$

$$\left[H{-}CH{=}C(O{-}H){-}CH_2CH_3\right] \longrightarrow \underset{\substack{\text{2-Butanone}\\\text{(Methyl ethyl ketone)}}}{H{-}CH_2{-}C({=}O){-}CH_2CH_3}$$

C. Replacement of Acetylenic Hydrogen by a Metal (Salt Formation). As was pointed out previously, the principal difference between the chemical behavior of the alkynes and that of the alkenes is due to the acidic character of the hydrogen atom bonded to an acetylenic carbon. Acetylenic hydrogen atoms can be replaced by metals to produce salts called acetylides (Sec. 3.17). Acetylene, when passed through ammoniacal solutions of cuprous chloride and silver nitrate, forms the acetylides of these metals.

$$H{-}C{\equiv}C{-}H + 2\ Cu(NH_3)_2Cl \longrightarrow \underset{\text{Copper acetylide}}{Cu{-}C{\equiv}C{-}Cu} + 2\ NH_4Cl + 2\ NH_3$$

$$H{-}C{\equiv}C{-}H + 2\ Ag(NH_3)_2NO_3 \longrightarrow \underset{\text{Silver acetylide}}{Ag{-}C{\equiv}C{-}Ag} + 2\ NH_4NO_3 + 2\ NH_3$$

Heavy metal salts of acetylene such as silver acetylide are extremely sensitive to shock when dry and may explode violently.

D. Oxidation of Acetylene (Combustion). Acetylene burns in an atmosphere of pure oxygen to produce extremely high temperatures (*ca.* 3000°C). This reaction makes possible the common oxyacetylene torch so useful for the welding and cutting of metals.

$$2\ H{-}C{\equiv}C{-}H + 5\ O_2 \longrightarrow 4\ CO_2 + 2\ H_2O + 619.7\ \text{kcal}$$

Summary, Part I

1. Unsaturated hydrocarbons with double bonds are called alkenes or olefins. The general formula for an acyclic alkene with one double bond is $\mathbf{C_nH_{2n}}$.
2. The double bond makes possible *cis,trans*-isomerism.
3. The availability of the π electrons causes alkenes to be more reactive than alkanes.
4. Alkenes are prepared by removing the elements of a simple molecule from adjacent carbon atoms by (a) the dehydration of an alcohol and (b) the dehydrohalogenation of an alkyl halide.
5. Reactions of the alkenes are principally those of addition. Markovnikov's rule is followed in the addition of hydrogen halides with the exception of HBr addition in the presence of peroxides.
   (a) Addition takes place with halogen, hydrogen, hydrogen halides, sulfuric acid, and hypochlorous acid.
   (b) Mild oxidation with $KMnO_4$ yields *cis*-diols called glycols.
   (c) Oxidation with ozone results in cleavage of the double bond to yield two carbonyl

   $$\left(\gt C{=}O\right)$$

   compounds.
   (d) Olefins may be polymerized to produce macromolecules called polymers.

6. A system of alternate single and double bonds is a conjugated system. Butadiene, the simplest example of a conjugated system, may add reagents by 1,2- or 1,4-addition.
7. The Diels–Alder reaction yields cyclohexenes by a concerted cycloaddition.
8. The terpenes are a class of natural products whose carbon skeletons are multiples of the C_5 isoprene unit. Many are "essential oils" used in perfumes and in flavoring agents.
9. Terpenes are classified according to the number of C_{10} "terpene" units present.

Summary, Part II

1. Unsaturated hydrocarbons with triple bonds are called acetylenes or alkynes. The general formula for an acyclic alkyne with one triple bond is $\mathbf{C_nH_{2n-2}}$.
2. The principal feature that distinguishes the alkynes from the alkenes is the acidic hydrogen on the triply bonded carbon atom.
3. Alkynes are prepared by (a) elimination of two simple molecules from adjacent carbon atoms and (b) replacement of the hydrogen on an existing alkyne by an alkyl group. Acetylene itself is prepared from calcium carbide.
4. The reactions of the alkynes are the following.
   (a) Addition (halogens, hydrogen, HCN, H_2O, acetic acid).
   (b) Oxidation (combustion), a highly exothermic reaction.
   (c) Salt formation. The sodium salt is an important intermediate. The heavy-metal salts are explosive when dry.

Supplementary Exercises

3.19 Write a structural formula and give another acceptable name for each of the following.

(a) propylene
(b) ethylene bromide
(c) isoprene
(d) isobutylene
(e) dimethylacetylene
(f) vinylacetylene

3.20 Draw the structural formulas for the compounds having the following IUPAC names.

(a) 2,5-dimethyl-3-hexene
(b) 2-ethyl-3-methyl-1-pentene
(c) *trans*-3,4-dichlorocyclohexene
(d) 1,3-cyclohexadiene
(e) 1,7-octadiene-4-yne
(f) 4-methyl-*cis*-2-pentene

3.21 Name the following structures according to IUPAC rules

(a) $CH_3CH{=}CH{-}CH_2{-}CH_2{-}CH_3$ 2 hexene
(b) $CH_3CH(Cl)CH{=}CHCH_3$ 4 chloro 2 pentene
(c) $CH_3{-}CH_2{-}C{\equiv}C{-}CH_3$ 2 pentyne
(d) $CH_3{-}C(Br){=}C(Br){-}CH_3$ 2,3 dibromo 2 butene

(e) $(CH_3)_2C{=}C(Br){-}CH_3$
(f) $(CH_3)_2CH{-}CH{=}CH_2$

(g) 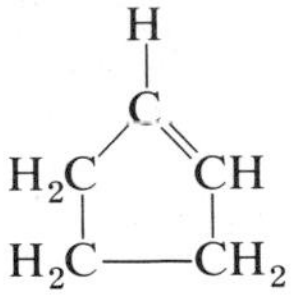
```
        H
        |
        C
      /   \\
  H2C       CH
    |       |
  H2C ----- CH2
```

(h)
```
       CH(CH3)
      /       \
  H2C          CH(CH3)
    |          |
  H2C -------- CH2
```

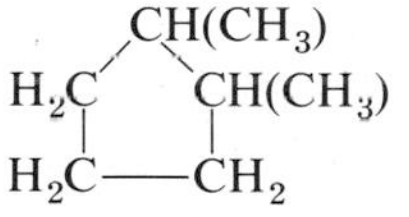

3.22 Draw twelve different structures for C_5H_{10}. Name each according to IUPAC Rules.

3.23 In each case tell why the following names are objectionable, and give a suitable one.

(a) 2,3-dimethylpentene
(b) 2-methyl-3-pentene
(c) 2-ethyl-3-hexene
(d) 3-methyl-4-heptene
(e) 4-methyl-2,4-hexadiene

3.24 Which structures in Exercise 3.21 may show *cis,trans*-isomerism?

3.25 Beginning with ethyl alcohol, CH_3CH_2OH, as your only starting material and using any other reagent you may require, show how you would prepare:

(a) ethene
(b) bromoethane
(c) 1-butyne
(d) ethylmagnesium bromide
(e) ethane
(f) acetylene
(g) 1,1,2-trichloroethane
(h) 1,2-dibromoethane
(i) vinyl chloride
(j) 2-chloroethanol
(k) 1,1-dibromoethane

3.26 Complete each of the following reactions by supplying the necessary but missing component. This could be a reactant, a catalyst, or a product.

(a) $(?) + Br_2 \text{ in } CCl_4 \longrightarrow CH_3CH(Br)CH_2Br$

(b) $CH_3{-}CH{=}CH_2 + (?) \longrightarrow CH_3CH(Br)CH_3$

(c) $CH_3{-}CH{=}CH_2 + H_2SO_4 \xrightarrow{(?)} CH_3CH(OH)CH_3$

(d) $CH_3{-}CH{=}CH_2 + HBr \xrightarrow{(?)} CH_3CH_2CH_2Br$

(e) $CH_3{-}C{\equiv}C{-}CH_3 + (?) \longrightarrow CH_3CH{=}CHCH_3$ (*cis*)

(f) $CH_3CH{=}CHCH_3 \xrightarrow{(?)} CH_3{-}CHD{-}CHD{-}CH_3$

(g) $CH_3CH{=}CHCH_3 + O_3 \xrightarrow{\text{then with Zn/H}_2\text{O}} (?)$

(h) cyclopentene + dilute $KMnO_4 \xrightarrow{5\%\ Na_2CO_3}$ (?)

(i) (?) $\xrightarrow[H_2SO_4]{H_2O,\ Hg^{2+}} [H_2C{=}CH(OH)] \longrightarrow CH_3CHO$

(j) α-pinene $\xrightarrow[Pt]{H_2}$ (?)

(k) β-selinene $\xrightarrow[CH_3OH]{O_3} \xrightarrow{CH_3SCH_3}$ (?)

(l) 1,3-Butadiene + HOCl $\longrightarrow$ (?)

3.27 For each of the structures below draw two isomeric alkyl halides that, on dehydrohalogenation, would produce the olefin shown. In each case predict which of the two halogen compounds would react more readily to produce the unsaturated compound. Give reasons for your choice.

(a) $CH_3CH{=}CH_2$
(b) $(CH_3)_2C{=}CHCH_3$
(c) $CH_3CH{=}CHCH_2CH_3$

3.28 A hydrocarbon, C_4H_8, neither decolorized a bromine–carbon tetrachloride solution nor reacted with HBr. When the hydrocarbon was heated to 200° with hydrogen in the presence of a nickel catalyst, a new hydrocarbon, C_4H_{10}, was formed. What was the original unknown?

3.29 Recalling that the alkynes and dienes are isomeric, with the type formula C_nH_{2n-2}, draw structures and assign names to two compounds having the formula C_5H_8, but with each showing only one of the following sets of properties:

Compound **A** adds two moles of bromine to yield a new substance, $C_5H_8Br_4$. Compound **A** forms a precipitate with ammoniacal $AgNO_3$. It may be hydrated in the presence of mercuric salts to yield a methyl ketone,

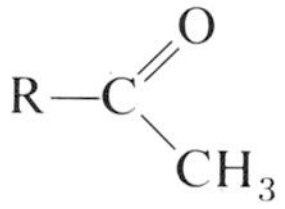

$$R{-}C({=}O){-}CH_3$$

Compound **B** also adds two moles of bromine to yield a substance of formula $C_5H_8Br_4$. Compound **B** is conjugated. It forms an ozonide which on reductive hydrolysis gives *three different* carbonyl-containing

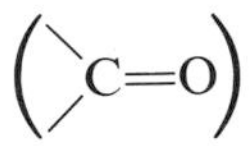

$$\left(>C{=}O\right)$$

structures. One of the latter has two carbonyl groups.

3.30 Predict the products (if any) of the following reactions.

(a) *cis*-2-pentene + $KMnO_4 \xrightarrow[dilute]{cold}$ (?)

(b) 2-methylpropene + $H_2O \xrightarrow{H_2SO_4}$ (?)

(c) 1-methylcyclohexene + HBr $\xrightarrow{\text{peroxides}}$ (?)

(d) *trans*-2-hexene + Br_2 $\longrightarrow$ (?)

(c) isoprene $\xrightarrow{\text{R—O—O—R}}$ (?)

(f) 2-methylpropene + BH_3 $\xrightarrow[\text{heat}]{H_2O_2,\ OH^-}$

(g) 2-butene + H_2 $\xrightarrow{\text{Pd}}$

(h) 2-butyne + $NaNH_2$ $\xrightarrow{\text{liq } NH_3}$

(i) 2-methylcyclohexanol + H_2SO_4 $\xrightarrow{\text{heat}}$

(j) 2-methylpropene + Br_2 + NaCl(aq) $\longrightarrow$

3.31 When ethene is bubbled through a methyl alcohol (CH_3OH) solution of bromine, not only is the expected 1,2-dibromoethane obtained, but also some 1-bromo-2-methoxyethane,

$$Br{-}CH_2CH_2{-}OCH_3$$

Account for this behavior in terms of Lewis acid–base theory.

3.32 Deduce the structure of a hydrocarbon that (1) adds only one mole of bromine, and (2) when treated with O_3 and subsequently hydrolyzed with Zn and dilute acid, gives

$$O{=}\overset{\displaystyle H}{\overset{|}{C}}CH_2CH_2CH_2CH_2\overset{\displaystyle H}{\overset{|}{C}}{=}O$$

as the only product.

3.33 Write complete equations for the following transformations. Include all conditions and reagents. If an intermediate step is missing, supply it.

(a) bromocyclohexane $\longrightarrow$ cyclohexene $\longrightarrow$ 1,2-cyclohexanediol
(b) acetylene $\longrightarrow$ 2-butyne $\longrightarrow$ butane
(c) 2-butene $\longrightarrow$ 2-butyne $\longrightarrow$ methyl ethyl ketone
(d) 1-butene $\longrightarrow$ 2-butene

3.34 The chlorination of propene under free-radical conditions (Sec. 2.11B) gives allyl chloride (3-chloro-1-propene) as the major product. Explain this result, which must have something to do with the stability of the intermediate free radical. (*Hint:* Review Sec. 1.14.)

3.35 Draw complete structural formulas for the following compounds, showing all carbon and hydrogen atoms.

(a) α-pinene
(b) camphor
(c) farnesol
(d) vitamin A

3.36 The inclusion of alkenes (olefins) in a gasoline blend improves its octane rating but makes the gasoline less stable in storage. Explain.

3.37 In Sec. 1.8 it was stated that the electronegativity of an element in a molecule depended to some extent on the state of hybridization. From your knowledge of the relative acidities of ethane, ethylene, and acetylene (Sec. 3.16), arrange the three hydrocarbons in order of decreasing electronegativity of their carbon atoms. Explain.

3.38 When 10 grams of a mixture of pentene and pentane was shaken with hydrogen over a platinum catalyst, 1120 cc of hydrogen (STP) were absorbed. What percent of the mixture is pentene?

4 Aromatic Hydrocarbons: Arenes

The aromatic hydrocarbons are ring hydrocarbons structurally related to benzene, C_6H_6. Although the formula of benzene indicates unsaturation, neither it nor its related compounds are olefinic in behavior. The aromatic hydrocarbons are found in the black, viscous, pitchlike material called coal tar, which is formed as a by-product when bituminous coal is converted to coke. Although coal tar itself is a vile-smelling substance, it is a treasure trove of aromatic hydrocarbons, some of which are of rather pleasant odor—hence the name ''aromatic.'' Though the term is not descriptive of many of these compounds, it has come into general usage to classify organic compounds that have as a common structural feature one or more 6-carbon benzene structures. Compounds in this category generally are named **arenes.** Most of these valuable chemicals are now obtainable in large volume from petroleum. From both sources, coal and petroleum, are harvested the basic materials for many pharmaceuticals, dyes, plastics, pesticides, explosives, and a host of other useful, everyday commodities.

4.1 Structure of Benzene

The discovery of benzene and the elucidation of its structure comprise an interesting and significant chapter in organic chemistry. This remarkable substance was first isolated in 1825 by Michael Faraday. He discovered benzene in the residual, oily condensate that collected in the illuminating gas lines of London. He established its empirical formula as (CH) and called it ''carburetted hydrogen.'' The molecular formula of benzene was subsequently established as C_6H_6 by Mitscherlich, who in 1834 prepared it by heating gum benzoin with lime. For the next thirty years, determination of the structure of benzene posed a problem that occupied the serious attention of the best scientific minds of the time.

Since the molecular formula for benzene provided only one hydrogen atom per carbon, the molecule was expected to be unsaturated and, therefore, very reactive. On the contrary, it appeared to be remarkably stable. Benzene neither decolorized dilute potassium permanganate (the Baeyer test for unsaturation) nor added bromine readily. One would expect an olefin to do both. When treated with bromine in the presence of bright sunlight, one gram molecular weight of benzene *added* three gram molecular weights of bromine to yield benzene hexabromide. Evidently each carbon in the benzene molecule has the potential for bonding to one more atom.

$$C_6H_6 + 3\ Br_2 \xrightarrow{\text{bright sunlight}} C_6H_6Br_6$$

When treated with bromine in the presence of iron, benzene formed a monobromo *substitution* product, a behavior remarkably different from that noted above.

$$C_6H_6 + Br_2 \xrightarrow{\text{Fe}} C_6H_5Br + HBr$$

The fact that only one bromobenzene and no isomeric products were obtained indicated to the German chemist Kekulé[1] that each hydrogen atom of benzene must be equivalent to every other hydrogen. In 1865 he proposed a structure for benzene that was in accord with its chemical behavior. The structure Kekulé proposed was a cyclic hexagonal planar structure of six carbon atoms with alternate double and single bonds. Each carbon atom was bonded to only one hydrogen atom. The Kekulé structure for benzene shown below was widely accepted by his contemporaries and is still used today.

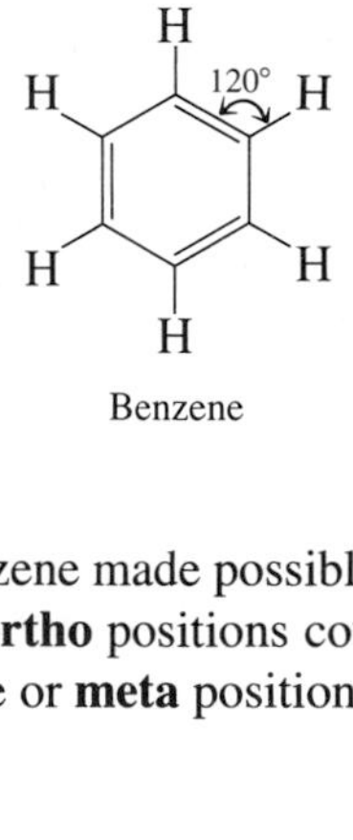

Benzene

The proposed structure for benzene made possible **three** isomeric disubstituted derivatives. In one isomer, adjacent or **ortho** positions could be occupied by the two substituents. In the second isomer, alternate or **meta** positions could be occupied, and in the third

[1]Friedrich August Kekulé (1829–1896), professor of chemistry successively at Heidelberg, Ghent, and finally at Bonn. Kekulé was an active researcher and a noted teacher but is best remembered for his formulation of the structure of benzene.

isomer, hydrogens bonded to the carbon atoms opposite or **para** to each other in the ring could be substituted.

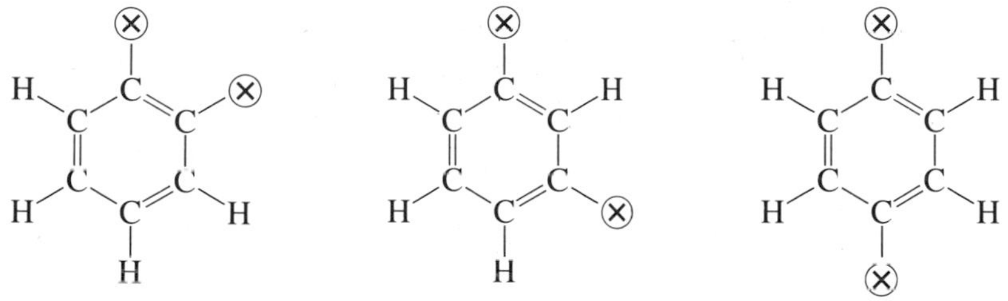

ortho positions

meta positions

para positions

If Kekulé's formula for benzene were correct, his critics argued, it would appear that **two** *ortho* isomers would be possible. In one *ortho* isomer, the carbon atoms holding the substituents would be separated by a double bond as shown below in structure (a). In the other *ortho* isomer, substituents would be separated by a single bond as shown in structure (b).

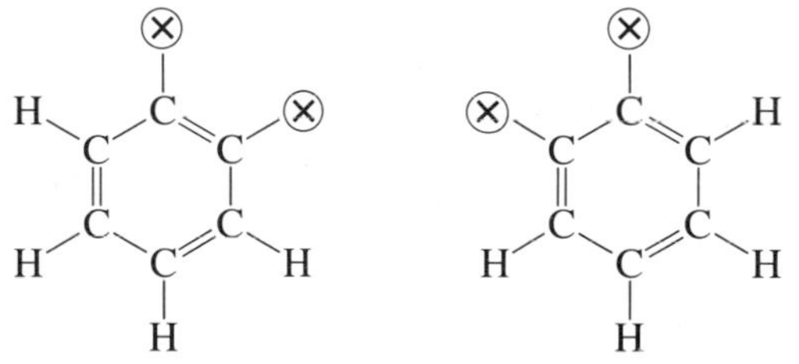

(a)

(b)

Kekulé reconciled the fact that two *ortho* isomers had not been found by stating that the system of alternate double and single bonds in the benzene ring was not a fixed system, but a dynamic one. He suggested that the structure of benzene could be that of either (a) or (b) and that the two structures were in equilibrium with each other.

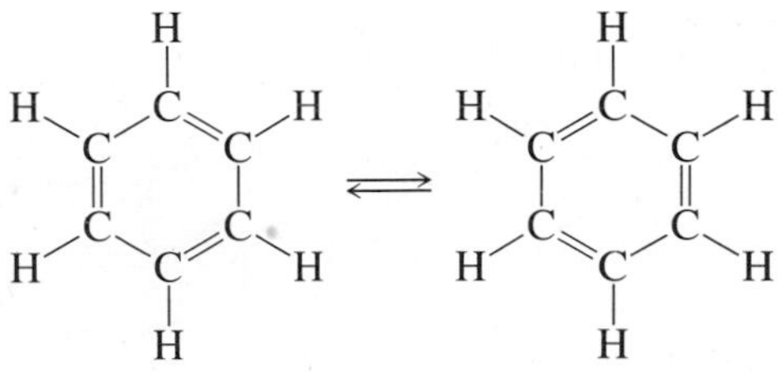

(a) $\rightleftharpoons$ (b)

Today we consider the structure of benzene to be that of a resonance hybrid intermediate between structures (a) and (b). Although the two preceding structures are contributing forms, the correct structure of benzene must lie somewhere between the following two extremes.

In formulating a molecular orbital representation of the benzene molecule, we form the planar, hexagonal framework by the overlap of sp^2 orbitals on the six carbon atoms as shown below. The six $2p$ orbitals, which appear to be the basis of the three double bonds between alternate carbon atoms in the Kekulé structures, combine instead to form six molecular orbitals that encompass the entire carbon framework. There are three bonding molecular orbitals, each occupied by two electrons called π electrons, and three empty antibonding orbitals. The lowest-energy orbital resembles that shown in Fig. 4.1. The next *two* lowest orbitals are of equal energy and also resemble Fig. 4.1, but each has a nodal plane perpendicular to the plane of the ring. Thus the π electrons are delocalized over the whole ring system. They appear as a cloud above and below the plane of the ring much like two hexagonal doughnuts with all six carbon and hydrogen atoms sandwiched in between. The bonding molecular orbitals and their relative energies can be represented schematically with the conventional symbols for $2p$ orbitals interconnected by lines to show overlap. Two common molecular models of benzene are shown in Fig. 4.2 on the opposite page.

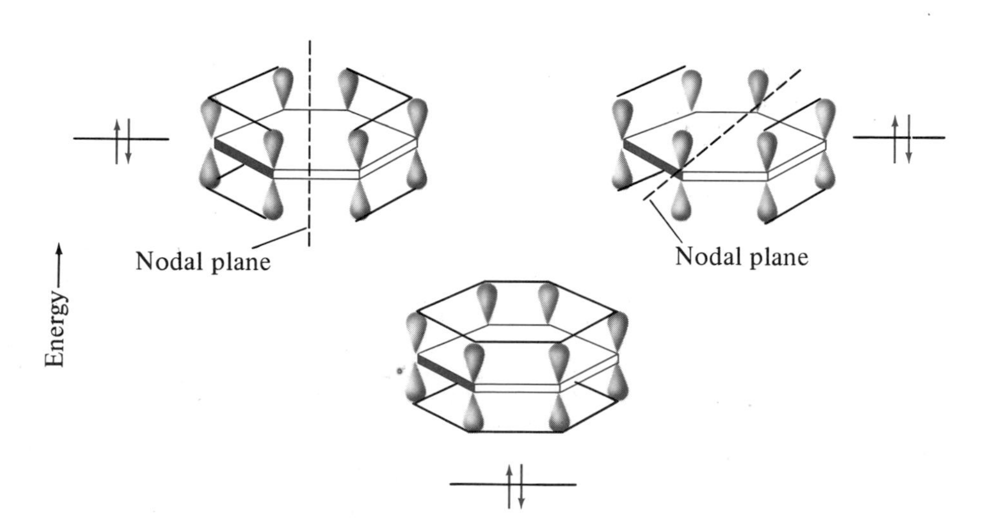

FIGURE 4.1 Structural Representation of Benzene Showing Overlap of Six π Electrons

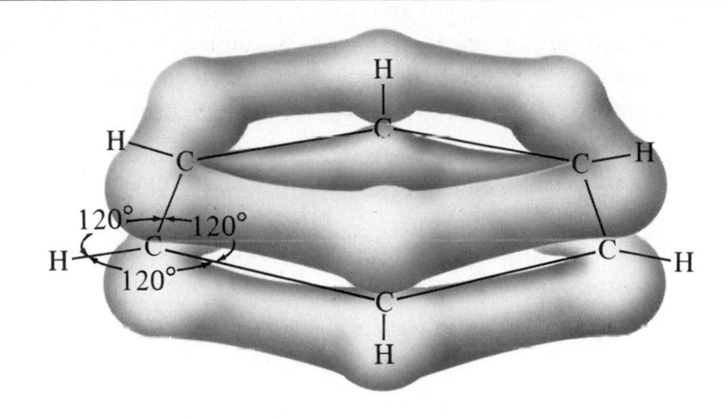

FIGURE 4.2 Molecular Models of Benzene: (a) Scale Model, (b) Ball and Stick Model

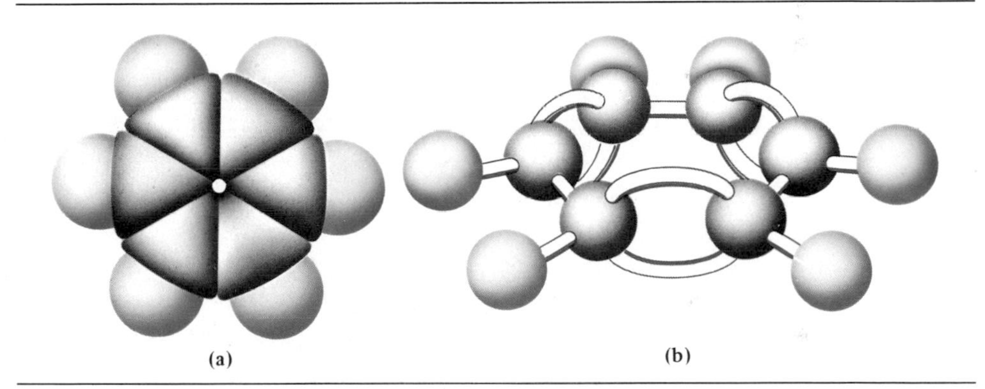

4.2 The Aromatic Sextet and Aromatic Character

To be classified as aromatic, a compound must meet the following requirements. It must be a cyclic structure in order for the delocalized π electrons to be in an orbital ring. It must be a planar structure in order to permit a maximum overlap of the p orbitals of all ring members to form a continuous, uninterrupted π system. Finally, if the ring is monocyclic, the number of p electrons must be equal to $4n + 2$, where n is an integer. The latter requirement is referred to as the Hückel Rule. If $n = 1$, we have the sextet of π electrons possessed by benzene. For other values of n the $4n + 2$ rule gives the necessary number of π electrons required for stability, and any cyclic, planar structure with this number of π electrons may exhibit aromaticity. The rule is a mathematical way of stating that monocyclic polyenes having 2, 6, 10, 14, 18, 22, . . . π electrons are likely to be aromatic, whereas those with 4, 8, 12, 16, 20, 24, . . . are likely to be nonaromatic or antiaromatic.

Aromaticity implies a high degree of stability. The π electrons above and below the plane of the molecule are able to participate in the formation of more than one bond. The results are strong bonds and a stable molecule. The additional stability possessed by benzene is called **stabilization** or **resonance energy,** and the benzene structure is said to be **stabilized by resonance.** The high degree of stability conferred upon the benzene structure by resonance is seen by comparing its heat of hydrogenation with that of an unsaturated 6-carbon ring structure in which little or no resonance occurs. For example, we might reasonably expect the heat of hydrogenation for a ring structure with three double bonds to be three times that of a compound such as cyclohexene, which has only one double bond. The heat evolved when one mole of cyclohexene is hydrogenated is 28.6 kcal.

$$\text{Cyclohexene} + H_2 \xrightarrow{\text{Ni}} \text{Cyclohexane} \qquad \Delta H = -28.6 \text{ kcal}$$

Cyclohexene   Cyclohexane

If benzene were simply a cyclohexatriene, its heat of hydrogenation should be three times 28.6 kcal or 85.8 kcal. Actually, the heat of hydrogenation of benzene is 49.8 kcal per mole.

$$\text{Benzene} + 3\,H_2 \xrightarrow{\text{Ni}} \text{Cyclohexane} \qquad \Delta H = -49.8 \text{ kcal.}$$

Benzene

85.8 kcal (expected heat) − 49.8 kcal (experimental heat) = 36 kcal (stabilization energy)

Benzene, therefore, gives up 36 kcal less energy than the expected amount because it must contain less energy. This means that benzene is more stable by 36 kcal than the hypothetical cyclohexatriene would be.

A frequently used graphic formula for benzene that embodies all the concepts of π bonding, resonance, and aromaticity is a regular hexagon with an inscribed circle.

The hexagon represents the six σ bonds between adjacent carbon atoms and the inner circle represents the π electron cloud. Each corner of the hexagon is understood to be occupied by one carbon with an attached hydrogen atom. All carbon and hydrogen atoms lie in the same plane, and the bonds between them describe angles of 120°. Although the hexagon–inscribed circle notation is a simple one to draw and use in writing the reactions of benzene, we will usually use the Kekulé representation in our discussion, with the same features being implied. Further, when any hydrogen atoms have been replaced, we will show only the substituents, with the understanding that the other hydrogens remain.

Exercise 4.1 Which of the following structures may be classified as *aromatic?* Which may be classified as *nonaromatic?* (a) cyclopropenyl cation; (b) cyclopentadienyl anion; (c) cyclobutadiene; (d) cycloheptatrienyl cation; (e) cyclohexene; (f) cycloöctatetraene.

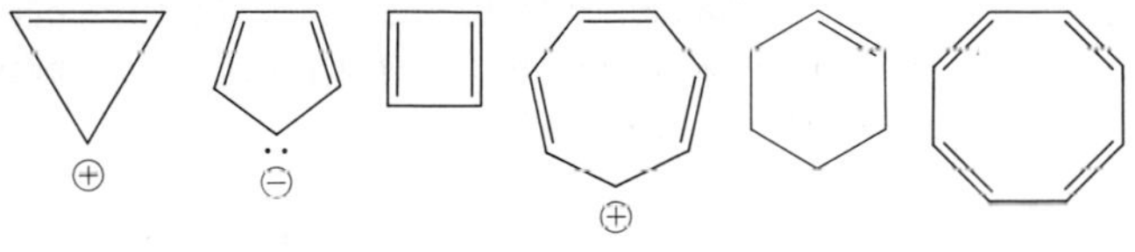

(a) (b) (c) (d) (e) (f)

4.3 Formulas and Nomenclature

The naming of aromatic compounds may be done in many ways. As in the aliphatic[2] or open-chain systems, aromatic compounds also have both trivial and systematic names. Compounds with one or more hydrogen atoms of benzene replaced by other atoms or groups may be named as substituted benzenes. Examples of such substitutive names are:

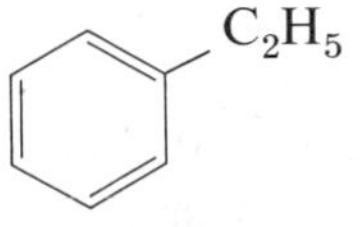

C_2H_5 Br NO_2

Ethylbenzene 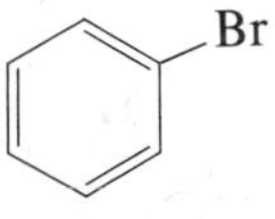 Bromobenzene 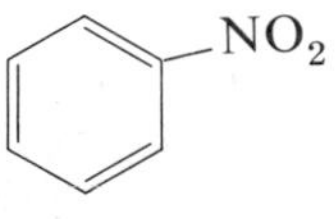 Nitrobenzene

Trivial or common names, usually of early origin, offer no clue to the nature of the substituent(s). However, such common names are so frequently used that they must be learned. When only two substituents appear on the benzene ring, their relative positions may be designated either by using the prefixes (*ortho* (*o*-), *meta* (*m*-), and *para* (*p*-) as described in Sec. 4.1), or by locating substituents on the ring by numbers. A few examples will illustrate these rules.

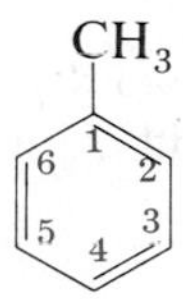

Toluene (Methylbenzene) | *o*-Xylene (1,2-Dimethylbenzene) | *m*-Xylene (1,3-Dimethylbenzene) | *p*-Xylene (1,4-Dimethylbenzene)

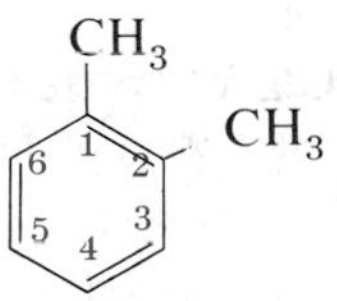

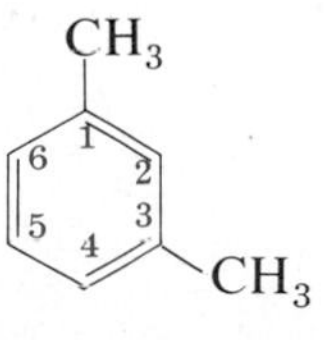

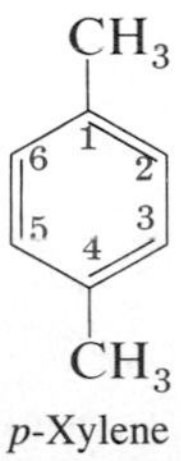

[2](Gr. *aleiphatos*, fat.) Many of the first open-chain structures studied were derived from fatty acids. In present usage aliphatic pertains to noncyclic carbon compounds, or to carbon compounds other than aromatic.

Names of substituents are listed in alphabetical order. If one of the substituents corresponds to a monosubstituted benzene with a common name, the substance is usually named as a derivative of that parent compound. When a common name is used, the group responsible for the name, e.g., methyl in the case of toluene or hydroxyl in the case of phenol, is considered to be on carbon 1. When a trivial name is not used, IUPAC rules require that the lowest number be given to the principal functional group, if a functional group is present. If not, for disubstituted compounds the lower number (1) is given to the group cited first in the name. If three or more substituents are present, the ring is numbered in such a way as to use the lowest possible numbers.

p-Chloronitrobenzene (1-Chloro-4-nitrobenzene)

2,4-Dinitrobenzoic acid

2-Bromo-4-nitrotoluene (2-Bromo-1-methyl-4-nitrobenzene)

4-Bromo-2-nitrotoluene (4-Bromo-1-methyl-2-nitrobenzene)

In 1972 the use of *o*-, *m*-, and *p*- was discontinued by *Chemical Abstracts*, the principal abstracting and indexing journal in the field of chemistry. In 1978 the use of most common names was discontinued. Since nonchemists and chemists alike use *Chemical Abstracts* to locate information on organic compounds, and since computer-based information retrieval systems are based largely on *Chemical Abstracts* nomenclature, it is important to be aware of the nomenclature used by this journal. The IUPAC rules still permit the use of *o*-, *m*-, and *p*-, as well as many common names. In this text we will continue to use common and IUPAC names, with *Chemical Abstracts* names that are different from these given in square brackets.

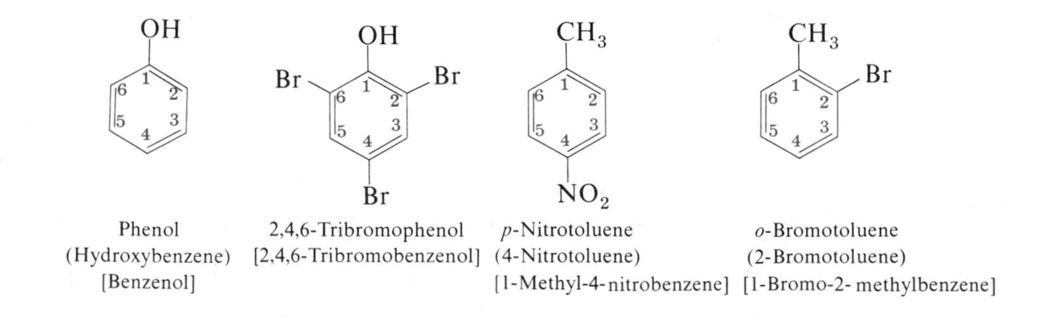

Phenol (Hydroxybenzene) [Benzenol]

2,4,6-Tribromophenol [2,4,6-Tribromobenzenol]

p-Nitrotoluene (4-Nitrotoluene) [1-Methyl-4-nitrobenzene]

o-Bromotoluene (2-Bromotoluene) [1-Bromo-2-methylbenzene]

The following are very common hydrocarbons in the aromatic series.

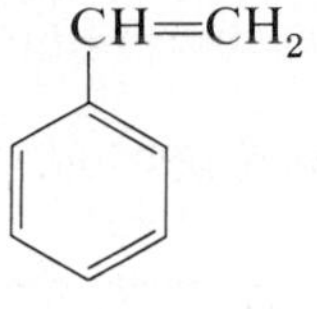

$CH{=}CH_2$ $CH_3{-}CH{-}CH_3$ CH_3 CH_3 CH_3

Styrene
[Ethenylbenzene]

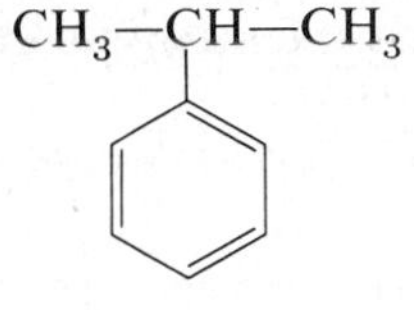

Cumene
[1-Methylethylbenzene]

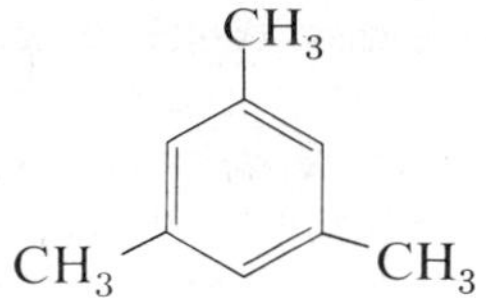

Mesitylene
(1,3,5-Trimethylbenzene)

Exercise 4.2 Cumene and mesitylene (above) have the molecular formula C_9H_{12}. Draw structures for and name six additional benzene derivatives with the same molecular formula.

4.4 Other Aromatic Hydrocarbons

Common aromatic hydrocarbons, other than the alkyl-substituted benzenes, include a number of polycyclic systems. In these systems the rings may be either condensed—that is, share two carbons in common—or be separate but joined through carbon–carbon bonds. Many of these hydrocarbons, like benzene, also are coal tar derivatives. The following structures, along with their numbering systems, are those that appear most frequently as part of polycyclic organic molecules.

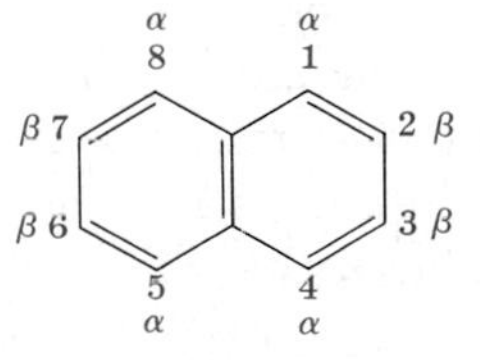

α α 8 1 β 7 2 β β 6 3 β 5 4 α α

Napthalene

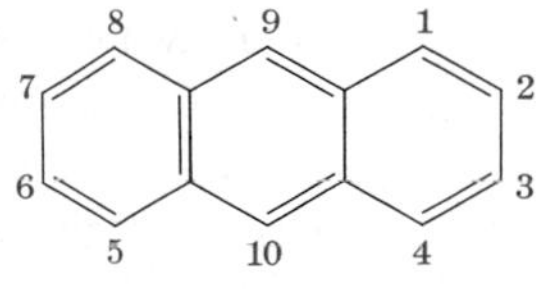

8 9 1 7 2 6 3 5 10 4

Anthracene

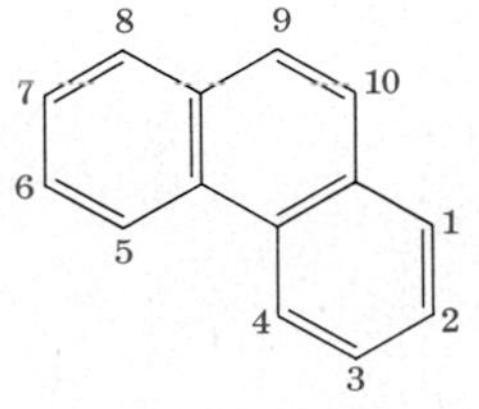

8 9 7 10 6 1 5 4 2 3

Phenanthrene

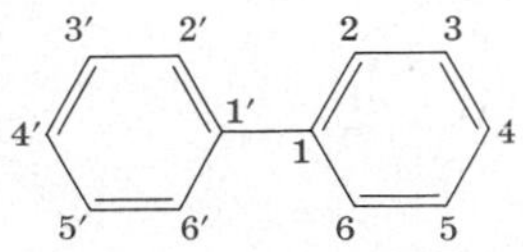

3′ 2′ 2 3 4′ 1′ 1 4 5′ 6′ 6 5

Biphenyl

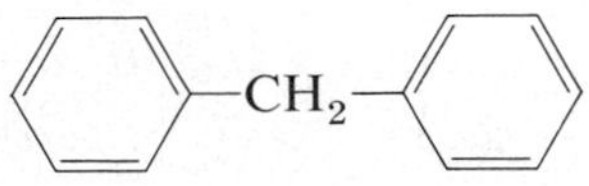

CH_2

Diphenylmethane

Exercise 4.3 Draw two additional structures for naphthalene other than the one shown above. Which structure is the most important contributor to the resonance hybrid?

Exercise 4.4 How many monosubstituted products are possible for naphthalene? How many disubstituted products?

4.5 Aromatic Groups

The removal of a hydrogen atom from an aromatic hydrocarbon, or arene, produces an **aryl group.** The abbreviation Ar is used for the aryl group. The name of the *phenyl* group, C_6H_5—, comes from *phene,* an early name for benzene. Common abbreviations for the phenyl group in structural formulas are Ph- and ϕ-, the former being much preferred. Table 4.1 gives the structures and names of the most commonly used aryl groups.

TABLE 4.1 Common Aryl Groups

Structure	Name of Group	Example of Usage	Name of Example
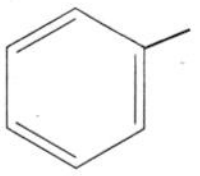 C_6H_5—	Phenyl	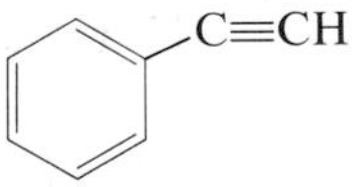 C_6H_5—C≡CH	Phenylacetylene [Ethynylbenzene]
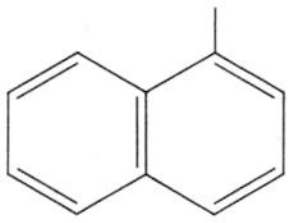 (1-naphthyl structure)	1-Naphthyl (α-Naphthyl)	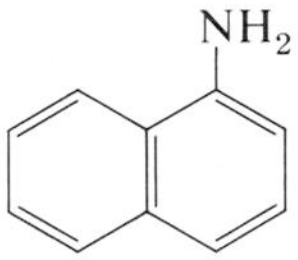 NH_2 (on naphthalene, position 1)	1-Naphthylamine (α-Naphthylamine)
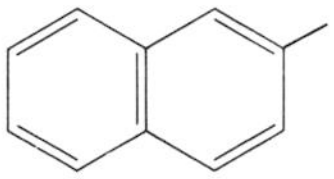 (2-naphthyl structure)	2-Naphthyl (β-Naphthyl)	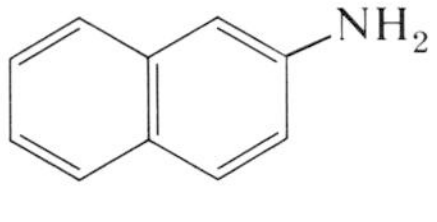 NH_2 (on naphthalene, position 2)	2-Naphthylamine (β-Naphthylamine)
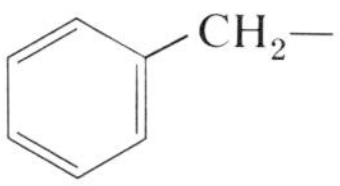 C_6H_5—CH_2—	Benzyl	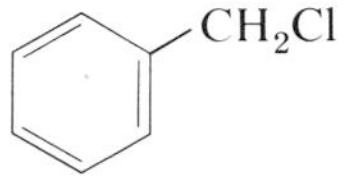 C_6H_5—CH_2Cl	Benzyl chloride [Phenylmethyl chloride]
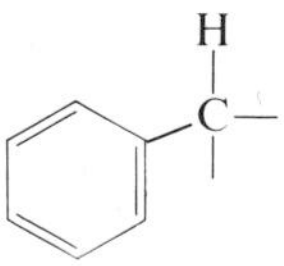 C_6H_5—CH<	Benzal	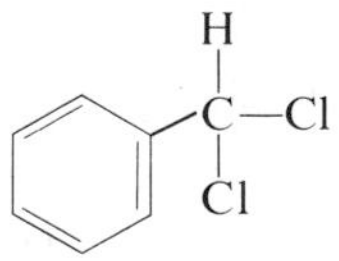 C_6H_5—CH(Cl)—Cl	Benzal chloride

Exercise 4.5 Draw structures for the following: (a) benzyl alcohol, (b) 1-phenylethanol, (c) triphenylmethane, (d) phenylmagnesium bromide, (e) 3-phenylhexane.

4.6 Reactions of the Aromatic Hydrocarbons

The reactions of benzene and other aromatic hydrocarbons are mainly reactions of substitution in the aromatic ring. Substitution reactions involving the aromatic ring, unlike those of the alkanes, are controllable and much more useful. The benzene ring is attacked in nearly every case by an electrophilic reagent that may be either a cation or a neutral but polarized molecule. The electron-deficient group, by sharing in an electron pair supplied at one of the ring carbons, first forms with the benzene ring a cationic intermediate, called a **sigma complex,** in which the positive charge is distributed among the remaining carbon atoms. In the second step a proton is lost to regenerate the more stable aromatic ring.

Catalysts are often necessary to generate the electrophile. Such catalysts, as you will see in the following sections, are often Lewis acids, each of which lacks an electron pair. The result of attracting a pair of electrons from the substituting reagent by the catalyst is the formation of either a positive ion or a polarized molecule. In either case the positive ion or the positive part of the molecule is the ring-attacking group. The mechanism of electrophilic aromatic substitution may be illustrated by the following.

STEP 1.

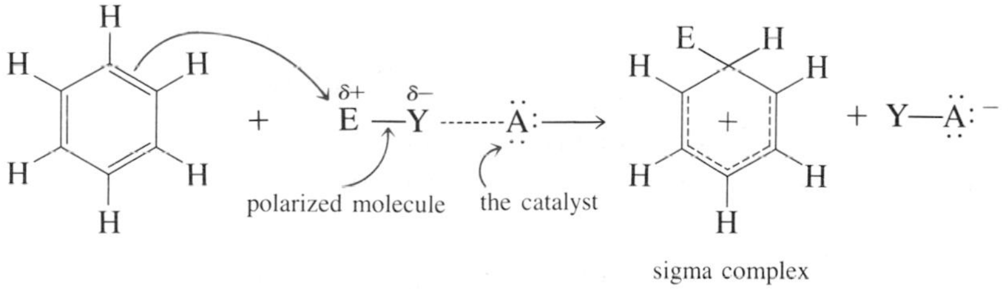

$$C_6H_6 + \overset{\delta+}{E}-\overset{\delta-}{Y}\cdots\ddot{\underset{..}{A}}: \longrightarrow \text{sigma complex} + Y-\ddot{\underset{..}{A}}:^-$$

polarized molecule   the catalyst

sigma complex

STEP 2.

E H

E E E

E

$$\longrightarrow C_6H_5E + H^+$$

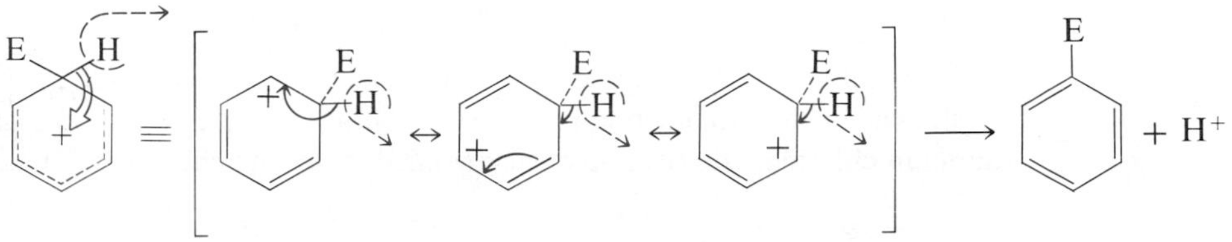

STEP 3. $H^+ + Y-\ddot{\underset{..}{A}}:^- \longrightarrow H-Y + \ddot{\underset{..}{A}}:$

The principal reactions of benzene are described in the following sections.

A. Halogenation. Halogenation of the aromatic ring may be illustrated by bromination. Bromine attacks the aromatic ring in a manner analogous to its attack upon the double bond of an olefin, except that a catalyst is necessary to assist in the polarization of the bromine molecule. Iron, usually in finely divided form, is used for this purpose. Presumably, ferric bromide, $FeBr_3$, is the actual catalyst, and this could form as shown in the first of the equations representing the bromination of benzene.[3]

$$3\,Br_2 + 2\,Fe \longrightarrow 2\,FeBr_3$$

Ferric bromide

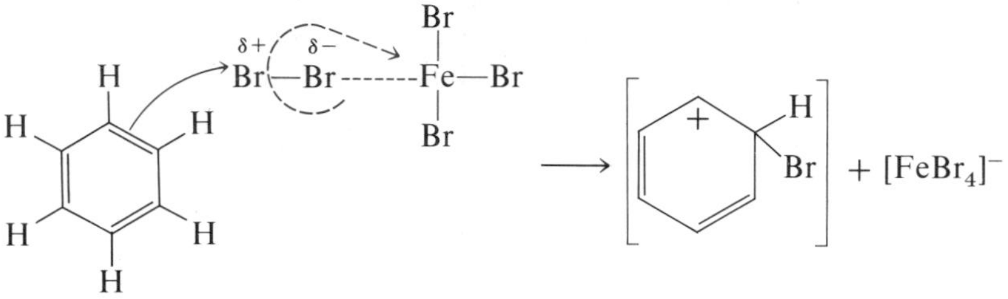

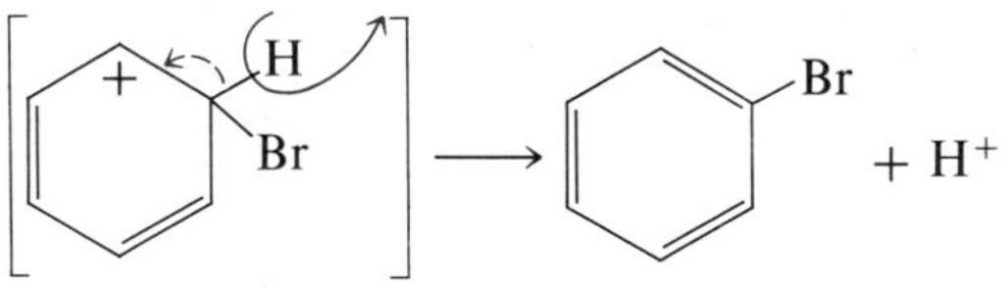

$$C_6H_6 + \overset{\delta+}{Br}\!-\!\overset{\delta-}{Br}\cdots FeBr_3 \longrightarrow [C_6H_6Br]^+ + [FeBr_4]^-$$

$$[C_6H_6Br]^+ \longrightarrow C_6H_5Br + H^+$$

Bromobenzene

$$H^+ + FeBr_4^- \longrightarrow FeBr_3 + HBr$$

Although we have indicated the bromination of benzene as a series of discrete steps, the net equation for the overall reaction may be written simply as

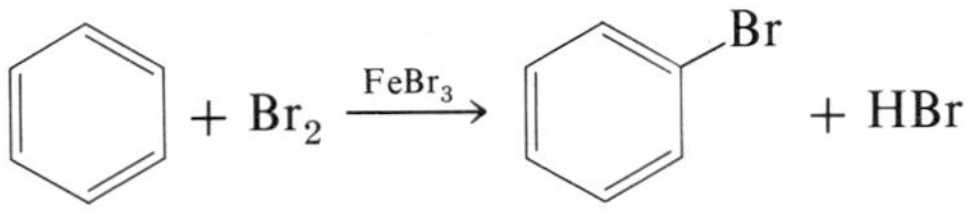

$$C_6H_6 + Br_2 \xrightarrow{FeBr_3} C_6H_5Br + HBr$$

Bromobenzene

Substitution of benzene by chlorine may also be accomplished in a manner similar to that shown for the bromination reaction. As before, a catalyst is necessary. Anhydrous aluminum chloride, $AlCl_3$, is an excellent catalyst for the chlorination reaction. If iron is

[3] In this and subsequent mechanisms for electrophilic aromatic substitution, only one resonance form, in brackets, is given for the cationic intermediate. The student should practice writing all three (or more) forms, referring to those above as necessary.

used, the catalyst presumably is $FeCl_3$.

$$C_6H_6 + Cl_2 \xrightarrow{AlCl_3} C_6H_5Cl + HCl$$

Chlorobenzene

B. Nitration of Benzene. Nitration of the aromatic ring is a very important reaction and produces a number of useful compounds unobtainable in any other way. Nitration is usually accomplished by treating benzene with a mixture of concentrated nitric and sulfuric acids. A mixture of these two acids is referred to as a nitrating mixture. The electrophilic species in the nitration of benzene appears to be the nitronium ion, $^+NO_2$, which may be formed from nitric acid by the action of sulfuric acid. The mechanism usually given for the nitration of benzene is illustrated by the following sequence of reactions.

$$HO{-}NO_2 + 2\,H_2SO_4 \longrightarrow 2\,HSO_4^- + H_3O^+ + {}^{\oplus}NO_2$$

Nitronium ion

$$C_6H_6 + {}^{\oplus}NO_2 \longrightarrow \left[C_6H_6NO_2^+\right] \longrightarrow C_6H_5NO_2 + H^+$$

Nitrobenzene

The net equation for the overall reaction may be written

$$C_6H_6 + HNO_3 \xrightarrow{H_2SO_4} C_6H_5NO_2 + H_2O$$

Nitrobenzene

C. Alkylation of Benzene. The simple alkyl benzenes such as toluene (methylbenzene) and the xylenes (dimethylbenzenes) are readily available from petroleum. On occasion, however, it may be necessary or desirable to introduce an alkyl group into the benzene ring. One method for accomplishing this substitution is a reaction known as the **Friedel–Crafts** reaction.[4] In the Friedel–Crafts reaction an alkyl group is attached to the ring by treating benzene with an alkyl halide in the presence of anhydrous aluminum chloride, $AlCl_3$. The electrophilic reagent that attacks the ring in the alkylation reaction appears to be a carbo-

[4]Charles Friedel (1832–1899), a French chemist known mainly for his discovery, in collaboration with an American chemist, James M. Crafts (1838–1917), of a convenient method of synthesis of aromatic ketones.

cation or incipient carbocation. Inasmuch as alkenes (equation 2) and alcohols (equation 3) are also capable of forming carbocations (Sec. 7.7), they too can be used to alkylate the benzene ring. The mechanism of the Friedel–Crafts reaction is illustrated in the following equations for the preparation of isopropylbenzene.

$$(1)\quad (CH_3)_2CHCl + AlCl_3 \longrightarrow \overset{\oplus}{C}H(CH_3)_2\cdots Cl\overset{(-)}{-}AlCl_3$$

Isopropyl chloride

$$C_6H_6 + \overset{\oplus}{C}H(CH_3)_2\cdots Cl\overset{(-)}{-}AlCl_3 \longrightarrow \left[\overset{+}{C_6H_6}\text{—}C(H)(CH_3)_2\right] + \overset{(-)}{AlCl_4}$$

$$\left[\overset{+}{C_6H_6}\text{—}CH(CH_3)_2\right] \longrightarrow C_6H_5\text{—}C(H)(CH_3)_2 + H^+$$

Isopropylbenzene (Cumene)

$$H^+ + \overset{(-)}{AlCl_4} \longrightarrow AlCl_3 + HCl$$

$$(2)\quad CH_3\text{—}CH\text{=}CH_2 + HCl + AlCl_3 \longrightarrow CH_3\text{—}\underset{\oplus}{C}H\text{—}CH_3 + \overset{(-)}{AlCl_4}$$

$$(3)\quad CH_3\text{—}CH(OH)\text{—}CH_3 + 2\,H_2SO_4 \longrightarrow CH_3\text{—}\underset{\oplus}{C}H\text{—}CH_3 + H_3\overset{+}{O} + 2\,HSO_4^-$$

Isopropyl alcohol

The isopropyl cation then combines with the aromatic ring in the same manner as indicated in the first equation above.

The Friedel–Crafts reaction, although useful, has the disadvantage of giving not only polysubstituted products, but also rearranged products if the group to be substituted into the benzene ring is unbranched and larger than ethyl. For example, if benzene is treated with *n*-propyl chloride and aluminum chloride, the product obtained is largely isopropylbenzene, or cumene. Only small amounts of the normal isomer are obtained. Formation of the more highly branched secondary carbocation, which may be accomplished by a rearrangement of a hydrogen atom and a pair of electrons as a hydride ion, appears to be favored because it represents a more stable intermediate than the primary carbocation. This type of migration is referred to as a 1,2-hydride shift.

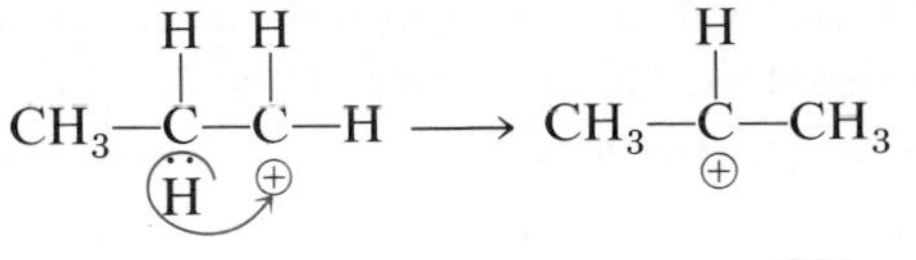

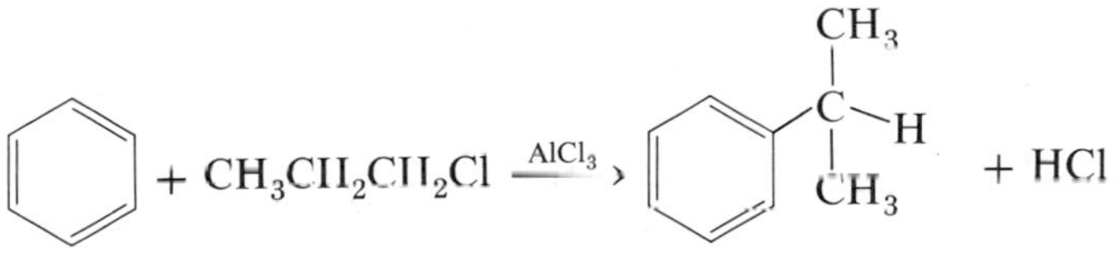

$$CH_3-CH(H)-CH_2^{\oplus} \longrightarrow CH_3-\overset{\oplus}{C}H-CH_3$$

$$C_6H_6 + CH_3CH_2CH_2Cl \xrightarrow{AlCl_3} C_6H_5CH(CH_3)_2 + HCl$$

Isopropylbenzene
(Cumene)

D. Sulfonation of Benzene. Benzenesulfonic acid is produced when benzene is heated with concentrated sulfuric acid. The ring-attacking group in the sulfonation reaction is not a positive ion as was the case in the previous reactions. Instead, it appears to be the electron-deficient sulfur trioxide molecule produced by the reaction of two molecules of sulfuric acid.

$$2\,H_2SO_4 \rightleftharpoons H_3O^+ + SO_3 + HSO_4^-$$

The sulfonation of aromatic compounds is a very important reaction, especially in the preparation of dyestuffs. Sulfonation of the aromatic ring is a convenient method for making aromatic compounds water-soluble, because many of the sulfonic acids form water-soluble metal salts, and many aromatic sulfonic acids are themselves quite soluble in water.

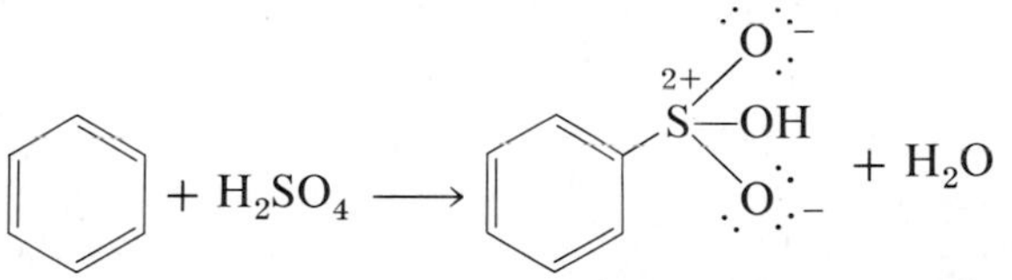

$$C_6H_6 + H_2SO_4 \longrightarrow C_6H_5S^{2+}(O^-)_2OH + H_2O$$

Benzenesulfonic acid

$$C_6H_5SO_3H + Na^+OH^- \longrightarrow C_6H_5SO_3^-Na^+ + H_2O$$

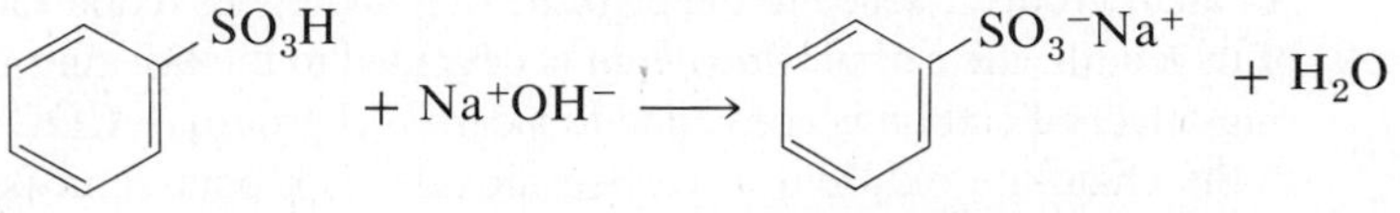

Sodium benzenesulfonate

E. Oxidation of Aromatic Hydrocarbons. Neither benzene nor polycyclic aromatic compounds are reactive to the usual oxidizing reagents such as potassium permanganate ($KMnO_4$) or potassium dichromate ($K_2Cr_2O_7$). However, the benzene ring can be ruptured and oxidized when treated with oxygen in the presence of vanadium pentoxide at high temperatures. The anhydride of maleic acid, a dicarboxylic acid (Sec. 11.8), may be produced in this manner.

$$\text{benzene} + 4\tfrac{1}{2}\,O_2 \xrightarrow[450^\circ C]{V_2O_5} \text{maleic anhydride} + 2\,CO_2 + 2\,H_2O$$

Maleic anhydride

Naphthalene also is oxidized under these same conditions. One benzene ring appears to facilitate the oxidation of the other. The anhydride of phthalic acid, an important industrial chemical used in the preparation of the glyptal resins (Sec. 11.4), is produced from naphthalene in this manner.

$$\text{Naphthalene} + 4\tfrac{1}{2}\,O_2 \xrightarrow[350^\circ C]{V_2O_5} \text{Phthalic anhydride} + 2\,CO_2 + 2\,H_2O$$

Naphthalene — Phthalic anhydride

Other reactions of naphthalene are similar to those shown for benzene. In nearly every instance the substituting group exhibits a predilection for the **alpha** positions (Sec. 4.4).

4.7 Oxidation of Alkyl Benzenes

An alkyl group attached to the benzene ring undergoes oxidation quite readily. Regardless of its length, the carbon *side chain* is degraded to the last ring-attached carbon atom. The ring-attached carbon is converted to a carboxyl group, **—COOH.** All other carbon atoms in the chain are oxidized to carbon dioxide. Hot potassium permanganate or potassium dichromate in sulfuric acid is usually used for the oxidation of side chains.

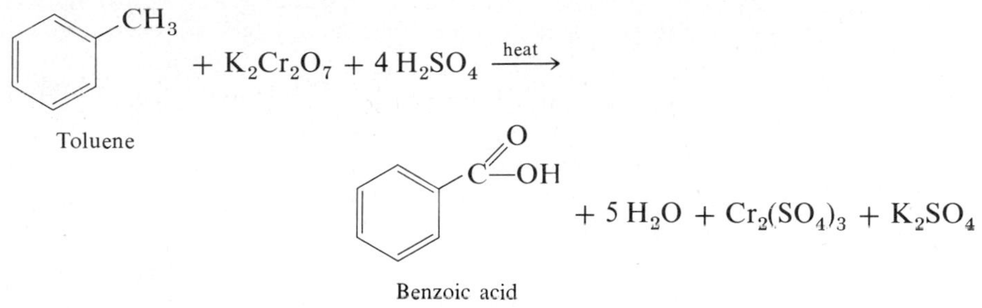

$$\text{Toluene} + K_2Cr_2O_7 + 4\,H_2SO_4 \xrightarrow{\text{heat}} \text{Benzoic acid} + 5\,H_2O + Cr_2(SO_4)_3 + K_2SO_4$$

Toluene

Benzoic acid

Exercise 4.6 Write an equation for the oxidation of isopropylbenzene (cumene) with alkaline potassium permanganate.

4.8 Directive Influence of Ring Substituents

A substituent already on the ring not only influences the facility with which a second group enters but also determines the position on the ring that the incoming group will occupy, that is, determines the **orientation** of further substitution. If a substituent has one or more pairs of nonbonded electrons, such as hydroxyl, (—$\ddot{O}$H), or amino, (—$\ddot{N}H_2$), the group may serve as an electron-pair donor and tend to stabilize the cationic intermediate formed preferentially at the *ortho* and *para* positions of the ring. Substitution at these sites is enhanced, and little, if any, substitution will occur at the *meta* position. Groups capable of orienting new substituents to the *ortho* and *para* positions in this manner are called **ortho-para directors.** Consider, for example, the electrophilic substitution of phenol and the cationic intermediates formed by the addition of the electrophile, E^+, to the *ortho*, *meta*, and *para* positions. Note that there are four contributing forms, including one with a positive charge on oxygen, for the *ortho* and *para* intermediates, but only three forms for the *meta* intermediate.

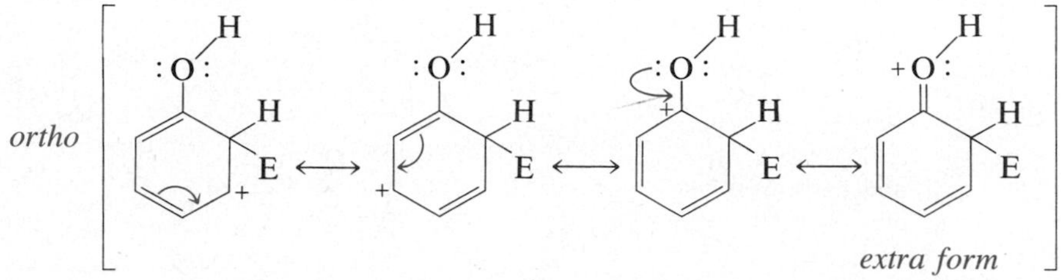

ortho

extra form

meta

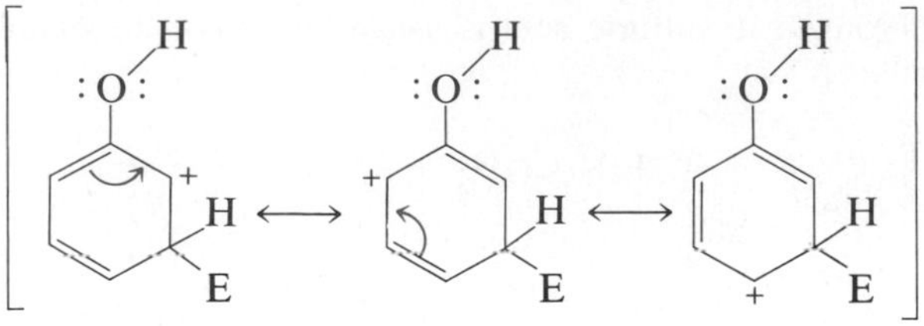

para

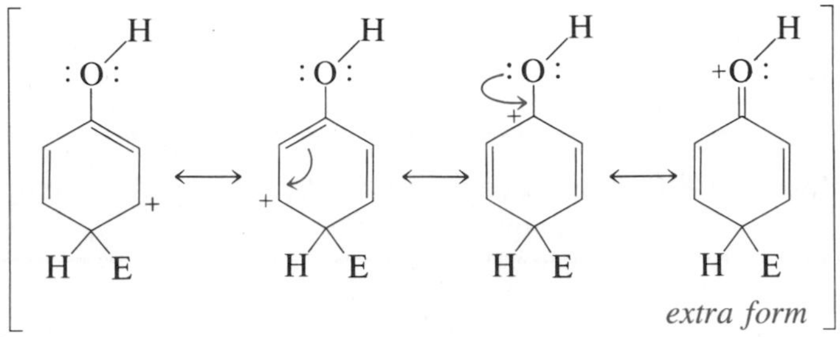

extra form

The stability of a charged species depends in part on the dispersal of the charge over as many atoms as possible, and the stability of a resonance hybrid depends upon the number of reasonable resonance forms that can be written. Both of these conditions are met better by the *ortho* and *para* intermediates than by the *meta* intermediate or by the sigma complex (p. 129) derived from unsubstituted benzene itself (three forms). Thus the hydroxyl group is not only an *o–p* director, but also has an activating effect on the ring, causing phenol to be more reactive than benzene. Indeed, phenol is so reactive that 2,4,6-tribromophenol may be prepared by simply shaking an aqueous solution of phenol with bromine water. No catalyst is required.

In contrast to the foregoing, an electron-attracting group tends to *deactivate* the ring and make it less reactive than benzene. Such substituents are usually groups in which the atom attached to the ring is multiply bonded to other atoms of greater electronegativity such as nitro:

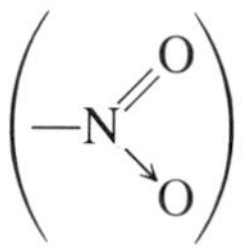

cyano: (—C≡N)

and carbonyl:

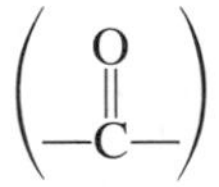

Substitution at the *ortho* and *para* positions will be inhibited because the cationic intermediates will be destabilized somewhat more than the intermediate for *meta* substitution. The nitro group is one of the most powerful electron-withdrawing groups and serves as a good model for other groups of this type.

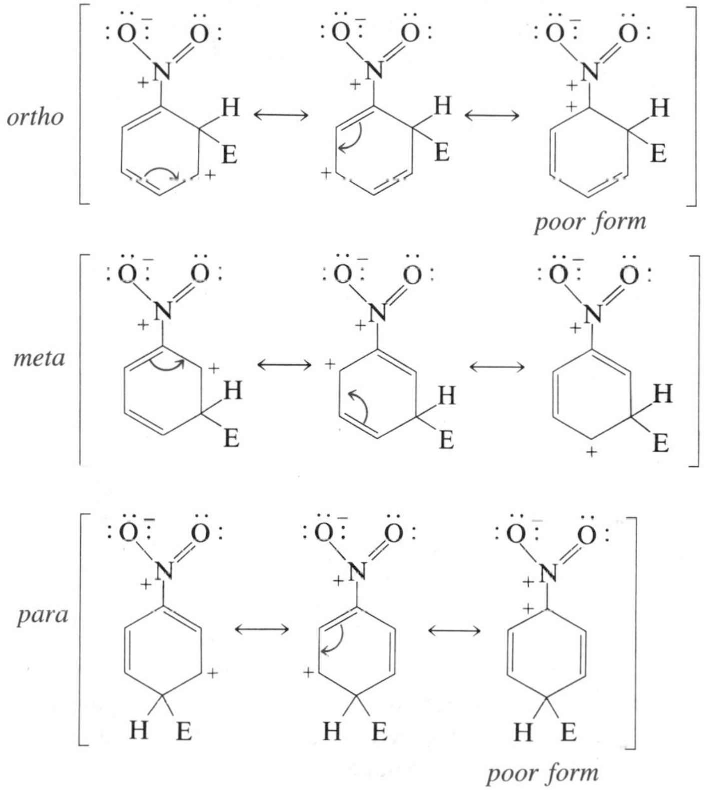

ortho

poor form

meta

para

poor form

All the contributing forms are destabilized by the electrostatic repulsion of the cationic positive charge and the strong dipole of the nitro group (Sec. 1.14).

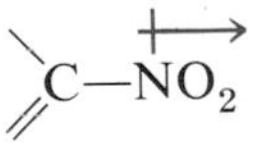

$$\text{C—NO}_2$$

Exercise 4.7 Draw structures for *p*-chloronitrobenzene and *p*-bromoaniline, then indicate the direction in which the electron density is increased.

The third contributing form for both the *ortho* and *para* cationic intermediates will be affected the most because the cation's positive charge is on the carbon atom attached to the strongly electron-withdrawing nitro group, that is, at the positive end of the nitro group dipole. Although all three of the cationic intermediates are destabilized relative to the intermediates for benzene, the *ortho* and *para* intermediates are destabilized the most. Thus an incoming substituent, if it bonds to the ring at all, must take next best and affix itself to one of the *meta* carbons. Groups capable of orienting other substituents to *meta* positions are called **meta directors.** The nitro group is a very strong electron attractor, deactivates the ring, and is a *meta* director. The deactivating power of the nitro group is sufficiently strong to prevent nitrobenzene from entering into a Friedel–Crafts reaction. Nitrobenzene, because of its unreactivity, is frequently used as the solvent for the Friedel–Crafts reaction. Table 4.2 lists the common *ortho–para* and *meta* directors in approximate order of diminishing directive power.

TABLE 4.2 Table of *ortho–para* and *meta* Directing Groups

Ortho–Para Directors (Activating)	Representative Compounds	Meta Directors (Deactivating)	Representative Compounds
—OH	Phenol, C_6H_5OH	$—\overset{+}{N}H_3$	Anilinium chloride, $C_6H_5\overset{+}{N}H_3Cl^-$
$—NH_2$	Aniline, $C_6H_5NH_2$	—N(→O)O	Nitrobenzene, $C_6H_5NO_2$
$—OCH_3$	Anisole, $C_6H_5OCH_3$	—C≡N	Benzonitrile, C_6H_5CN
—N(H)—C(=O)—CH_3	Acetanilide, $C_6H_5N(H)C(=O)CH_3$	—S(→O)(→O)—OH	Benzenesulfonic acid, $C_6H_5SO_3H$
—O—C(=O)—CH_3	Phenyl acetate, $C_6H_5O—C(=O)—CH_3$	—C(=O)—CH_3	Acetophenone, $C_6H_5C(=O)—CH_3$
$—CH_3$	Toluene, $C_6H_5CH_3$	—C(=O)—H	Benzaldehyde, $C_6H_5C(H)=O$
—Cl, —Br, —I* (deactivating)*	Bromobenzene, C_6H_5Br	—C(=O)—OH	Benzoic acid, $C_6H_5C(=O)—OH$

*The halogens are *ortho–para* directing by virtue of their nonbonded electron pairs, but they are also electron-attracting because of their high electronegativity. The second influence destabilizes all three intermediates and results in ring deactivation.

In planning the synthesis of an aromatic compound with two or more substituents on the ring, it is essential to consider the orientation effects of the substituents in order to determine which substituent should be introduced first, second, and so on. For example, if the objective is to prepare *m*-bromonitrobenzene, the nitro group must be introduced first, as is shown in the following equations.

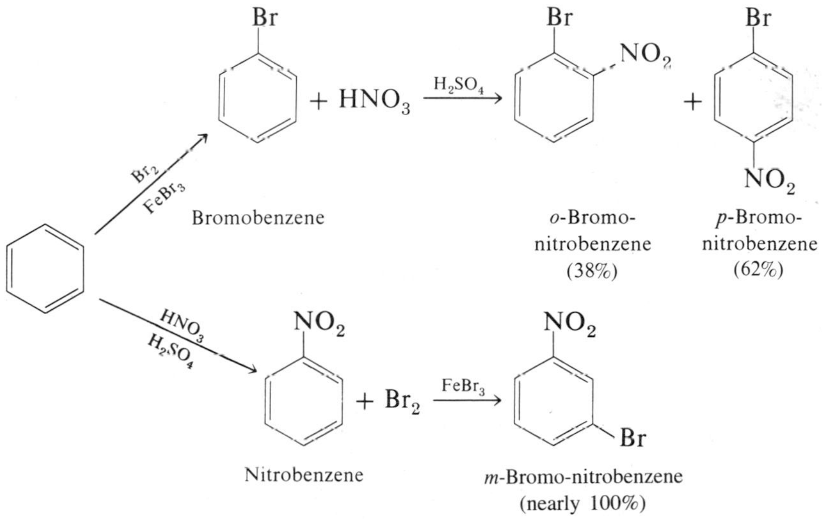

Br; + HNO_3 $\xrightarrow{H_2SO_4}$ Br, NO_2 + Br, NO_2

Br_2, $FeBr_3$

Bromobenzene

o-Bromo-nitrobenzene (38%)

p-Bromo-nitrobenzene (62%)

HNO_3, H_2SO_4

NO_2 + Br_2 $\xrightarrow{FeBr_3}$ NO_2, Br

Nitrobenzene

m-Bromo-nitrobenzene (nearly 100%)

PROBLEM IN SYNTHESIS

Through what sequence of reactions could *p*-nitrobenzoic acid be prepared?

Solution The orientation of the nitro group to the *para* position necessitates beginning with an *o–p* director. Our choice of starting material is therefore toluene, which is inexpensive and readily available. We cannot begin with benzoic acid because the carboxyl group is a *meta* director. After separation of *p*-nitrotoluene from its *ortho* and *meta* isomers (which will also be formed), we will oxidize the methyl group to carboxyl. The nitro group is not oxidizable.

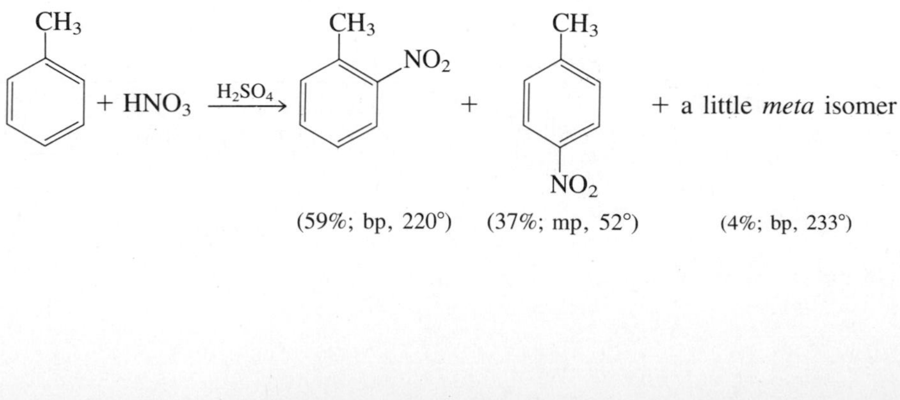

CH_3 + HNO_3 $\xrightarrow{H_2SO_4}$ CH_3, NO_2 + CH_3, NO_2 + a little *meta* isomer

(59%; bp, 220°) (37%; mp, 52°) (4%; bp, 233°)

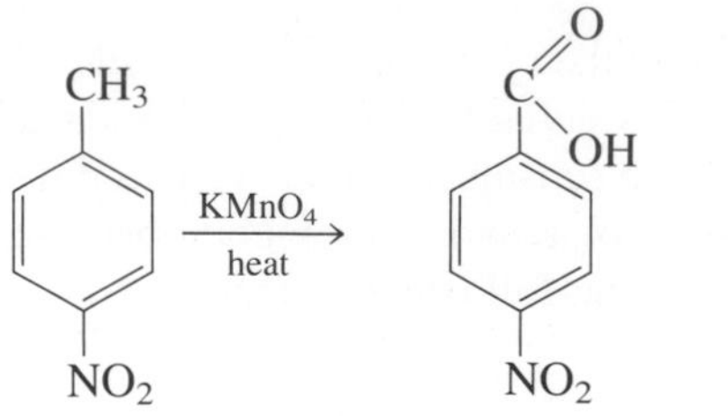

CH_3 (p-nitrotoluene) $\xrightarrow[\text{heat}]{KMnO_4}$ p-$NO_2C_6H_4COOH$

p-Nitrobenzoic acid

Exercise 4.8 Ring substitution of aniline takes place almost exclusively at the *ortho* and *para* positions when carried out in neutral or slightly basic solution. When substitution of aniline is attempted in an acidic solution, the incoming group seeks the *meta* position. Explain.

4.9 Heterocyclic and Heteroaromatic Compounds

The ring systems of the cyclic compounds studied up to this point have been those in which only carbon atoms are joined together. Such ring systems are called **carbocyclic.** Rings that contain, in addition to carbon, one or more other atoms in the ring are called **heterocyclic** (Gr. *heteros,* other). The **hetero atoms** that occur most frequently in heterocyclic rings are nitrogen, sulfur, and oxygen. Heterocyclic compounds, like carbocyclic compounds, may be saturated, unsaturated, or aromatic. Natural products containing one or more heterocyclic rings in their molecular structures are so numerous, and their properties so diverse, that we can consider only a few of the most important systems in this text. The names and structures of the five- and six-membered heterocyclic ring systems most frequently encountered in natural products or in the chemistry laboratory are given in Table 4.3. You will note that numbering of the ring atoms begins with a hetero atom.

TABLE 4.3 Common Heterocyclic Ring Systems

Structure	Name	Occurs in These Examples
(A) *Five-membered Rings*		
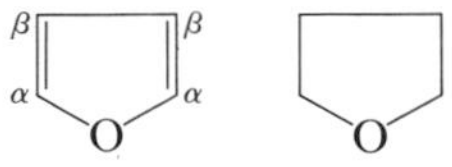 (a) (β, α, O) (b) (O)	(a) Furan (b) Tetrahydrofuran	Furfural Morphine
 (β 4, 3 β, α 5, 2 α, 1 S)	Thiophene	

TABLE 4.3 *Continued*

Structure	Name	Occurs in These Examples
(A) *Five-membered Rings (continued)*		
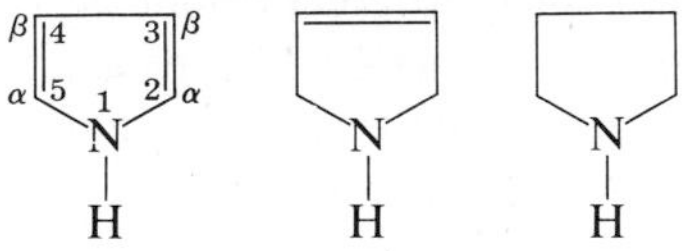 (a) (b) (c)	(a) Pyrrole (b) Pyrroline (c) Pyrrolidine	Vitamin B_{12} Chlorophyll-a Heme Cytochrome P450 Proline Nicotinc
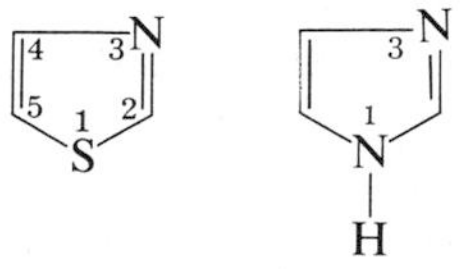 (a) (b)	(a) Thiazole (b) Imidazole	Penicillins Vitamin B_1 Histidine
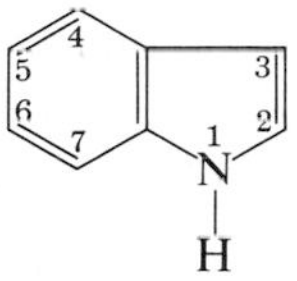	Indole	Tryptophan Strychnine Reserpine Serotonin
(B) *Six-membered Rings*		
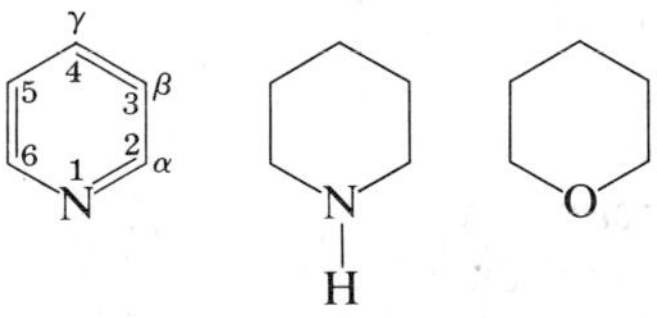 (a) (b) (c)	(a) Pyridine (b) Piperidine (c) Tetrahydropyran	NAD, Nicotine Vitamin B_{12} Quinine Morphine Cocaine, Coniine Reserpine Marijuana
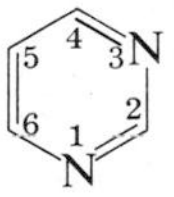	Pyrimidine	Vitamin B_1 Barbiturates Nucleic acids
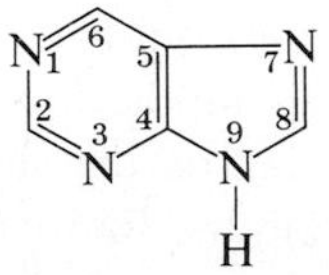	Purine	CoA enzyme NAD enzyme Nucleic acids
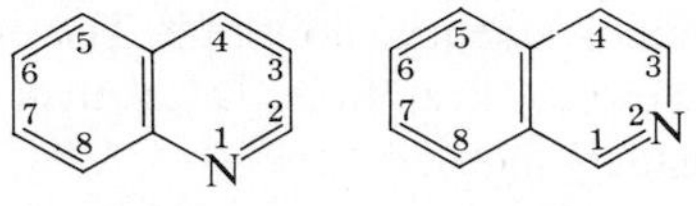 (a) (b)	(a) Quinoline (b) Isoquinoline	Curare

Other ring atoms (in rings with but one hetero atom) are often designated in nonsystematic nomenclature as α, β, and γ. The α-atom is the ring member adjacent to the hetero atom. The chemistry of the saturated and unsaturated heterocyclic systems differs very little from that of related acyclic compounds; therefore, this chemistry will be described along with that of the acyclic analogs in the following chapters. The rich chemistry of the **heteroaromatic** compounds may appear to be quite different from that of aromatic compounds containing only carbon in the ring; however, the differences are often differences in degree, not in kind. Thus, heteroaromatic compounds may be more or less aromatic than benzene, and their reactions will reflect this difference.

Consider the two heterocyclic systems, pyridine and pyrrole. Both have one nitrogen atom in the ring, but pyridine is a 6-ring and pyrrole is a 5-ring system. In pyridine the nitrogen atom is sp^2 hybridized, as are all the carbon atoms. Each carbon atom and the nitrogen contribute one electron to the six-electron π system. The unshared pair on the nitrogen atom lies in an sp^2 orbital *in the plane of the ring*. Since this orbital cannot overlap with the $2p$ orbitals making up the π system, the unshared pair cannot be a part of the π system; therefore, there are only six, not eight, π electrons in pyridine. We conclude that pyridine is a hybrid of two Kekulé-like structures, and we predict that the molecule is likely to be aromatic. Its resonance energy has been estimated to be about 32 kcal/mole, not greatly different from that of benzene itself. Pyridine, mainly isolated from coal tar, is used in the laboratory and in industry as a solvent and as an organic base in reactions requiring basic catalysis.

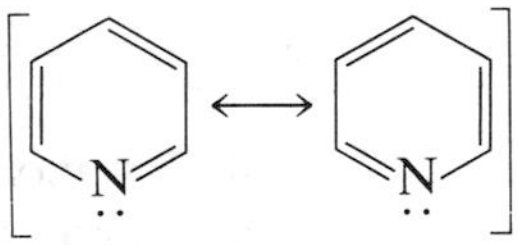

Pyridine

The molecular structure of pyrrole is quite different from the aromatic structures we have previously discussed (see Exercise 4.1). In pyrrole it may appear, at first, that the nitrogen atom would be like the nitrogen atom in ammonia, sp^3 hybridized with the bond pair to the hydrogen atom in a sigma orbital resulting from sp^3–$1s$ overlap, with the unshared pair in an sp^3 orbital, and that these orbitals should be directed away from the nitrogen at angles somewhat above and below the plane of the ring, respectively. However, the actual state of hybridization of the nitrogen atom in any molecule will be that yielding the most stable molecule—in this case, sp^2. Let us consider the result of sp^2 hybridization of the nitrogen atom. The bond pair to the hydrogen atom will be in an sp^2–$1s$ sigma orbital lying in the plane of the ring, whereas the unshared pair will be in a $2p$ orbital that can overlap easily with the four $2p$ orbitals on the four carbon atoms. Thus, the six electrons required for an aromatic sextet may be derived from *two* $2p$ electrons from the nitrogen atom and four $2p$ electrons from the four carbon atoms. In drawing the resonance forms for this system, we observe that the forms are not exactly equivalent and that there is charge separation in a neutral molecule. Therefore, we are not surprised that the resonance energy of pyrrole, 21 kcal/mole, is less than that of benzene and that pyrrole is said to be "less aromatic" than benzene.

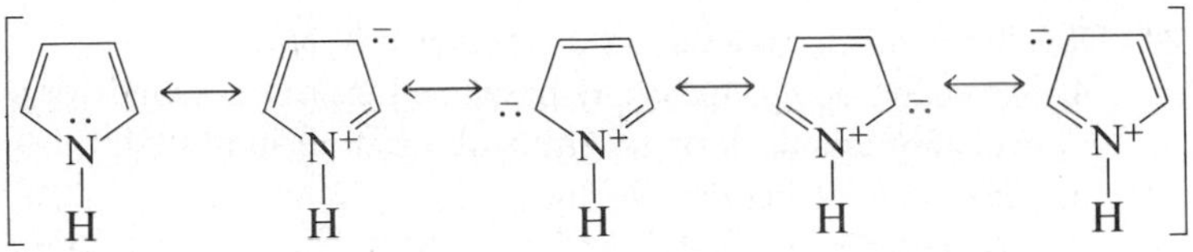

Pyrrole

Thiophene, furan, and pyrrole are more reactive toward electrophiles than is benzene, with substitution tending to take place at the 2-position of the ring and a relative order of reactivity of pyrrole > furan > thiophene ≫ benzene. Although much more reactive than benzene, thiophene is closer to benzene in chemical behavior than are the other two 5-ring heterocycles, and replacement of a benzene ring by a thiophene ring in a medicinal product often yields a new drug with similar pharmacological properties. Pyridine is much less reactive than benzene in electrophilic substitution reactions, with substitution occurring almost exclusively at the 3-position. The very low reactivity of pyridine in electrophilic substitution is probably due to the reaction of the lone-pair electrons on the pyridine nitrogen with a proton or the electrophile (which is a Lewis acid) to place a positive charge on the nitrogen atom. Thus, the transition state for an electrophilic substitution reaction will have *two* positive charges on the ring, and the repulsion between the charges will destabilize the transition state. Electrophilic substitution reactions of furan and pyrrole must be carried out under very mild conditions and those of pyridine under very drastic conditions.

Exercise 4.9 The product from the nitration of thiophene under very mild conditions is a mixture of 2-nitrothiophene and 3-nitrothiophene in a ratio of 14:1. Explain this result by drawing the resonance forms for the sigma complexes (cationic intermediates). See Sec. 4.6.

Exercise 4.10 The nitration of pyridine at 300°C with nitric acid and sulfuric acid (plus some potassium nitrate and iron powder!) yields only about 20% of 3-nitropyridine. Explain this result by drawing the resonance forms for the sigma complexes (cationic intermediates) for substitution at the 2-, 3-, and 4-positions. *Hint:* Start with the reaction of the lone-pair electrons with a proton (i.e., protonation of the ring nitrogen).

Summary

1. Aromatic hydrocarbons are derived from coal tar and petroleum. They are cyclic, relatively stable systems containing one or more six-carbon benzenoid structures.
2. Aromatic hydrocarbons may be homologs of benzene or polycyclic systems. The latter may be either fused or isolated rings.

3. The benzene structure is a resonance hybrid.
4. Aromatic compounds may be named by trivial names or as derivatives of benzene. Relative positions of substituents must be indicated.
5. Reactions of benzene include
   (a) halogenation
   (b) nitration
   (c) alkylation (Friedel–Crafts)
   (d) sulfonation
   (e) oxidation of the side chain.
6. Substituents already on the ring govern the ease of further substitution and possess a directive influence upon the ring position that entering groups assume.
7. *Ortho–para* directors (except the halogens) activate the ring and enhance further substitution; *meta* directors deactivate the ring and inhibit further substitution.
8. Heterocyclic compounds are cyclic compounds containing one or more atoms other than carbon in the cyclic system.
9. Heterocyclic compounds may be saturated, unsaturated, or aromatic. If the latter, they are called heteroaromatic.

Supplementary Exercises

4.11 Assign an acceptable name to each of the following:

(a) Br, Br, Br, OH

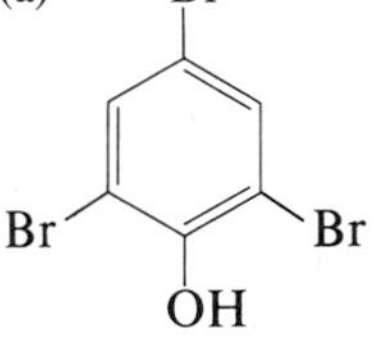

(b) NO_2, NO_2

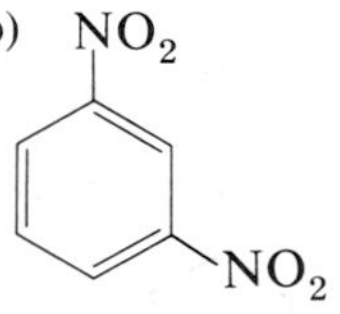

(c) $-CH{=}CH_2$

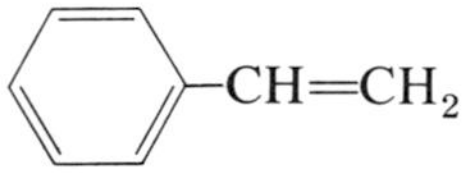

(d) $-CH_2{-}C{\equiv}N$

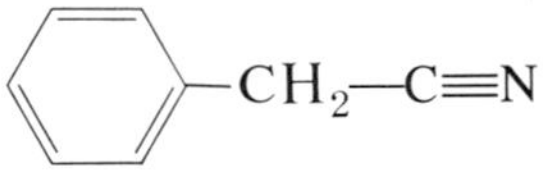

(e) Br— —Br

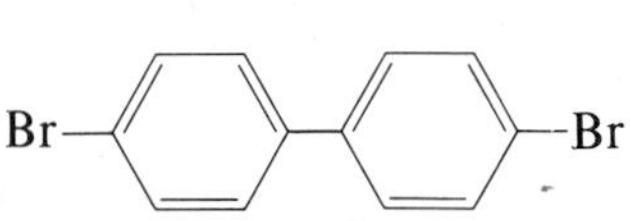

(f) C, H

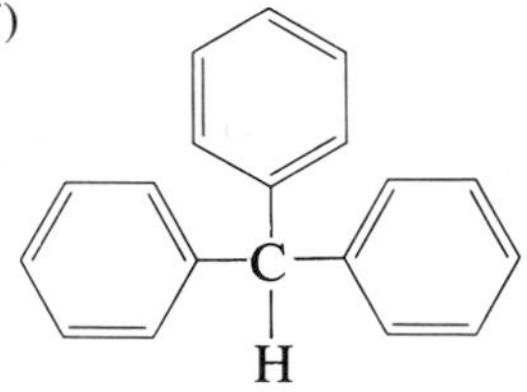

(g) SO_3H

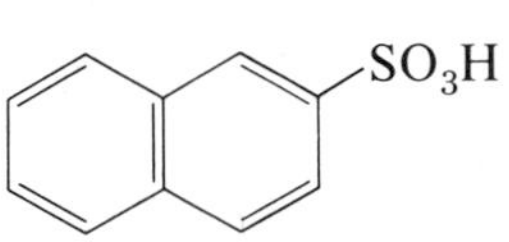

4.12 Draw structural formulas for the following useful substances:

(a) T.N.T. (2,4,6-trinitrotoluene, a high explosive)
(b) Salicylic acid (*o*-hydroxybenzoic acid, a pharmaceutical)
(c) Pentachlorophenol (a wood preservative and fungicide)
(d) *p*-Dichlorobenzene (a moth repellent and larvicide)
(e) *p*-Ethoxyacetanilide (an antipyretic and analgesic)

4.13 Draw structures for all isomeric homologs of benzene with the formula $C_{10}H_{14}$. Assign an acceptable name to each.

4.14 In the bromination of benzene, iron filings were used as a "carrier." Explain by what mechanism iron could promote the formation of an electrophilic ring-seeking bromine atom, when halogens are by nature such strong electronegative elements.

Do you think $AlCl_3$ could serve as a catalyst in the bromination reaction?

4.15 Write structural formulas for:

(a) A compound, C_8H_{10}, which can give only one theoretically possible monobromo ring substitution product.
(b) A compound, C_9H_{12}, which can give only one theoretically possible mononitro ring substitution product.
(c) Four compounds, $C_{10}H_{14}$, which can give only one mononitro ring substitution product.

4.16 An unidentified liquid is thought to be either benzene, cyclohexene, or cyclohexane. What simple chemical test will identify it?

4.17 Using toluene or benzene as your only aromatic organic starting material and any other reagents you may require, devise synthetic routes that will lead to the following products.

(a) *o*-nitrobenzoic acid
(b) *m*-bromonitrobenzene
(c) *p*-bromobenzoic acid
(d) *p*-toluenesulfonic acid
(e) cumene (isopropylbenzene)
(f) benzyl chloride
(g) 1-bromo-4-nitrobenzene

4.18 Show the probable mono-nitrated products when each of the following is treated with a nitrating mixture.

(a) OH, CH_3
(b) CH_3
(c) H O, N—C—CH_3
(d) H O, N—C
(e) OH
(f) CH_3, Br
(g) O, C—OH
(h) O, C, O
(i) NO_2
(j) NO_2, CH_2

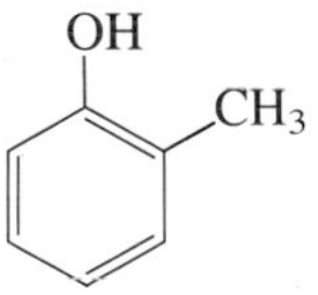

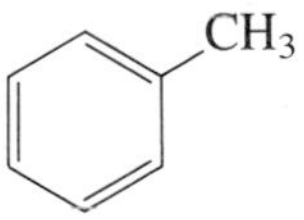

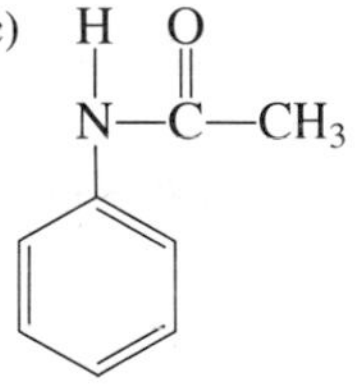

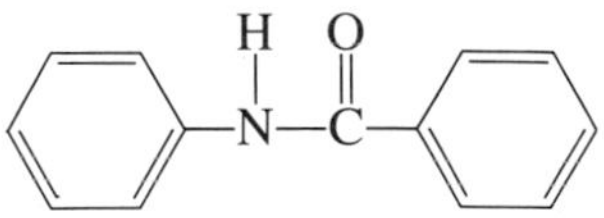

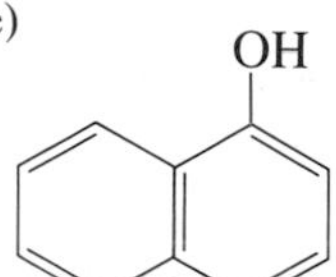

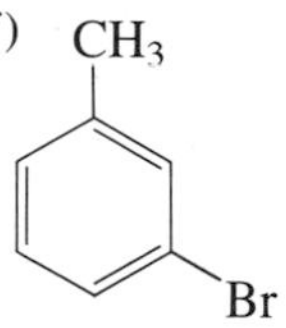

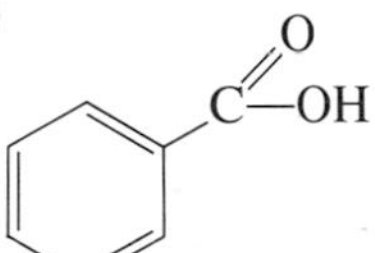

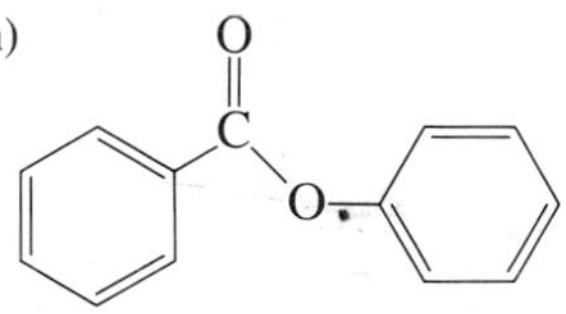

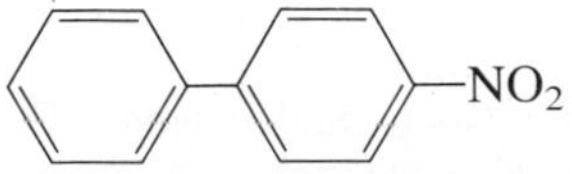

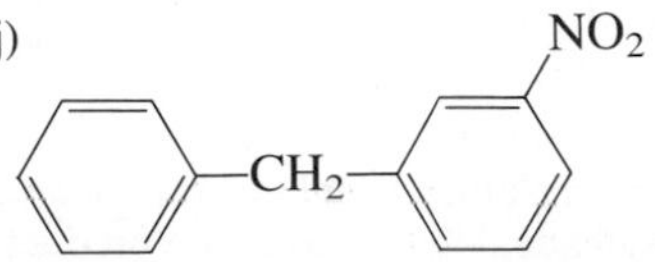

4.19 Complete the following reactions showing the principal products, if any, expected in each.

(a) *o*-Nitrotoluene + $Br_2 \xrightarrow{Fe}$

(b) Benzene + 3 $H_2 \xrightarrow{Ni}$

(c) Toluene + $Cl_2 \xrightarrow{sunlight}$

(d) Nitrobenzene + 2-Chloropropane $\xrightarrow{AlCl_3}$

(e) Bromobenzene + Mg $\xrightarrow[ether]{anhydrous}$

(f) 3-Methylpyridine + hot $KMnO_4 \longrightarrow$

(g) Benzoic acid + $Br_2 \xrightarrow{Fe}$

(h) Acetanilide + $Br_2 \longrightarrow$

(i) *m*-Xylene + $Br_2 \xrightarrow{Fe}$

(j) Thiophene + $Cl_2 \longrightarrow$

4.20 Starting with benzene, toluene, ethylbenzene, and any alkyl halides of your choice, show how the following compounds could be synthesized. The review of earlier chapters may be helpful. You may use as many steps as are necessary.

(a) 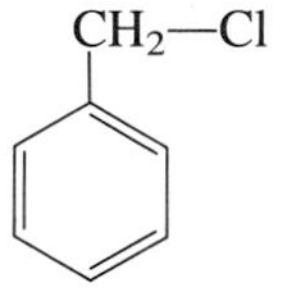 C_6H_5–CH_2—Cl (benzene ring bearing CH_2—Cl)

(b) 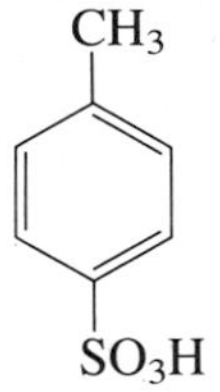 CH_3 and SO_3H on a benzene ring (para)

(c) COOH and NO_2 on a benzene ring (meta)

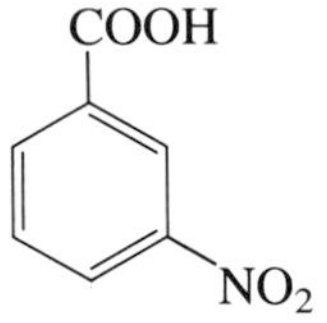

(d) 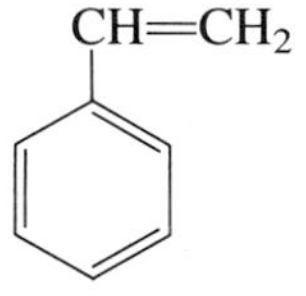 CH=CH_2 on a benzene ring

(e) 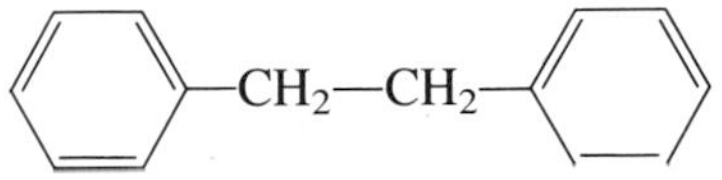 C_6H_5—CH_2—CH_2—C_6H_5

(f) 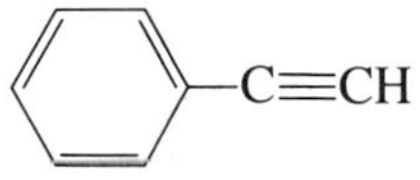 C_6H_5—C≡CH

(g) 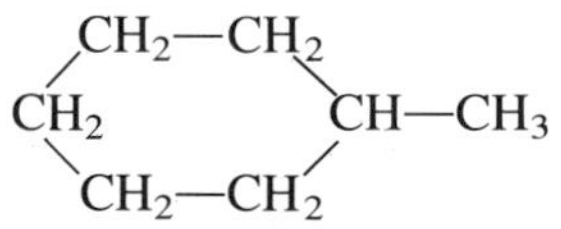 Ring: CH_2—CH_2, CH_2, CH—CH_3, CH_2—CH_2 (methylcyclohexane)

(h) 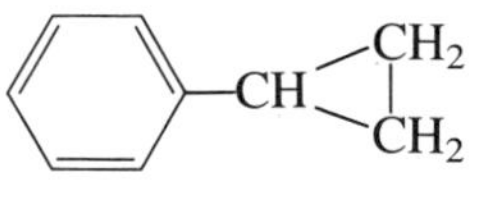 C_6H_5—CH< (ring with CH_2—CH_2) (cyclopropylbenzene)

4.21 A hydrocarbon with the formula C_8H_{10} yielded two monobromo derivatives. On strong oxidation it gave an acid identical to the oxidation product of naphthalene. What was the original hydrocarbon?

4.22 Draw the more important resonance structures for the following compounds. In each example you should have five forms (including the two Kekulé forms) because π electrons from the ring may be delocalized into the substituent or from the substituent into the ring. In the former case the ring will have a net positive charge, and in the latter, a net negative charge. If done properly, there will be a consistent pattern for the location of such charges around the ring.

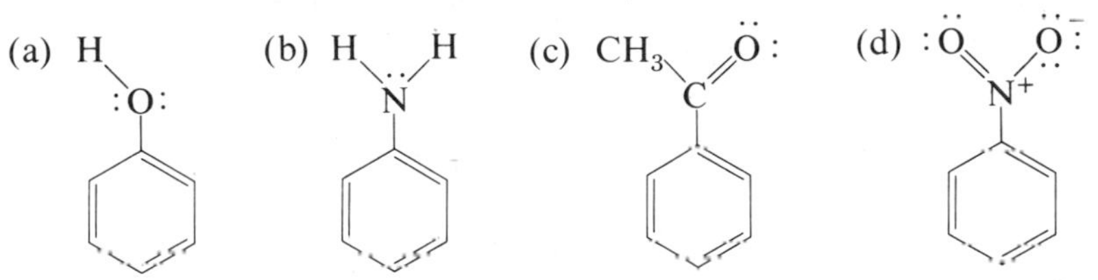

(a) Phenol (b) Aniline (c) Acetophenone (d) Nitrobenzene

4.23 Draw the principal resonance forms for benzyl radical and benzyl cation. You should have a total of five forms for each, including the Kekulé forms (see Exercise 4.22). Would you expect either or both of these reactive intermediates to be more stable than primary alkyl radicals and cations?

Benzyl radical Benzyl cation

4.24 The methyl group of toluene is an *o-p* director and activates the ring toward electrophilic substitution. Draw the resonance forms for the cationic intermediates in the electrophilic substitution (by the general reagent E^+) at all three positions. Suggest an explanation for the *o-p* direction and ring activation. (*Hint:* Look at each contributing form carefully to see whether or not it might be more stable than the others, recalling what you have learned about the stability of *simple* carbocations (Sec. 3.8A).)

4.25 The bromination of ethylbenzene under free-radical conditions gives exclusively 1-bromo-1-phenylethane. Explain this result. (*Hint:* Review Sec. 2.11B and Exercise 4.23.)

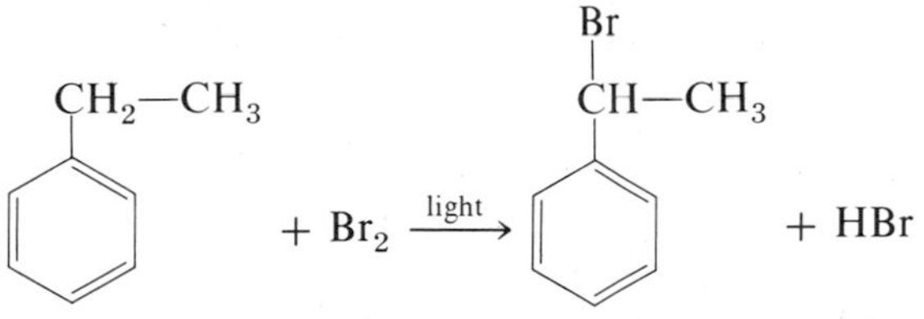

$C_6H_5CH_2{-}CH_3 + Br_2 \xrightarrow{\text{light}} C_6H_5CHBr{-}CH_3 + HBr$

4.26 When benzene is warmed with 1-chloro-2-methylpropane in the presence of anhydrous $AlCl_3$, the only product obtained is *tert*-butylbenzene. How do you account for this (Sec. 3.8A)?

4.27 As you saw in Exercise 4.1, aromaticity is not limited to compounds with six-membered rings. The important factor in the stabilization of cyclic structures is the delocalization of π electrons over the entire ring. Using this concept and the drawing of resonance structures as a means of visualizing delocalization, explain the aromaticity of the following compounds:

(a) 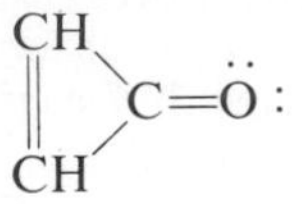

(b) 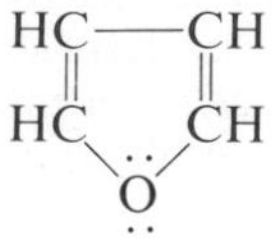

(c) 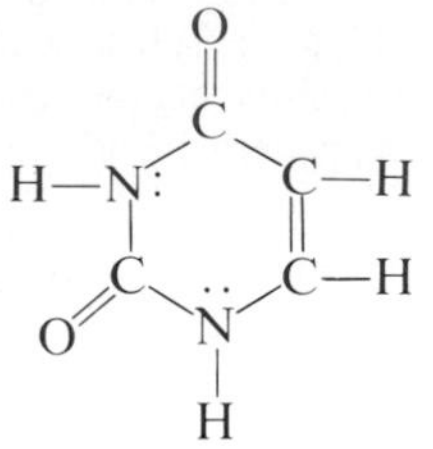

4.28 Write the structural formulas for the following compounds.

(a) 3-methylpyridine (β-picoline)
(b) 8-hydroxyquinoline
(c) 3-methylindole (skatole)
(d) 2-propylpiperidine (coniine)
(e) 4-phenylpyridine

4.29 The heterocyclic system porphine is an important structural unit in hemoglobin, chlorophyll, and cytochrome. Although porphine appears to be made up of four pyrrole rings, it is a heteroaromatic ring system that meets the requirements of the Hückel rule. Locate the 18-π-electron *monocyclic* ring that is responsible for the aromaticity by tracing it with a colored pen or pencil.

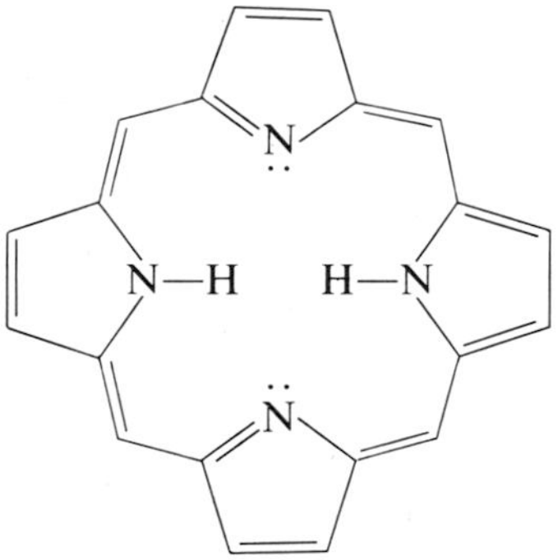

4.30 The dehydrohalogenation of 1-phenyl-2-chloropropane yields only 1-phenylpropene and none of the isomeric 3-phenylpropene. Why would the first product be formed to the exclusion of the second? (*Hint:* See Sec. 3.4.)

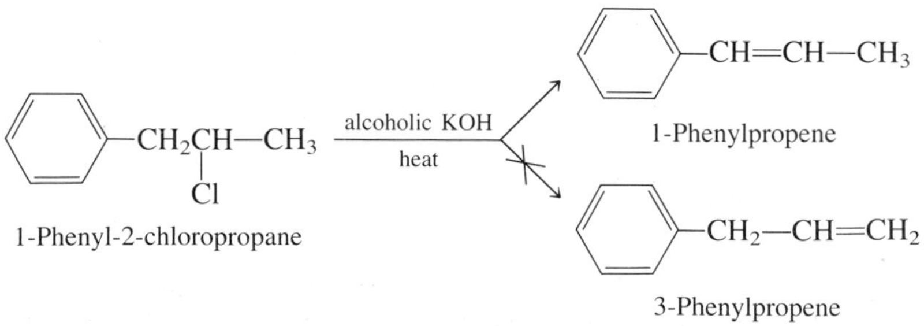

5 Stereoisomerism

The possibility of isomerism was first discussed in Section 1.6, where it was pointed out that two different structures could be drawn to represent the molecular formula C_4H_{10}. These two structural variations represented two *different* molecular species—namely, *n*-butane and isobutane. Because these two hydrocarbons are different compounds, they have different physical and chemical properties. We observed that the substitution reactions of the alkanes and the arenes, as well as the addition reactions of the alkenes, could result in the formation of isomeric products. As we continue our study of organic chemistry, we realize that isomerism is an important aspect of it, and that if we understand *why* isomeric products are formed, then we also will understand *how* organic molecules react. The kind of isomers we will consider in the following sections are molecular structures in which the same atoms and groups that are joined together in one isomer are similarly joined in the other. The atomic linkages are the same in both isomers, but the spatial arrangement of the atoms and groups is different. Such isomers are called **stereoisomers** (Gr. *stereos,* solid). Our definition of stereoisomers does not include such spatial variations as those allowed 1,2-dichloroethane in its staggered and eclipsed forms (Sec. 2.6) nor those for cyclohexane in its boat and chair forms (Sec. 2.9). Such spatial arrangements are interconverted by the expenditure of relatively small amounts of energy, and this may be accomplished by the simple expedient of rotation about single bonds. Such different arrangements, you will recall, are referred to as conformations. Stereoisomers, on the other hand, have different **configurations** and cannot be interconverted without breaking bonds and rearranging groups.

5.1 Optical Isomerism (Mirror-Image Isomerism). Polarized Light

Optical isomerism (or **mirror-image isomerism**) gets its name from the unique properties of certain isomeric *pairs* of compounds called **enantiomers** (Gr. *enantios,* opposite; *meros,* part). Enantiomers have the same chemical properties and the same physical properties—with one notable exception. They differ in the direction in which each is able to rotate a beam of **plane-polarized light.**[1] The degree of rotation is the same for each enantiomer, but the direction of rotation is opposite.

In order to understand the phenomenon of ''optical activity,'' it would be helpful to review the nature of plane-polarized light. Ordinary white light exhibits an electromagnetic wave motion in which waves of varying lengths are vibrating in all possible planes at right angles to the path of the ray. Monochromatic light, used in the measurement of optical activity, is light of only one wavelength. It is usually produced in one of two ways: (1) all unwanted wavelengths of ordinary light are removed by means of a colored filter, or (2) light of one wavelength is generated from a special source such as a sodium or a mercury lamp. Monochromatic light, like ordinary light, also consists of waves vibrating in all possible planes at right angles to the path of propagation.

Certain substances such as tourmaline crystals, polaroid, or specially prepared prisms, called Nicol prisms, act as screens when light is passed through them. Waves vibrating in only one plane pass through such special screens and all those vibrating in other planes are either rejected or absorbed. Such specially filtered light is said to be *plane-polarized*.

We can illustrate the action of such a **polarizer** best with the Nicol prism. The Nicol prism is an ingenious device made of calcite, $CaCO_3$. Calcite is a clear crystal with the shape of a rectangular rhombohedron and exhibits an unusual optical property called *birefringence,* or double refraction. A ray of light entering the crystal is refracted or bent in two slightly different directions to produce two rays. To form the Nicol prism, the crystal is cut in a plane diagonally through the obtuse angles and perpendicular to the two end faces. The cut surfaces are then polished and recemented with Canada balsam. After entering the crystal and striking the cemented surface, one of the refracted rays vibrating in one plane is transmitted while others are reflected out of the crystal. Figure 5.1 illustrates graphically how a beam of monochromatic light vibrating in more than one plane might appear (a) before and (b) after it had passed through a Nicol prism polarizer.

A maximum transmission of plane-polarized light may be observed through a second prism, called the **analyzer,** only when the latter is oriented in the same optical plane as the polarizer. On the other hand, when the analyzer is rotated, the intensity of the emergent beam, as seen by the viewer, is gradually diminished. A point of minimum transmission is reached when the second prism has been rotated through an angle of 90° from the point of maximum transmission. The phenomenon can be illustrated graphically by using two polaroid lenses (Fig. 5.2).

An **optically active** compound is one that is capable of rotating the plane of polarized light. The compound is said to be **dextrorotatory** (L. *dexter,* right) when the plane of polarized light is rotated to the right, or clockwise. A dextrorotatory substance is always indicated by the positive (+) sign. A compound is described as **levorotatory** (L. *laevus,* left) when the direction of rotation is to the left, or counterclockwise. In this case the

[1]There are several types of isomers, which may or may not show optical activity, included under the broad terms, optical isomerism or mirror-image isomerism. Only enantiomers show equal but opposite optical rotation. Enantiomers may vary in their rates of reaction with other optically active substances. This is especially true in biological systems. Thus one enantiomer may be attacked by a bacterium, the other not. Again, one of a pair of enantiomers may have a hormonal activity far greater than that of the other. These differences are due to the spatial requirements of each reactant in combining with or attacking the other.

FIGURE 5.1 Polarization: (a) A Beam of Monochromatic Light Vibrating in All Planes, (b) A Beam of Plane-Polarized Light

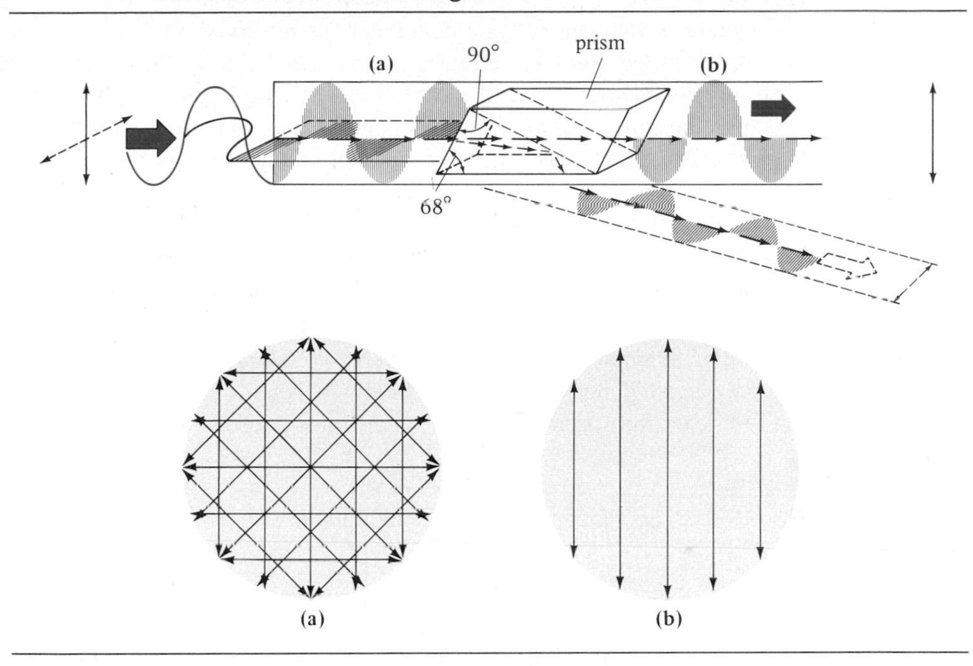

FIGURE 5.2 Transmission: (a) Maximum Transmission of Plane-Polarized Light, (b) Plane-Polarized Beam Blocked

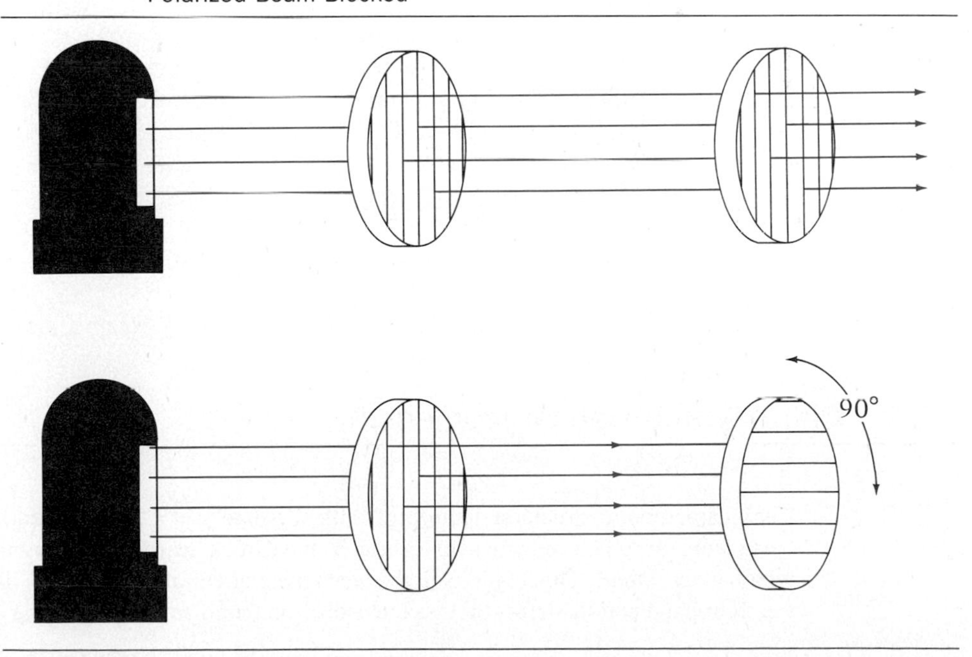

negative (−) sign is used to describe its rotation. The angle of rotation, α, is measured in degrees.

Optical activity in organic compounds is observed and measured by means of an instrument called a **polarimeter** (Fig. 5.3). The value of the specific rotation, $[\alpha]$, depends upon the length of the tube containing the sample solution, measured in decimeters, and upon the concentration of the solution, measured in grams per milliliter. If the sample is a pure liquid, its density is used in place of the concentration.

$$[\alpha] = \frac{\text{observed rotation in degrees}}{\text{length of sample tube in dm} \times \text{concentration (g/mL)}}$$

In order to standardize the specific rotation values of optically active substances, both the temperature at which observations are made and the light source employed are indicated. For example, the term $[\alpha]_D^{25°} = +54°$ indicates that the specific rotation of an optically active compound is 54° to the right when the measurement is made at 25°C and the D line of the spectrum (sodium light source, 5893 Å) is used.

FIGURE 5.3 Schematic Diagram of the Polarimeter, Showing Components

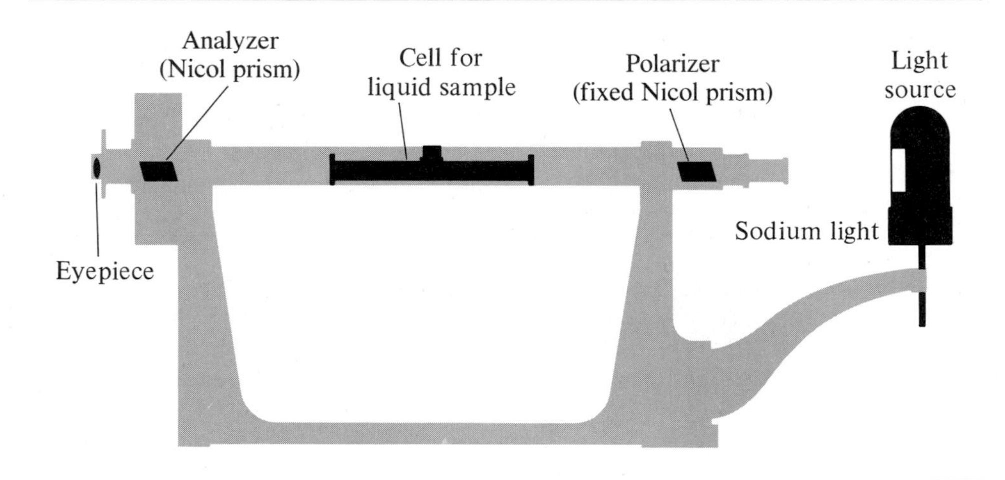

5.2 Chirality and Optical Isomerism

The characteristic structural feature of optical isomers is a property called **chirality** or "handedness." The term has its origin in the Greek word *cheir* (rhymes with spire), which means *hand*. The relationship between pairs of chiral molecules is like that between the right hand and the left—that is, one isomer is the mirror image of the other and is not

FIGURE 5.4 Nonsuperimposable Mirror Images

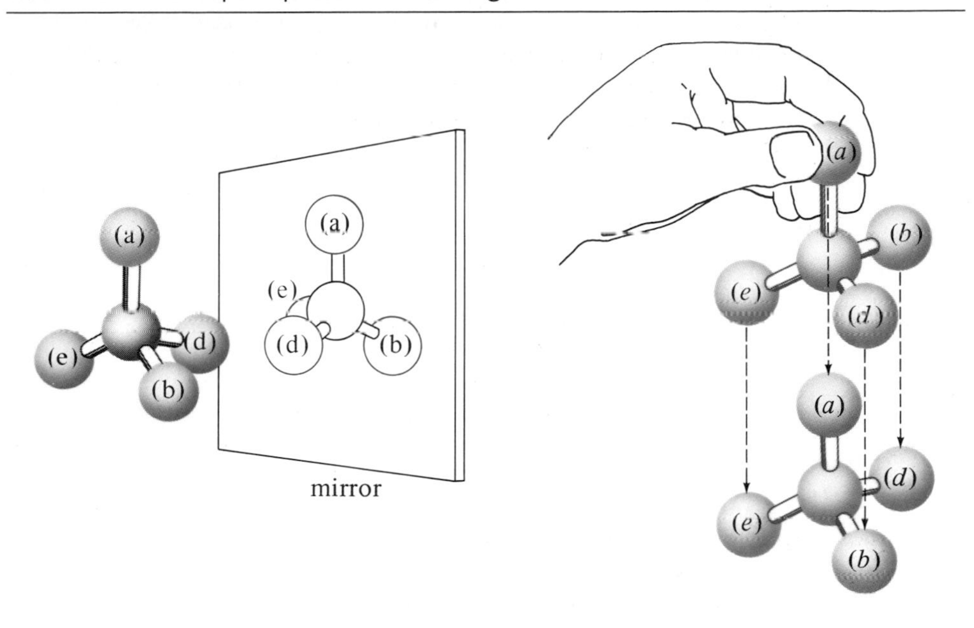

superimposable upon the other. By nonsuperimposable we mean that one isomer cannot be positioned above the other with all elements of the isomer above coinciding with those of the isomer below. The mirror-image relationship of a pair of enantiomers is shown in Fig. 5.4.

All molecules have mirror images, but those whose mirror images are superimposable are identical and are said to be **achiral** or to lack chirality. For example, the mirror image of 2-chloropropane need only be turned 180° about the C—H axis to reveal that the molecule and its image are one and the same.

A common feature of most chiral molecules is the presence of a carbon atom that is bonded to *four different groups*. Such a carbon has been referred to by organic chemists for many years as an asymmetric carbon atom, and was often indicated in a formula by placing an asterisk near it. For example, in the formula for *sec*-butyl chloride, the number 2 carbon atom is indicated as being asymmetric.

```
     H  H  H  H
     |  |  |* |
  H—C—C—C—C—H
     |  |  |  |
     H  H  Cl H
```

sec-Butyl chloride

Labelling a carbon atom as asymmetric can be confusing, because only molecules and objects can be asymmetric—not atoms. A carbon atom bonded to four dissimilar groups is better described as a **chiral center** or a point of asymmetry. All molecules that contain only *one* such carbon are indeed asymmetric and chiral; however, molecules containing more than one such carbon may be chiral or achiral. We will illustrate this point in the following section.

Exercise 5.1 Which of the following objects may exist in enantiomeric forms? (a) glove, (b) shoe, (c) cap, (d) axe, (e) wood screw, (f) a pair of pliers, (g) a pair of scissors, (h) a golf club, (i) a coil spring, (j) a shovel

Exercise 5.2 Which of the following are chiral molecules? (a) 2-bromopropane, (b) 1-bromo-2-methylpropane, (c) 1,2-dibromopropane, (d) 1-chloro-2-propanol

5.3 Compounds Containing More Than One Chiral Center

(Case A) Two Unlike Chiral Centers. The total number of optical isomers possible when a molecule contains n unlike chiral centers is 2^n. This is a statement of the **van't Hoff rule.**[2] The rule may be illustrated by considering the stereoisomerism possible for a molecule

[2] J. H. van't Hoff (1852–1911), Dutch physical chemist, was one of the first to recognize asymmetry in a compound in which four different groups were attached to a carbon atom. He postulated that the four different groups in such an asymmetric structure could have two possible spatial arrangements.

with a configuration like that exhibited by the ball and stick model of 2-bromo-3-chloro-butane.

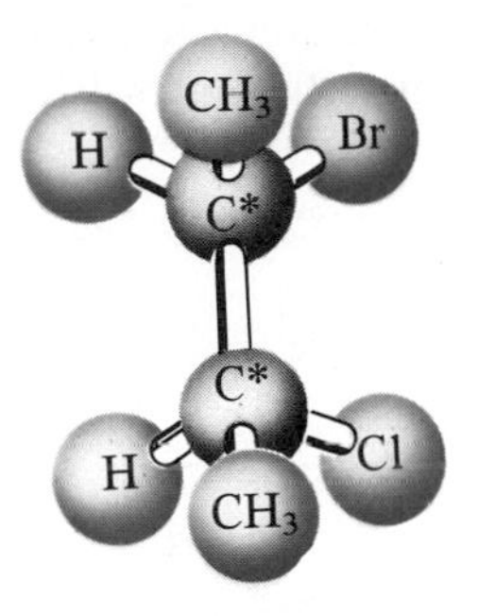

Designating the upper carbon α and the lower one β in structures IV to VII, we can show that two configurations are possible for the four groups bonded to each carbon atom. Structures IV and V represent a pair of enantiomers, as do VI and VII. Although structures V and VI (also structures V and VII, IV and VI, and IV and VII) are optically active stereoisomers, they are not mirror images. Stereoisomers that are not mirror images are called **diastereoisomers.**

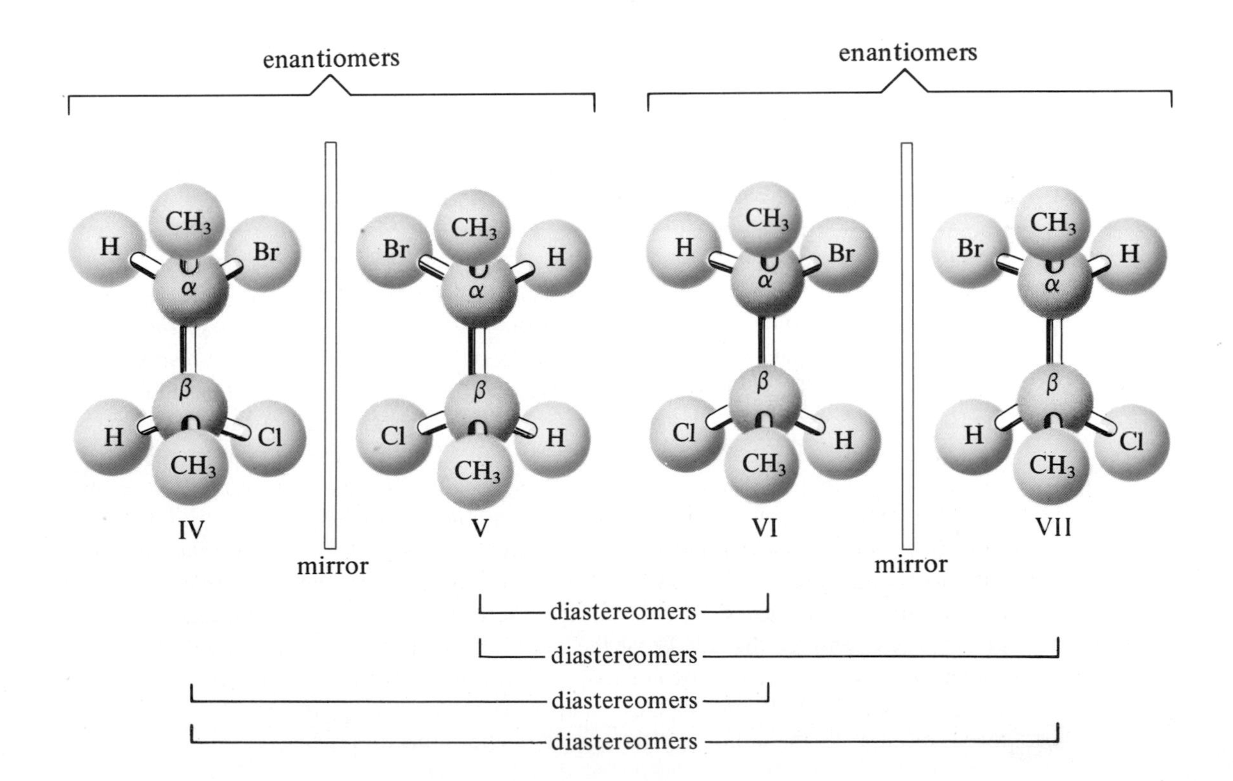

(Case B) Two Similar Chiral Centers. 2,3-Dibromobutane is an example of a compound having two chiral centers that are similar—that is, both carbons are attached to the same four groups. A compound with two similar chiral centers will exist in three stereoisomeric forms. Two will be optically active enantiomers, and the third an optically inactive form called a **meso isomer.** Structures VIII and IX are nonsuperimposable mirror images.

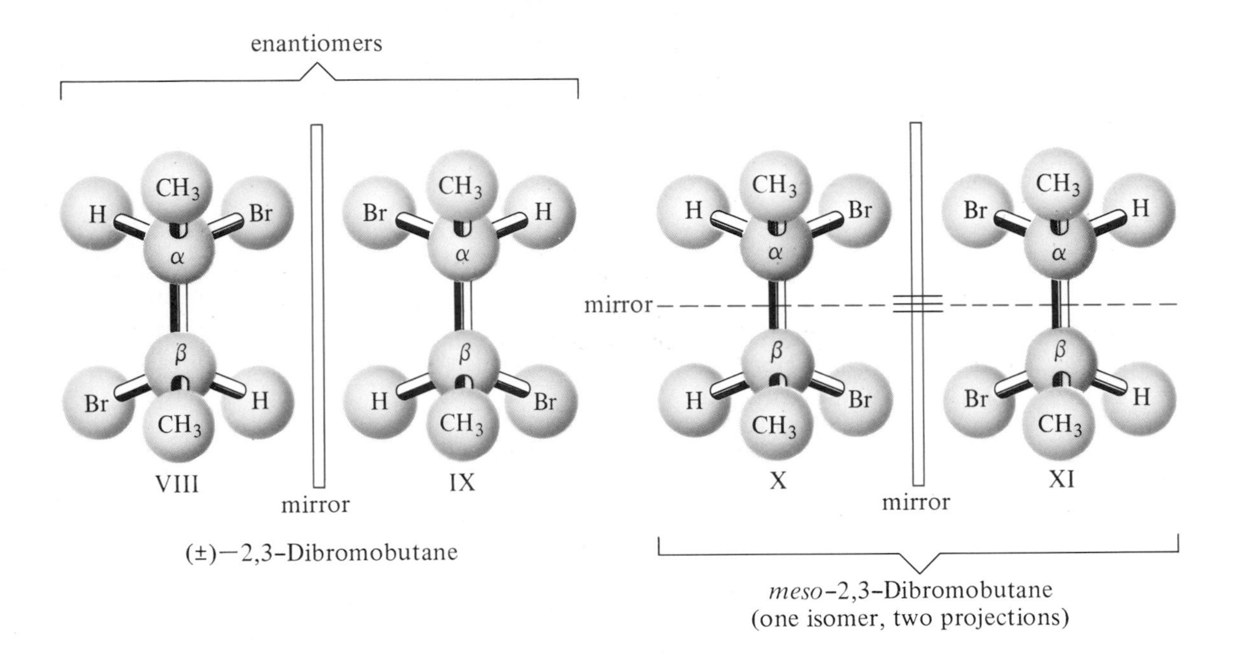

Structures X and XI, although mirror images, can be made superimposable by rotating one of them 180° in the plane of the paper. Thus, X and XI do not represent two different isomers but two different representations of the same isomer. To emphasize this, an equality sign (≡) is drawn between the structures. The meso form of 2,3-dibromobutane has physical properties unlike those of the optically active (+) and (−) forms.

Although we have explained why there is only one meso isomer corresponding to structures X and XI, we have not explained why this isomer is optically inactive, even though there are two chiral atoms in the structure. The lack of optical activity is best explained by drawing a second mirror plane through the center of the molecule. Now it should be apparent that the upper half of the molecule is the mirror image of the lower half. Thus, when plane-polarized light passes through the molecule, the light is rotated to the left by one half of the molecule and an equal amount to the right by the other half, leaving no net rotation. We say that the meso isomer is optically inactive by *internal compensation*. The easiest way to detect a meso isomer is through the use of a mirror plane passed through the middle of the molecule. If a mirror plane passed through the center of a molecule having two or more chiral centers divides the molecule into mirror-image halves, the molecule is a meso isomer. The plane may pass through a bond or even bisect an atom if the atom is at the center of the molecule.

We can illustrate this last point by examining the structure of *cis*-1,2-dimethylcyclopentane.

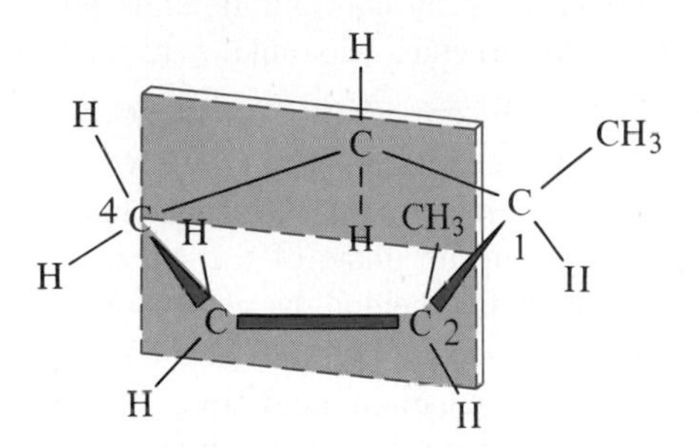

cis–1,2–Dimethylcyclopentane (a meso form)

When a mirror plane is passed *between* carbons 1 and 2 and *through* carbon 4, the symmetry of the molecule is apparent. *cis*-1,2-dimethylcyclopentane is optically inactive even though carbons 1 and 2 are bonded to four dissimilar groups.

Exercise 5.3 Did the addition of bromine to cyclopentene (Sec. 3.7) produce a pair of optical isomers?

Exercise 5.4 How many enantiomers and meso forms does 3-chloro-2,4-dibromopentane have?

5.4 Relative Configurations and Nomenclature of Enantiomers and Diastereomers

The structural formula we draw for an organic molecule should describe the actual structure of the molecule as unambiguously as possible. This means that our structural formula must show the stereochemical configuration of the molecule. Often we can fulfill this requirement by using perspective (dashed line–wedge) formulas (Sec. 1.12 and Exercise 2.7), or Newman projections (Sec. 2.6). However, we must also be able to indicate the stereochemistry of the molecule in the name we assign to it. The prefixing of a (+) or (−) to the name of an enantiomer or diastereomer tells us only that it is dextro- or levorotatory based on a polarimetric measurement in the laboratory. These prefixes do not tell us anything about the configuration or absolute structure of the molecule. However, organic chemists have devised, and are continuing to devise, ways of drawing the structures of stereoisomers and of naming them in such a way as to correctly indicate their stereochemistry. We will examine two conventions currently being widely used for the nomenclature of enantiomers and diastereomers.

A. The D,L-Convention. Because the **absolute,** or true, structures of optical isomers were unknown to the early chemists working with optically active compounds, Fischer[3] decided to relate as many configurations as possible to that of a standard structure, one whose absolute structure was unknown but was to be arbitrarily defined by him. His choice for the standard structure was that of glyceraldehyde, $HOCH_2\overset{*}{C}H(OH)CHO$ (for reasons that will become apparent when we study the sugars in Sec. 14.4). In the Fischer projection of glyceraldehyde, the carbon chain is drawn vertically with only the asymmetric carbon atom in the plane of the paper. Both the carbonyl and the hydroxymethyl groups are drawn as if behind the plane—the carbonyl group (—CHO) at the top and the hydroxymethyl group ($—CH_2OH$) at the bottom. The hydroxyl group and the hydrogen atom attached to the carbon atom are drawn as if in *front* of the plane of the paper—the hydroxyl group to the right and the hydrogen to the left (Fig. 5.5(a)). Flattening out the structure produces the projection formulas shown in Fig. 5.5(b). This configuration was arbitrarily designated as the D-configuration of glyceraldehyde by Fischer and is identified by a small capital D. Its mirror-image enantiomer, with the opposite configuration, is identified as belonging to the L series.

FIGURE 5.5 (a) Perspective and (b) Fischer Projection Formulas of D(+)-Glyceraldehyde

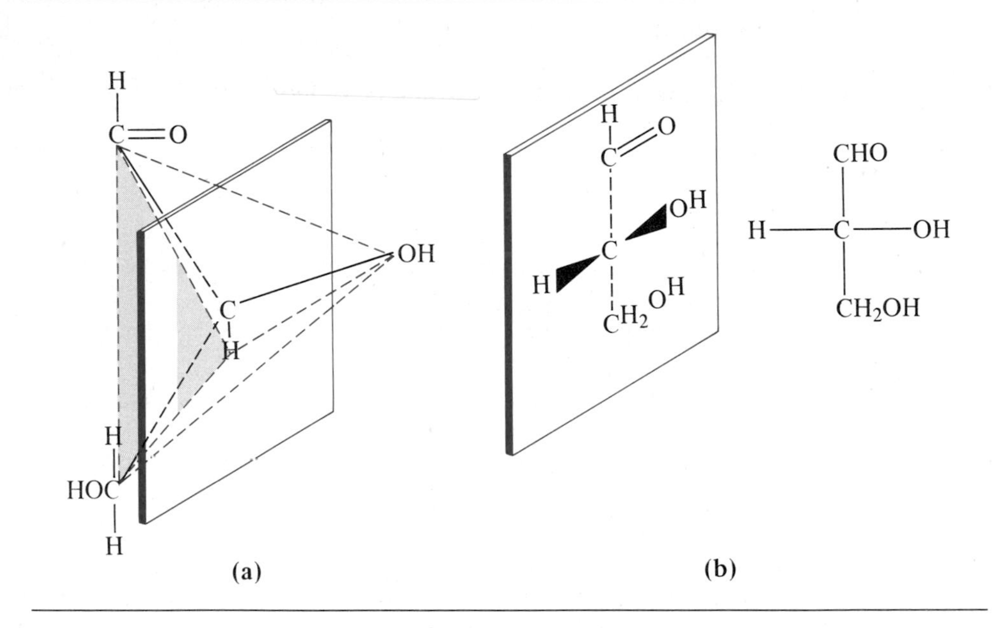

[3]Emil Fischer (1852–1919), professor of chemistry, University of Berlin. Winner of the Nobel Prize in Chemistry, 1902.

The structure of any other optically active compound of the type R—CHX—R′ is drawn with the carbon chain

$$\begin{array}{c} \text{R} \\ | \\ -\text{C}- \\ | \\ \text{R}' \end{array}$$

in the vertical direction, with the *lowest numbered* carbon atom at the top. If the X group (usually —OH, —NH_2, or halogen) is on the right, the relative configuration is D- (dee); if it is on the left, the configuration is L- (ell). If there is more than one X-group attached to the chain, the relative configuration is determined by the configuration of the *highest numbered* X group.

The designations "D" and "L" refer to configuration only and are not to be interpreted as "dextro" and "levo." The latter terms refer to the direction of rotation. If the direction of rotation is specified in the name of an enantiomer, either with or without the "D" or "L", the designations (+) for the dextrorotatory isomer and (−) for the levorotatory isomer should be used. Occasionally the abbreviations *d* and *l* are used for dextro and levo, respectively. Although the D-form of glyceraldehyde was arbitrarily chosen as the dextrorotatory isomer without a knowledge of its absolute or true configuration, the choice was a fortuitous one; for in 1951, with the use of modern analytical methods, the dextro isomer was definitely established as being the D-form. It is possible, you see, for an optically active substance to have the L-configuration and still be dextrorotatory.

Exercise 5.5 The very useful organic acid, tartaric acid (Sec. 11.6), exists in three stereoisomeric forms, two of which are enantiomers. The L-isomer is a by-product of wine making, and the D-isomer is about 5% by weight of the *Bauhini* plant, which is common in central Africa. Draw the Fischer projections for the three stereoisomers and label the D- and L-forms. What label do we put on the third form?

$$HO_2C—CHOH—CHOH—CO_2H$$

Tartaric acid

B. The (*R*),(*S*)-Convention. Unfortunately, the D,L-system is not applicable to many molecules of interest to the organic chemist; therefore, the IUPAC has adopted a system called the (*R*),(*S*)- or the Cahn–Ingold–Prelog convention. This system is capable of specifying the absolute configuration of any optical isomer. The "right-handed" enantiomer of a pair of optical isomers is designated in this system as the (*R*)-form, and the "left-handed" enantiomer as the (*S*)-form. These prefixes are derived from the Latin *rectus*, right, and *sinister*, left. To use this system we must assign a priority to each of the four atoms or groups attached to the chiral carbon atom. The group of lowest priority is placed in the position most remote from us as we view the chiral carbon atom. The remaining three groups are assigned positions in a *clockwise* order of decreasing priority for the (*R*)-isomer, and in a *counterclockwise* order of decreasing priority for the (*S*)-isomer. The detailed rules for assigning priorities to atoms and groups are beyond the scope of this text; however, there are two basic rules:

(1) Higher atomic number is of higher priority than lower atomic number, and higher atomic mass over lower atomic mass, if the atomic numbers are the same.
(2) Priorities of groups are determined by working outward from the point of attachment to the chiral carbon atom using rule (1) until a point of difference is reached, treating double and triple bonds as two or three of the same atom single bonds. Thus

$$>C{=}O \quad \text{is treated as} \quad >C(-O)(-O)$$

and

$$>C{=}C< \quad \text{as} \quad >C(-C{<})(-C{<})$$

The priority sequence for the most common atoms and groups is as follows.

(*High priority*) $-I$, $-Br$, $-Cl$, $-SO_2R$, $-SOR$, $-SR$, $-SH$, $-F$,

$-O-C(=O)-R$, $-OR$, $-OH$, $-NO_2$, $-NH-C(=O)-R$, $-NR_2$, $-NHR$, $-NH_2$,

$-CCl_3$, $-C(=O)-Cl$, $-C(=O)-OR$, $-C(=O)-OH$, $-C(=O)-NH_2$, $-C(=O)-R$, $-C(=O)-H$,

$-C{\equiv}N$, $-CR_2-OH$, $-CH(OH)-R$, $-CH_2OH$, C_6H_5-, $-CR_3$, $-CHR_2$,

$-CH_2R$, $-CH_3$, $-D$, $-H$ (*Low priority*)

In molecules with more than one chiral center, an (*R*)-substituent has higher priority than an (*S*)-substituent.

The proper placement of all groups is quite easy if one simply visualizes the chiral carbon atom as the hub of a three-spoked steering wheel with the atom of lowest priority attached to the steering column. The (*R*)- and (*S*)-enantiomers of 2-chlorobutane are shown.

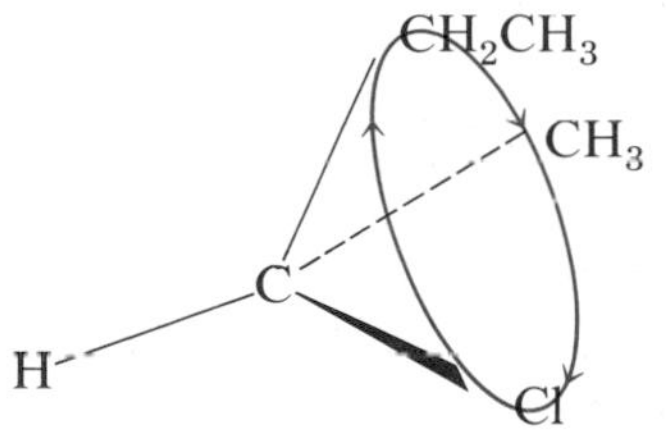

(*R*)-
(*R*)-2-Chlorobutane

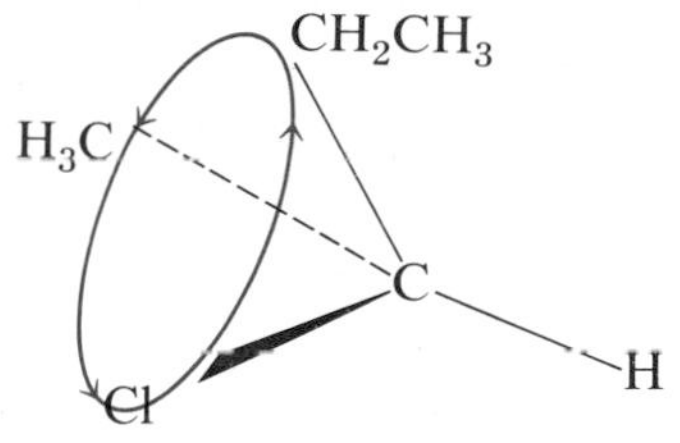

(*S*)-
(*S*)-2-Chlorobutanc

Exercise 5.6 Show that for 2-bromo-3-chlorobutane (Sec. 5.3) the following configurational assignments are correct: IV, (*R*)-2-bromo-(*S*)-3-chlorobutane; V, (*S*)-2-bromo-(*R*)-3-chlorobutane; VI, (*R*)-2-bromo-(*R*)-3-chlorobutane; VII, (*S*)-2-bromo-(*S*)-3-chlorobutane. (*Hint:* View the carbon atom designated α carefully, with the lowest priority group farthest from the eye. Then proceed from the highest priority group to the next in either a clockwise or counterclockwise direction to arrive at the (*R*)- or (*S*)-designation. Repeat the procedure with the β-carbon.)

5.5 Framework or "Zigzag" Structures and Stereochemistry

As was pointed out in Secs. 1.6 and 3.13, framework representations of organic structures are becoming increasingly popular among organic chemists seeking to simplify the writing of structural formulas, particularly those of complex molecules of the type found in important natural products and medicinal chemicals. Many of these molecules are chiral; therefore, rules will be needed to guide us in drawing framework structures for the various stereoisomers. Until such rules are agreed upon, the following guidelines will suffice for most purposes. The principal carbon chain of an acyclic molecule is drawn in a ''zigzag'' fashion. If there is a functional group at either end of the chain, it is added in the same way. Then, any substituents of known stereochemistry are placed on the zigzag chain, using a bold line for groups that project above the plane of the chain and a dashed line for groups that project below the plane of the chain. If there are two different substituents attached to a carbon atom of the chain, their stereochemistry is shown by using the solid wedge–dashed line symbolism described in Sec. 1.7. Hydrogen atoms are not shown; however, a hydrogen atom attached to a chiral center can be considered to project in a direction opposite to that of the other substituent on the chiral center. These simple guidelines will be better understood by studying the following examples.

(*S*)-2-Bromopentane

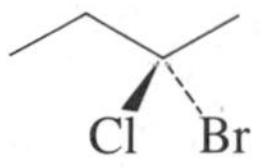

(*S*)-2-Bromo-2-chlorobutane

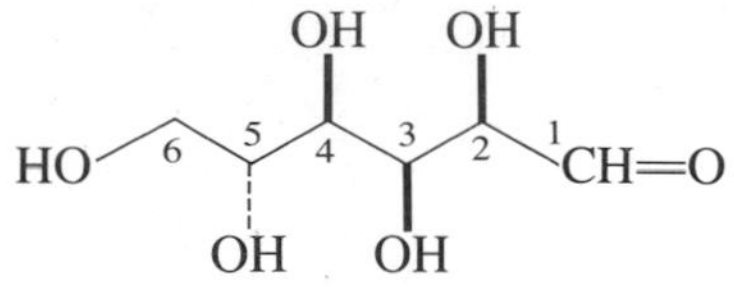

D-Glucose
(the (2*R*,3*S*,4*R*,5*R*)-configuration)

The reason zigzag structures are popular with organic chemists and students is that it is very easy to determine the configuration at chiral centers in such structures *if there is a hydrogen atom at the chiral center*. Consider the structure of D-glucose, remembering that the unwritten hydrogen atoms on chiral centers 2, 3, and 4 point away from you and the hydrogen atom on chiral center 5 points toward you. About each chiral center we draw an arc from the highest priority group to the next highest priority group and finally to the third highest priority group. The unshown hydrogen atoms are the lowest priority groups in all cases. For the substituents attached to the chain by bold lines: if the arc is clockwise, the configuration is (*R*)-; if the arc is counterclockwise, the configuration is (*S*)-. For substituents attached to the chain by dashed lines, the assignments are just the reverse of the above.

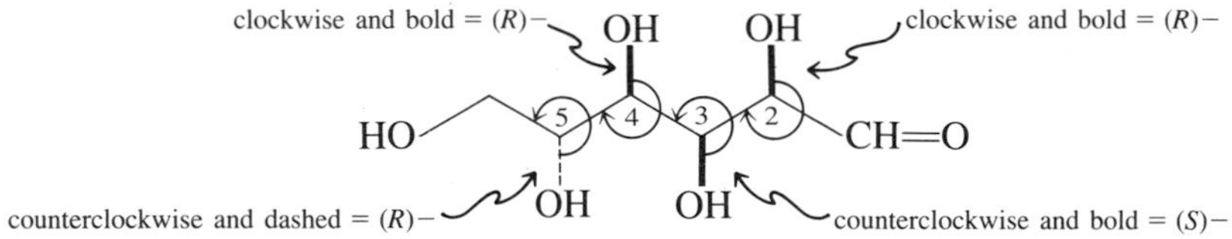

clockwise and bold = (*R*)−
clockwise and bold = (*R*)−
counterclockwise and dashed = (*R*)−
counterclockwise and bold = (*S*)−

5.6 Asymmetric Synthesis

Life as we know it depends on chirality. The carbohydrates and proteins as well as numerous small molecules involved in the living process, such as some of the terpenes of Sec. 3.13, are chiral molecules. Most of these exist as a single enantiomer, even when many enantiomers are possible. Simple calculations have suggested that it could hardly be otherwise. In their simplest forms, proteins are polymers made from various combinations of about 20 different amino acids (Chap. 15), and all but one of these amino acids have a chiral center. A protein having 100 chiral centers would not be unusual; many have more than this number. However, such a protein could have as many as 2^{100}, or 10^{30}, enantiomers. If only *one molecule* of each of these enantiomers were placed in a box the size of a football field, the box would be filled to a depth of about six feet. The duplication of cells and structures, metabolic and catabolic processes, and the efficient use of biological energy resources would be made prohibitively complex if so many enantiomers had to be dealt with. Thus, order is achieved (in part) by restricting most biological molecules to only *one* of the many possible enantiomers existing in either the (*R*)- *or* the (*S*)-form! In our hypothetical protein molecule, every amino acid unit, with but one possible exception, would have the (*S*)-configuration. How can this selectivity be achieved? It is the result of building small molecules and assembling these into large molecules on the surfaces of *chiral* biological catalysts, called enzymes, by **asymmetric synthesis,** the selective synthesis of only one of a possible two or more stereoisomers.

Molecules undergoing reactions catalyzed by enzymes are bound to the enzyme surface by various combinations of van der Waals forces (Sec. 2.10), electrostatic forces, hydrogen bond formation (Sec. 7.3), and covalent bond formation. Asymmetric synthesis can be achieved if the enzyme will bind substrate (reactant) in such a way that (1) only one group on a chiral carbon atom in the substrate is exposed to attack by the "reagent," which may also be bound to the enzyme, or (2) only one side or face of a double bond in the substrate is exposed to attack. One way of achieving this specificity of binding is to require that three of the four groups on a chiral carbon atom be bound to three specific sites on the enzyme surface. We will illustrate this type of binding with the chiral molecule, lactic acid. Assume that the three binding sites are in a cavity in the enzyme surface just large enough to accommodate the lactic acid molecule, and that the three specific binding sites are represented by the labeled circles arranged in the triangular pattern shown. Only the (*R*)-isomer can bind to all three of the binding sites and leave the H-atom exposed to attack by a reagent. The (*S*)-isomer can bind to two sites at best, and if it does, the H-atom is in the cavity, where it cannot be readily attacked by the reagent.

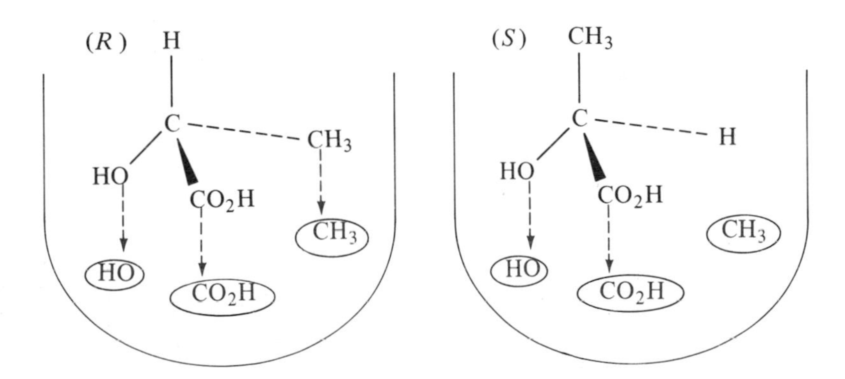

As a result of the specificity of their enzyme systems, many organisms can use only one enantiomer in their metabolic processes. The classical example of this phenomenon is the discovery by Louis Pasteur over 100 years ago that when the green mold *Penicillium glaucum* is grown in a medium containing both (+)- and (−)-tartaric acid, the mold uses the (+)-tartaric acid preferentially, allowing recovery of the unconsumed (−)-enantiomer.

It is now popular in organic chemistry to refer to the total number of all pure enantiomers that exist on earth at a given time as the **chiral pool.** The chiral pool is our shared resource from which we draw when we need chiral molecules for chemical or biological purposes, and to which we contribute when we make new enantiomerically pure substances. Most contributions to the chiral pool are made by plants and animals through their enzymatic processes, that is, by asymmetric syntheses catalyzed by enzymes. In the laboratory, asymmetric syntheses are much less common and are usually syntheses in which a new chiral center is introduced into a molecule already having at least one such center in the molecule; that is, a molecule taken from the chiral pool.

Most laboratory syntheses on molecules with no chiral center lead to a 1:1 mixture of the two possible enantiomers. Consider, for example, 2-chlorobutane, which, as we have seen, is a chiral molecule, because there are four different groups attached to the second carbon atom. The preparation of 2-chlorobutane by the direct chlorination of *n*-butane leads to an optically inactive product, because an equal number of (*R*)- and (*S*)-isomers are produced. Such 50-50 mixtures of equal parts of enantiomorphs are called **racemates,** *d,l*-mixtures, or (±)-mixtures. The optical activity of one half of the isomers in the mixture in nullified by that of the other half, and the mixture is said to be optically inactive by what is called **external compensation.**

It is not difficult to understand why equal amounts of enantiomeric forms of 2-chlorobutane result when one considers the planar structure of the hydrocarbon free radical that forms as one of the reaction intermediates (Sec. 2.11). Obviously there is an equal opportunity for the chlorine molecule to react with the free radical from either side. Since Avogadro's number (the number of molecules in one mole) is about 6×10^{23}, the number of molecules involved is enormous, and the probability of the chlorine molecule's approaching from either side is exactly one-half. Employing the convention for indicating bond positions used in Exercise 2.7, we can show two reaction sequences graphically to illustrate the synthesis of racemic 2-chlorobutane.

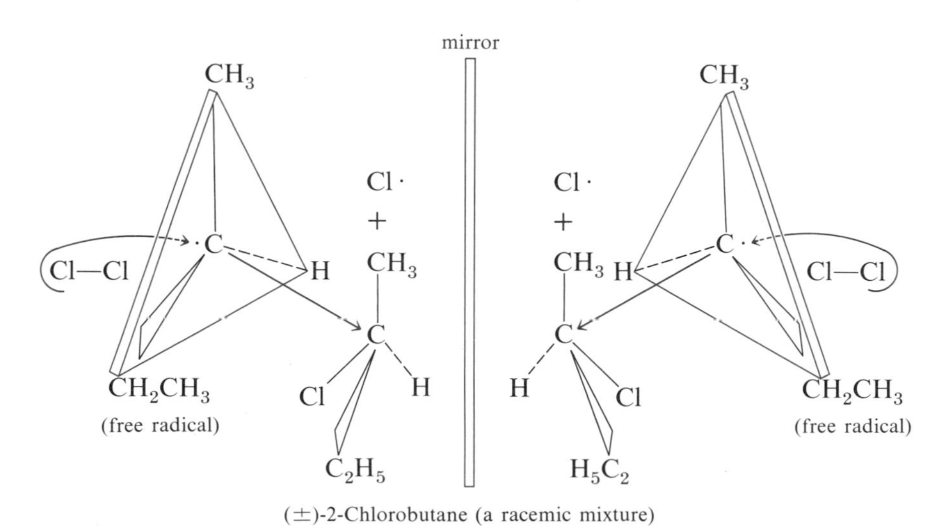

(±)-2-Chlorobutane (a racemic mixture)

Although chemists are not quite ready to compete with enzymes in the realm of asymmetric synthesis, some small successes are being achieved—and after all, the enzymes have had a few million years' head start. For example, the usual catalytic reduction of pyruvic acid over a *heterogeneous* catalyst leads to racemic lactic acid. On the other hand, pyruvic acid is reduced by yeast to D-lactic acid. Lactic acid isolated from muscle tissue is L-lactic acid.

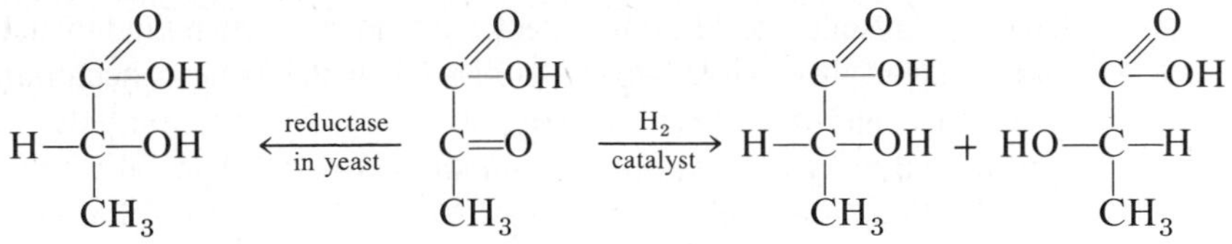

D-Lactic acid ← (reductase, in yeast) Pyruvic acid → (H_2, catalyst) (±)-Lactic acid

By using a *chiral, homogeneous* catalyst, it is now possible to hydrogenate the double bond of α-acetamidocinnamic acids to yield mixtures containing as much as 39 parts of one enantiomer to only 1 part of the other. In the example shown, hydrolysis of the product yields the important amino acid, L-dopa, which is used in the treatment of Parkinson's disease.

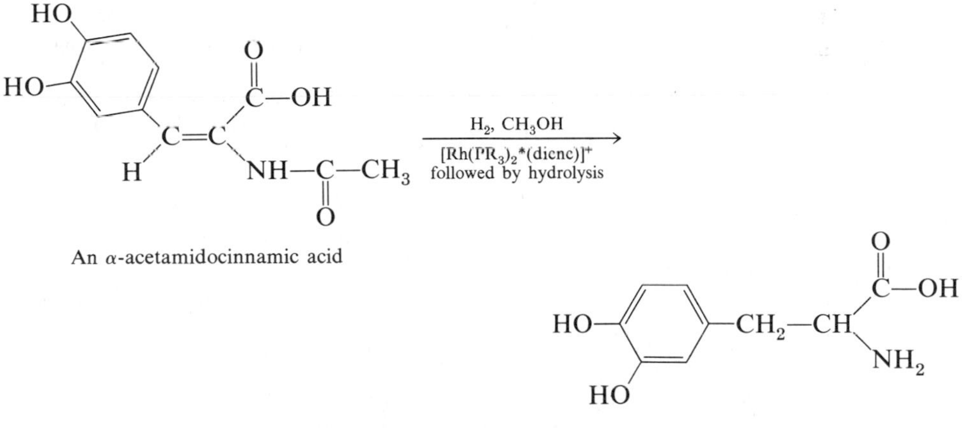

An α-acetamidocinnamic acid → (H_2, CH_3OH, $[Rh(PR_3)_2{}^*(diene)]^+$, followed by hydrolysis) L-Dopa

$(PR_3)^*{}_2$ = a chiral diphosphine
(diene) = 1,5-cyclooctadiene

Exercise 5.7 Draw a Fischer projection formula for L-dopa and show that this configuration is also (*S*).

5.7 Resolution of Optical Isomers

The separation of a racemic mixture into its optically active (*R*)- and (*S*)-forms is called **resolution.** Resolution is a common method used by chemists and biochemists to contribute to the chiral pool. Several techniques have been employed to effect such separations.

A. Mechanical Separation. The first resolution of a racemic mixture was the historic accomplishment of Pasteur in 1848. He observed that the crystalline tartaric acid salts (Sec. 12.5), when precipitated from solution, appeared to consist of "right- and left-handed" crystals. He carefully selected a number of each kind by using a pair of tweezers and a magnifying glass. Separate solutions of the same concentration were prepared of each form, and their rotations were observed. The rotation values of the two solutions were found to be alike *but of opposite direction*. Mechanical resolutions such as the one done by Pasteur are seldom possible. Other methods of resolution are usually necessary.

B. Preparation and Separation of Diastereoisomers. Reaction of a racemic mixture with an optically active reagent results in the preparation of diastereoisomers. The physical properties of diastereoisomers are different, and they may be separated by crystallization, distillation, or some other technique. An example of such chemical resolution is the preparation of salts from a racemic acid and a *levo*-base.

$$\underset{\text{Enantiomers}}{\left\{\begin{array}{c}\textit{dextro}\text{-acid}\\ \textit{levo}\text{-acid}\end{array}\right\}} + \textit{levo}\text{-base} \longrightarrow \underset{\text{Diastereoisomers}}{\left\{\begin{array}{c}\textit{dextro}\text{-acid}\cdot\textit{levo}\text{-base salt}\\ \textit{levo}\text{-acid}\cdot\textit{levo}\text{-base salt}\end{array}\right\}}$$

Each diastereoisomer, having physical properties different from those of the other, may be separated by fractional crystallization. Subsequent treatment with a mineral acid liberates the optically active organic acid.

$$\textit{dextro}\text{-acid}\cdot\textit{levo}\text{-base salt} + \text{HCl} \longrightarrow \textit{dextro}\text{-acid} + \textit{levo}\text{-base}\cdot\text{HCl}$$
$$\textit{levo}\text{-acid}\cdot\textit{levo}\text{-base salt} + \text{HCl} \longrightarrow \textit{levo}\text{-acid} + \textit{levo}\text{-base}\cdot\text{HCl}$$

C. Biochemical Resolution. Because certain microorganisms preferentially attack one enantiomer to the exclusion of the other when allowed to "feed" on a racemic mixture, one enantiomer is destroyed, leaving the other in solution. Pasteur's use of *Penicillium glaucum* in the resolution of racemic tartaric acid (Sec. 5.5) was the first example of this technique. Biochemical resolutions as preparative procedures, however, are limited in their application.

5.8 *cis,trans*-Isomerism

As was shown in Secs. 2.8 and 3.2, *cis,trans*-isomerism is a form of stereoisomerism that results when the rotation of one carbon atom with respect to another is restricted by the presence of a double bond, or when such rotation is prevented by the incorporation of the atoms in a ring. Several examples of this type of isomerism have been given in the previous chapters. Although the terms *cis* and *trans* will continue to be used for simple alkenes, they have been replaced by the IUPAC for more complex structures with the more general and less ambiguous designations (*E*)- (from the German *entgegen,* across) and (*Z*)- (from the German *zusammen,* together). In the (*E*),(*Z*)-system (no pun intended), the two groups attached to each end of the double bond are assigned priorities exactly as in the (*R*),(*S*)-system. If the two higher priority groups are on the same side of the double bond, the configuration is (*Z*); if on the opposite sides, the configuration is (*E*).

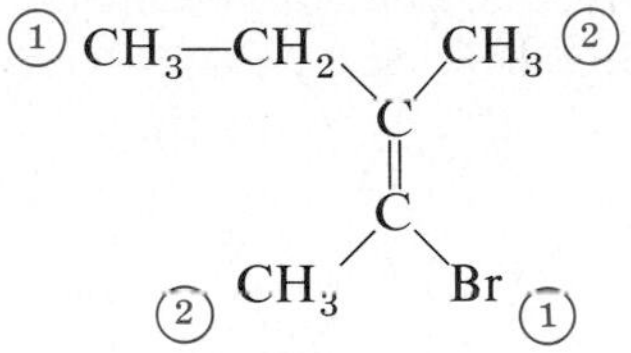

Relative priorities are encircled.

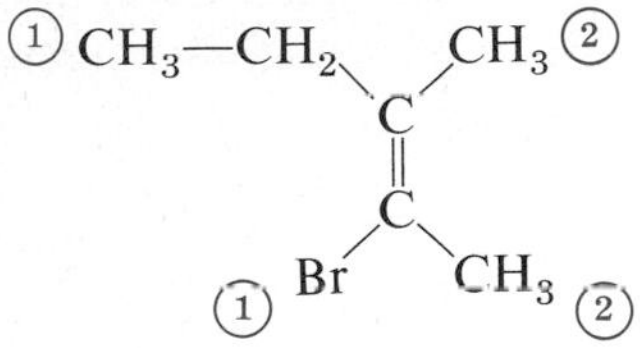

(*E*)-2-Bromo-3-methyl-2-pentene

(*Z*)-2-Bromo-3-methyl-2-pentene

For the cycloalkanes, the *cis* and *trans* designations of configuration have been retained. For flexible ring systems, such as cyclohexane, the ring may be treated as if it were planar for purposes of assigning configuration.

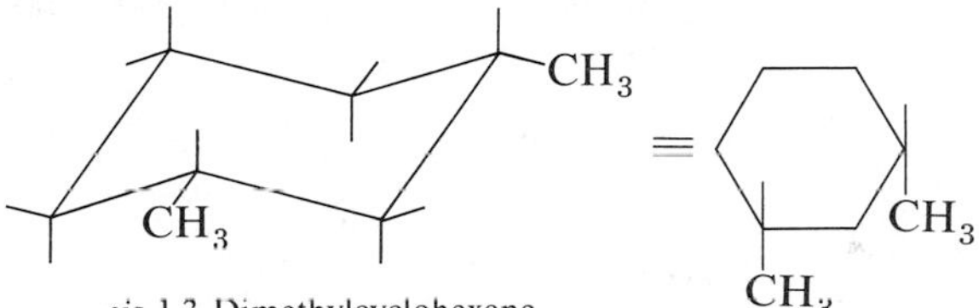

cis-1,3-Dimethylcyclohexane

Unlike enantiomers, *cis,trans*-isomers are diastereoisomers and therefore have widely different physical properties and often vary in their chemical behavior. Measurable differences in these properties allow us to distinguish between *cis* and *trans* forms rather easily. For example, *cis*-1,2-dichloroethene has a measurable **dipole moment,**[4] whereas *trans*-1,2-dichloroethene has a zero dipole moment. The dipole moment of a molecule with polar bonds, such as the C—Cl bond, will be the vector sum of the individual bond (dipole) moments. These bond moments cancel each other in the *trans*-isomer but add to give a net dipole moment in the *cis*-isomer.

(polar)

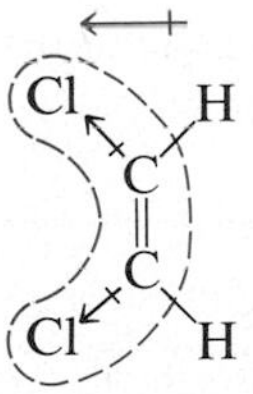

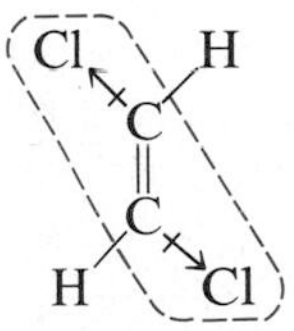

(nonpolar)

cis-1,2-Dichloroethene
bp 60°
$\mu = 1.85\ D$

trans-1,2-Dichloroethene
bp 48°
$\mu = 0$

[4]The dipole moment of a compound measures the concentration of positive and negative charges in different parts of the molecule. It is defined by the expression $\mu = e \times d$, where e is the charge (in electrostatic units) and d is the distance between the centers of positive and negative charge in angstroms (Å). Dipole moments are measured with a device called a dipolimeter and are expressed in Debye units, *D*.

The chemical behavior of *cis*- and *trans*-isomers may also vary sufficiently to enable us to distinguish one from the other. For example, the two carboxyl groups of maleic acid are on the same side of the double bond and thus properly oriented in space to permit the easy removal of a water molecule, when heated, to give maleic anhydride. Its *trans*-isomer, fumaric acid, cannot form the anhydride.

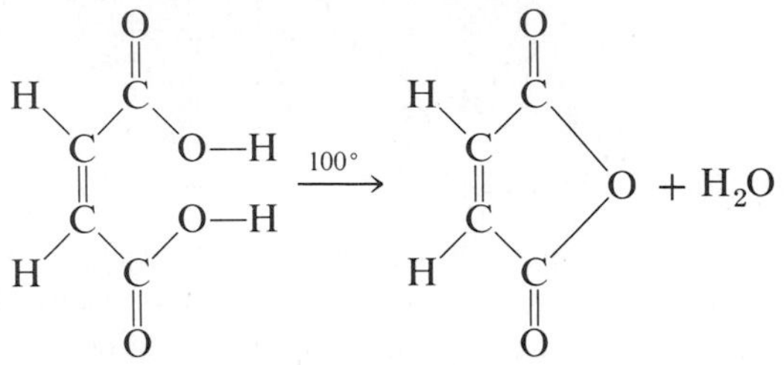

$\xrightarrow{100°}$ + H_2O

Maleic acid (*cis*-Butenedioic acid) mp 130°

Maleic anhydride

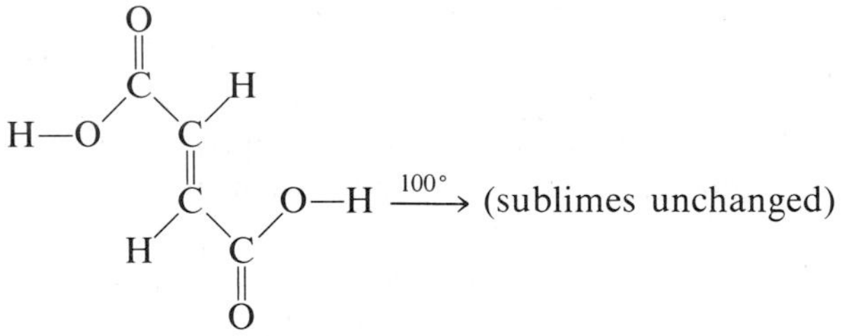

$\xrightarrow{100°}$ (sublimes unchanged)

Fumaric acid (*trans*-Butenedioic acid) mp 270°

Exercise 5.8 Natural rubber and its isomer, gutta-percha, are polymers of isoprene. Natural rubber has a *cis*-configuration; gutta-percha has a *trans*-configuration. Draw partial structures of each natural product.

Summary

1. Isomers that have similar structural formulas but different configurations (spatial arrangements) are called **stereoisomers.**
2. Stereoisomers may be either enantiomers or *cis,trans*-isomers.
3. An **optically active** compound is one capable of rotating the plane of polarized light. Compounds that are **chiral** are optically active. Chirality is the property of ''handed-

ness.'' A chiral center is usually a carbon atom with four different atoms or groups bonded to it.

4. Plane-polarized light is light in which all rays are vibrating in a single plane.
5. Optical isomers that are mirror images, or enantiomers, rotate plane-polarized light to the same degree, but in opposite directions.
6. The direction of rotation, if to the right, is designated **dextro** (+); if to the left, **levo** (−).
7. The configurations of some optically active compounds may be related to that of glyceraldehyde and designated as D- or L-forms.
8. Absolute configurations of optically active compounds may be designated as (*R*)- or (*S*)-forms.
9. Optical activity can be measured by means of an instrument called a **polarimeter.**
10. A synthesis that produces a compound containing a chiral center usually results in a 50-50 mixture of (*R*)- and (*S*)-forms. Such a mixture is called a **racemic mixture.** Racemic mixtures are optically inactive.
11. A molecule with *n different* chiral carbon atoms can have 2^n optical isomers.
12. A molecule with *similar* chiral carbon atoms can have, in addition to optically active forms, one or more **meso** forms that are optically inactive.
13. **Diastereoisomers** are stereoisomers that are not enantiomers.
14. Separation of a mixture of enantiomers is called **resolution.** Resolution can be accomplished mechanically, chemically, or biologically.
15. The configurations of *cis,trans*-isomers may be designated as (*E*)- or (*Z*)-forms.

Supplementary Exercises

5.9 Predict the number and kind of stereoisomers possible for each of the following.

(a) 2-Butenoic acid, $CH_3CH{=}CH{-}COOH$
(b) 1,2-Dimethylcyclopentane
(c) 2,3-Butanediol, $CH_3CH(OH)CH(OH)CH_3$
(d) 3-Bromocyclohexene
(e) Glucose, $HOCH_2(CHOH)_4CHO$

5.10 Indicate which structures in the following sets are identical and which are enantiomers.[5]

	(a)	(b)	(c)	(d)	(e)
top	COOH	OH	H	CH_3	H
left	H_3C	H	HO	HO	H_3C
right	OH	CH_3	CH_3	H	COOH
bottom	H	COOH	COOH	COOH	OH

[5]The chiral carbon atom is often omitted in drawing Fischer projections, and other simplifying abbreviations are sometimes used (Sec. 14.4).

5.11 How many optically active forms are possible for each of the following compounds?

(a) *sec*-Butylbenzene
(b) 2-Bromo-1-methylcylcopentane
(c) *trans*-4-Bromo-1-methylcyclohexane
(d) 1,2-Dibromopentane
(e) 1,2-Dibromobenzene
(f) 1-Bromo-2-chloropropane

(g) NO_2 CO_2H Cl NO_2

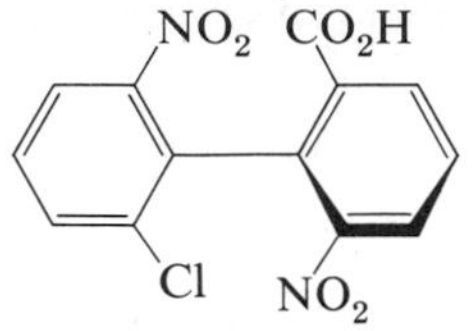

Rings perpendicular to each other; rotation around central bond restricted.

2-Carboxy-2′-chloro-6,6′-dinitrobiphenyl

5.12 Could the following compounds be resolved into (+) and (−) forms?

(a) $CH_3CH_2CH(OH)CH_3$

(b) $CH_3CH(Br)CH_3$

(c) CH_3, Br—C, H, C=C, H, H, H

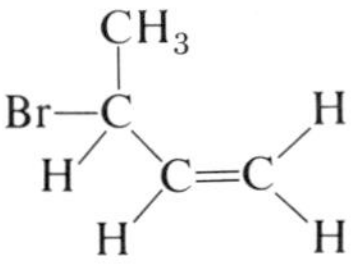

(d) CH_3, H, CH_3, H

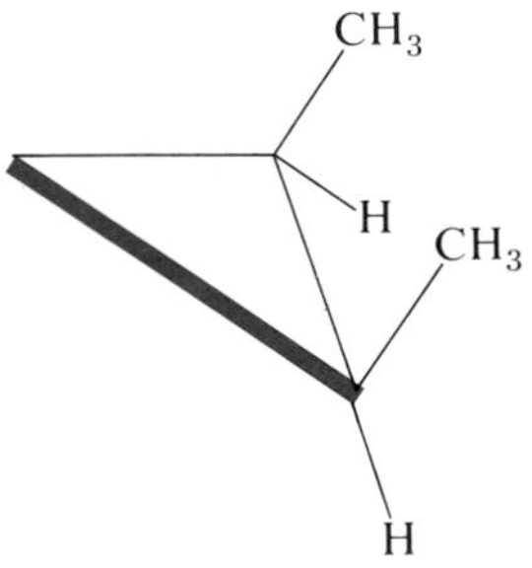

(e) H, —C—CH_3, Cl

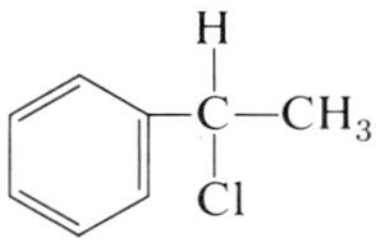

5.13 Draw (*R*)- and (*S*)-configurations for compound (c) in Exercise 5.12.

5.14 Draw Fischer projection formulas and zigzag formulas for

(a) H, —CH_2—C—COOH, NH_2

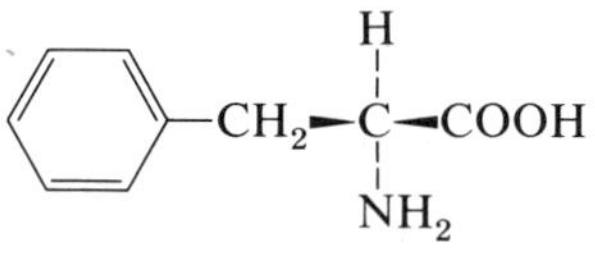

Phenylalanine

(b) H, CH_3—C—COOH, OH

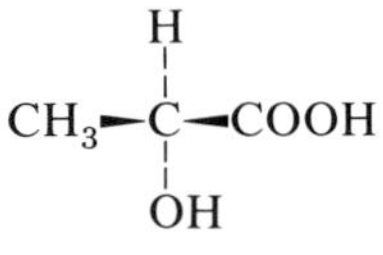

Lactic acid

Identify each as D or L; as (*R*) or (*S*).

5.15 Draw stereochemical formulas for three unsaturated compounds of formula C_5H_9Br that would show optical activity. Do any of these also have *cis* and *trans* forms?

5.16 Complete the following reactions. Are the principal products resolvable into optically active forms?

(a) 1-Butene + HBr ⟶
(b) 2-Butene + HBr ⟶
(c) Cyclopentene + HBr ⟶
(d) Cyclopentene + Br_2 ⟶
(e) 2-Butene + HOCl ⟶
(f) Propylene + HOCl ⟶
(g) Propylene + cold, dilute $KMnO_4$ ⟶

5.17 What stereoisomers are possible when bromine is added to *trans*-2-butene? What stereoisomers are possible when bromine is added to *cis*-2-butene? (*Hint:* See Section 3.7. Also draw three-dimensional structures or Fischer projections for reactants and products.)

5.18 An optically active liquid (A), C_4H_9Cl, when heated with alcoholic KOH, lost its optical activity. The product, an optically inactive gas (B), C_4H_8, was bubbled through bromine–carbon tetrachloride solution to give a new substance (C), $C_4H_8Br_2$, that could not be resolved into (+) and (−) forms. Write the reactions that lead to products (B) and (C) and show the stereochemistry involved by drawing three-dimensional formulas.

5.19 Draw the Fischer projection of the compound whose Newman projection is given below.

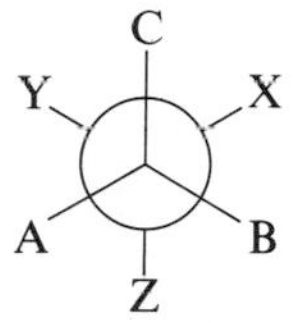

C
Y X
A B
Z

Properly done, the answer will provide a useful device for interconverting Fischer and Newman projections.

5.20 Label each of the double bonds in the following naturally occurring substances as (*E*) or (*Z*). Label each chiral center as (*R*) or (*S*).

(a) O COOH HO OH

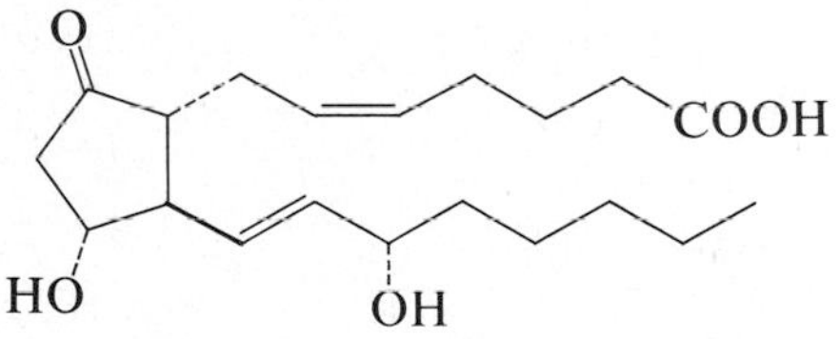

Prostaglandin PGE_2 (a biochemical regulator)

(b) $CH_3(CH_2)_3$, H, C=C, H, H, C=C, $(CH_2)_7CO_2H$, H, C=C, H, H

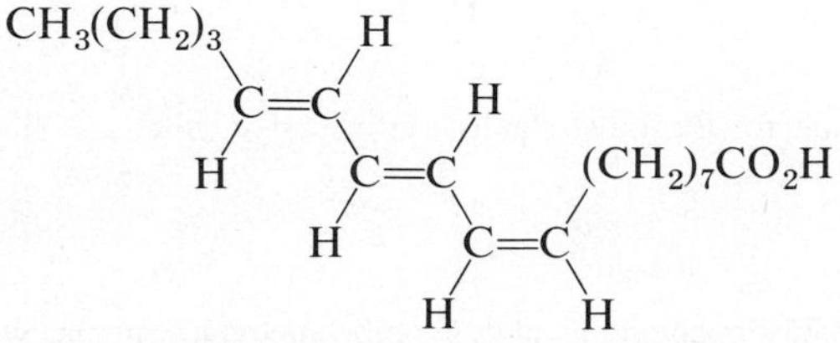

Eleostearic acid (a component of linseed oil)

5.21 Determine which, if any, of the following is a *meso* isomer. (See footnote 5.)

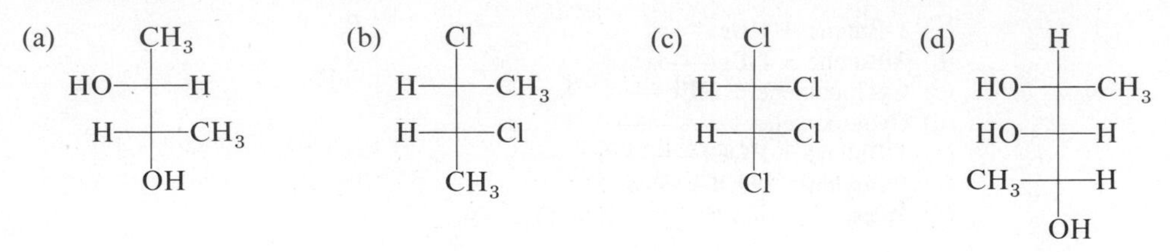

5.22 Label each of the chiral centers in the following compounds as (*R*) or (*S*). (See footnote 5.)

(a) CH_3; H—OH; H—Cl; CH_3

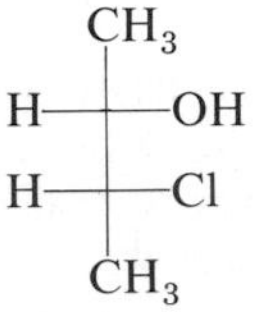

(b) CH_3; H—Cl; Cl—H; CH_3

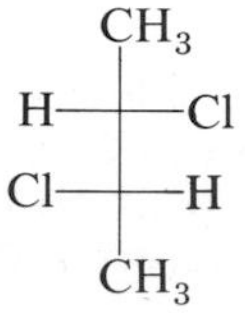

(c) CH_3; H—OH; H—OH; CH_3

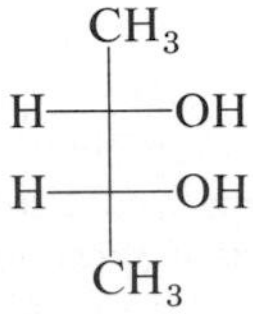

(d) CH_3; H—OH; H—Cl; H—OH; CH_3

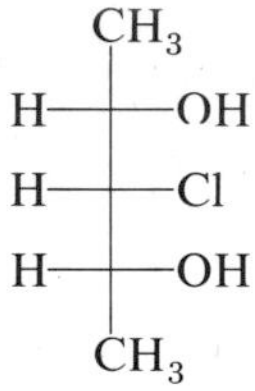

(e) CH=O; H—OH; H—OH; H—OH; H—OH; CH_2OH

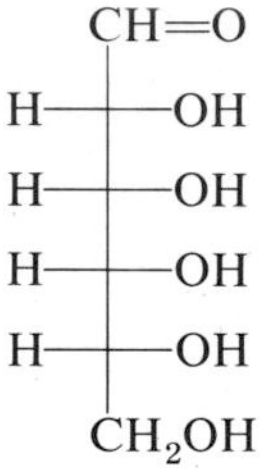

5.23 Draw the Fischer projection of (*R*)-2-chlorobutane. Which of the following structures are also representations of (*R*)-2-chlorobutane? Which are representations of (*S*)-2-chlorobutane? (See footnote 5.)

(a) H; CH_3—CH_2CH_3; Cl

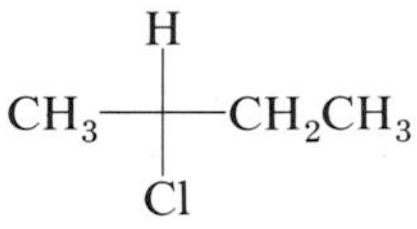

(b) CH_2CH_3; H—Cl; CH_3

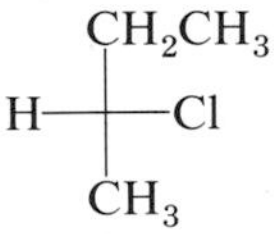

(c) CH_3, H, Cl on C—C bearing H, H, CH_3

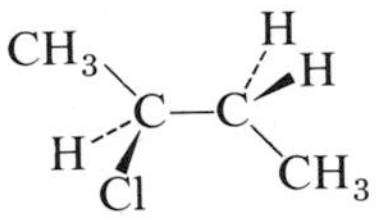

(d) CH_3, H, H on C—C bearing CH_3, H, Cl

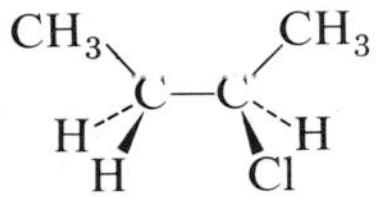

(e) H; H, CH_3 (front); CH_3, Cl; H

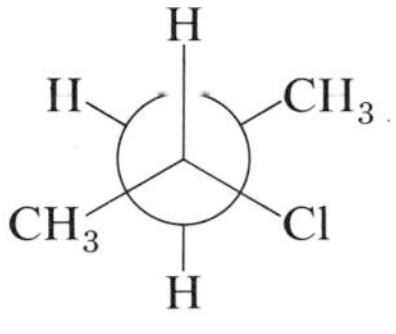

(f) Cl; H, H; CH_3, H; CH_3

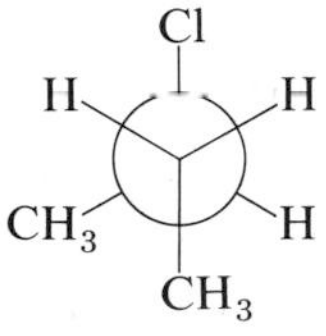

5.24 Draw a Newman projection for the most stable conformer of *meso*-2,3-dibromobutane. Identify each chiral carbon as (*R*) or (*S*).

5.25 If we identify 1,2-dibromocyclopentane as (1*R*,2*S*)-dibromocyclopentane, would it still be necessary to prefix *trans* to the name? Explain.

5.26 Using the structural formulas given in Supplementary Exercise 1.26, locate and circle all of the chiral carbon atoms in the molecules (a), (c), and (d).

5.27 Calculate the number of stereoisomers that are possible for D-glucose (formula in Section 5.5).

5.28 A solution of 0.90 g of limonene (Sec. 3.13) in sufficient methanol (CH_3OH) to make 50 ml of solution gave an observed rotation, α, of $+1.91°$ in a 10-cm polarimeter tube at 20°C. Calculate $[\alpha]_D^{20°}$.

5.29 Draw skeletal structures (not showing individual carbon or hydrogen atoms) for the *cis,cis-*, *trans,cis-*, *trans,trans-*, and *cis,trans-*isomers of 2,4-hexadiene. Which, if any, of the four structures are identical? Label each isomer according to the (*E*),(*Z*)-system.

5.30 In Sec. 3.6 the catalytic hydrogenation of 1,2-dimethylcyclopentene was described to give *cis*-1,2-dimethylcyclopentane. Does *cis*-1,2-dimethylcyclopentane exist in enantiomeric forms?

5.31 Identify the chiral carbon atoms in cholesterol and calculate the number of enantiomers possible.

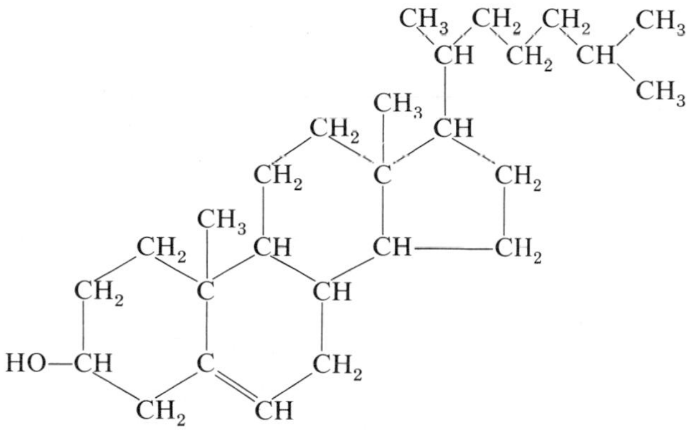

Cholesterol

5.32 How many stereoisomers are possible for ricinoleic acid?

$$CH_3{-}(CH_2)_5{-}CH(OH){-}CH_2{-}CH{=}CH{-}(CH_2)_7COOH$$

Ricinoleic acid

5.33 Could the simple procedure described in Section 5.5 for determining the absolute configuration of zigzag structures with hydrogen atoms at each chiral center be extended to cyclic molecules if we were to use the same convention of bold and dashed lines? Verify your answer by labeling each of the chiral centers in the following compounds as (*R*) or (*S*).

(a) (b)

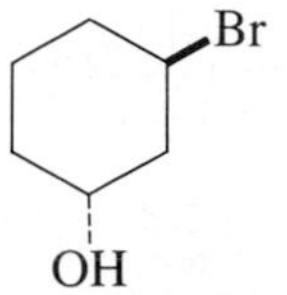

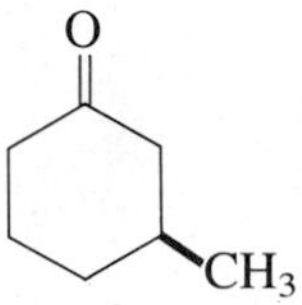

6 Organic Halogen Compounds

The earth's crust (including the oceans) contains inorganic halides of every kind in great abundance. In contrast to this, there are relatively few naturally occurring organic compounds in which halogen atoms are covalently bonded to carbon. Yet the importance of this family of substances is so great that many synthetic routes lead to end products only by way of some halogen intermediate. Many organic halogen compounds are synthesized, therefore, simply to be used as chemical reagents. Others are prepared as useful commodities in their own right and are employed as propellants in aerosol sprays, and as refrigerants, pesticides, dry cleaners, plastics, and anesthetics.

We found double and triple bonds in the alkenes and alkynes to be the structural features that bestowed a characteristic reactivity upon these compounds. We shall now study for the first time a functional group composed of a single atomic species—the covalently bound halogen.

6.1 Classification and Nomenclature

All organic halogen compounds may be classified as either aliphatic or aromatic. If aliphatic, they are alkyl halides that may be classified further according to structure—that is, they are either primary, secondary, or tertiary alkyl halides. This classification is based upon the nature of the carbon atom to which the halogen atom is bonded (*see* Sec. 2.3).

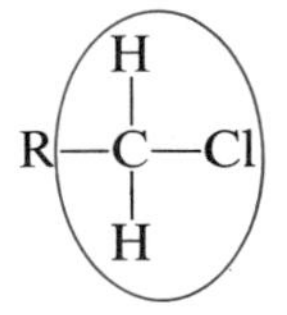

A primary alkyl halide

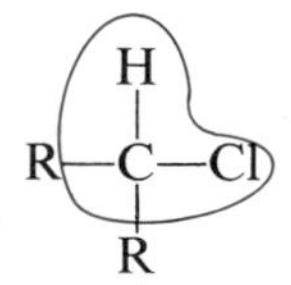

A secondary alkyl halide

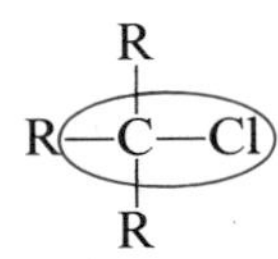

A tertiary alkyl halide

The nomenclature of the halogen compounds is not difficult to learn. Common names are frequently used for the simpler members of this family. In such names the alkyl group to which the halogen is attached is given first, and this is then followed by the name of the halogen. The halogen (as halide) is a separate word. Alkyl halides of five carbons or greater and the aromatic halogen compounds are named more conveniently as substituted hydrocarbons according to IUPAC rules. The following examples illustrate each kind of nomenclature.

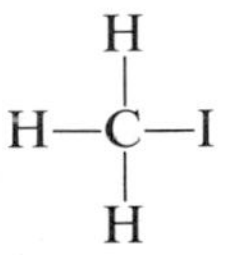

CH_3I — Methyl iodide

$(CH_3)_3C{-}Cl$ — *tert*-Butyl chloride

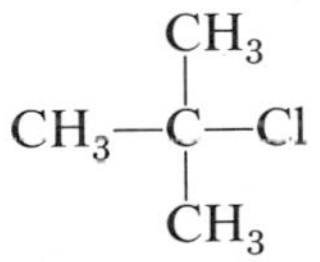

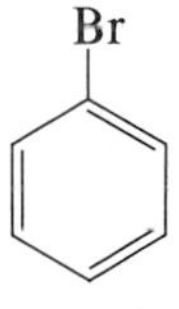

Bromobenzene

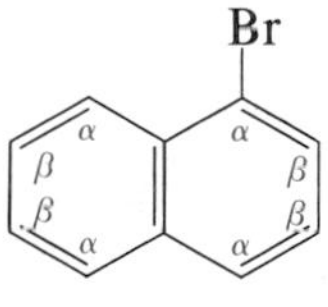

1-Bromonaphthalene (α-Bromonaphthalene)

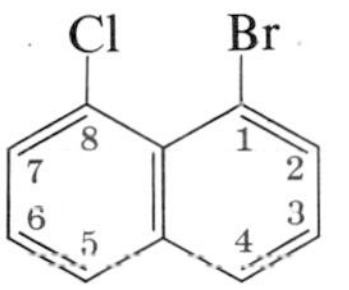

1-Bromo-8-chloronaphthalene

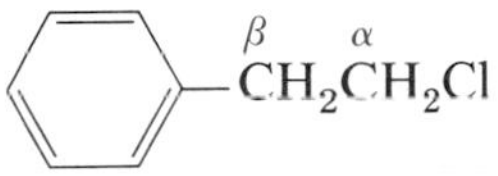

$C_6H_5{-}\overset{\beta}{C}H_2\overset{\alpha}{C}H_2Cl$ — β-Phenylethyl chloride (1-Chloro-2-phenylethane)

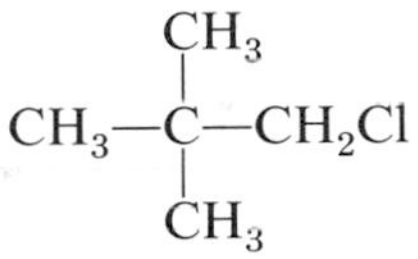

$(CH_3)_3C{-}CH_2Cl$ — Neopentyl chloride (1-Chloro-2,2-dimethylpropane)

Exercise 6.1 Eight structures may be drawn for the molecular formula $C_5H_{11}Cl$. Classify each as a primary, secondary, or tertiary chloride. Name each according to the IUPAC system. Which of the compounds may exist as d,l-forms?

6.2 Preparation of Organic Halogen Compounds

A. Preparation by Direct Halogenation of Hydrocarbons. The aromatic halogen compounds in which the halogen is attached to the ring may be prepared by electrophilic halogenation (Sec. 4.6A). The halogenation of alkanes by a free-radical chain process (Sec. 2.11) is difficult to control and produces a mixture of isomers and polyhalogenated products. The chlorination of methane, for example, produces all of the possible mono- and polysubstituted methanes. For this reason direct halogenation of alkanes is seldom used as a means of preparing alkyl halides in the laboratory, although it is used in some industrial processes.

Chlorination of the methyl group of toluene under free-radical conditions by light catalysis also results in the production of varying amounts of polysubstituted derivatives and is an industrial, not a laboratory, process.

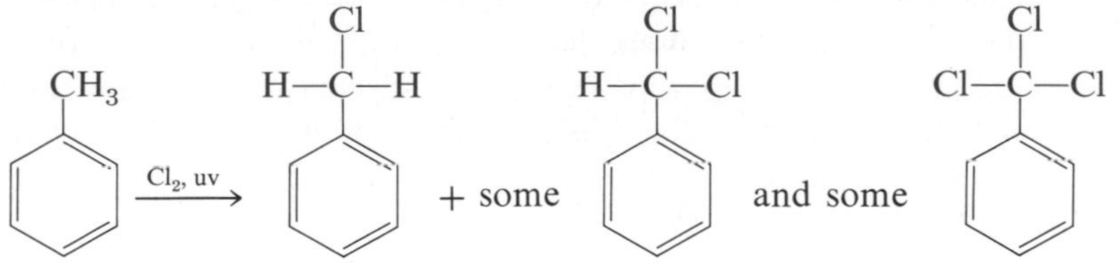

$$C_6H_5CH_3 \xrightarrow{Cl_2,\ uv} C_6H_5CH_2Cl + \text{some } C_6H_5CHCl_2 \text{ and some } C_6H_5CCl_3$$

Toluene | Benzyl chloride | Benzal chloride | Benzotrichloride

With side chains longer than methyl, chlorine reacts with alkylbenzenes to give a mixture of isomeric halides. However, bromine, which reacts relatively slowly with alkanes, reacts readily with alkylbenzenes under free-radical conditions to give only the product in which the bromine atom is attached to the carbon atom next to the ring, called the benzylic carbon or **benzylic position.** The reason for this highly selective attack is that an odd electron on the benzylic carbon may be delocalized into the ring, as may be seen from the resonance structures for the benzyl radical. This resonance interaction stabilizes the free-radical intermediate and favors its formation. We shall see this same phenomenon manifested throughout chemical behavior—namely, the reaction that leads to a stabilized intermediate, or product, is often the favored reaction.

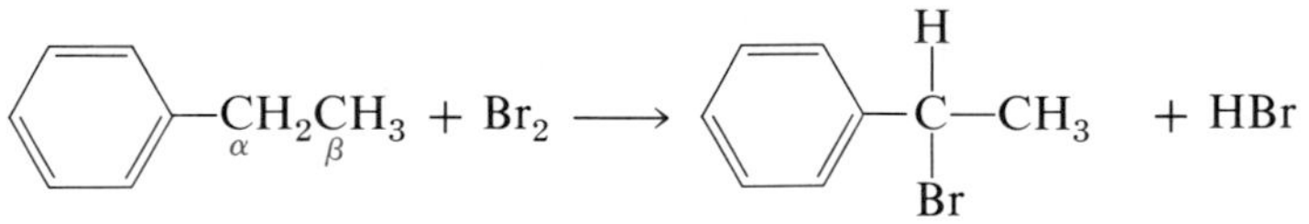

$$C_6H_5\underset{\alpha}{CH_2}\underset{\beta}{CH_3} + Br_2 \longrightarrow C_6H_5CH(Br)CH_3 + HBr$$

Ethylbenzene | α-Phenylethyl bromide (1-Bromo-1-phenylethane)

In general, then, the relative reactivity of hydrogens in the free-radical reactions is benzylic > tertiary > secondary > primary.

Exercise 6.2 If you have not already done so (Exercise 4.23), draw the resonance forms for the benzyl radical.

$$C_6H_5\dot{C}H_2$$

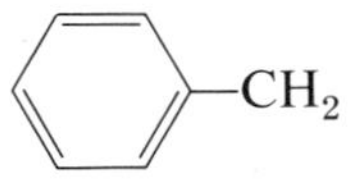

[*Hint:* Use one-electron shifts rather than two-electron shifts to derive the forms, as shown below with fishhooks (single-barbed arrows) for the allyl radical.]

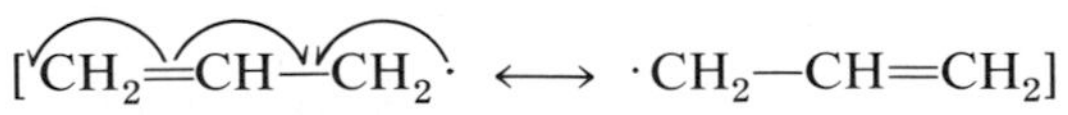

$$[CH_2{=}CH{-}CH_2\cdot \longleftrightarrow \cdot CH_2{-}CH{=}CH_2]$$

B. Preparation from Alcohols. A laboratory preparation of an alkyl halide may begin with an alcohol as starting material. Treatment of the appropriate alcohol with any of the following reagents results in replacement of the hydroxyl group (—OH), by halide.

(a) concentrated hydrobromic acid, HBr.
(b) concentrated hydriodic acid, HI.
(c) concentrated hydrochloric acid, $HCl + ZnCl_2$.
(d) phosphorus tribromide, phosphorus triiodide, or phosphorus pentachloride.
(e) thionyl chloride, $SOCl_2$.

Two examples of the preparation of organic halogen compounds from alcohols by the use of these reagents are given below.

$$3\ CH_3CH_2OH + PBr_3 \longrightarrow 3\ CH_3CH_2Br + H_3PO_3$$

Ethyl alcohol → Ethyl bromide

$$C_6H_5\text{—CH(OH)—}CH_3 + SOCl_2 \longrightarrow C_6H_5\text{—CH(Cl)—}CH_3 + HCl + SO_2$$

1-Phenylethanol → 1-Chloro-1-phenylethane

Thionyl chloride, $SOCl_2$, is a superior chlorinating reagent because the by-products formed are gases and escape from the reaction mixture.

C. Preparation from Olefins. Halogen acids (HBr, HCl, HI) will add to olefins to yield alkyl halides. The mode of addition follows Markovnikov's rule, except for the addition of hydrogen bromide in the presence of air, light, or peroxides. The mechanisms for both modes of addition were shown in Sec. 3.8A.

$$CH_3\text{—CH=}CH_2 \xrightarrow[\text{no light}]{HBr} CH_3\text{—CHBr—}CH_3$$

Propylene (Propene) → Isopropyl bromide (2-Bromopropane)

$$CH_3\text{—CH=}CH_2 \xrightarrow[\text{light}]{HBr} CH_3\text{—}CH_2\text{—}CH_2\text{—Br}$$

Propylene (Propene) → *n*-Propyl bromide (1-Bromopropane)

D. Preparation via Halogen Exchange. Certain halides, in some instances, are best prepared from other halides by an exchange. This method is well suited to the preparation of iodo and fluoro compounds. Sodium iodide, unlike sodium chloride or sodium bromide, is soluble in anhydrous acetone. If a primary or secondary alkyl chloride is dissolved in a solution of sodium iodide in acetone, the halogens exchange. The equilibrium favors formation of the alkyl iodide because the sodium chloride that forms is not soluble in acetone, and precipitates.

$$\underset{\text{Primary or secondary}}{R{-}Cl} + Na^{+}I^{-} \xrightarrow{\text{acetone}} R{-}I + NaCl$$

Fluorine compounds often are prepared via a halogen exchange through the use of inorganic fluorides, rather than by reactions utilizing fluorine or hydrogen fluoride, because the latter chemicals are extremely reactive and difficult to handle.

$$\underset{\text{Methyl bromide}}{2\ CH_3Br} + \underset{\text{Mercurous fluoride}}{Hg_2F_2} \longrightarrow \underset{\text{Fluoromethane}}{2\ CH_3F} + \underset{\text{Mercurous bromide}}{Hg_2Br_2}$$

Exercise 6.3 If you were given propene and benzene as your only organic starting materials and any other inorganic reagents that you might require, how would you prepare (a) isopropyl chloride, (b) *n*-propyl bromide, (c) *n*-propyl iodide, and (d) 2-chloro-2-phenylpropane?

6.3 Polyhalogen Compounds

Halogenated products prepared from methane become increasingly dense, higher boiling, and nonflammable as halogen successively replaces the four hydrogen atoms. Polychlorinated methanes may be prepared by the direct halogenation of methane (Sec. 2.11) or, as in the case of carbon tetrachloride, by the reaction of chlorine with carbon disulfide.

$$CS_2 + 2\ Cl_2 \longrightarrow CCl_4 + 2\ S$$

The chlorine atoms of carbon tetrachloride, in turn, can be exchanged for fluorine in the preparation of the fluorinated methanes. For example,

$$CCl_4 + 2\ SbF_3 \longrightarrow \underset{\substack{\text{Dichlorodifluoromethane} \\ \text{(Freon-12)}}}{CCl_2F_2} + 2\ SbF_2Cl$$

Carbon tetrachloride is an excellent solvent for oils, waxes, fats, and greases and has been used as a dry cleaner. Although it is nonflammable, it is very volatile, and repeated inhalation of its vapor can cause cumulative liver damage. Carbon tetrachloride as a fire-extinguishing agent has been replaced largely by liquefied carbon dioxide, bromotrifluoromethane ($CBrF_3$), and dibromodifluoromethane (CBr_2F_2). The fluorinated methanes may be used safely in confined spaces such as submarines, aircraft, boats, and automobiles.

The trihalomethanes—chloroform, bromoform, and iodoform—may be prepared by treating acetone or ethyl alcohol with alkaline solutions of the halogens (the haloform reaction), as is described in Sec. 8.8.

$$CH_3{-}CH_2{-}OH + 4\,Cl_2 + 6\,NaOH \longrightarrow$$

$$\underset{\text{Chloroform}}{HCCl_3} + \underset{\text{Sodium formate}}{H{-}C({=}O){-}O^-Na^+} + 5\,NaCl + 5\,H_2O$$

At present chloroform is prepared industrially by the chlorination of methane. It is a heavy, nonflammable, volatile liquid. Medicinally, it found some use at one time as an inhalation anesthetic and in cough medicines. It is now considered not only too toxic for anesthesia but also possibly carcinogenic. **Halothane,** $CF_3{-}CHClBr$, **enflurane,** $CHFCl{-}CF_2{-}O{-}CHF_2$, and **isoflurane,** $CF_3{-}CHCl{-}O{-}CHF_2$, are effective, relatively nontoxic inhalation anesthetics. Iodoform, CHI_3, a yellow powder with a characteristic pungent odor, has found some application in medicine as a topical antiseptic.

The fluorochloro compounds, marketed under the name of **"Freons,"** are extremely inert, nontoxic, noncorrosive, low-boiling liquids and gases. They are widely used as refrigerants and have also been used extensively as propellants in aerosol sprays of all kinds. Unfortunately, it has been suggested, on the basis of theoretical studies, that the widespread release of the Freons into the upper atmosphere from such use may result in an eventual depletion of the atmosphere's ozone layer—thought to protect us from some of the harmful effects of the sun's ultraviolet radiation. In the United States the use of Freons in pressurized cans for the dispensing of various substances (from shaving cream to paint) was prohibited by law in 1979 for all but "essential uses."

Polyfluoro- and polychloroethylenes are used as the structural units in several very useful plastics. One of these, "Teflon," is a polyfluoroethylene.

$$x\,F{-}C(F){=}C(F){-}F \xrightarrow[\text{45–50 atm.}]{\text{catalysts,}} \left({-}CF_2{-}CF_2{-}\right)_x$$

$(x = 5{,}000\text{–}20{,}000)$

"Teflon"

Teflon is a waxy plastic completely inert to nearly all reagents and ideal for gasket materials and valve packings where corrosive chemicals and high temperatures are encountered. Its great resistance to high temperatures suggested its use as nose cone material for missiles and spacecraft. Teflon is widely used as a liner in frying pans and muffin tins, and on other utensils and tools to provide nonsticking surfaces.

Vinyl chloride ($CH_2{=}CHCl$, Sec. 6.6) is the organic halogen compound manufactured in greatest amount in the United States, some 6.5 billion pounds/year, principally for the preparation of the important polymer, polyvinyl chloride, used in the manufacture of leather substitutes, tubes and pipes, packaging films, etc. Although the polymer is safe, unless decomposed by excessive exposure to high temperatures, the monomer, vinyl chloride, poses a serious health problem because repeated inhalation can cause a rare form of cancer. Polymerization of vinyl chloride with vinylidene chloride (1,1-dichloroethene, $CH_2{=}CCl_2$) produces a copolymer with properties superior to those of the homopolymer, polyvinyl chloride, for some uses. The popular packaging wrap, "Saran Wrap," is an example of such a copolymer.

Many polychlorinated hydrocarbons are effective pesticides, but unfortunately these substances are "hard" insecticides; that is, they are very resistant to biological degradation. Because of the persistent and damaging effects they have had on wildlife, especially birds and fish, the indiscriminate use of polychlorinated insecticides in agriculture has been severely criticized in recent years. DDT, which is estimated to have saved 75,000,000 lives from the ravages of malaria and other mosquito-borne diseases, but which is also a hard insecticide, was banned in 1972 for agricultural use in the United States. The "hard" pesticides aldrin and chlordane, as well as a derivative of aldrin, dieldrin (not shown), have been banned for all but limited use in termite control and a few other applications because of the higher persistence of the substances in the environment and a presumptive link with cancer in certain rodents.

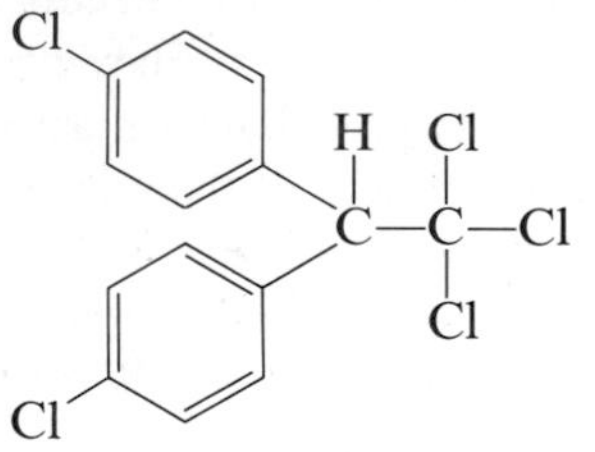

*Di*chloro*di*phenyl*tri*chloroethane (DDT)

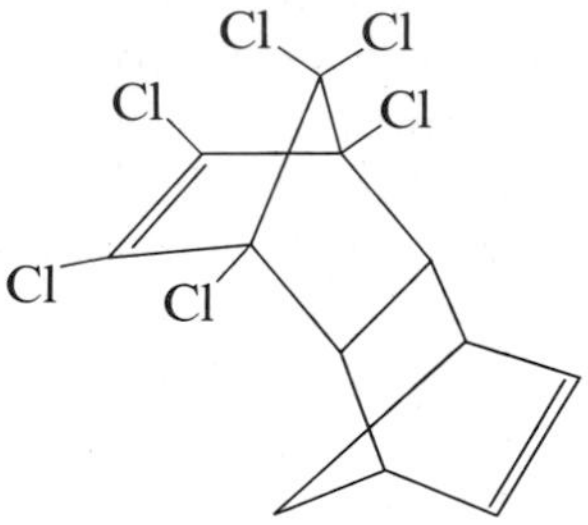

Aldrin

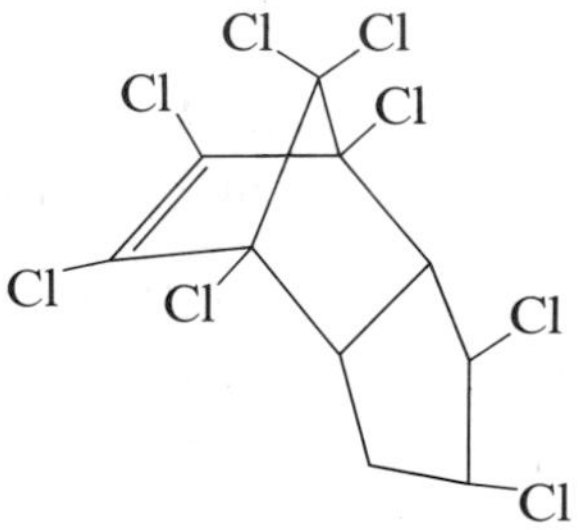

Chlordane

For many years woolen blankets and clothing in storage were protected by the use of "mothballs," originally composed of naphthalene. Later naphthalene was replaced for this purpose by *p*-dichlorobenzene. Several polychlorobiphenyls (PCB's) have been used in large electrical transformers because of their good dielectric properties and their outstanding thermal stability even at temperatures up to 700°C. Although not introduced deliberately into the environment, PCB's have made their way into some rivers and lakes from leaks and improper disposal of wastes. Because of the high level of toxicity of the PCB's, fish showing traces of these materials are considered inedible.

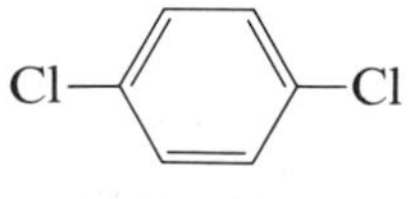

p-Dichlorobenzene

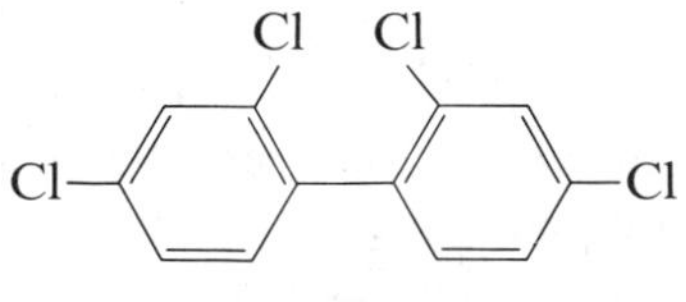

A PCB

1,2-Dibromoethane, 1,3-dichloropropene, and 1,2-dibromo-3-chloropropane have been used as soil fumigants for the control of nematodes.

Although it may appear to you at this stage that most organic halogen compounds are toxic, the student of chemistry must maintain the proper perspective. Thorough toxicological studies have been carried out on relatively few organic compounds. In many instances, only qualitative rather than quantitative evaluations of the hazards associated with the use of chemicals are available. On the basis of the principle "better safe than sorry," many chemicals are being labeled "toxic" or "cancer suspect" on the basis of preliminary evidence, pending thorough investigation. Thus, the following halides are just a sample of those on the list[1] of cancer-suspect substances: methyl iodide, *sec*-butyl chloride, *tert*-butyl bromide, 1,2-dibromoethane, trichloroethylene ($ClCH{=}CCl_2$), chloromethyl methyl ether ($Cl{-}CH_2{-}O{-}CH_3$), and 2-chloroethyl ethyl ether ($Cl{-}CH_2{-}CH_2{-}O{-}CH_2CH_3$).

6.4 Reaction Kinetics. Molecularity

The study of the reactions of alkyl halides and how they take place has been of great importance in the development of our understanding of reaction mechanisms. This understanding enables us to better classify and establish order among the very large number of organic reactions known to us, to reduce our reliance on rote memory of reaction types, and to invent or predict new reactions when we need them. Much of what we have learned is based on the measurement of **reaction kinetics,** which is the study of the rates of chemical reactions and the factors influencing them. As we saw in Section 1.12, the rate of a chemical reaction depends on the activation energy of the slowest step, the *rate-determining step,* because at a given temperature only a fraction of the reactant species present will have energies greater than the activation energy and be able to get over the energy barrier. Remember that the reactant species may be molecules, ions, or radicals, depending on the nature of the reaction being studied. To increase the rate of reaction, we must increase the number of such species that have the necessary energy. One way of doing this is to raise the temperature; another is to increase the concentration of one or more of the reactants needed for the transition state.

If the reaction involves only *one* reactant species, A, in the transition state for the rate-determining step, we say that the reaction is **unimolecular.** In general, we find that if all other conditions are held constant, the rate of a unimolecular reaction can be described by the simple mathematical equation:

$$\text{Rate of reaction} = k[\text{A}]$$

where [A] is the concentration of A present at a given instant, and k is the **rate constant** of the reaction. The value of k depends on the reaction being studied. A reaction with this rate equation is said to be **first order.**

[1] *Registry of Toxic Effects of Chemical Substances,* Superintendent of Documents, U.S. Government Printing Office, Washington, D. C. 20402. Most chemistry libraries have one or more reference volumes on the hazards of chemicals, and most chemical companies attempt to label the more hazardous chemicals as such in their catalogs, e.g., the *Aldrich Catalog/Handbook of Fine Chemicals,* Aldrich Chemical Company.

If the reaction involves *two* reactant species, A and B, in the transition state for the rate-determining step, we say that the reaction is **bimolecular.** For a bimolecular reaction, generally the rate equation will be:

$$\text{Rate of reaction} = k[\text{A}][\text{B}]$$

and the reaction is said to be **second order.** There are higher-order reactions and very complex rate equations, especially in biochemistry, but we need not be concerned with them here. The **molecularity** of a reaction is determined by the number of molecules, or other species, that are involved in the transition state for the rate-determining step, while the **kinetic order** of a reaction depends on the number of concentration terms in the rate equation. In special cases the numbers associated with the kinetic order and the reaction molecularity may appear to be different, but again we need not be concerned with these special cases in this text.

Chemists use reaction kinetics as a tool to aid them in establishing mechanisms of reactions. Actually, mechanisms are rarely, if ever, considered to have been proven correct beyond all doubt. We can only gather as much evidence as possible about a reaction, and if all of this evidence supports one proposed mechanism to the exclusion of others, the mechanism may be said to be accepted or established. Thus, if we study the kinetics of a reaction and find it to be a second-order reaction, this suggests that the reaction is bimolecular and has a mechanism in which two molecules, ions, or radicals are involved in the transition state. We will then seek other supporting evidence, such as the behavior of enantiomers when subjected to the reaction under study, as will be illustrated in the next section.

Exercise 6.4 Consider the reaction, A + B $\rightarrow$ [*transition state*] $\rightarrow$ C + D, which is known to be bimolecular and whose kinetics follow the second-order rate equation given above. Assume that at the start of the reaction the concentrations of A and B are equal; that is, [A] = [B]. (a) When the concentrations of A and B have dropped to one-half of their original values, will the rate of the reaction be (1) one-half of the original, (2) one-fourth of the original, or (3) the same as the original rate? Explain your answer. (b) Draw the potential energy–reaction coordinate diagram for the reaction.

6.5 Reactions of the Alkyl Halides

The reactions of the alkyl halides are principally of two types—**substitution** and **elimination.** The halogen atom in a substitution reaction is displaced by an electron-rich *nucleophilic* reagent (Sec. 1.13). Such nucleophilic reagents are Lewis bases and often are negative ions. Substitution reactions may take place by two different mechanisms. In one of these the rate of reaction depends upon the concentration of *one* reacting species, and the reaction is called a *substitution, nucleophilic, unimolecular,* or simply S_N1. In the other substitution mechanism, the rate of reaction depends upon the concentration of *two* reacting species, and the reaction is called a *substitution, nucleophilic, bimolecular,* or S_N2.

In S_N1 reactions the rate-determining step is the formation of a carbocation. Any factors such as steric effects (Sec. 7.4B), resonance stabilization (Sec. 1.14), or inductive effects (Sec. 9.3) that help to promote the formation of the carbocation will increase the rate of reaction. The reaction rate may be increased further through the assistance of a polar solvent, which can solvate both cation and anion. The ionic charges are then dispersed over solvent molecules as well as the ions, thereby assisting in ion formation. An example of a reaction that proceeds largely by way of the S_N1 mechanism is the hydrolysis of *tert*-butyl chloride in an 80% aqueous alcoholic solution (below).

The intermediate carbocation in the S_N1 reaction is planar (Sec. 1.12). The attacking nucleophile, water, may approach the positive carbon atom from either the front or the back. The product of hydrolysis of *tert*-butyl chloride by attack from either side is *tert*-butyl alcohol. Racemization will result if a chiral alkyl halide is hydrolyzed in an S_N1

manner. Let us illustrate this point by using one of the enantiomers of α-phenylethyl chloride:

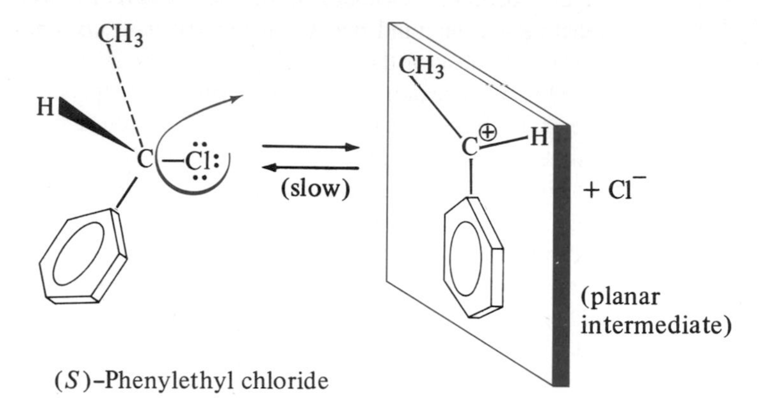

(*S*)-Phenylethyl chloride

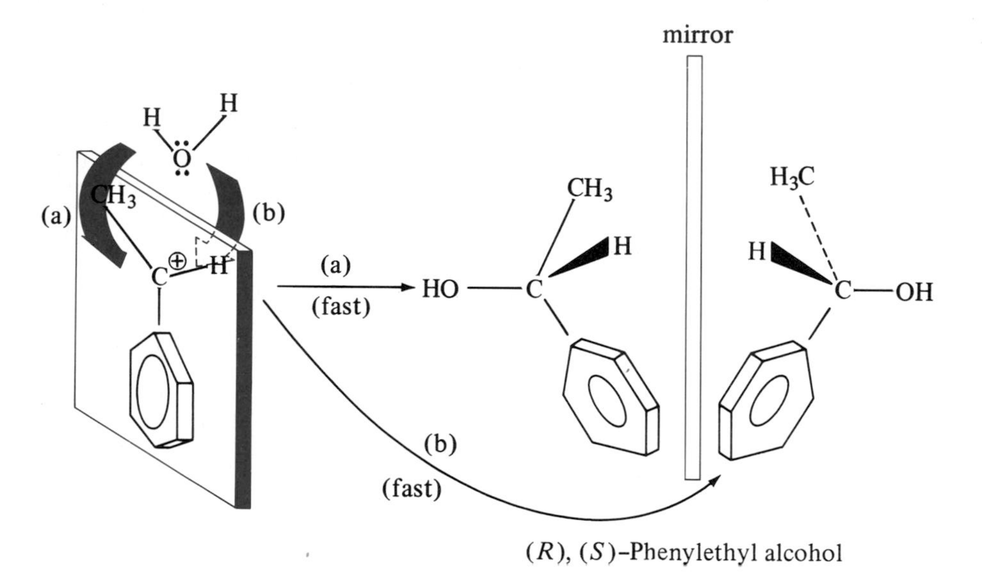

(*R*), (*S*)-Phenylethyl alcohol

Regardless of which enantiomer of α-phenylethyl chloride is hydrolyzed, the product will be either a racemic mixture or a partially racemized mixture. Racemization may not be complete to the extent that an equal number of (*R*)- and (*S*)-enantiomers of the product are formed, because the departing anion (chloride, in this case) may still be near enough to one face of the carbocation intermediate to interfere with the approach of the nucleophile from that side. The isomeric product resulting from the backside approach will predominate in those examples where racemization is not complete. However, racemization is characteristic of all S_N1 reactions.

Let us now return to the *tert*-butyl chloride hydrolysis reaction and see what has happened to the remainder of our starting material. In order to account for the other 17%, we must consider another possible reaction route. If, instead of combining with the nucleophilic reagent, the carbocation loses a proton (Route II) in the fast step, then the product of the reaction is an alkene, and the reaction mechanism is referred to as E1. As in the S_N1 mechanism, the number in the abbreviation refers to unimolecular. The E indicates elimination. Because of the competition between substitution and elimination, more than one product often results.

Route II (E1 mechanism)

$$(CH_3)_2C^{+}\text{—}CH_3 + H_2\ddot{O}: \xrightarrow{(fast)} (CH_3)_2C{=}CH_2 + H_3\overset{+}{O}$$

The S_N2 reaction mechanism may be illustrated by the alkaline hydrolysis of *sec*-butyl chloride to *sec*-butyl alcohol. In this case in the slow step of the reaction sequence, both *sec*-butyl chloride and hydroxide ion are involved. The hydroxide ion displaces the chloride ion, not by a head-on approach to the face of the carbon atom bonded to the chlorine atom, but by an attack from the rear. In making this backside approach, a new bond forms between the carbon atom and the hydroxyl group as the bond between the carbon atom and chlorine atom is broken. Thus, the reaction proceeds through a transition state in which the carbon atom is partially bonded to both the hydroxyl group and the chlorine atom. Transition states are often shown in brackets to emphasize that they are a transient, high-energy species.

As the transition state collapses to the products, the groups bonded to the carbon atom change their relative positions as shown, and the configuration becomes exactly opposite to that of the starting material. Such a reversal of configuration is referred to as a **Walden inversion** and invariably occurs in an S_N2 reaction. Thus, if we were to begin with (*S*)-*sec*-butyl chloride, we would obtain as our product (*R*)-*sec*-butyl alcohol. Indeed, it was through such stereochemical studies that the S_N2 mechanism was established.

Route I (S_N2 mechanism)

$$H\ddot{O}:^- + C(CH_3)(H)(C_2H_5)\text{—}\ddot{C}l: \longrightarrow \left[HO\cdots C(CH_3)(H)(C_2H_5)\cdots \ddot{C}l:\right] \longrightarrow HO\text{—}C(CH_3)(H)(C_2H_5) + Cl^-$$

partial bonds

(*S*)-*sec*-Butyl chloride (transition state) (*R*)-*sec*-Butyl alcohol

It should be pointed out that there is also an E2 reaction path that may enter into competition with the S_N2 mechanism through the elimination of a β-hydrogen atom. You will remember the dehydrohalogenation reactions used in the preparation of the alkenes (Sec. 3.4) as examples of such β-eliminations.

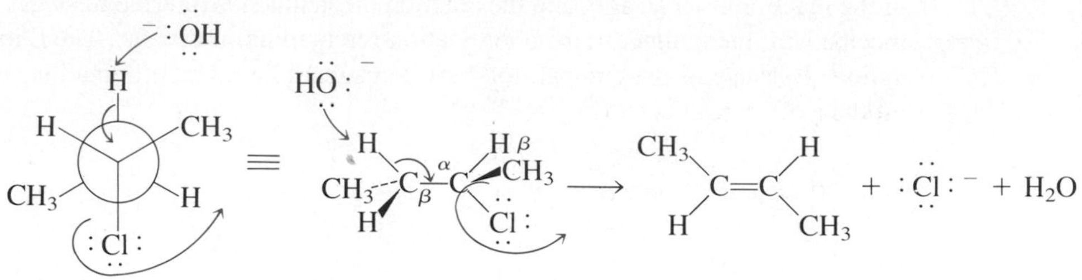

Newman projection sec-Butyl chloride trans-2-Butene

$$\longrightarrow \text{CH}_3\text{CH}{=}\text{CHCH}_3 + :\ddot{\text{Cl}}:^- + \text{H}_2\text{O}$$

The stereochemistry of E2 elimination is such that in the transition state the four atoms of the alkyl halide that are involved (hydrogen, two carbons, and halogen) must lie in the same plane, and the hydrogen and halogen must be *anti* to each other (Sec. 2.6). This arrangement is called anticoplanar. As shown, *trans*-2-butene is the principal product, although smaller amounts of both 1-butene and *cis*-2-butene are formed. The relative amounts of the three isomers will depend on the nature of the base and of the halogen—iodide, bromide, or chloride. As was pointed out in Sec. 3.4, the most stable alkene usually predominates, and the most stable alkene is generally that with the most substituents on the double bond. These observations suggest that the transition state for E2 elimination is stabilized by the same structural factors that stabilize alkenes.

EXAMPLE

From which diastereomer of 1-chloro-2-methylcyclohexane will the following be obtained as the principal product by dehydrohalogenation?

(a) 1-methylcyclohexene?

(b) 3-methylcyclohexene?

Solution Both (a) and (b) are the result of a β-elimination. Therefore, in each case we must draw the structures of the substituted cyclohexanes with the halogen and the hydrogen to be eliminated in an anticoplanar arrangement. We can do this by drawing the chair forms of each with the hydrogen to be eliminated and the chlorine on axial bonds and *trans* to each other.

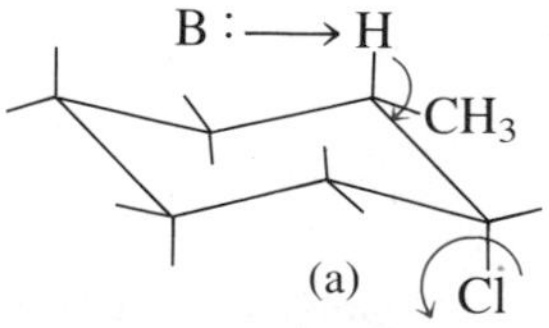

B:⟶H CH₃ (a) Cl

cis-1-chloro-2-methylcyclohexene

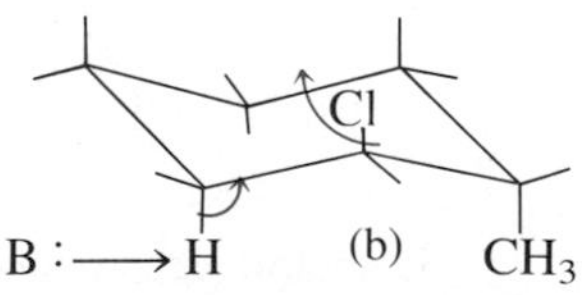

Cl B:⟶H (b) CH₃

trans-1-chloro-2-methylcyclohexene

Exercise 6.5 The dehydrohalogenation of 2-chlorobutane produces *trans*-2-butene as the predominant product as was illustrated above, but a small amount of *cis*-2-butene is also formed. Use a Newman projection to show the conformation of 2-chlorobutane that would yield *cis*-2-butene.

Which of the four reaction mechanisms predominates depends upon the concentration of reagents, the structures of the alkyl halides involved, the solvent used, and the nature of the nucleophile. Unbranched, primary halides react with good nucleophiles principally by the S_N2 route and *not* by S_N1. Only with strong, bulky bases do they give E2 reactions. Tertiary halides react, in the absence of strong base, by S_N1 and E1 routes, but elimination usually predominates, and tertiary halides are normally not used for substitution reactions. In the presence of strong bases, tertiary halides give good yields of elimination products by the E2 route. Secondary halides are intermediate in behavior and are less predictable. However, with good nucleophiles the S_N2 mechanism is favored, and with strong bases good yields of elimination products are formed by the E2 mechanism. The order of reactivity in elimination reactions is *tertiary* > *secondary* > *primary*.

The more important nucleophilic substitutions of alkyl halides are illustrated by the following general equations. The nucleophile in each case is underlined in color.

$$R{-}X + Na^+\underline{OH^-} \longrightarrow R{-}OH + Na^+X^-$$

Sodium hydroxide → An alcohol

$$R{-}X + Na^+\underline{OR^-} \longrightarrow R{-}OR + Na^+X^-$$

Sodium alkoxide → An ether

$$R{-}X + Na^+\underline{SH^-} \longrightarrow R{-}SH + Na^+X^-$$

Sodium hydrosulfide → A thio alcohol

$$R{-}X + Na^+\underline{CN^-} \longrightarrow R{-}CN + Na^+X^-$$

Sodium cyanide → An alkyl cyanide (nitrile)

$$R{-}X + Na^+\underline{NH_2^-} \longrightarrow R{-}NH_2 + Na^+X^-$$

Sodium amide → An amine

$$R{-}X + Na^+\underline{:C{\equiv}C{-}H} \longrightarrow R{-}C{\equiv}C{-}H + Na^+X^-$$

Sodium acetylide → An acetylene

$$R{-}X + \underline{:NH_3} \longrightarrow R\overset{+}{N}H_3 + X^-$$

Ammonia → An ammonium ion

$$R{-}X + \underline{:PR'_3} \longrightarrow R\overset{+}{P}R'_3 + X^-$$

A phosphonium ion

A generalization helpful in selecting the proper halogen compound to employ in a reaction is that the iodides are more reactive than the bromides, and the bromides are more reactive than the chlorides. Unfortunately, their cost parallels this order.

Exercise 6.6 What would be the stereochemical result if (*R*)-2-Chloro-3-methylbutane were heated with alcoholic KOH and the resulting product treated with bromine?

6.6 Reactions of Aryl, Vinyl, and Allyl Halides

The displacement reactions of the alkyl halides discussed in the previous section are not applicable to the aryl halides unless, of course, halogen is in the side chain. Halogen held to a doubly bonded carbon such as that found in chlorobenzene is unreactive. The ring halogen is not sufficiently labile to engage in displacement reactions similar to those illustrated for the alkyl halides. Only when the *ortho* and/or the *para* positions of the aromatic ring are occupied by strong electron-attracting groups can the ring halogen be displaced by a nucleophile. Electron-attracting groups in the *ortho* or *para* positions diminish the electron density at the halogen-attached ring carbon and make it susceptible to nucleophilic attack. For example, 2,4-dinitrochlorobenzene may be hydrolyzed in a potassium hydroxide solution to produce 2,4-dinitrophenol. Under the same conditions chlorobenzene fails to react.

$$\text{2,4-}(NO_2)_2C_6H_3Cl + KOH \longrightarrow \text{2,4-}(NO_2)_2C_6H_3OH + KCl$$

2,4-Dinitrochlorobenzene 2,4-Dinitrophenol

The ease with which the fluorine atom is displaced from 2,4-dinitrofluorobenzene has a practical application in peptide chemistry (Sec. 15.8).

Vinyl chloride, $CH_2{=}CH{-}Cl$, also has a chlorine joined to an unsaturated carbon atom and, like chlorobenzene, is rather unreactive. The electrons of chlorine can interact with the π-system of the carbon–carbon double bond to shift *away from* the chlorine, as shown in the following resonance structure. As a result, the chlorine atom is held somewhat closer to the carbon, the carbon–halogen bond is strengthened, and displacement of chlorine is made more difficult. Chlorine on the benzene ring represents a similar system, except that in this case the electrons of chlorine can interact with the π electrons of the benzene ring.

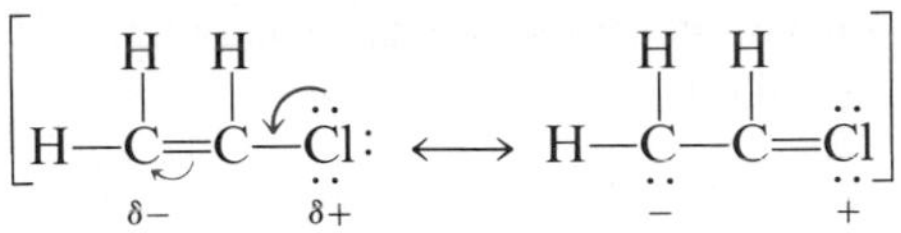

$$\left[\mathrm{H{-}\overset{H}{\underset{\delta-}{C}}{=}\overset{H}{C}{-}\underset{\delta+}{\ddot{C}l}:} \longleftrightarrow \mathrm{H{-}\overset{H}{\underset{-}{\ddot{C}}}{-}\overset{H}{C}{=}\underset{+}{\ddot{C}l}} \right]$$

Resonance forms of vinyl chloride

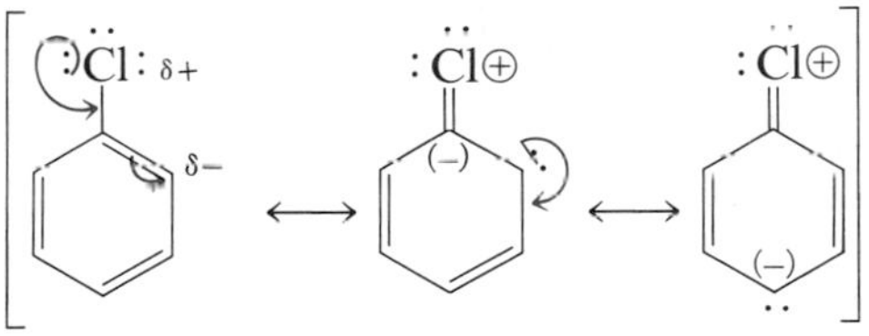

Contributing forms to the chlorobenzene structure

In contrast to vinyl halides, **allyl halides,** $CH_2{=}CH{-}CH_2X$, are extremely reactive. In allylic halides the halogen is on the carbon next to the double bond. A different factor now comes into play. The allyl cation that remains, once the halide ion has separated, is stabilized by resonance (Sec. 1.14).

$$CH_2{=}CH{-}CH_2{-}Cl \longrightarrow [CH_2{=}CH{-}\overset{\oplus}{C}H_2 \longleftrightarrow \overset{\oplus}{C}H_2{-}CH{=}CH_2] + Cl^-$$

Benzyl chloride, which has a similar allylic structure, reacts in a manner similar to that of allyl chloride.

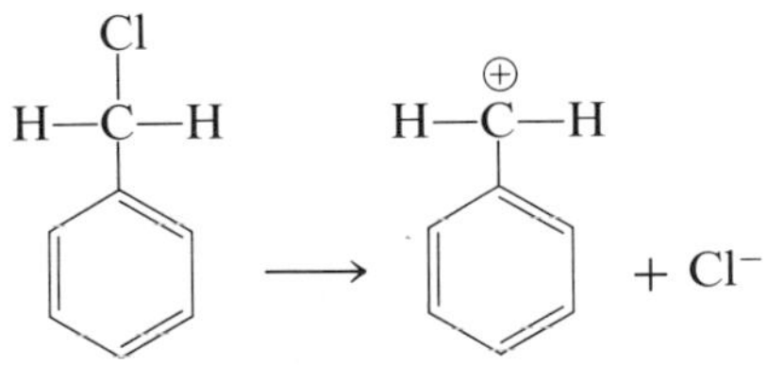

Benzyl chloride

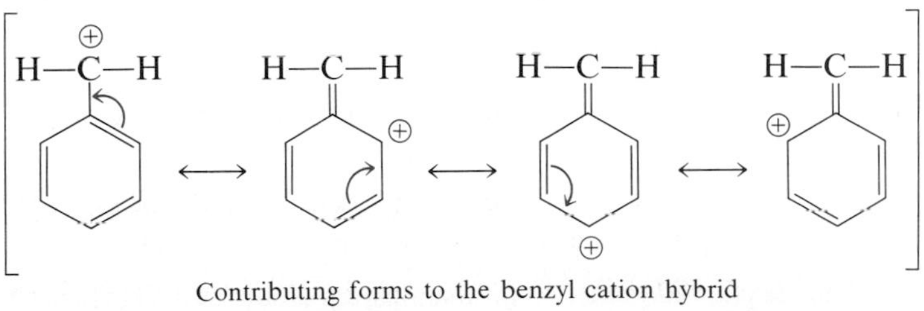

Contributing forms to the benzyl cation hybrid

Halogens on carbon atoms other than that next to the ring (the benzylic carbon) behave like simple alkyl halides.

Halogen compounds that react through ionic intermediates (tertiary, allyl, and benzyl) readily precipitate silver halide when warmed with alcoholic silver nitrate. This is a convenient test for highly reactive alkyl halides. Vinyl and aryl halides do not react under these conditions.

We now can generalize further the order of reactivity of halogen compounds in S_N1 reactions as follows:

$$\text{tertiary} > \text{allylic} > \text{benzylic} > \text{secondary} \gg \text{primary} > \text{vinyl}$$

For S_N2 reactions the order is:

$$\text{benzylic} > \text{allylic} > \text{primary} > \text{secondary} \gg \text{tertiary} > \text{vinyl}$$

Exercise 6.7 Which type of reaction mechanism, S_N1, S_N2, E1, or E2, is illustrated by each of the following reactions?

(a) $$CH_2{=}CH{-}CH_2Cl + AgNO_3 + C_2H_5OH \longrightarrow \underline{AgCl} + CH_2{=}CH{-}CH_2OC_2H_5 + HNO_3$$

(b) $$(CH_3)_3CCl + KOH \xrightarrow{\text{alcohol}} (CH_3)_2C{=}CH_2$$

(c) $$CH_3CH_2CH_2CH_2Cl + NaI \xrightarrow{\text{acetone}} CH_3CH_2CH_2CH_2I + NaCl$$

(d) $$CH_3{-}CH{=}CHCl + \overset{+}{Na}\overset{-}{N}H_2 \longrightarrow CH_3{-}C{\equiv}C{-}H + NH_3 + NaCl$$

6.7 Reactions of Halogen Compounds with Metals

The reaction of an alkyl halide with magnesium to form the important Grignard reagents has already been discussed (Sec. 2.13A); however, these reagents are sufficiently important that we will examine their formation and properties in somewhat greater detail. It is thought that the formation of Grignard reagents proceeds by the following mechanism.

$$R{-}X + Mg \longrightarrow R\cdot + X{-}Mg\cdot$$

$$R\cdot + X{-}Mg\cdot \longrightarrow R{-}Mg{-}X$$

The alkyl halide reacts with magnesium atoms at the surface of the metal to form an alkyl radical and an MgX· radical still tightly bound to the surface. Then the alkyl radical reacts with the MgX· radical to form the Grignard reagent, R—Mg—X. Although it is the custom to draw the Grignard reagent as though the metal were covalently bonded to the alkyl group, metals are electropositive, and in many organometallic compounds the carbon–metal bonds have a high degree of ionic character. Thus, dipolar resonance structures can be important contributors to the structures of organometallic compounds.

$$[R{-}M \longleftrightarrow R{:}^{-}M^{+}] \qquad (M = \text{metal})$$

The importance of the dipolar form will depend in part on the electronegativity (Sec. 1.8) of the metal: the more electropositive (less electronegative) the metal, the more ionic the bond. Organomagnesium compounds, such as Grignard reagents, are polar but still partially covalent. Organolithium compounds are much more ionic and are best regarded as lithium salts of the corresponding carbanions. Magnesium compounds have another interesting property: the metal does not have a full octet of electrons in the valence shell; therefore, in Grignard reagents the magnesium attempts to complete its octet by coordination with whatever electron donors are present, usually halide ions or solvent ether molecules. Thus, in solution the Grignard reagent may consist of the following species in proportions that will vary with the concentration.

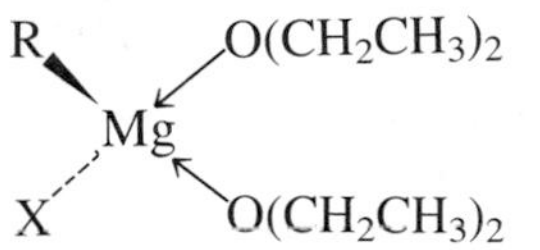

R, $O(CH_2CH_3)_2$, Mg, X, $O(CH_2CH_3)_2$

Monomeric Grignard reagent

R, X, $O(CH_2CH_3)_2$, Mg, Mg, $(CH_3CH_2)_2O$, X, R

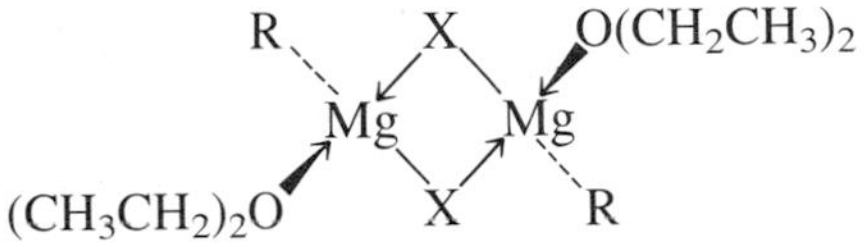

Dimeric Grignard reagent

With the exception of methyl iodide, the bromo compounds are usually employed for the preparation of Grignard reagents. Grignard reagents may be prepared from both alkyl and aryl halides. An example of a frequently prepared aryl Grignard reagent is phenylmagnesium bromide.

$$C_6H_5{-}Br + Mg \xrightarrow[\text{ether}]{\text{anhydrous}} C_6H_5{-}MgBr$$

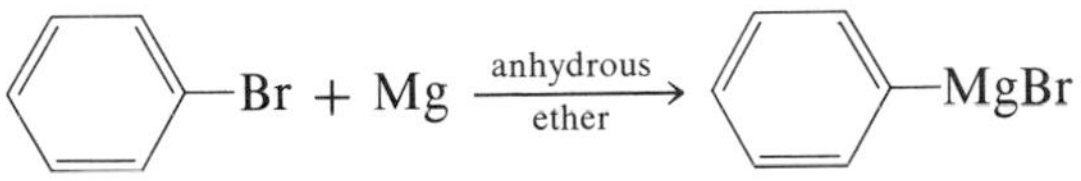

Bromobenzene → Phenylmagnesium bromide

The Grignard reagent is one of the most useful of all organometallic compounds and plays a very important role in numerous synthetic procedures. The various applications of the Grignard reagent are illustrated in later sections.

A few Grignard reagents require special procedures. Vinylmagnesium bromide is made from the halide, but in tetrahydrofuran (Sec. 7.9) rather than ethyl ether. Acetylenic Grignard reagents are made by an acid–base reaction between a terminal alkyne (a carbon acid) and any simple Grignard reagent (a carbanionlike base).

$$CH_2{=}CH{-}Br + Mg \xrightarrow[\text{tetrahydrofuran}]{} CH_2{=}CH{-}MgBr$$

Vinyl bromide → Vinylmagnesium bromide

$$H{-}C{\equiv}C{-}H + CH_3CH_2{-}MgBr \longrightarrow H{-}C{\equiv}C{-}MgBr + CH_3{-}CH_3$$

Ethyne + Ethylmagnesium bromide → Ethynylmagnesium bromide + Ethane

The preparation of organolithium reagents and lithium dialkylcuprates was described in Sec. 2.13. Further aspects of their chemistry will be discussed in Sec. 8.12.

Summary

1. Halogen compounds may be classified as:
   (a) Aliphatic (R—X) or aromatic (Ar—X).
   (b) Primary, secondary, or tertiary halogen compounds.
   (c) Allylic, benzylic, and vinylic.
   (d) Iodo, bromo, chloro, or fluoro compounds.
2. Halogen compounds can be prepared by
   (a) direct halogenation (most useful in the preparation of aromatic halogen compounds).
   (b) replacement of the hydroxyl group (—OH) of alcohols.
   (c) addition of HX to alkenes (Markovnikov's rule is followed).
   (d) exchange for other covalently bound halogens. Most useful in the preparation of iodo and fluoro compounds.
   (e) the haloform reaction that yields trihalomethanes.
3. Reactions of the halogen compounds include
   (a) displacement of the halogen atom as halide by another negative group—i.e., hydroxy (OH^-), alkoxy (OR^-), amino (NH_2^-), mercapto (SH^-), cyano (CN^-), acetylide ($H—C{\equiv}C^-$).
   (b) elimination (dehydrohalogenation) to form alkenes.
   (c) reactions with metals: preparation of the Grignard reagent.
4. Important polychloro compounds include chloroform, carbon tetrachloride, and polyvinyl chloride.
5. The polyfluoromethanes, ethanes, and cyclobutanes known as "Freons" are useful refrigerants. "Teflon" (polyfluoroethylene) is a very useful inert, heat-resistant plastic.
6. The polyhalogenated compounds, such as DDT, aldrin, and chlordane, are powerful pesticides but are not easily removed from the biosphere.

Supplementary Exercises

6.8 Name each of the following compounds.

(a) $CH_2{=}CH—CH_2Br$

(b) CH_3MgI

(c) $CH_3—\underset{Br}{\overset{H}{C}}—CH_3$

(d) HCI_3

(e) Cl / NO_2

(f) Cl / CH_3

```
        F   F
        |   |
    F—C—C—F
(g)     |   |
    F—C—C—F
        |   |
        F   F
```

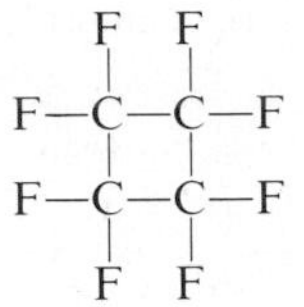

(h) 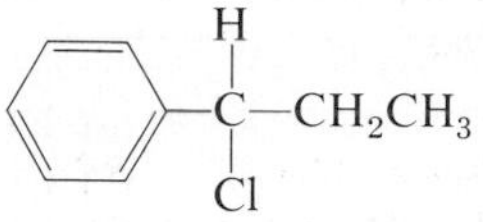 C_6H_5–CH(Cl)–CH_2CH_3 (i) 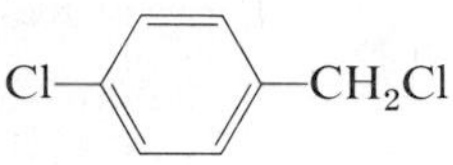 Cl–C_6H_4–CH_2Cl

6.9 Write structures for and assign acceptable names to all isomers with the molecular formula (a) C_4H_9Br, (b) $C_4H_8Cl_2$. Are any of the structures you have drawn those of optically active compounds?

6.10 Indicate the principal organic products (if any) given by the following reactions.

(a) *tert*-C_4H_9OH + HCl $\longrightarrow$
(b) $CH_3CH_2CH{=}CH_2$ + HBr $\xrightarrow{\text{peroxide}}$
(c) Bromobenzene + Mg $\xrightarrow{\text{anhydrous ether}}$
(d) Product of (a) + alcoholic KOH $\longrightarrow$
(e) Product of (b) + aqueous KOH $\longrightarrow$
(f) Product of (e) + I_2 + NaOH $\longrightarrow$
(g) Chlorobenzene + alcoholic KOH $\longrightarrow$
(h) Benzoic acid + Br_2 $\xrightarrow{\text{Fe}}$
(i) *cis*-2-Butene + Br_2 $\longrightarrow$
(j) 2,4-Dinitrochlorobenzene + NH_3 $\longrightarrow$
(k) *n*-C_4H_9Cl + NaI $\xrightarrow{\text{acetone}}$
(l) Ethylbenzene + Cl_2 $\xrightarrow{\text{sunlight}}$
(m) *n*-Butyl chloride + benzene $\xrightarrow{AlCl_3}$
(n) Ethyl bromide + sodium acetylide $\longrightarrow$
(o) Ethyl bromide + sodium methoxide $\longrightarrow$
(p) (*R*)-2-Bromobutane + NaOH $\xrightarrow{S_N2}$
(q) Isopropylbenzene + Cl_2 $\xrightarrow{\text{uv light}}$
(r) 2-Methyl-2-butene + HBr $\longrightarrow$
(s) 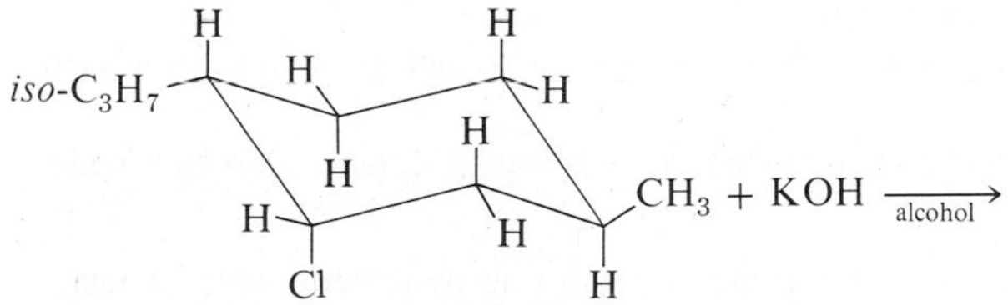 *iso*-C_3H_7–(cyclohexane chair bearing H, H, H, H, H, H, H, Cl, H, CH_3, H) + KOH $\xrightarrow[\text{alcohol}]{}$

6.11 Show how the following transformations can be effected.

(a) isopropyl alcohol → isopropyl bromide → propane
(b) benzene → ethylbenzene → styrene
(c) isopropyl bromide → *n*-propyl bromide
(d) 1,2-dibromopropane → 2,2-dibromopropane
(e) toluene → *p*-bromotoluene → *p*-bromobenzoic acid

6.12 Give the structure of the major product expected, if any, when *n*-propyl chloride is treated with each of the following reagents.

(a) NaOH, H_2O
(b) Mg, anhydrous ether
(c) cold, concentrated H_2SO_4
(d) benzene, anhydrous $AlCl_3$
(e) KI, acetone
(f) $CH_3—C≡C:^-Na^+$
(g) KOH, C_2H_5OH
(h) NH_3

6.13 Complete the table below by making a comparison between S_N1 and S_N2 reaction mechanisms with respect to each of the characteristics in the left column.

	S_N1	S_N2
(a) Molecularity	______	______
(b) Stereochemistry	______	______
(c) Relative rates of reaction of primary, secondary, tertiary alkyl halides	______	______
(d) Effect of doubling concentration of nucleophile	______	______
(e) Effect of increasing polarity of solvent	______	______

6.14 What simple chemical test, i.e., test tube reaction, will serve to distinguish between:

(a) cyclohexene and chlorobenzene
(b) α-phenylethyl chloride and β-phenylethyl chloride
(c) chlorobenzene and 2,4-dinitrochlorobenzene
(d) *p*-bromotoluene and benzyl bromide

6.15 A halogen compound contained 58% bromine. When the compound was heated with alcoholic potassium hydroxide, a combustible gas evolved. The gas, when hydrogenated in the presence of a catalyst, absorbed one mole of H_2. On combustion the gas united with oxygen in a ratio of 1:6 by volume (STP) to give carbon dioxide and water as the only products. When this same gas was subjected to chemical oxidation with hot, concentrated $KMnO_4$, only acetic acid, CH_3COOH, was obtained as the oxidation product. Give the formula for the original compound.

6.16 An analysis of an organic compound established its formula as C_8H_9Cl. What would be its structure and name if it had shown the following behavior:

(a) It was optically active, precipitated AgCl when warmed with an alcoholic silver nitrate solution, and on oxidation yielded benzoic acid.
(b) It was optically inactive, precipitated AgCl when warmed with alcoholic silver nitrate, and on oxidation yielded phthalic acid, $C_6N_4(COOH)_2$.
(c) It was optically inactive and on oxidation yielded *p*-chlorobenzoic acid.

6.17 When *cis*-3-methylcyclopentyl chloride is heated with potassium hydroxide, only *trans*-3-methylcyclopentanol is formed.

(a) Draw stereochemical structures for both starting material and product.
(b) Did this displacement proceed by the S_N1 or the S_N2 reaction pathway? Explain.

6.18 When *tert*-butyl bromide is hydrolyzed in an ethanolic solution, three products result. What are they and which of the three would you expect to predominate? Why?

6.19 When ethyl bromide is heated with a solution of potassium hydroxide in ethyl alcohol, the principal product is ethyl ether.

$$CH_3—CH_2—Br + KOH \xrightarrow{CH_3—CH_2—OH} CH_3—CH_2—O—CH_2—CH_3$$

Explain this result. What does this result tell you about the nucleophilic species actually present in alcoholic solutions of potassium hydroxide?

6.20 The reaction of *trans*-2-chlorocyclohexanol, $C_6H_{11}ClO$, with aqueous sodium hydroxide solution gives a product having the formula $C_6H_{10}O$. Suggest a structure for this product and a mechanism for its formation. (*Hint:* The preceding exercise may be of some help.)

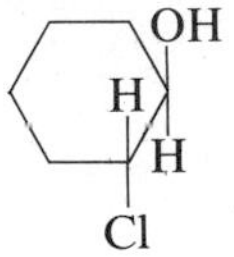

trans-2-chlorocyclohexanol

6.21 The base-catalyzed E2 elimination of *one* molecule of hydrogen bromide from (*R*,*R*)-1,2-dibromo-1,2-diphenylethane gives only 1-bromo-*trans*-1,2-diphenylethene, while (*R*,*S*)-1,2-dibromo-1,2-diphenylethane gives 1-bromo-*cis*-1,2-dephenylethene. Explain these results.

6.22 One mechanism for the substitution reactions of aryl halides is an *addition-elimination* mechanism. First the nucleophile is added, followed by the elimination of halide. This bimolecular mechanism is thought to involve an intermediate sigma complex, one of whose resonance forms is shown. Write structures for the important resonance forms of the sigma complex that would be formed from 2,4-dinitro-1-chlorobenzene and hydroxide ion. Using these structures, explain the greater reactivity of 2,4-dinitro-1-chlorobenzene compared to chlorobenzene.

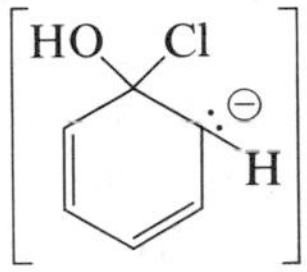

6.23 Halides of the type shown below show little tendency to react by S_N1, S_N2, E1, or E2 reactions. Suggest an explanation for the failure of S_N1 and S_N2 substitutions. (*Hint:* Looking at a model may help.)

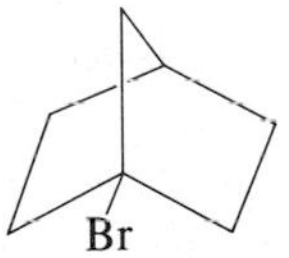

+ good nucleophile $\xrightarrow[S_N2]{S_N1 \text{ or}}$ no reaction

***6.24** An organic compound, C_3H_6ClI, gave an nmr spectrum with a triplet at δ 3.6, another triplet at δ 3.3, and a quintet at δ 2.2. All coupling constants were approximately 7 Hz, and the integral indicated that each multiplet represented two hydrogen atoms. Suggest a structure for the compound.

***6.25** An organic compound, $C_5H_8Br_4$, gave an nmr spectrum in which there was only one peak, a singlet, at δ 3.3. Suggest a structure for the compound.

6.26 From the discussion of S_N1 and S_N2 reactions of alkyl halides, it should be clear that the S_N1 reaction involves the sequence, alkyl halide $\rightarrow$ transition state $\rightarrow$ carbocation $\rightarrow$ transition state $\rightarrow$ product, whereas the S_N2 reaction involves the sequence, alkyl halide $\rightarrow$ transition state $\rightarrow$ product. Based on these sequences, draw approximate reaction coordinate diagrams (Section 1.12) for S_N1 and S_N2 reactions.

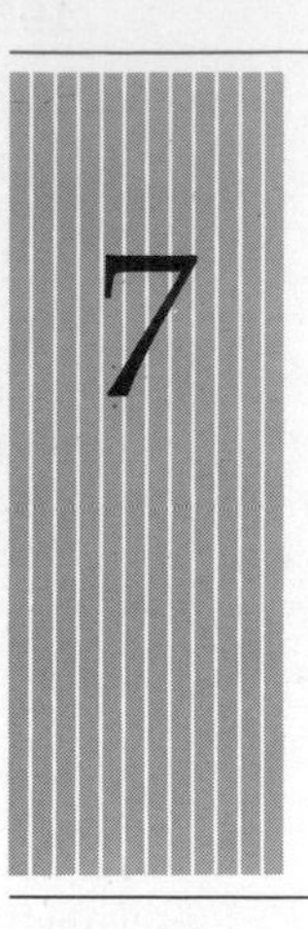

7 Alcohols, Phenols, and Ethers

Alcohols, phenols, and ethers represent three classes of oxygen-containing compounds in which the oxygen atom is singly bonded (—O—) to two other atoms. The oxygen atom bridges carbon to hydrogen in both the alcohols and the phenols. In ethers the oxygen is the bridge between two carbon atoms. All three classes of compounds provide us with a great number of useful products. These include germicides, antifreeze agents, pharmaceuticals, explosives, solvents, anesthetics, and plastics.

7.1 Structure of the Alcohols and Phenols

Alcohols and phenols may be considered as hydroxyl-substituted hydrocarbons of the general formulas R—OH and Ar—OH, respectively. The **hydroxyl group** (—OH) is the functional group that characterizes both alcohols and phenols. Compounds that have hydroxyl groups joined to carbon atoms of alkyl groups are alcohols. Compounds that have hydroxyl groups joined to carbon atoms of the aromatic ring compounds are phenols. Both types may be illustrated by the following general formulas and specific examples.

$R—\ddot{O}:—H$	$CH_3—OH$	$CH_3CH_2—OH$
An alcohol	Methyl alcohol	Ethyl alcohol

$Ar—\ddot{O}:—H$	$C_6H_5—OH$	$C_{10}H_7—OH$
A phenol	Phenol	β-Naphthol

7.2 Classification and Nomenclature of Alcohols and Phenols

The alcohols, like the alkyl halides, may be classified as *primary, secondary,* or *tertiary* according to the number of hydrocarbon groups attached to the carbon atom bearing the hydroxyl group. However, the nomenclature of the alcohols is somewhat more extensive than that encountered in other families of substances. **Common,** or **trivial, names** usually are employed for the simpler members having one to four carbon atoms. Such names are formed simply by naming the alkyl group bonded to the hydroxyl function, followed by the word *alcohol.* The following examples are illustrations of common names.

CH_3—OH
Methyl alcohol
(A primary alcohol)

CH_3—CH(CH_3)—OH
Isopropyl alcohol
(A secondary alcohol)

CH_3—C(CH_3)(CH_3)—OH
tert-Butyl alcohol
(A tertiary alcohol)

The IUPAC system of nomenclature is better adapted to naming the more complex members of the alcohol family, for which prefixes such as *secondary* and *tertiary* have little significance. Alcohols are named according to IUPAC rules by selecting and naming the longest carbon chain *including the hydroxyl group*. The terminal **e** of the parent alkane is replaced by **ol.** As before, the chain is numbered to confer upon the functional group the smallest number, even if carbon–carbon double bonds or triple bonds are present. If more than one hydroxyl group appears on the chain, prefixes such as **di, tri,** etc., are used. Alkyl side chains and other groups are named and their positions indicated. It would be profitable to consider a few examples.

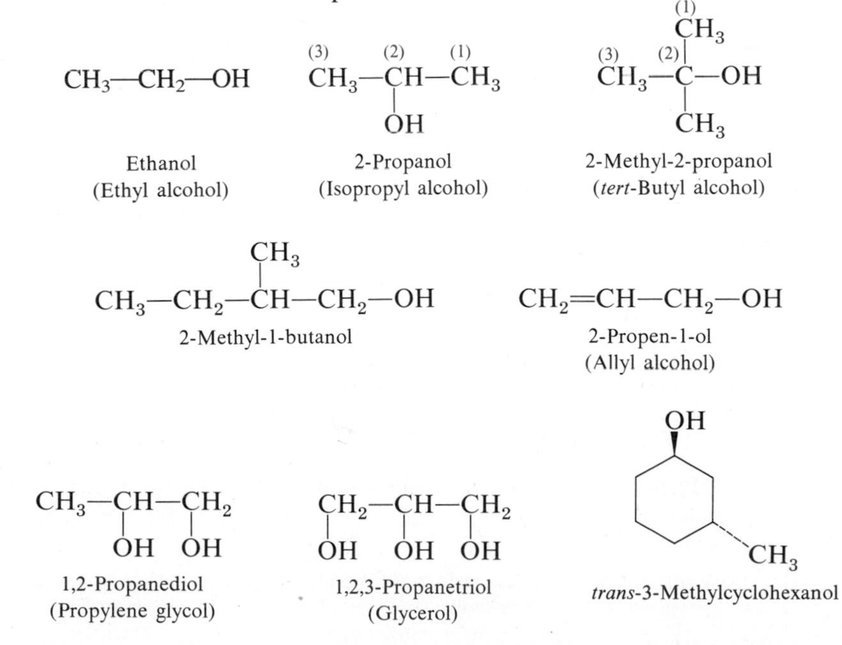

Ethanol
(Ethyl alcohol)

2-Propanol
(Isopropyl alcohol)

2-Methyl-2-propanol
(*tert*-Butyl alcohol)

2-Methyl-1-butanol

2-Propen-1-ol
(Allyl alcohol)

1,2-Propanediol
(Propylene glycol)

1,2,3-Propanetriol
(Glycerol)

trans-3-Methylcyclohexanol

EXAMPLE

Name the following structure according to IUPAC Rules.

$$CH_3CH_2CH_2{-}\overset{\displaystyle CH_2OH \atop |}{\underset{| \atop \displaystyle CH_2OH}{C}}{-}CH_3$$

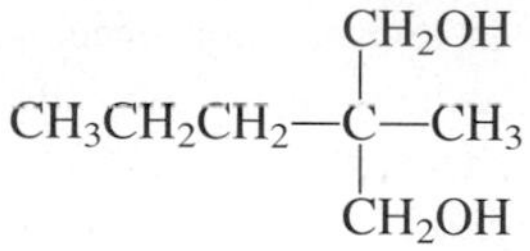

Solution We select the longest carbon chain *containing the functional grouping*. The name is 2-methyl-2-propyl-1,3-propanediol.

The suffix **ol** is generic for compounds that contain hydroxyl groups. Although names such as cres**ol,** glycer**ol,** and cholester**ol** contain no clues to their structures, such names do indicate that each contains one or more hydroxyl groups.

A third system of nomenclature is called a **derived system.** According to this system, alcohols are considered to be derivatives of the simplest one—that is, hydroxy methane, or **carbinol.** Groups attached to the carbon atom bearing the hydroxyl are named as one word with the suffix carbinol. This system of nomenclature is seldom used except to name the phenyl-substituted methanols.

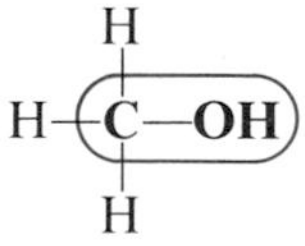

Carbinol (Methyl alcohol) Diphenylcarbinol 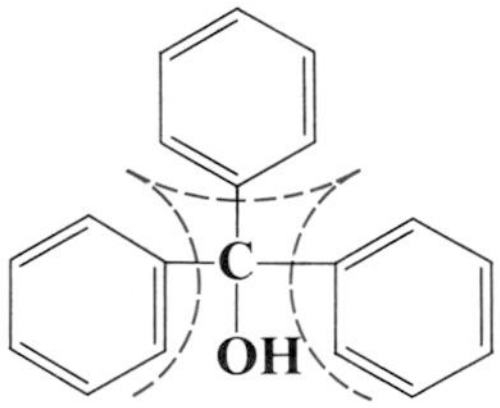 Triphenylcarbinol

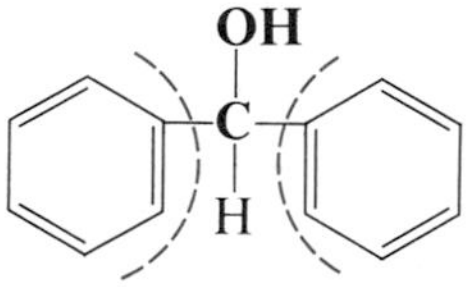

Exercise 7.1 Draw structures for eight alcohols with a molecular formula $C_5H_{12}O$ and name them according to IUPAC rules. Classify each as primary, secondary, or tertiary.

Exercise 7.2 Which of the eight alcohols in Exercise 7.1 are chiral?

Phenols are commonly named as derivatives of the parent substance, and simplest member of the family, **phenol.** Other substituents on the phenol ring are located by number or by *ortho, meta,* or *para* designations. In *Chemical Abstracts* substituted phenols are named as derivatives of benzenol; however, at present these names are not commonly used elsewhere and will not be used in this text (except in square brackets to illustrate the system).

p-Bromophenol [4-Bromobenzenol]

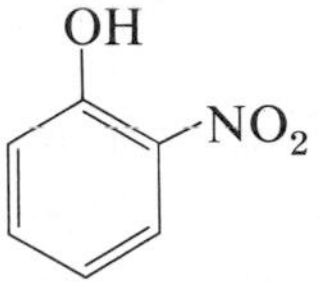

o-Nitrophenol [2-Nitrobenzenol]

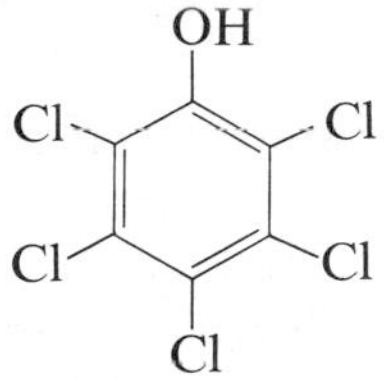

Pentachlorophenol [2,3,4,5,6-Pentachlorobenzenol]

You may have observed that, in the IUPAC system, some functional groups, such as the double and triple bonds, are named only with suffixes and others, such as the halogens and nitro compounds, only with the prefixes. Most of the other functional groups can be named with either prefixes or suffixes. If only one such functional group is present, the suffix name is used. When more than one functional group in a molecule can be named with either a prefix or a suffix, the functional group with the "highest priority" is named with its suffix name, and all others are named with their prefix names. The order of priorities is not the same as that used in the (*R*),(*S*)- and (*E*),(*Z*)-systems of nomenclature, but one that evolved with organic chemistry. For the purposes of this text the following priority listing (or that inside the front cover) should suffice:

carboxylic acids > aldehydes > ketones > alcohols > phenols > amines

The prefix name for the alcohols and phenols is **hydroxy.** The same *prefix* names and priorities are often employed in both the IUPAC and the common systems of nomenclature.

A number of important phenols and alcohols have nonsystematic common names, whose use, although discouraged, persists. Examples of these and the hydroxy system of nomenclature follow.

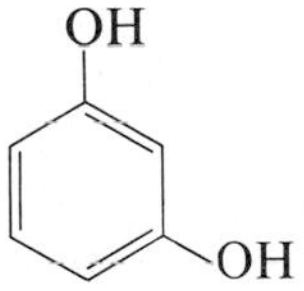

Resorcinol [1,3-Benzenediol]

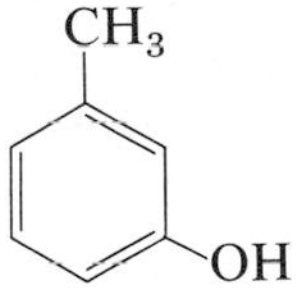

m-Cresol (3-Methylphenol)

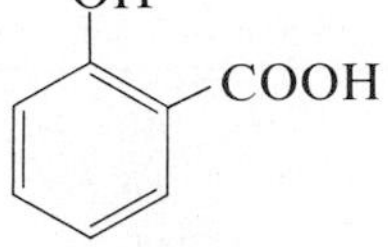

Salicylic acid (*o*-Hydroxybenzoic acid)

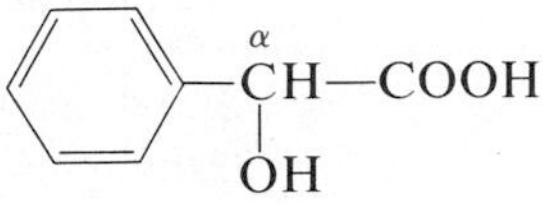

Mandelic acid (α-Hydroxyphenylacetic acid)

7.3 Structure and Properties of the Alcohols and Phenols. Hydrogen Bonding. Acidity

Molecules in which hydrogen is bonded to a highly electronegative element such as fluorine, oxygen, or nitrogen may exhibit a dipole, the positive end of which is a relatively exposed hydrogen nucleus. Because of such exposure, a strong attractive force develops between the hydrogen atom of one molecule and the electronegative element of a neighboring molecule. Water in the solid and liquid states exhibits such a dipole, the hydrogen atoms of one molecule being attracted to the oxygen atoms of another. This type of intermolecular attraction between hydrogen and a donor atom is called **hydrogen bonding,** or **H-bonding,** and is represented between molecules by dotted lines. Liquids in which H-bonding occurs between molecules are called *associated* liquids and have properties unlike those of ''normal'' liquids.

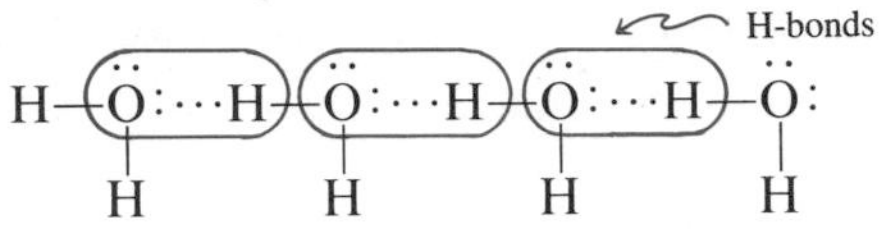

H-bonds

H—Ö:···H—Ö:···H—Ö:···H—Ö:
  | | | |
  H H H H

The alcohol molecule, like water, is highly polar; and though the alcohol molecule has but one hydrogen attached to oxygen, it also is associated through hydrogen bond formation. Molecules capable of H-bonding are attracted to each other by the electrostatic interaction of their dipoles and the affinity of their hydrogens for a second electronegative atom.

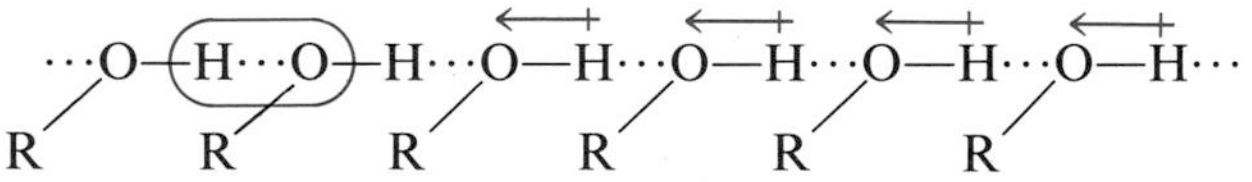

···O—H···O—H···O—H···O—H···O—H···O—H···
R R R R R R

To pass from the liquid into the vapor form, a molecule of alcohol, like one of water, must receive sufficient energy to overcome the attraction of its neighbors. Hydrogen bonding thus explains why the boiling point of water is relatively high for such a low-molecular weight molecule and also explains why the alcohols have considerably higher boiling points than their nonhydroxylic isomers. For example, the boiling point of ethyl alcohol, C_2H_5OH, is 78°C, and that of its isomer, dimethyl ether, $CH_3—O—CH_3$ (incapable of H-bonding), is −24.9°C. Comparisons made in Table 7.1 illustrate how H-bonding, more than an increase in molecular weight, affects the boiling points of liquids.

TABLE 7.1 Boiling Points of Some Substituted Ethanes

Compound	Molecular Wt.	bp, °C
Ethyl alcohol	46	78.4
Ethyl chloride	64.5	12.2
Ethyl bromide	109	38.0
Ethyl iodide	156	72.4

The lower members of the alcohol family are like water in yet another respect. Inasmuch as the hydroxyl group comprises much of the molecular structure, the lower members of the alcohol family are miscible with water in all proportions. The alcohol molecule becomes more and more oil-like in character as the hydrocarbon segment becomes larger, and its solubility in water diminishes markedly.

Phenol is a colorless, low-melting solid that is moderately soluble in cold water and totally soluble in hot water. Other phenols and polyhydroxy benzenes have similar properties but are usually much less soluble in water, their properties being influenced by their substituents and molecular weight. Although the alcohols have about the same or somewhat less acidity than water, phenols are much more acidic than water, but much less acidic than the mineral acids such as sulfuric and hydrochloric acids. Phenols are also much less acidic than the organic acids such as acetic. In fact, on the acidity scale, phenol is about halfway between ethyl alcohol and acetic acid (Table 7.2). Because of their acidity, phenols react readily with aqueous hydroxide bases to form salts called phenoxides. Alcohols do not react with hydroxide bases.

$$C_6H_5OH + NaOH \longrightarrow C_6H_5O^-Na^+ + H_2O$$

Sodium phenoxide

TABLE 7.2 Acidities of Some Acids

Compound	Formula	Acid Ionization Constant
Hydrochloric acid	HCl	1.6×10^{2}
Acetic acid	CH_3COOH	1.8×10^{-5}
Phenol	C_6H_5OH	1.0×10^{-10}
Water	HOH	2.0×10^{-16}
Ethyl alcohol	CH_3CH_2OH	1.3×10^{-16}

Relating the relative acidities of organic acids to their structures is not always easy, because the structures of acids affect their acidity in several, sometimes conflicting, ways. Furthermore, acidity depends not only on structure but also on the effects of solvent molecules. However, one thing is certain: the stronger the acid, the weaker its conjugate base, and vice versa. Often we can gain a better understanding of the relative acidities of two types of acids by comparing the strengths of their conjugate bases. We can do this by studying the effects of structure on the stability of the conjugate bases; that is, the anions, relative to the stability of the acids. Structural effects that influence acidity include resonance effects, other electronic effects, stereochemical effects, and steric effects. Consider, for example, resonance effects. Resonance that stabilizes the anion of an acid more than it stabilizes the acid itself results in a stronger acid than would otherwise be expected. Thus, phenol and its conjugate base, phenoxide ion, are stabilized by the following resonance structures:

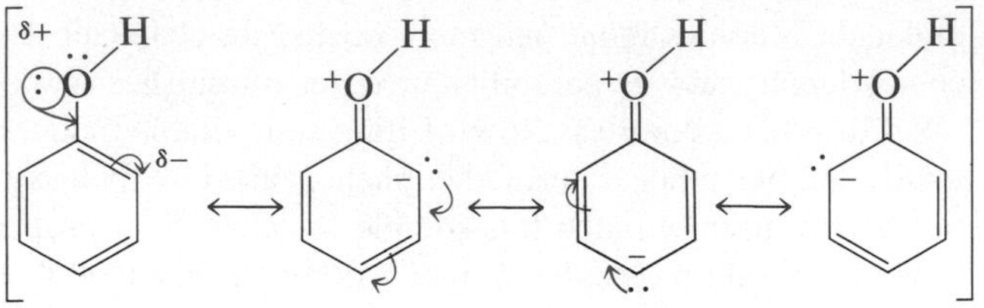

Resonance forms of phenol

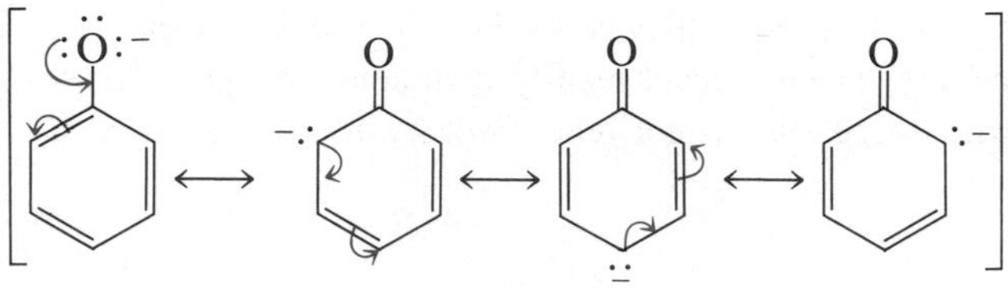

Resonance forms of phenoxide ion

Resonance stabilizes phenoxide ion somewhat more than it does phenol, because there is no separation of positive and negative charges in the phenoxide resonance forms (Sec. 1.14). Separation of charges requires energy to keep them apart and is a destabilizing effect. There is no resonance stabilization of either an alcohol or its conjugate base, the alkoxide ion. Therefore, relative to their respective acids, the phenoxide ion is more stable than the alkoxide ion. We conclude from this that phenoxide is more stable, less basic, than alkoxide ion and that phenol is a stronger acid than an alcohol. This is seen somewhat more easily in the following simple energy plot, where we have arbitrarily put phenol and the alcohol at the same level, since we are only interested in the differences in stability between each acid and its anion.

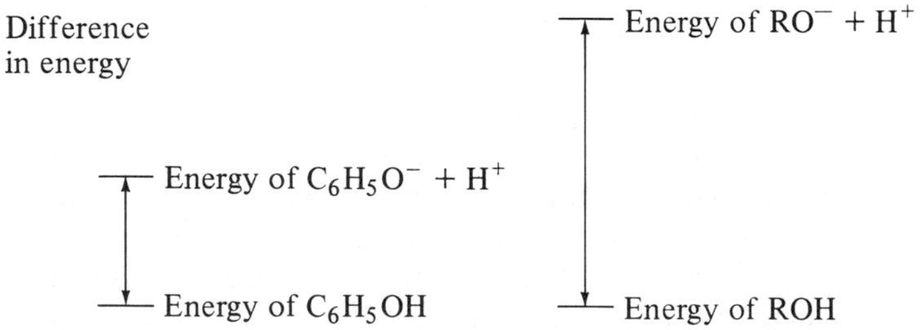

Difference in energy

Energy of $RO^- + H^+$

Energy of $C_6H_5O^- + H^+$

Energy of C_6H_5OH

Energy of ROH

Among the alcohols, the higher alcohols tend to be less acidic than methanol and ethanol, mainly because the alkyl groups of the larger alkoxide ions shield the negative charges of the anions and prevent them from being stabilized by solvation. Solvation distributes a negative charge and makes it less concentrated. The alkoxide ions of the higher alcohols are also destabilized somewhat by another electronic effect called the inductive effect, which we will discuss in Chapter 9.

Phenols are not sufficiently acidic to react with sodium carbonate, but will react with the strong bases sodium and potassium hydroxide to yield the phenoxide ions of these metals. This behavior allows other stronger organic acids that react with sodium carbonate to be separated as water-soluble salts from a mixture that might also include phenol. If strong electron-withdrawing groups are substituted in the *ortho* or *para* position of phenol, the substituted phenol is strongly acidic.

Exercise 7.3 Which of the following compounds would be most acidic? Least acidic? Why? (a) ethyl alcohol, (b) *tert*-butyl alcohol, (c) phenol, (d) *o*-nitrophenol, (e) *o*-cresol, (f) *p*-chlorophenol, (g) 2,4-dinitrophenol. Arrange in order of diminishing acidity.

7.4 Preparation of the Alcohols

Before we consider general methods for the preparation of the alcohols, let us discuss briefly the very important first two alcohols of the aliphatic series.

Methanol (methyl alcohol or ''wood alcohol'') was made prior to 1923 by the destructive distillation of hardwoods. Since then, synthetic methanol has been made from synthesis gas (Sec. 2.18), carbon monoxide, and hydrogen under high pressure in the presence of catalysts.

$$CO + 2\,H_2 \xrightarrow[260°,\ 150\ \text{atm.}]{\text{Cu catalyst}} \underset{\text{Methanol}}{CH_3OH}$$

About 8.5 billion pounds of methanol were made in the United States in 1985. Since procedures are known for the conversion of methanol into alkenes and gasoline feedstocks, methanol may become one of our most important organic chemicals in the years ahead.

Methanol is a very poisonous substance which, if taken internally, causes visual impairment, complete blindness, or death. Death from the ingestion of as little as 30 ml of methanol has been reported. It is to prevent such tragedies that the less suggestive IUPAC name, methanol, is used by the chemical industry, instead of methyl alcohol or wood alcohol.

Ethanol is sometimes called **grain alcohol** because starch from grain, when hydrolyzed to sugars and fermented by enzymes, produces ethanol and carbon dioxide. Starch from any source is a suitable starting material. The fermentation of sugar by yeast is a reaction practiced since antiquity and is the basis for the production of alcoholic spirits and for the leavening action required in the baking process.

$$\underset{\text{Starch}}{2\,(C_6H_{10}O_5)_n} + n\,H_2O \xrightarrow[\text{in malt}]{\text{diastase}} \underset{\text{Maltose}}{n\,C_{12}H_{22}O_{11}}$$

$$C_{12}H_{22}O_{11} + H_2O \xrightarrow{\text{maltase}} \underset{\text{Glucose}}{2\,C_6H_{12}O_6}$$

$$C_6H_{12}O_6 \xrightarrow{\text{zymase}} \underset{\text{Ethanol}}{2\,C_2H_5OH} + 2\,CO_2$$

Inasmuch as the second step in the preceding sequence produces glucose, ethanol also may be prepared from this simple sugar directly. Grape juice, a rich source of glucose, will ferment to produce a wine with a maximum alcoholic content of approximately 12% by volume. The alcoholic content of liquors is usually designated as **proof spirit,** 100 proof indicating an alcoholic content of 50% by volume. The term "proof spirit" supposedly had its origin in an early and rather crude analytical procedure for determining the alcoholic content of whiskey. Whiskey of high alcoholic content, when poured onto a small mound of gunpowder and ignited, would burn with a flame sufficiently hot to ignite the powder. This was "proof" of spirit content. If the gunpowder failed to ignite, the presence of too much water was indicated, and the powder became too wet to burn.

Ethanol used in the laboratory for solvent purposes is seldom pure alcohol. It is usually a mixture of 95% alcohol and 5% water. Ninety-five per cent represents the maximum purity obtainable when alcohol is distilled because this is the constant-boiling composition. A constant-boiling mixture of liquids, called an **azeotrope,** cannot be separated by fractional distillation. In order to obtain **absolute,** or 100% pure, ethanol the water must be removed by methods other than fractionation. One method is to distill a ternary mixture composed of alcohol, water, and benzene. These three liquids, in a composition of 18.5%, 7.4%, and 74.1%, respectively, also form an azeotrope with a constant-boiling temperature of 64.85°. Therefore, if sufficient benzene is added to 95% alcohol and the mixture is distilled, the water is removed in the distillate along with benzene and some alcohol, but pure alcohol is left in the still pot.

In the laboratory the last traces of water may be removed by chemical combination. Calcium oxide, CaO, for example, reacts with water to produce calcium hydroxide, $Ca(OH)_2$, but does not react with the alcohol. Alcohol "dried" in this manner can be recovered by a simple distillation.

Alcoholic beverages have always been a prime source of tax revenue for most governments. Ethanol, when used as a solvent or as a reagent, is tax free. In order to prevent the use of tax-free alcohol for purposes other than scientific or industrial, the government requires that it be **denatured.** Denatured alcohol is alcohol rendered unfit for beverages by the addition of substances repugnant and difficult to remove. The reason for the government's vigilance over the manufacture and use of alcohol may be illustrated clearly by the following simple economics. The list price of 95% (190 proof) ethanol in 1985 in bulk lots was about two dollars per *gallon* when purchased as a tax-free reagent, but about $60.00 per gallon when included in alcoholic beverages, the difference being due largely to excise taxes.

A. Preparation of Alcohols from Alkenes. Large quantities of ethanol are being made by the fermentation of sugars found in blackstrap molasses or, increasingly, by the fermentation of grain or plant wastes. Fermentation alcohol is used largely for beverage or motor fuel ("gasohol") purposes. The synthesis of ethanol from ethylene is the dominant industrial process, about 500 million gallons being made by this route in 1985. Alkenes can be converted to alcohols by hydration, which, in this case, is the reverse of the reaction we employed to produce an alkene from an alcohol (Sec. 3.8C). In practice, the elements of water are added across the double bond indirectly. Sulfuric acid is added first, and the alkyl sulfuric acid thus formed, when diluted with water and heated, undergoes hydrolysis to produce the alcohol according to the following equations, which illustrate the preparation of ethanol from ethylene.

$$\mathrm{H{-}CH{=}CH{-}H} + \mathrm{H^{+}\ \overset{(-)}{O}SO_3H} \longrightarrow \mathrm{CH_3{-}CH_2{-}OSO_3H}$$

Ethyl sulfuric acid

$$\mathrm{CH_3{-}CH_2{-}OSO_3H} + \mathrm{H_2O} \longrightarrow \mathrm{CH_3{-}CH_2{-}OH} + \mathrm{H_2SO_4}$$

Ethanol

Other members of the alcohol family can be prepared from alkenes by the same process.

$$\mathrm{R{-}CH{=}CH_2} + \mathrm{\overset{+}{H}\ HSO_4^-} \longrightarrow \mathrm{R{-}CH(OSO_3H){-}CH_3} \xrightarrow{\mathrm{H_2O}} \mathrm{R{-}CH(OH){-}CH_3}$$

A secondary alcohol

$$\mathrm{R_2C{=}CH_2} \xrightarrow{\mathrm{H_2SO_4}} \mathrm{R_2C(OSO_3H){-}CH_3} \xrightarrow{\mathrm{H_2O}} \mathrm{R_2C(OH){-}CH_3}$$

A tertiary alcohol

The addition of ionic reagents to isolated carbon–carbon double bonds follows Markovnikov's rule (Sec. 3.8A). The alcohol that results from the hydration of the alkene is always the more highly branched of the two possibilities. An "anti-Markovnikov" hydration of an alkene to an alcohol can be accomplished by the hydroboration reaction discussed in Sec. 3.8D.

Exercise 7.4 By what sequence of reactions might one convert 2-methyl-2-propanol to 2-methyl-1-propanol?

B. Hydrolysis of Alkyl Halides. Alcohols may be prepared by the hydrolysis of alkyl halides, but the reaction is of limited usefulness because alkene formation by *elimination* of halogen acid competes with the *substitution* reaction. Alkyl halides are hydrolyzed in a neutral, rather than an alkaline, alcoholic solution in order to minimize the formation of alkenes, but some elimination may result even when a neutral hydrolysis is carried out. This is especially true when highly branched tertiary alkyl halides are hydrolyzed. The **steric effects** provided by bulky groups, when attached to the same carbon atom that bears the halogen, not only inhibit a nucleophilic attack from the rear (S_N2), but appear to have a crowding effect upon each other. Strain, due to such crowding, may be relieved by the departure of the halide ion. The intermediate, tertiary carbocation now can be attacked by the solvent either at the face of the carbon atom, or at a β hydrogen. You will recognize the first choice as the S_N1, the second as the E1 reaction (Sec. 6.5). The elimination route appears to be favored, especially if an alkaline hydrolysis with a strong base is attempted. When alkene formation is not possible, the hydrolysis of an alkyl halide results in the formation of an alcohol. A useful application of this method is the synthesis of benzyl alcohol from benzyl chloride.

$$C_6H_5CH_3 \xrightarrow{Cl_2,\ uv} C_6H_5CH_2Cl \xrightarrow{Na^+OH^-} C_6H_5CH_2OH$$

Toluene — Benzyl chloride — Benzyl alcohol

C. Addition of Grignard Reagents to Aldehydes and Ketones. The addition of Grignard reagents of the form RMgX or ArMgX to aldehydes or to ketones (Chapter 8) provides one of the best routes leading to the preparation of alcohols. An alkyl or aryl magnesium halide adds to the carbonyl group of an aldehyde or a ketone to produce a mixed salt of divalent magnesium. The organic group forms a new carbon–carbon bond at the carbonyl carbon atom, *thereby lengthening the carbon chain*. The electropositive MgX^+ group is usually drawn as attached to the newly formed alkoxide oxygen atom. However, it must be remembered that this bond is ionic rather than covalent. The mode of addition is illustrated by

$$\overset{\delta+}{>C}=\overset{\delta-}{O} + \overset{\delta-}{R}\,\overset{\delta+}{MgX} \longrightarrow -\overset{|}{C}(OMgX)(R)$$

The addition product is treated with dilute hydrochloric acid, or a saturated solution of ammonium chloride, and is smoothly converted into an alcohol and a magnesium dihalide.

$$-\overset{|}{C}(OMgX)(R) + H^+Cl^- \longrightarrow -\overset{|}{C}(OH)(R) + MgClX$$

This reaction is a versatile one by which alcohols of any class may be produced. Of the aldehydes, formaldehyde alone yields a primary alcohol when treated with a Grignard reagent. Other aldehydes give secondary alcohols, and ketones yield tertiary alcohols. General equations for the preparation of alcohols of each class from Grignard reagents are as follows:

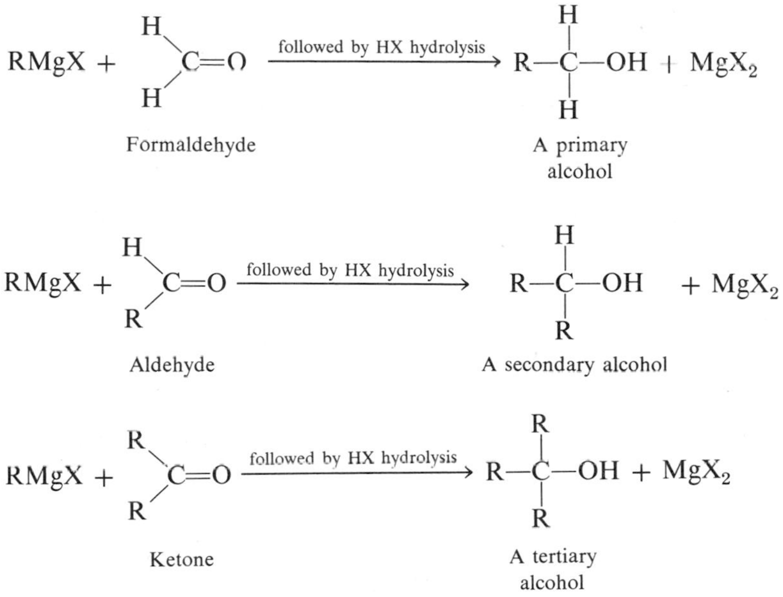

$$\mathrm{RMgX} + \underset{\text{Formaldehyde}}{\mathrm{H_2C{=}O}} \xrightarrow{\text{followed by HX hydrolysis}} \underset{\text{A primary alcohol}}{\mathrm{R{-}CH_2{-}OH}} + \mathrm{MgX_2}$$

$$\mathrm{RMgX} + \underset{\text{Aldehyde}}{\mathrm{RHC{=}O}} \xrightarrow{\text{followed by HX hydrolysis}} \underset{\text{A secondary alcohol}}{\mathrm{R{-}CH(R){-}OH}} + \mathrm{MgX_2}$$

$$\mathrm{RMgX} + \underset{\text{Ketone}}{\mathrm{R_2C{=}O}} \xrightarrow{\text{followed by HX hydrolysis}} \underset{\text{A tertiary alcohol}}{\mathrm{R{-}CR_2{-}OH}} + \mathrm{MgX_2}$$

An inspection of the structure of the desired alcohol will always indicate which Grignard reagent and which carbonyl compound may be combined to produce it. For example, the preparation of 3-methyl-2-butanol, a secondary alcohol, suggests an aldehyde for the carbonyl compound. Which aldehyde shall we use, and from what shall we prepare our Grignard reagent? By inspection it may be seen that the portion of the structure indicated by the broken line can have its origin in acetaldehyde. The portion indicated by the solid line could come from isobutyraldehyde. In either case, the carbon bonded to the oxygen atom in the starting carbonyl compound will be bonded to the oxygen atom of the hydroxyl group in the product alcohol.

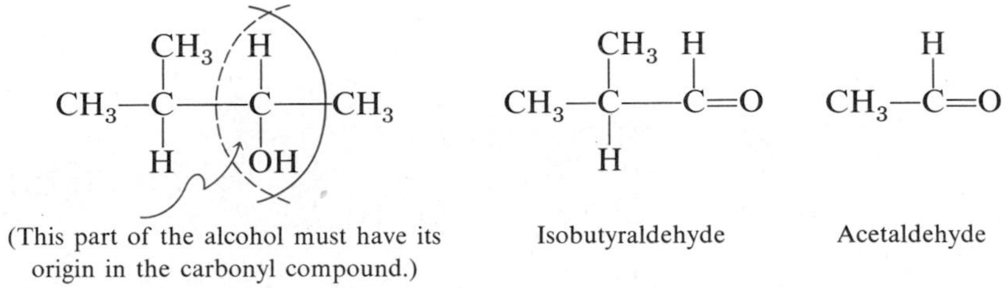

$$\mathrm{CH_3{-}CH(CH_3){-}CH(OH){-}CH_3}$$

(This part of the alcohol must have its origin in the carbonyl compound.)

$$\underset{\text{Isobutyraldehyde}}{\mathrm{CH_3{-}CH(CH_3){-}CH{=}O}} \qquad \underset{\text{Acetaldehyde}}{\mathrm{CH_3{-}CH{=}O}}$$

If acetaldehyde is our choice for the carbonyl group, it follows that the Grignard reagent used with it must be prepared from an isopropyl halide. On the other hand, if isobutyraldehyde is selected, the Grignard reagent must be prepared from methyl halide. The first set of reagents is more appealing than the second and may be used in the following manner.

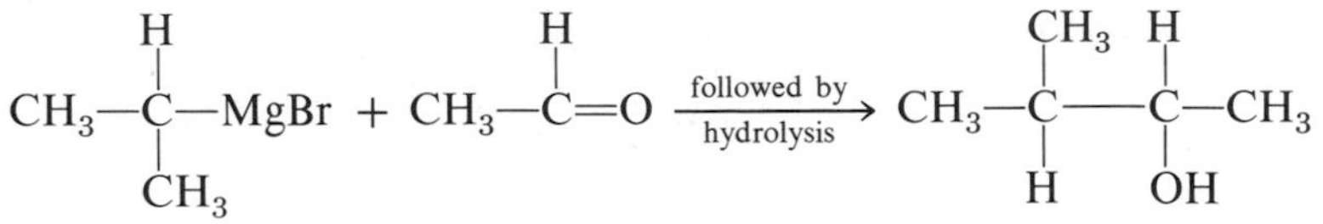

$$CH_3-\underset{CH_3}{\overset{H}{C}}-MgBr + CH_3-\overset{H}{C}=O \xrightarrow[\text{hydrolysis}]{\text{followed by}} CH_3-\underset{H}{\overset{CH_3}{C}}-\underset{OH}{\overset{H}{C}}-CH_3$$

Isopropylmagnesium bromide — Acetaldehyde — 3-Methyl-2-butanol (a secondary alcohol)

The synthesis of the isomeric, tertiary five-carbon alcohol, 2-methyl-2-butanol, must begin with a ketone.

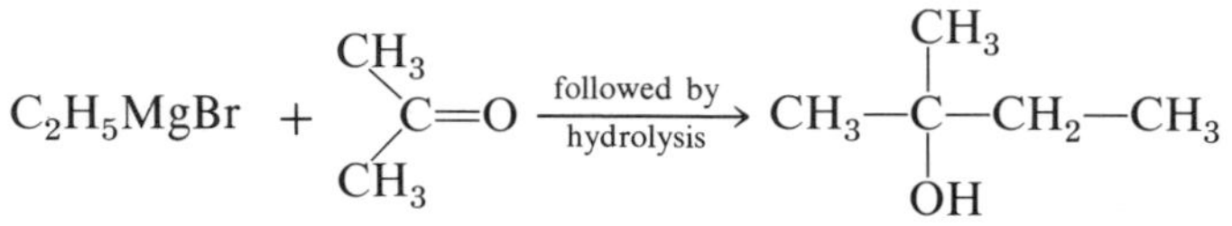

$$C_2H_5MgBr + (CH_3)_2C=O \xrightarrow[\text{hydrolysis}]{\text{followed by}} CH_3-\underset{OH}{\overset{CH_3}{C}}-CH_2-CH_3$$

Ethylmagnesium bromide — Acetone — 2-Methyl-2-butanol (a tertiary alcohol)

Another synthetic route leading to primary alcohols is the reaction of a Grignard reagent with ethylene oxide (Sec. 7.13). The reaction has the added advantage of lengthening the carbon chain by *two* carbons.

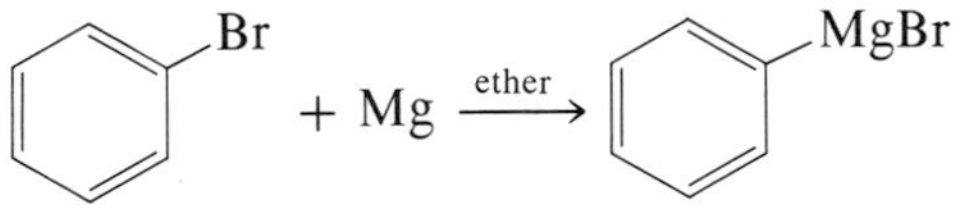

$$C_6H_5Br + Mg \xrightarrow{\text{ether}} C_6H_5MgBr$$

Phenylmagnesium bromide

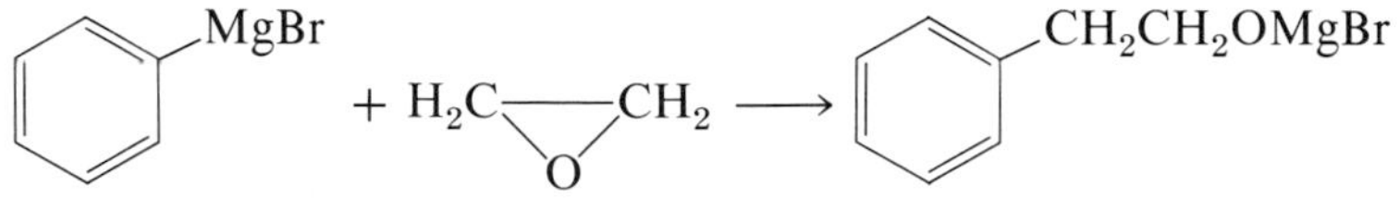

$$C_6H_5MgBr + H_2C\overset{O}{-}CH_2 \longrightarrow C_6H_5CH_2CH_2OMgBr$$

Ethylene oxide

$$C_6H_5CH_2CH_2OMgBr \xrightarrow{H_2O} C_6H_5-CH_2CH_2OH + HOMgBr$$

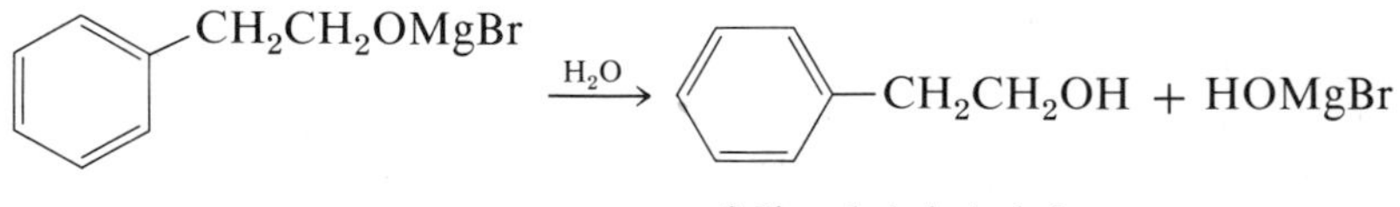

β-Phenylethyl alcohol (2-Phenylethanol)

This result should be contrasted with that obtained from the reaction of the same Grignard reagent with formaldehyde, which lengthens the chain by *one* carbon atom.

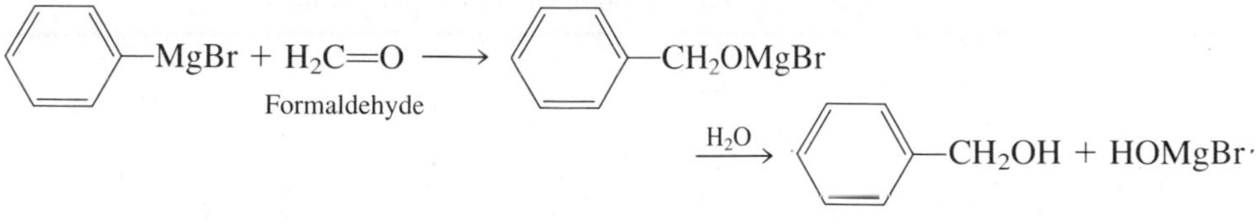

$$C_6H_5\text{—}MgBr + \underset{\text{Formaldehyde}}{H_2C\text{=}O} \longrightarrow C_6H_5\text{—}CH_2OMgBr \xrightarrow{H_2O} \underset{\text{Benzyl alcohol}}{C_6H_5\text{—}CH_2OH} + HOMgBr$$

The addition of Grignard reagents to carbonyl compounds and to ethylene oxide is very important in synthesis, because it involves the creation of new carbon–carbon bonds and allows us to build complex molecules from simple ones. Generally, smaller molecules are easier and less expensive to obtain than more complex ones.

EXAMPLE

What sets of reagents would you require for a Grignard preparation of (a) 2-methyl-1-butanol, and (b) 3-methyl-1-butanol?

Solution The structure of 2-methyl-1-butanol is that of a primary alcohol.

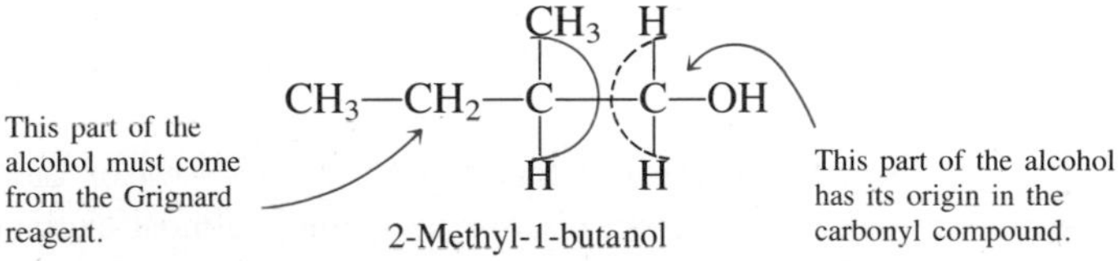

$$CH_3\text{—}CH_2\text{—}\overset{CH_3}{\underset{H}{C}}\text{—}\overset{H}{\underset{H}{C}}\text{—}OH$$

This part of the alcohol must come from the Grignard reagent.

2-Methyl-1-butanol

This part of the alcohol has its origin in the carbonyl compound.

(a) The only carbonyl compound with two hydrogen atoms on the carbonyl carbon is formaldehyde. The remaining hydrocarbon portion must be provided by *sec*-butylmagnesium halide.

(b) In the structure of 3-methyl-1-butanol we again have a primary alcohol, but in this case we have a choice of reagents.

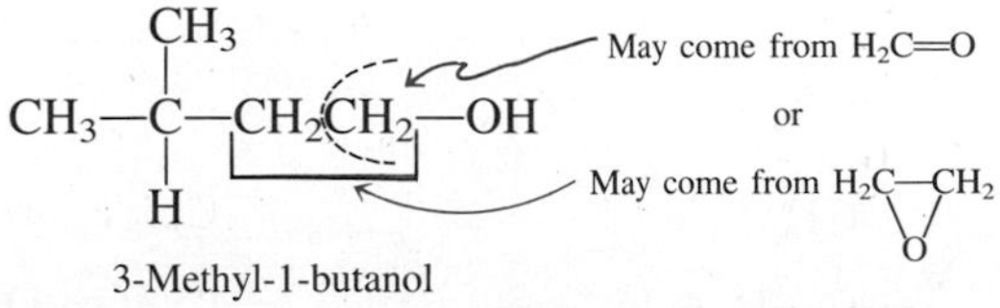

$$CH_3\text{—}\overset{CH_3}{\underset{H}{C}}\text{—}CH_2CH_2\text{—}OH$$

May come from $H_2C\text{=}O$

or

May come from $H_2C\text{—}CH_2$ (O bridging, ethylene oxide)

3-Methyl-1-butanol

If formaldehyde is our choice for one reagent, the Grignard reagent must be isobutylmagnesium halide; if ethylene oxide is our choice for one reagent, the other will have to be isopropylmagnesium halide.

Exercise 7.5 Beginning with propylene, ethylene oxide, and any other inorganic reagents that you might require, show how you might prepare (a) isopropyl alcohol, (b) *n*-propyl-alcohol, (c) 2,3-dimethyl-2-butanol, (d) 1-pentanol.

7.5 Preparation of the Phenols

Phenols cannot be prepared from aryl halides by displacement reactions unless the ring is substituted with powerful electron-withdrawing groups (Sec. 6.6). Simple aryl halides react under conditions suitable only for industrial processes by an elimination-addition mechanism thought to involve an unstable **benzyne** intermediate. For example, phenol was formerly produced from chlorobenzene by treatment with aqueous alkali at a very high temperature and pressure. Hydrogen chloride is eliminated to give benzyne, which then adds water to yield phenol. Phenol is converted by the alkali to sodium phenoxide, but may be recovered by acidification with mineral acid.

$$C_6H_5Cl \xrightarrow[300°C,\ 200\ atm.]{\text{dilute NaOH}} [C_6H_4] \xrightarrow{H-OH} C_6H_5OH \xrightarrow{NaOH} C_6H_5O^-Na^+ \xrightarrow{HCl} C_6H_5OH$$

Benzyne — Sodium phenoxide — Phenol

At present the starting material for about 90% of the phenol made industrially is cumene (isopropylbenzene). Cumene is made almost exclusively from benzene and propene *via* a Friedel–Crafts reaction. Air, when forced through cumene in the presence of a trace of base, produces a hydroperoxide. The latter, on treatment with acid, rearranges and decomposes into phenol and the important by-product acetone.

$$C_6H_6 + CH_3{-}CH{=}CH_2 \xrightarrow{AlCl_3} C_6H_5{-}CH(CH_3)_2$$

Cumene

$$C_6H_5{-}CH(CH_3)_2 + O_2 \longrightarrow C_6H_5{-}C(CH_3)_2{-}O{-}OH \xrightarrow{H_2SO_4} C_6H_5OH + (CH_3)_2C{=}O$$

Cumene hydroperoxide — Acetone

Phenol may be prepared in the laboratory by the fusion of sodium benzenesulfonate (Sec. 4.6D) with sodium hydroxide. The sodium phenoxide produced by this fusion is converted to free phenol by acid treatment.

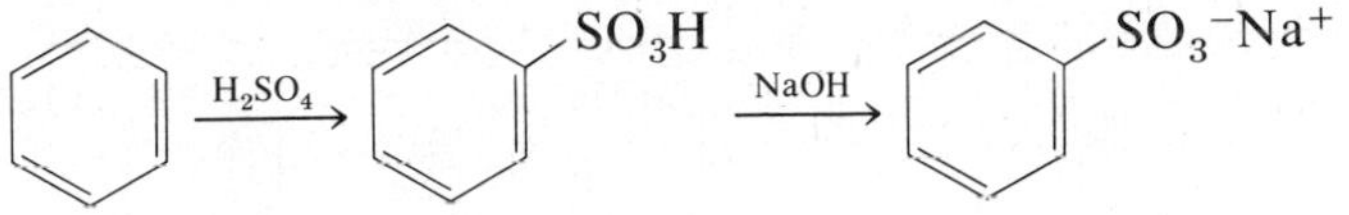

$$C_6H_6 \xrightarrow{H_2SO_4} C_6H_5SO_3H \xrightarrow{NaOH} C_6H_5SO_3^-Na^+$$

Benzenesulfonic acid

Sodium benzenesulfonate

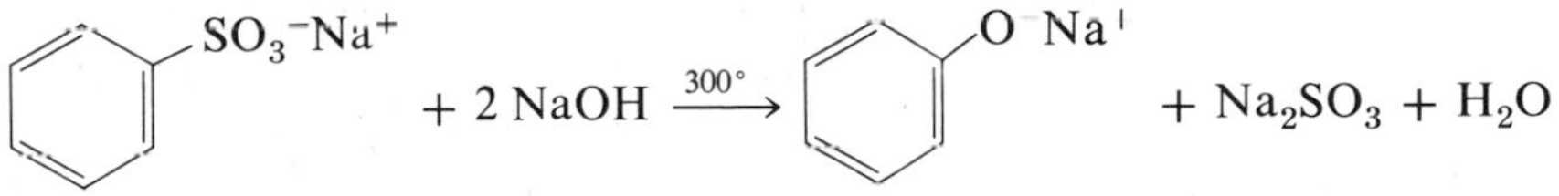

$$C_6H_5SO_3^-Na^+ + 2\,NaOH \xrightarrow{300^\circ} C_6H_5O^-Na^+ + Na_2SO_3 + H_2O$$

Sodium phenoxide

$$C_6H_5O^-Na^+ + H^+ \longrightarrow C_6H_5OH + Na^+$$

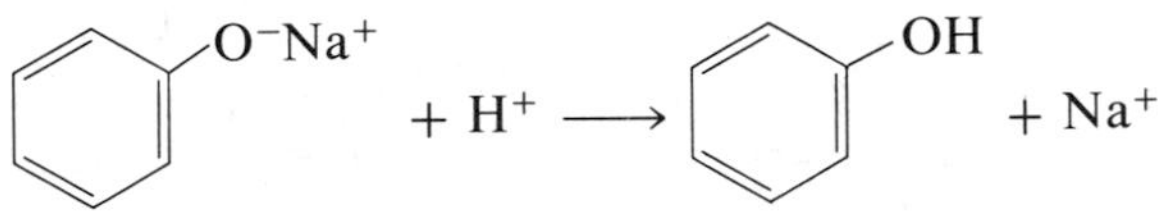

Phenol

7.6 Dihydric Phenols and Quinones

The three dihydric phenols can be prepared by standard procedures but, because of their industrial importance, they are often made by special methods.

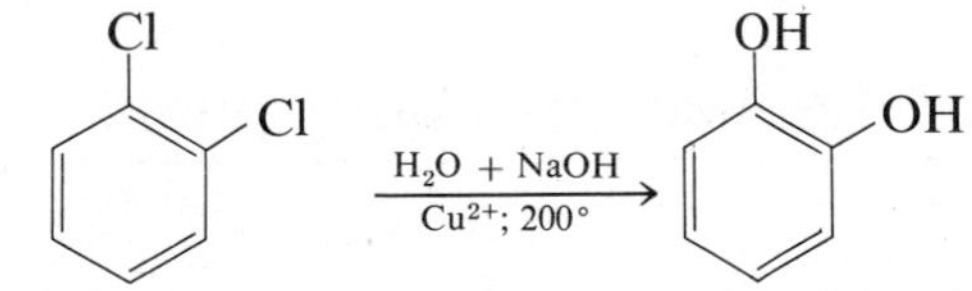

$$o\text{-}C_6H_4Cl_2 \xrightarrow[Cu^{2+};\ 200^\circ]{H_2O\ +\ NaOH} o\text{-}C_6H_4(OH)_2$$

o-Dichlorobenzene

Catechol

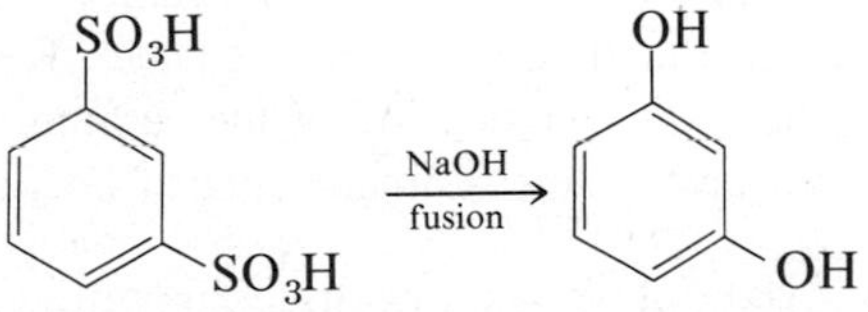

$$m\text{-}C_6H_4(SO_3H)_2 \xrightarrow[fusion]{NaOH} m\text{-}C_6H_4(OH)_2$$

1,3-Benzenedisulfonic acid

Resorcinol

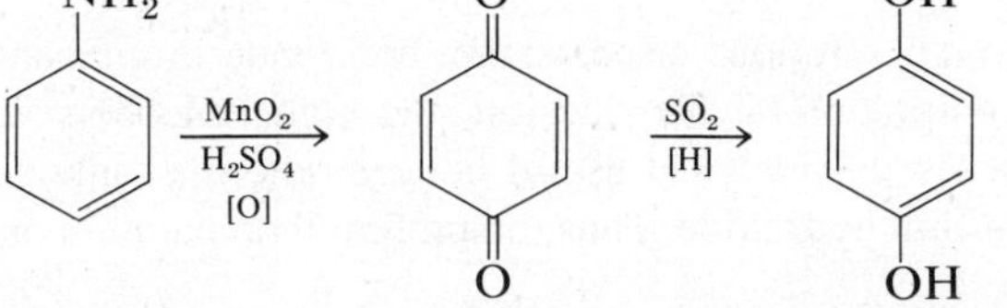

$$C_6H_5NH_2 \xrightarrow[H_2SO_4 \\ [O]]{MnO_2} O{=}C_6H_4{=}O \xrightarrow[[H]]{SO_2} p\text{-}C_6H_4(OH)_2$$

Aniline

p-Benzoquinone

Hydroquinone

Both 1,2- and 1,4-dihydroxybenzene derivatives are characterized by easy oxidation to **quinones** and the easy reduction of the quinones back to the dihydric phenols.

$$\text{Catechol} \underset{Na_2S_2O_4}{\overset{Ag_2O}{\rightleftharpoons}} o\text{-Benzoquinone}$$

$$\text{Hydroquinone} \underset{Na_2S_2O_4}{\overset{Na_2Cr_2O_7,\ H_2SO_4}{\rightleftharpoons}} p\text{-Benzoquinone}$$

Dihydric and polyhydric phenols and their derivatives, as well as the quinones, are widely distributed in nature. The oxidation-reduction reactions of hydroquinone and quinone derivatives are important in certain biochemical processes. The quinones are not aromatic compounds, as may be readily seen by application of the Hückel rule (Sec. 4.2); rather, they are α,β-unsaturated ketones (Sec. 8.12) and show the reactions expected of such substances.

7.7 Reactions of the Alcohols and Phenols

The reactions of the alcohols and phenols are of the following types:

(A) Reactions that result in O—H bond cleavage, RO⌇H
(B) Reactions that result in C—O bond cleavage, R⌇OH
(C) Reactions that result in oxidation of the carbinol carbon
(D) Reactions that involve the aromatic ring of phenols.

A-1. Cleavage of the O—H Bond. Salt Formation. The aliphatic alcohols are weaker acids than water; therefore, the following equilibrium lies well to the left in aqueous solution.

$$R{-}OH + HO^-K^+ \rightleftharpoons R{-}O^-K^+ + H{-}OH$$

Although solutions of potassium hydroxide in methanol or ethanol contain considerable concentrations of alkoxide ion, the higher alcohols are less acidic and for all practical purposes do not yield useful concentrations of alkoxides on treatment with sodium or potassium hydroxide. Thus the preferred preparation of the alkoxides is the reaction of the alcohol with active metals such as sodium or potassium, which takes place with the liberation of hydrogen gas.

$$2\ RO{-}H + 2\ Na \longrightarrow 2\ RO^-Na^+ + H_2$$

Sodium alkoxide

Primary alcohols exhibit a greater reactivity on treatment with sodium metal than do secondary alcohols, and the latter show a greater reactivity than tertiary alcohols. Of the alkali metals, potassium is more reactive toward any class of alcohols than is sodium. Thus the reaction of potassium metal with *tert*-butyl alcohol gives potassium *tert*-butoxide, a very powerful base, widely used in the laboratory and in industry.

The liberation of hydrogen gas from an unknown liquid is sometimes used as a test for alcohols. Of course, the liquid to be tested must be free of water.

A-2. Cleavage of the O—H Bond. Ether Formation. The alkoxides or phenoxides of the alkali metals are strong bases (nucleophiles) that easily enter into S_N2 displacements of halogen from alkyl halides. This reaction, referred to as the **Williamson ether synthesis,** is best used to prepare mixed ethers—that is, ethers in which the two groups bridged by the oxygen atom are not the same.

$$(CH_3)_2CHO^-Na^+ + CH_3I \xrightarrow{S_N2} (CH_3)_2CH{-}O{-}CH_3 + NaI$$

Isopropyl methyl ether

If one of the groups in the ether is to be a branched structure, that part of the ether should have its origin in the alkoxide. If the alkyl halide is highly branched, elimination, rather than substitution, results.

$$CH_3O^-Na^+ + (CH_3)_2CHI \xrightarrow{E2} CH_3CH{=}CH_2 + CH_3OH + NaI$$

Propylene

A-3. Ester Formation. Alcohols react with carboxylic acids (Chapter 9) to produce esters. The reaction is catalyzed by strong mineral acids and, as indicated by the double arrow in the equation, is reversible.

$$R{-}\overset{\overset{\large O}{\|}}{C}{-}OH + H{-}OR \rightleftharpoons R{-}\overset{\overset{\large O}{\|}}{C}{-}OR + H_2O$$

Acid Alcohol Ester

The **esterification** reaction is discussed in greater detail in Section 9.6B.

B. Reactions that Result in C—O Bond Cleavage. The reactions of alcohols in which the carbon–oxygen bond is cleaved fall into two familiar classes, substitution and elimination. The mechanisms for these reactions are similar to those for the S_N1, S_N2, E1, and E2 substitution and elimination reactions of alkyl halides (Sec. 6.5).

The E2 mechanism for the acid-catalyzed dehydration of ethyl alcohol to ethylene and the E1 mechanism for the acid-catalyzed dehydration of *tert*-butyl alcohol to isobutylene were described in detail in Sec. 3.3. The important feature of each of these mechanisms and of the S_N1 and S_N2 mechanisms of alcohols is the role of the catalyst. We can illustrate this role by comparing the S_N2 reaction of (a) ethyl bromide with potassium iodide and (b) the S_N2 reaction of ethyl alcohol with aqueous hydroiodic acid.

$$\text{(a)}\quad K^+ + I^- + \underset{\displaystyle CH_3}{\underset{|}{C}}H_2{-}Br \longrightarrow I{-}\underset{\displaystyle CH_3}{\underset{|}{C}}H_2 + K^+ + Br^-$$

$$\text{(b)}\quad CH_3CH_2{-}OH + H^+ + I^- \longrightarrow CH_3CH_2{-}OH_2^+ + I^-$$

$$I^- + \underset{\displaystyle CH_3}{\underset{|}{C}}H_2{-}OH_2^+ \longrightarrow I{-}\underset{\displaystyle CH_3}{\underset{|}{C}}H_2 + H_2O$$

In substitution and elimination reactions, the stability of the **leaving group** that is being displaced or eliminated is very important. The bond to this group is being broken and the rate of the reaction will depend in part upon the ease of this bond-breaking. If the leaving group is reasonably stable, the bond to it will be more easily broken, the energy of the transition state will be lower, and the reaction will proceed more rapidly. Generally, good leaving groups tend to be the conjugate bases of strong acids or small neutral molecules such as H_2O and its organic derivatives, NH_3 and its organic derivatives, and N_2. Since the leaving group is gaining electron density when it is displaced or eliminated, it should be a stable anion or neutral molecule. Thus, in reaction (a), the leaving group is the stable, weakly basic bromide ion, the conjugate base of the strong acid HBr. In reaction (b), the leaving group is the neutral water molecule. If in reaction (b) the alcohol had not been protonated on the oxygen atom, the displacement would have required the removal of hydroxide ion, a strong base. Strong bases are not good leaving groups. Protonation of the hydroxyl group to form the alkyloxonium ion changes the leaving group from hydroxide ion to water and facilitates both substitution and elimination. The use of acid catalysis to protonate hydroxyl and alkoxyl groups for the purpose of creating good leaving groups is a fairly common practice in organic chemistry.

Exercise 7.6 Write mechanisms for the S_N2 reaction of *n*-butyl alcohol with concentrated hydrobromic acid to yield *n*-butyl bromide, and the S_N1 reaction of *tert*-butyl alcohol with cold concentrated hydrobromic acid to yield *tert*-butyl bromide.

Reactions of the foregoing types proceeding via E1 or S_N1 mechanisms *may* be complicated by the facility with which some carbocations rearrange to more stable carbocations. Rearrangements, when they occur, are often of the following types:

$$CH_3-\underset{H}{\overset{CH_3}{C}}-\underset{OH}{\overset{H}{C}}-CH_3 \xrightleftharpoons{H^+Br^-} CH_3-\underset{H}{\overset{CH_3}{C}}-\underset{\overset{+}{O}H_2}{\overset{H}{C}}-CH_3 \xrightleftharpoons{-H_2O} CH_3-\underset{H}{\overset{CH_3}{C}}-\overset{H}{\underset{+}{C}}-CH_3$$

3-Methyl-2-butanol

secondary carbocation

$$\Updownarrow$$

$$CH_3-\underset{Br}{\overset{CH_3}{C}}-\underset{H}{\overset{H}{C}}-CH_3 \xleftarrow{Br^-} CH_3-\underset{+}{\overset{CH_3}{C}}-\underset{H}{\overset{H}{C}}-CH_3$$

2-Bromo-2-methylbutane

tertiary carbocation

$$CH_3-\underset{CH_3}{\overset{CH_3}{C}}-\underset{OH}{\overset{H}{C}}-CH_3 \xrightleftharpoons{H_2SO_4} CH_3-\underset{CH_3}{\overset{CH_3}{C}}-\underset{\overset{+}{O}H_2}{\overset{H}{C}}-CH_3 \xrightleftharpoons{-H_2O} CH_3-\underset{CH_3}{\overset{CH_3}{C}}-\overset{H}{\underset{+}{C}}-CH_3$$

3,3-Dimethyl-2-butanol

secondary carbocation

$$\Updownarrow$$

$$CH_3-\overset{CH_3}{C}=\underset{CH_3}{C}-CH_3 \xleftarrow{-H^+} CH_3-\underset{+}{\overset{CH_3}{C}}-\underset{CH_3}{\overset{H}{C}}-CH_3$$

2,3-Dimethyl-2-butene

tertiary carbocation

In each of the examples a group on a carbon next to the positively charged carbon migrated to the positive carbon, bringing with it a pair of electrons and creating a new, more stable carbocation. Rearrangements of this type are common in carbocation chemistry.

EXAMPLE

When neopentyl bromide is heated with ethyl alcohol, ethyl neopentyl ether is not obtained, but 2-ethoxy-2-methylbutane and 2-methyl-2-butene are found in the reaction mixture. Show a mechanism whereby each product is formed.

Solution Although neopentyl bromide is a primary alkyl halide, the bulkiness of the tertiary butyl group attached to the α-carbon prevents an easy back-side approach for an S_N2 type of displacement. Therefore, a slow dissociation of alkyl halide into a carbocation and bromide ion must occur, followed by an immediate rearrangement of the primary carbocation into a more stable tertiary carbocation. The tertiary carbocation then enters one of two reaction pathways.

$$CH_3-\overset{\overset{CH_3}{|}}{\underset{\underset{CH_3}{|}}{C}}-\overset{\overset{H}{|}}{\underset{\underset{H}{|}}{C}}-Br \longrightarrow CH_3-\overset{\overset{CH_3}{|}}{\underset{\underset{CH_3}{|}}{C}}-\overset{\overset{H}{|}}{\underset{\underset{H}{|}}{C^+}} + Br^-$$

$$CH_3-\overset{\overset{CH_3}{|}}{\underset{\underset{CH_3}{|}}{C}}-\overset{\overset{H}{|}}{\underset{\underset{H}{|}}{C^+}} \longrightarrow CH_3-\overset{\overset{CH_3}{|}}{\underset{+}{C}}-CH_2CH_3 \xrightarrow[S_N1]{C_2H_5OH} CH_3-\overset{\overset{CH_3}{|}}{\underset{\underset{OC_2H_5}{|}}{C}}-CH_2CH_3$$

Rearrangement

2-Ethoxy-2-methyl-butane

$$\xrightarrow[E1]{-H^+} CH_3-\overset{\overset{CH_3}{|}}{C}=CHCH_3$$

2-Methyl-2-butene

Exercise 7.7 Two reaction routes are open to the carbocation that results when 1-butanol is treated with strong acid. Which route leads to the principal product?

B-1. Replacement of the Hydroxyl Group by Acid Anions. The reactions of alcohols with phosphorus halides and thionyl chloride, $SOCl_2$, were reviewed under the preparation of the alkyl halides (Sec. 6.2B). Most phenols do *not* undergo these reactions.

Treatment of an aliphatic alcohol with the **Lucas reagent** (a solution of zinc chloride in concentrated hydrochloric acid) produces an alkyl chloride.

$$R-OH + HCl \xrightarrow{ZnCl_2} R-Cl + H_2O$$

The alkyl chloride, when formed in this manner, is insoluble in the reagent and either produces a cloudy appearance or forms two layers. The Lucas reagent is used to distinguish between primary, secondary, and tertiary alcohols because a tertiary alcohol reacts immediately, the secondary after a few minutes, and the primary only when heated for some hours.

An alcohol treated with concentrated sulfuric acid at room temperature produces an alkyl hydrogen sulfate.

$$C_2H_5(OH + H)O-SO_2-OH \longrightarrow C_2H_5-O-SO_2-OH + H_2O$$

Ethyl hydrogen sulfate

The alkyl hydrogen sulfates still have one acidic hydrogen remaining and are capable of forming salts. The sodium salts of the long-chain alkyl hydrogen sulfates are excellent detergents (Sec. 9.6A).

Alcohols react with nitric acid to yield alkyl nitrates. Perhaps the most widely used alkyl nitrate is glyceryl trinitrate (commonly called "nitroglycerine"), the principal explosive ingredient in dynamite.

B-2. Cleavage of the C—O Bond. Dehydration. Ethanol may, in effect, be dehydrated to an olefin when its hydrogen sulfate ester is heated to 150° or higher.

$$CH_3CH_2{-}OSO_3H \xrightarrow{150^\circ} H_2C{=}CH_2 + H_2SO_4$$

If the hydrogen sulfate ester is heated with an excess of alcohol at a temperature lower than 150°, the bisulfate ion is displaced by alcohol to produce a simple ether.

$$CH_3CH_2{-}\ddot{O}H + CH_3CH_2{-}OSO_3H \xrightleftharpoons{130^\circ} {}^{-}OSO_3H + CH_3CH_2{-}\overset{H}{\overset{|}{O^{\oplus}}}{-}CH_2CH_3$$

$$CH_3CH_2{-}\overset{H}{\overset{|}{O^{\oplus}}}{-}CH_2CH_3 + {}^{-}OSO_3H \rightleftharpoons \underset{\text{Diethyl ether}}{CH_3CH_2OCH_2CH_3} + H_2SO_4$$

C. Oxidation. When oxidized, primary and secondary alcohols give organic products containing the same number of carbon atoms. Primary alcohols can be oxidized to carboxylic acids (Sec. 9.5A) and secondary alcohols to ketones (Sec. 8.4A). Tertiary alcohols are not oxidizable without a rupture of carbon–carbon bonds. Reagents usually employed for the oxidation of alcohols are potassium dichromate in combination with concentrated sulfuric acid, or a hot, alkaline solution of potassium permanganate.

$$3\,\underset{\text{Ethanol}}{CH_3CH_2OH} + 2\,\underset{\text{(orange)}}{K_2Cr_2O_7} + 8\,H_2SO_4 \longrightarrow$$

$$3\,\underset{\text{Acetic acid}}{CH_3{-}C({=}O){-}OH} + 2\,K_2SO_4 + 2\,\underset{\text{(green)}}{Cr_2(SO_4)_3} + 11\,H_2O$$

The color change in this oxidation reaction of ethanol is the basis for the breath analysis of suspected inebriated drivers. The intensity of the color change is directly proportional to the alcoholic content in a measured volume of exhaled breath.

$$3\,\underset{\text{2-Propanol}}{CH_3{-}CH(OH){-}CH_3} + 2\,KMnO_4 \longrightarrow$$

$$3\,\underset{\text{Acetone}}{CH_3{-}\overset{O}{\overset{\|}{C}}{-}CH_3} + 2\,MnO_2 + 2\,KOH + 2\,H_2O$$

D. Ring Substitution in Phenols. The hydroxyl group of phenols strongly activates the ring toward electrophilic substitution (Sec. 4.8). The hydroxyl group serves as an electron-pair donor, stabilizing the cationic intermediates for substitution at positions *ortho-* and *para-* to the hydroxyl group. Thus, when phenol is nitrated with dilute HNO_3, both the *ortho* and the *para* isomeric nitrophenols are formed.

$$2\ C_6H_5OH + 2\ HNO_3\ (\text{dilute}) \xrightarrow{25\,°C} o\text{-}O_2NC_6H_4OH + p\text{-}O_2NC_6H_4OH + 2\ H_2O$$

o-Nitrophenol *p*-Nitrophenol

The *ortho* form can be separated from its *para* isomer by steam distillation. The *ortho* isomer is able to form an H-bond within itself, or *intra*molecularly, to produce a **chelated ring** (Gr. *chele*, claw). *Para* isomers must hydrogen-bond *inter*molecularly to form an associated compound not volatile with steam.

Intramolecular H-Bonding (Chelation)

Intermolecular H-Bonding (Association)

Phenol, when nitrated directly with concentrated nitric acid, undergoes oxidation. For this reason the highly explosive 2,4,6-trinitrophenol, or picric acid, is obtained through a synthesis that begins with chlorobenzene. The first product, 2,4-dinitrochlorobenzene, is then easily hydrolyzed to 2,4-dinitrophenol (Sec. 6.6) and the nitration continues to give picric acid in good yield.

$$C_6H_5Cl + 2\ HNO_3 \xrightarrow{H_2SO_4} 2,4\text{-}(NO_2)_2C_6H_3Cl + 2\ H_2O$$

Chlorobenzene

$$\text{2-chloro-1,4-dinitrobenzene (Cl, } NO_2, NO_2) + Na^+OH^- \longrightarrow \text{(OH, } NO_2, NO_2) + NaCl$$

$$\text{(OH, } NO_2, NO_2) + HNO_3 \xrightarrow{H_2SO_4} \text{(OH, } O_2N, NO_2, NO_2) + H_2O$$

Picric acid
(2,4,6-Trinitrophenol)

Many organic substances form crystalline molecular complexes with picric acid. Butesin picrate, the picrate of *n*-butyl-*p*-aminobenzoate, is a useful surface anesthetic for the treatment of burns. Picric acid, a bright yellow compound, also has been used as a direct dye on silk and wool.

Phenol, when treated with sulfuric acid, yields both *ortho* and *para* phenolsulfonic acids. The *ortho* isomer predominates at low temperatures, the *para* at high temperatures.

$$\text{(OH)} + 2\,H_2SO_4 \longrightarrow \text{(OH, } SO_3H) + \text{(OH, } SO_3H) + 2\,H_2O$$

o-Phenolsulfonic acid *p*-Phenolsulfonic acid

Phenol is very easily brominated. 2,4,6-Tribromophenol can be prepared simply by shaking an aqueous phenol solution with a saturated solution of bromine in water.

$$\text{(OH)} + 3\,Br_2 \xrightarrow{H_2O} \text{(OH, Br, Br, Br)} + 3\,HBr$$

2,4,6-Tribromophenol

The catalytic reduction of phenol produces cyclohexanol.

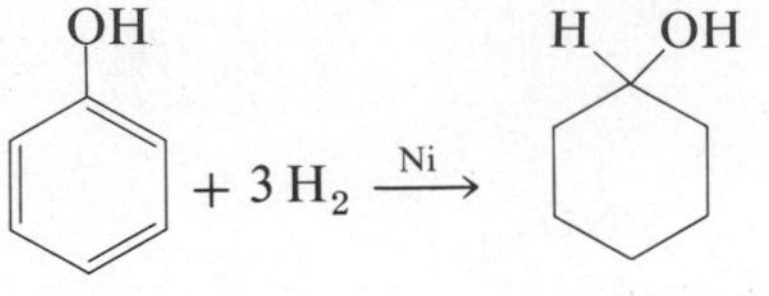

$$C_6H_5OH + 3\,H_2 \xrightarrow{Ni} C_6H_{11}OH$$

Cyclohexanol

7.8 Alcohols and Phenols in Nature, Medicine, and Industry

Mono-, di-, and polyhydric alcohols and phenols are quite common in nature, and numerous examples of such naturally occurring alcohols and phenols are given in later chapters. Many of these examples have fairly complex structures, but even relatively simple alcohols and phenols play important roles in nature. For example, a number of the **pheromones** fall into these classes. The pheromones are substances secreted by animals and insects for the purpose of intraspecies communication. They may be used for such diverse purposes as marking a trail, warning of danger, attracting the opposite sex, or calling assembly. The use of natural sex attractants, or their synthetic analogs, to lure insects into traps may prove to be a relatively safe alternative to the use of chemical pesticides in the control of selected insect species, for it is known that as few as 30 molecules, or perhaps even fewer, will attract certain moths. Thus the concentrations of chemicals required are incredibly small. Some typical pheromones are shown.

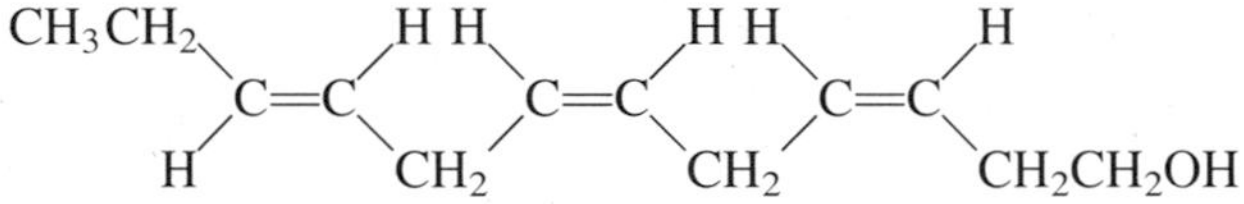

$CH_3CH_2CH{=}CHCH_2CH{=}CHCH_2CH{=}CHCH_2CH_2OH$

Trailmarker for termites

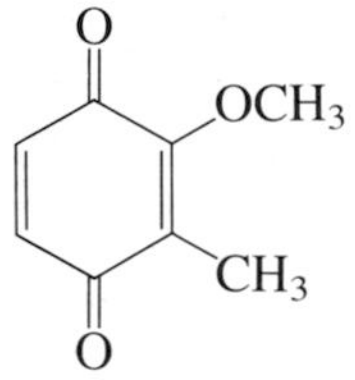

Chemical defense for millipedes

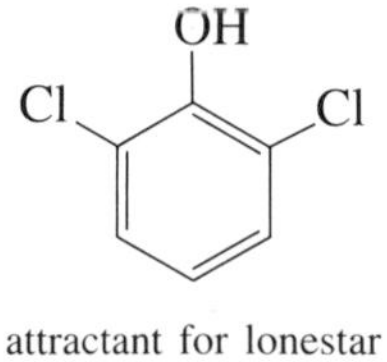

Sex attractant for lonestar tick

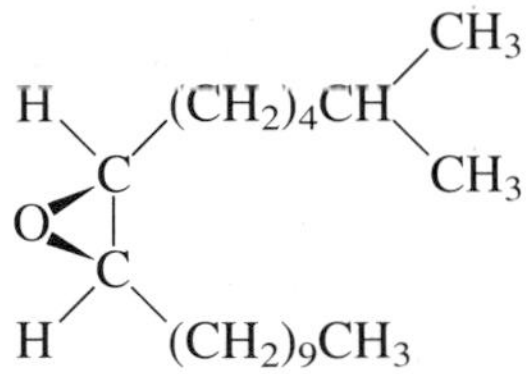

Disparlure, sex attractant for the gypsy moth

Exercise 7.8 Disparlure is used to control gypsy moths in hardwood trees. Label the two carbon atoms in its three-membered ring as (*R*)- or (*S*)-. The enantiomer of disparlure is only 1/100th as effective in insect control. Draw its structure.

Dihydric phenols and the quinones serve in other important capacities in biochemical processes. Examples are the K vitamins (Sec. 17.7) and coenzyme Q, the latter being involved in electron transport in the mitochondrial respiratory chain. The catechol moiety is represented in nature in the urushiols, the active ingredient in the allergenic oil of poison ivy.

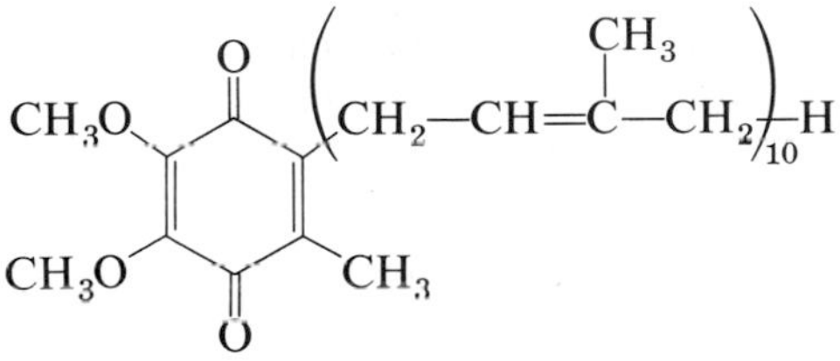

CH_3O, CH_3O, CH_3, $\left(CH_2{-}CH{=}\overset{CH_3}{\overset{|}{C}}{-}CH_2\right)_{10}{-}H$

Mammalian coenzyme Q

OH, OH, $(CH_2)_7{-}CH{=}CH{-}(CH_2)_5{-}CH_3$

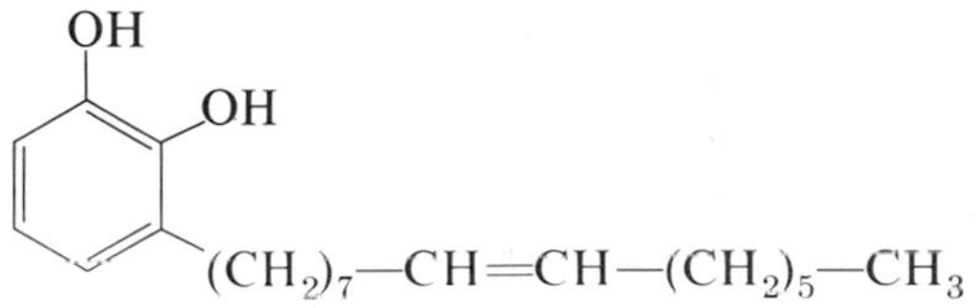

A urushiol

Two very important terpene alcohols are farnesol (Sec. 3.13) and Vitamin A (Sec. 17.7). The former is the biological precursor of **cholesterol,** a steroidal alcohol that is present to some extent in all animal tissue, but is found principally in the brain and spinal cord (Sec. 17.10). Normal humans have about 0.3 pound of cholesterol per 100 pounds of body weight. Cholesterol is perhaps better known to the public for its role in arteriosclerosis (''hardening of the arteries'') and heart disease.

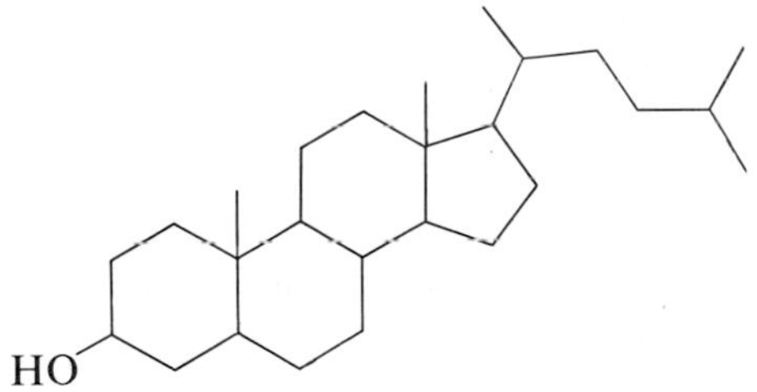

HO

Cholesterol

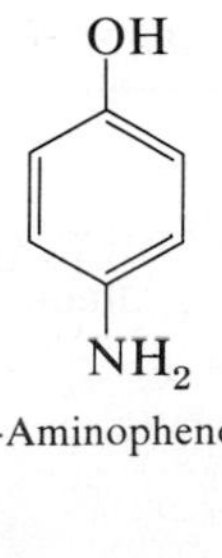

OH, NH_2

p-Aminophenol

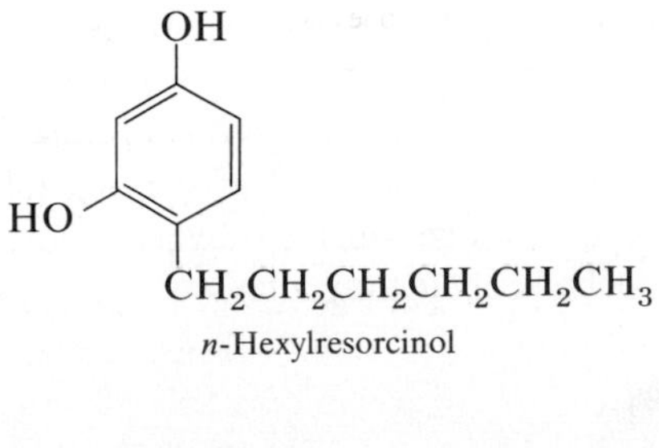

OH, HO, $CH_2CH_2CH_2CH_2CH_2CH_3$

n-Hexylresorcinol

Phenols of many types are of great industrial importance. The principal commercial application of phenol is in the manufacture of phenol-formaldehyde resins of the **Bakelite** type (Sec. 8.11); however, one type of **nylon** is still made with phenol as a starting material. Hydroquinone and related substances (particularly *p*-aminophenol) are important photographic developers because of their ability to function as reducing agents (Sec. 7.6). Important synthetic pharmaceuticals, such as **aspirin** (Sec. 11.6), as well as naturally occurring medicinal products, such as **adrenalin (epinephrine,** Sec. 13.7), are often phenolic compounds. Phenol itself is one of the oldest of disinfectants. All phenolic compounds appear to have a germicidal power that is enhanced by the presence of alkyl groups on the ring. It appears that the optimum size of the side chain for maximum germicidal activity is six carbons, making ***n*-hexylresorcinol** a very fine antiseptic. The killing power that an antiseptic has against microorganisms is measured against that of phenol. The germicidal efficiency of an antiseptic is measured in terms of an arbitrary unit called a **phenol coefficient.** For example, a germicide in a 1% solution that kills an organism in the same length of time as that required for a 5% phenol solution is assigned a phenol coefficient of 5.

Chlorine-substituted phenols are especially active against bacteria and fungi. **Pentachlorophenol,** for example, is an excellent fungicide and wood preservative and is widely used to protect against "dry rot" and termites. The chlorine-substituted biphenyl phenolic compound 2,2′-dihydroxy-3,3′,5,5′,6,6′-hexachlorodiphenylmethane, popularly known as "hexachlorophene," has a phenol coefficient of about 125. In dilute form it has been used extensively in the manufacture of germicidal soaps, some toothpastes, and many deodorants, but products containing it have recently come under scrutiny as possibly causing systemic damage, especially when used on infants. The sale of products containing hexachlorophene is now restricted.

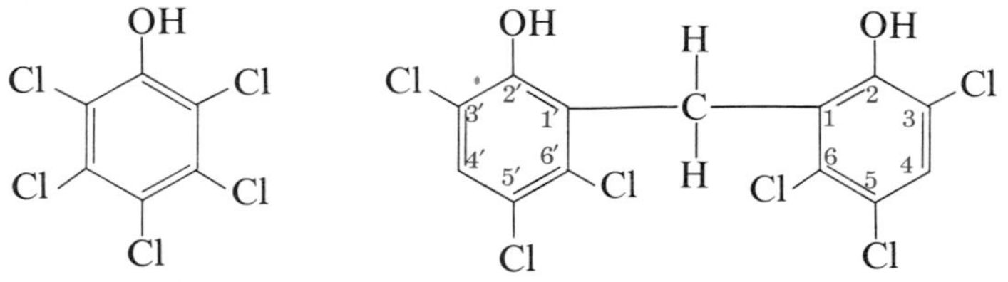

Pentachlorophenol

Hexachlorophene

Phenol and its homologs are toxic substances and have a caustic action on animal tissue. Care must be exercised in handling phenols; direct contact or inhalation of their vapors should be avoided.

7.9 Structure and Nomenclature of Ethers

Ethers are compounds of the general formula R—O—R, or R—O—Ar, in which an oxygen bridge joins two hydrocarbon groups. Although the ethers are isomeric with the alcohols, their properties are vastly different.

Ethers may be named by the IUPAC system by choosing one of the alkyl groups bonded to the oxygen as the parent chain and treating the other alkyl group as an **alkoxy** substituent. For mixed ethers the larger alkyl group will be chosen as the parent chain. Ethers may also be named by the common nomenclature system by naming both alkyl groups attached to the oxygen as separate words and adding ''ether'' as a separate word. If both groups are alike, the prefix ''di'' is employed. Sometimes the ''di'' prefix is omitted; however, for clarity it is now considered desirable to include it. The following examples illustrate these rules.

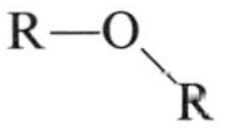

R—O—R
A simple ether

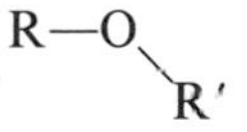

R—O—R'
A mixed ether

$CH_3—O—CH_3$
Dimethyl ether
(Methyl ether)

$C_2H_5—O—C_2H_5$
Diethyl ether
(Ethyl ether)

$CH_3—O—C_2H_5$
Methyl ethyl ether

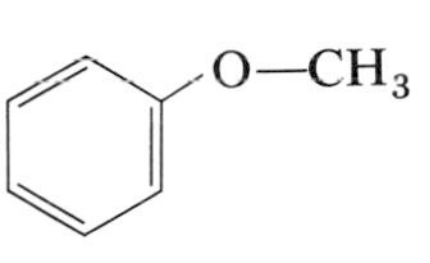

$C_6H_5—O—CH_3$
Methyl phenyl ether
(Anisole)

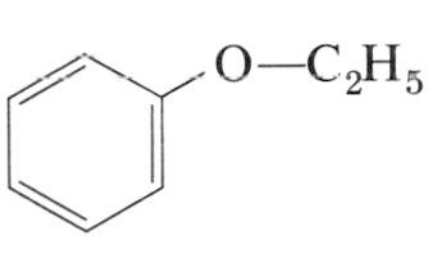

$C_6H_5—O—C_2H_5$
Ethoxybenzene
(Phenetole)

$CH_3—O—CH_2CH_2—O—CH_3$
1,2-Dimethoxyethane
(DME)

Heterocyclic ethers are named as derivatives of the parent heterocyclic system or are given common names, as shown in the following examples.

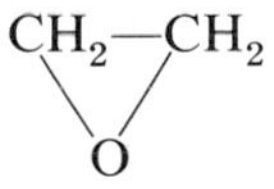

$CH_2—CH_2$, O
Ethylene oxide
(Oxirane)

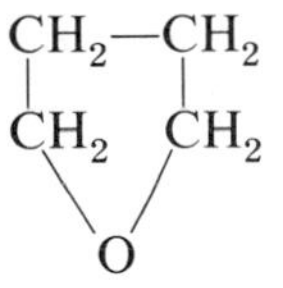

$CH_2—CH_2$, CH_2 CH_2, O
Tetrahydrofuran
(THF)

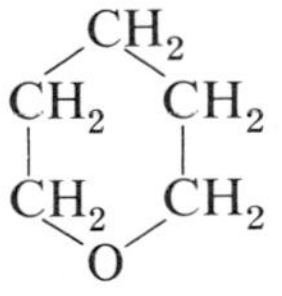

CH_2, CH_2 CH_2, CH_2 CH_2, O
Tetrahydropyran

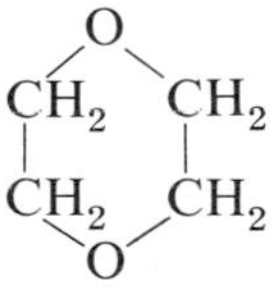

O, CH_2 CH_2, CH_2 CH_2, O
1,4-Dioxane
(Dioxane)

7.10 Properties of Ethers

Ethers boil at much lower temperatures than do the alcohols from which they are derived because the oxygen atom now is attached only to carbon. Thus, H-bonding between many molecules is not possible. The boiling points of the ethers closely parallel those of the alkanes of the same molecular weight. You will note that the methylene group,

—(CH_2)—, has almost the same formula weight (14) as an oxygen atom (16). Diethyl ether, with a molecular weight of 74, boils at 35°C; *n*-pentane, with a molecular weight of 72, boils at 36°C. The similarity may be illustrated also with di-*n*-propyl ether and *n*-heptane.

$$CH_3CH_2CH_2—O—CH_2CH_2CH_3 \qquad CH_3CH_2CH_2CH_2CH_2CH_2CH_3$$

Di-*n*-propyl ether M.W., 102; bp 91°C — *n*-Heptane M.W., 100; pb 98°C

The lower members of the aliphatic ethers are highly volatile and very flammable. Diethyl ether, the most important member of the family, is both an excellent organic solvent and a fine general anesthetic. Its high flammability, however, presents a hazard in the laboratory and in the operating room. The vapor of diethyl ether is heavier than air and has an annoying tendency to flow along the top of a laboratory bench and become ignited by a burner some distance away.

Diethyl ether, while not completely immiscible with water, is an excellent solvent to employ in extraction procedures involving aqueous solutions.[1]

7.11 Preparation of Ethers

A simple ether can be produced by the elimination of a molecule of water from two molecules of alcohol. The mechanism for this reaction is shown in Section 7.7B2. In practice, a mixture of the alcohol and sulfuric acid is heated to a temperature of approximately 140°C. An additional volume of alcohol is then added as the ether distills.

$$C_2H_5\boxed{—OH + H}—O—C_2H_5 \xrightarrow{H_2SO_4} C_2H_5—O—C_2H_5 + H_2O$$

Ethyl alcohol — Diethyl ether

Inasmuch as concentrated sulfuric acid may cause the elimination of a molecule of water intramolecularly, conditions must be carefully controlled to minimize this competing reaction. The several courses of reaction open to a mixture of alcohol and sulfuric acid have already been illustrated (Sec. 7.7B). The sulfuric acid method is employed for the preparation of simple ethers. A method applicable to the preparation of mixed ethers involves the reaction of a metallic alkoxide and an alkyl halide (Williamson's synthesis). The choice of reagents to employ in a Williamson ether synthesis, as was indicated in Section 7.7A-2, must be made with some care.

[1]At 25°C diethyl ether is soluble in water to the extent of 6%, and water is soluble in ether to the extent of approximately 1.5%. Ether dissolved in water, however, is easily removed by distillation.

Mixed ethers of the alkyl-aryl type may also be prepared by the Williamson method. The methyl and ethyl ethers of phenol are usually prepared by treating sodium phenoxide or phenol with the appropriate alkyl iodide or dialkyl sulfate in the presence of a base.

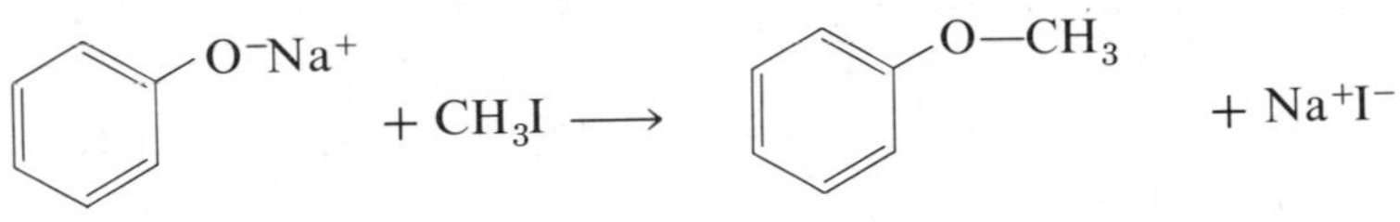

$$C_6H_5O^-Na^+ + CH_3I \longrightarrow C_6H_5\text{—}O\text{—}CH_3 + Na^+I^-$$

Methyl phenyl ether (Anisole)

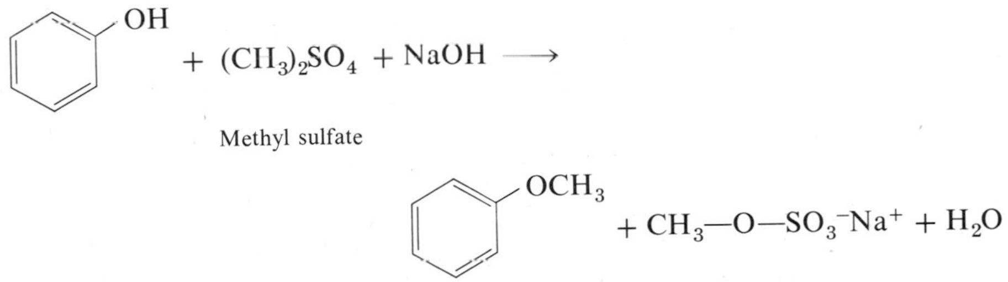

$$C_6H_5OH + (CH_3)_2SO_4 + NaOH \longrightarrow$$

Methyl sulfate

$$C_6H_5OCH_3 + CH_3\text{—}O\text{—}SO_3^-Na^+ + H_2O$$

Methyl phenyl ether (Anisole) — Sodium methyl sulfate

Exercise 7.9 A student wishes to prepare ethyl *tert*-butyl ether as a laboratory project. What reagents should be chosen for this synthesis? Why?

7.12 Reactions of the Ethers

The ethers represent an extremely stable group of compounds, and their reactions are few. They are soluble in strong mineral acids because of the formation of **oxonium** salts. The ethereal oxygen provides unshared electron pairs for bond formation with an acid. Solubility in sulfuric acid is thus a convenient method for distinguishing ethers from hydrocarbons and alkyl halides.

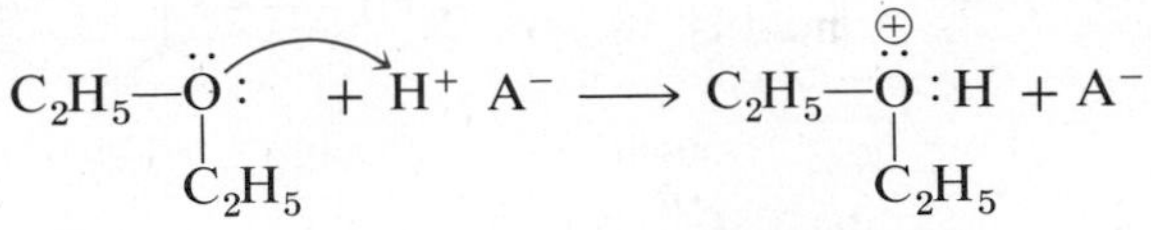

$$C_2H_5\text{—}\ddot{\underset{|}{O}}{:}(C_2H_5) + H^+\ A^- \longrightarrow C_2H_5\text{—}\overset{\oplus}{\ddot{O}}{:}H(C_2H_5) + A^-$$

Oxonium ion

Ethers are resistant to attack by the usual chemical oxidizing agents; yet anhydrous ethyl ether, when exposed repeatedly to air over long periods of time, forms a highly explosive peroxide.

$$CH_3{-}CH_2{-}O{-}CH_2{-}CH_3 + O_2 \xrightarrow[\text{radical reaction}]{\text{slow free}} CH_3{-}CH_2{-}O{-}\underset{}{\overset{OOH}{\overset{|}{C}H}}{-}CH_3 +$$

A hydroperoxide

$$\begin{array}{c} CH_3{-}CH_2{-}O{-}CH{-}CH_3 \\ | \\ O \\ | \\ O \\ | \\ CH_3{-}CH_2{-}O{-}CH{-}CH_3 \end{array}$$

A peroxide

Such ether peroxides are extremely dangerous, and anhydrous ether should be tested before it is distilled. In one test a sample of the ether is treated with an acidified solution of potassium iodide. If peroxides are present in the ether they will oxidize the iodide ion, I^-, to molecular iodine, I_2, yielding the characteristic brown color of iodine. Peroxides in ether may be removed by washing the ether with a ferrous sulfate solution. The peroxides oxidize ferrous ions to ferric and are thus eliminated.

Other reactions of the ethers involve a cleavage of the carbon–oxygen linkage. Concentrated hydriodic acid is an excellent reagent to employ for this cleavage. Each molecule of ether cleaved produces one equivalent of alkyl iodide and one equivalent of alcohol. If hydroiodic acid is used in excess, the alcohol initially formed is also converted to an alkyl iodide.

$$R{-}O{-}R' + HI \longrightarrow RI + R'OH$$

$$R{-}\ddot{\underset{..}{O}}{-}R' + HI \longrightarrow R{-}\overset{\oplus}{\ddot{O}}{:}H \;\; I^{\ominus} \xrightarrow[\text{(excess)}]{HI} RI + R'I + H_2O$$
$$\qquad\qquad\qquad\qquad\quad \diagdown R'$$

If one group of an ether is alkyl, such as methyl or ethyl, and the other group is aryl, such as phenyl, then the iodide of one of the alkyl groups is usually one product of the cleavage, the other being a phenol.

$$C_6H_5{-}OCH_3 + HI \xrightarrow{100°C} C_6H_5{-}OH + CH_3I$$

Anisole

Cleavage of ethers may also be accomplished by the use of concentrated (48%) hydrobromic acid.

Exercise 7.10 A student storeroom assistant one morning found that the following labels had come undone from their bottles and were lying on the floor: Ethyl alcohol, Ethyl bromide, Ethyl ether.

What simple test could the student have performed on a sample from each of the bottles in order to properly relabel them?

7.13 Epoxides, Polyhydric Alcohols

The simplest heterocyclic ethers are those with a three-membered ring, the **oxiranes** (IUPAC name), which are often called **epoxides** or **olefin oxides.** The epoxides perform important functions in nature but are perhaps best known as valuable industrial intermediates, being relatively easy to prepare by the **epoxidation** of alkenes with oxygen or peroxy compounds, or by the cyclization of a chlorohydrin by an internal Williamson reaction. These same reactions may be easily carried out in the laboratory.

Ethylene oxide is prepared industrially by the silver-catalyzed oxidation of ethylene with oxygen (obtained by the fractionation of air).

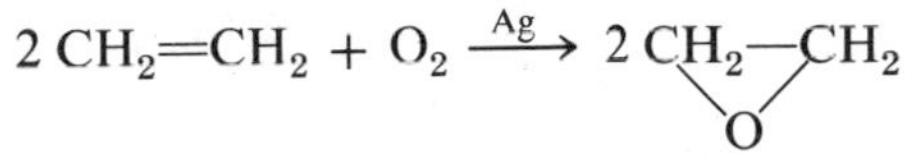

$$2\,CH_2{=}CH_2 + O_2 \xrightarrow{Ag} 2\,\underset{\diagdown\,O\,\diagup}{CH_2{-}CH_2}$$

Ethylene — Ethylene oxide (Oxirane)

This procedure works well only with ethylene; therefore, the higher epoxides must be made by one of the alternative routes cited. Although simple epoxides can be made easily from most alkenes by treatment with an alkyl hydroperoxide (R—O—O—H) or a peroxycarboxylic acid

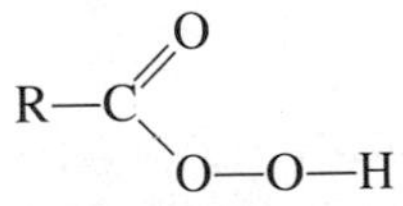

$$R{-}C({=}O){-}O{-}O{-}H$$

we will illustrate direct epoxidation with an example of an exciting, new asymmetric synthesis of epoxides derived from allylic alcohols, which have the structural unit

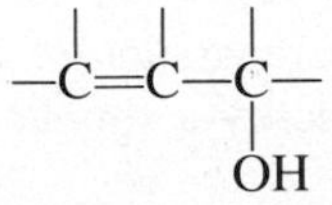

$$-\overset{|}{C}{=}\overset{|}{C}-\underset{OH}{\overset{|}{\underset{|}{C}}}-$$

The epoxidation of a double bond by either of the two peroxy compounds named is stereospecific in that both carbon-oxygen bonds of the resulting oxirane are formed at the same time. Because the oxygen is added across the double bond in a *syn* manner from either below or above the plane of the double bond, enantiomers result when the reaction is applied to an allylic alcohol:

$$\mathrm{H(CH_2OH)C{=}CH} \xrightarrow[\mathrm{CH_2Cl_2}]{\mathrm{R{-}C(=O){-}OOH}} \text{epoxide (O above)} + \text{epoxide (O below)}$$

As was explained in Sec. 5.6, chemists have discovered only a few asymmetric syntheses and none as simple, versatile, and inexpensive as the **Sharpless epoxidation.** The synthesis requires an allylic alcohol, titanium tetraisopropoxide, an ester of one of the two enantiomers of tartaric acid (Exercise 5.5), and *t*-butyl hydroperoxide. With the D-tartrate the oxygen is added from above the plane of the double bond, and with the L-tartrate, from below.

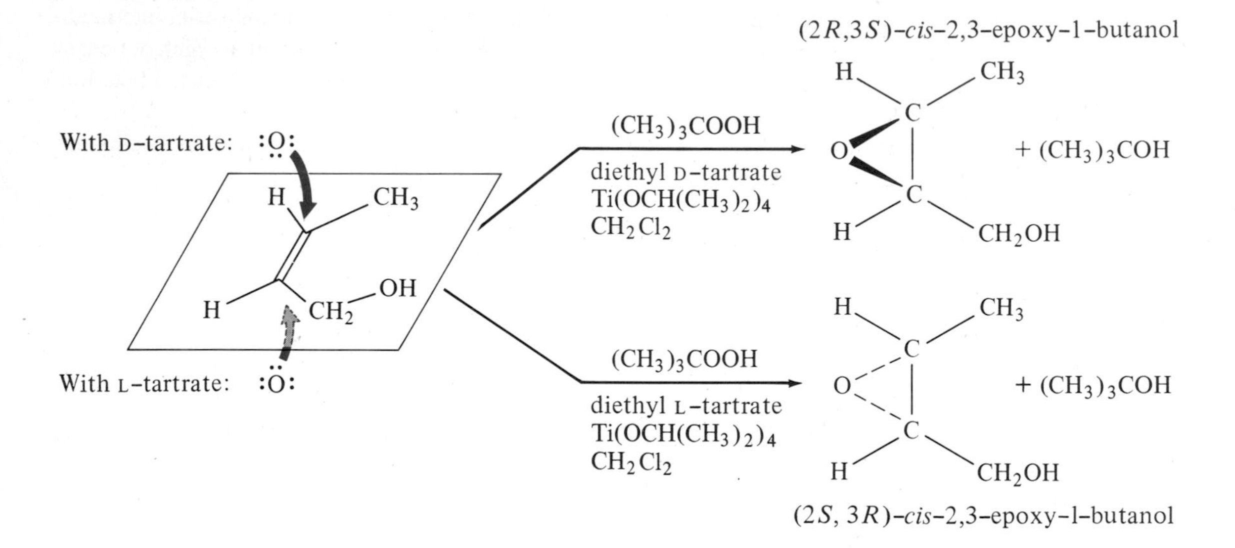

Although this reaction was first reported in 1980, it has already been used in numerous syntheses of antibiotics, insect pheromones, and families of sugar molecules. Its use resulted in the reduction of the price per gram of disparlure (Sec. 7.8) from $2000 to $250.

The use of the internal Williamson reaction may be illustrated by the current industrial process for the preparation of **epichlorohydrin.** About half of the propylene oxide manufactured is made by a similar cyclization.

$$\underset{\text{Propylene}}{\mathrm{CH_3{-}CH{=}CH_2}} + \mathrm{Cl_2} \xrightarrow{600\,^\circ\mathrm{C}} \underset{\text{Allyl chloride}}{\mathrm{Cl{-}CH_2{-}CH{=}CH_2}} + \mathrm{HCl}$$

$$\mathrm{Cl{-}CH_2{-}CH{=}CH_2 + \overset{(Cl_2+H_2O)}{HO^-Cl^+}} \longrightarrow$$

$$\underset{\text{(70\% 1,3-Dichloro-2-propanol + 30\% 2,3-Dichloro-1-propanol)}}{\mathrm{Cl{-}CH_2{-}\underset{OH}{CH}{-}\underset{Cl}{CH_2}}} \xrightarrow[90°]{Ca(OH)_2} \underset{\text{Epichlorohydrin (2-Chloromethyloxirane)}}{\mathrm{Cl{-}CH_2{-}CH{-}CH_2} \text{ (with O bridging the last two carbons)}}$$

Note that in the first step of the reaction, the chlorine replaces a hydrogen of the methyl group rather than adding to the double bond. This occurs because, at the high temperature of the reaction, substitution by a free-radical process involving the fairly stable allyl radical (Exercise 3.34) predominates over addition, and about 85% of the product is allyl chloride.

Like the other ethers, oxiranes undergo carbon–oxygen bond cleavage when treated with acids. However, because of the highly strained three-membered ring (as discussed in Sec. 2.9), much milder acidic conditions may be employed than with acyclic ethers. Furthermore, the ring may be opened by treatment with aqueous base, a reaction not shown by simple ethers. These reactions are generally used to convert epoxides into 1,2-diols, or **glycols.** Thus, **ethylene glycol,** the principal component in automobile antifreeze solutions, is made from the acid-catalyzed cleavage of ethylene oxide at moderate temperatures (or with water alone at higher temperatures and pressures), a reaction that involves the S_N2 attack of water on the protonated epoxide with an alcohol as the leaving group.

$$\mathrm{CH_2{-}CH_2\ (\text{with O bridging}) + H_2O \xrightarrow{HCl} \underset{OH}{CH_2}{-}\underset{OH}{CH_2}}$$

Ethylene glycol

$$\mathrm{H{-}\ddot{\underset{..}{O}}{-}H \;+\; CH_2{-}CH_2\ (\text{bridged by } {:}\overset{+}{O}{-}H) \longrightarrow \mathrm{H{-}\overset{..+}{O}{-}H \text{ on } CH_2{-}CH_2 \text{ with } {:}\ddot{O}H}}$$

Ethylene glycol is miscible with water in all proportions. Its great solubility in water, along with its high boiling point (197.5°C), makes it an excellent antifreeze, as it is not readily lost by evaporation. **Propylene glycol,** $CH_3CH(OH)CH_2OH$, is prepared from propylene oxide in a manner similar to that used for the preparation of ethylene glycol. Propylene glycol is much less toxic than ethylene glycol and is used as an emulsifying and softening agent in cosmetics; however, its principal application is in the preparation of polyurethane polymers.

Glycerol, or glycerine as it is commonly called, is a trihydroxy alcohol, or **triol.** Glycerol is prepared industrially from the base-catalyzed hydrolysis and ring cleavage of epichlorohydrin.

$$\text{Cl—CH}_2\text{—}\underset{\diagdown\,\text{O}\,\diagup}{\text{CH—CH}_2} + 2\,\text{H}_2\text{O} \xrightarrow{\text{NaOH}} \underset{\text{OH}}{\underset{|}{\text{CH}_2}}\text{—}\underset{\text{OH}}{\underset{|}{\text{CH}}}\text{—}\underset{\text{OH}}{\underset{|}{\text{CH}_2}}$$

Glycerol
(1,2,3-Propanetriol)

About half of the glycerol manufactured in the United States is obtained as a by-product in the manufacture of soaps from animal and vegetable fats and oils (Sec. 9.6A). Glycerol is a viscous, sweet-tasting liquid and, as might be expected with three hydroxyl groups present, is miscible with water in all proportions. It is an excellent humectant, or moisture-retaining agent, and is used in the tobacco and cosmetic industries. About 4% of the glycerol production is used for the preparation of ''nitroglycerine,'' which is converted to the explosive ''dynamite'' by adsorption on silica gel. Glycerol is also used in the production of plastics, synthetic fibers, and surface coatings (Sec. 11.4).

$$\begin{array}{l}\text{CH}_2\text{—OH}\\ |\\ \text{CH—OH}\\ |\\ \text{CH}_2\text{—OH}\end{array} + \text{HO—NO}_2 \xrightarrow{\text{H}_2\text{SO}_4} \begin{array}{l}\text{CH}_2\text{—O—NO}_2\\ |\\ \text{CH—O—NO}_2\\ |\\ \text{CH}_2\text{—O—NO}_2\end{array} + 3\,\text{H}_2\text{O}$$

Glycerol — Glyceryl trinitrate ("Nitroglycerine")

Glyceryl trinitrate is also used to promote the rapid relaxation of smooth muscles, especially of the blood vessels of the heart, where it functions as a vasodilator of short duration for the relief of angina pectoris.

7.14 Uses of Ethers

Diethyl ether was once used as the principal general inhalation anesthetic. It has been largely replaced by halothane and the halogenated ethers, enflurane and isoflurane (Sec. 6.3), which are less hazardous. In the laboratory diethyl ether is still a popular solvent, particularly for the preparation of Grignard reagents, although its use is avoided whenever possible in industrial applications because of its potential hazards. Tetrahydrofuran (commonly called ''THF'') and 1,4-dioxane are not only good organic solvents, but they are also miscible with water in all proportions. In addition, tetrahydrofuran is a good solvent for Grignard reagents and other anionic species. Dioxane is used as a solvent in a number of biological applications. Methyl *t*-butyl ether has been proposed as an additive to motor fuels to improve their octane ratings.

An interesting new class of ethers that has attracted considerable attention in recent years is the **crown ethers,** which are macrocyclic polymers of ethylene glycol. These ethers are given names of the form *m*-crown-*n*, where *m* is the number of atoms in the macrocyclic ring and *n* is the number of oxygens. The most common of these ethers is 18-crown-6, which has the unusual property of being able to solvate a potassium cation. The metallic cation is in a highly polar cavity, but the exterior framework is essentially hydrocarbon in nature; thus the ionic complex is soluble in nonpolar organic solvents. A spectacular demonstration of this property is the extraction of potassium permanganate from aqueous solution into a benzene solution containing 18-crown-6. For other cations, crown ethers with smaller or larger central cavities are required. Naturally occurring cyclic compounds of similar structure, or with some of the ether oxygens replaced by nitrogen, apparently are involved in the transport of ions across biological membranes.

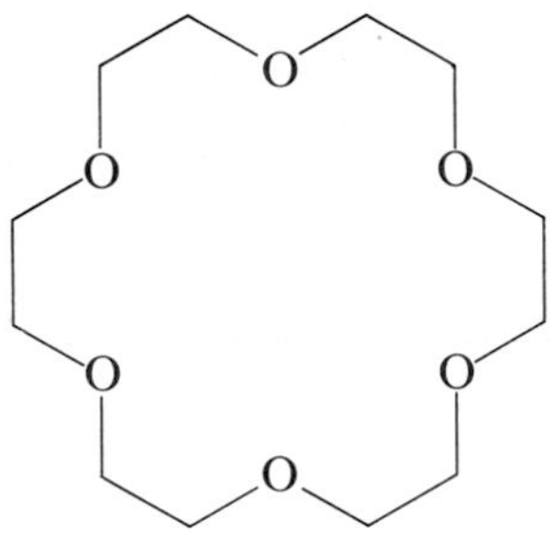

MnO_4^-

18-Crown-6

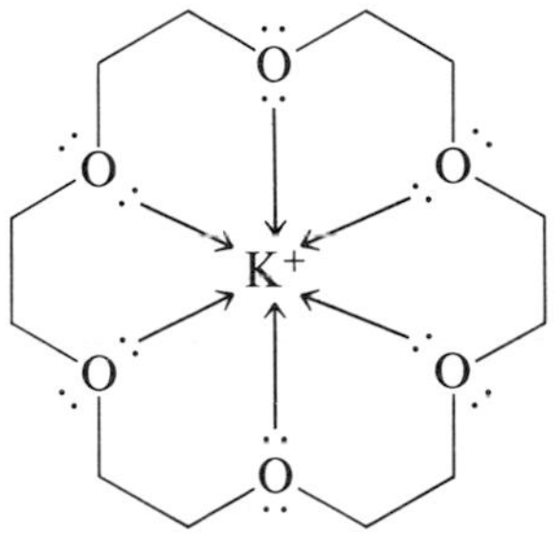

18-Crown-6 complex

The herbicides, 2,4-dichlorophenoxyacetic acid and 2,4,5-trichlorophenoxyacetic acid, popularly known as **2,4-D** and **2,4,5-T,** respectively, are ethers derived from halogenated phenols. The use of 2,4,5-T has been questioned because, if improperly manufactured, it may become contaminated with **dioxin,** one of the most toxic chlorine-containing compounds known.

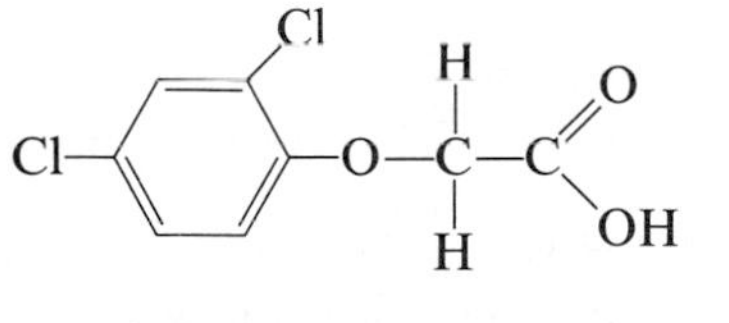

2,4-Dichlorophenoxyacetic acid
(2,4-D)

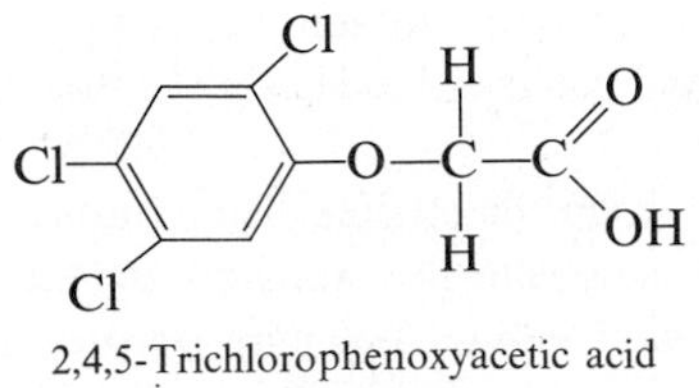

2,4,5-Trichlorophenoxyacetic acid
(2,4,5-T)

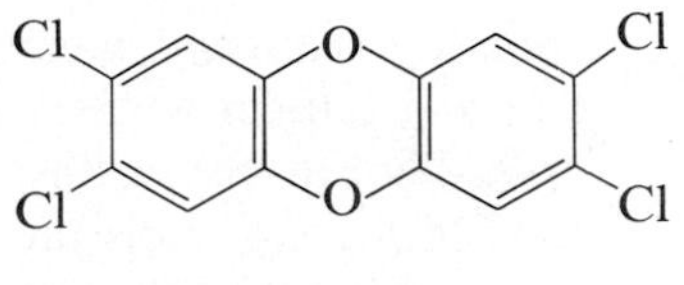

2,3,7,8-Tetrachlorodibenzo-*p*-dioxin
(Dioxin)

7.15 Organic Sulfur Compounds

The divalent sulfur compounds can be regarded as the sulfur analogs of the corresponding oxygen compounds. This relationship is shown in the parallelism of their nomenclature and, to a certain extent, their chemistry. The rules of nomenclature for the thiols or mercaptans (R—SH), sulfides or thioethers (R—S—R), and disulfides (R—S—S—R) will be apparent from the examples given. Note that the prefix *thio-* indicates that sulfur has replaced oxygen in an organic compound and that *alkylthio-* is the sulfur equivalent of *alkoxy-*. The —SH group is called the *sulfhydryl* group but is commonly referred to as the *mercapto* group. The latter term is used as a prefix if other groups are also present in a compound.

$CH_3—CH_2—SH$
Ethanethiol
(Ethyl mercaptan)

$HS—CH_2—CH_2—SH$
1,2-Ethanedithiol

$CH_3—CH=CH—CH_2—SH$
2-Butene-1-thiol

$CH_3CH_2—S—CH_2CH_3$
Diethyl sulfide
((Ethylthio)ethane)

$C_6H_{11}—S—CH_3$
Cyclohexyl methyl sulfide
((Methylthio)cyclohexane)

$C_6H_5—SH$
Thiophenol

$CH_3CH_2—S—S—CH_2CH_3$
Diethyl disulfide
((Ethyldithio)ethane)

$HS—CH_2—CH_2—OH$
2-Mercaptoethanol
(β-Mercaptoethyl alcohol)

Unlike the alcohols, the thiols show little tendency to form strong hydrogen bonds with the oxygen of water and are therefore less soluble than the corresponding alcohols. As nonassociated liquids, their boiling points also are much lower. For example, the boiling point of methyl mercaptan, CH_3SH, is 6° while that of its oxygen analog, methanol, is 65°.

Perhaps the most outstanding property of the lower-molecular weight thiols and sulfides is their disagreeable odor. The defensive mechanism of the skunk, when threatened, is to spray all about him with a mixture of butyl mercaptans. The odor we detect in a gas leak is not that of natural or L.P. gas, both of which are odorless, but that of a small amount of methyl sulfide added as an odorant and safety precaution. Certain mercaptans are also to be found in onions and garlic. Like H_2S, the lower molecular weight mercaptans and sulfides are very toxic.

Although the sulfhydryl group has been listed along with other nucleophiles (Sec. 6.5), the thiols are not satisfactorily prepared by the S_N2 reaction of sodium hydrosulfide (NaSH) with an alkyl halide, but may be prepared in better yield by the alkylation of the highly nucleophilic thiourea followed by basic hydrolysis.

$$(H_2N)_2C{=}S + R{-}Br \longrightarrow (H_2N)_2C^{+}{-}S{-}R\ Br^-$$

Thiourea — S-Alkylisothiouronium bromide

$$(H_2N)_2C^{+}{-}S{-}R\ Br^- \xrightarrow[\text{2. } H^+]{\text{1. } H_2O + NaOH} R{-}SH + (NH_2CN)_x$$

Thiol — Polymer

To prepare thiols from halides, which do not easily undergo S_N2 substitution, an alternative route is the treatment of alkyl or aryl Grignard reagents with sulfur.

$$C_6H_5{-}MgBr + S \xrightarrow{\text{ether}} C_6H_5{-}S{-}MgBr \xrightarrow[H^+]{H_2O} C_6H_5{-}SH$$

Phenylmagnesium bromide — Thiophenol

Sulfides are prepared by a variation of the Williamson synthesis.

$$CH_3CH_2{-}SH + NaOH \longrightarrow CH_3CH_2{-}S^-Na^+ + H_2O$$

Ethanethiol

$$CH_3CH_2{-}S^-Na^+ + CH_3{-}Br \longrightarrow CH_3CH_2{-}S{-}CH_3 + NaBr$$

Ethyl methyl sulfide

The greater acidity of ethanethiol (pK_a 10.5) relative to that of ethanol (pK_a 15.9) permits the preparation of the alkylthio (mercaptide) anion by treatment with aqueous sodium hydroxide rather than with metallic sodium.

The thiols and the disulfides form a redox system that is extremely important in biological systems. In the laboratory, thiols are readily oxidized to disulfides with *mild* oxidizing agents such as iodine, and disulfides are reduced to thiols with reagents such as lithium aluminum hydride or lithium metal in ammonia.

$$2\,R{-}SH \underset{LiAlH_4}{\overset{I_2}{\rightleftarrows}} R{-}S{-}S{-}R$$

Thiol — Disulfide

In nature the disulfide bond serves to bind together different parts of the same protein chain or two or more different protein chains (Sec. 15.7), and is involved in the functions of the coenzyme lipoic acid.

$$\text{(S–S ring: } CH_2{-}CH_2{-}CH\text{)}{-}CH_2CH_2CH_2CH_2{-}COOH$$

Lipoic acid

The ease with which the disulfide bond can be reduced and the sulfhydryl group oxidized provides the basis for the ''permanent'' wave when the reactions are applied to *keratin*, the fibrous protein of hair (Sec. 15.9).

Summary

1. Alcohols may be classified as *primary, secondary,* or *tertiary;* ethers are classified as simple or mixed.
2. Alcohols associate through H-bond formation. H-bonding accounts for their abnormally high boiling points. Ethers are incapable of H-bonding and are not associated. They are low boiling, good solvents, and rather inert chemically.
3. Phenols generally do not give the same reactions as do the aliphatic alcohols. Phenols have an acidic hydrogen. The hydroxyl group in phenols is a strong *ortho-para* director. Phenols are used in germicides, herbicides, fungicides, plastics, dyes, explosives, and many other useful everyday commodities.
4. **Nomenclature:**
   (A) Alcohols
      (a) (IUPAC) The suffix ''ol'' replaces terminal ''e'' of corresponding alkanes.
      (b) (Common) Group attached to —OH is named and followed by the word alcohol.
      (c) (Derived) All groups attached to the hydroxylated carbon are named as substituents of carbinol, e.g., CH_3OH.
   (B) Phenols
      (a) (IUPAC) Are named as hydroxy benzenes.
      (b) (Common) Are named as phenol derivatives.
   (C) Ethers
      (a) (IUPAC) Are named as alkoxy alkanes.
      (b) (Common) Are named simply as ethers. Both hydrocarbon groups are named.
5. **Preparation:**
   (A) Alcohols can be prepared by one of the following methods:
      (a) Synthetic methanol is prepared from CO and H_2.
      (b) Ethanol is prepared by fermentation of sugars.
      (c) Alcohols, in general, are prepared by the hydration of olefins (via H_2SO_4 or B_2H_6).
      (d) Hydrolysis of alkyl halides.
      (e) Addition of Grignard reagents to aldehydes and ketones.
   (B) Phenol is prepared from
      (a) Chlorobenzene.
      (b) Cumene (isopropylbenzene).
      (c) Benzenesulfonic acid.
   (C) Ethers are prepared by one of the following methods:
      (a) Simple ethers from alcohols by the sulfuric acid method.
      (b) Mixed ethers can be obtained via the Williamson synthesis.

6. **Reactions:** The reactions of alcohols, phenols, and ethers may be summarized as follows:
   (A) Alcohols
      (a) Cleavage of the O—H bond to produce alkoxides, ethers, and esters.
      (b) Cleavage of the C—O bond to produce alkyl halides, sulfates, and nitrates.
      (c) Dehydration to produce alkenes.
      (d) Oxidation of primary alcohols to acids; secondary alcohols to ketones.
   (B) Phenols
      (a) Replacement of H by reaction with a strong base.
      (b) Reduction of ring by catalytic hydrogenation.
      (c) Ring substitution at *o*- and *p*-positions.
   (C) Ethers
      (a) Formation of oxonium salts in strong acids.
      (b) Formation of peroxides.
      (c) Cleavage with HI or HBr.
7. Divalent sulfur compounds may be considered to be sulfur analogs of the alcohols, ethers, and peroxides. IUPAC nomenclature uses the suffix thiol to name the —SH group of thio alcohols (mercaptans). Sulfide and disulfide nomenclature parallels that of the ethers, the prefix alkylthio corresponding to alkoxy.
8. Preparation of sulfur compounds
   (a) Thiols are prepared from alkyl halides by successive treatment with thiourea and aqueous base or by treatment of a Grignard reagent with sulfur.
   (b) Sulfides are prepared by a Williamson synthesis using an alkyl halide and an alkali metal mercaptide.
   (c) Disulfides are prepared by oxidation of thiols with iodine.
9. The thiols and disulfides form a redox system that is important in biological systems.

Supplementary Exercises

7.11 Assign an acceptable name to each of the following compounds:

(a) $(CH_3)_2CH—O—CH(CH_3)_2$

(b) $(CH_3)_2CH—CHOH—CHCl—CH_3$

(c) $CH_3CH_2CH(OH)CH_3$

(d) $(CH_2CH_2O)_5$ (cyclic)

(e) $CH_3CH_2C(CH_3)_2OH$

(f) 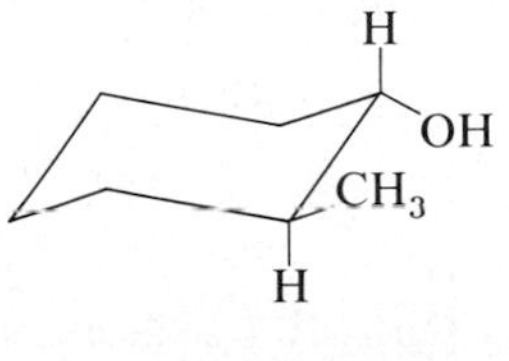 [cyclohexane chair: one carbon bearing H and OH; adjacent carbon bearing CH_3 and H]

(g) $CH_3CH(OH)CH(OH)CH_3$

(h)  HO—C_6H_4—CH_3 (para)

(i) 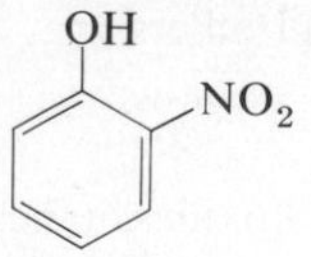 OH, NO_2 (ring structure)

(j) 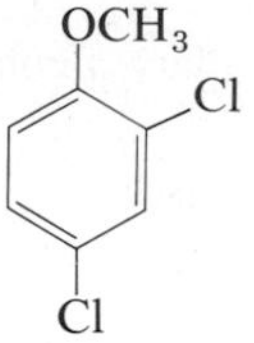 OCH_3, Cl, Cl (ring structure)

(k) 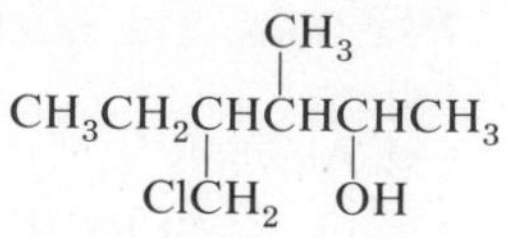 $CH_3CH_2CHCHCHCH_3$ with CH_3, $ClCH_2$, OH substituents

(l) 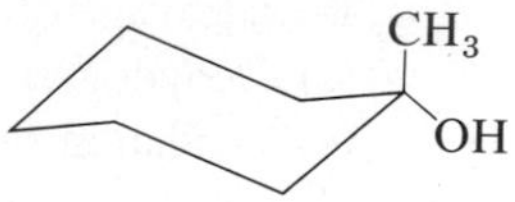 CH_3, OH (cyclohexane structure)

(m) 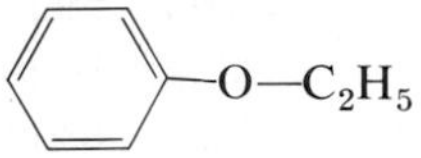 $-O-C_2H_5$ (ring structure)

7.12 Write structural formulas for the following:

(a) 4-methyl-1-naphthol
(b) 2-phenyl-2-propanol
(c) *p*-chlorobenzyl alcohol
(d) 2,6-di-*tert*-butylphenol
(e) 1-phenylethanol
(f) 2-butene-1-ol
(g) dibenzyl ether
(h) *trans*-3-phenylcyclopentanol
(i) potassium *m*-methylphenoxide
(j) sodium isopropoxide

7.13 Using ethyl alcohol as your only organic starting material, write equations showing how you would prepare the following.

(a) ethylene
(b) ethyl bromide
(c) acetylene
(d) 1-butyne
(e) ethylene chlorohydrin
(f) ethylene glycol
(g) ethylene oxide
(h) acetic acid
(i) iodoform
(j) *n*-butyl alcohol

7.14 Complete the following reactions.

(a) $CH_3-CH=CH_2 + HOCl \longrightarrow$

(b) $(CH_3)_2CHMgCl$ + ethylene oxide $\xrightarrow{\text{followed by hydrolysis}}$

(c) $(CH_3)_3COH + HCl \longrightarrow$

(d) *n*-propyl alcohol + $PI_3 \xrightarrow{\text{heat}}$

(e) 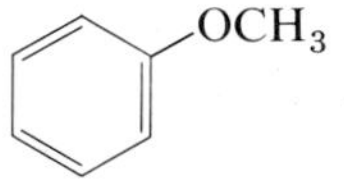 (benzene ring)$-OCH_3$ + HI $\xrightarrow{\text{heat}}$

(f) C_2H_5OH + sodium metal $\xrightarrow{\text{room temperature}}$

(g) *n*-propyl alcohol + $K_2Cr_2O_7 + H_2SO_4 \xrightarrow{\text{heat}}$

(h) 1-butene + $(BH_3)_2 \xrightarrow{H_2O_2,\ ^-OH}$

(i) 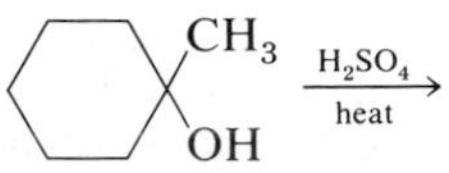 $C_6H_{10}(CH_3)(OH) \xrightarrow[\text{heat}]{H_2SO_4}$

(j) 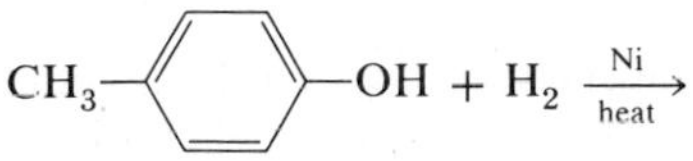 $CH_3-C_6H_4-OH + H_2 \xrightarrow[\text{heat}]{\text{Ni}}$

(k) 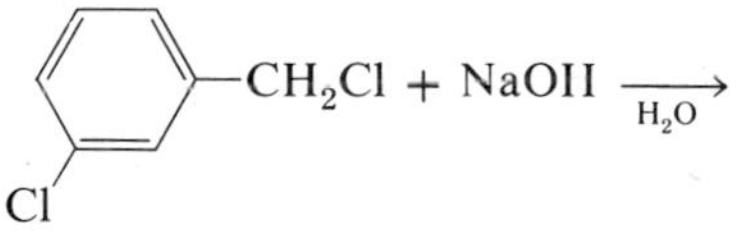 $Cl-C_6H_3$ (ring) $-CH_2Cl + NaOH \xrightarrow[H_2O]{}$

(l) 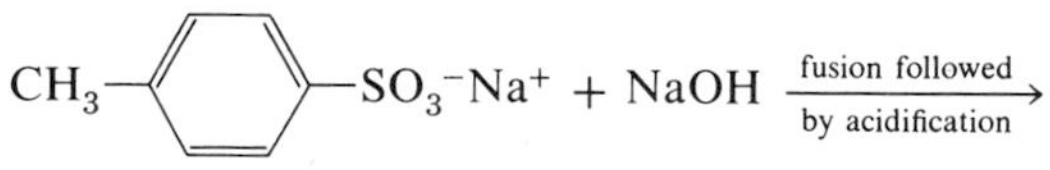 $CH_3-C_6H_4-SO_3^-Na^+ + NaOH \xrightarrow[\text{by acidification}]{\text{fusion followed}}$

(m) 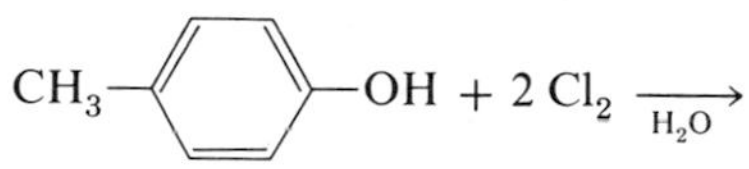 $CH_3-C_6H_4-OH + 2\,Cl_2 \xrightarrow[H_2O]{}$

(n) $CH_3CH_2-SH + KOH \longrightarrow$

(o) $CH_3CH_2-S-S-CH_2CH_3 \xrightarrow{LiAlH_4}$

(p) $CH_3CH_2-SH \xrightarrow{[O]}$

7.15 Show two different reaction pathways whereby each of the following may be converted to the products shown.

(a) propene $\longrightarrow$ 1-propanol
(b) 1-butanol $\longrightarrow$ 1-bromobutane
(c) styrene $\longrightarrow$ styrene oxide

7.16 Arrange the compounds in the following series: (A) in an increasing order of solubility in water, (B) in an increasing order of reactivity toward Lucas reagent, and (C) in an increasing order of reactivity with sodium metal.

(a) 1-butanol
(b) 2-butanol
(c) 2-methyl-2-propanol
(d) 2-methyl-1-propanol
(e) ethanol
(f) 2,3-dimethyl-2-butanol

7.17 Give a simple chemical test that would distinguish one from the other in each of the following pairs of compounds.

(a) *sec*-butyl alcohol and 1-hexyne
(b) isopropyl alcohol and isopropyl ether
(c) ethyl ether and *n*-pentane
(d) *tert*-butyl alcohol and *n*-butyl alcohol
(e) *sec*-butyl alcohol and *n*-butyl alcohol
(f) *p*-cresol and benzyl alcohol

7.18 Draw the structures of both *cis* and *trans*-cyclopentane-1,2-diol. The *trans* isomer has a boiling point somewhat higher than that found for the *cis* isomer. Does this higher boiling point indicate a greater degree of association for the *trans* isomer? Why would the *cis* isomer be less inclined to form H-bonds with its neighbors?

7.19 No mechanism was given for the reaction of epoxides under conditions of basic catalysis. Write a mechanism for the reaction of ethylene oxide with aqueous sodium hydroxide to form ethylene glycol. The leaving group is not a particularly good one. Why do you suppose that this reaction goes so well with one of the poorer leaving groups when simple ethers are not cleaved under similar conditions?

7.20 Write equations showing how propylene oxide (2-methyloxirane) would be expected to react with the following substances. In some cases isomeric products could form. Write the structures of both and indicate which you think would be the major product.

(a) aqueous sodium hydroxide
(b) ethylmagnesium bromide (followed by hydrolysis)
(c) hydrogen bromide (in a nonpolar solvent)
(d) sodium methoxide in methanol (followed by acidification)

7.21 Suggest structures for the compounds given letter names.

(a) An optically active compound A, $C_6H_{14}O$, reacts with sodium dichromate in acidic solution to form a new optically active compound B, $C_6H_{12}O$.
(b) Compound C, C_7H_8O, is soluble in aqueous sodium hydroxide solution. Hydrogenation of C with hydrogen and nickel gives D, $C_7H_{14}O$, which is insoluble in aqueous sodium hydroxide but does react with sodium metal with the liberation of hydrogen. Oxidation of D with sodium dichromate in acidic solution gives E, $C_7H_{12}O$. Compound E cannot be resolved into two enantiomeric forms.
(c) Compound F, C_9H_{10}, decolorizes both aqueous potassium permanganate solution and bromine in carbon tetrachloride solution. Ozonolysis of F yields G, C_8H_8O, and H, CH_2O. Treatment of F with diborane followed by oxidative workup with alkaline hydrogen peroxide yields I, $C_9H_{12}O$. Compound I can be resolved into two enantiomeric forms.

7.22 The cleavage of cyclohexene oxide with either aqueous acid or base gives only one of the two possible stereoisomeric glycols. Predict whether this will be the *cis*- or *trans*-glycol. Explain. (*Hint:* Sec. 3.7 and 7.13.)

7.23 Write equations showing how each of the following conversions may be made. Show all intermediate steps.

(a) propene ⟶ 1-propanol
(b) neopentyl alcohol ⟶ 2-methyl-2-butene
(c) 3,3-dimethyl-2-butanol ⟶ 2-chloro-2,3-dimethylbutane
(d) benzene ⟶ β-phenylethyl alcohol
(e) 3-methyl-1-butanol ⟶ 2-methyl-2-butene
(f) aniline ⟶

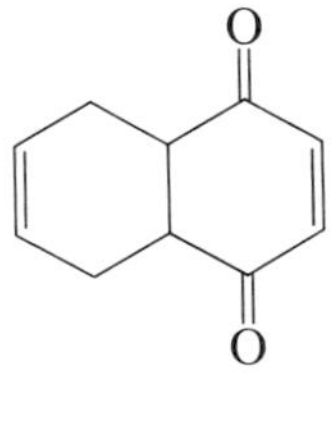

(g) 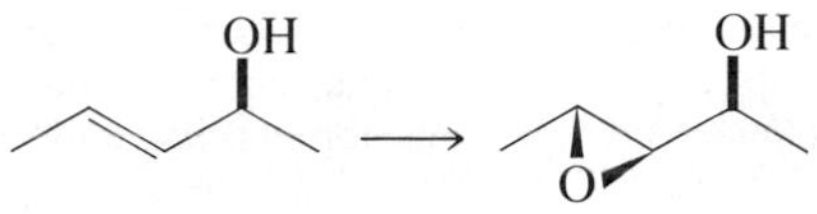 [structure: OH-bearing alkene] ⟶ [epoxide with OH]

(h) 2-methylcyclopentanol ⟶ 1-methylcyclopentanol
(i) styrene ⟶ 4-phenyl-2-methyl-2-butanol

7.24 One of the classic rearrangements of carbocation chemistry is the *pinacol–pinacolone rearrangement*.

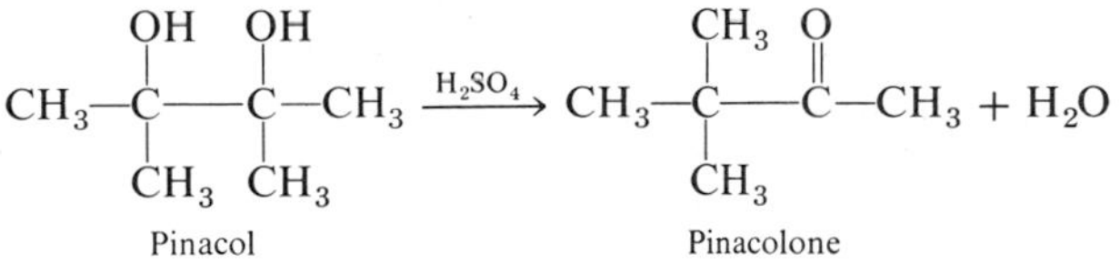

$$CH_3-\underset{CH_3}{\overset{OH}{C}}-\underset{CH_3}{\overset{OH}{C}}-CH_3 \xrightarrow{H_2SO_4} CH_3-\underset{CH_3}{\overset{CH_3}{C}}-\overset{O}{\overset{\|}{C}}-CH_3 + H_2O$$

Pinacol — Pinacolone

Suggest a mechanism for this rearrangement. (*Hint:* In acid-*catalyzed* reactions a proton usually is added in the first step and removed in the last step.)

7.25 The product in each of the following acid-catalyzed reactions results from a typical reaction of a carbocation intermediate. Suggest a sequence of steps (mechanism) that could lead to each product.

(a) [cyclohexanol with OH, H, CH_3, CH_3] $\xrightarrow{H^+}$ [cyclohexene with CH_3, CH_3]

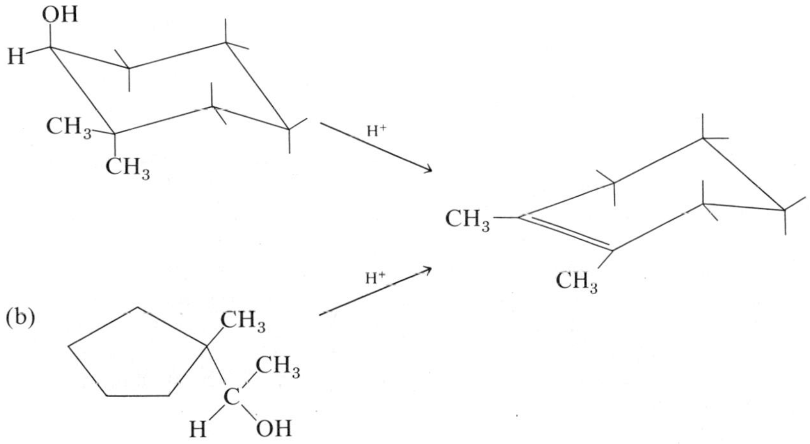

(b) [cyclopentane bearing CH_3 and C(H)(OH)CH_3] $\xrightarrow{H^+}$ [same cyclohexene product with CH_3, CH_3]

(c) $C_6H_5-CH=CH_2 \xrightarrow{H^+}$ [indane with C(H)(CH_3), CH_2, C(H)(phenyl)]

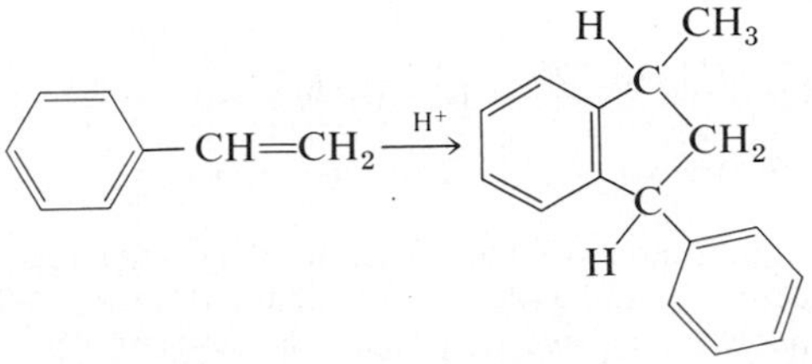

7.26 In the biosynthesis of cholesterol from squalene, the triterpene lanosterol is an important intermediate. Apparently the immediate precursor of lanosterol is squalene oxide. Suggest a mechanism for the formation of lanosterol from squalene oxide. The reaction is acid-catalyzed (H^+) and involves rearrangements of the type discussed in Sec. 7.7B.

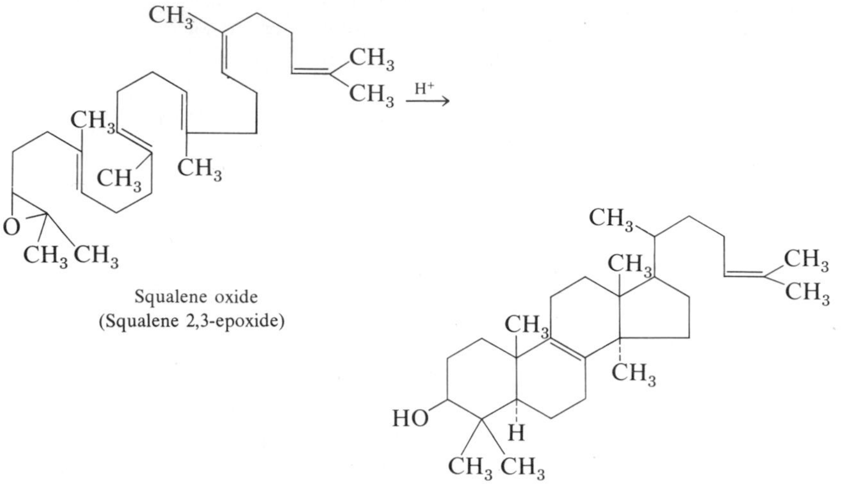

Squalene oxide
(Squalene 2,3-epoxide)

Lanosterol

7.27 The manufacturers of glycerol would like to avoid the use of chloro compounds (because of the relatively useless calcium and sodium chlorides formed). One possible alternative would be the epoxidation of allyl alcohol; thus what is needed is a chlorine-free route to allyl alcohol. One such route is the acid-catalyzed rearrangement of propylene oxide. Write a mechanism for the rearrangement of propylene oxide to allyl alcohol with aqueous acid.

***7.28** An optically active compound, $C_8H_{10}O$, gave an nmr spectrum consisting of a three-proton doublet at δ 1.33, a one-proton singlet at δ 2.75, a one-proton quartet at δ 4.67, and a five-proton singlet at δ 7.20. What are the name and structure of the compound?

***7.29** The infrared spectrum of a compound containing only carbon, hydrogen, and oxygen (molecular weight, 98) showed a broad, very strong band at 3400 cm^{-1}, a weak band at 2110 cm^{-1}, and the usual C—H bands at about 3000 cm^{-1}, but no other absorption above 1500 cm^{-1}. The nmr spectrum of the compound consisted of a one-proton singlet at δ 3.33, a one-proton singlet at δ 2.32, a two-proton quartet at δ 1.67, a three-proton singlet at δ 1.45, and a three-proton triplet at δ 0.95. What is the structure of the compound?

7.30 The reaction of sulfides with periodic acid (HIO_4) yields sulfoxides:

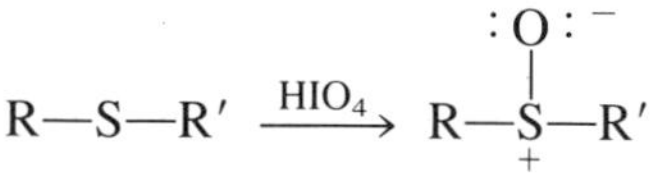

$$\underset{\text{Sulfide}}{\text{R—S—R}'} \xrightarrow{HIO_4} \underset{\text{Sulfoxide}}{\text{R—}\overset{\text{:}\ddot{O}\text{:}^-}{\underset{+}{S}}\text{—R}'}$$

If the two alkyl substituents on the sulfur atom are different, the resultant sulfoxide is found to be resolvable into a pair of enantiomeric sulfoxides (mirror-image isomers). What kind of structure must the sulfoxides have to allow the existence of optical isomerism?

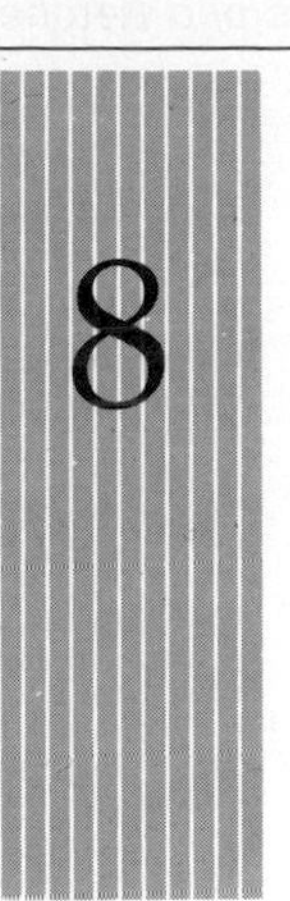

8 Aldehydes and Ketones

The aldehydes and ketones are compounds that contain the **carbonyl** group—a carbon–oxygen double bond,

$$-\overset{|}{C}=O$$

The carbonyl is one of the most frequently encountered and, from the standpoint of synthetic organic chemistry, one of the most useful of the functional groups. Compounds that contain one or more carbonyl groups also are widely distributed in nature. For the most part, the aldehydes and ketones are of pleasant odor and are responsible for the active principles in a number of delightful-smelling, natural substances. For this reason, certain aldehydes and ketones are used as perfumes and as flavoring agents.

8.1 Structure of Aldehydes and Ketones. The Carbonyl Group

The aldehydes and ketones are often referred to collectively as carbonyl compounds; however, the two families differ in structure and in properties. If the carbonyl group is attached to two hydrogens or to one hydrogen and one alkyl or aryl group, the compound is an aldehyde. If the carbonyl group is attached to two alkyl groups, to two aryl groups, or to one alkyl and one aryl group, the compound is a ketone. The various structural possibilities are illustrated in the following examples.

$$R-\overset{\overset{H}{|}}{C}=O$$

Aliphatic aldehyde

$$Ar-\overset{\overset{H}{|}}{C}=O$$

An aromatic aldehyde

$H-\overset{\overset{\large H}{|}}{C}=O$ — Formaldehyde

$CH_3-\overset{\overset{\large H}{|}}{C}=O$ — Acetaldehyde

$C_6H_5-\overset{\overset{\large H}{|}}{C}=O$ — Benzaldehyde

$R-\overset{\overset{\large O}{\|}}{C}-R$ — A simple aliphatic ketone

$R-\overset{\overset{\large O}{\|}}{C}-R'$ — A mixed aliphatic ketone

$Ar-\overset{\overset{\large O}{\|}}{C}-Ar$ — An aromatic ketone

$CH_3-\overset{\overset{\large O}{\|}}{C}-CH_3$ — Dimethyl ketone (Propanone)

$CH_3-\overset{\overset{\large O}{\|}}{C}-CH_2CH_3$ — Methyl ethyl ketone (2-Butanone)

$C_6H_5-\overset{\overset{\large O}{\|}}{C}-C_6H_5$ — Diphenyl ketone (Benzophenone)

$Ar-\overset{\overset{\large O}{\|}}{C}-R$ — An aliphatic-aromatic ketone

$C_6H_5-\overset{\overset{\large O}{\|}}{C}-CH_3$ — Methyl phenyl ketone (Acetophenone)

The state of hybridization of the carbonyl carbon atom is like that of the doubly bonded carbon atoms in the alkenes (Sec. 3.2), approximately sp^2. Thus, the carbonyl carbon forms three σ bonds with its three attached atoms, and these bonds lie in a plane separated by bond angles of approximately 120°. The remaining $2p$ orbital on the carbon atom overlaps with one $2p$ orbital on the oxygen atom to form a π bond between these atoms above and below the plane in which these atoms lie. The remaining orbitals of the oxygen atom hold two pairs of nonbonding (unshared) electrons. The nature of the bonding and the planarity of the carbonyl group are illustrated with structural formulas for formaldehyde.

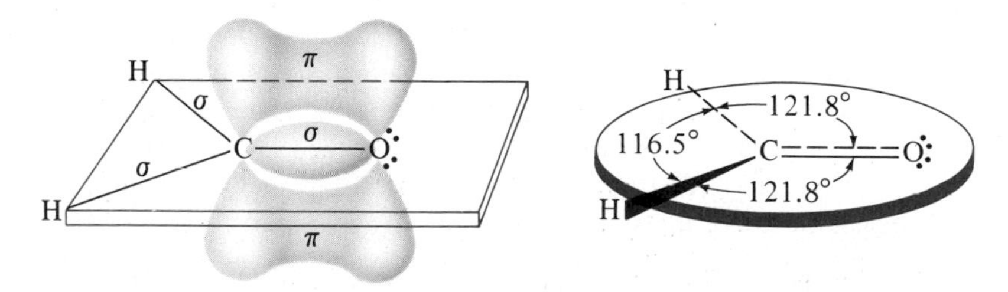

Since the oxygen atom is much more electronegative than the carbon atom (Sec. 1.8), the electrons in the double bond, especially the more mobile π electrons, will be drawn

toward the oxygen atom, causing the carbon–oxygen bond to be highly polarized. The oxygen will be electron-rich and the carbon will be electron-deficient, as shown by the resonance structures for the carbonyl group.

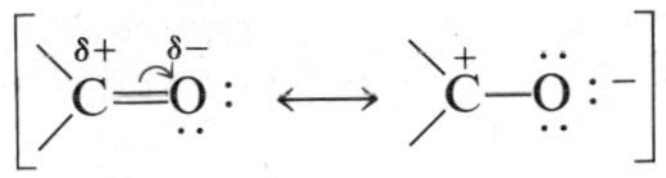

$$\left[\overset{\delta+}{>C} = \overset{\delta-}{\ddot{O}}: \longleftrightarrow \overset{+}{>C} - \ddot{O}:^- \right]$$

Because of this polarity the carbonyl compounds have fairly high dipole moments (Sec. 5.8) and react with bases and nucleophiles at the electron-deficient carbon atom and with acids and electrophiles at the weakly basic oxygen atom.

8.2 Nomenclature

Aldehydes are assigned common names derived from those of the carboxylic acids into which they are convertible by oxidation.

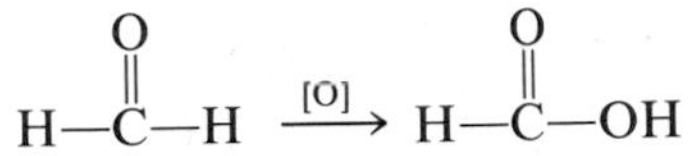

$$H-\overset{O}{\overset{\|}{C}}-H \xrightarrow{[O]} H-\overset{O}{\overset{\|}{C}}-OH$$

Formaldehyde — Formic acid

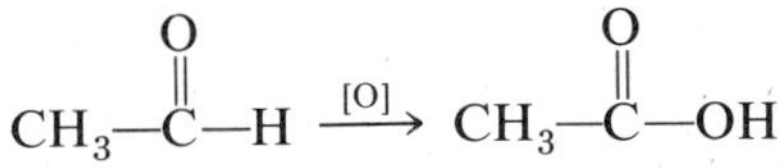

$$CH_3-\overset{O}{\overset{\|}{C}}-H \xrightarrow{[O]} CH_3-\overset{O}{\overset{\|}{C}}-OH$$

Acetaldehyde — Acetic acid

Common names for the simple ketones are formed by naming both groups attached to the carbonyl carbon atom, then adding the word *ketone*. Trivial names, long in use, also are commonly employed. The IUPAC system of nomenclature follows established rules. The longest chain including the carbonyl carbon is named after the parent hydrocarbon with *-al* added as a suffix to designate an *al*dehyde. The carbonyl carbon atom of an aldehyde is always number one in the carbon chain and takes precedence over other functional groups that may be present. The carbonyl carbon atom of acyclic ketones, on the other hand, may appear at any point in the chain and must be located by number. In cyclic ketones, cycloalkanones, and cycloalkenones, the carbonyl group is always number one. The suffix *-one* is used to designate a ket*one*. The examples illustrate these rules.

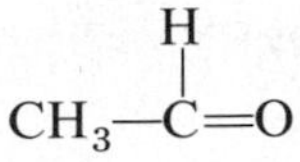

$$CH_3-\overset{H}{\overset{|}{C}}=O$$

Ethanal
(Acetaldehyde)

$$CH_3-\underset{CH_3}{\underset{|}{\overset{H}{\overset{|}{C}}}}-\overset{H}{\overset{|}{C}}=O$$

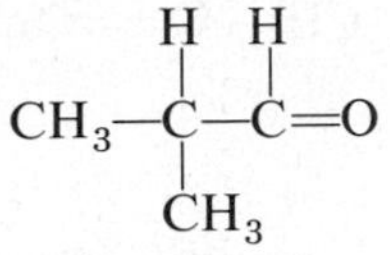

2-Methylpropanal
(Isobutyraldehyde)

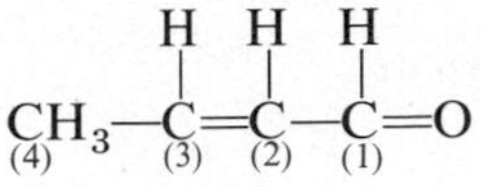

$$\underset{(4)}{CH_3}-\underset{(3)}{\overset{H}{\overset{|}{C}}}=\underset{(2)}{\overset{H}{\overset{|}{C}}}-\underset{(1)}{\overset{H}{\overset{|}{C}}}=O$$

2-Butenal
(Crotonaldehyde)

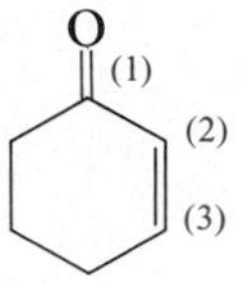

(cyclohexene ring with C=O at (1), double bond between (2) and (3))

2-Cyclohexenone

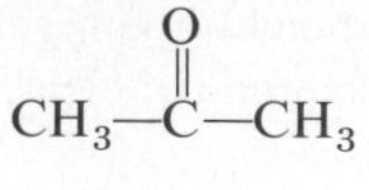

$$CH_3-\overset{O}{\overset{\|}{C}}-CH_3$$

Propanone
(Acetone)

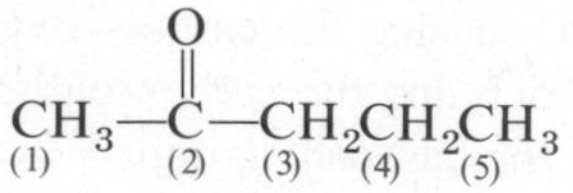

$$\underset{(1)}{CH_3}-\underset{(2)}{\overset{O}{\overset{\|}{C}}}-\underset{(3)}{CH_2}\underset{(4)}{CH_2}\underset{(5)}{CH_3}$$

2-Pentanone
(Methyl *n*-propyl ketone)

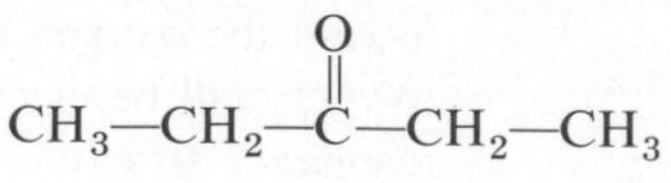

$$CH_3-CH_2-\overset{O}{\overset{\|}{C}}-CH_2-CH_3$$

3-Pentanone
(Diethyl ketone)

If the carbonyl group is attached to a benzene ring, ketones are given IUPAC names based on the names of the carboxylic acids from which they could be considered to be derived.

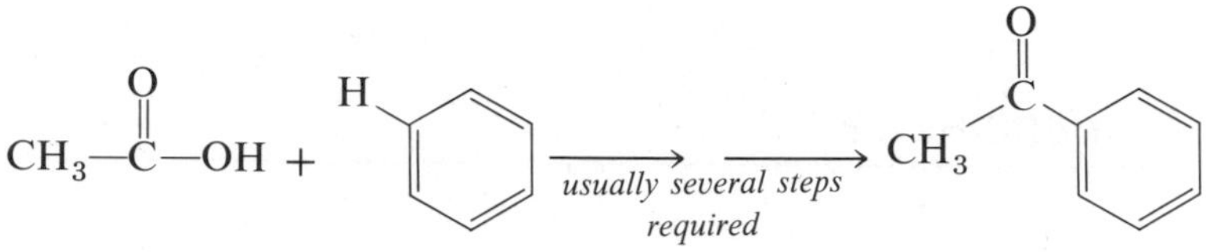

$$CH_3-\overset{O}{\overset{\|}{C}}-OH + C_6H_5-H \xrightarrow{\text{usually several steps required}} CH_3-\overset{O}{\overset{\|}{C}}-C_6H_5$$

Acetic acid → Acetophenone

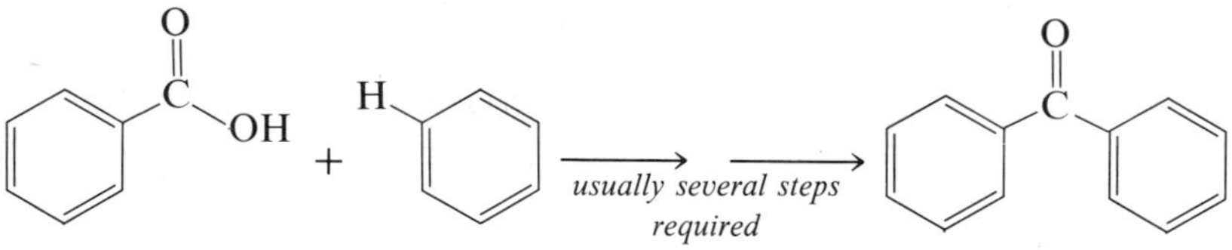

$$C_6H_5-\overset{O}{\overset{\|}{C}}-OH + C_6H_5-H \xrightarrow{\text{usually several steps required}} C_6H_5-\overset{O}{\overset{\|}{C}}-C_6H_5$$

Benzoic acid → Benzophenone

Usually the name is formed by replacing the *ic* or *oic acid* of the acid name by *ophenone* (exception: propionic acid → propiophenone).

In many naturally occurring substances, both *-al* and *-one* suffixes frequently are employed in nonsystematic names to indicate the presence of aldehyde or ketone functions. The stem of the name often indicates the source of the substance. For example, *civetone,* a ketone found in a glandular secretion of the civet cat, has a cyclic structure. Civetone in a very dilute solution has a pleasant odor and is used in perfumery. *Citral,* an aldehyde found in oil of citrus fruits, is an unsaturated aldehyde used as a flavoring agent. The structures of both natural carbonyl compounds are shown following.

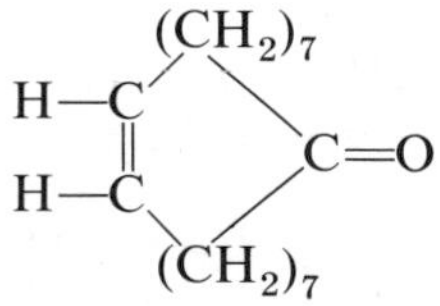

$$\text{H}-C=C-\text{H ring: } -CH=CH-(CH_2)_7-C(=O)-(CH_2)_7-$$

Civetone (secretion of civet cat)
(9-Cycloheptadecene-1-one)

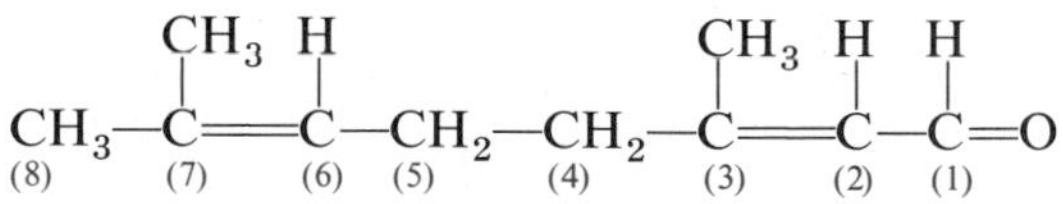

$$\underset{(8)}{CH_3}-\underset{(7)}{\overset{CH_3}{\overset{|}{C}}}=\underset{(6)}{\overset{H}{\overset{|}{C}}}-\underset{(5)}{CH_2}-\underset{(4)}{CH_2}-\underset{(3)}{\overset{CH_3}{\overset{|}{C}}}=\underset{(2)}{\overset{H}{\overset{|}{C}}}-\underset{(1)}{\overset{H}{\overset{|}{C}}}=O$$

Citral (oil of lemon)
(3,7-Dimethyl-2,6-octadienal)

EXAMPLE

Name the following structure according to IUPAC Rules.

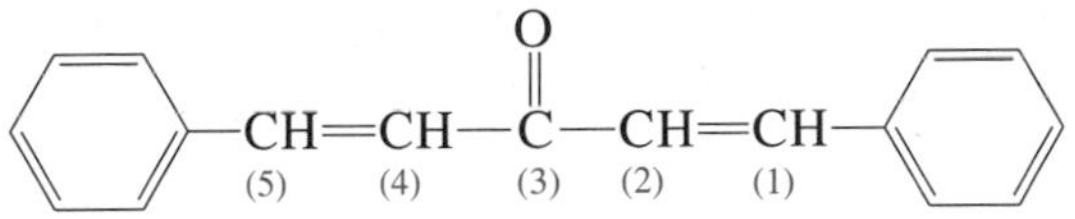

$$\text{C}_6\text{H}_5\text{—}\underset{(5)}{\text{CH}}\text{=}\underset{(4)}{\text{CH}}\text{—}\underset{(3)}{\overset{\overset{\text{O}}{\|}}{\text{C}}}\text{—}\underset{(2)}{\text{CH}}\text{=}\underset{(1)}{\text{CH}}\text{—C}_6\text{H}_5$$

Solution This ketone is a symmetrical structure; therefore, we may number the carbon chain from either end. Carbons 1 and 5 bear phenyl groups, and double bonds appear between carbons 1 and 4. Carbon 3 is designated the *one*. The name is 1,5-diphenyl-1,4-pentadien-3-one.

Exercise 8.1 The open-chain aldehydes and ketones are isomeric. Draw structures for seven carbonyl compounds with the molecular formula $C_5H_{10}O$. Assign IUPAC names to each.

Exercise 8.2 Assign acceptable names to the following:

(a) 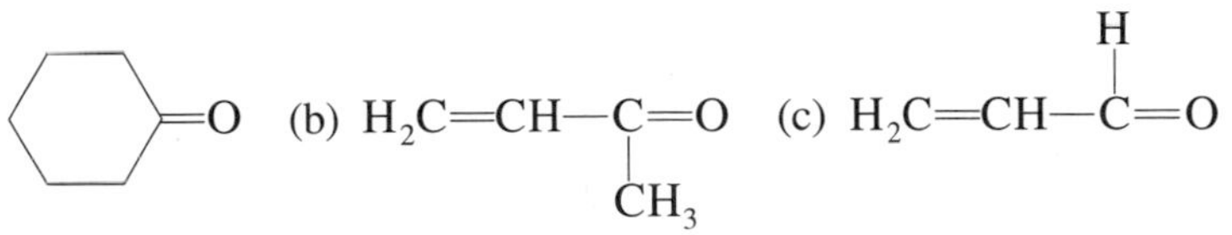 cyclohexane=O (b) $H_2C\text{=}CH\text{—}\underset{\underset{CH_3}{|}}{C}\text{=}O$ (c) $H_2C\text{=}CH\text{—}\overset{\overset{H}{|}}{C}\text{=}O$

8.3 Properties

With the exception of formaldehyde, a gas, the lower-molecular weight aldehydes and ketones are liquids that have higher boiling points than either the alkanes or alkenes, but lower boiling points than the alcohols of approximately the same molecular weight (Table 8.1). The interactions holding the molecules of the alkanes and alkenes together are the very weak van der Waals forces. The aldehydes and ketones are higher boiling because their molecular attraction is due to somewhat stronger dipole–dipole interactions, the slightly positive carbonyl carbon atom of one molecule being attracted to the slightly negative oxygen atom of another. The alcohols are the highest boiling because of even stronger hydrogen-bonding interactions. The lower-molecular weight aldehydes and ketones are soluble in water, in part because of hydrogen bonding between the carbonyl

TABLE 8.1 Boiling Points of Aldehydes and Ketones Compared to Those of Alcohols with the Same Carbon Chain

Compound bp, °C	Formula	Compound bp, °C	Formula
Formaldehyde (−21)	$CH_2{=}O$	Methyl alcohol (64.6)	$CH_3{-}OH$
Acetaldehyde (20.2)	$CH_3{-}CH{=}O$	Ethyl alcohol (78.3)	$CH_3{-}CH_2{-}OH$
Propionaldehyde (48.8)	$CH_3CH_2CH{=}O$	*n*-Propylalcohol (97.8)	$CH_3CH_2CH_2{-}OH$
Acetone (56.1)	$CH_3{-}C({=}O){-}CH_3$	Isopropyl alcohol (82.5)	$CH_3{-}CH(OH){-}CH_3$
n-Butyraldehyde (75.7)	$CH_3{-}CH_2CH_2{-}CH{=}O$	*n*-Butyl alcohol (117.7)	$CH_3CH_2CH_2CH_2{-}OH$
Methyl ethyl ketone (79.6)	$CH_3{-}CH_2{-}C({=}O){-}CH_3$	*sec*-Butyl alcohol (99.5)	$CH_3{-}CH_2{-}CH(OH){-}CH_3$

oxygen atom and the hydrogen atoms of the water molecule. The carbonyl–carbonyl and carbonyl–water interactions are illustrated in the following structures.

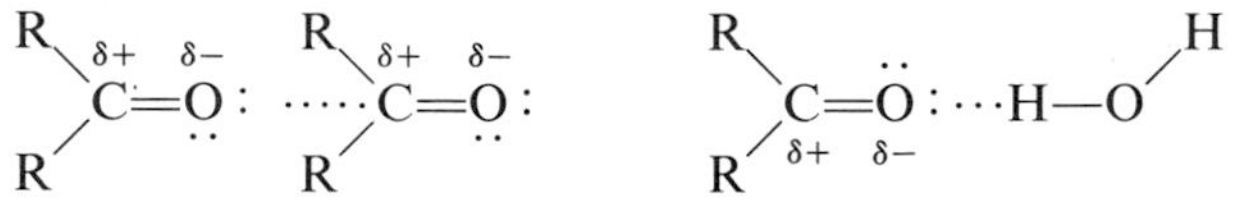

$$R_2\overset{\delta+}{C}{=}\overset{\delta-}{\ddot{O}}: \cdots\cdots R_2\overset{\delta+}{C}{=}\overset{\delta-}{\ddot{O}}: \qquad R_2\overset{\delta+}{C}{=}\overset{\delta-}{\ddot{O}}:\cdots H{-}O{-}H$$

The aldehydes and ketones, except the lower members containing up to four carbon atoms, are practically insoluble in water. The lower members of the aldehyde family have sharp, irritating odors, but the higher-molecular weight members and nearly all members of the ketone family are fragrant. As was pointed out in the introductory section, certain of these are used in perfumery and as flavoring agents.

8.4 Preparation of Aldehydes and Ketones

A-1. Oxidation of Alcohols (Dehydrogenation). The word *aldehyde* is a composite name originally given to this family of compounds to describe them as products of *al*cohol *dehyd*rogenation. The dehydrogenation of primary and secondary alcohols is an industrial method for the preparation of aldehydes and ketones, respectively. Dehydrogenation is accomplished by passing the alcohol vapors over a heated copper or silver catalyst.

$$R\!-\!\overset{\displaystyle H}{\underset{\displaystyle H}{C}}\!-\!O\!-\!H \xrightarrow[300\text{–}600^\circ C]{Cu} H_2 + R\!-\!\overset{\displaystyle H}{C}\!=\!O \qquad R\!-\!\overset{\displaystyle R}{\underset{\displaystyle H}{C}}\!-\!O\!-\!H \xrightarrow[300\text{–}600^\circ C]{Cu} H_2 + R\!-\!\overset{\displaystyle R}{C}\!=\!O$$

Chemical oxidation of a primary alcohol is one of the most direct methods for the preparation of the simple aldehydes. Formaldehyde and acetaldehyde, perhaps the most important members of the aliphatic series of aldehydes, may be produced by oxidation of methyl and ethyl alcohols, respectively. The oxidation is generally done in air with the help of a catalyst.

$$2\,H\!-\!\overset{\displaystyle H}{\underset{\displaystyle H}{C}}\!-\!OH + O_2\,(\text{air}) \xrightarrow[250^\circ C]{Cu,} 2\;\underset{\text{Formaldehyde}}{H\!-\!\overset{\displaystyle H}{C}\!=\!O} + 2\,H_2O$$

Aldehydes are especially susceptible to further oxidation in the presence of water. Therefore, an aldehyde formed by the oxidation of an alcohol will be converted directly into an acid unless some provision is made for its removal, or unless the oxidation is done under anhydrous conditions with a special reagent.

$$\underset{\text{Primary alcohol}}{R\!-\!\overset{\displaystyle H}{\underset{\displaystyle H}{C}}\!-\!OH} \xrightarrow{[O]} \underset{\text{Aldehyde}}{R\!-\!\overset{\displaystyle H}{C}\!=\!O} \xrightarrow[\substack{H_2O\\ \text{fast}}]{[O]} \underset{\text{Acid}}{R\!-\!\overset{\displaystyle O}{\overset{\|}{C}}\!-\!OH}$$

$$\underset{\substack{\text{Ethyl alcohol}\\ (\text{bp, } 78.3^\circ C)}}{3\,CH_3CH_2OH} + Cr_2O_7^{2-} + 8\,H^+ \xrightarrow{50^\circ C} \underset{\substack{\text{Acetaldehyde}\\ (\text{bp, } 20.8^\circ C)}}{3\,CH_3\overset{\displaystyle H}{C}\!=\!O} + 2\,Cr^{3+} + 7\,H_2O$$

$$3\,CH_3\!-\!\overset{\displaystyle H}{C}\!=\!O + Cr_2O_7^{2-} + 8\,H^+ \longrightarrow \underset{\substack{\text{Acetic acid}\\ (\text{bp, } 118.1^\circ C)}}{3\,CH_3\!-\!\overset{\displaystyle O}{\overset{\|}{C}}\!-\!OH} + 2\,Cr^{3+} + 4\,H_2O$$

The boiling points of the aldehydes are not only much lower than those of the alcohols from which they may be obtained (Table 8.1), but are, as illustrated in the example just given, also considerably lower than the boiling points of the corresponding acids they would yield on further oxidation. This difference in physical properties is fortunate, for it makes possible the removal of the lower molecular weight aldehydes as they are formed by simple distillation, and prevents further oxidation. For less volatile aldehydes, various complexes between chromium trioxide and the heterocyclic amine, pyridine, may be used as oxidants. The most popular complex is a complex between chromium trioxide, pyridine, and hydrogen chloride, called pyridinium chlorochromate or PCC. However, for acid-sensitive compounds pyridinium dichromate (PDC) is better. These complexes are used in methylene chloride solution.

$$CH_3(CH_2)_4-\underset{H}{\overset{H}{C}}-OH \xrightarrow[\text{or } (C_5H_5NH)_2^+Cr_2O_7^= \text{ (PDC)},\ CH_2Cl_2]{CrO_3(C_5H_5N)HCl \text{ (PCC)}} CH_3(CH_2)_4-C(=O)H$$

1-Hexanol Hexanal

Ketones also may be prepared by oxidation methods using secondary alcohols as starting material. Ketones, unlike the aldehydes, are not easily oxidized because carbon–carbon bonds must be broken, and can be made in high yield with several oxidizing agents, including PCC and PDC.

$$CH_3-\underset{OH}{\overset{H}{C}}-CH_3 \xrightarrow{[O]} CH_3-\underset{\|}{\underset{O}{C}}-CH_3$$

2-Propanol (Isopropyl alcohol) 2-Propanone (Acetone)

A-2. Oxidation of Alkenes (Ozonolysis). The oxidation of alkenes as a possible route to carbonyl compounds is discussed in some detail in Sec. 3.9B.

B. Hydration of Alkynes. Acetaldehyde and certain ketones may be made by hydration of alkynes. This reaction is discussed in Sec. 3.18B.

C. Friedel–Crafts Acylation. Certain aromatic aldehydes may be prepared by the direct introduction of the aldehyde group into the benzene nucleus through a modified Friedel–Crafts reaction known as **formylation.** Formyl chloride,

$$H-\overset{O}{\overset{\|}{C}}-Cl$$

is an unstable compound and decomposes to carbon monoxide and hydrogen chloride. A combination of these two gases in the presence of anhydrous aluminum chloride accomplishes the same result as if the acid chloride were used. The reaction is known as the Gatterman–Koch reaction.

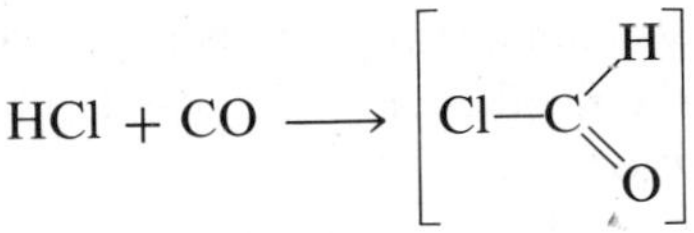

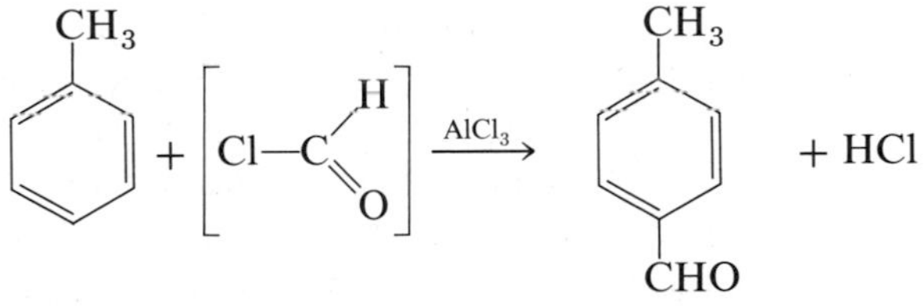

$$HCl + CO \longrightarrow \left[Cl{-}C{\overset{H}{\underset{O}{\lessgtr}}} \right]$$

$$\text{Toluene} + \left[Cl{-}C(H){=}O \right] \xrightarrow{AlCl_3} p\text{-Tolualdehyde} + HCl$$

Toluene

p-Tolualdehyde

Aromatic ketones also may be prepared by the Friedel–Crafts acylation reaction using either acid chlorides or acid anhydrides. In either case, an acyl group,

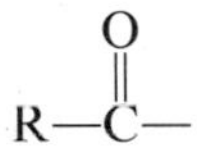

$$R{-}\overset{\overset{\displaystyle O}{\|}}{C}{-}$$

is attached directly to the aromatic ring. The method is illustrated in the preparation of acetophenone using acetyl chloride, the acid chloride of acetic acid.

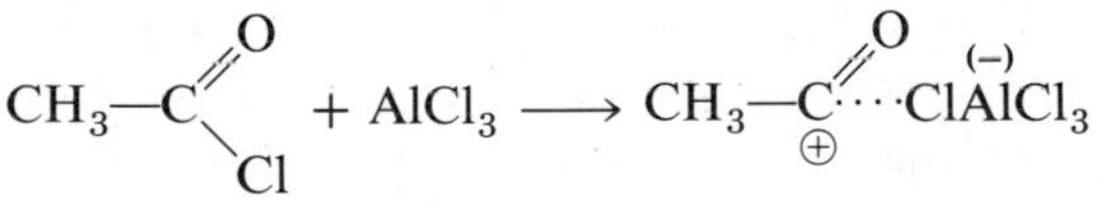

$$CH_3{-}C(=O){-}Cl + AlCl_3 \longrightarrow CH_3{-}\overset{\oplus}{C}(=O)\cdots Cl\overset{(-)}{Al}Cl_3$$

Acetyl chloride

$$C_6H_6 + \overset{\oplus}{C}(CH_3)(=O)\cdots Cl\overset{(-)}{Al}Cl_3 \longrightarrow C_6H_5{-}\overset{\overset{\displaystyle O}{\|}}{C}{-}CH_3 + HCl + AlCl_3$$

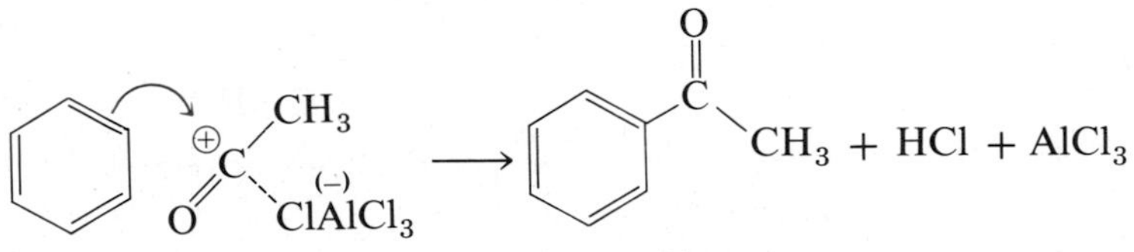

Acetophenone

Exercise 8.3 The Friedel–Crafts acylation reaction, unlike the alkylation reaction, produces only a monosubstituted product. Explain.

D. Industrial Procedures. **Formaldehyde** is manufactured from methanol by two processes: (1) dehydrogenation using a silver or copper catalyst, or (2) oxidation using a zinc-chromium or iron-molybdenum oxide catalyst. Although the silver catalyst is expensive, the dehydrogenation is so cleverly designed that no silver is lost and the catalyst is easily regenerated when necessary. Furthermore, the hydrogen evolved may be burned to generate some of the required heat of reaction (in which instance the process is called oxidative dehydrogenation).

About 80% of the **acetone** produced in the United States results as a coproduct in the preparation of phenol (Sec. 7.5). Most of the remainder comes from the dehydrogenation of isopropyl alcohol. In other countries the direct oxidation of propylene with oxygen or air, catalyzed by a mixture of palladium and cuprous chlorides (the Wacker Process), is an important alternative. Some acetone is still obtained as a by-product in the Weizmann fermentation of carbohydrates. The Weizmann fermentation, a process discovered by Chaim Weizmann[1] during World War I, is a bacterial fermentation. Sugar from either corn mash or blackstrap molasses is inoculated with the bacterium *Clostridium acetobutylicum*. Under anaerobic (''without air'') conditions the fermentation produces *n*-butyl alcohol, acetone, and ethyl alcohol in approximate yields of 60%, 30%, and 10%, respectively.

Acetone is used principally as a solvent and as an intermediate in organic synthesis.

Acetaldehyde, which was at one time prepared by the hydration of acetylene, is now made largely from ethylene by the Wacker process or, to a much lesser extent, by dehydrogenation of ethanol using copper or silver catalysts. Acetaldehyde is used principally in the production of acetic acid and acetic anhydride. However, these substances now are being made increasingly from methanol and carbon monoxide, causing a dramatic decline in the manufacture of acetaldehyde.

Furfuraldehyde, $C_5H_4O_2$, usually called **furfural,** is a very important industrial aldehyde obtained from agricultural wastes such as corn cobs and oat hulls. It is a colorless liquid when freshly distilled, but on exposure to air it becomes oxidized to a deep brown or black liquid. Treatment of furfural with a mineral acid results in a ring cleavage. In neutral or basic solutions furfural gives all the reactions of benzaldehyde. Furfural is used in the petroleum industry for the refining of lubricants and as a starting material for certain synthetic resins. Since six tons of corn cobs are required to produce one ton of furfural, the high cost of collection and transportation, and the relatively modest yield, have limited the use of furfural in the United States.

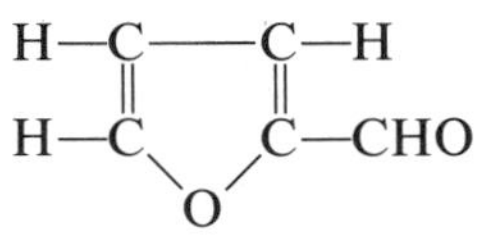

```
H—C———C—H
   ‖     ‖
H—C     C—CHO
    \   /
      O
```

Furfuraldehyde
(Furfural)

[1] Chaim Weizmann, 1874–1952. University of Manchester, England. First president of Israel.

8.5 Chemical Characteristics of Carbonyl Compounds

The principal chemical characteristic of the carbonyl compounds is the highly polar carbonyl group. Thus, in the addition reactions illustrated in the following sections, bases and nucleophiles (Sec. 1.13) attack the electron-deficient carbonyl carbon atom, and acids and electrophiles (Sec. 1.13) attack the basic oxygen atom. Therefore, with most reagents carbonyl addition reactions show the same overall course: addition of the negative nucleophilic part of the reagent to the carbon atom and addition of the positive electrophilic part of the reagent to the oxygen atom.

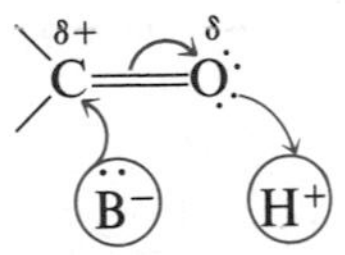

A carbonyl group attached to an aromatic ring behaves as an electron-withdrawing group and deactivates the ring toward electrophilic substitution. The ring is made more resistant than benzene to such substitution, and the carbonyl group directs incoming groups to the *meta* position when substitution does take place. These effects may be attributed to resonance interaction between the carbonyl group and the aromatic ring. One result of this interaction is a weakening of the positive charge on the carbonyl carbon atom through dispersal of the charge into the ring. As a result, carbonyl groups attached to aromatic rings are less reactive in addition reactions than are aliphatic and alicyclic carbonyl compounds.

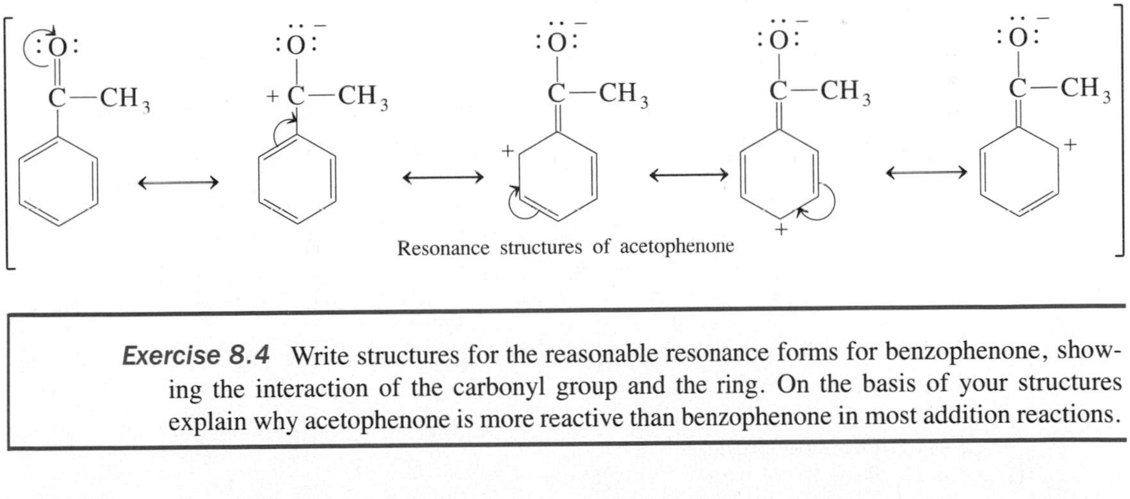

Resonance structures of acetophenone

Exercise 8.4 Write structures for the reasonable resonance forms for benzophenone, showing the interaction of the carbonyl group and the ring. On the basis of your structures explain why acetophenone is more reactive than benzophenone in most addition reactions.

Another important characteristic of carbonyl compounds is the acidity of hydrogen atoms on carbon atoms *alpha* to the carbonyl group (called α-hydrogens). We first encountered carbon acids in the alkynes (Sec. 3.15). Acetone is about 100,000 times stronger as an acid than acetylene. Because of the reactivity of the α-hydrogens, aldehydes and ketones may exist in solution as equilibrium mixtures of two isomeric forms, a **keto** form and an **enol** form.

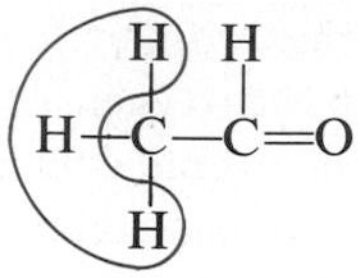

$$\mathrm{H{-}C(H)(H){-}C(H){=}O}$$

Acetaldehyde
(three α-hydrogens)

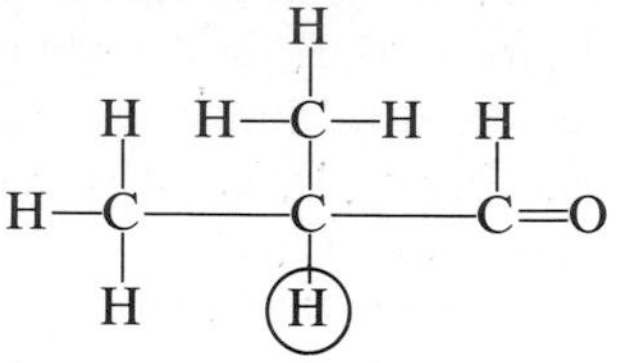

$$\mathrm{H{-}C(H)(H){-}C(H)(CH_3){-}C(H){=}O}$$

Isobutyraldehyde
(one α-hydrogen)

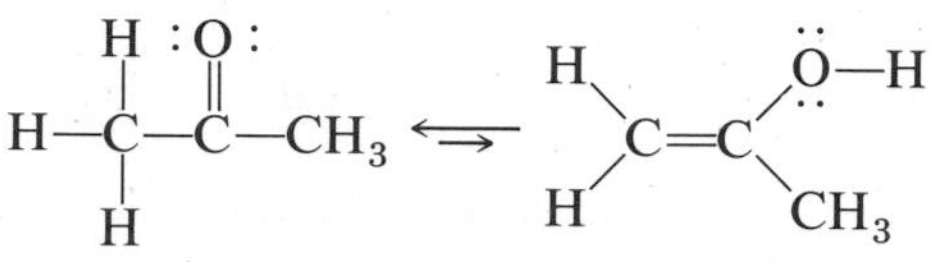

$$\mathrm{H{-}C(H)(H){-}C(=O){-}CH_3 \rightleftharpoons (H)_2C{=}C(OH)CH_3}$$

Keto form of acetone — Enol form of acetone

This type of isomerism is called **tautomerism,** and the isomers are known as **tautomers.** For most simple carbonyl compounds, very little enol is present at equilibrium, but the equilibrium is established very rapidly. Therefore, enols are important intermediates in many of the reactions of carbonyl compounds. Enols may be stabilized by conversion of the hydroxyl group to ethers or esters (Sec. 10.4). An example is the important, high-energy metabolic intermediate, phosphoenolpyruvate (PEP), shown in the acid form.

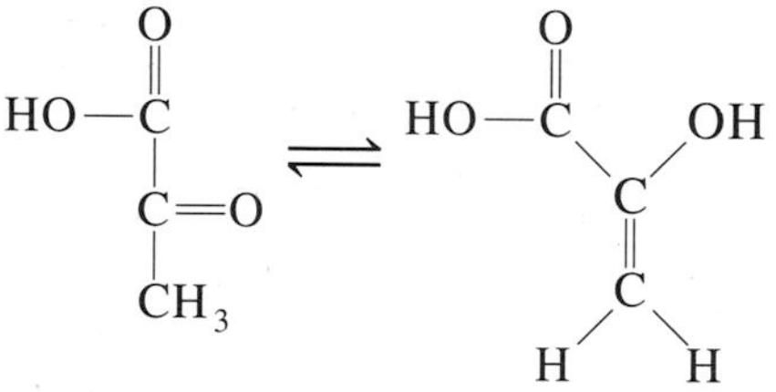

$$\mathrm{HO{-}C(=O){-}C(=O){-}CH_3 \rightleftharpoons HO{-}C(=O){-}C(OH){=}CH_2} \qquad \mathrm{HO{-}C(=O){-}C(O{-}P(=O)(OH)_2){=}CH_2}$$

Pyruvic acid — Pyruvic acid enol 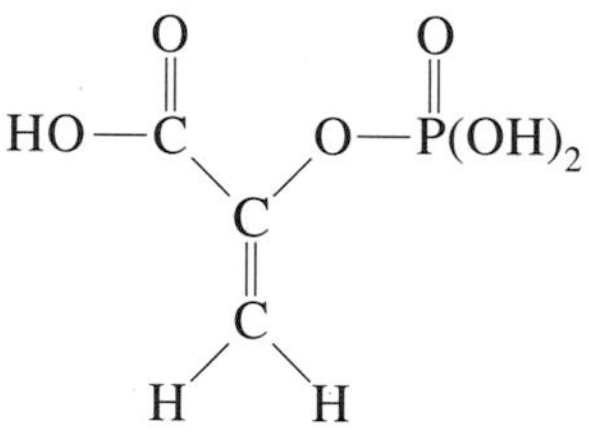 — Phosphoric acid ester of the enol

Enolization is catalyzed by both acids and bases as is shown briefly in the following equations:

Base-catalyzed enolization:

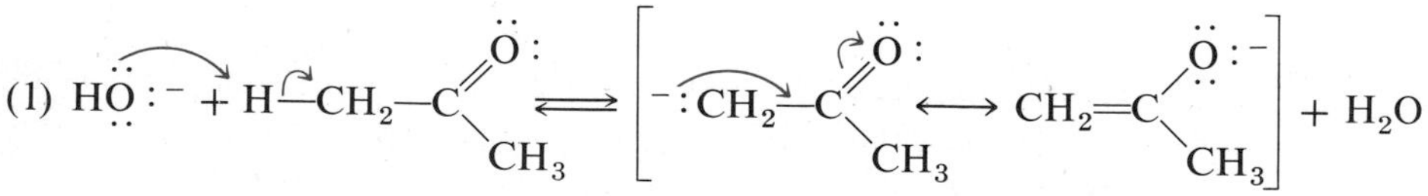

$$\text{(1)}\quad \mathrm{HO{:}^- + H{-}CH_2{-}C(=O)CH_3 \rightleftharpoons \left[{}^-{:}CH_2{-}C(=O)CH_3 \longleftrightarrow CH_2{=}C(O{:}^-)CH_3\right] + H_2O}$$

Enolate anion

$$(2)\ \left[{}^{-}\!:\!CH_2{-}C(=\ddot{O}:)CH_3 \longleftrightarrow CH_2{=}C(\ddot{O}:^{-})CH_3 \right] + H{-}\ddot{O}H \rightleftharpoons CH_2{=}C(\ddot{O}{-}H)CH_3 + :\ddot{O}H^{-}$$

Acid-catalyzed enolization:

$$(1)\ H{-}CH_2{-}C(=\ddot{O}:)CH_3 + H^{+} \rightleftharpoons \left[H{-}CH_2{-}C(=\dot{O}^{+}{-}H)CH_3 \longleftrightarrow H{-}CH_2{-}\overset{+}{C}(\ddot{O}{-}H)CH_3 \right]$$

Protonated acetone (oxonium ion)

$$(2)\ \left[H{-}CH_2{-}C(=\dot{O}^{+}{-}H)CH_3 \longleftrightarrow H{-}CH_2{-}\overset{+}{C}(\ddot{O}{-}H)CH_3 \right] \rightleftharpoons CH_2{=}C(\ddot{O}{-}H)CH_3 + H^{+}$$

The mechanism for base-catalyzed enolization helps us to explain the acidity of the α-hydrogen atoms in aldehydes and ketones. As was stated in Section 7.3, greater acid strength depends on lower basicity—that is, greater stability—of the conjugate base of the acid, where the conjugate base is simply the anion formed from the acid. When an α-hydrogen is removed from an aldehyde or ketone, the anion formed is not a simple carbanion but is a resonance-stabilized hybrid anion. The stabilization of the anion by resonance is responsible for the greater acidity of acetone relative to acetylene or to other simple hydrocarbons that cannot form resonance-stabilized anions. Since oxygen is more electronegative than carbon, the resonance structure with the charge on oxygen is more important than the resonance structure with the charge on carbon. This means that in the resonance hybrid the negative charge is largely on the oxygen atom. For this reason chemists usually draw the enolate anion as

$$CH_2{=}C(:\ddot{O}:^{-}){-}CH_3$$

However, as was stated in Section 3.11, resonance hybrids may show the reactions that would be expected of more than one contributing resonance structure. Thus, enolate ions can and do behave as carbanions, as will be demonstrated in the next section.

Exercise 8.5 2,4-Pentanedione, $CH_3COCH_2COCH_3$, is a much stronger acid than acetone. It is converted almost quantitatively to its anion in aqueous sodium hydroxide. Explain the acidity of this compound by drawing the resonance structures of its enolate anion. (*Hint:* The central carbon atom is the most acidic.)

8.6 Replacement of α-Hydrogen by Halogen (Haloform Reaction)

Aldehydes and ketones are much more readily halogenated than the alkanes. The reaction, which is ionic rather than free radical in nature, invariably results in replacement of an α-hydrogen to give an α-halo ketone and is catalyzed by both acids and bases. Acid-catalyzed halogenation proceeds through the enol form of the aldehyde or ketone and tends to stop after the introduction of one halogen atom.

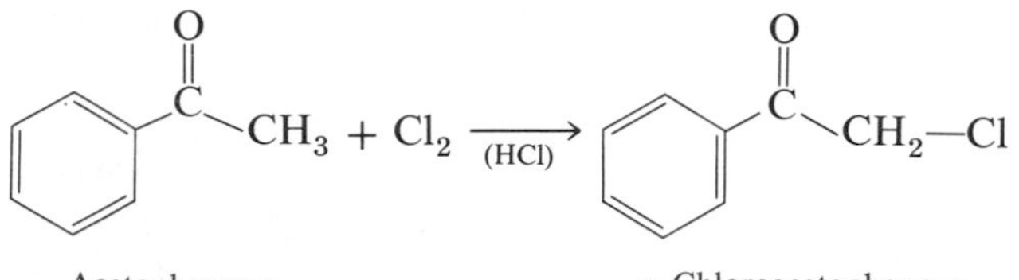

$$C_6H_5\text{—}\overset{\overset{\large O}{\|}}{C}\text{—}CH_3 + Cl_2 \xrightarrow[(HCl)]{} C_6H_5\text{—}\overset{\overset{\large O}{\|}}{C}\text{—}CH_2\text{—}Cl$$

Acetophenone

α-Chloroacetophenone
(Phenacyl chloride)

Phenacyl chloride is a potent **lachrymator** or tear gas and is used widely in law enforcement. It is a constituent of the self-defense agent, Mace.

Base-catalyzed halogenation proceeds through the enolate ion as shown in the following equation.

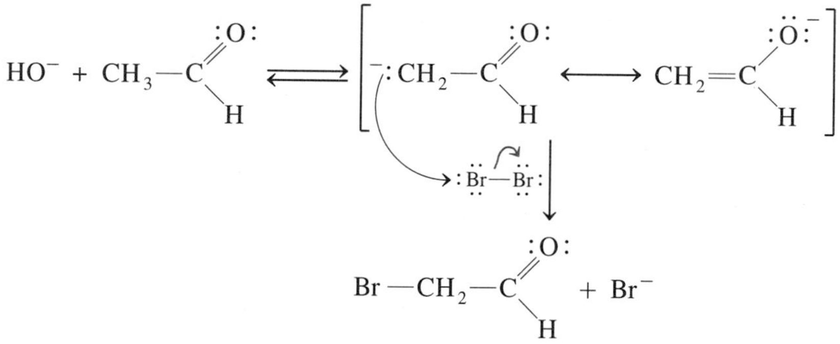

$$HO^- + CH_3\text{—}CHO \rightleftharpoons \left[{}^-:CH_2\text{—}CH{=}O \longleftrightarrow CH_2{=}CH\text{—}O^- \right]$$

$$\xrightarrow{:Br\text{—}Br:} Br\text{—}CH_2\text{—}CHO + Br^-$$

This type of halogenation tends to give polyhalogenation products. Furthermore, when acetaldehyde or a methyl ketone is warmed with an alkaline solution of chlorine, bromine, or iodine, the product is chloroform, bromoform, or iodoform, respectively. This reaction is called the **haloform reaction** and appears to take place in two stages. In the first stage the three hydrogens on the α-carbon are successively replaced by halogen, each additional hydrogen more easily than the one before because of the strong electron-withdrawing effect of the halogen atom(s) that makes the remaining hydrogen atom(s) more acidic.

This new electronic effect is called an **inductive effect** and can be explained as follows. When atoms of different electronegativities are bonded, there will be a shift of electrons toward the more electronegative atom and away from the less electronegative atom (Sec. 1.8), resulting in the formation of a dipole (Sec. 5.8). The electrostatic effect of this dipole is called the inductive effect. The effect of the chlorine atoms on the acidity of the α-hydrogen atoms of an aldehyde or ketone can be explained by studying the structure of the enolate ion:

$$\left[\mathrm{Cl} \leftrightarrows \ddot{\mathrm{C}}^{-}(\mathrm{H})-\mathrm{C}(=\ddot{\mathrm{O}})\mathrm{R} \longleftrightarrow \mathrm{Cl} \leftrightarrows \mathrm{C}(\mathrm{H})=\mathrm{C}(-\ddot{\mathrm{O}}:^{-})\mathrm{R} \right]$$

Since the chlorine atom tends to draw the electrons away from the α-carbon atom, the negative charge on the carbon atom is dispersed partially onto the chlorine atom. Delocalization or dispersal of charge is a stabilizing effect; therefore, the anion is stabilized. Thus, the effect of the chlorine can be said to enhance the acidity of the α-hydrogen by stabilizing the enolate ion formed by the removal of the α-hydrogen by base.

Now let us examine the two stages of the haloform reaction.

Stage 1: Halogenation

$$\underset{\text{Acetaldehyde}}{CH_3-C(=O)H} + 3\,Cl_2 + 3\,NaOH \longrightarrow Cl_3C-C(=O)H + 3\,H_2O + 3\,NaCl$$

Stage 2: Cleavage

$$Cl_3C-\overset{\delta+}{C}(=\overset{\delta-}{\ddot{O}}:)H + {}^{-}:\ddot{O}H \xrightarrow{\text{addition}} Cl_3C-C(-:\ddot{O}:^{-})(H)-\ddot{O}H \xrightarrow{\text{elimination}} Cl_3C:^{-} + H-C(=\ddot{O}:)\ddot{O}H$$

$$Cl_3C:^{-} + H-C(=\ddot{O}:)\ddot{O}H \longrightarrow \underset{\text{Chloroform}}{Cl_3C-H} + \underset{\text{Formate ion}}{H-C(=\ddot{O}:)\ddot{O}:^{-}}$$

The first stage is polyhalogenation via the enolate ion. The second stage is a cleavage by base as the result of an **addition-elimination mechanism.** We will find this type of mechanism common in the reactions of the carbonyl group of both acids and acid derivatives (Chapter 9).

A structural requirement for any compound that gives a positive haloform reaction is that it contain an **acetyl group,**

$$CH_3-\overset{O}{\overset{\|}{C}}-$$

or one oxidizable to an acetyl group. The first requirement is met by acetaldehyde and all methyl ketones. The second requirement is met by all methyl carbinols of the structure

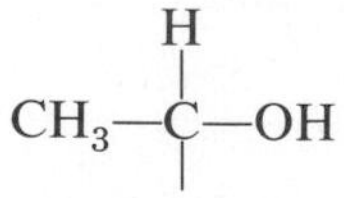

$$CH_3-\overset{\displaystyle H}{\overset{|}{\underset{|}{C}}}-OH$$

Of the primary alcohols, ethanol alone gives the haloform reaction because it is oxidizable to acetaldehyde. The haloform reaction is useful not only as a preparative method for the haloforms (Sec. 6.3), but also as a diagnostic test for the presence of the groupings indicated. In practice, a solution of iodine is added to an unknown compound in an aqueous alkaline solution. A positive reaction will yield iodoform, CHI_3, a bright yellow solid which may be identified by its sharp pungent odor and its melting point. Chloroform and bromoform are liquids.

Exercise 8.6 Of the two structures shown, which carbonyl carbon atom will be more readily attacked by the hydroxide ion? Why?

$$CH_3-\overset{\displaystyle O}{\overset{\|}{C}}-CH_3 \qquad CH_3-\overset{\displaystyle O}{\overset{\|}{C}}-CCl_3$$

8.7 Addition Reactions

A. Addition of Hydrogen Cyanide. The elements of hydrogen cyanide, HCN, add to aldehydes and ketones to yield **cyanohydrins** when the reaction is carried out with a basic catalyst.

$$HCN + \overset{-}{O}H \longrightarrow H_2O + \overset{-}{C}N$$

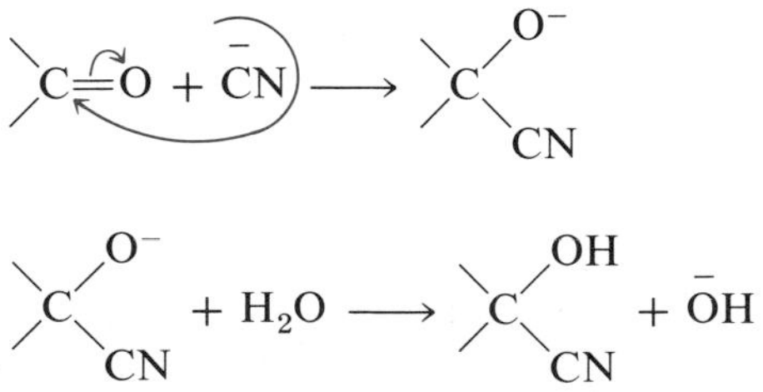

$$\rangle C{=}O + \overset{-}{C}N \longrightarrow \rangle C\langle^{O^-}_{CN}$$

$$\rangle C\langle^{O^-}_{CN} + H_2O \longrightarrow \rangle C\langle^{OH}_{CN} + \overset{-}{O}H$$

Hydrogen cyanide, a highly toxic gas, is rarely handled directly, but is produced at the site of a reaction by treating sodium or potassium cyanide with a mineral acid. However, the amount of acid used must never be enough to react with all the cyanide ion; otherwise the basic conditions favorable to the reaction will be lacking.

$$(CH_3)_2C{=}O + NaCN + H_2SO_4 \xrightarrow{10\text{–}20^\circ C} (CH_3)_2C(OH)CN + NaHSO_4$$

Acetone → Acetone cyanohydrin

Aldehydes, aliphatic methyl ketones, and cyclic ketones form cyanohydrins. Inasmuch as these addition compounds can be hydrolyzed to acids (Sec. 9.5B), the cyanohydrins are valuable intermediates in organic synthesis. The conversion of acetaldehyde to lactic acid is an example.

$$CH_3{-}CH{=}O + HCN \longrightarrow CH_3{-}CH(OH){-}CN$$

Acetaldehyde → Acetaldehyde cyanohydrin

$$CH_3{-}CH(OH){-}CN + HCl + 2\,H_2O \longrightarrow CH_3{-}CH(OH){-}C(=O){-}OH + NH_4Cl$$

Lactic acid (a hydroxy acid)

B. Addition of Water and Alcohols. The aldehydes and ketones show little tendency to form *stable* hydrates. In a few compounds in which the carbonyl group is attached to other strong electron-attracting groups, a hydrate sufficiently stable to be isolated is formed. One of the best examples of such a compound is trichloroacetaldehyde, or **chloral,** which forms a stable hydrate used medicinally as a soporific.

$$Cl_3C{-}CH{=}O + H_2O \longrightarrow Cl_3C{-}CH(OH){-}OH$$

Chloral → Chloral hydrate

On the other hand, alcohols will add to aldehydes in the presence of acid catalysts. An unstable addition product called a **hemiacetal,** formed in the first stage of the reaction, reacts with a second molecule of alcohol to yield a stable compound called an **acetal.** The mechanism of hemiacetal and acetal formation proceeds through the formation of oxonium ions and is analogous to the S_N1 reaction whereby ethers were produced from alcohols (Sec. 7.7B).

STEP 1.

$$R'-\overset{H}{\overset{|}{C}}=O \underset{A^-}{\overset{H^+}{\rightleftharpoons}} R'-\overset{H}{\overset{|}{C}}=\overset{+}{O}H \underset{-ROH}{\overset{ROH}{\rightleftharpoons}} R'-\underset{\underset{H}{\underset{|}{R-O^+}}}{\overset{H}{\overset{|}{C}}}-OH \underset{H^+}{\overset{A^-}{\rightleftharpoons}} R'-\underset{OR}{\underset{|}{\overset{H}{\overset{|}{C}}}}-OH$$

A hemiacetal (unstable)

$$R'-\underset{OR}{\underset{|}{\overset{H}{\overset{|}{C}}}}-OH \underset{A^-}{\overset{H^+}{\rightleftharpoons}} R'-\underset{OR}{\underset{|}{\overset{H}{\overset{|}{C}}}}-\overset{H}{\overset{|}{\underset{+}{O}}}H \underset{+H_2O}{\overset{-H_2O}{\rightleftharpoons}} R'-\overset{H}{\overset{|}{C}}=\overset{+}{O}R$$

STEP 2.

$$R'-\overset{H}{\overset{|}{C}}=\overset{+}{O}R \underset{-ROH}{\overset{ROH}{\rightleftharpoons}} R'-\underset{\underset{H}{\underset{|}{R-O^+}}}{\overset{H}{\overset{|}{C}}}-OR \underset{H^+}{\overset{A^-}{\rightleftharpoons}} R'-\underset{OR}{\underset{|}{\overset{H}{\overset{|}{C}}}}-OR$$

An acetal (stable)

As is indicated by the double arrows, the reaction is reversible, and the acetal, when hydrolyzed in an acid solution, readily regenerates the original alcohol and aldehyde.

The equilibrium for the formation of higher molecular weight acetals and most **ketals** is unfavorable. Therefore, the water formed is removed by azeotropic distillation (Sec. 7.4) to push the equilibrium to the right. Cyclic acetals and ketals and 5- and 6-membered rings form very readily and are widely employed in carbohydrate chemistry. Acetone is the ketone most frequently used with carbohydrates, giving a cyclic ketal known as an isopropylidene derivative.

$$CH_3-\overset{O}{\overset{\|}{C}}-CH_3 + HO-CH_2-CH_2-OH \overset{H^+}{\rightleftharpoons} \text{(cyclic ketal: } CH_2-CH_2 \text{ joined through two O atoms to } C(CH_3)_2\text{)}$$

Acetone Ethylene glycol A cyclic ketal

Exercise 8.7 Label the following compounds as acetals, hemiacetals, or neither.

(a) 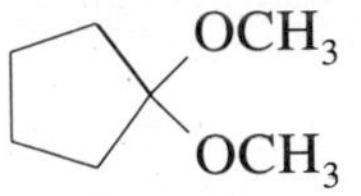 OCH_3 OCH_3 (b) 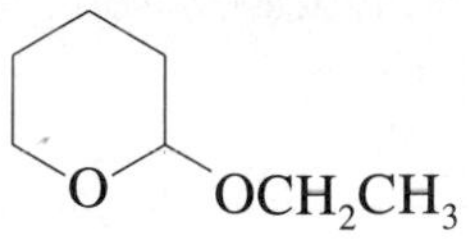 O OCH_2CH_3

(c) 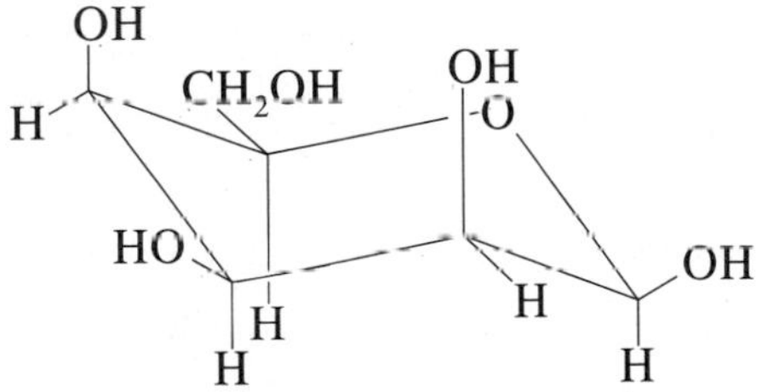 OH H CH_2OH OH O HO OH H H H H (d) 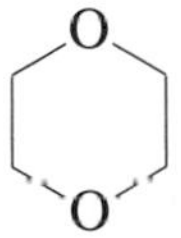 O O (e) 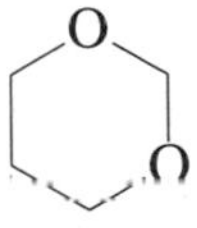 O O

B. Addition of Grignard Reagents. The addition of Grignard reagents (Sec. 6.7) of the form RMgX or ArMgX to aldehydes or to ketones as a route to the alcohols has already been presented in detail in Sec. 7.4C.

Exercise 8.8 Using propene and any inorganic reagents that you might need, show how you could obtain the necessary Grignard reagent and carbonyl compound needed to prepare 2,3-dimethyl-2-butanol.

8.8 Addition Reactions with Loss of Water. Condensations

Reactions in which two reactants combine with the loss of a molecule of water or some other neutral molecule are called **condensations.**

A. Condensation with Ammonia Derivatives. Certain derivatives of ammonia that contain the primary amino group, $—NH_2$, add to aldehydes and ketones to form tetrahedral intermediates. The initial addition product loses the elements of water (condensation) to form a carbon–nitrogen double bond.

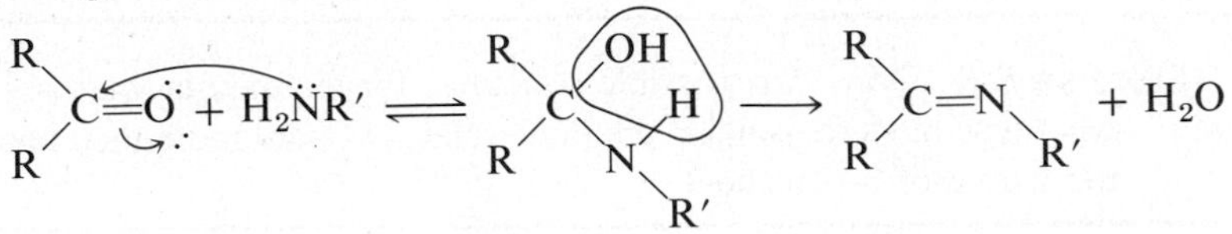

$$R_2C{=}\ddot{O}: + H_2\ddot{N}R' \rightleftharpoons R_2C(OH)(NHR') \longrightarrow R_2C{=}NR' + H_2O$$

Many of these condensation products are crystalline solids with sharp melting points. For this reason they are frequently employed for the preparation of aldehyde and ketone derivatives needed in identification work. Ammonia derivatives commonly used and their condensation products follow.

Carbonyl Compound		Ammonia Derivative		Condensation Product
$R_2C{=}O$	+	$H_2N{-}OH$	$\xrightarrow{H^+}$	$R_2C{=}N{-}OH + H_2O$
A ketone*		Hydroxylamine		An oxime
$R_2C{=}O$	+	$H_2N{-}NH{-}C_6H_5$	$\xrightarrow{H^+}$	$R_2C{=}N{-}NH{-}C_6H_5 + H_2O$
A ketone		Phenylhydrazine		A phenylhydrazone
$R_2C{=}O$	+	$H_2N{-}NH{-}C({=}O){-}NH_2$	$\xrightarrow{H^+}$	$R_2C{=}N{-}NH{-}C({=}O){-}NH_2 + H_2O$
A ketone		Semicarbazide		A semicarbazone

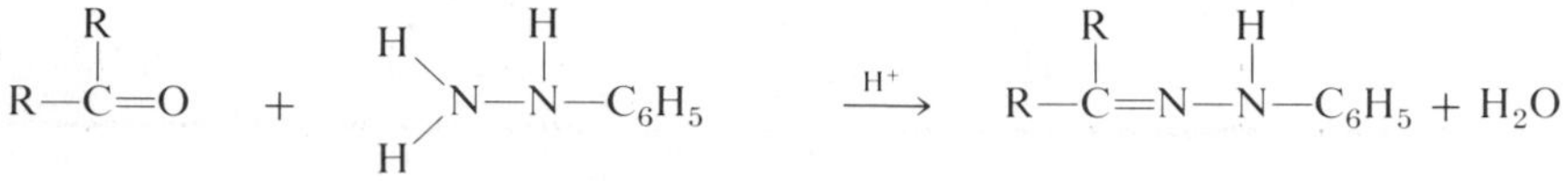

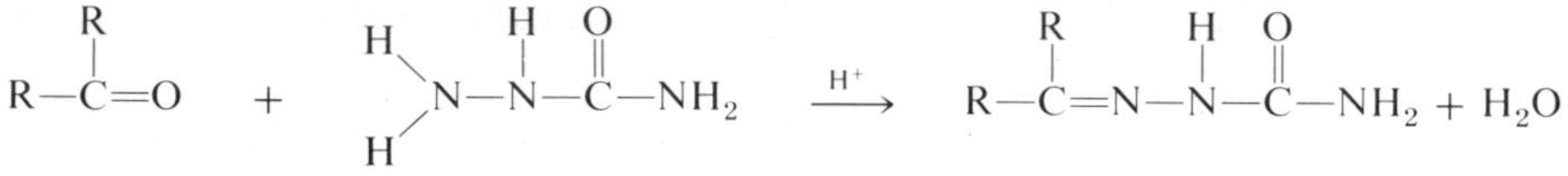

*If an R were H, the compound would be an aldehyde.

The carbonyl derivative formed in each case simply is designated as the **aldoxime,** the **ketoxime,** the **phenylhydrazone,** or the **semicarbazone** of the carbonyl compound from which it was prepared.

$$CH_3{-}CH{=}O + H_2N{-}OH \longrightarrow CH_3{-}CH{=}N{-}OH + H_2O$$

Acetaldoxime (mp, 47°C)

$$(CH_3CH_2)_2C{=}O + H_2NOH \longrightarrow (CH_3CH_2)_2C{=}N{-}OH + H_2O$$

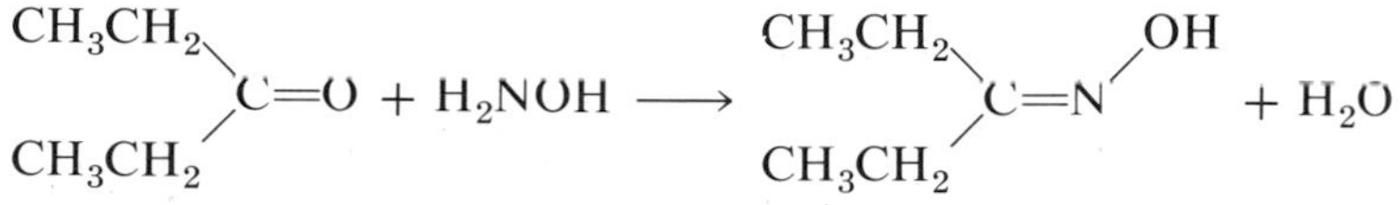

Diethylketoxime (mp, 69°C)

Exercise 8.9 Draw two possible structures for the oxime obtainable from 2-butanone. To what type of stereoisomers are these related? Could more than one structure be drawn for the oxime of 3-pentanone?

Frequently, the higher the molecular weight of a carbonyl derivative, the higher its melting point. Thus, 2,4-*dinitrophenylhydrazine* is often used in place of phenylhydrazine for the preparation of carbonyl derivatives.

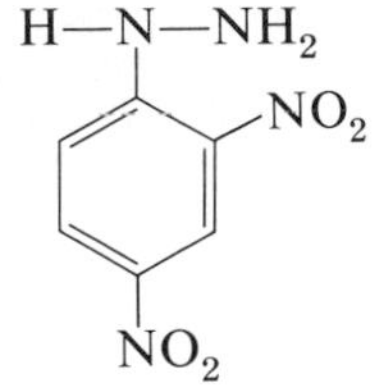

$$H{-}N{-}NH_2 \text{ (attached to benzene ring bearing } NO_2 \text{ at positions 2 and 4)}$$

2,4-Dinitrophenylhydrazine

B. The Aldol Condensation. When an enolate ion adds to another molecule of aldehyde or ketone, the reaction is called an **aldol[2] condensation.** An aldol condensation involving acetaldehyde reacting with itself in the presence of a basic catalyst is shown as an example. Because the aldol condensation is reversible, the reaction often gives better results if it is followed by one with a more favorable equilibrium such as dehydration, or by an essentially irreversible reaction. The reversibility of the aldol condensation is important in biological processes since it is involved in both the biosynthesis and degradation of carbohydrates, with enzymes called *aldolases* serving as the catalysts.

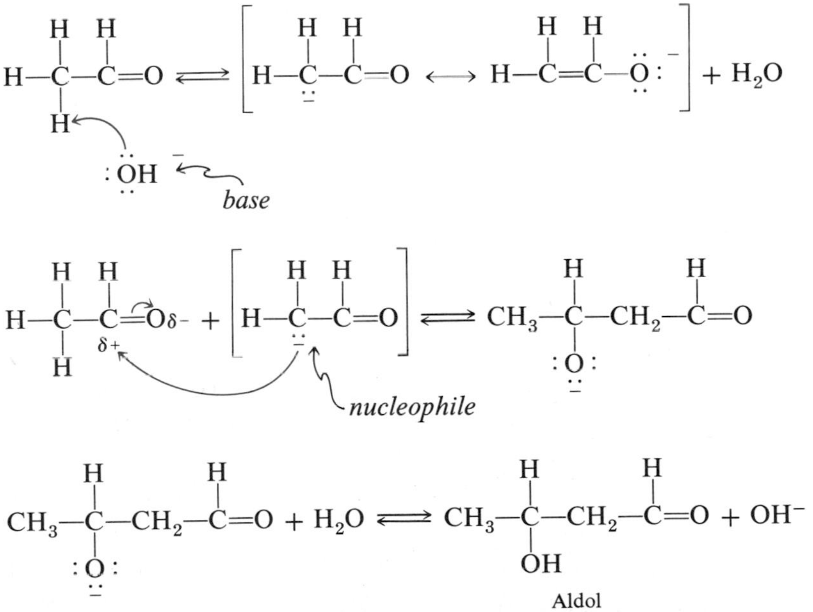

$$H_3C{-}CH{=}O \;\rightleftharpoons\; \left[H_2\ddot{C}^{-}{-}CH{=}O \;\longleftrightarrow\; H_2C{=}CH{-}\ddot{O}:^{-} \right] + H_2O$$

$$:\ddot{O}H^{-} \text{ (base)}$$

$$H_3C{-}\overset{\delta+}{C}H{=}O^{\delta-} + \left[H_2\ddot{C}^{-}{-}CH{=}O \right] \text{ (nucleophile)} \;\rightleftharpoons\; CH_3{-}\underset{:\ddot{O}:^{-}}{\overset{H}{C}}{-}CH_2{-}\overset{H}{C}{=}O$$

$$CH_3{-}\underset{:\ddot{O}:^{-}}{\overset{H}{C}}{-}CH_2{-}\overset{H}{C}{=}O + H_2O \;\rightleftharpoons\; CH_3{-}\underset{OH}{\overset{H}{C}}{-}CH_2{-}\overset{H}{C}{=}O + OH^-$$

Aldol
(3-Hydroxybutanal or
β-Hydroxybutyraldehyde)

[2]Aldol, a composite word for *ald*ehyde + alcoh*ol*.

Under more vigorous conditions aldols lose water to give ***α,β*-unsaturated carbonyl compounds.**

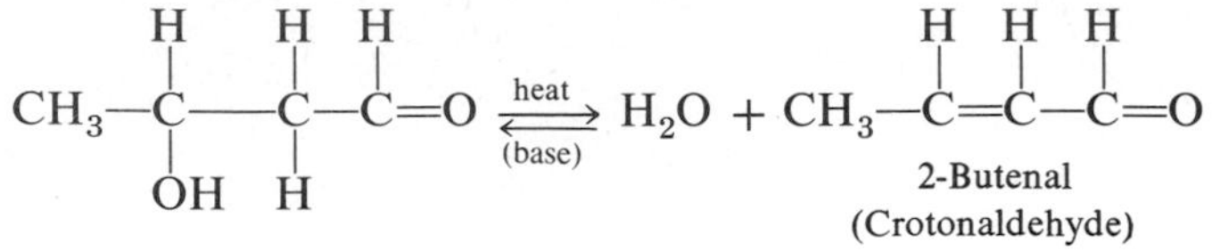

$$CH_3-\underset{OH}{\overset{H}{C}}-\underset{H}{\overset{H}{C}}-\overset{H}{C}=O \xrightleftharpoons[\text{(base)}]{\text{heat}} H_2O + CH_3-\overset{H}{C}=\overset{H}{C}-\overset{H}{C}=O$$

2-Butenal
(Crotonaldehyde)

Ketones containing α-hydrogen also are capable of an aldol-type condensation.

An aldehyde that lacks an α-hydrogen can enter into a mixed aldol condensation with another aldehyde that has α-hydrogens only by serving as the carbanion acceptor. For example, benzaldehyde reacts with acetaldehyde to produce cinnamaldehyde, an α,β-unsaturated aromatic aldehyde used as a flavoring agent.

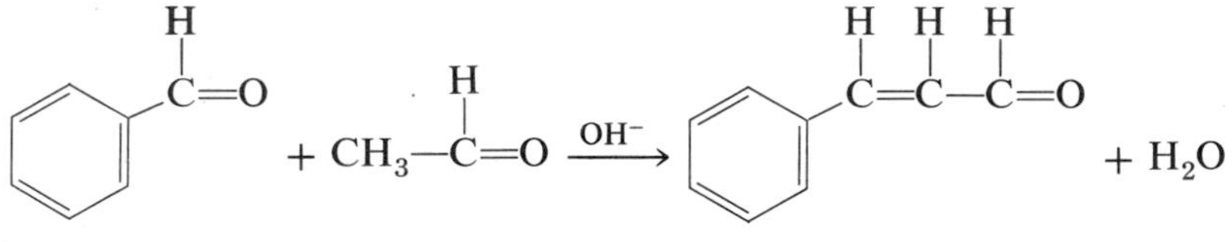

$$C_6H_5-\overset{H}{C}=O + CH_3-\overset{H}{C}=O \xrightarrow{OH^-} C_6H_5-\overset{H}{C}=\overset{H}{C}-\overset{H}{C}=O + H_2O$$

Benzaldehyde Cinnamaldehyde

EXAMPLE

What condensation products result when a mixture of acetaldehyde and butyraldehyde is warmed with dilute sodium hydroxide?

Solution Acetaldehyde has three α-hydrogens and butyraldehyde has two. The catalyst will recognize little difference between the two aldehydes in preparing the carbanion. We will thus have a reaction mixture composed of four different condensation products.

$$CH_3-CH=CH-CHO$$

2-Butenal

$$CH_3-CH=\overset{C_2H_5}{C}-CHO$$

2-Ethyl-2-butenal

$$CH_3CH_2CH_2CH=CH-CHO$$

2-Hexenal

$$CH_3CH_2CH_2CH=\overset{C_2H_5}{C}-CHO$$

2-Ethyl-2-hexenal

Exercise 8.10 The following compounds can be made by the aldol condensation, perhaps followed by another reaction. Write the equations for the preparation of these substances.

(1) 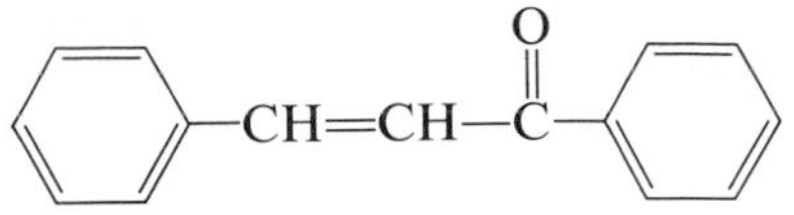 $C_6H_5-CH{=}CH-\overset{\overset{\displaystyle O}{\|}}{C}-C_6H_5$

(2) 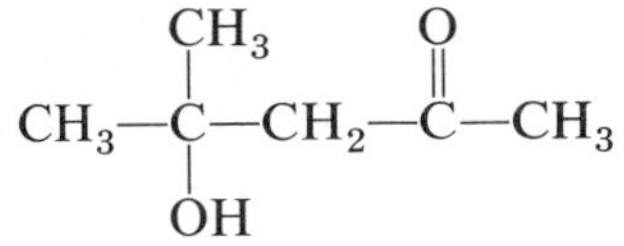 $CH_3-\underset{\underset{\displaystyle OH}{|}}{\overset{\overset{\displaystyle CH_3}{|}}{C}}-CH_2-\overset{\overset{\displaystyle O}{\|}}{C}-CH_3$

(3) 3-methylcyclohex-2-en-1-one (ring with =O and CH_3)

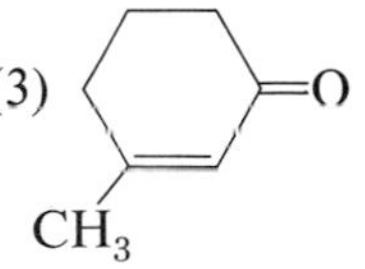

C. The Wittig Reaction. A very important and useful synthesis of alkenes known as the **Wittig reaction**[3] involves the reaction between an aldehyde or ketone and a phosphorus **ylide.** An ylide is an internal salt formed by the removal of a proton by a strong base from a carbon atom adjacent to a positively charged heteroatom such as phosphorus. Four reactants are required for the Wittig reaction: a ketone such as 2-butanone, an alkyl halide such as methyl bromide, an organic phosphine such as the relatively inexpensive triphenylphosphine, and a powerful base such as sodamide. In the first step of the reaction sequence the alkyl halide undergoes an S_N2 displacement by the phosphine to give the stable methyltriphenylphosphonium salt.

$$\text{(1)}\quad \underset{\text{Triphenylphosphine}}{(C_6H_5)_3P:} + CH_3-\ddot{\underset{..}{Br}}: \longrightarrow \underset{\text{Methyltriphenylphosphonium bromide}}{CH_3-\overset{+}{P}(C_6H_5)_3Br^-}$$

$$\text{(2)}\quad \underset{\text{Sodamide}}{Na^+NH_2:^-} + H-CH_2-\overset{+}{P}(C_6H_5)_3Br^- \longrightarrow$$

$$\underset{\text{Methylenetriphenylphosphorane (An ylide)}}{\left[:\overset{-}{C}H_2-\overset{+}{P}(C_6H_5)_3 \longleftrightarrow CH_2{=}P(C_6H_5)_3\right]} + NaBr$$

[3]Georg Wittig (1897–), Professor Emeritus of Chemistry, University of Heidelberg, West Germany. Shared the Nobel Prize in Chemistry for 1979.

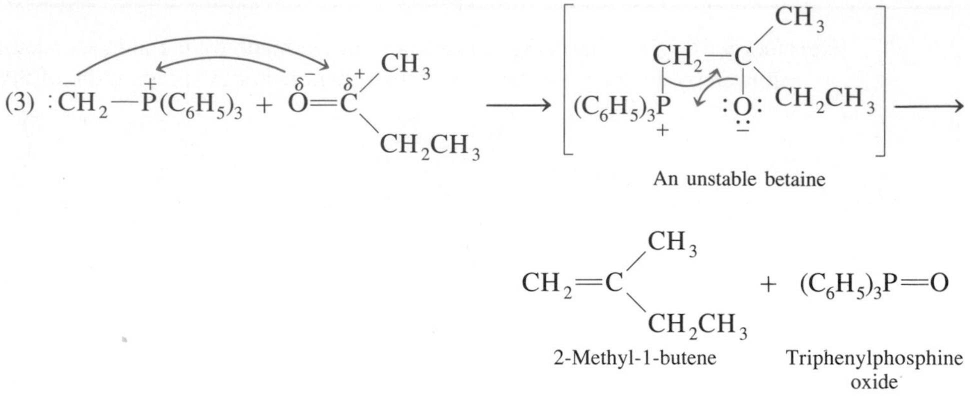

$$(3)\ :\overset{-}{C}H_2-\overset{+}{P}(C_6H_5)_3 + \overset{\delta^-}{O}=\overset{\delta^+}{C}(CH_3)(CH_2CH_3) \longrightarrow \left[(C_6H_5)_3\overset{+}{P}-CH_2-C(CH_3)(CH_2CH_3)-\ddot{O}:^- \right] \longrightarrow$$

An unstable betaine

$$CH_2=C(CH_3)(CH_2CH_3) + (C_6H_5)_3P=O$$

2-Methyl-1-butene Triphenylphosphine oxide

Because of the positive charge on the phosphorus, the hydrogen atoms on the methyl group are now sufficiently acidic that one of them can be removed in the second step of the reaction by treatment with the powerful base, sodamide, to give the ylide, shown here as a resonance hybrid. In the final step of the sequence the ylide, reacting as a carbanion, adds to the carbonyl group of 2-butanone to give an unstable **betaine.** As shown in the equation the betaine loses a molecule of the very stable triphenylphosphine oxide, to yield the alkene, 2-methyl-1-butene. Most aldehydes and ketones can be used in this reaction sequence, and alkyl halides that react well by the S_N2 mechanism can be substituted for methyl bromide. The reaction is so popular that "instant Wittig reagents" are now being sold for use in the laboratory, including student laboratories. These are mixtures of solid, dry alkyltriphenylphosphonium salts and sodamide. The mixture is reasonably stable in the dry state, but when mixed with tetrahydrofuran, quickly forms the ylide, which can then be treated with the desired aldehyde or ketone to give the alkene.

More than one combination of reagents *may* give the same alkene. To decide what choices are available, divide the alkene into two parts at the double bond. One part will come from the aldehyde or ketone and the other from the alkyl halide, as shown in the following example.

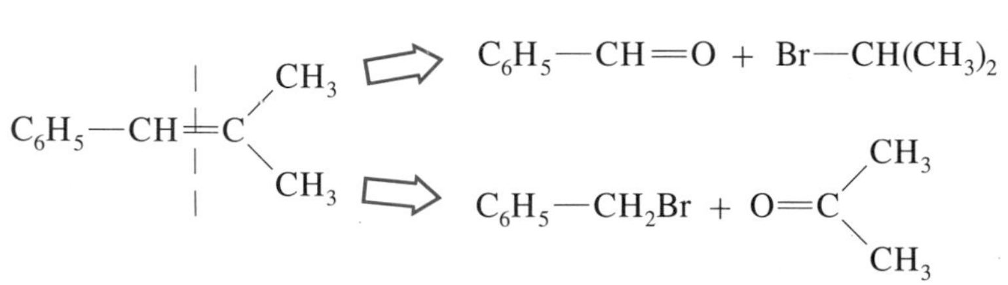

$$C_6H_5-CH=C(CH_3)_2 \Rightarrow C_6H_5-CH=O + Br-CH(CH_3)_2$$

$$C_6H_5-CH=C(CH_3)_2 \Rightarrow C_6H_5-CH_2Br + O=C(CH_3)_2$$

In the example given, either combination of reactants would be acceptable; however, generally the combination involving the better S_N2 halide should be used.

The Wittig reaction is run on a very large scale industrially, and much of the synthetic vitamin A manufactured today is made by a reaction sequence involving the Wittig reaction.

Exercise 8.11 The following compounds can be made by the Wittig Reaction. Write the equations for the preparation of these substances. Explain your choice of reactants.

(a) $CH_2{=}CH{-}C(CH_3)_3$ (b) $CH_2{=}$cyclohexylidene

8.9 Oxidation Reactions

Oxidation by Tollens' Reagent and Fehling's and Benedict's Solutions. Aldehydes are so easily oxidized that even the mildest of oxidizing reagents will serve to bring about their conversion to acids. Ketones, on the other hand, are fairly resistant to oxidation. The oxidation of ketones, when forced by the use of strong oxidizing reagents and heat, results in a rupture of carbon–carbon bonds to produce acids. The reaction probably involves the cleavage of the carbon–carbon double bond in the enol form of the ketone.

$$CH_3{-}\overset{O}{\overset{\|}{C}}{-}CH_2CH_3 \xrightarrow[\text{heat}]{KMnO_4,\ H^+} CH_3{-}C\overset{O}{\underset{OH}{<}} \quad \text{and/or} \quad CH_3{-}CH_2{-}C\overset{O}{\underset{OH}{<}}$$

The ease with which oxidation takes place provides a simple method for distinguishing between aldehydes and ketones. Mild oxidizing agents may be used for this purpose. **Tollens' reagent,** an ammoniacal solution of silver oxide, $Ag(NH_3)_2OH$, **Fehling's solution,** an alkaline solution of cupric ion complexed with sodium potassium tartrate (Rochelle salt), and **Benedict's solution,** an alkaline solution of cupric ion complexed with sodium citrate, are three reagents commonly used to detect the presence of an aldehyde group. When Tollens' reagent is used to oxidize an aldehyde, the silver ion is reduced to the metallic form and, if the reaction is carried out in a clean test tube, deposits as a mirror.

$$R{-}\overset{H}{\overset{|}{C}}{=}O + 2\,Ag(NH_3)_2OH \longrightarrow R{-}\overset{O}{\overset{\|}{C}}{-}O^-NH_4^+ + 2\,Ag + H_2O + 3\,NH_3$$

When Fehling's and Benedict's solutions are used to oxidize an aldehyde, the complexed deep blue cupric ion is reduced to red cuprous oxide.

$$R{-}\overset{H}{\overset{|}{C}}{=}O + 2\,Cu(OH)_2 + NaOH \longrightarrow R{-}\overset{O}{\overset{\|}{C}}{-}O^-Na^+ + Cu_2O + 3\,H_2O$$

Aromatic aldehydes react with Tollens' reagent but do not react with either Fehling's or Benedict's solutions. A means of distinguishing between aliphatic and aromatic aldehydes is thus provided by this difference in reactivity with the two types of reagents.

Exercise 8.12 How might you use simple test tube reactions to distinguish between:

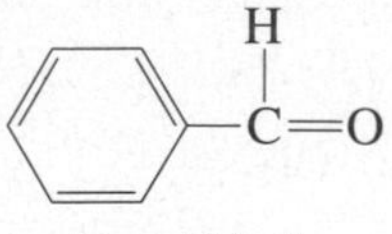

$$C_6H_5\text{—}\overset{\displaystyle H}{\overset{|}{C}}\text{=O} \qquad C_6H_5\text{—}CH_2\text{—}\overset{\displaystyle H}{\overset{|}{C}}\text{=O} \qquad C_6H_5\text{—}\overset{\displaystyle H}{\overset{|}{C}}\text{=}\overset{\displaystyle H}{\overset{|}{C}}\text{—}\overset{\displaystyle H}{\overset{|}{C}}\text{=O}$$

Benzaldehyde 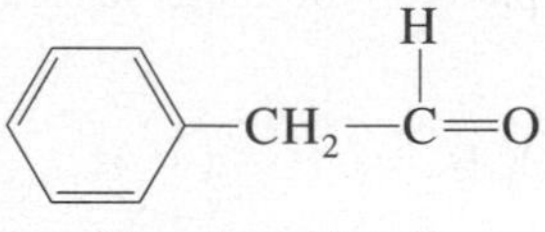 Phenylacetaldehyde 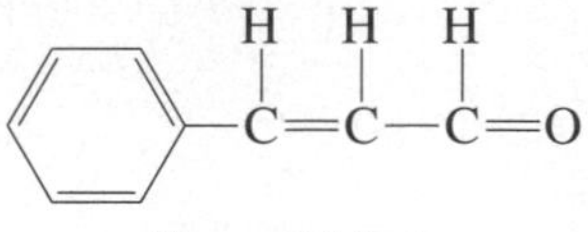 Cinnamaldehyde

8.10 Reduction

The carbonyl group of aldehydes and ketones may be reduced to primary and secondary alcohols, respectively. This transformation can be accomplished either catalytically by hydrogenation, or by means of a chemical reducing agent such as lithium aluminum hydride, $LiAlH_4$.

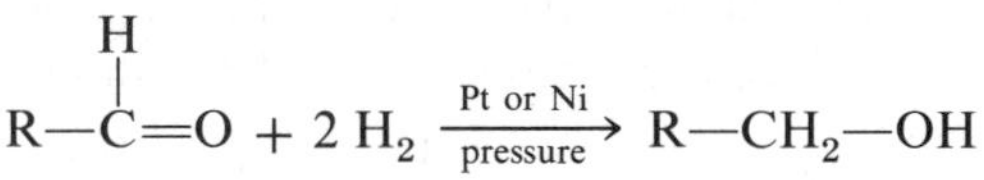

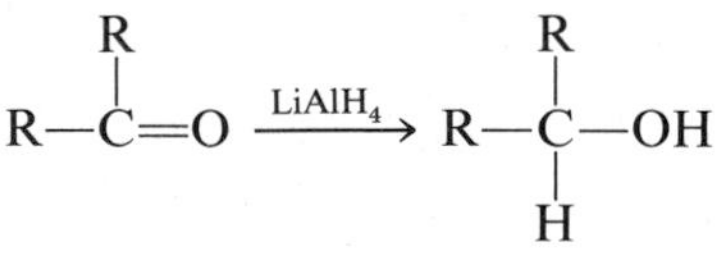

$$R\text{—}\overset{\displaystyle H}{\overset{|}{C}}\text{=O} + 2\,H_2 \xrightarrow[\text{pressure}]{\text{Pt or Ni}} R\text{—}CH_2\text{—OH}$$

$$R\text{—}\overset{\displaystyle R}{\overset{|}{C}}\text{=O} \xrightarrow{LiAlH_4} R\text{—}\underset{\displaystyle H}{\underset{|}{\overset{\displaystyle R}{\overset{|}{C}}}}\text{—OH}$$

Another reducing agent specific for the reduction of carbonyl groups is sodium borohydride, $NaBH_4$. This reagent is less reactive than $LiAlH_4$ and may be used in water or alcoholic solutions. Lithium aluminum hydride, on the other hand, reacts violently with such hydroxylic solvents.

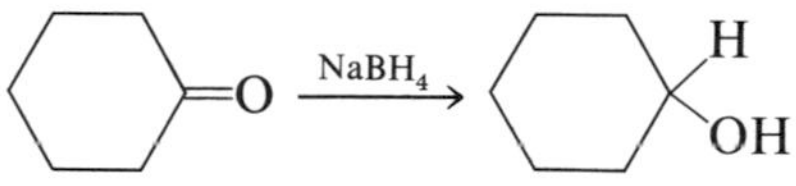

$$C_6H_{10}\text{=O} \xrightarrow{NaBH_4} C_6H_{10}(H)(OH)$$

Cyclohexanone Cyclohexanol

Exercise 8.13 Show how cyclohexanone may be converted to (a) 1,6-hexanediol, (b) *cis*-1,2-dihydroxycyclohexane.

The carbonyl group of a ketone can be reduced to a methylene group when refluxed with concentrated hydrochloric acid in the presence of amalgamated zinc. The reaction, known as the **Clemmensen reduction,** provides a method for converting carbonyls to hydrocarbons and is often used as a sequel to the Friedel–Crafts acylation of the aromatic ring. The net result is the monosubstitution of the ring by an unbranched alkyl group—an objective not always possible in the Friedel–Crafts alkylation.

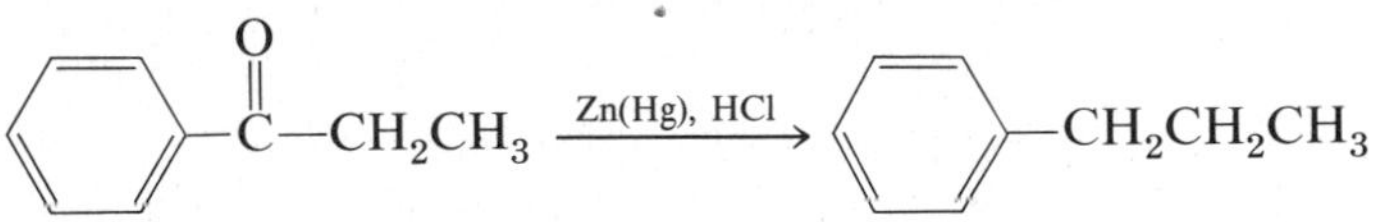

$$C_6H_5-\overset{O}{\overset{\|}{C}}-CH_2CH_3 \xrightarrow{Zn(Hg),\ HCl} C_6H_5-CH_2CH_2CH_3$$

Propiophenone (Ethyl phenyl ketone) → *n*-Propylbenzene

8.11 Polymerization of Aldehydes

Acetaldehyde, as you have noted, is rather unstable toward oxidation. This inherent instability, along with a boiling point of only 20°C, makes acetaldehyde a difficult compound to store and use. Fortunately, when treated with acid at a low temperature, acetaldehyde undergoes self-addition to give the cyclic trimer,[4] **paraldehyde** (bp, 125°C).

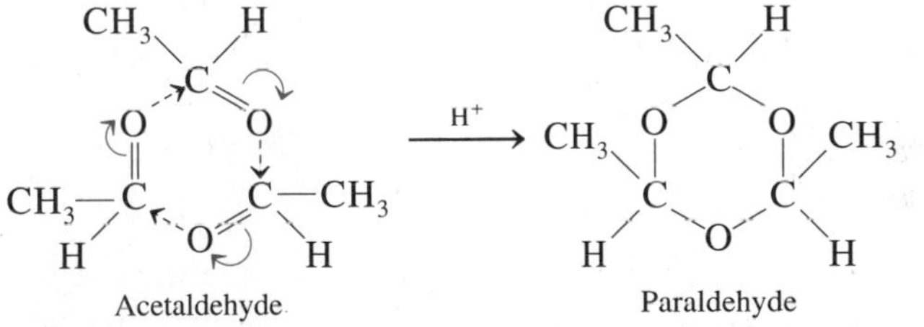

$$3\,CH_3CHO \xrightarrow{H^+} (CH_3CHO)_3$$

Acetaldehyde → Paraldehyde

In this form the aldehyde not only is stable to oxidation but no longer is easily lost by evaporation. Paraldehyde, when warmed, depolymerizes to regenerate acetaldehyde. Acetaldehyde, when warmed in concentrated alkaline solution, appears to undergo repeated aldol condensation accompanied by dehydration and polymerization to yield viscous, resinous products of undetermined structure. Acetaldehyde (as paraldehyde) has been used medicinally as a soporific, but its most important use is as a starting material in many organic syntheses.

Formaldehyde is not marketed in its gaseous form, but either as **formalin,** a 37–40% aqueous solution, or as the polymer, **paraformaldehyde,** $HO(CH_2O)_nH$, with n having an average value of 30. Paraformaldehyde is an amorphous white solid prepared by slowly

[4]Monomer (Gr. *mono,* one; *meros,* part). The simplest structural unit in a polymer. A trimer is composed of three molecules of monomer.

evaporating formalin under reduced pressure. The polymerization involves the self-addition of many molecules of formaldehyde.

$$H_2C{=}O + H_2O \rightleftharpoons [HOCH_2OH]$$

$$[HOCH_2OH] + n\,HCHO \longrightarrow HO\left(\begin{array}{c}H\\|\\C\\|\\H\end{array}-O\right)_{n+1}H$$

Paraformaldehyde

Depolymerization of paraformaldehyde, as in the case of paraldehyde, is brought about by heating. This ready change of state from solid to gaseous allows formaldehyde to be easily stored and used.

A high molecular weight linear polymer named ''Delrin'' has been prepared from formaldehyde. It is a remarkable plastic and shows promise of becoming a very useful one for structural materials where strength and resiliency are important.

Formaldehyde is perhaps the most important member of the aldehyde family. Its industrial importance lies principally in its ability to **copolymerize**[5] with phenol, with urea [$(H_2N)_2C{=}O$], and with melamine ($C_3H_6N_6$, a cyclic triamino compound) to produce hard, electrical nonconducting, infusible resins of the ''Bakelite'' and ''Melmac'' type. Structural units of these useful plastics are shown.

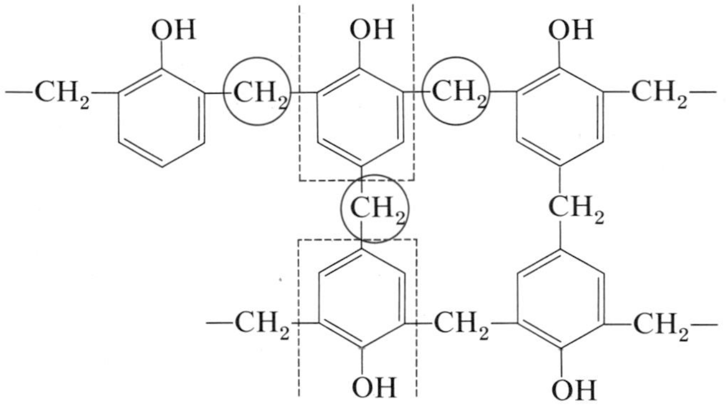

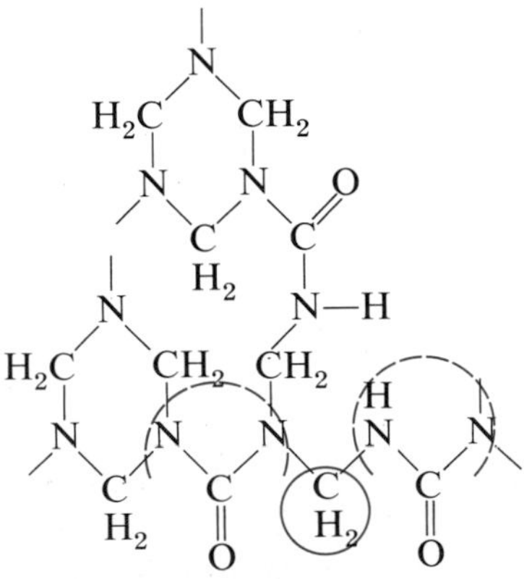

Structural units in
(a) Bakelite and
(b) urea-formaldehyde resins

[5]Copolymerization: a reaction in which two or more unlike monomers polymerize with each other.

Formaldehyde reacts with proteins to harden them and makes them less susceptible to putrefaction. For this reason it is used in the preservation of biological specimens and in embalming agents. Formaldehyde is toxic to insects and many microorganisms, and in the form of paraformaldehyde "candles" is conveniently used as a fumigant.

8.12 α,β-Unsaturated Carbonyl Compounds

When a carbonyl group is conjugated with a double bond, as in the α,β-unsaturated aldehydes and ketones, the conjugated system tends to show both normal 1,2- and conjugate 1,4-addition reactions (Sec. 3.11). Thus hydrogen bromide adds in the 1,4- manner to methyl vinyl ketone.

$$\overset{\delta+}{\underset{④}{CH_2}}{=}\underset{③}{CH}-\underset{②}{\overset{\delta+}{C}}(={\overset{\delta-}{:O:}}①)-CH_3 + H^+Br^- \longrightarrow$$

Methyl vinyl ketone

$$\left[CH_2{=}CH-C(={}^{+}\ddot{O}H)-CH_3 \longleftrightarrow \overset{+}{C}H_2-CH{=}C(-:\ddot{O}H)-CH_3\right] + Br^-$$

two of the three principal forms

$$Br^- + \left[\overset{+}{C}H_2-CH{=}C(-:\ddot{O}H)-CH_3\right] \longrightarrow$$

$$\left[\underset{Br}{\underset{|}{CH_2}}-CH{=}C(-:\ddot{O}H)-CH_3\right] \rightleftarrows \underset{Br}{\underset{|}{CH_2}}-\underset{H}{\underset{|}{CH}}-C(={:O:})-CH_3$$

unstable enol 4-Bromo-2-butanone

There are two important observations that should be made concerning additions to conjugated systems. First, the numbers 1,2- and 1,4- do not refer to the carbon chain numbering schemes used in nomenclature. They refer to the conjugated system; and, as shown for methyl vinyl ketone, the numbering *usually* begins with the hetero atom, oxygen. Second, although the addition of hydrogen bromide takes place in a 1,4- manner, because the intermediate enol is unstable and rearranges to the ketone, the final product *appears* to be that formed by *anti*-Markovnikov 3,4-addition. Thus the terms 1,2- and 1,4- refer to the *mechanism* of addition, *not* to the structures of the final products. In 1,2-addition, the addition occurs on two adjacent atoms. In 1,4-addition the mechanism requires addition across a 1,4-system.

Nucleophilic reagents, such as organometallic reagents or carbanions, may add 1,2- or 1,4- or both, depending on the nature of the conjugated system and the reagent. The 1,4-addition of water to α,β-unsaturated carbonyl systems is a common step in biochemical reactions (Figure 12.1). One of the many useful 1,4-additions is that of a cuprate (Sec. 2.13), as illustrated in the following example.

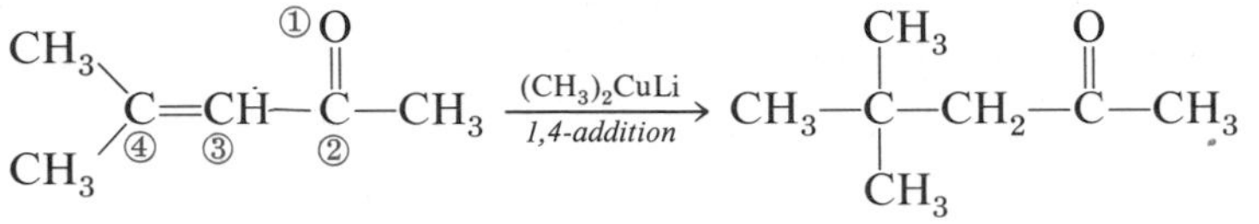

$$(CH_3)_2\overset{④}{C}{=}\overset{③}{C}H{-}\overset{②}{C}(=\overset{①}{O}){-}CH_3 \xrightarrow[\text{1,4-addition}]{(CH_3)_2CuLi} CH_3{-}C(CH_3)_2{-}CH_2{-}C(=O){-}CH_3$$

Mesityl oxide (4-Methyl-3-penten-2-one) → 4,4-Dimethyl-2-pentanone (Methyl neopentyl ketone)

Exercise 8.14 Draw the important resonance structures for methyl vinyl ketone.

Summary

1. **Structure**
The functional group of both the aldehydes and ketones is the carbonyl group,

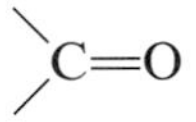

$$>C{=}O$$

The general formula for the aldehydes is

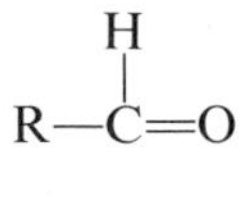

$$R{-}\overset{\overset{\large H}{|}}{C}{=}O$$

for the ketones,

$$R{-}\overset{\overset{\large R}{|}}{C}{=}O$$

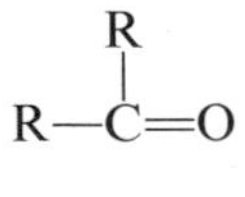

2. **Nomenclature**
IUPAC nomenclature uses the suffix **-al** to name an aldehyde and **one** to designate a ketone. Systematic nomenclature is governed by IUPAC rules previously outlined.

3. **Properties**
   (A) The simple aldehydes and ketones are nonassociated, low-boiling liquids.
   (B) Only aldehydes and ketones of low molecular weight are soluble in water.
   (C) The simple aldehydes have sharp, irritating odors.
   (D) The aromatic aldehydes and nearly all the ketones are fragrant.
   (E) The electron-withdrawing effect of the carbonyl oxygen is transmitted to the α-carbon atom and bestows acidic properties upon the hydrogen atoms bonded to it.
4. **Preparation**
   (A) General methods for the preparation of the aldehydes are as follows:
      (a) The catalytic dehydrogenation or oxidation of primary alcohols.
      (b) The hydrolysis of geminal dihalides (must be 1,1-dihalides).
   (B) Special methods of preparation of aldehydes are as follows:
      (a) The Gatterman–Koch reaction for aromatic aldehydes.
      (b) The hydration of acetylene for preparing acetaldehyde.
   (C) General methods for the preparation of the ketones are as follows:
      (a) The oxidation of secondary alcohols.
      (b) The hydration of alkynes other than acetylene.
   (D) Special methods for the preparation of ketones are as follows:
      (a) The Friedel–Crafts acylation reaction for aromatic ketones.
      (b) The Weizmann fermentation for acetone.
5. **Reactions of Carbonyl Compounds**
   (A) Oxidation.
      Aldehydes are easily oxidized to acids even by mild oxidizing reagents; ketones generally cannot be oxidized without a rupture of the carbon chain.
   (B) Addition of nucleophilic reagents—i.e., bases, electron-pair donors.
      (a) The following reagents give "straight" addition products.
         (i) HCN adds to aldehydes and ketones to form cyanohydrins.
         (ii) Hydrogen adds to the carbonyl group of aldehydes and ketones to produce primary and secondary alcohols, respectively.
         (iii) Grignard reagents, when treated with aldehydes, lead to the preparation of *secondary* alcohols; when treated with ketones, Grignard reagents lead to the preparation of *tertiary* alcohols. The only aldehyde capable of forming a primary alcohol with a Grignard reagent is formaldehyde.
         (iv) Water adds to the carbonyl group of aldehydes and ketones to give unstable hydrates. Alcohols add to give stable acetals and ketals.
      (b) Certain nitrogen-containing nucleophilic reagents add to aldehydes and ketones to produce unstable addition products. Subsequent loss of water (condensation) produces a carbon–nitrogen double bond. *Hydroxylamine* reacts with aldehydes and ketones to yield *oximes;* hydrazine and its derivatives give the corresponding *hydrazones*. *Semicarbazide* reacts to produce *semicarbazones*. All three types of derivatives are useful in the identification of "unknowns."
   (C) Reactions of α-Hydrogen.
      (a) An aldol condensation is the result when an aldehyde or a ketone with

α-hydrogen is treated with a dilute base. The aldol readily loses water, yielding an α,β-unsaturated carbonyl.

(b) Acetaldehyde and methyl ketones undergo replacement of all three α-hydrogen atoms, when warmed with an alkaline solution of the halogens, to yield *haloforms* and salts of acids.

(D) The Wittig reaction.
Aldehydes and ketones react with phosphorus ylides to give alkenes.

(E) Polymerization.
Formaldehyde is capable of self-addition to yield polymers of varying molecular weights. Acetaldehyde trimerizes to paraldehyde. Formaldehyde copolymerizes with phenol (Bakelite), urea, and melamine (Melmac). These are useful plastics with desirable thermal and electrical properties.

(F) Conjugate addition.
α,β-Unsaturated aldehydes and ketones may undergo 1,2- or 1,4- addition or both.

Supplementary Exercises

8.15 Assign an acceptable name to each of the following compounds.

(a) 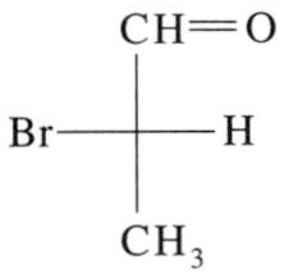 Fischer projection: CH=O (top), Br (left), H (right), CH_3 (bottom)
D or L ? (*R*) or (*S*)?

(b) 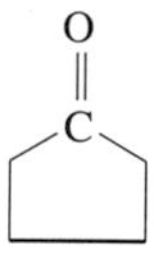 Cyclopentanone (C=O in five-membered ring)

(c) Benzene ring bearing OH and $C(CH_3)=O$ on adjacent carbons

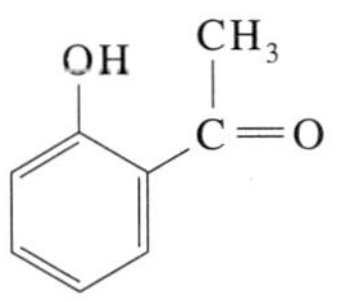

(d) 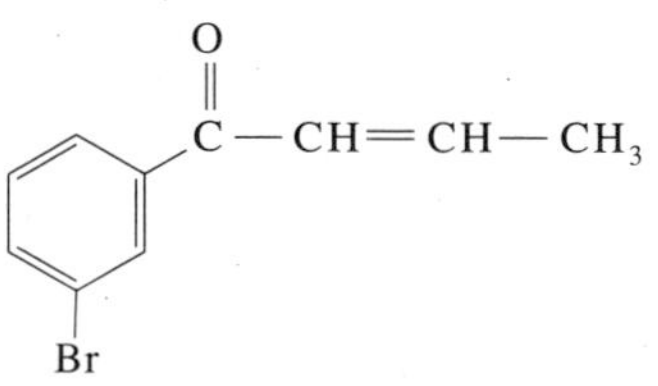 Benzene ring bearing $C(=O)-CH=CH-CH_3$ and Br (meta)

(e) 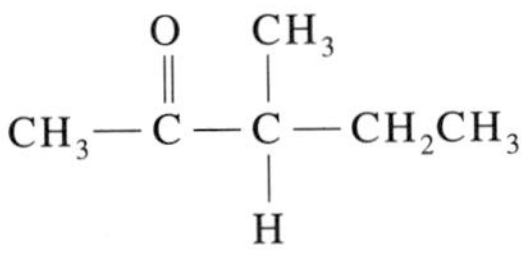 $CH_3-C(=O)-CH(CH_3)-CH_2CH_3$

(f) 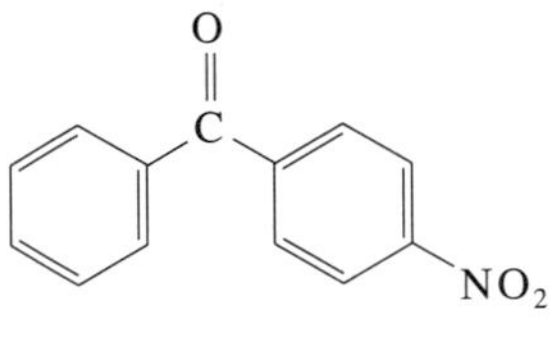 Phenyl–C(=O)–phenyl with NO_2 in the para position

(g) 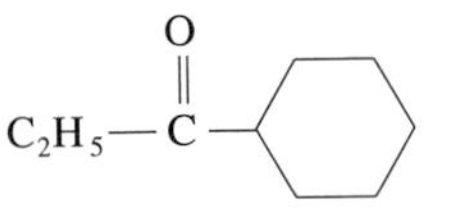 $C_2H_5-C(=O)-$cyclohexyl

(h) 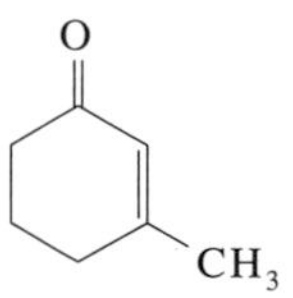 Cyclohexenone with CH_3 on the β-carbon

(i) 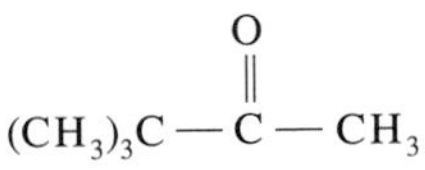 $(CH_3)_3C-C(=O)-CH_3$

(j) 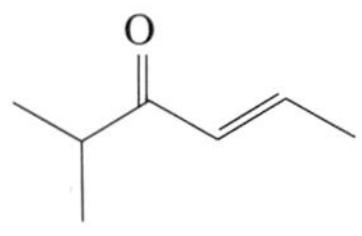 Skeletal structure: isopropyl ketone with a propenyl group
(*E*) or (*Z*)?

(k) 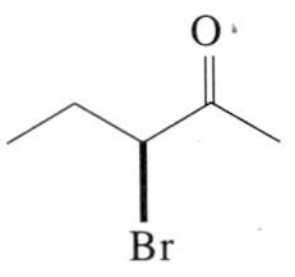 Skeletal structure: methyl ketone with Br on the α-carbon bearing an ethyl group
(*R*) or (*S*)?

(l) 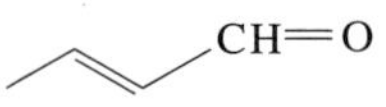 Skeletal structure: $CH_3CH=CH-CH=O$
(*E*) or (*Z*)?

8.16 Write structures for the following.

(a) 3-methylpentanal
(b) diisopropyl ketone
(c) *trans*-3,4-dimethylcyclohexanone
(d) 2-octanone
(e) α-bromoacetophenone
(f) cinnamaldehyde
(g) *p*-hydroxybenzaldehyde
(h) 4-phenyl-3-buten-2-one

8.17 Complete the following reactions by supplying the necessary but missing component. This could be either a reactant, a catalyst, or a product.

(a) (?) + I_2 in aqueous KI + NaOH $\longrightarrow$ $CH_3COO^-Na^+ + CHI_3$

(b) Acetone + (?) $\longrightarrow$ $(CH_3)_2C(OH)CN$

(c) (?) + $C_2H_5OH \xrightarrow{H^+}$ [cyclopentane ring bearing two OC_2H_5 groups on the same carbon]

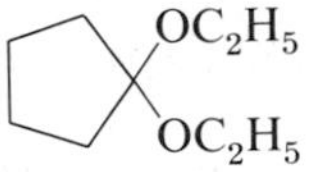

(d) (?) + $C_2H_5MgBr \xrightarrow[\text{hydrolysis}]{\text{followed by}} (C_2H_5)_2C(OH)CH_3$

(e) Acetophenone + $H_2NOH \xrightarrow{H^+}$ (?)

(f) *n*-Butyraldehyde + $CH_2{=}P(C_6H_5)_3 \longrightarrow$ (?)

(g) Propionaldehyde $\xrightarrow{10\%\ \text{NaOH}}$ (?)

(h) Benzaldehyde + (?) $\longrightarrow C_6H_5{-}CH_2OH$

(i) Methanol $\xrightarrow{\text{CuO, heat}}$ (?)

(j) (?) + $Ag(NH_3)_2OH \longrightarrow \underline{Ag} + CH_3COO^-NH_4^+$

8.18 What product would you expect from the reaction of cyclohexanone with each of the following reagents?

(a) benzaldehyde + aqueous NaOH
(b) $NaBH_4$ in methanol
(c) NaCN and aqueous H_2SO_4
(d) phenylhydrazine + H^+
(e) ethylene glycol + H^+
(f) $Ph_3P{=}$[cyclohexylidene]
(g) Br_2 in acetic acid
(h) $CH_3CH_2{-}MgBr$ in ether followed by $H_2O + H^+$
(i) $H_2N{-}OH + H^+$
(j) zinc amalgam + HCl
(k) $CH_3C{\equiv}C^-Na^+$ in liquid NH_3

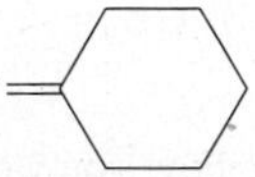

8.19 Write structures for the products formed when each of the following is treated with ethylmagnesium bromide, followed by acid hydrolysis.

(a) $H-\overset{H}{\overset{|}{C}}=O$

(b) $CH_3-\overset{H}{\overset{|}{C}}=O$

(c) $CH_3-\overset{CH_3}{\overset{|}{C}}=O$

(d) $C_2H_5-\overset{CH_3}{\overset{|}{C}}=O$

(e) 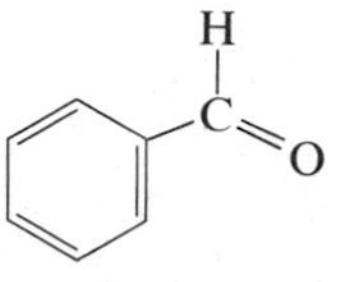 $C_6H_5-\overset{H}{\overset{|}{C}}=O$

(f) 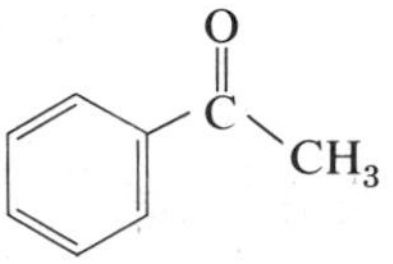 $C_6H_5-\overset{O}{\overset{\|}{C}}-CH_3$

(g) 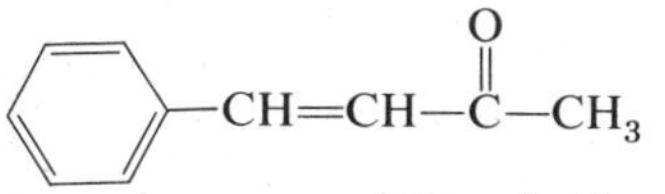 $C_6H_5-CH=CH-\overset{O}{\overset{\|}{C}}-CH_3$

(two products)

8.20 Which of the following compounds will give a positive iodoform reaction?

(a) $(CH_3)_2CHOH$

(b) $C_2H_5-\overset{CH_3}{\overset{|}{C}}=O$

(c) $CH_3-\underset{CH_3}{\underset{|}{\overset{CH_3}{\overset{|}{C}}}}-OH$

(d) 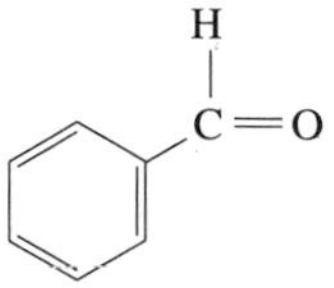 $C_6H_5-\overset{H}{\overset{|}{C}}=O$

(e) CH_3CH_2OH

(f) 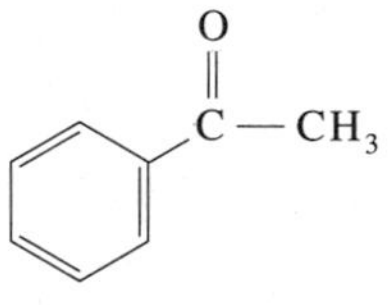 $C_6H_5-\overset{O}{\overset{\|}{C}}-CH_3$

(g) $CH_3CH_2-\underset{H}{\underset{|}{\overset{OH}{\overset{|}{C}}}}-CH_3$

(h) $CH_3CH_2-\overset{H}{\overset{|}{C}}=O$

(i) $CH_3CH_2-\overset{O}{\overset{\|}{C}}-CH_2CH_3$

(j) 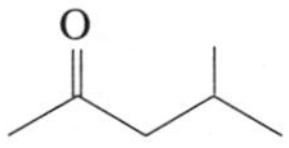 $CH_3-\overset{O}{\overset{\|}{C}}-CH_2-CH(CH_3)_2$

(k) CH_3OH

(l) $CH_3-\overset{H}{\overset{|}{C}}=O$

8.21 Using acetylene as your only organic starting material, show how you would synthesize each of the following compounds. You may use whatever inorganic reagents you consider necessary.

(a) acetaldehyde
(b) ethyl bromide
(c) iodoform
(d) acetic acid
(e) 2-butanol
(f) methyl ethyl ketone
(g) 2-butenol
(h) propionic acid, CH_3CH_2COOH

8.22 Identify the aldehyde or ketone that will react with ethylmagnesium bromide to produce

(a) 2-butanol
(b) 3-methyl-3-pentanol
(c) *n*-propyl alcohol
(d) 2-methyl-2-butanol
(e) 3-methyl-3-octanol

8.23 Outline sequences of reactions that would be suitable for bringing about the following transformations. Use any other reagents required.

(a) 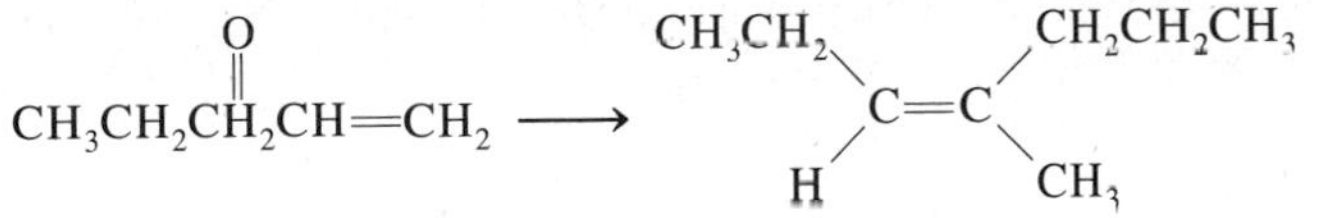 $CH_3CH_2\overset{O}{\overset{\|}{C}}H_2CH{=}CH_2 \longrightarrow$ $(CH_3CH_2)(H)C{=}C(CH_2CH_2CH_3)(CH_3)$

(b) $CH_3CH_2\overset{O}{\overset{\|}{C}}CH_2CH_3 \longrightarrow CH_3CH_2{-}\underset{CH_2-CH_3}{C}{=}\overset{CH_3}{C}{-}CH_3$

(c) $CH_3CH_2CH_2CH{=}CH_2 \longrightarrow CH_3CH_2CH_2C(=O)OH$

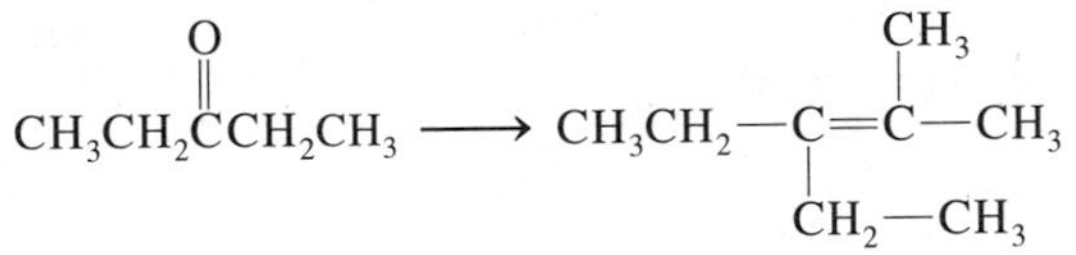

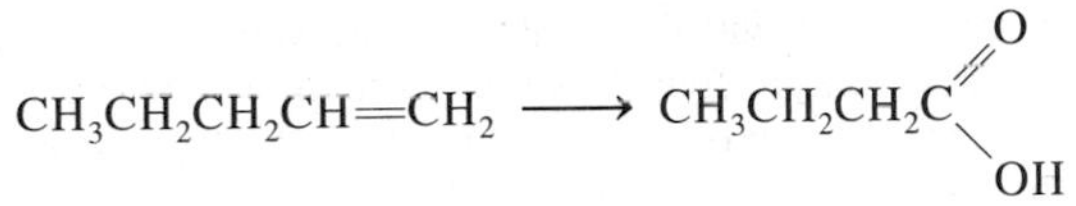

(d)  $HC{\equiv}CH \longrightarrow CH_3CH_2CH_2CH_2\overset{O}{\overset{\|}{C}}CH_3$

(e) 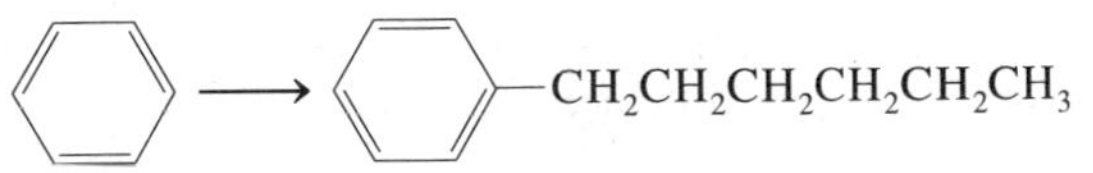 [benzene] $\longrightarrow$ [benzene]$-CH_2CH_2CH_2CH_2CH_2CH_3$

(f) 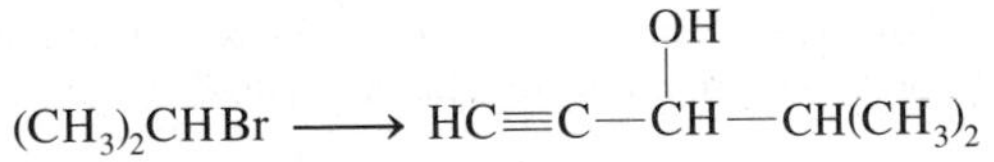 $(CH_3)_2CHBr \longrightarrow HC{\equiv}C{-}\overset{OH}{\overset{|}{C}H}{-}CH(CH_3)_2$

(g) 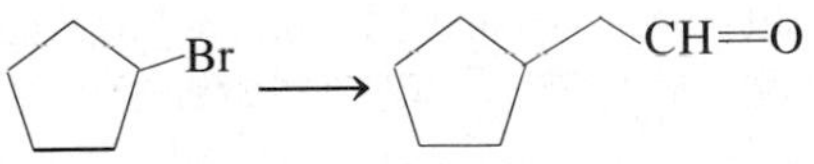 [cyclopentyl]$-Br \longrightarrow$ [cyclopentyl]$-CH_2CH_2CH{=}O$

8.24 Indicate simple test tube reactions that would serve to distinguish between members of the following pairs of carbonyl compounds without the need for taking boiling points or melting points, or preparing solid derivatives.

(a) acetaldehyde and acetone
(b) 2-octanone and benzaldehyde
(c) 2,2-dimethylpropanal and 2-butanone
(d) pentanol and cyclopentanone
(e) acetophenone and benzaldehyde

8.25 An optically active compound, $C_6H_{12}O$, reacted with 2,4-dinitrophenylhydrazine to give a red precipitate and gave a positive haloform test. Write the structural formula for the compound and give its name.

8.26 Compound A, $C_5H_{12}O$, when refluxed with potassium dichromate and sulfuric acid, was converted to B, $C_5H_{10}O$. Compound B formed a derivative with 2,4-dinitrophenylhydrazine but did not give a positive haloform reaction. Give the structure and name of the original compound.

8.27 An alkene, C_6H_{12}, after ozonolysis and subsequent hydrolysis, yielded two products. One of these gave a positive iodoform reaction but a negative Tollens' test. The other product gave a positive Tollens' test but a negative iodoform reaction. What are the structure and name of the alkene?

8.28 What would be the structure of the principal product if natural rubber were subjected to ozonolysis? What would be its name?

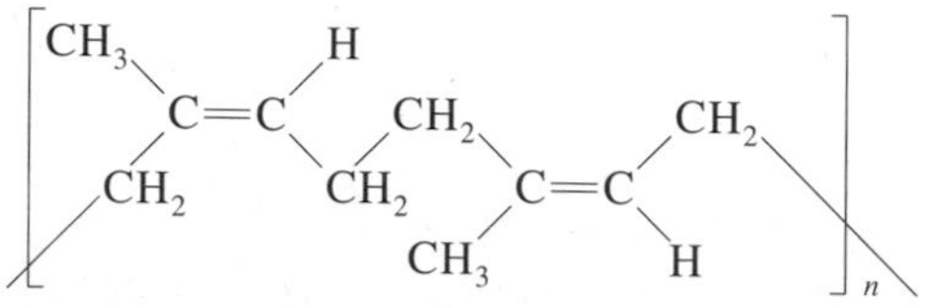

$$\left[\begin{array}{c} \text{CH}_3\diagdown\quad\diagup\text{H} \\ \text{C}=\text{C} \\ \diagup\text{CH}_2\quad\diagdown\text{CH}_2-\text{CH}_2\diagdown\quad\diagup\text{CH}_2\diagdown \\ \text{C}=\text{C} \\ \text{CH}_3\diagup\quad\diagdown\text{H} \end{array}\right]_n$$

a segment of the natural rubber structure

8.29 *Vanillin,* $C_8H_8O_3$, is the flavoring agent extracted from the vanilla bean. Six carbon and three hydrogen atoms are part of the aromatic ring. What is the nature of the ring substituents if vanillin (a) reacts with hydroxylamine, (b) gives a silver mirror with Tollens' reagent, (c) yields methyl iodide when treated with concentrated hydriodic acid, and (d) is soluble in sodium hydroxide solution? Propose a structure for vanillin.

***8.30** An organic compound, $C_5H_{10}O$, showed strong absorption at 1700 cm^{-1} in the ir region. With only this much information it is possible to draw seven different structures that fit the above molecular formula, but only one will be in agreement with the nmr spectrum of the compound, which consisted of only a quartet at δ 2.45 and a triplet at δ 0.97. The ratio of the areas was quartet:triplet = 2:3. What is the name of the compound in question?

***8.31** A colorless liquid, $C_6H_{12}O_3$, absorbed strongly in the ir at 1720 cm^{-1} and gave the following nmr spectrum: a 1-proton triplet at δ 4.71, a 6-proton singlet at δ 3.29, a 2-proton doublet at δ 2.64, and a 3-proton singlet at δ 2.11. Suggest a structure for the compound.

***8.32** Periodic acid, HIO_4, is a reagent used to cleave the carbon–carbon bond in 1,2-glycols, 1,2-dicarbonyl compounds, and α-hydroxy carbonyl compounds. In the course of the cleavage, hydroxyl groups are converted to carbonyl groups and carbonyl groups to carboxylic acid groups, as shown.

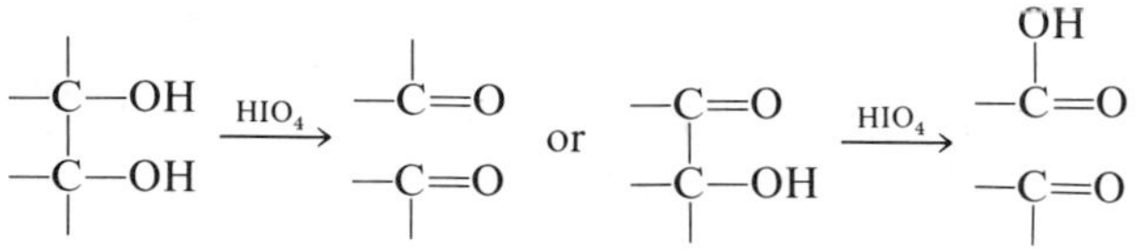

$$\begin{array}{c} | \\ -\text{C}-\text{OH} \\ | \\ -\text{C}-\text{OH} \\ | \end{array} \xrightarrow{HIO_4} \begin{array}{c} | \\ -\text{C}=\text{O} \\ \\ -\text{C}=\text{O} \\ | \end{array} \quad \text{or} \quad \begin{array}{c} -\text{C}=\text{O} \\ | \\ -\text{C}-\text{OH} \\ | \end{array} \xrightarrow{HIO_4} \begin{array}{c} \text{OH} \\ | \\ -\text{C}=\text{O} \\ \\ -\text{C}=\text{O} \\ | \end{array}$$

When compound A, $C_6H_{14}O_2$, was treated with HIO_4, it gave a single compound B, C_3H_6O. The latter showed a strong ir absorption band at 1720 cm^{-1} and reacted immediately with a solution of iodine in aqueous sodium hydroxide solution to produce iodoform. When compound A was treated with sulfuric acid, a new compound C, $C_6H_{12}O$, resulted, which also gave a positive iodoform reaction. The nmr spectrum of C showed only two singlets at δ 1.10 and δ 2.50 in a proton ratio of 3:1. What are the structures and names of compounds A, B, and C?

8.33 The important insect control agent and pheromone, disparlure (Sec. 7.8), may be synthesized in four steps from *cis*-2-tridecen-1-ol as shown below. Write equations for this synthesis, supplying the missing reagents and any organic reactants required.

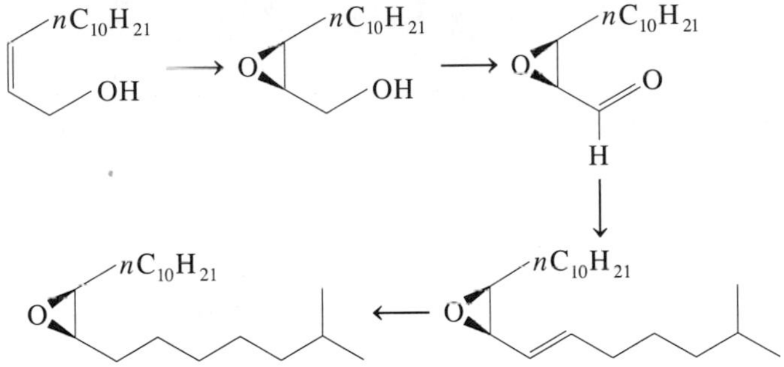

8.34 Devise a synthesis for the starting material in the previous exercise, 2-tridecen-1-ol, beginning with *n*-decyl bromide, $CH_3(CH_2)_9Br$, and using whatever other organic and inorganic reagents that may be required. (*Hint:* See Sec. 6.7.)

8.35 In Section 8.6 we explained the greater acidity of the α-hydrogens in chloroacetaldehyde using an argument based on the stabilizing effect of the chlorine atom on the enolate anion. Strictly speaking, we also should have considered the effect of the chlorine atom on the stability of chloroacetaldehyde, because only effects that stabilize the anion more than the acid from which it is derived will increase acid strength (Sec. 7.3). Complete the argument by describing the effect of the chlorine atom on the stability of chloroacetaldehyde.

9 Carboxylic Acids

Substances that contain a **carboxyl group,**

$$-C\begin{matrix}{\nearrow}^{O} \\ {\searrow}_{OH}\end{matrix}$$

(also written simply as —COOH, or as —CO_2H), make up a large family of compounds known as **carboxylic acids.** The carboxyl group represents the highest oxidation state of a carbon atom when bonded to another carbon and is a group found as a part of the structure of many natural products. It is not surprising to learn that acids occur naturally when one considers that we live in an oxidizing atmosphere. Many carboxylic acids have been known for a long time. For example, acetic acid, CH_3COOH, the principal component of vinegar, from which it takes its name (L. *acetum,* vinegar), has been used for centuries as a condiment and food preservative. It is frequently referred to in the Bible. A number of acids play vital roles in body functions and are important intermediates in the metabolic processes. A large number of acids, or their derivatives, are useful everyday commodities.

9.1 Formulas and Nomenclature

Carboxylic acids of the aliphatic series frequently are referred to as ''fatty'' acids because those containing an even number of carbon atoms, four or greater, exist in a combined form with glycerol as fats and oils. Acids that contain but one carboxyl group (monocarboxylic) have the general formula R—COOH or Ar—COOH.

$H{-}C(=O){-}OH$ $CH_3{-}C(=O){-}OH$ $CH_3CH_2CH_2{-}C(=O){-}OH$

Formic acid Acetic acid Butanoic acid

The carboxyl group in aromatic acids is attached directly to the benzene ring. However, the phenyl group, C_6H_5—, may appear as a substituent on any carbon atom of an aliphatic acid. Such acids should be classified as aryl-substituted aliphatic acids because their reactions essentially are those given by acids of the aliphatic series.

$C_6H_5{-}COOH$ $Br{-}C_6H_4{-}COOH$ $C_6H_5{-}CH_2{-}COOH$

Benzoic acid 4-Bromobenzoic acid Phenylacetic acid

Many of the carboxylic acids had been known before rules for systematic nomenclature were devised. As a result, they are often called by their common names. Many of these have their origin in Greek or Latin, and their names often indicate the original source of the acid. There are no easy rules for remembering common names; they simply must be learned. IUPAC nomenclature follows general rules. The final **e** of the hydrocarbon stem is replaced by **oic,** followed by the word **acid.** The names, formulas, and derivations of some of the more common carboxylic acids are given in Table 9.1. Names appearing in boldface are preferred in present day chemical literature.

TABLE 9.1 Some Common Carboxylic Acids

Name			
(Common)	(IUPAC)	Formula	Derivation
Formic	Methanoic	H—COOH	L. *formica,* ant
Acetic	Ethanoic	CH_3COOH	L. *acetum,* vinegar
Propionic	**Propanoic**	CH_3CH_2COOH	Gr. *protos,* first; *pion,* fat
n-Butyric	**Butanoic**	$CH_3CH_2CH_2COOH$	L. *butyrum,* butter
n-Valeric	**Pentanoic**	$CH_3(CH_2)_3COOH$	L. *valere,* powerful
Caproic	**Hexanoic**	$CH_3(CH_2)_4COOH$	L. *caper,* goat
Caprylic	**Octanoic**	$CH_3(CH_2)_6COOH$	L. *caper,* goat
Capric	**Decanoic**	$CH_3(CH_2)_8COOH$	L. *caper,* goat
Lauric	**Dodecanoic**	$CH_3(CH_2)_{10}COOH$	Laurel
Palmitic	**Hexadecanoic**	$CH_3(CH_2)_{14}COOH$	Palm oil
Stearic	**Octadecanoic**	$CH_3(CH_2)_{16}COOH$	Gr. *stear,* tallow
Benzoic	Benzenecarboxylic acid	$C_6H_5{-}COOH$	(gum benzoin)

The carbon atom of the carboxyl group is always carbon 1 in systematic nomenclature, other numbers locating the position of substituents. Greek letters, α, β, γ, δ, etc., are used to locate substituents in common names. The α-carbon in an acid is always the carbon atom adjacent, or joined, to the carboxyl group.

$$\underset{(5)}{\overset{\delta}{C}}-\underset{(4)}{\overset{\gamma}{C}}-\underset{(3)}{\overset{\beta}{C}}-\underset{(2)}{\overset{\alpha}{C}}-\underset{(1)}{C}(=O)OH$$

In a few cases certain acids are more conveniently named as derivatives of acetic acid or as alkanecarboxylic acids. The following examples will illustrate the various methods for naming substituted fatty acids.

$\underset{(3)}{\overset{\beta}{CH_3}}-\underset{(2)}{\overset{\alpha}{CH}}(Br)-\underset{(1)}{COOH}$

2-Bromopropanoic acid
(α-Bromopropionic acid)

$\underset{(4)}{\overset{\gamma}{CH_3}}-\underset{(3)}{\overset{\beta}{CH}}(OH)-\underset{(2)}{\overset{\alpha}{CH_2}}-\underset{(1)}{COOH}$

3-Hydroxybutanoic acid
(β-Hydroxybutyric acid)

$CH_2=CH-CH_2COOH$

3-Butenoic acid
(Vinylacetic acid)

$CH_3-C(CH_3)_2-COOH$

Trimethylacetic acid
(2,2-Dimethylpropanoic acid)
(Pivalic acid)

$C_6H_5-CH_2-C(=O)OH$

Phenylacetic acid

$C_6H_{11}-COOH$

Cyclohexanecarboxylic acid

Aromatic acids are usually designated by common names, or named as derivatives of the parent acid—benzoic, C_6H_5COOH. Substituents on the benzene ring are designated either by number, or by the prefixes *ortho-* (*o*-), *meta-* (*m*-), or *para-* (*p*-). The ring carbon atom that bears the carboxyl group is always number 1 when an aromatic acid is named as a substituted benzoic acid.

$o\text{-}HOC_6H_4COOH$

Salicylic acid
(*o*-Hydroxybenzoic acid)

$2,6\text{-}(CH_3)_2C_6H_3COOH$

2,6-Dimethylbenzoic acid

$C_6H_5-CH(OH)-COOH$

Mandelic acid
(α-Hydroxyphenylacetic acid)

Exercise 9.1 Assign acceptable names to the following: (a) $CH_3-CH=CH-COOH$ (b) $CH_3-CH(NH_2)-COOH$ (c) $BrCH_2CH_2CH_2COOH$ (d) $(CH_3)_2CHCH_2COOH$

9.2 Acetic Acid

Acetic acid, although classified as one of the fatty acids, is not a component acid of fats and oils. As stated earlier, it is the acid found in vinegar—a 5% solution of acetic acid. Synthetic acetic acid brought to this dilution is "white" vinegar, whereas "brown" vinegar is the natural product of apple juice fermentation. Vinegars may also be made from other fruit juices. The fermentation process produces ethyl alcohol as the first product. Certain enzymes, if present, then catalyze the further oxidation of ethyl alcohol to acetic acid.

$$\underset{\text{Fruit sugar}}{C_6H_{12}O_6} \xrightarrow[\text{of yeast}]{\text{enzymes}} 2\,C_2H_5OH + 2\,CO_2$$

$$C_2H_5OH + O_2\ (\text{air}) \xrightarrow[\text{acetobacter}]{\text{enzymes of}} CH_3{-}C(=O){-}OH + H_2O$$

Pure acetic acid, mp 16.7°C, is referred to as "glacial" acetic acid because it appears as an icelike solid at lower than room temperature. Acetic acid is by far the most important of the monocarboxylic acids.

Exercise 9.2 Vinegar has a density of 1.0055 g/mL. If this represents a 5% (by weight) acetic acid solution, what is the concentration of vinegar in terms of molarity?

Exercise 9.3 The density of acetic acid is 1.0491. How many mL of acetic acid would you need to dilute to prepare a liter of "white" vinegar? *Answer:* 47.9 mL

9.3 Acidity and Structure

The carboxylic acids are much weaker acids than the mineral acids (HCl, H_2SO_4, HNO_3), but they are more acidic than the phenols. The carboxylic acids do not show a strong tendency to dissociate into protons and their corresponding acid anions. Indeed, the equilibrium between the ionized and unionized forms of a carboxylic acid lies far to the left. The reaction of acetic acid with water may be used as an illustration.

$$\underset{\text{Acetic acid}}{CH_3{-}C(=O){-}OH} + H_2O \rightleftharpoons \underset{\text{Acetate ion}}{CH_3{-}C(=O){-}O^-} + \underset{\text{Hydronium ion}}{H_3\overset{+}{O}}$$

Acetic acid in a 0.1 M aqueous solution has been found to dissociate only to the extent of 1.34% at room temperature. Statistically, this means that less than two molecules of acetic acid out of every hundred undergo ionization at any one time. The hydronium ion concentration may be related to that of other components in the acid solution by means of an equilibrium constant, K_a, called an ionization constant. A small numerical value for K_a indicates a weak acid in which the bulk of the acid molecules remain in the undissociated form.

$$K_a = \frac{\text{concentration of hydronium ion} \times \text{concentration of acetate ion}}{\text{concentration of undissociated acetic acid}^1}$$

$$= \frac{[H_3O^+][CH_3COO^-]}{[CH_3COOH]}$$

$$= \frac{(0.1 \times 0.0134)^2}{0.1 \times (1 - 0.0134)} = 1.8 \times 10^{-5}$$

An acid stronger than acetic acid would be dissociated to a greater extent. Thus, for such an acid the numerator in the equation for K_a would be larger, and the denominator would be smaller, leading to a value of K_a larger than that for acetic acid. Therefore, the stronger the acid, the larger its ionization constant, K_a.

Because most organic acids are relatively weak, their ionization constants are very small numbers expressed in negative powers of 10. Thus, it is convenient and common practice to convert these rather cumbersome numbers into numbers easier to handle. This is done by taking the *negative* logarithm of K_a and calling the result the pK_a value for the acid.

$$pK_a = -\log K_a$$

Thus for acetic acid

$$pK_a = -\log (1.8 \times 10^{-5}) = 4.74$$

Note that the *smaller* the value of pK_a, the *stronger* the acid will be.

Exercise 9.4 Calculate the ionization constants for the following acids whose pK_a values are given in parentheses: (a) trifluoroacetic acid (0.23), (b) phenol (9.99), (c) propanoic acid (4.87).

Although the tendency of carboxylic acids to dissociate is but slight, two factors play a part in the process. The first factor, **resonance stabilization,** helps to explain why the

[1]The concentration of water has been omitted from this expression for K_a because the number of water molecules involved in the formation of hydronium ions is negligible with respect to the total number of water molecules present. It may be assumed that the concentration of water remains nearly constant.

carboxylic acids are acids at all. The gain in stability afforded through resonance of the acid anion serves as the driving force that helps to promote the ionization process.

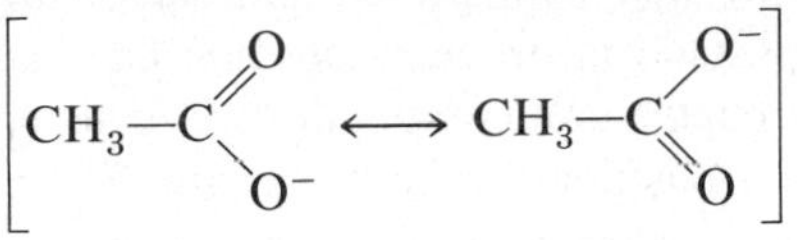

$$\left[CH_3{-}C\begin{matrix}{=}O \\ {-}O^- \end{matrix} \longleftrightarrow CH_3{-}C\begin{matrix}{-}O^- \\ {=}O \end{matrix} \right]$$

(a) (b)

The resonance forms shown for the acetate ion represent two extremes. The actual structure of the acetate ion appears to be that of a hybrid intermediate between (a) and (b) and is more correctly represented by the structure below.

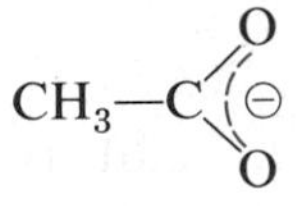

$$CH_3{-}C\begin{matrix}O \\ \ominus \\ O\end{matrix}$$

The second factor is an **inductive effect.** Electrophilic groups on the hydrocarbon portion of an acid have the effect of promoting ionization. Such groups withdraw electrons from the carboxyl group, promote the departure of the positive proton, and help stabilize the anion. For example, chloroacetic acid, $Cl{-}CH_2COOH$, is far more acidic than acetic acid due to the inductive effect of the chlorine atom.

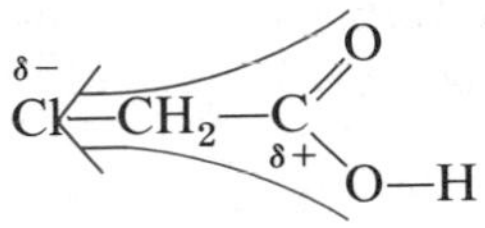

$$\overset{\delta-}{Cl}{\leftarrow}CH_2{-}\overset{\delta+}{C}\begin{matrix}{=}O \\ {-}O{-}H\end{matrix}$$

A second chlorine atom increases the acidity still further, and three chlorine atoms on the α-carbon produce an acid about 13,000 times as strong as acetic acid based on the relative K_a values. The inductive effects diminish when electron-withdrawing groups are farther removed from the carboxyl group and are strongest when such substituents are positioned on the α carbon. Inductive effects also vary with the electronegativities of the elements bonded to the α-carbon atom. Reference to Table 9.2 will show that fluoroacetic acid is nearly four times as strong an acid as iodoacetic acid.

Alkyl groups tend to be mild electron-donating groups relative to hydrogen. Therefore, the inductive effects of such electron-releasing groups on the α-carbon atom destabilize the anion by intensifying the negative charge, thus making proton departure more difficult and the acid weaker.

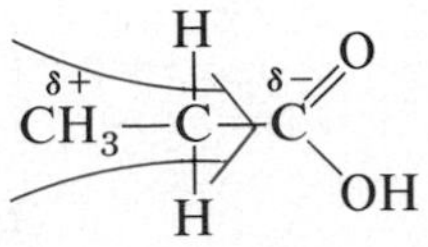

$$\overset{\delta+}{CH_3}{-}\underset{H}{\overset{H}{C}}{\rightarrow}\overset{\delta-}{C}\begin{matrix}{=}O \\ {-}OH\end{matrix}$$

In the aromatic series, ring-activating groups are electron-releasing, while ring-deactivating groups are electron-withdrawing (Sec. 4.8). Therefore, when ring-deactivating groups are substituted in the *ortho* or *para* positions of benzoic acid, the substituted acid is stronger than benzoic acid because electrons are drawn away from the carboxyl group, promoting the loss of a proton and stabilizing the carboxylate anion. Ring-activating groups substituted in the *para* position have an acid-weakening effect. Table 9.2 compares the strengths of a number of carboxylic acids and illustrates the effects of substituents and their positions on acid strength. The table is divided in such a way as to focus your attention on meaningful comparisons.

TABLE 9.2 Relative Strengths of Some Organic Acids (25°C)

Name of Acid	Structure	pK_a
Water*	H—OH	15.74
Formic	H—COOH	3.75
Acetic	CH_3COOH	4.76
Monochloroacetic	$ClCH_2COOH$	2.86
Dichloroacetic	$Cl_2CHCOOH$	1.30
Trichloroacetic	Cl_3CCOOH	0.64
Chloroacetic	$ClCH_2COOH$	2.86
Bromoacetic	$BrCH_2COOH$	2.90
Iodoacetic	ICH_2COOH	3.18
Fluoroacetic	FCH_2COOH	2.59
Propanoic	CH_3CH_2COOH	4.87
2-Chloropropanoic	$CH_3CH(Cl)COOH$	2.80
3-Chloropropanoic	$ClCH_2CH_2COOH$	4.00
Benzoic	C_6H_5COOH	4.21
p-Chlorobenzoic	Cl—C_6H_4—COOH	3.99
p-Nitrobenzoic	O_2N—C_6H_4—COOH	3.44
p-Methoxybenzoic	CH_3O—C_6H_4—COOH	4.49

*While water does not belong in the above category, it is included for reference value. The value of pK_a is given for the equation: $K = [H^+][OH^-]/[H_2O]$, where $[H_2O]$ is 55.5 moles per liter.

Exercise 9.5 Without consulting Table 9.2, arrange the acids in the following set in order of diminishing acidity: (a) acetic acid, (b) benzoic acid, (c) propanoic acid, (d) chloroacetic acid, (e) bromoacetic acid.

9.4 Properties

The lower-molecular weight members of the aliphatic series of carboxylic acids are liquids with sharp or disagreeable odors. Those with four to ten carbon atoms are particularly obnoxious. The odors of rancid butter, limburger cheese, and stale sweat vividly exemplify this unpleasant property. The higher members are waxlike solids and almost odorless. The boiling points of fatty acids increase regularly by an approximate 20°C increment per methylene unit. The abnormally high boiling points of the fatty acids are accounted for by hydrogen bond formation between acid molecules. Hydrogen bonding also accounts for the fact that molecular weight measurements reveal the fatty acids to be largely ''dimeric,'' or double, molecules. Both linear and cyclic dimers are present, the latter predominating.

$$\mathrm{R{-}C}\begin{matrix}\mathrm{\diagup\!\!\!\diagup O\cdots H{-}O\diagdown} \\ \mathrm{\diagdown O{-}H\cdots O\diagup\!\!\!\diagup}\end{matrix}\mathrm{C{-}R}$$

Cyclic dimer

Carboxylic acids, with the exception of the first five members of the aliphatic series, are not very soluble in water. The aromatic acids usually are crystalline solids, also sparingly soluble in cold water.

9.5 Preparation of Acids

A number of straight-chain aliphatic acids and some aromatic acids are available as natural products. Others can be prepared by one of the following methods.

A. Oxidation Methods. Direct oxidation of primary alcohols provides one of the most direct routes to the corresponding aliphatic acids. Potassium or sodium dichromate in combination with concentrated sulfuric acid is frequently used to bring about this change.

$$3\,\mathrm{R{-}CH_2OH} + 2\,\mathrm{Cr_2O_7^{2-}} + 16\,\mathrm{H^+} \longrightarrow 3\,\mathrm{R{-}C}\begin{matrix}\mathrm{\diagup\!\!\!\diagup O} \\ \mathrm{\diagdown OH}\end{matrix} + 4\,\mathrm{Cr^{3+}} + 11\,\mathrm{H_2O}$$

The benzene carboxylic acids are obtained by the oxidation of alkyl derivatives of benzene. The benzene ring itself appears to be rather resistant to oxidation but facilitates the oxidation of the carbon atom attached to it. Should more than one alkyl group be attached to the ring, all become oxidized to carboxyl groups. Thus, toluene and other monoalkylated benzenes yield benzoic acid, and the three isomeric xylenes, $C_6H_4(CH_3)_2$, yield the corresponding phthalic acids.

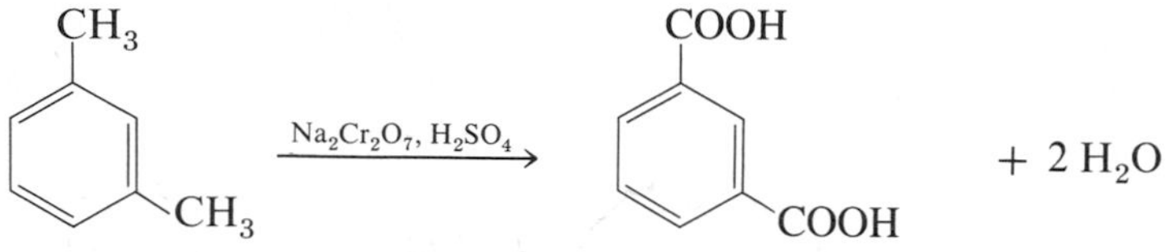

$$C_6H_4(CH_3)_2 \xrightarrow{Na_2Cr_2O_7,\ H_2SO_4} C_6H_4(COOH)_2 + 2\ H_2O$$

m-Xylene

1,3-Benzenedicarboxylic acid (Isophthalic acid)

Although the length of the side chain on the benzene nucleus may vary, it is degraded to the last or ring-attached carbon atom. All other carbon atoms in the side chain are oxidized to carbon dioxide.

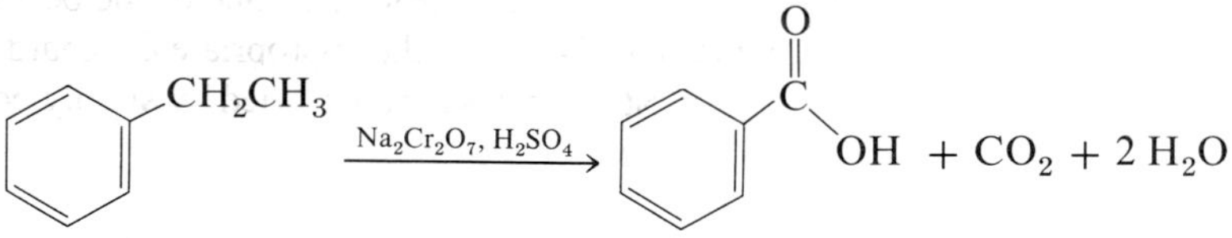

$$C_6H_5CH_2CH_3 \xrightarrow{Na_2Cr_2O_7,\ H_2SO_4} C_6H_5C(=O)OH + CO_2 + 2\ H_2O$$

Ethylbenzene

Benzoic acid

Exercise 9.6 Balance the equation for the oxidation of ethylbenzene.

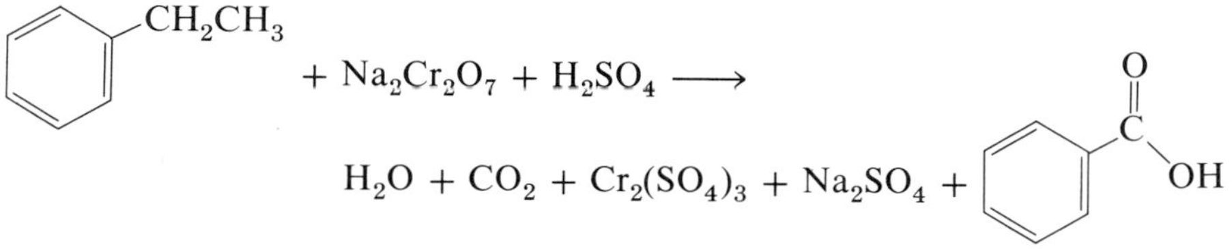

$$C_6H_5CH_2CH_3 + Na_2Cr_2O_7 + H_2SO_4 \longrightarrow H_2O + CO_2 + Cr_2(SO_4)_3 + Na_2SO_4 + C_6H_5C(=O)OH$$

Naphthalene, obtainable from coal tar, is a source of phthalic acid. Phthalic anhydride, formed when phthalic acid is heated in excess of 200°C, is an important industrial material used in the manufacture of glyptal resins for surface coatings (Sec. 11.4).

$$\text{Naphthalene} + 9\,[O] \xrightarrow[470°C]{V_2O_5\ \text{air}} \text{Phthalic anhydride} + 2\,H_2O + 2\,CO_2$$

Naphthalene   Phthalic anhydride

$$\text{Phthalic anhydride} + H_2O \longrightarrow \text{Phthalic acid}$$

Phthalic acid
(1,2-Benzenedicarboxylic acid)

B. The Carbonation of Grignard Reagents. Grignard reagents are especially useful for the preparation of acids, either aliphatic or aromatic. One of the best general methods for the preparation of carboxylic acids is to treat the appropriate Grignard reagent with anhydrous carbon dioxide. In the laboratory this is accomplished by simply pouring the ether solution of the Grignard reagent onto "dry ice."

$$\overset{\delta-}{R}-\overset{\delta+}{MgX} + O{=}\overset{\delta+}{C}{=}\overset{\delta-}{O} \longrightarrow R-C(=O)-O^{-}\ {}^{+}MgX$$

$$R-C(=O)-OMgX + HX \longrightarrow R-C(=O)-OH + MgX_2$$

A very important feature of the carbonation of Grignard reagents with carbon dioxide is that the reaction increases the length of the carbon chain by one carbon.

EXAMPLE

Show how trimethylacetic acid might be prepared from a Grignard reagent by carbonation with CO_2 followed by hydrolysis.

Solution

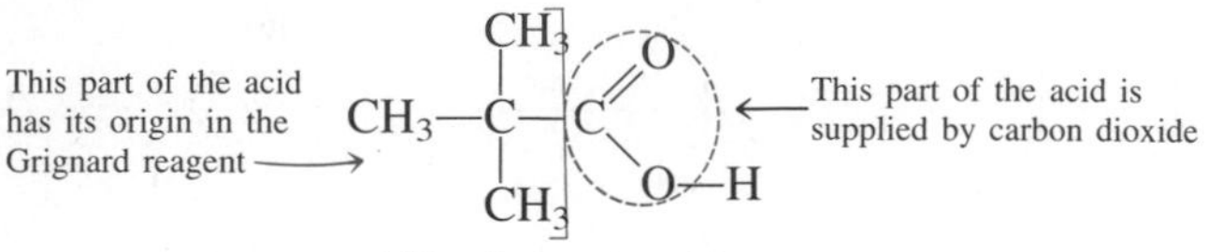

This part of the acid has its origin in the Grignard reagent ⟶

$$CH_3-\overset{CH_3}{\underset{CH_3}{C}}-C(=O)-O-H$$

⟵ This part of the acid is supplied by carbon dioxide

Trimethylacetic acid

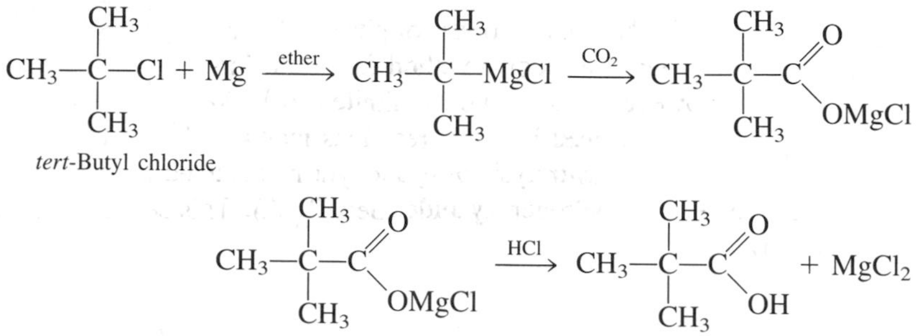

$$CH_3-\overset{CH_3}{\underset{CH_3}{C}}-Cl + Mg \xrightarrow{\text{ether}} CH_3-\overset{CH_3}{\underset{CH_3}{C}}-MgCl \xrightarrow{CO_2} CH_3-\overset{CH_3}{\underset{CH_3}{C}}-C(=O)OMgCl$$

tert-Butyl chloride

$$CH_3-\overset{CH_3}{\underset{CH_3}{C}}-C(=O)OMgCl \xrightarrow{HCl} CH_3-\overset{CH_3}{\underset{CH_3}{C}}-C(=O)OH + MgCl_2$$

Exercise 9.7 Show how an alkyl bromide can be converted to an acid having *two* more carbons in the carbon chain by a sequence involving only one Grignard reaction and an oxidation.

C. Hydrolysis of Nitriles. Nitriles may be prepared from the S_N2 reaction of alkyl halides with potassium cyanide, as shown below for the synthesis of propanenitrile.

$$CH_3CH_2-Br + KCN \longrightarrow CH_3CH_2C\equiv N$$

Propanenitrile
(Propionitrile)

Simple nitriles are named according to the acids they yield on hydrolysis. Thus, methyl cyanide, CH_3CN, is named *acetonitrile;* phenyl cyanide, C_6H_5CN, *benzonitrile,* etc. In the IUPAC system nitriles are named by adding *nitrile* as a suffix to the hydrocarbon stem of the same length. Acetonitrile with a total of two carbons is ethanenitrile.

Nitriles may be hydrolyzed to acids having the same number of carbon atoms by refluxing with aqueous acid or alkali.

Acid hydrolysis

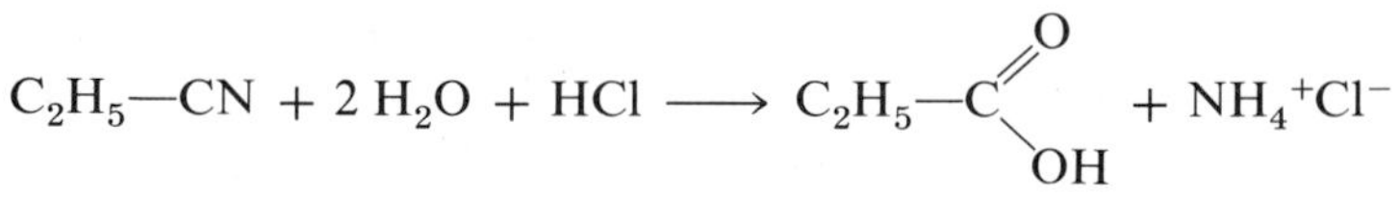

$$C_2H_5-CN + 2\,H_2O + HCl \longrightarrow C_2H_5-C(=O)OH + NH_4^+Cl^-$$

Propanoic acid
(Propionic acid)

Alkaline hydrolysis

$$C_2H_5{-}CN + 2\,H_2O + NaOH \longrightarrow C_2H_5{-}C(=O)O^-Na^+ + NH_3 + H_2O$$

Sodium propanoate
(Sodium propionate)

The formation of acids by the hydrolysis of nitriles obtained from the alkyl halides of one less carbon atom provides a second method for increasing the length of the carbon chain by one carbon; however, this method is limited to halides that react well by the S_N2 mechanism. Cyanides formed by other reactions may also be hydrolyzed to acids. One important example is the hydrolysis of cyanohydrins obtained from the reaction of aldehydes and ketones with hydrogen cyanide (Sec. 8.7A). This sequence yields α-hydroxy acids as shown below.

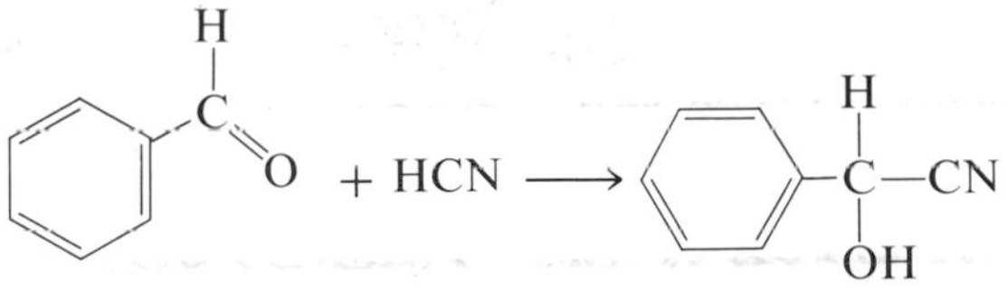

$$C_6H_5{-}CHO + HCN \longrightarrow C_6H_5{-}CH(OH){-}CN$$

Benzaldehyde

Benzaldehyde cyanohydrin (Mandelonitrile)

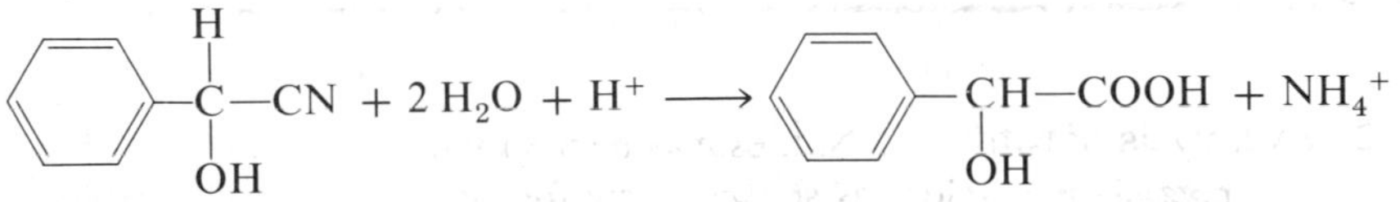

$$C_6H_5{-}CH(OH){-}CN + 2\,H_2O + H^+ \longrightarrow C_6H_5{-}CH(OH){-}COOH + NH_4^+$$

Mandelic acid

Exercise 9.8 The attempted preparation of trimethylacetic acid by treatment of *tert*-butyl bromide with potassium cyanide followed by hydrolysis produced 2-methylpropene instead. Explain.

9.6 Reactions of the Carboxylic Acids

The reactions of the carboxylic acids may be divided into five categories: (1) salt formation or the replacement of the acidic hydrogen, (2) nucleophilic addition at the carboxyl carbon, (3) replacement of α hydrogen in the hydrocarbon segment, (4) reduction of the carboxyl group, and (5) decarboxylation in which CO_2 is expelled and the carbon chain is shortened. An example of each type of reaction will be illustrated.

A. Preparation of Salts. Soaps. The carboxylic acids, although feebly acidic when compared to mineral acids, are relatively strong when compared to water (Table 9.2). They may be neutralized quantitatively by carbonates, bicarbonates, and hydroxide bases to produce salts and water.

$$2\,R{-}C(=O)OH + Na_2CO_3 \longrightarrow 2\,R{-}C(=O)O^-Na^+ + CO_2 + H_2O$$

$$R{-}C(=O)OH + NaHCO_3 \longrightarrow R{-}C(=O)O^-Na^+ + CO_2 + H_2O$$

$$R{-}C(=O)OH + NaOH \longrightarrow R{-}C(=O)O^-Na^+ + H_2O$$

Salts of organic acids are named as one would name salts of inorganic acids—that is, the cation is named first, followed by the name of the acid anion. The latter is derived by dropping the *ic* of the acid and adding *ate*. For example,

$$CH_3{-}C(=O)O^-Na^+$$

Sodium acetate

$$CH_3{-}(CH_2)_{16}{-}C(=O)O^-Li^+$$

Lithium stearate

Exercise 9.9 Draw structures for: (a) ammonium benzoate, (b) potassium formate, (c) calcium acetate, (d) sodium palmitate.

The reaction of sodium or potassium hydroxide with natural fats produces the alkali metal salts of the long-chain fatty acids twelve to eighteen carbons in length. These salts are called **soaps,** and the process whereby they are formed is called **saponification.** Many of the synthetic detergents (*syndets*) also are acid salts but differ from soaps principally in the anionic part of the molecule. Soaps are carboxylates, whereas synthetic detergents are mainly sulfates or sulfonates. Some detergents are ammonium salts, and others are non-ionic molecules. However, the molecular structures of synthetic detergents and soaps are similar in that both types of cleansing agents are characterized by the presence of a long, nonpolar, hydrophobic (water-hating) hydrocarbon tail and a polar, hydrophilic (water-loving) head. Examples of all four types of detergents are shown.

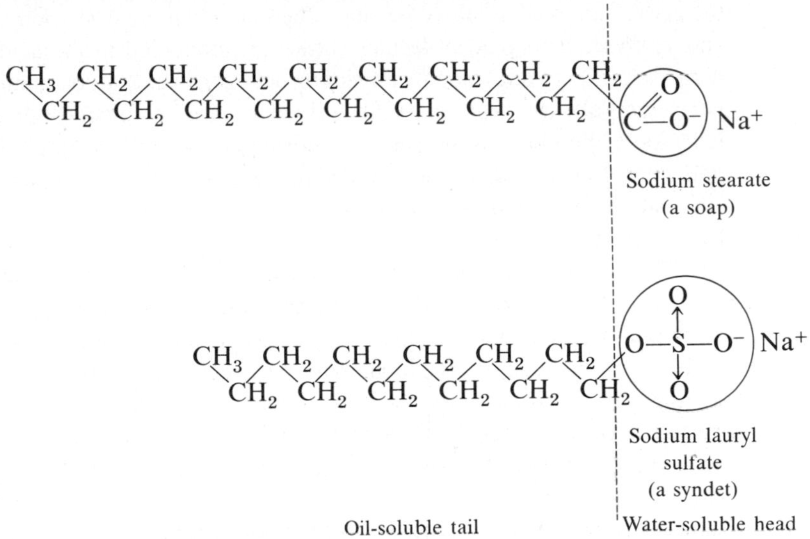

$$CH_3CH_2CH_2CH_2CH_2CH_2CH_2CH_2CH_2CH_2CH_2CH_2CH_2CH_2CH_2CH_2CH_2C(=O)O^- \; Na^+$$

Sodium stearate
(a soap)

$$CH_3CH_2CH_2CH_2CH_2CH_2CH_2CH_2CH_2CH_2CH_2CH_2-O-S(\rightarrow O)_2-O^- \; Na^+$$

Sodium lauryl
sulfate
(a syndet)

Oil-soluble tail | Water-soluble head

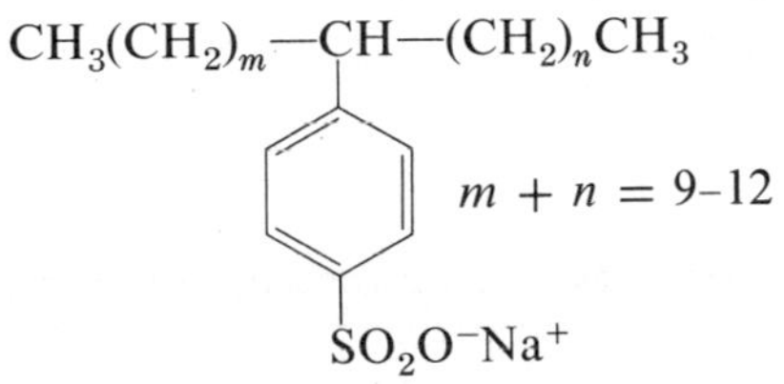

$$CH_3(CH_2)_m-\underset{\underset{SO_2O^-Na^+}{|}}{\underset{C_6H_4}{|}}{CH}-(CH_2)_nCH_3 \qquad m + n = 9\text{–}12$$

An alkyl benzene sulfonate (Anionic)

$$R-\overset{+}{N}(CH_3)_3Cl^-$$

(Cationic)
$R = C_{16}$

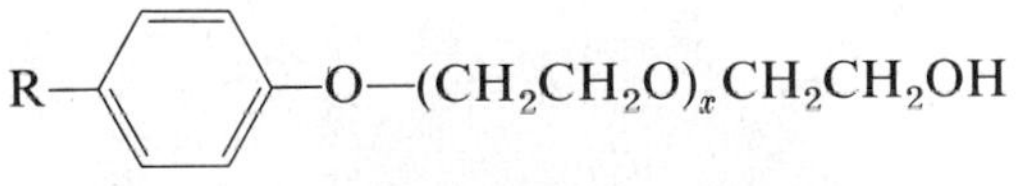

$$R-C_6H_4-O-(CH_2CH_2O)_xCH_2CH_2OH$$

(Nonionic)
$R = C_8\text{—}C_{10}$; $x = 8 - 12$

Soaps, when used in hard water, precipitate as insoluble calcium, magnesium, and iron salts of the long-chain fatty acids, but synthetic detergents are not precipitated in the presence of these ions. Solubility in hard water is one of the principal advantages that a synthetic detergent has over a soap. One the other hand, the large-scale use of syndets in this country within the past several decades has presented a serious disposal problem. Many of these detergents have an aromatic ring in the hydrocarbon segment of the molecule. This type is generally classified as alkyl benzene sulfonate (ABS) detergents.

The straight-chain fatty acid salts with an even number of carbon atoms are biodegradable—that is, easily degraded by microorganisms. Unfortunately, some of the early

ABS types of detergents had highly branched alkyl side chains that resisted degradation, and the detergents reappeared quite active and sudsy in some freshwater sources. Present day ABS derivatives have the biodegradable structure shown, with branching confined to the carbon atom adjacent to the ring. The long-chain hydrocarbon segments of both soap and synthetic detergent molecules shown are embodied in the acid anion portion of sodium salts. This does not mean that a detergent, like a soap, must be a sodium or a potassium salt. On the contrary, the oil-soluble tail or "business end" of some detergents is cationic. Solubilizing power, not sudsing action, is the principal criterion for a good detergent. For this reason certain detergents have been developed, especially for use in automatic clothes and dish washers, that are neither anionic nor cationic but are simply long-chain polar molecules.

The mechanics of detergent action seem to be the same for both soaps and syndets. In the removal of soil from clothing and grease from dishes, the hydrocarbon segment of the detergent molecule dissolves in the oily or greasy layer (like dissolves like). The oily layer is then dislodged as small globules by the mechanical action of rubbing, tumbling, or stirring. Once loosened, each globule becomes emulsified, or suspended in water, because the polar end of the detergent molecule is attracted to the polar water molecule. Such tiny droplets of "oil-in-water" emulsion do not coagulate, or condense, into huge globules because the electrical nature of the outer film surrounding each droplet is the same. A particle that has adjacent to its surface a double layer of charges of opposite sign is called a **micelle.** One micelle thus tends to repel the other. The role that a detergent or a soap molecule plays in the removal of oil is illustrated (in part) by Fig. 9.1.

FIGURE 9.1 The Solvating Action of Soap on a Droplet of Water

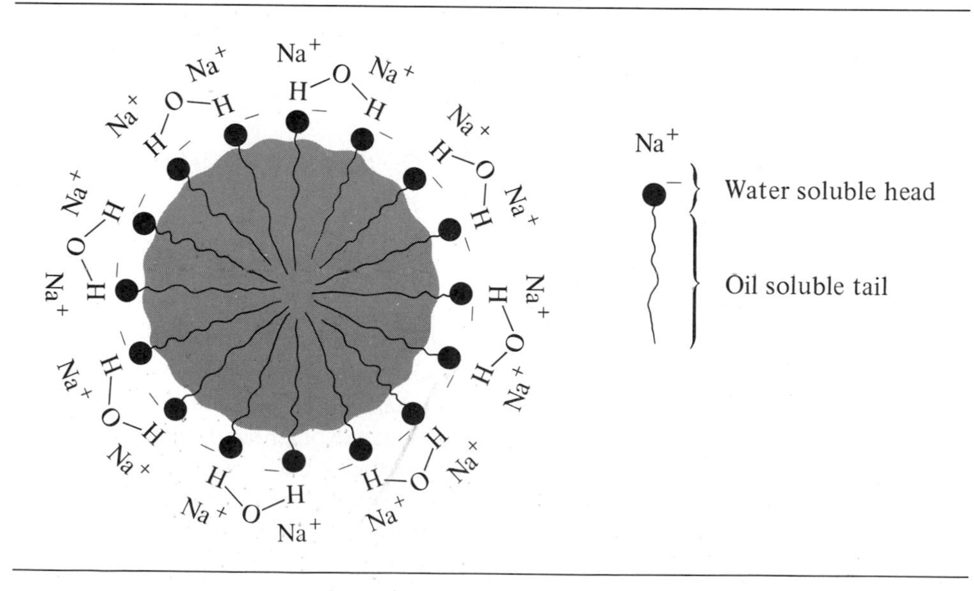

Acid salts other than soaps are to be found in a number of everyday commodities. Lithium stearate and certain heavy metal salts, when blended with oils, form lubricating grease. Calcium propanoate, $(CH_3CH_2COO)_2Ca$, is a commonly used additive in bread to retard spoilage. Certain of the zinc salts of the higher fatty acids also are excellent fungicides and are used in the treatment of skin diseases such as athlete's foot. One of these, zinc undecylenate from 10-undecenoic acid, is particularly effective. Its structure is shown.

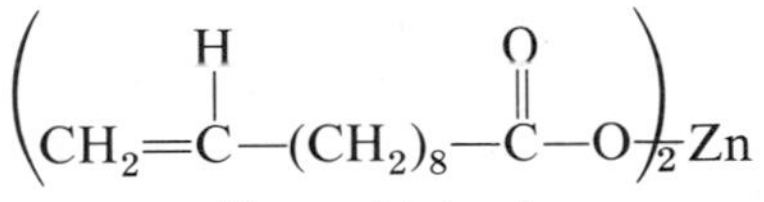

$$\left(CH_2{=}\overset{H}{\overset{|}{C}}{-}(CH_2)_8{-}\overset{O}{\overset{\|}{C}}{-}O\right)_2Zn$$

Zinc undecylenate

The cupric salts of naphthenic acids, which are cyclic, nonaromatic acids derived from petroleum, are used as wood preservatives.

B. Nucleophilic Addition. Preparation of Esters. The reaction of an acid with an alcohol in the presence of a mineral acid catalyst produces an ester. A direct esterification of alcohols and acids in this manner is known as the **Fischer esterification.**

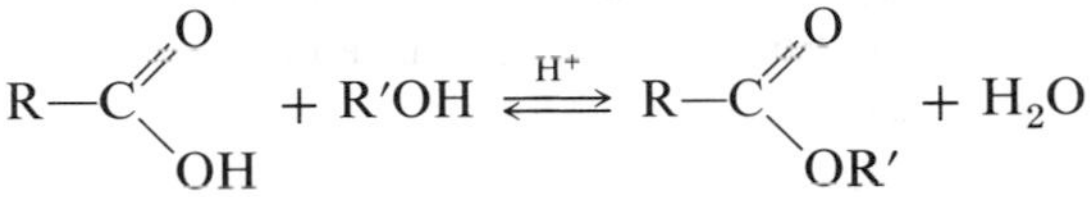

$$R{-}C({=}O){-}OH + R'OH \underset{}{\overset{H^+}{\rightleftharpoons}} R{-}C({=}O){-}OR' + H_2O$$

Direct esterification is a reversible reaction, and at equilibrium appreciable amounts of unreacted acid and alcohol may be present. Thus, if we start with equimolar quantities of acetic acid and ethyl alcohol and allow them to react until equilibrium is established, two-thirds of the acid and the alcohol will have been converted to ester, and one-third will remain as unreacted acid and alcohol.

$$K_e = \frac{\text{ethyl acetate} \times \text{water}}{\text{acetic acid} \times \text{ethyl alcohol}} = \frac{(2/3)^2}{(1/3)^2} = 4$$

The principle of LeChatelier may be applied to shift the equilibrium to the right and promote esterification, either by increasing the concentration of the alcohol or the acid or by removing the ester or the water as it is formed. The reverse reaction, hydrolysis, on the other hand, can be made complete only when carried out in an alkaline solution.

Esters are named in a manner similar to that used in naming the salts of the carboxylic acids, in that the group attached to the oxygen is named first.

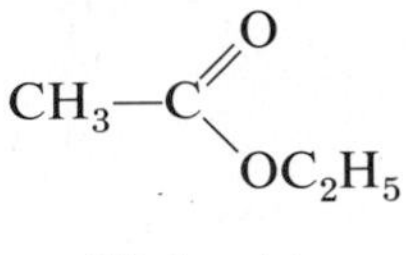

$$CH_3{-}C({=}O){-}OC_2H_5 \qquad C_6H_5{-}C({=}O){-}OCH_3$$

Ethyl acetate 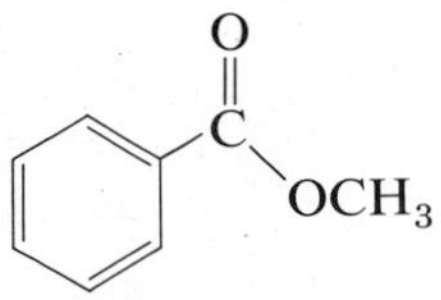 Methyl benzoate

The following mechanism has been proposed for the acid-catalyzed esterification of acids.

$$CH_3{-}C{\overset{O}{\underset{OH}{\diagup\!\!\diagdown}}} + H^+ \rightleftharpoons CH_3{-}C(=O){-}\overset{\oplus}{O}H_2$$

$$CH_3{-}C{\overset{O}{\underset{OH}{\diagup\!\!\diagdown}}} + H^+ \rightleftharpoons \left[CH_3{-}C(=\overset{\oplus}{O}H){-}OH \longleftrightarrow CH_3{-}C(OH)(=\overset{\oplus}{O}H) \longleftrightarrow CH_3{-}\overset{\oplus}{C}(OH)(OH) \right]$$

The acid is protonated on both oxygens, but only the oxonium ion formed by protonation on the carbonyl oxygen is involved in the reaction with alcohol.

$$\left[CH_3{-}\overset{\oplus}{C}(OH)(OH) \right] + \underset{H}{\overset{|}{O}}{-}C_2H_5 \rightleftharpoons CH_3{-}C(OH)(OH){-}\underset{H}{\overset{|}{O}}{-}C_2H_5$$

$$CH_3{-}C(OH)(OH){-}\overset{\oplus}{O}(H){-}C_2H_5 \rightleftharpoons CH_3{-}C(OH)(\overset{\oplus}{O}H_2){-}O{-}C_2H_5$$

$$CH_3{-}C(OH)(\overset{\oplus}{O}H_2){-}O{-}C_2H_5 \rightleftharpoons$$

$$H_2O + \left[CH_3{-}\overset{\oplus}{C}(OH){-}O{-}C_2H_5 \longleftrightarrow CH_3{-}C(=\overset{\oplus}{O}H){-}O{-}C_2H_5 \right]$$

$$\left[CH_3{-}C(=\overset{\oplus}{O}H){-}O{-}C_2H_5 \right] \rightleftharpoons H^+ + CH_3{-}C{\overset{O}{\underset{OC_2H_5}{\diagup\!\!\diagdown}}}$$

In the esterification reaction, the oxygen atom in the alkoxy group of the ester is derived from the alcohol, and the oxygen atom in the molecule of water formed is derived from the acid. Thus, when alcohols containing the ^{18}O isotope are used in the esterification reaction, the **labeled** oxygen is found exclusively in the ester and not in the water.

$$CH_3C(=O)OH + CH_3CH_2{}^{18}OH \xrightarrow{H^+} CH_3C(=O){}^{18}O{-}CH_2CH_3 + H_2O$$

Exercise 9.10 (a) If a Fischer esterification of acetic acid with ethanol were carried out using 1 mole of acid and 10 moles of alcohol, how many moles of ethyl acetate would appear in the equilibrium mixture? (b) How many moles of ester would appear in the equilibrium mixture had we started with 5 moles of the acid and 1 mole of the alcohol?
Answer: (a) 0.97; (b) 0.945

C. Replacement of α-Hydrogen. Hydrogen in the α-position may be replaced by halogen and provide an easy route to α-halogenated acids and indirectly to a number of other substituted carboxylic acids. In practice, the α-bromoacids usually are prepared. The appropriate acid is treated with red phosphorus and bromine to produce the acid bromide as the initial product. Acid bromides form enols much more readily than do carboxylic acids. Therefore, in the example that follows the bromine will attack selectively the α-carbon of acetyl bromide rather than another molecule of acetic acid to yield α-bromoacetyl bromide. After bromination, α-bromoacetyl bromide transfers its acetyl bromine to an unreacted molecule of acetic acid to give another molecule of acetyl bromide.

$$6\,CH_3C(=O)OH + 2\,P + 3\,Br_2 \longrightarrow 6\,H{-}CH_2{-}C(=O)Br + 2\,P(OH)_3$$

Acetyl bromide

$$CH_3{-}C(=O)Br + Br_2 \longrightarrow BrCH_2{-}C(=O)Br + HBr$$

α-Bromoacetyl bromide

$$Br{-}CH_2{-}C(=O)Br + CH_3{-}C(=O)OH \rightleftharpoons Br{-}CH_2{-}C(=O)OH + CH_3{-}C(=O)Br$$

α-Bromoacetic acid

The reaction thus continues until all the original acid is converted to the α-bromo derivative, and the phosphorus tribromide need be present only in catalytic amounts. The above sequence of reactions is known as the Hell–Volhard–Zelinsky reaction.

Exercise 9.11 What simple test tube reaction would serve to distinguish between bromoacetic acid and ethyl bromide?

D. Reduction of Carboxylic Acids. Carboxylic acids may be reduced with lithium aluminum hydride or with diborane but *not* with sodium borohydride.

$$4\,R{-}C(=O){-}OH + 3\,LiAlH_4 \xrightarrow[\text{ether}]{\text{anhydrous}} (R{-}CH_2{-}O)_4AlLi + 4\,H_2$$

$$(R{-}CH_2{-}O)_4AlLi \xrightarrow[H^+]{H_2O} 4\,R{-}CH_2{-}OH$$

$$3\,R{-}C(=O){-}OH + B_2H_6 \longrightarrow (R{-}CH_2{-}O)_3B + H_3BO_3 \xrightarrow[H^+]{H_2O} 3\,R{-}CH_2{-}OH$$

E. Decarboxylation. Carboxylic acids will undergo decarboxylation with a loss of carbon dioxide and a degradation of the carbon skeleton when their silver salts are treated with halogen.

$$R{-}C(=O){-}O^-Ag^+ + Br_2 \xrightarrow{CCl_4} R{-}Br + AgBr\downarrow + CO_2$$

Strong heating of monocarboxylic acids usually results in a fragmentation of the carbon chain. However, β-ketoacids or acids bearing a second carboxyl group on the α-carbon atom decarboxylate smoothly simply on heating.

$$\underset{\text{Acetoacetic acid}}{CH_3{-}C(=O){-}CH_2{-}(COO)H} \xrightarrow{\text{heat}} \underset{\text{Acetone}}{CH_3{-}C(=O){-}CH_3} + CO_2$$

$$\underset{\substack{\text{Malonic acid}\\ \text{(Propanedioic acid)}}}{HOOC{-}CH_2{-}(COO)H} \xrightarrow{\text{heat}} \underset{\text{Acetic acid}}{CH_3{-}COOH} + CO_2$$

Exercise 9.12 The decarboxylation of silver salts of acids appears to take place by way of a free-radical mechanism. Propose a reaction mechanism that would explain the loss of CO_2 and the formation of R—X from the intermediate R—COOBr.

9.7 Properties of the Aromatic Acids

The reactions of the aromatic carboxylic acids are, in general, very similar to those of the aliphatic acids. The carboxyl group, you will recall, is a *meta*-director and prevents other substituents from entering *ortho* positions. *Ortho, para*-substituted benzoic acids, therefore, must be prepared indirectly. For example, *p*-nitrobenzoic acid is not prepared from benzoic acid but is obtained by the oxidation of *p*-nitrotoluene.

$$CH_3C_6H_4NO_2 + 3\,[O] \xrightarrow[\text{heat}]{K_2Cr_2O_7,} HOOCC_6H_4NO_2 + H_2O$$

p-Nitrotoluene — *p*-Nitrobenzoic acid

The synthesis of an organic compound requires not only the use of the proper reactions but also the use of these reactions in the proper order or sequence. Sometimes such routes are lengthy and circuitous but may be the only ones leading to the desired product. At other times the orientation of substitution may be such that devising a synthesis is relatively simple.

Summary

1. **Structure**
   The carboxyl group, —COOH, is the functional group of the carboxylic acids. The aliphatic acids are referred to as *fatty acids* because many appear in a combined form with glycerol as fats.
2. **Nomenclature of Acids**
   The carboxylic acids may be named in two different ways. In common names substituents in the hydrocarbon segment of an acid are located by Greek letters; in systematic nomenclature, by number. In the case of aromatic acids, substituents are located either by number or by *ortho, meta,* and *para* prefixes.
3. **Properties of the Acids**
   The carboxylic acids are weak acids. They are capable of associating (H-bonding) and are high-boiling liquids or crystalline solids sparingly soluble in water. H-bonding also makes possible the formation of double molecules or dimers.
4. **Preparation of the Acids**
   The carboxylic acids may be prepared by any of the following methods.
   (a) Oxidation of primary alcohols, methyl ketones (haloform reaction), or side chains on the aromatic nucleus.
   (b) Hydrolysis of nitriles.
   (c) Carbonation of Grignard reagents.

5. **Reactions**

Aliphatic and aromatic acids give reactions that are almost identical. Any differences are of degree rather than of kind. These reactions include:

(a) Replacement of the ionizable hydrogen by another cation (salt formation). The alkali metal salts of the long-chain acids are called soaps.

(b) Replacement of the hydroxyl by an alkoxyl, —OR, to yield **esters,**

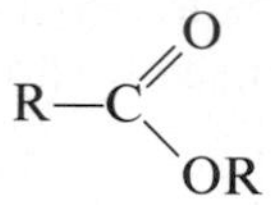

$$R-C(=O)-OR$$

(c) Replacement of hydrogen in the hydrocarbon segment.

(d) Reduction of the carboxyl group to $—CH_2OH$.

(e) Degradation of the carbon chain by decarboxylation.

Supplementary Exercises

9.13 Draw structural formulas for each of the following compounds.

(a) pentanoic acid
(b) 2,3-dimethylhexanoic acid
(c) ethyl bromoacetate
(d) methyl salicylate
(e) pentanenitrile
(f) γ-phenylbutyric acid
(g) butyronitrile
(h) *p*-ethoxybenzoic acid
(i) sodium trifluoroacetate
(j) methyl isobutyrate

9.14 Assign an acceptable name to each of the following.

(a) 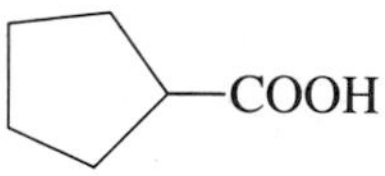 cyclopentane–COOH

(b) 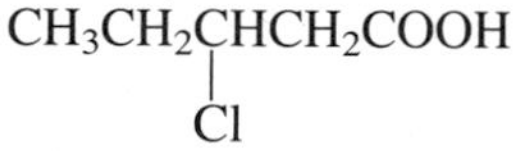 $CH_3CH_2CHCH_2COOH$ (Cl on C-3)

(c) $CH_3CH_2CH_2CHCOOH$ (CH_3 on C-2)

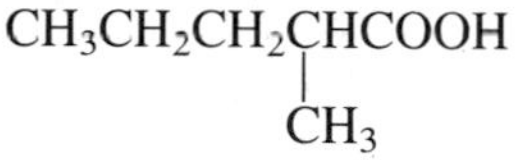

(d) 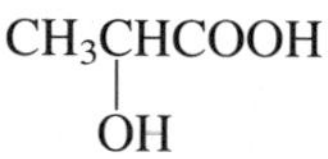 $CH_3CHCOOH$ (OH on C-2)

(e) 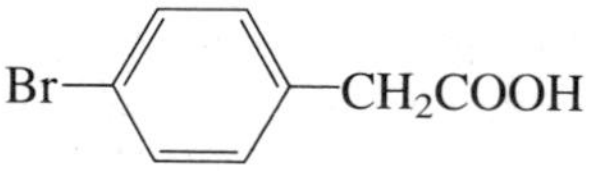 Br–C₆H₄–CH_2COOH (para)

(f) 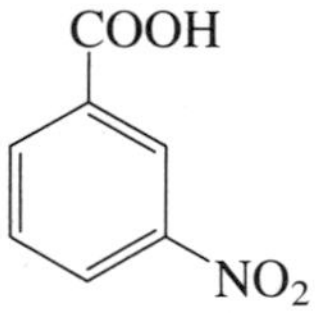 COOH on benzene ring with NO_2 meta

(g) 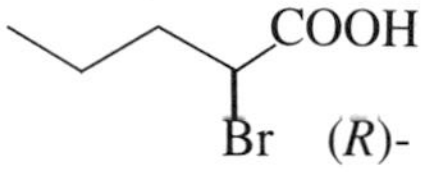 skeletal chain with COOH and Br on the same carbon (*R*)- or (*S*)-?

(h)  skeletal chain with Cl (dashed) and CH_3 (wedge), COOH (*R*)- or (*S*)-?

(i) 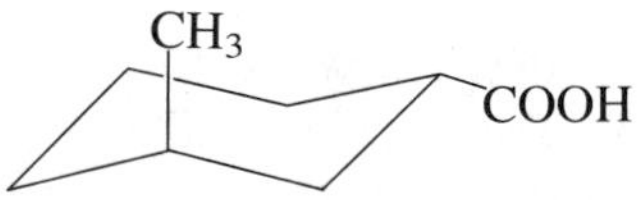 cyclohexane with CH_3 and COOH *cis* or *trans*?

9.15 What product would you expect from the reaction of 3-methylbutanoic acid with the following reagents?

(a) silver hydroxide, followed by Br_2 in CCl_4
(b) $LiAlH_4$ in ether, followed by water and H^+
(c) Br_2 and PBr_3
(d) NaOH in water
(e) cyclohexanol plus a trace of H^+

9.16 Write the resonance structures for benzoic acid, benzoate anion, *p*-nitrobenzoic acid, and *p*-nitrobenzoate anion. Use these resonance structures to explain why *p*-nitrobenzoic acid is a stronger acid than benzoic acid.

9.17 Identify each lettered product in the following reaction sequences.

(a) $CH_3CN \xrightarrow[\text{reflux}]{Na^+OH^-} (A) + (B);$

$A + HCl \longrightarrow CH_3COOH + NaCl$

(b) $n\text{-}C_4H_9\text{—}Br \xrightarrow{NaCN} (A) \xrightarrow[HCl]{H_2O} (B) \xrightarrow{LiAlH_4} (C)$

(c) $C_2H_5Br \xrightarrow[\text{ether}]{\text{Mg, anhydrous}} (A) \xrightarrow[\text{(2) hydrolysis}]{\text{(1) } CO_2} (B) \xrightarrow{PBr_3} (C) \xrightarrow{CH_3CH_2COOH} (D)$

(d) $C_6H_5\text{—}CH(CH_3)_2 + K_2Cr_2O_7 + H_2SO_4 \longrightarrow (A)$

9.18 Calculate the pK_a values for the following acids whose acid ionization constants are given in parentheses following their names: (a) methoxyacetic acid (2.9×10^{-4}), (b) 3-butenoic acid (4.5×10^{-5}), (c) 3-butynoic acid (4.8×10^{-4}).

9.19 Without referring to a table of acidity constants, arrange the acids in the following set in an order of diminishing acidity: (a) benzoic acid, (b) *p*-nitrobenzoic acid, (c) phenylacetic acid, (d) *p*-methoxybenzoic acid.

9.20 Using acetic acid as your only carbon-containing starting material and any other inorganic reagents you may need, show how you might synthesize the following compounds. (*Note:* The product of one reaction may be used as starting material for the preparation of another.)

(a) ethanol
(b) ethyl bromide
(c) 2-butanol
(d) iodoform
(e) propanoic acid
(f) acetone
(g) isopropyl alcohol
(h) 2-bromopropane
(i) bromoacetic acid
(j) 2-methylpropanoic acid

9.21 Outline a procedure for the removal of acetic acid (bp, 118°) from a mixture that also contains *n*-butyl alcohol (bp, 117°) and phenol.

9.22 Using a minimum number of steps, outline a reaction sequence for the conversion of a carboxylic acid to the next higher homolog. Use a readily available acid as your starting acid.

9.23 Which of the preparative methods discussed in Section 9.5 should be used to convert 1,2-dibromoethane into succinic acid (butanedioic acid), $HOOC\text{—}(CH_2)_2\text{—}COOH$?

9.24 Using $Ba^{14}CO_3$ as your source of carbon-14, outline a procedure for the preparation of $CH_3CH_2{}^{14}COOH$.

9.25 Beginning with toluene as your only organic starting material, outline a reaction or reaction sequence that would lead to the preparation of the following compounds.

(a) *p*-bromotoluene
(b) 4-methylbenzoic acid
(c) *p*-bromophenylacetic acid
(d) terephthalic acid (p-$C_6H_4(CO_2H)_2$)
(e) benzyl phenyl ketone (Hint: *see* Section 8.12)

9.26 An unknown acid was believed to be either *o*-bromobenzoic acid or 2,4-dibromobenzoic acid. A 0.2412-g sample of the acid neutralized 12.4 ml of 0.098 *N* sodium hydroxide solution. Which acid was it?

9.27 The following compounds have the molecular formula $C_9H_{10}O_2$, and the structural feature

$$-C(=O)-O-$$

in common. Draw a structure for each isomer that is consistent with the following sets of data:

Compound A: A sweet-smelling liquid that was insoluble in water and in cold dilute base. Strong oxidation of the compound produced benzoic acid. Alkaline hydrolysis produced benzyl alcohol as one product.
Compound B: A white, odorless solid that was insoluble in water but soluble in dilute sodium hydroxide. Vigorous oxidation produced phthalic acid.

9.28 Outline sequences of reactions that would be suitable for bringing about the following transformations.

(a) $C_6H_{10}{=}CH_2 \longrightarrow C_6H_{11}{-}CH_2COOH$ (methylenecyclohexane → cyclohexylacetic acid)

(b) $CH_3CH{=}C(CH_3)_2 \longrightarrow CH_3CH_2C(CH_3)_2{-}COOH \longrightarrow CH_3CH_2C(CH_3)_2{-}C(=O){-}CH_3$

(Hint: *see* Section 8.12)

(c) $CH_3CH_2CH_2CH_2COOH \longrightarrow CH_3CH_2CH_2COOH$

***9.29** Which of the compounds in Exercise 9.27 would be expected to have an nmr spectrum consisting of a 3-proton singlet at δ 2.00, a 2-proton singlet at δ 5.05, and a 5-proton singlet at δ 7.29?

***9.30** An organic compound, $C_6H_{12}O_2$, gave an ir spectrum with strong absorption at 1740 and 1715 cm^{-1} and an nmr spectrum consisting of a 9-proton singlet at δ 1.08, a 2-proton singlet at δ 2.20, and a 1-proton singlet at δ 11.24. Propose a structure for the compound.

***9.31** An organic compound, $C_6H_{11}BrO_2$, showed strong absorption in the ir at 1733 cm^{-1} and gave an nmr spectrum consisting of a 3-proton triplet at δ 1.19, a 6-proton singlet at δ 1.92, and a 2-proton quartet at δ 4.20. Propose a structure for the compound.

***9.32** Propose a structure for an acid, $C_8H_8O_2$, that would give an nmr spectrum consisting of three singlets in a 5:2:1 ratio.

10 Carboxylic Acid Derivatives

The term ''acid derivative'' is used to describe four general classes of compounds in which some other atom or group has replaced the hydroxyl group of the acid, and from which the original acid can be regained by hydrolysis. In the following sections are outlined a number of the more general reactions by which the acid derivatives are prepared. Following these are the specific reactions of the derivative. It soon will become apparent that in a number of instances a reaction involving one acid derivative is a route leading to the preparation of another.

10.1 General Considerations

The acid derivatives have the general structure

$$R{-}C(=O){-}Z$$

where Z = X (halogen), R′O (alkoxy), RCOO (carboxy), and R'_2N (amido). Although the members of the four classes differ from each other in many ways, much of their chemistry is quite similar. Indeed, the principal differences in many of their reactions are the abilities of the anions Z^- to function as good leaving groups (Sec.7.7B). The four classes of compounds and their anionic leaving groups are as follows:

Acid halides (acyl halides)	$Z^- = X^-$	(halide ion)
Acid anhydrides	$Z^- = RCOO^-$	(carboxylate ion)
Esters	$Z^- = R'O^-$	(alkoxide ion)
Amides	$Z^- = R'_2N^-$	(amide anion)

Recalling that good leaving groups are generally either anions derived from strong acids or neutral molecules, we can arrange the four anions as follows:

$$X^- > RCOO^- \gg R'O^- > R'_2N^-$$

The last two of these are poor leaving groups, but they do function as such in the acid derivative series. Of course, if they are protonated to form the alcohol, R′OH, or the amine, R'_2NH, they become better leaving groups because they can then leave as neutral molecules. Of the four, only X^- (halide) is a truly good leaving group; therefore, the acid halides are the most reactive of the acid derivatives. Many of the reactions of the acid derivatives with nucleophiles have similar mechanisms, differing only as necessary to persuade the leaving group to leave. A very approximate overview of the mechanism of these reactions follows.

$$R{-}C(=O)Z + :Nu^- \rightleftharpoons R{-}C(O^-)(Nu)Z \rightleftharpoons R{-}C(=O)Nu + :Z^-$$

Nucleophile · Tetrahedral intermediate · Leaving group

The nucleophile adds to the carbon of the carbonyl group, forming a tetrahedral intermediate. Departure of the leaving group yields a new acid derivative. Of course, if $:Nu^-$ is a carbanion, we will get an aldehyde or ketone rather than a new acid derivative. The details of the mechanism will vary, depending on whether or not acid or base catalysis is employed to overcome any deficiencies of the leaving group. The mechanism may be classified as an **addition-elimination mechanism,** that is, the addition of one nucleophile followed by the elimination of another. We have already encountered such a mechanism in the haloform reaction (Sec. 8.6).

10.2 Acid Halides

A. Preparation of Acid Halides. The acid halides are also referred to as the **acyl halides.** The name of an acyl group,

$$R{-}C(=O){-}$$

is derived from that of an aliphatic acid by dropping the *ic* ending of the acid and adding *yl*. The aromatic acid halides or aroyl halides, are formed and named in the same manner. The following examples will illustrate the usage of the term as a group name.

$CH_3C(=O)OH$ — Acetic acid (Ethanoic acid)

$CH_3C(=O)Cl$ — Acetyl chloride (Ethanoyl chloride)

$CH_3C(=O)CH_2C(=O)CH_3$ — Acetylacetone (2,4-Pentanedione)

$C_6H_5C(=O)OH$ — Benzoic acid

$C_6H_5C(=O)Cl$ — Benzoyl chloride

$C_6H_5C(=O)C_6H_4COOH$ — *o*-Benzoylbenzoic acid (Benzophenone-*o*-carboxylic acid)

Acid halides other than the chlorides can be made, but the chlorides are more easily and economically prepared. The acyl chlorides are very reactive compounds and are usually employed when acyl halides are required. They are easily prepared by the reaction of the appropriate acid with either phosphorus trichloride, phosphorus pentachloride, or thionyl chloride. Thionyl chloride as a chlorinating reagent is superior to the phosphorus halides for two reasons: (1) the by-products of the reaction are gases and are easily removed, and (2) the reagent is a low-boiling liquid. The second advantage permits any excess reagent to be easily removed by distillation.

$$C_6H_5C(=O)OH + SOCl_2 \longrightarrow C_6H_5C(=O)Cl + SO_2 + HCl$$

Benzoic acid, Thionyl chloride, Benzoyl chloride

B. Reactions of the Acid Halides. The acid halides are low-boiling liquids of irritating odor. They are extremely reactive compounds because the inductive effect of the halogen atom further diminishes the electron density on the carbonyl carbon atom. Furthermore, as was pointed out earlier, the weakly basic chloride ion is an excellent leaving group. Thus, attack on the carbonyl carbon by nucleophiles (Sec. 1.13) is enhanced.

$$Nu{:} \rightarrow R\overset{\delta+}{C}(=O^{\delta-})Cl$$

$$Nu{:} = H_2\ddot{O}{:},\ R\ddot{O}H{:},\ H{-}\ddot{N}H_2 \text{ or } R{-}C(=O)\ddot{O}{:}^-$$

The reaction of an acyl halide with water is called **hydrolysis** and results in the displacement of halogen by hydroxyl to reform the original organic acid.

$$H_2\ddot{O}: + R-C(=\ddot{O}:)-\ddot{C}l: \rightleftharpoons H_2\overset{+}{O}-C(-\ddot{O}:^-)(R)-\ddot{C}l: \longrightarrow H_2\overset{+}{O}-C(=\ddot{O}:)-R + :\ddot{C}l:^-$$

$$H-\overset{\oplus}{\ddot{O}}(H)-C(=O)-R \longrightarrow HO-C(=O)-R + H^+$$

The net equation may be shown as

$$R-C(=O)-Cl + HOH \longrightarrow R-C(=O)-OH + HCl$$

The reaction of an acyl halide with an alcohol is called **alcoholysis** and results in the displacement of halogen by an alkoxy group, —OR, to produce an **ester** (Sec. 10.4A).

$$R-C(=O)-Cl + HOR' \longrightarrow \underset{\text{An ester}}{R-C(=O)-OR'} + HCl$$

The reaction of an acyl halide with ammonia is called **ammonolysis** and results in the displacement of halogen by an amino group, $—NH_2$, to produce an **amide** (Sec. 10.5A).

$$R-C(=O)-Cl + 2\,NH_3 \longrightarrow \underset{\text{An amide}}{R-C(=O)-NH_2} + NH_4Cl$$

Acyl halides react with salts of acids to yield acid **anhydrides.**

$$\underset{\text{Sodium acetate}}{CH_3-C(=O)-O^-Na^+} + \underset{\text{Acetyl chloride}}{CH_3-C(=O)-Cl} \longrightarrow \underset{\text{Acetic anhydride}}{CH_3-C(=O)-O-C(=O)-CH_3} + Na^+Cl^-$$

Exercise 10.1 Hydrolysis of an acyl halide is irreversible; that is, we cannot make an acyl halide by the reaction of an acid with hydrogen chloride. Explain this using the mechanism written for hydrolysis of an acyl halide.

10.3 Acid Anhydrides

The structures of the acid anhydrides consist of two acyl groups connected by an oxygen atom. Both simple and mixed anhydrides are known. However, mixed anhydrides are relatively uncommon. To name an anhydride, add the word *anhydride* to the common or IUPAC name(s) of the acid(s), as illustrated for acetic anhydride below.

A. Preparation of Acid Anhydrides. The preparation of an acid anhydride is a reaction that has the net effect of removing a molecule of water from between two molecules of the acid.

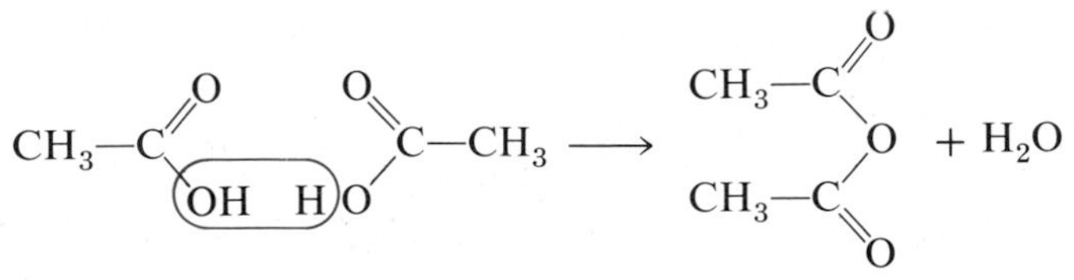

$$CH_3-C(=O)-OH + HO-C(=O)-CH_3 \longrightarrow (CH_3-C(=O))_2O + H_2O$$

Acetic anhydride
(Ethanoic anhydride)

Direct dehydration, however, is seldom practiced. Dehydration usually is accomplished indirectly by reaction between the sodium salt of the acid with its acid chloride. Two acyl groups are thus bridged by an oxygen atom by a method much like that employed in the Willilamson ether synthesis (Sec. 7.11). Acetic anhydride is by far the most important acid anhydride. Industrially acetic anhydride is prepared from the very reactive "anhydride," ketene, $CH_2{=}C{=}O$. Ketene is prepared by the high-temperature dehydration of acetic acid.

$$\underset{\text{Acetic acid}}{CH_3-COOH} \xrightarrow[700^\circ]{AlPO_4} H_2O + \underset{\text{Ketene}}{CH_2{=}C{=}O}$$

Acetic acid adds to ketene to produce acetic anhydride.

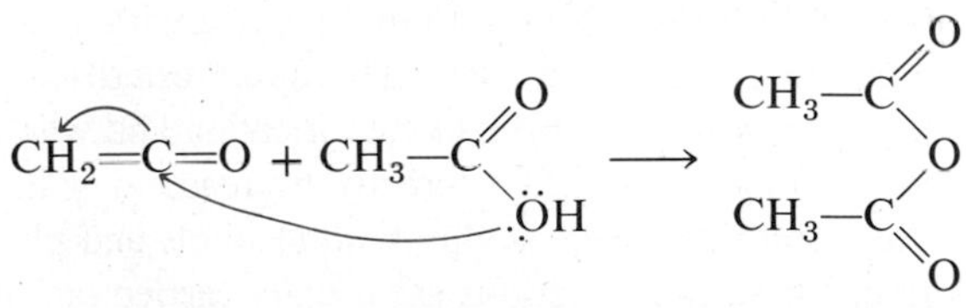

$$\underset{\text{Ketene}}{CH_2{=}C{=}O} + \underset{\text{Acetic acid}}{CH_3-C(=O)-\ddot{O}H} \longrightarrow \underset{\text{Acetic anhydride}}{(CH_3-C(=O))_2O}$$

B. Reactions of the Acid Anhydrides. The reactions of the acid anhydrides with water, alcohol, and ammonia parallel those already shown for the acyl halides. However, anhydrides are somewhat less reactive than acyl halides because the acyloxy group,

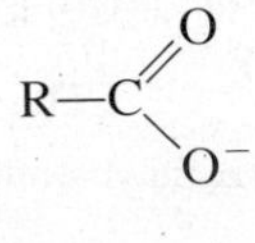

$$R-C(=O)-O^-$$

is less electron-withdrawing and activates the carbonyl group less than chlorine. Also, the acetate ion is a poorer leaving group. As shown in the following reactions, at least one molecule of the organic acid is a product of each reaction.

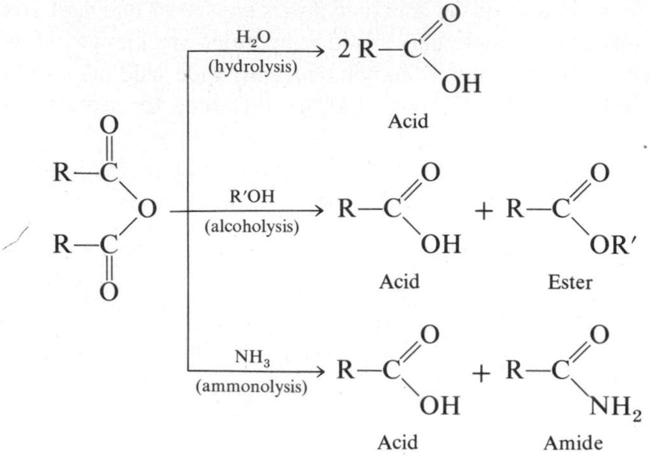

$$(RCO)_2O \xrightarrow[\text{(hydrolysis)}]{H_2O} 2\,R{-}C(=O){-}OH \quad \text{(Acid)}$$

$$(RCO)_2O \xrightarrow[\text{(alcoholysis)}]{R'OH} R{-}C(=O){-}OH \;(\text{Acid}) + R{-}C(=O){-}OR' \;(\text{Ester})$$

$$(RCO)_2O \xrightarrow[\text{(ammonolysis)}]{NH_3} R{-}C(=O){-}OH \;(\text{Acid}) + R{-}C(=O){-}NH_2 \;(\text{Amide})$$

Acid anhydrides may be used in place of acyl halides as acylating reagents in the Friedel–Crafts reaction (Sec. 8.4C).

10.4 Esters. The Claisen Condensation

Esters of carboxylic acids are named in the same manner as carboxylate salts: the name of the alkyl group attached to the oxygen atom is given first, followed by the name of the carboxylate group (the acid name minus *ic* plus *ate*). Several examples are given in the following sections and in Section 9.6B.

A. Preparation of Esters. The reaction of an acid with an alcohol in the presence of a mineral acid catalyst produces an ester. The direct esterification of acids with alcohols in this manner is known as the Fischer esterification and was covered in detail in Section 9.6B. Phenols cannot be esterified directly by reaction with a carboxylic acid.

Acyl chlorides react readily with alcohols and phenols to yield esters and hydrogen chloride. These esterifications are usually carried out in the presence of an organic base, such as pyridine or triethylamine (Section. 13.1), which serves to neutralize the hydrogen chloride formed in the reaction. This reaction, unlike the Fischer esterification, is not reversible.

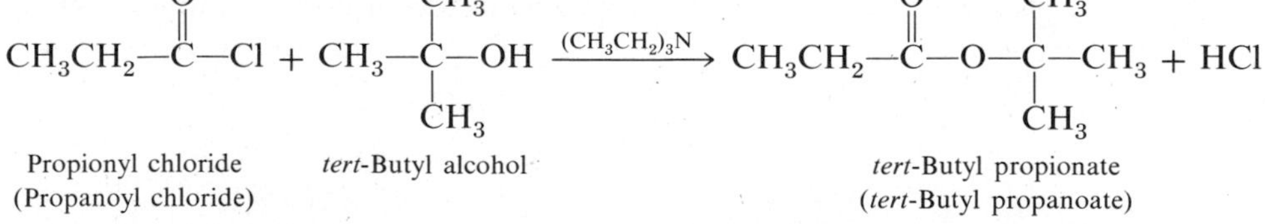

$$CH_3CH_2{-}C(=O){-}Cl + CH_3{-}C(CH_3)_2{-}OH \xrightarrow{(CH_3CH_2)_3N} CH_3CH_2{-}C(=O){-}O{-}C(CH_3)_2{-}CH_3 + HCl$$

Propionyl chloride (Propanoyl chloride) | *tert*-Butyl alcohol | *tert*-Butyl propionate (*tert*-Butyl propanoate)

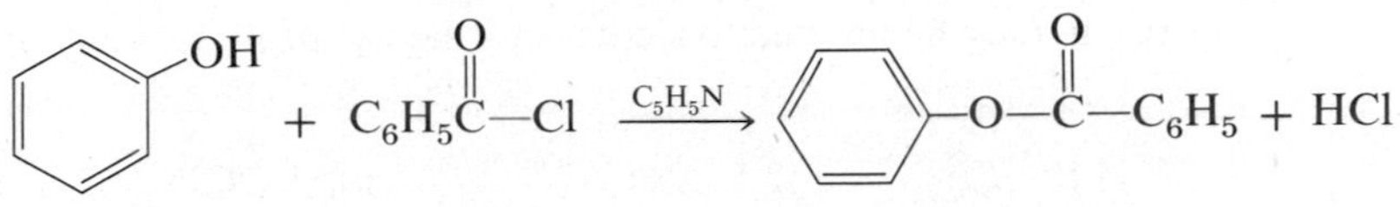

$$C_6H_5OH + C_6H_5\overset{O}{\overset{\|}{C}}-Cl \xrightarrow{C_5H_5N} C_6H_5-O-\overset{O}{\overset{\|}{C}}-C_6H_5 + HCl$$

Benzoyl chloride — Phenyl benzoate

Esters also may be prepared by heating the silver salt of an acid with an alkyl halide. This reaction likewise is not reversible, in part because the AgI precipitates from solution.

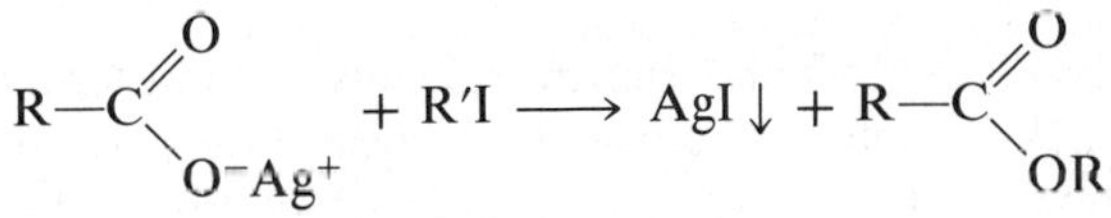

$$R-C(=O)O^-Ag^+ + R'I \longrightarrow AgI\downarrow + R-C(=O)OR'$$

B. Reactions of the Esters. The acid hydrolysis of esters has been shown to be an equilibrium reaction. In order to hydrolyze an ester irreversibly—that is, to have it quantitatively reform the acid (as a salt) and the alcohol—hydrolysis must be carried out in an alkaline solution. Under these conditions the acid component of the ester is converted to the salt of the carboxylic acid.

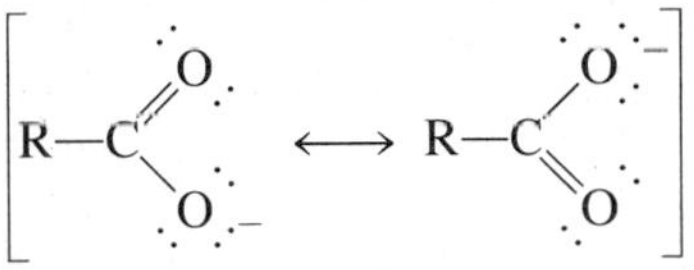

$$\left[R-C(=\ddot{O}:)-\ddot{O}:^- \longleftrightarrow R-C(-\ddot{O}:^-)=\ddot{O}:\right]$$

The highly resonance-stabilized carboxylate ion carries a full negative charge. This ion is a nucleophile and will repel rather than attract other nucleophilic species; therefore, reversal of hydrolysis will not occur. Alkaline hydrolysis of an ester is called **saponification** because soaps are prepared by the alkaline hydrolysis of fats and oils, esters of glycerol (Sec. 9.6A). As an example of saponification, the alkaline hydrolysis of ethyl acetate is shown below. Note that recovery of the organic acid requires that the carboxylate salt be treated with mineral acid.

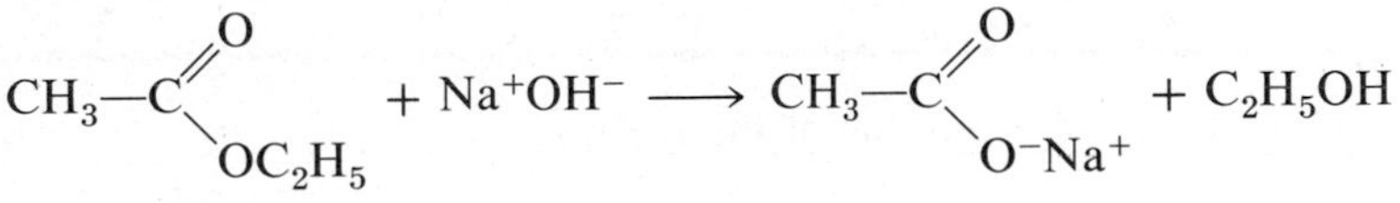

$$CH_3-C(=O)OC_2H_5 + Na^+OH^- \longrightarrow CH_3-C(=O)O^-Na^+ + C_2H_5OH$$

Ethyl acetate — Sodium acetate

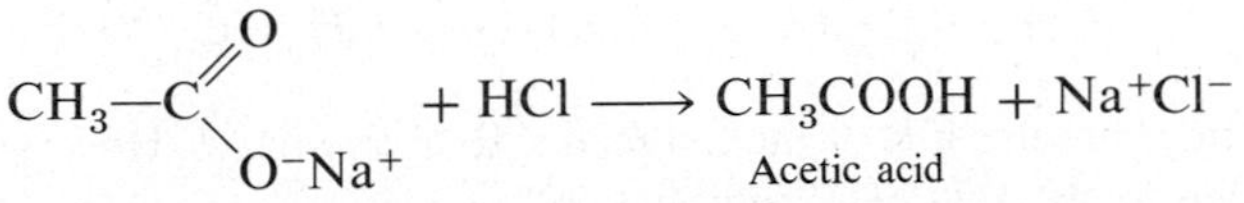

$$CH_3-C(=O)O^-Na^+ + HCl \longrightarrow CH_3COOH + Na^+Cl^-$$

Acetic acid

Ammonolysis of an ester forms an alcohol and an **amide.**

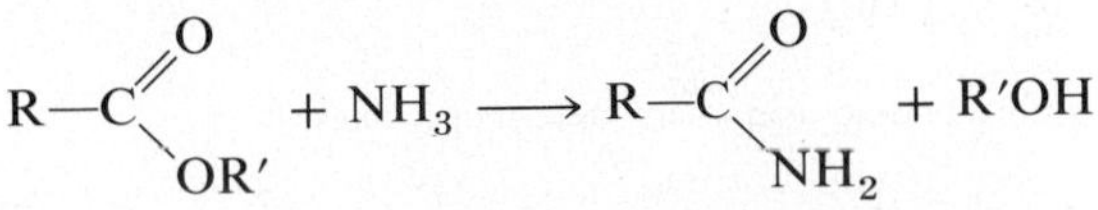

$$R-C(=O)OR' + NH_3 \longrightarrow R-C(=O)NH_2 + R'OH$$

An amide

Esters may be reduced to alcohols by the use of lithium aluminum hydride, $LiAlH_4$.

$$4\,R-C(=O)-OR' + 2\,LiAlH_4 \xrightarrow[\text{ether}]{\text{anhydrous}} LiAl(OCH_2R)_4 + LiAl(OR')_4$$

$$LiAl(OCH_2R)_4 + 4\,HCl \longrightarrow LiCl + AlCl_3 + 4\,RCH_2OH$$

$$LiAl(OR')_4 + 4\,HCl \longrightarrow LiCl + AlCl_3 + 4\,R'OH$$

Esters react with Grignard reagents to form tertiary alcohols[1] in which two of the three alkyl groups come from *the Grignard reagent*. In the tertiary alcohols made by the reaction of a ketone with a Grignard reagent, two of the three alkyl groups come from *the ketone* and the third from the Grignard reagent (Sec. 7.4C).

$$R-C(=\ddot{O}\!:)-OR' + \overset{\delta-}{R''}\overset{\delta+}{MgX} \longrightarrow R-C(R'')(:\ddot{O}:^{-}\,\overset{+}{MgX})-O-R'$$

$$R-C(R'')(:\ddot{O}:^{-}\,\overset{+}{MgX})-\ddot{O}R' \longrightarrow R(R'')C=O + R'-O^{-}\,\overset{+}{MgX}$$

$$R(R'')C=O + R''MgX \longrightarrow R''(R)(R'')C-OMgX$$

$$R-C(R'')_2-OMgX + HX \longrightarrow R-C(R'')_2-OH + MgX_2$$

Exercise 10.2 The rate of the reaction

$$CH_3-C(=O)-OR + NaOH \longrightarrow CH_3-C(=O)-O^-Na^+ + ROH$$

is 125 times greater if R = methyl than if R = *tert*-butyl. How would you account for this difference in the rate of saponification?

[1]Exception: Formic acid esters yield *sec*-alcohols.

C. Properties and Uses of Esters. The esters are liquids of rather pleasant odor. A number are responsible for the odors of certain fruits and other plant parts. For example, isopentyl acetate has the odor of bananas, isopentyl valerate that of apples, and butyl butyrate that of pineapples. Methyl salicylate (Sec. 11.6), "oil of wintergreen," has a mint-like odor. Many of the esters alone or as blends are used in artificial flavorings and in perfumes. The esters are excellent solvents for lacquers and plastics.

D. The Claisen Condensation. The α-hydrogen atoms of an ester such as ethyl acetate are sufficiently acidic to react with a strong base such as sodium ethoxide to produce the ester enolate anion. The ester anion is a strong nucleophile, which then may attack the carbonyl carbon atom of a second ester molecule. Elimination of ethoxide ion results in the formation of the β-keto ester, ethyl acetoacetate in the case of ethyl acetate. The reaction, illustrated in the following sequence, is called the **Claisen condensation**[2] and bears a resemblance to the aldol condensation (Sec. 8.8B) except that it is an addition-elimination rather than a simple addition reaction.

$$\underset{\text{Ethoxide ion}}{C_2H_5O^-} + H\text{—}CH_2\text{—}C(=\ddot{O}\!:)OC_2H_5 \rightleftharpoons$$

$$C_2H_5OH + \left[:\bar{C}H_2\text{—}C(=\ddot{O}\!:)OC_2H_5 \longleftrightarrow CH_2\text{=}C(\text{—}\ddot{O}\!:^-)OC_2H_5\right]$$

Ester enolate anion

$$CH_3\text{—}\overset{\delta+}{C}(=\overset{\delta-}{O})OC_2H_5 + \left[:\bar{C}H_2\text{—}C(=O)OC_2H_5\right] \rightleftharpoons CH_3\text{—}C(O^-)(OC_2H_5)\text{—}CH_2\text{—}C(=O)OC_2H_5$$

$$CH_3\text{—}C(\ddot{O}\!:^-)(OC_2H_5)\text{—}CH_2\text{—}C(=O)OC_2H_5 \rightleftharpoons C_2H_5O^- + \underset{\text{Ethyl acetoacetate}}{CH_3\text{—}C(=O)\text{—}CH_2\text{—}C(=O)OC_2H_5}$$

[2] Ludwig Claisen (1851–1930) was professor of chemistry at the University of Kiel.

The Claisen condensation is also an important biological reaction involved in the degradation and biosynthesis of the fatty acids (Sec. 12.6), although the conditions for the biochemical reactions are quite different from those employed in the laboratory.

The β-keto esters are useful in a number of organic syntheses that lead to ketones and carboxylic acids that are difficult or impossible to obtain by other methods. The β-keto esters such as ethyl acetoacetate are stronger acids than either the ketones or the esters and are easily converted into fairly stable enolate anions. These anions can be used as nucleophiles in a number of important reactions, including the S_N2 substitution of alkyl halides and conjugate addition to α,β-unsaturated carbonyl compounds (Sec. 8.12). The greater acidity of the β-keto esters is also reflected in the greater stability of the enol form relative to the keto form. The stability of the enol tautomer may be attributed in part to an intramolecular H-bond.

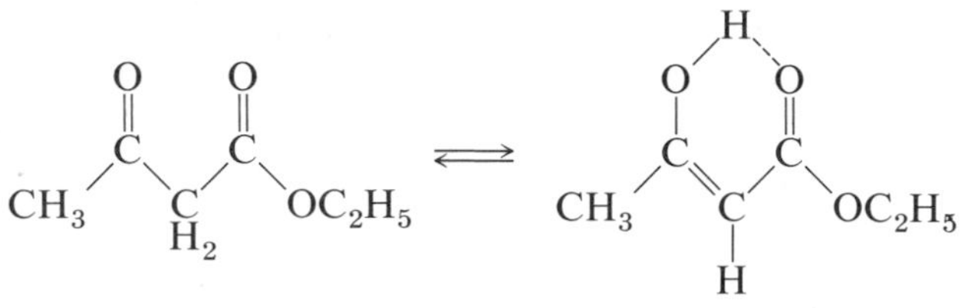

$$CH_3-\overset{O}{\overset{\|}{C}}-CH_2-\overset{O}{\overset{\|}{C}}-OC_2H_5 \rightleftharpoons CH_3-\overset{O-H}{\overset{|}{C}}=\underset{H}{\underset{|}{C}}-\overset{O}{\overset{\|}{C}}-OC_2H_5$$

Keto form
In hexane: 51%
In water: 90%
Reacts with $C_6H_5NHNH_2$

Enol form
In hexane: 49%
In water: 10%
Reacts with bromine and gives a red color with $FeCl_3$ *(a test reaction for phenols and enols)*

The readiness with which β-keto *esters* may be alkylated, combined with the ease with which β-keto *acids* may be decarboxylated, makes β-keto esters useful intermediates in a number of organic syntheses. One of the more important of these syntheses is the **acetoacetic ester synthesis,** which consists of the following steps: (1) conversion of ethyl acetoacetate to its enolate anion by treatment with a strong base, (2) alkylation of the enolate on the α carbon atom between the carbonyl groups using a good S_N2 halide, (3) hydrolysis of the alkylated β-keto ester with *dilute* base or with mineral acid to form the β-keto acid, and (4) decarboxylation of the β-keto acid to yield a substituted acetone derivative,

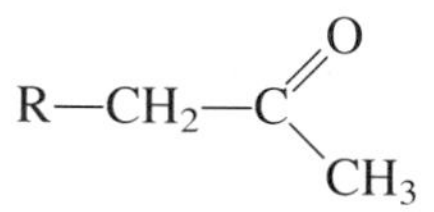

$$R-CH_2-\overset{O}{\overset{\|}{C}}-CH_3$$

The alkyl group R is derived from the alkyl halide used in the alkylation step. This sequence is illustrated in the following example.

In planning an acetoacetic ester synthesis, we need only to draw the structure of the ketone we want and by inspection decide which alkyl halide we shall need.

EXAMPLE

Prepare 2-hexanone via the acetoacetic ester synthesis.

This part of the product has its origin in the acetoacetic ester

$$CH_3{-}\overset{\overset{\displaystyle O}{\|}}{C}{-}CH_2 \;\Big|\; CH_2CH_2CH_3$$

This part of the product must come from the alkyl halide

Solution STEP 1. Prepare the carbanion from acetoacetic ester.

$$CH_3{-}\overset{\overset{\displaystyle O}{\|}}{C}{-}\underset{\underset{\displaystyle H}{|}}{\overset{\overset{\displaystyle H}{|}}{C}}{-}\overset{\overset{\displaystyle O}{\|}}{C}{-}OC_2H_5 + C_2H_5O^-Na^+ \longrightarrow$$

$$CH_3{-}\overset{\overset{\displaystyle O}{\|}}{C}{-}\underset{\ddot{}}{\overset{\overset{\displaystyle H}{|}}{C}}{-}\overset{\overset{\displaystyle O}{\|}}{C}{-}OC_2H_5 + Na^+ + C_2H_5OH$$

STEP 2. Displace the halogen of the alkyl halide.

$$CH_3{-}\overset{\overset{\displaystyle O}{\|}}{C}{-}\underset{\underset{\displaystyle H}{|}}{\overset{..}{\bar{C}}}{-}COOC_2H_5 + CH_3CH_2CH_2Br \longrightarrow CH_3{-}\overset{\overset{\displaystyle O}{\|}}{C}{-}\underset{\underset{\displaystyle H}{|}}{\overset{\overset{\displaystyle C_3H_7}{|}}{C}}{-}COOC_2H_5 + Br^-$$

STEP 3. Saponify the substituted acetoacetic ester.

$$CH_3{-}\overset{\overset{\displaystyle O}{\|}}{C}{-}\underset{\underset{\displaystyle H}{|}}{\overset{\overset{\displaystyle C_3H_7}{|}}{C}}{-}COOC_2H_5 + Na^+OH^- \longrightarrow CH_3{-}\overset{\overset{\displaystyle O}{\|}}{C}{-}\underset{\underset{\displaystyle H}{|}}{\overset{\overset{\displaystyle C_3H_7}{|}}{C}}{-}COO^-Na^+ + C_2H_5OH$$

STEP 4. Neutralize the acid salt with HCl and heat.

$$CH_3{-}\overset{\overset{\displaystyle O}{\|}}{C}{-}\underset{\underset{\displaystyle H}{|}}{\overset{\overset{\displaystyle C_3H_7}{|}}{C}}{-}COO^-Na^+ + HCl \longrightarrow CH_3{-}\overset{\overset{\displaystyle O}{\|}}{C}{-}\underset{\underset{\displaystyle H}{|}}{\overset{\overset{\displaystyle C_3H_7}{|}}{C}}{-}COOH + NaCl$$

$$CH_3{-}\overset{\overset{\displaystyle O}{\|}}{C}{-}\underset{\underset{\displaystyle CH_2CH_2CH_3}{|}}{\overset{\overset{\displaystyle H}{|}}{C}}{-}\overset{\overset{\displaystyle O}{\|}}{C}{-}OH \xrightarrow{\text{heat}} CH_3{-}\overset{\overset{\displaystyle O}{\|}}{C}{-}CH_2CH_2CH_2CH_3 + CO_2$$

2-Hexanone

Exercise 10.3 In step (2) of the previous reaction sequence a tertiary alkyl halide may not be used. Why?

Exercise 10.4 Which parts of the acetoacetic ester synthesis sequence would we need to repeat, and what alkyl halide would be required to prepare 3-methyl-2-hexanone?

Exercise 10.5 How could the 2-hexanone we prepared via the acetoacetic ester synthesis be converted into pentanoic acid?

10.5 Amides

A. Preparation of Amides. Amides are conveniently prepared in the laboratory by the ammonolysis of acyl halides (Sec. 10.2B) or acid anhydrides.

Amides are named after their parent acids by replacing *ic* or *oic* with *amide*.

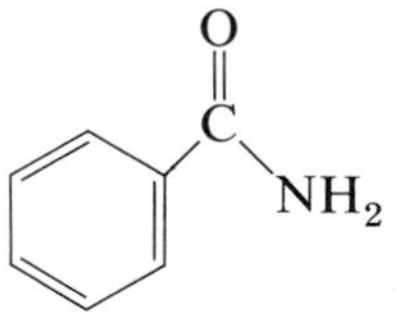

$$C_6H_5-C(=O)-NH_2 \qquad CH_3-CH_2-C(=O)-NH_2$$

Benzamide 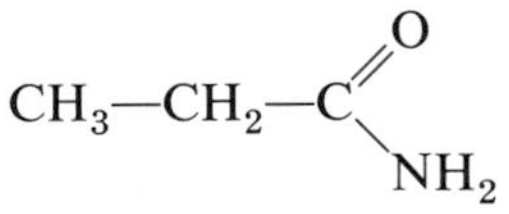   Propionamide (Propanamide)

With the exception of formamide,

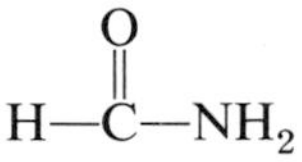

$$H-\overset{O}{\overset{\|}{C}}-NH_2$$

the amides are solids, with sharp melting points. This property makes them very useful derivatives for the identification of acids.

B. Reactions of the Amides. The amides can be hydrolyzed in acid or in alkaline solution. Hydrolysis carried out in an acid solution produces the free organic acid and an ammonium salt. Hydrolysis carried out in a basic solution produces the free base (NH_3) and the salt of the organic acid. Both types of procedures are illustrated by the following reaction equations.

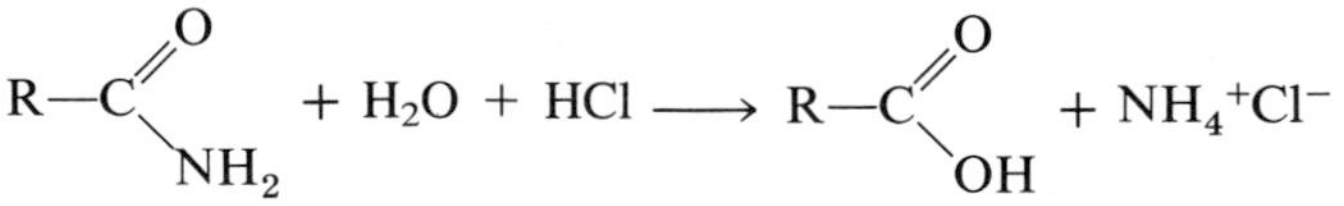

$$R-C(=O)-NH_2 + H_2O + HCl \longrightarrow R-C(=O)-OH + NH_4^+Cl^-$$

Acid hydrolysis

$$R-C(=O)-NH_2 + \overset{+}{Na}\,\overset{-}{O}H \longrightarrow R-C(=O)-O^-Na^+ + NH_3$$

Alkaline hydrolysis

Exercise 10.6 What simple test will serve to distinguish benzamide (mp, 128°) from benzoic acid (mp, 122°)?

Summary

1. Derivatives of the carboxylic acids are compounds in which the hydroxyl group of the acid has been replaced, but which, on hydrolysis, regenerate the original acid. The derivatives of both aliphatic and aromatic acids are prepared in the same manner and give reactions that are almost identical. Any differences are of degree rather than of kind. These include:

   (a) Replacement of the hydroxyl by halide to yield **acyl** or **aroyl halides,**

   $$R-C(=O)-X \quad \text{or} \quad Ar-C(=O)-X$$

   (b) Replacement of the hydroxyl by an alkoxyl, —OR, to yield **esters,**

   $$R-C(=O)-OR$$

   (c) Replacement of the hydroxyl by an amino group to yield **amides,**

   $$R-C(=O)-NH_2$$

   (d) Replacement of the hydroxyl by carboxylate,

   $$-O-C(=O)-R'\text{, to yield } \textbf{anhydrides,} \quad R-C(=O)-O-C(=O)-R'$$

2. Esters may be reduced to alcohols. Reduction is accomplished directly by the use of $LiAlH_4$. The ester is reduced to produce two alcohols.
3. Esters react with Grignard reagents to give tertiary alcohols in which two of the three alkyl groups originate in the Grignard reagent. (Exception: Formates give secondary alcohols).
4. The Claisen condensation is a reaction that results in the preparation of a β-keto ester. The β-keto esters are important intermediates in organic synthesis.

Supplementary Exercises

10.7 Draw structural formulas for each of the following compounds.

(a) butanamide
(b) ethyl bromacetate
(c) butanoyl chloride
(d) phenylacetamide
(e) N,N-dimethylacetamide
(f) bromoacetyl bromide
(g) methyl 2,3-dimethylhexanoate
(h) methyl salicylate
(i) ethyl α-propylacetoacetate
(j) benzoic anhydride
(k) ethyl phenylacetate
(l) methyl dodecanoate
(m) ethyl cyclobutanecarboxylate
(n) isopropyl 2-methylpropanoate

10.8 Give an acceptable name for each of the following.

(a) $H-\overset{\overset{O}{\|}}{C}-OC_2H_5$ (b) $C_6H_5-\overset{\overset{O}{\|}}{C}-NH_2$ (c) $(CH_3)_2CHCH_2\overset{\overset{O}{\|}}{C}-Cl$

(d) [benzene ring with Br, bearing $C(=O)OCH_3$] (e) [cyclohexane ring bearing $C(=O)OCH_2$] (f) [cyclohexane ring bearing $O-C(=O)CH_3$]

(g) [benzene ring–CH_2–$C(=O)NH_2$] (h) [skeletal structure with $C(=O)Cl$] (i) [skeletal ester structure with O]

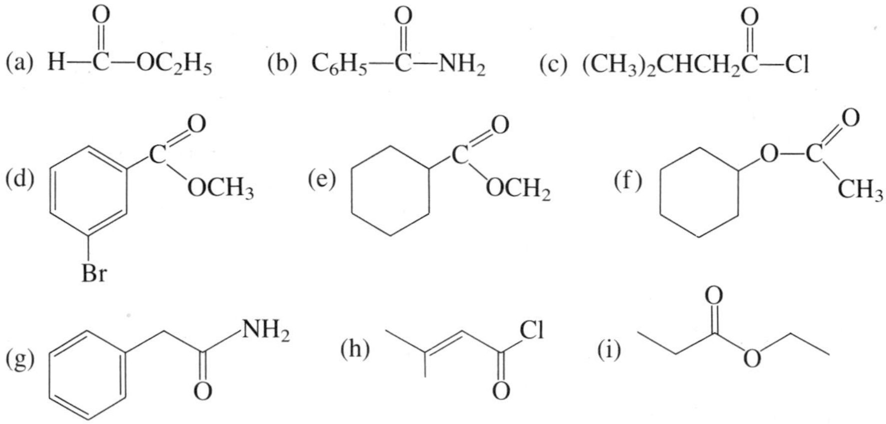

10.9 The pK_a of acetoacetic ester is about 11, whereas the pK_a of acetone is about 19. Therefore, acetoacetic ester is much the stronger acid of the two. The greater acid strength of acetoacetic ester has been attributed to greater resonance stabilization of its anion. Compare the resonance stabilization of the carbanions of acetone and acetoacetic ester by drawing the principal contributing forms for each anion.

10.10 Show how cyclohexanecarboxylic acid could be converted into the following compounds.

(a) cyclohexanecarbonyl chloride
(b) cyclohexanecarboxamide
(c) isopropyl cyclohexanecarboxylate
(d) sodium cyclohexanecarboxylate
(e) cyclohexanemethanol
(f) 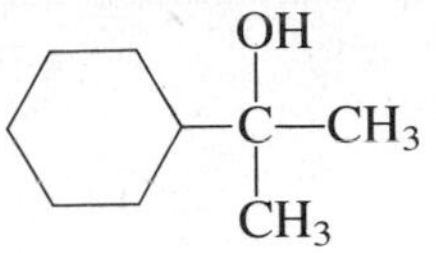 cyclohexyl–C(OH)(CH_3)–CH_3

10.11 Write a mechanism for the base-catalyzed hydrolysis of ethyl acetate and explain why the reaction is irreversible.

10.12 How would you proceed to prepare a mixed anhydride of the formula

$$R-\overset{O}{\overset{\|}{C}}-O-\overset{O}{\overset{\|}{C}}-R',$$

where $R \neq R'$? Illustrate with an example.

10.13 Why is saponification of an ester with sodium hydroxide a better method of hydrolysis than treatment with sulfuric acid?

10.14 Which product in each of the following reactions will contain the labeled oxygen atom, ^{18}O?

(a) Benzoic acid + $CH_3{}^{18}OH \xrightarrow{H_3\overset{+}{O}}$

(b) Acetyl chloride + $CH_3{}^{18}OH \xrightarrow{\text{base}}$

(c) $CH_3-\overset{O}{\overset{\|}{C}}-{}^{18}OC_2H_5 + NaOH \xrightarrow[\Delta]{}$

(d) $C_6H_5-\overset{O}{\overset{\|}{C}}-{}^{18}OC_2H_5 + H_3\overset{+}{O} \xrightarrow[\Delta]{}$

10.15 Write equations to show how the following reaction sequences might be carried out. You may use any readily available reagents you require.

(a) Ethyl bromide $\rightarrow$ propanoic acid $\rightarrow$ ethyl propanoate $\rightarrow$ 2-methyl-2-butanol
(b) Ethanol $\rightarrow$ ethyl acetate $\rightarrow$ ethyl acetoacetate $\rightarrow$ 2-pentanone $\rightarrow$ butanoic acid

10.16 Illustrate a reaction between an ester and a Grignard reagent that could be used to prepare a *secondary* alcohol.

10.17 A neutral compound, $C_5H_{10}O_2$, was hydrolyzed to yield two new products, A and B. Compound B gave a positive iodoform test reaction, reacted readily with the Lucas reagent, and on treatment with concentrated H_2SO_4 gave propylene. Give the name and structure of the original compound.

10.18 A liquid, $C_7H_{14}O_2$, on hydrolysis gave compounds A and B. Compound B gave a positive iodoform reaction, and when treated with PBr_3 formed an alkyl bromide. A Grignard reagent, when prepared from the alkyl bromide and caused to react with carbon dioxide, produced an acid identical to compound A. Give the structure and name of the original compound.

10.19 A student was issued a sweet-smelling liquid as an unknown for identification and logically assumed it to be an ester. A boiling point determination of 77° led the student to believe the compound to be either ethyl acetate (bp,77°) or methyl propanoate (bp,79°). If you were this student, what would be your next step toward making a positive identification of the compound?

***10.20** If the compound in question in Exercise 10.19 gave the nmr spectrum, (a), which of the two isomers would the unknown be? Which isomer gives the nmr spectrum shown in (b)?

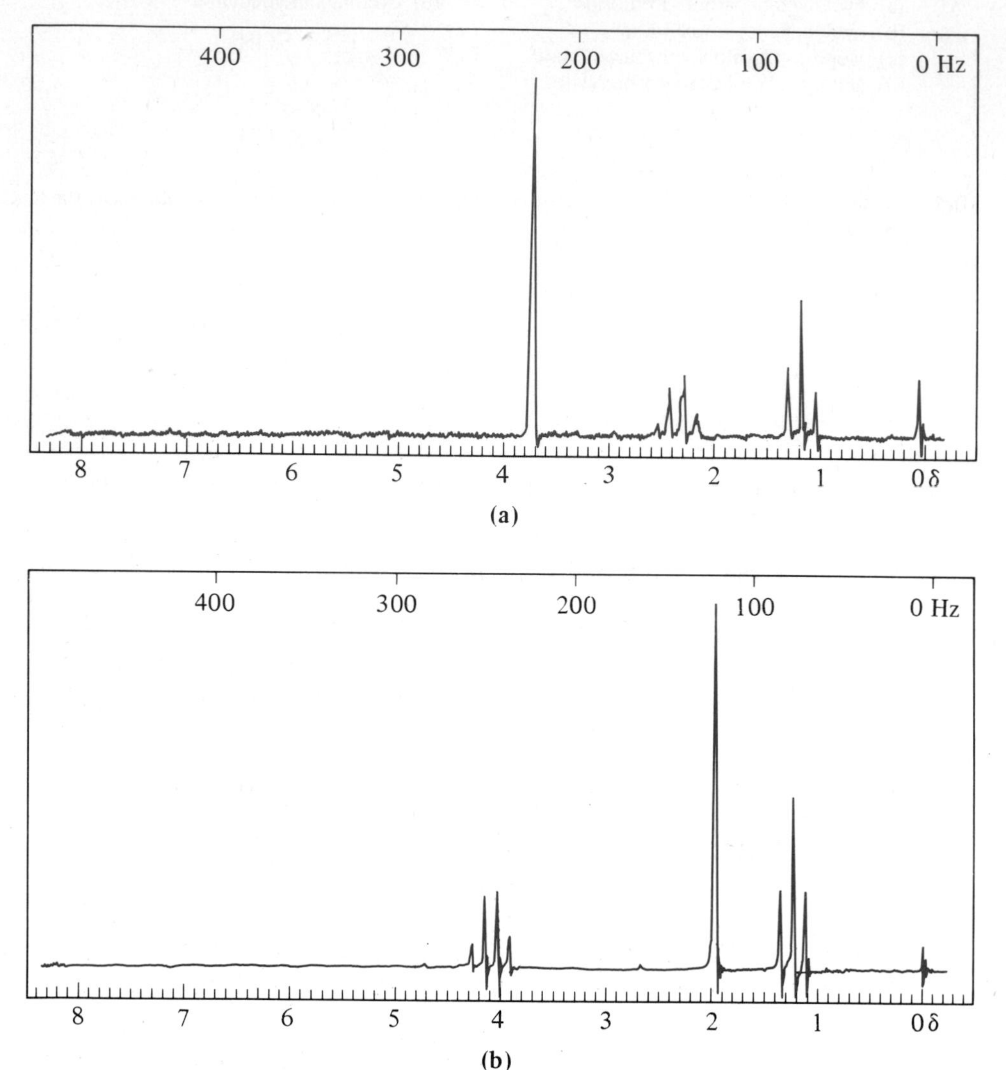

***10.21** The correct identity of two liquids in unlabeled bottles is in question. One liquid is thought to be propyl acetate (bp,101.6°) and the other ethyl propanoate (bp,99.1°). Draw structures for both esters and tell how an nmr spectrum of each would make an unequivocal identification.

10.22 The conjugate addition of a good carbon nucleophile, such as the enolate anion of ethyl acetoacetate, to an α,β-unsaturated carbonyl compound is an important synthesis of 1,5-dicarbonyl compounds, called the Michael reaction. Predict the product of the following Michael reaction.

$$CH_3-\overset{\displaystyle O}{\overset{\|}{C}}-CH_2-\overset{\displaystyle O}{\overset{\|}{C}}-OCH_2CH_3 + CH_2{=}CH-\overset{\displaystyle O}{\overset{\|}{C}}-CH_3 \xrightarrow{NaOCH_2CH_3} \xrightarrow[HCl]{H_2O} \xrightarrow[heat]{-CO_2}$$

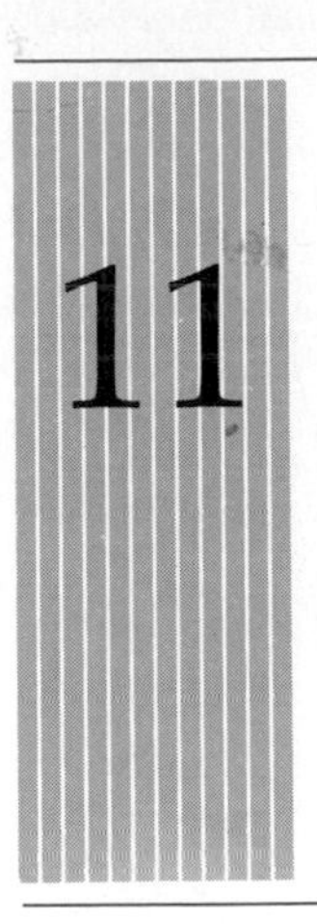

11 Bifunctional Acids

Compounds discussed in previous chapters usually had but one functional group. Occasionally, as in the case of the diols and the dihalides, more than one functional group was encountered, but in these instances the groups were of the same kind. In the present chapter the compounds considered are primarily acids but have, in addition to the carboxyl, another functional group. The chemical behavior of a bifunctional acid, for the most part, is the behavior characteristic of each group separately. However, the dual nature of bifunctional acids confers special properties upon these compounds. Such changes in properties are particularly significant when the functional groups are in close proximity to each other.

Dicarboxylic Acids

11.1 Nomenclature and Properties

The aliphatic dicarboxylic acids are naturally occurring, colorless, crystalline solids often referred to by common names usually of Latin or Greek derivation. Such names frequently indicate a natural source of the acid. IUPAC nomenclature follows established rules. One simply adds the suffix *dioic* to the parent hydrocarbon of the same number of carbon atoms. For example, oxalic acid (HOOC—COOH), the simplest member of the dicarboxylic acid series, is named **ethanedioic acid.** Malonic acid (HOOC—CH_2—COOH), the next member of the family, is named **propanedioic acid,** etc. Table 11.1 lists the common names and structures, along with the acidic properties, of the dicarboxylic acids containing 2–10 carbon atoms.

TABLE 11.1 Dicarboxylic Acids

Name	Formula	mp (°C)	pK_1 (25°)	pK_2 (25°)
Acetic*	CH_3COOH	16.6	4.76	
Oxalic	HOOC—COOH	187	1.27	4.27
Malonic	HOOC—CH_2—COOH	135 (dec.)	2.86	5.70
Succinic	HOOC—$(CH_2)_2$—COOH	185	4.21	5.64
Glutaric	HOOC—$(CH_2)_3$—COOH	97.5	4.34	5.42
Adipic	HOOC—$(CH_2)_4$—COOH	151	4.43	5.41
Pimelic	HOOC—$(CH_2)_5$—COOH	105	4.50	5.42
Suberic	HOOC—$(CH_2)_6$—COOH	142	4.52	5.41
Azelaic	HOOC—$(CH_2)_7$—COOH	106	4.53	5.40
Sebacic	HOOC—$(CH_2)_8$—COOH	134	4.70	5.42
Phthalic	$C_6H_4(COOH)_2$ (1,2-benzenedicarboxylic acid)	191	2.95	5.41
Maleic	HOOC—CH=CH—COOH (both H on same side)	130	1.94	6.22
Fumaric	HOOC—CH=CH—COOH (H on opposite sides)	287	3.02	4.38

*Included for reference value.

The first two members of the dicarboxylic acid family are much stronger acids than acetic acid. The presence of a second carboxyl group adjacent to, or near, another appears to withdraw electrons from the first. This inductive effect becomes weaker if the carboxyl functions are further separated by intervening methylene groups. An electron withdrawal on the part of one carboxyl group tends to promote ionization of the other carboxyl group by stabilizing the negatively charged monoanion. Ionization of the second carboxyl group is more difficult because of electrostatic repulsion in the dianion between the two negatively charged carboxylate groups that tends to destabilize the dianion. The value of the second ionization constant, for this reason, is generally much smaller than that of the first.

11.2 Preparation and Reactions of the Dicarboxylic Acids

Dicarboxylic acids usually can be prepared by adapting methods used for the preparation of the monocarboxylic acids. Functional groups that are easily convertible to carboxyl groups generally provide a route to the dicarboxylic acids. For example, the oxidation of unsaturated acids and the hydrolysis of nitriles are methods that are frequently used. The preparation of malonic and azelaic acids by such methods is illustrated in the following reactions.

Oxidation of an unsaturated acid

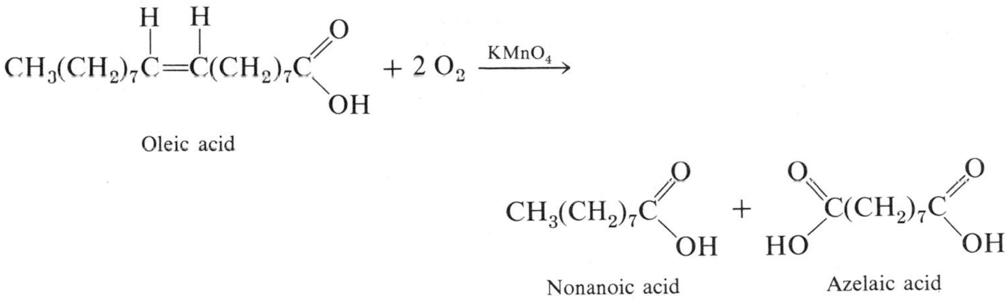

$$CH_3(CH_2)_7\overset{H}{\overset{|}{C}}{=}\overset{H}{\overset{|}{C}}(CH_2)_7C(=O)OH + 2\,O_2 \xrightarrow{KMnO_4}$$

Oleic acid

$$CH_3(CH_2)_7C(=O)OH + HO(O=)C(CH_2)_7C(=O)OH$$

Nonanoic acid Azelaic acid

Hydrolysis of a nitrile

$$ClCH_2COOH + NaHCO_3 \longrightarrow ClCH_2COO^-Na^+ + H_2O + CO_2$$

Chloroacetic acid; Sodium chloroacetate

$$ClCH_2COO^-Na^+ + K^+CN^- \longrightarrow N{\equiv}C{-}CH_2COO^-Na^+ + KCl$$

Sodium cyanoacetate

$$N{\equiv}C{-}CH_2COO^-Na^+ + 2\,HCl + 2\,H_2O \longrightarrow HOOC{-}CH_2{-}COOH + NH_4Cl + NaCl$$

Malonic acid

Exercise 11.1 Beginning with succinic acid, assign IUPAC names to the dicarboxylic acids.

The derivatives of the dicarboxylic acids are, in general, the same as those produced from the monocarboxylic acids and include salts, esters, amides, and acid halides. The behavior of the dicarboxylic acids, when heated, is unique, however, and deserves special mention.

Oxalic acid, when strongly heated, undergoes decomposition according to the following equation.

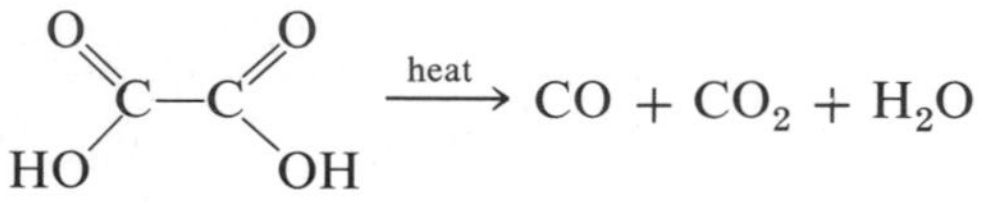

$$HOOC{-}COOH \xrightarrow{\text{heat}} CO + CO_2 + H_2O$$

Oxalic acid

Malonic acid is so easily decomposed by heat that it is decarboxylated at its melting point (137°C).

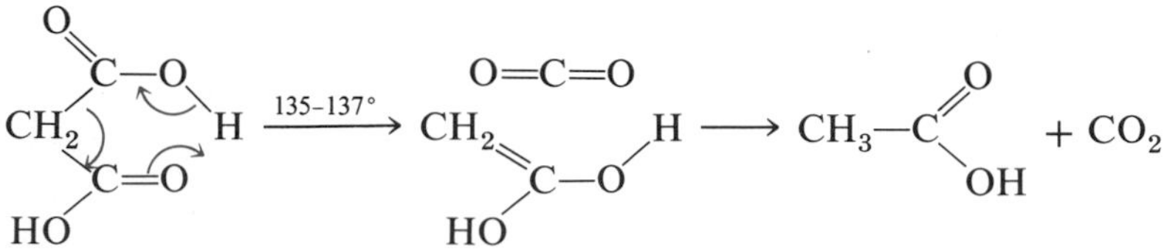

$$HO{-}C(=O){-}CH_2{-}C(=O){-}O{-}H \xrightarrow{135\text{–}137°} O{=}C{=}O + CH_2{=}C(OH){-}O{-}H \longrightarrow CH_3{-}C(=O){-}OH + CO_2$$

Malonic acid | Enol of acetic acid | Acetic acid

Succinic and glutaric acids, when heated, lose water to produce their corresponding cyclic anhydrides.

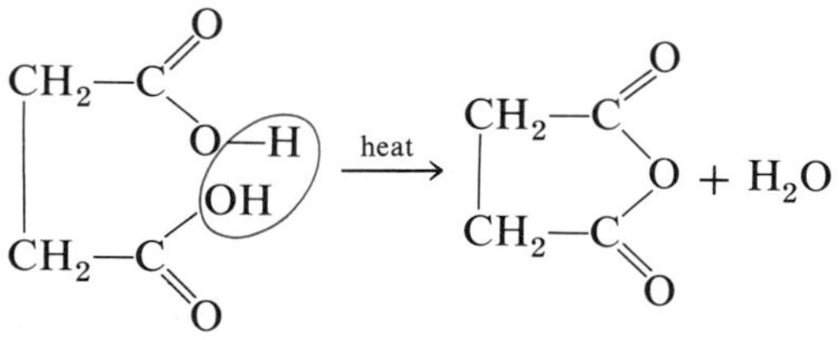

$$\begin{array}{l} CH_2{-}C(=O){-}O{-}H \\ | \\ CH_2{-}C(=O){-}OH \end{array} \xrightarrow{\text{heat}} \begin{array}{l} CH_2{-}C(=O) \\ | \quad\quad\quad\;\; \backslash O \\ CH_2{-}C(=O) \end{array} + H_2O$$

Succinic acid | Succinic anhydride

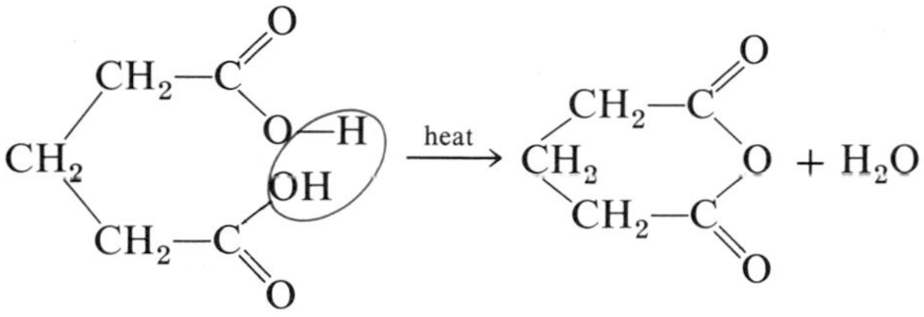

$$CH_2\begin{array}{l} \diagup CH_2{-}C(=O){-}O{-}H \\ \diagdown CH_2{-}C(=O){-}OH \end{array} \xrightarrow{\text{heat}} CH_2\begin{array}{l} \diagup CH_2{-}C(=O) \\ \diagdown CH_2{-}C(=O) \end{array}\!\!\!\!\!\! \diagdown\!\!\!\!\diagup O + H_2O$$

Glutaric acid | Glutaric anhydride

Exercise 11.2 Why is it not possible to convert a halogen acid into a Grignard reagent and then, by carbonation of the Grignard reagent with anhydrous CO_2, form a second carboxyl group?

11.3 Malonic Ester Synthesis

Substituted malonic acids, like malonic acid, are easily decarboxylated. The ease with which carbon dioxide can be split out from only one carboxyl group of malonic acid makes it a very useful acid in a number of organic syntheses. A synthesis that involves a substituted malonic acid begins with its ethyl ester, diethyl malonate, or ''malonic ester'' as it usually is called. The methylene carbon of ethyl malonate, as in ethyl acetoacetate, is flanked on both sides by electron-withdrawing carbonyl groups. This dual influence causes the α-hydrogens to be much more acidic (pK_a 13) than those in a simple ester such as ethyl acetate (pK_a 24.5). Thus, these hydrogens are easily removed by a basic reagent, such as sodium ethoxide. The resonance-stabilized anion of sodiomalonic ester, when formed, is nucleophilic and reacts readily with alkyl halides.

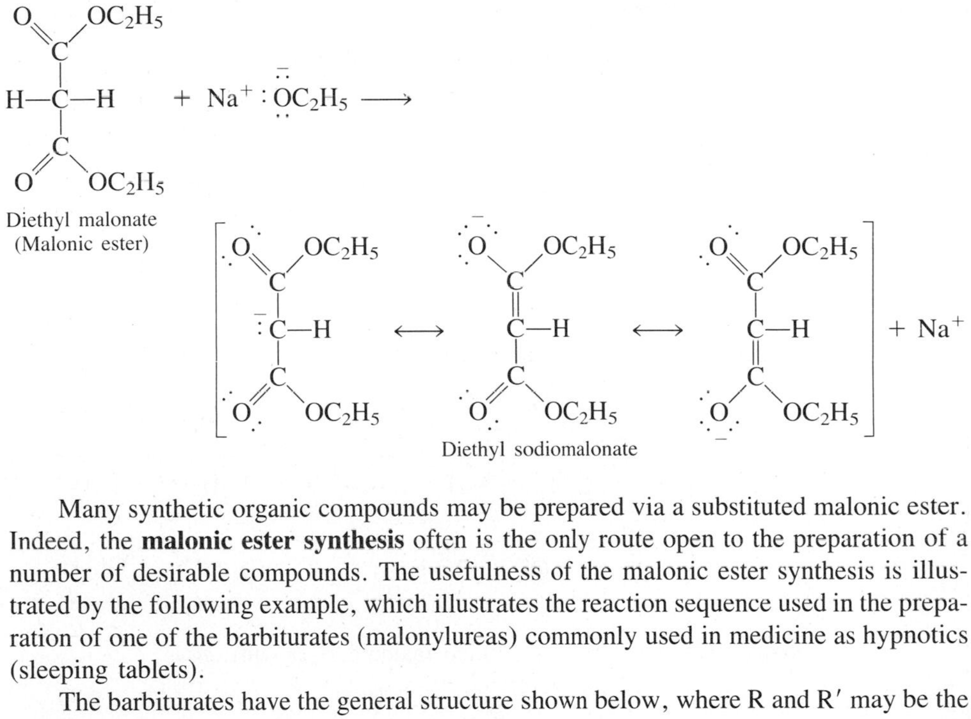

$$CH_2(COOC_2H_5)_2 + Na^+ \, {}^-:\!OC_2H_5 \longrightarrow$$

Diethyl malonate (Malonic ester)

$$\left[{}^-:CH(COOC_2H_5)_2 \longleftrightarrow (C_2H_5O)(^-O)C{=}CH{-}COOC_2H_5 \longleftrightarrow C_2H_5OOC{-}CH{=}C(O^-)(OC_2H_5) \right] + Na^+$$

Diethyl sodiomalonate

Many synthetic organic compounds may be prepared via a substituted malonic ester. Indeed, the **malonic ester synthesis** often is the only route open to the preparation of a number of desirable compounds. The usefulness of the malonic ester synthesis is illustrated by the following example, which illustrates the reaction sequence used in the preparation of one of the barbiturates (malonylureas) commonly used in medicine as hypnotics (sleeping tablets).

The barbiturates have the general structure shown below, where R and R′ may be the same or different.

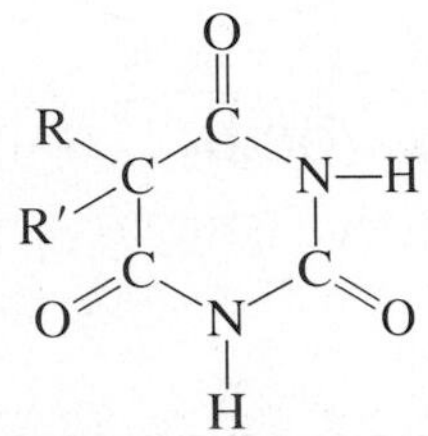

[Structure: barbiturate ring — C(R)(R′) flanked by two C=O groups, each bonded to N–H, with the two N–H groups joined by a C=O]

EXAMPLE

Show the different steps required in the preparation of 5,5-diethylbarbituric acid *(Barbital)* via the "malonic ester synthesis."

Solution Step 1. As in the acetoacetic ester synthesis, we first must prepare the carbanion.

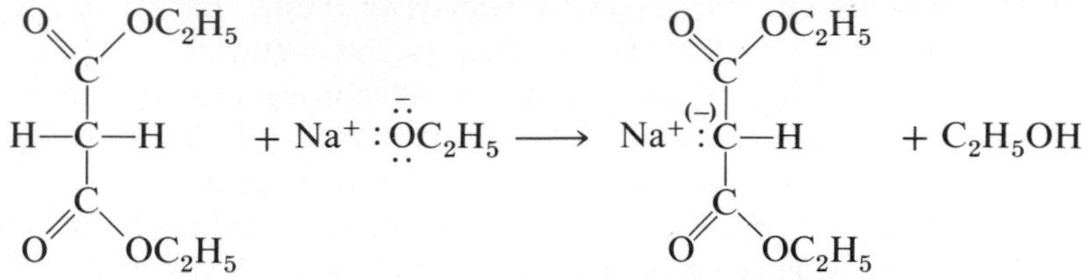

$$H_2C(COOC_2H_5)_2 + Na^+ :\ddot{O}C_2H_5^- \longrightarrow Na^+ :\overset{(-)}{C}H(COOC_2H_5)_2 + C_2H_5OH$$

Diethyl malonate (Malonic ester) — Diethyl sodiomalonate

Step 2. Displace the iodide of ethyl iodide (S_N2).

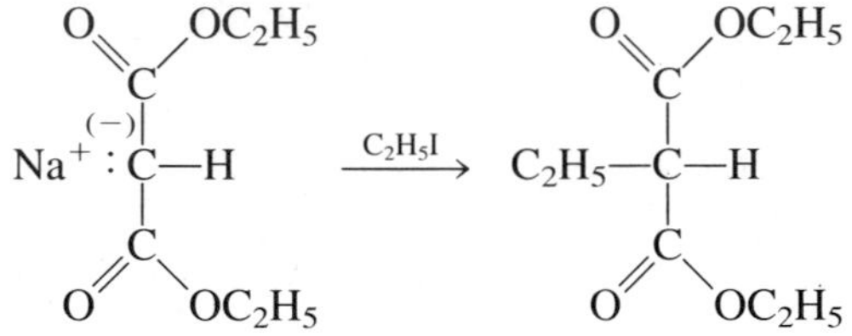

$$Na^+ :\overset{(-)}{C}H(COOC_2H_5)_2 \xrightarrow{C_2H_5I} C_2H_5\text{—}CH(COOC_2H_5)_2$$

Diethyl ethylmalonate

Step 3 and Step 4. Repeat Steps 1 and 2.

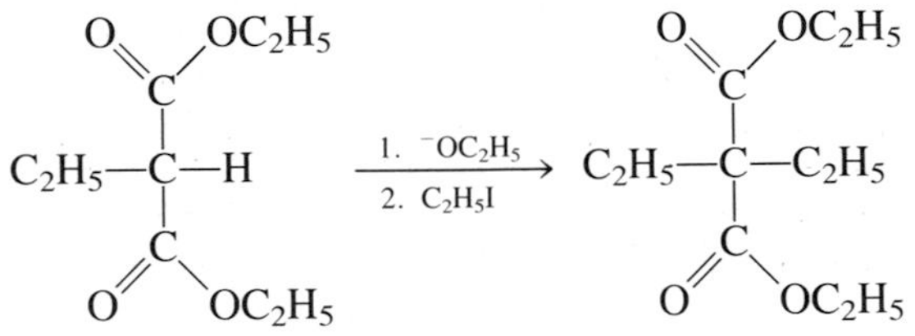

$$C_2H_5\text{—}CH(COOC_2H_5)_2 \xrightarrow[2.\ C_2H_5I]{1.\ ^-OC_2H_5} (C_2H_5)_2C(COOC_2H_5)_2$$

Diethylmalonic ester

Step 5. Condense the substituted malonic ester with urea.

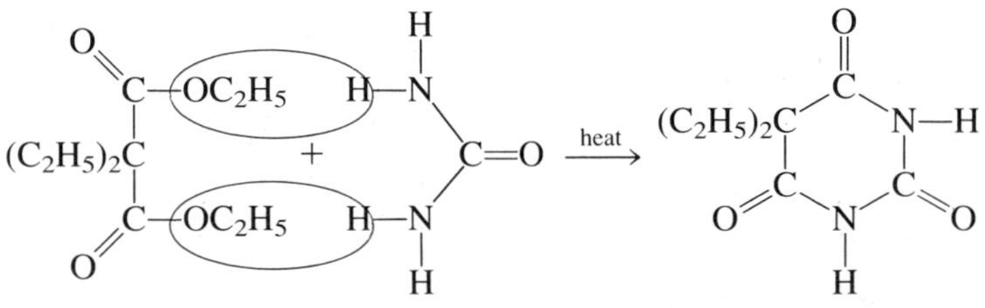

$$(C_2H_5)_2C(COOC_2H_5)_2 + H_2N\text{—}CO\text{—}NH_2 \xrightarrow{heat} (C_2H_5)_2C\text{(CO—NH)}_2C\text{=O}$$

Urea — 5,5-Diethylbarbituric acid (Barbital)

Sodium pentothal is a thiobarbiturate that is used for the induction of general anesthesia by intravenous injection. It is prepared by condensing a substituted malonic ester with thiourea. Sodium pentothal is used in veterinary medicine for euthanasia.

$$CH_3CH_2CH_2CH(CH_3)\text{—}C(C_2H_5)\text{ ring: }C(=O)\text{—}N^-\,Na^+\text{—}C(=S)\text{—}N(H)\text{—}C(=O)$$

Sodium pentothal

In the laboratory the malonic ester synthesis is widely used for the synthesis of substituted acetic acids,

$$R\text{—}CH_2\text{—}CO_2H \quad \text{and} \quad \begin{matrix} R \\ R' \end{matrix}\!\!>CH\text{—}CO_2H$$

where R and R′ are derived from the alkyl halides used to alkylate malonic ester, and the acetic acid moiety is derived from malonic acid by decarboxylation. To illustrate this useful procedure, all we need to do is interrupt the preparation of 5,5-diethylbarbituric acid after Step 2 in our example synthesis, hydrolyze the ester, and heat in order to prepare butanoic acid.

$$C_2H_5\text{—}CH(COOC_2H_5)_2 \xrightarrow[HCl]{H_2O} C_2H_5\text{—}CH(COOH)_2 \xrightarrow[heat]{-CO_2} C_2H_5\text{—}CH_2\text{—}COOH$$

Ethylmalonic ester — Ethylmalonic acid — Butanoic acid

> **Exercise 11.3** Show by means of equations how the malonic ester synthesis could be used to prepare 2-ethyl-5-methylhexanoic acid.
>
> **Exercise 11.4** If our intermediate disubstituted malonic ester prepared in Exercise 11.3 had not been hydrolyzed to the acid, but instead at this stage had been condensed with urea, the resulting product would have been the barbiturate *Amytal*. Draw the structure of *Amytal*.

11.4 Other Important Dicarboxylic Acids

Adipic acid, used in large quantities for the production of nylon, is an important industrial chemical. One commercial method for the production of adipic acid uses cyclohexane as starting material. Cyclohexane is oxidized to adipic acid according to the following reaction.

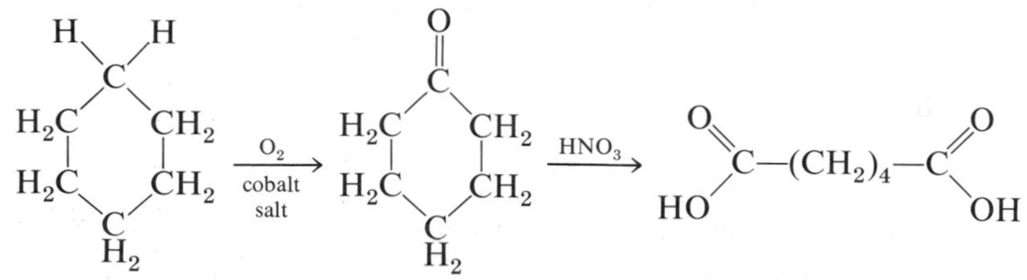

$$\text{Cyclohexane} \xrightarrow[\text{cobalt salt}]{O_2} \text{Cyclohexanone} \xrightarrow{HNO_3} HO{-}\overset{O}{\overset{\|}{C}}{-}(CH_2)_4{-}\overset{O}{\overset{\|}{C}}{-}OH$$

Cyclohexane Cyclohexanone Adipic acid

When adipic acid is heated with hexamethylenediamine, $H_2N{-}(CH_2)_6{-}NH_2$, nylon 66 is formed. The 66 designation means that this nylon (there are others) has two six-carbon units.

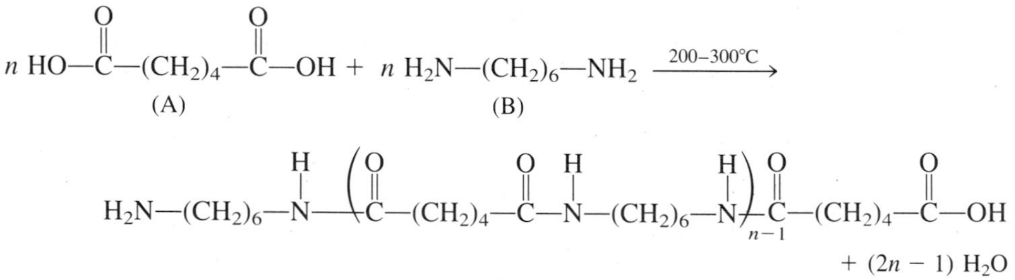

$$n\ HO{-}\overset{O}{\overset{\|}{C}}{-}(CH_2)_4{-}\overset{O}{\overset{\|}{C}}{-}OH + n\ H_2N{-}(CH_2)_6{-}NH_2 \xrightarrow{200\text{–}300°C}$$

(A) (B)

$$H_2N{-}(CH_2)_6{-}\overset{H}{\overset{|}{N}}{-}\left(\overset{O}{\overset{\|}{C}}{-}(CH_2)_4{-}\overset{O}{\overset{\|}{C}}{-}\overset{H}{\overset{|}{N}}{-}(CH_2)_6{-}\overset{H}{\overset{|}{N}}\right)_{n-1}{-}\overset{O}{\overset{\|}{C}}{-}(CH_2)_4{-}\overset{O}{\overset{\|}{C}}{-}OH$$

$$+ (2n - 1)\ H_2O$$

Nylon 66 (n = 450–500)

Phthalic acid, or 1,2-benzenedicarboxylic acid, is one of the most useful aromatic dicarboxylic acids. Phthalic acid is obtained by the vigorous oxidation of naphthalene (Secs. 9.5A and 4.6E). The *para* isomer, terephthalic acid, is prepared by the oxidation of the methyl groups of *p*-xylene.

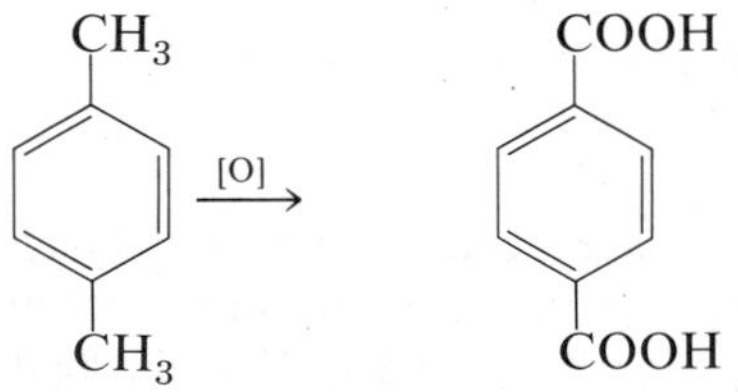

$$CH_3{-}C_6H_4{-}CH_3 \xrightarrow{[O]} HOOC{-}C_6H_4{-}COOH$$

p-Xylene

Terephthalic acid
(1,4-Benzene dicarboxylic acid)

Phthalic and terephthalic acids, when esterified with polyhydric alcohols, produce high-molecular weight polyesters. The reaction of phthalic acid either with ethylene glycol or with glycerol produces the **glyptal resins** (*gly*cerol + *phthal*ic acid) that are now widely used as synthetic auto finishes. The synthetic fiber known as "Dacron" or "Terylene" is a polyester of ethylene glycol and terephthalic acid. Dacron is not produced by the direct esterification of terephthalic acid, but rather by an ester interchange or **transesterification** between methyl terephthalate and ethylene glycol.

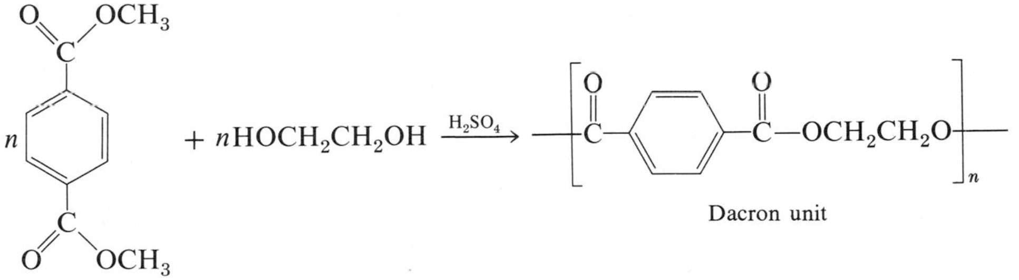

$$n\ CH_3OOC{-}C_6H_4{-}COOCH_3 + n\,HOCH_2CH_2OH \xrightarrow{H_2SO_4} \left[{-}\overset{O}{\overset{\|}{C}}{-}C_6H_4{-}\overset{O}{\overset{\|}{C}}{-}OCH_2CH_2O{-}\right]_n$$

Dacron unit

The same polyester, when produced in the form of a film, is marketed under the trade name "Mylar."

Hydroxy and Halogen Acids

11.5 Structure and Nomenclature

Certain of the hydroxy acids (see Table 11.2) are very common naturally occurring substances and are usually referred to by common names. The halogen acids, on the other hand, are mostly synthetic and are named systematically. Hydroxy and halogen substituents on the carbon chain can be located either by numbers (IUPAC) or by Greek letters (common names).

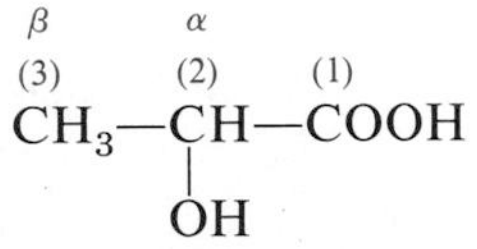

$$\overset{\beta}{\underset{(3)}{CH_3}}{-}\overset{\alpha}{\underset{(2)}{CH}}(OH){-}\underset{(1)}{COOH}$$

2-Hydroxypropanoic acid
α-Hydroxypropionic acid
(Lactic acid)

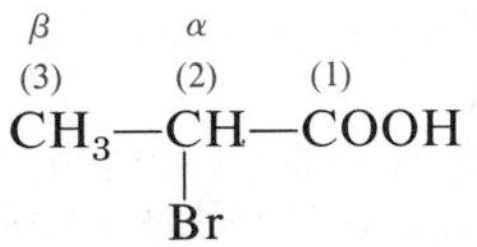

$$\overset{\beta}{\underset{(3)}{CH_3}}{-}\overset{\alpha}{\underset{(2)}{CH}}(Br){-}\underset{(1)}{COOH}$$

2-Bromopropanoic acid
α-Bromopropionic acid

TABLE 11.2 Naturally Occurring Hydroxy Acids

Common Name	Structure	pK_n (25°)
Glycolic (*Glycerol*)	$HOCH_2COOH$	3.82
Lactic (L. *lactis,* milk)	$CH_3-CH(OH)-COOH$	3.86
Malic (L. *malus,* apple)	$HO-C(H)(CH_2COOH)-COOH$	pK_1 3.40 pK_2 5.05
Tartaric	$HO-C(H)(COOH)-C(H)(OH)-COOH$	pK_1 3.22 pK_2 4.81
Citric (L. *citrum,* lemon, lime)	$HO-C(CH_2-COOH)_2-COOH$	pK_1 3.13 pK_2 4.76 pK_3 6.40
Salicylic (L. *salix,* willow)	o-HOC_6H_4COOH	pK_1 3.00 pK_2 12.38
Mandelic (Gr. *mandel,* almond)	$C_6H_5-CH(OH)-C(=O)OH$	3.41

11.6 Preparation and Reactions of Substituted Acids

The chemistry of substituted acids is interrelated. A method that leads to the preparation of one substituted acid usually involves a reaction of another acid already bearing a different replaceable substituent.

Acids that have a halogen atom on the α-carbon atom are important starting materials for a number of substituted acids. Such α-halogen acids are easily prepared via the Hell–

Volhard–Zelinsky reaction (Sec. 9.6C). The reactions exhibited by halogen acids are similar in nature to those reviewed for the alkyl halides. For example, dehydrohalogenation with alcoholic potassium hydroxide converts a halogen acid to an unsaturated acid.

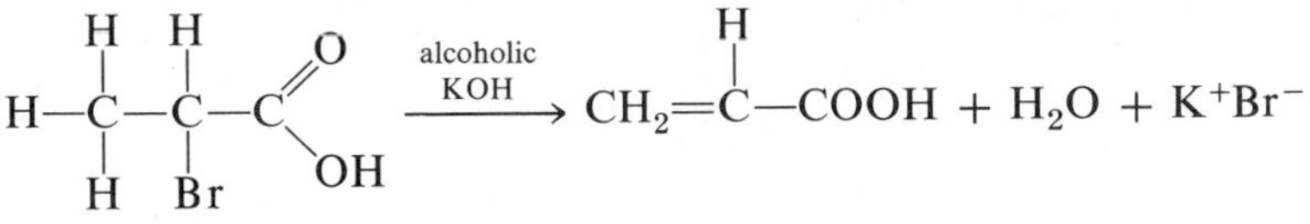

$$\underset{\alpha\text{-Bromopropionic acid}}{H{-}\underset{H}{\overset{H}{C}}{-}\underset{Br}{\overset{H}{C}}{-}C(=O)OH} \xrightarrow[\text{KOH}]{\text{alcoholic}} \underset{\text{Acrylic acid}}{CH_2{=}\overset{H}{C}{-}COOH} + H_2O + K^+Br^-$$

The addition of hydrogen bromide to acrylic acid will not reform α-bromopropionic acid but instead will yield β-bromopropionic acid by 1,4-addition (Sec. 8.12).

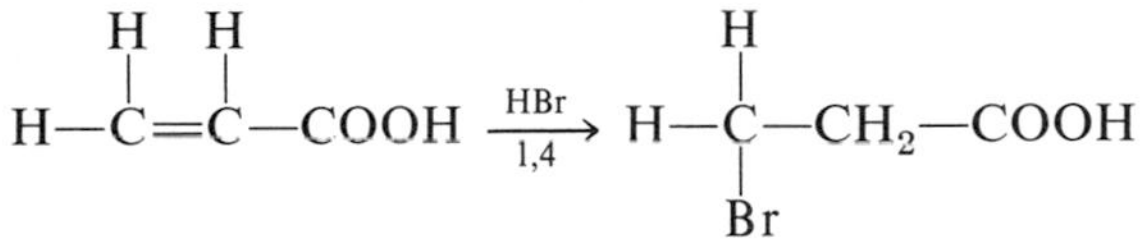

$$\underset{\text{Acrylic acid}}{H{-}\overset{H}{C}{=}\overset{H}{C}{-}COOH} \xrightarrow[1,4]{\text{HBr}} \underset{\beta\text{-Bromopropionic acid}}{H{-}\underset{Br}{\overset{H}{C}}{-}CH_2{-}COOH}$$

Hydrolysis of a halogen acid with dilute aqueous alkali produces the corresponding hydroxy acid.

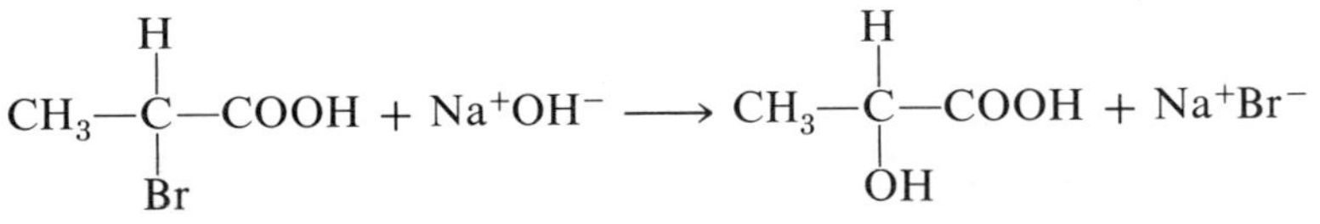

$$CH_3{-}\underset{Br}{\overset{H}{C}}{-}COOH + Na^+OH^- \longrightarrow \underset{\text{Lactic acid}}{CH_3{-}\underset{OH}{\overset{H}{C}}{-}COOH} + Na^+Br^-$$

Hydroxy acids also may be prepared by the hydrolysis of cyanohydrins (Sec. 8.7A).

The hydroxy acids, when heated strongly, show a tendency to lose water. Strong heating converts α-hydroxy acids into cyclic diesters known as lactides. In the intermolecular action of two molecules of α-hydroxy acid to form a lactide, each molecule supplies both an alcohol and an acid function for the esterification of the other.

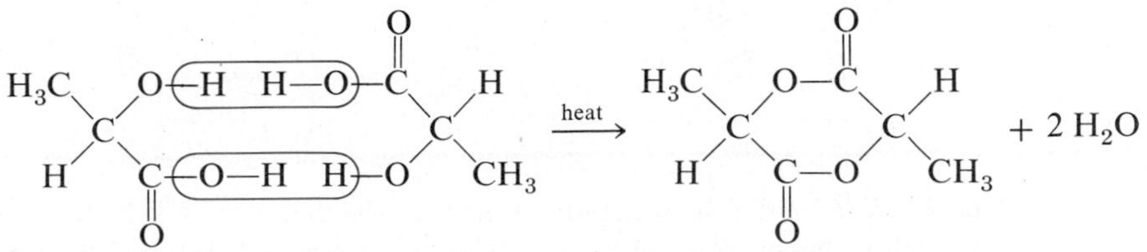

$$\underset{\text{Lactic acid (two molecules)}}{2\ CH_3CH(OH)COOH} \xrightarrow{\text{heat}} \underset{\text{A lactide}}{(CH_3CH{-}O{-}CO)_2} + 2\,H_2O$$

The β-hydroxy acids, when heated strongly, form α,β-unsaturated acids,

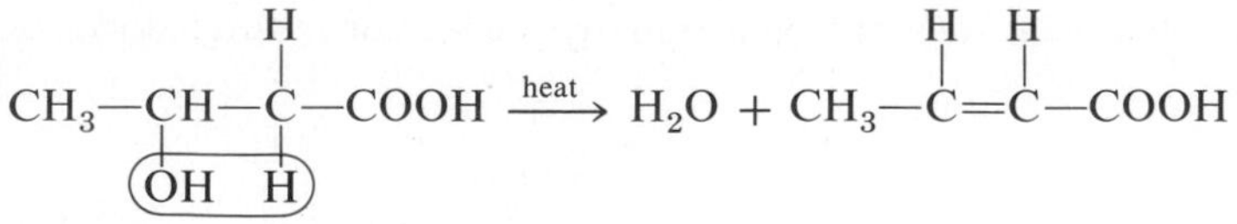

$$CH_3-\underset{\boxed{OH}}{CH}-\overset{H}{\underset{\boxed{H}}{C}}-COOH \xrightarrow{\text{heat}} H_2O + CH_3-\overset{H}{C}=\overset{H}{C}-COOH$$

β-Hydroxybutyric acid

2-Butenoic acid
(Crotonic acid)

The γ- and the δ-hydroxy acids can react intramolecularly to produce five- and six-atom cyclic inner esters known as **γ- and δ-lactones.**

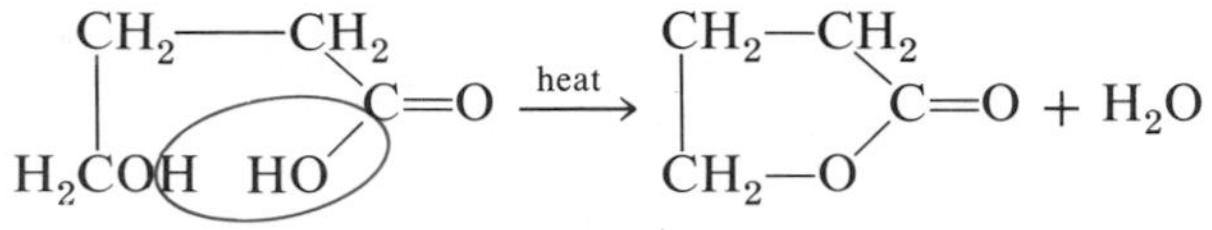

$$\begin{matrix} CH_2 - CH_2 \\ | \quad\quad \searrow \\ \quad\quad\quad C=O \\ | \quad\quad \nearrow \\ H_2COH \quad HO \end{matrix} \xrightarrow{\text{heat}} \begin{matrix} CH_2-CH_2 \\ | \quad\quad \searrow \\ \quad\quad C=O \\ | \quad\quad \nearrow \\ CH_2-O \end{matrix} + H_2O$$

γ-Hydroxybutyric acid

γ-Butyrolactone

If, instead of the hydroxyl group, the amino group, ($-NH_2$), is a substituent in the γ, δ, or ϵ position, heating produces analogous compounds in which the nitrogen atom becomes part of the ring. Such compounds are called **lactams.**

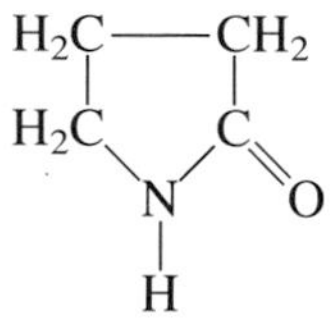

$$\begin{matrix} H_2C - CH_2 \\ | \quad\quad | \\ H_2C \quad\quad C=O \\ \searrow \ \swarrow \\ N \\ | \\ H \end{matrix}$$

γ-Butyrolactam

(ring with positions β, α, γ, δ, ϵ; $C=O$ and NH in ring)

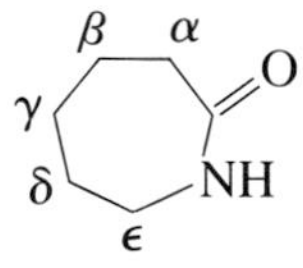

ϵ-Caprolactam

Tartaric acid has a useful culinary application. The half-salt, potassium hydrogen tartrate, called "cream of tartar" is blended with an equivalent amount of sodium bicarbonate (baking soda) to form the active ingredients in "baking powder." When the acid-base mixture is moistened, the resulting reaction releases carbon dioxide, which causes the dough to rise.

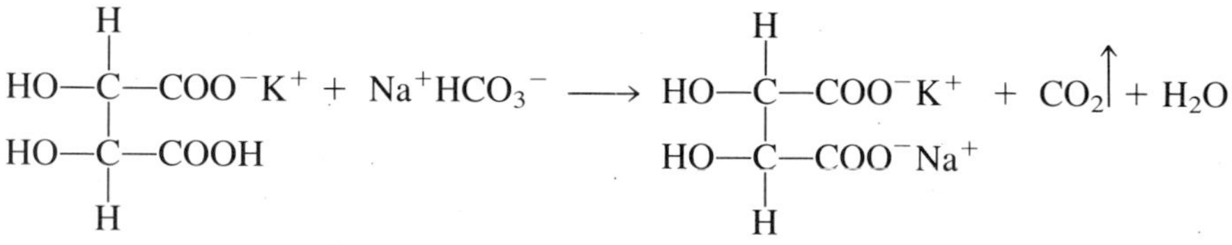

$$\begin{matrix} H \\ | \\ HO-C-COO^-K^+ \\ | \\ HO-C-COOH \\ | \\ H \end{matrix} + Na^+HCO_3^- \longrightarrow \begin{matrix} H \\ | \\ HO-C-COO^-K^+ \\ | \\ HO-C-COO^-Na^+ \\ | \\ H \end{matrix} + CO_2\uparrow + H_2O$$

Exercise 11.5 Write the equation for the reaction that provides the leavening action required for baking when sour milk and baking soda are used in combination.

One of the most important of the aromatic hydroxy acids, perhaps, is salicylic, or *o*-hydroxybenzoic, acid. The importance of salicylic acid and its derivatives lies in the fact that they are widely used as antipyretics and analgesics.

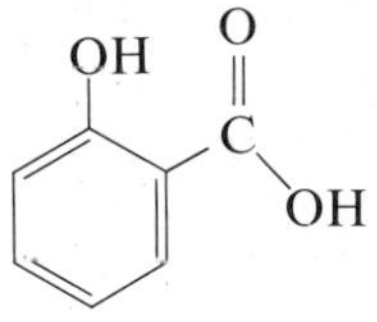

Salicylic acid

Salicylic acid is manufactured commercially in large quantities by heating sodium phenoxide with carbon dioxide under high pressure. The free acid is obtained from its sodium salt by treatment with mineral acid.

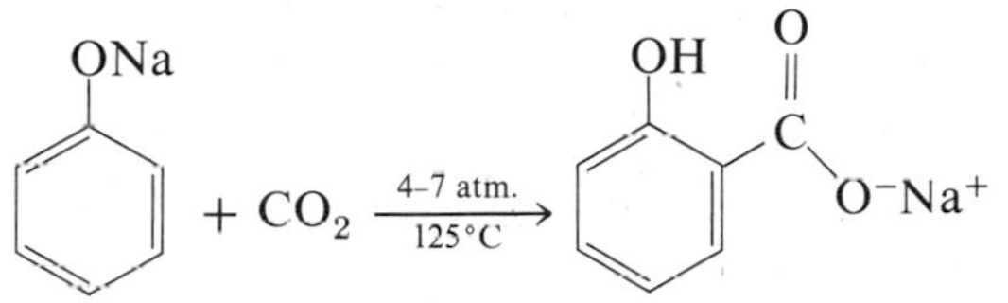

ONa + CO_2 $\xrightarrow[125°C]{4\text{–}7\text{ atm.}}$ OH, C(=O)O^-Na^+

Sodium phenoxide — Sodium salicylate

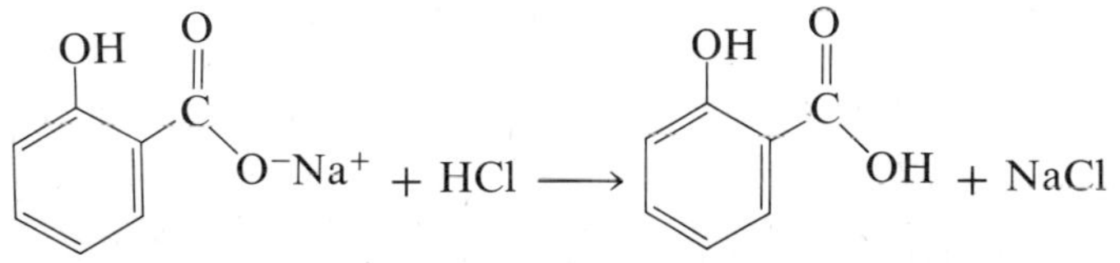

OH, C(=O)O^-Na^+ + HCl ⟶ OH, C(=O)OH + NaCl

Salicylic acid

Salicylic acid is a stronger acid than benzoic due to the presence of the adjacent phenolic group. The ionization of salicylic acid is promoted because the salicylate ion, once formed, is stabilized by H-bonding between the phenolic hydrogen and one or the other of the oxygen atoms of the carboxylate group.

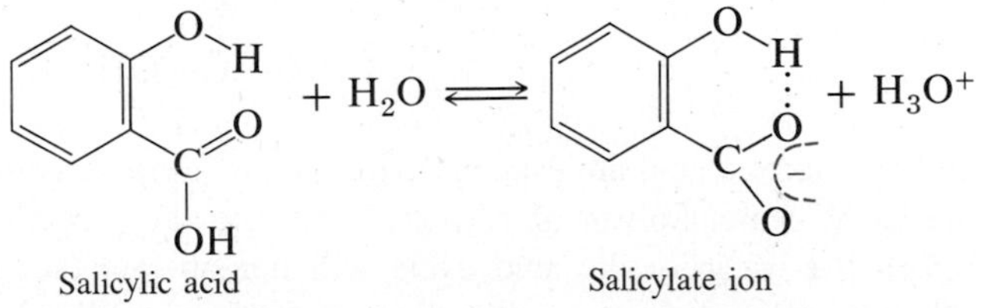

+ H_2O ⇌ + H_3O^+

Salicylic acid — Salicylate ion

Salicylic acid is the starting material for the preparation of several very useful esters. In some of these the carboxyl group is involved; in others only the phenolic group. Esters formed from only the carboxyl group of salicylic acid are called *salicylates*. Methyl salicylate, a flavoring agent widely used in confections and in toothpastes, and referred to as *oil of wintergreen,* is an example of such an ester. It is prepared by the direct esterification of salicylic acid by methyl alcohol.

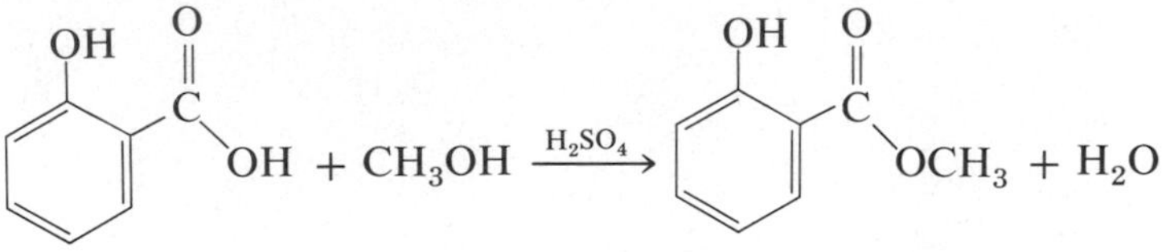

$$\text{HO-C}_6\text{H}_4\text{-COOH} + CH_3OH \xrightarrow{H_2SO_4} \text{HO-C}_6\text{H}_4\text{-C(=O)OCH}_3 + H_2O$$

Methyl salicylate
(Oil of wintergreen)

Aspirin is by far the best example of an ester involving only the phenolic group of salicylic acid. Aspirin, or acetylsalicylic acid, is prepared by acetylation of the hydroxy group by acetic anhydride.

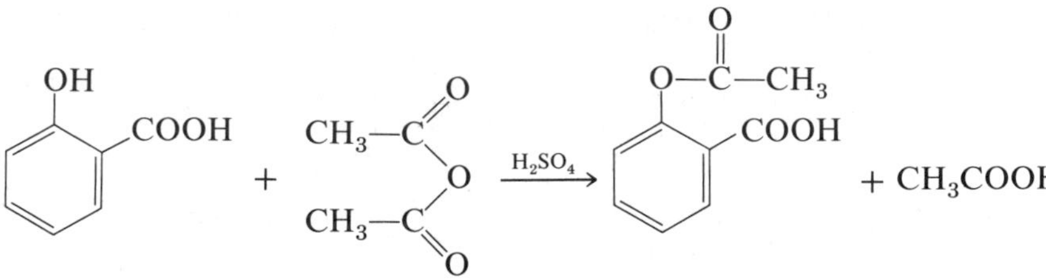

$$\text{HO-C}_6\text{H}_4\text{-COOH} + (CH_3CO)_2O \xrightarrow{H_2SO_4} \text{CH}_3\text{C(=O)-O-C}_6\text{H}_4\text{-COOH} + CH_3COOH$$

Acetic anhydride

Acetyl salicylic acid
(Aspirin)

11.7 Keto Acids

Certain of the keto acids play vital roles as metabolic intermediates in biological oxidation and reduction reactions. Pyruvic acid,

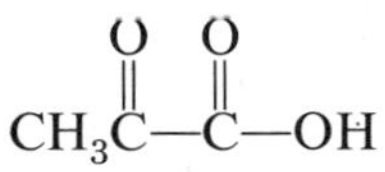

$$CH_3\overset{O}{\overset{\|}{C}}-\overset{O}{\overset{\|}{C}}-OH$$

and acetoacetic acid are especially important. Pyruvic acid is a principal intermediate in the aerobic metabolism of carbohydrates and the precursor to oxalacetic acid, without which the tricarboxylic acid cycle will not operate (see Fig. 14.1). Inspection of the scheme outlined for the metabolism of fatty acids (Sec. 12.6) shows the CoA ester of acetoacetic acid also as an intermediate in the metabolic process. The abnormal metabolism of fatty acids by diabetics, if uncontrolled, releases an excess of acetoacetic acid into the bloodstream. The tendency of β-keto acids to decarboxylate results in the release of acetone. The accumulation of acetoacetic acid and acetone in the blood may reach concentrations sufficient to cause death.

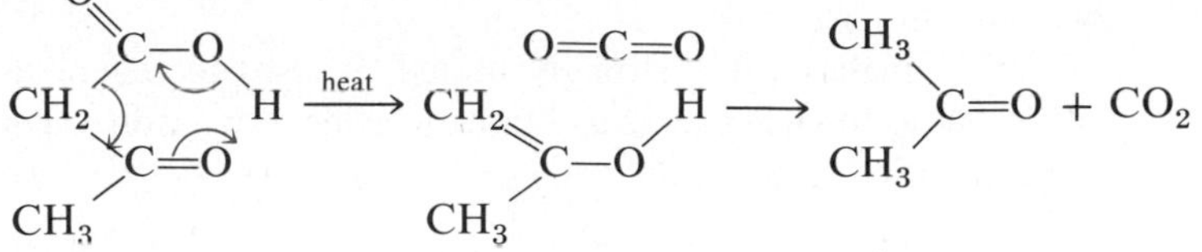

$$\text{Acetoacetic acid } (CH_3-CO-CH_2-CO-OH) \xrightarrow{\text{heat}} \text{Enol form of acetone } (CH_3-C(OH)=CH_2) + O=C=O \longrightarrow (CH_3)_2C=O + CO_2$$

Acetoacetic acid — Enol form of acetone — Acetone

> ***Exercise 11.6*** Compare the mechanisms given for the decarboxylation of malonic acid and of acetoacetic acid, and suggest a reason for the experimental observation that β-keto acids such as acetoacetic acid decarboxylate much more readily when heated than do malonic acid and its derivatives.

The readiness with which a β-keto acid decarboxylates, when heated, makes such acids useful intermediates in a number of organic syntheses. The α-hydrogens of ethyl acetoacetate are even more acidic than those of malonic ester; therefore, they may be replaced in the same manner used for malonic ester, as described in Section 10.4D.

11.8 Unsaturated Acids. *cis,trans*-Isomerism

The only unsaturated acids that we will consider in this chapter are the α,β-unsaturated acids. Other unsaturated acids and derivatives will be described in Chapter 12. The simplest α,β-unsaturated acid is propenoic acid, $CH_2{=}CH{-}CO_2H$, commonly called acrylic acid. It and its derivatives are important industrial materials used both in the manufacture of polymers and plastics and as starting materials for the synthesis of other useful products. Acrylonitrile, $CH_2{=}CH{-}C{\equiv}N$, is prepared by the catalytic oxidation of propene in the presence of ammonia and is used to prepare polyacrylonitrile, from which the synthetic fibers known as "Orlon" or "Acrilan" are made. Acrylonitrile is also used in the manufacture of nylon (Sec. 11.4) and of ABS plastic (Sec. 3.11). In 1984 2.2×10^9 pounds of acrylonitrile were produced in the United States!

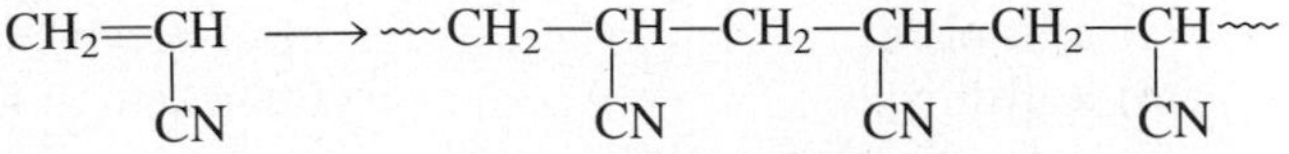

$$CH_2{=}CH(CN) \longrightarrow \sim CH_2{-}CH(CN){-}CH_2{-}CH(CN){-}CH_2{-}CH(CN)\sim$$

Acrylonitrile — Polyacrylonitrile (Orlon) (Acrilan)

The methyl ester of 2-methylpropenoic acid is known as methyl methacrylate and is the monomer from which the clear, glasslike thermoplastic called Lucite or Plexiglass is formed. Methyl methacrylate is prepared by the dehydration and esterification of acetone cyanohydrin (Sec. 8.7).

When malic acid is strongly heated, it loses the elements of water from adjacent carbon atoms to yield maleic and fumaric acids—two isomeric α,β-unsaturated dicarboxylic acids (Sec. 5.8).

$$2\ \mathrm{HO{-}CH(COOH){-}CH_2{-}COOH} \xrightarrow{\text{heat}} H_2O + \mathrm{(H)(COOH)C{=}C(H)(COOH)} + \mathrm{(H)(COOH)C{=}C(COOH)(H)} + H_2O$$

Malic acid

Maleic acid (*cis*-Butenedioic acid) mp, 130°C

Fumaric acid (*trans*-Butenedioic acid) mp, 287°C

When heated, maleic acid loses water to form an anhydride. Fumaric acid, however, is incapable of anhydride formation.

$$\mathrm{H{-}C{-}C(=O){-}OH \;\; \| \;\; H{-}C{-}C(=O){-}O{-}H} \xrightarrow{300^\circ C} \mathrm{H{-}C{-}C(=O) \;\; \| \;\; H{-}C{-}C(=O)}\!>\!O + H_2O$$

Maleic acid

Maleic anhydride

Maleic anhydride is an important industrial material that is manufactured by a combined oxidation and dehydrogenation of butane using a phosphorus and vanadium catalyst. Maleic anhydride is one of the most powerful dienophiles known (Sec. 3.12). When either maleic or fumaric acid is hydrogenated, succinic acid is produced.

Summary

1. The bifunctional acids contain, in addition to a carboxyl group, one or more other functions. These may be
   (a) a halogen
   (b) a hydroxyl
   (c) a carbon–carbon double bond
   (d) a second carboxyl group
   (e) a keto group
2. The chemical behavior of bifunctional acids, generally, is either that of the monocarboxylic acids or that usually associated with the second functional group.
3. The presence of a second functional group in an acid molecule may lead to cyclization, unsaturation, loss of carbon dioxide, or polymerization.

4. Dicarboxylic acids may be prepared by the same general methods used for the preparation of monocarboxylic acids. Such methods include
   (a) the oxidation of unsaturated acids
   (b) hydrolysis of nitriles
5. Dicarboxylic acids usually are referred to by common names.
6. Adipic acid is used in the manufacture of "Nylon." Phthalic acid is used in the manufacture of *glyptal* resins, and terephthalic acid is used in the manufacture of "Dacron."
7. The chemistry of the halogen and hydroxy acids is interrelated.
8. Halogen acids usually are prepared by
   (a) the direct halogenation by the use of halogen and red phosphorus—e.g., the Hell–Volhard–Zelinsky reaction
   (b) the action of an inorganic halide upon a hydroxy acid
   (c) the oxidation of a halohydrin
   (d) the addition of hydrogen halide to an unsaturated acid
9. Hydroxy acids may be prepared by
   (a) the alkaline hydrolysis of the corresponding halogen acid
   (b) the hydrolysis of a cyanohydrin
10. Malic acid, when dehydrated, produces a pair of isomeric, unsaturated dicarboxylic acids called **fumaric** and **maleic.** These acids are *cis,trans*-isomers.

Supplementary Exercises

11.7 Write structural formulas for each of the following.

(a) adipyl chloride
(b) methyl succinate
(c) terephthalic acid
(d) ethylmalonic acid
(e) ethyl cyanoacetate
(f) δ-valerolactone
(g) α-methylglutaric acid
(h) aspirin

11.8 Name each of the following structures.

(a) 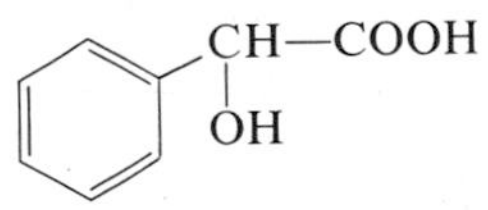 C_6H_5—CH(OH)—COOH

(b) Cl—CH_2—C(=O)—OH

(c) H(HOOC)C=C(COOH)H

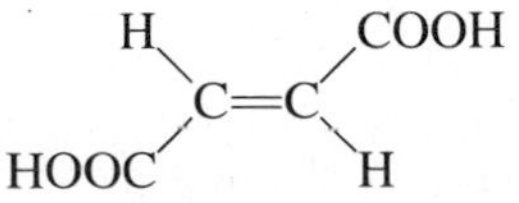

(d) 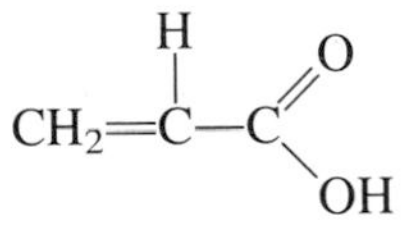 CH_2=CH—C(=O)—OH

(e) 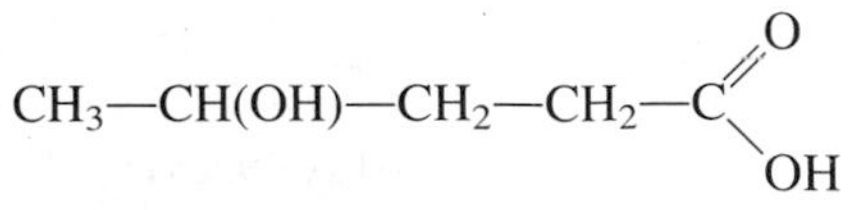 CH_3—CH(OH)—CH_2—CH_2—C(=O)—OH

(f) 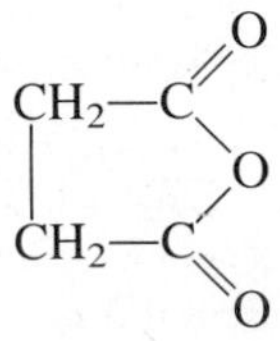 CH_2—C(=O)—O—C(=O)—CH_2 (cyclic: the two CH_2 groups bonded to each other)

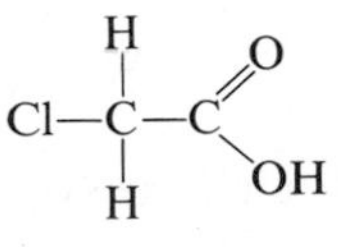

(g) 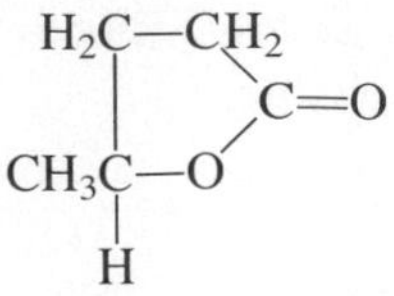 H_2C-CH_2 / $CH_3C(H)-O$ / $C=O$ (ring)

(l) 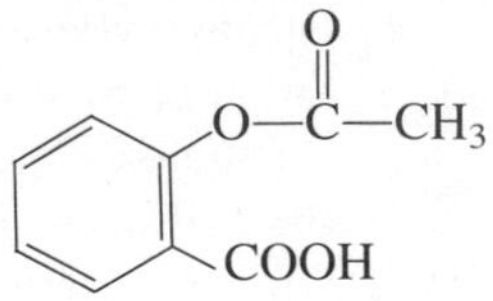 o-(CH_3COO)C_6H_4COOH (aryl ring with $O-\overset{O}{\overset{\|}{C}}-CH_3$ and COOH)

(h) $C_2H_5O-\overset{O}{\overset{\|}{C}}-C(C_2H_5)_2-\overset{O}{\overset{\|}{C}}-OC_2H_5$

(m) 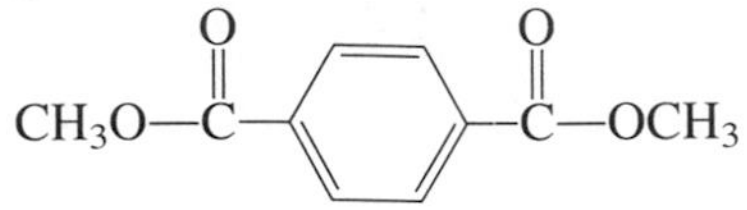 $CH_3O-\overset{O}{\overset{\|}{C}}-C_6H_4-\overset{O}{\overset{\|}{C}}-OCH_3$

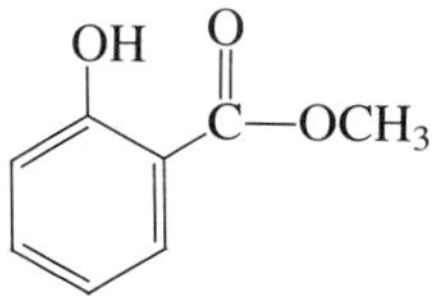

(i) $CH_2{=}CH-\overset{O}{\overset{\|}{C}}-OCH_3$

(j) $HOOC-CH_2-\underset{CH_3}{\overset{CH_3}{C}}-CH_2-CH_2-COOH$

(n) 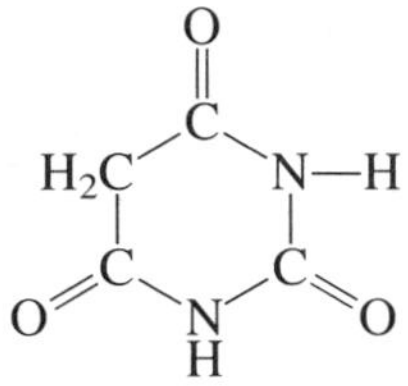 H_2C, $C{=}O$, N—H, $C{=}O$, N(H), $C{=}O$ (six-membered ring)

(k) OH, $\overset{O}{\overset{\|}{C}}-OCH_3$ (ortho-substituted benzene ring)

(o) H, COOH, H, OH (cyclohexane chair)

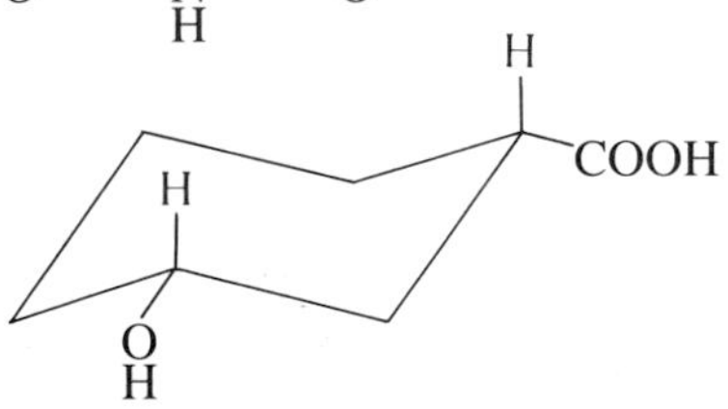

11.9 Complete the following reaction sequences and identify each lettered product.

(a) $H_2C{=}CH_2 \xrightarrow{HOCl} (A) \xrightarrow{[O]} (B) \xrightarrow[2.\ NaCN]{1.\ NaOH} (C) \xrightarrow{H_3O^+,\ heat} (D)$

(b) 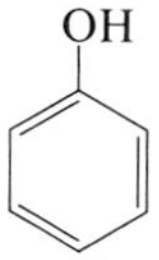 $C_6H_5OH \xrightarrow{NaOH} (A) \xrightarrow[2.\ Heat,\ pressure]{1.\ CO_2} (B) \xrightarrow{H_3O^+} (C) \xrightarrow[H_2SO_4]{(CH_3\overset{O}{C})_2O} (D)$

(c) $CH_3-\overset{H}{\overset{|}{C}}{=}O \xrightarrow{HCN} (A) \xrightarrow{H_3O^+,\ reflux} (B) \xrightarrow{heat} (C)$

(d) $CH_3CH_2CH_2COOH \xrightarrow[2.\ H_2O]{1.\ Red\ P,\ Br_2} (A) \xrightarrow{alcoholic\ KOH} (B) \xrightarrow{HBr} (C)$

(e) $CH_2(COOC_2H_5)_2 \xrightarrow[2.\ C_6H_5OCH_2CH_2Br]{1.\ C_2H_5O^-Na^+} (A) \xrightarrow[heat]{48\%\ HBr} (B) + (C) + CO_2$

$(B) + FeCl_3 \longrightarrow$ red color

(f) civetone $\xrightarrow[2.\ H_2O\ +\ HCl]{1.\ KMnO_4\ +\ H_2O,\ heat} (A)$

(g) fumaric acid + $Br_2 \longrightarrow (A)$

11.10 Arrange the following in an order of increasing acidity.

(a) salicylic acid
(b) chloroacetic acid
(c) oxalic acid
(d) malonic acid
(e) mandelic acid

11.11 Explain why the pK_a for the ionization of the first carboxyl group of maleic acid is *smaller* than that for fumaric acid, and why the pK_a for the ionization of the second carboxyl group of maleic acid is *larger* than that for fumaric acid.

11.12 Which of the following acids could exist as *cis,trans*-isomers? Which as optical isomers?

(a) acrylic acid
(b) 2-butenoic acid
(c) malic acid
(d) 1,2-cyclopentanedicarboxylic acid
(e) ricinoleic acid

11.13 Suggest a simple chemical test for distinguishing between

(a) aspirin and phenyl salicylate
(b) acetyl chloride and chloroacetic acid
(c) maleic and malonic acid

11.14 Beginning with ethyl alcohol as your only organic starting material, and any other reagents you might require, show how you might prepare:

(a) acetic acid
(b) ethyl acetate
(c) ethyl acetoacetate
(d) β-hydroxybutyric acid
(e) 2-butenoic acid
(f) malonic acid
(g) 2-pentanone
(h) *n*-butyric acid

11.15 A compound, $C_4H_8O_3$, responded to a series of tests as follows:

(a) Water $\rightarrow$ aqueous solution acid to litmus
(b) $Na_2Cr_2O_7 + H_2SO_4$ + heat $\rightarrow$ blue-green coloration
(c) Strong heating $\rightarrow C_4H_6O_2$
(d) Product of (c) + dilute $KMnO_4 \rightarrow$ decolorization
(e) $I_2 + NaOH \rightarrow$ yellow solid
(f) Rotated plane-polarized light

What was the compound? Write equations for reactions (a) through (e).

11.16 Show how the following compounds could be prepared from malonic ester. Use any inorganic reagents and other organic reactants you might require.

(a) $CH_3CH(CH_3)CH_2CH_2CH_2COOH$ (i.e., $CH_3\underset{|}{C}HCH_2CH_2CH_2COOH$ with CH_3 on the second carbon)

(b) $HOOCCH_2CH_2COOH$

(c) 5-methylbarbituric acid

(d) $C_6H_5CH_2CH(CH_3)COOH$ (drawn as $C_6H_5CH_2$ and CH_3 attached to CHCOOH)

11.17 Write a sequence of equations describing the synthesis of methyl methacrylate (Sec. 11.8) from acetone.

11.18 Nylon 66 is one of the most common of the nylon polymers; however, nylon 6 is also made in considerable quantity. Assuming that the single six means that there is only one six-carbon unit in the chain, rather than two, suggest a structure for nylon 6. (*Hint:* See Sec. 11.6.)

12 Lipids

Natural products that are soluble in ether, chloroform, carbon tetrachloride, and other water-immiscible organic solvents, but insoluble in water, are known as **lipids.** The lipids are important constituents of all plant and animal tissue. About 40–50% of most membranes is composed of lipids of various types. In this chapter our attention will be focused on the saponifiable lipids derived from the fatty acids: the fats, oils,[1] and waxes. The important biochemical regulators, the prostaglandins, as well as the phospholipids, are also derived from the fatty acids and are included in our discussion. In addition to the fats and oils (complex lipids), there are other water-insoluble natural substances—the steroids, hormones, and fat-soluble vitamins—which are known as simple lipids and are discussed in Chapter 17. Not only do the edible fats and oils make up approximately 40% of the American diet, but they also provide the raw material for the preparation of numerous important commodities.

12.1 Waxes

Waxes are esters of long-chain, unbranched fatty acids and long-chain ''fatty'' alcohols. Both the acid and the alcohol that combine to form a wax may be sixteen to thirty carbons in length. The general formula of a wax is essentially that of a simple ester,

$$R{-}\overset{\displaystyle O}{\overset{/\!/}{C}}{-}O{-}R'$$

[1]The term oils, as used in the present chapter, refers to glycerides (see Sec. 12.2) that are liquid at room temperature, and *not* to mineral oils or petroleum products.

Both plants and animals produce natural waxes. Waxes usually are mixtures of esters and contain, in addition, small amounts of free acids, alcohols, and even hydrocarbons. The waxes melt over a wide range of temperature (35–100°C), have a satiny "waxy" feel, and are very insoluble in water. On the other hand, they are quite soluble in many organic solvents and may be compounded into a number of useful everyday commodities. Such wax solutions generally are used as protective coatings. They are often protective coatings in nature, too.

Beeswax is the material with which the bee encases its honey in the "honeycomb." It is largely ceryl myristate,

$$C_{13}H_{27}\overset{O}{\overset{\|}{C}}{-}O{-}C_{26}H_{53}$$

along with some esters of cerotic acid, $C_{25}H_{51}COOH$, and a few per cent of hydrocarbons. It melts between 62 and 65°C and is used in the preparation of polishes, candles, and paper coatings.

Lanolin used in emollients and lotions has its origin in wool wax.

Fats and Oils

12.2 Structure and Composition of Fats and Oils

Fats and oils differ from waxes in that they are glycerides, or esters of glycerol, a trihydroxy alcohol (Sec. 7.13).

$$\begin{array}{l} H \qquad\quad O \\ | \qquad\quad\; \| \\ H-C-O-C-R \\ | \qquad\quad O \\ | \qquad\quad\; \| \\ H-C-O-C-R \\ | \qquad\quad O \\ | \qquad\quad\; \| \\ H-C-O-C-R \\ | \\ H \end{array}$$

A simple glyceride

A simple glyceride is one in which all R groups in the previous general formula are identical. If R in the general formula represents an aliphatic group, C_nH_{2n+1}, then the number of carbons in the group usually is an odd number from 3 to 17. Such chains are saturated. If R is an unsaturated alkyl group of the form C_nH_{2n-1}, C_nH_{2n-3}, or C_nH_{2n-5}, then n usually is 17.

$$\begin{array}{l}
\quad\ \ \mathrm{H} \qquad\ \mathrm{O} \\
\quad\ \ | \qquad\quad \| \\
\mathrm{H{-}C{-}O{-}C{-}(CH_2)_{14}{-}CH_3} \\
\quad\ \ | \qquad\quad\ \mathrm{O} \\
\quad\ \ | \qquad\quad \| \\
\mathrm{H{-}C{-}O{-}C{-}(CH_2)_{14}{-}CH_3} \\
\quad\ \ | \qquad\quad\ \mathrm{O} \\
\quad\ \ | \qquad\quad \| \\
\mathrm{H{-}C{-}O{-}C{-}(CH_2)_{14}{-}CH_3} \\
\quad\ \ | \\
\quad\ \ \mathrm{H}
\end{array}$$

Glyceryl tripalmitate (a simple glyceride)

The natural fats and oils are generally not simple glycerides. The three acid residues produced when fats and oils are hydrolyzed usually vary not only in length but also in the degree of unsaturation. The principal structural difference between oils and fats lies in the degree of saturation of the acid residues, and this accounts for the differences in both the physical and chemical properties of these two classes of glycerides.

The fats are glyceryl esters in which long-chain *saturated* acid components predominate. They are solids or semisolids and are principally animal products. **Lauric,** $CH_3(CH_2)_{10}COOH$; **palmitic,** $CH_3(CH_2)_{14}COOH$; and **stearic,** $CH_3(CH_2)_{16}COOH$ are the most important acids obtained by the hydrolysis of fats. Such long-chain carboxylic acids usually are called *fatty acids* because they are obtained from fats. They are, for the most part, insoluble in water but soluble in organic solvents. It is entirely possible that all three of the fatty acids named above could be present in a mixed glyceride such as that shown by the formula below.

$$\begin{array}{l}
\quad\ \ \mathrm{H} \qquad\ \mathrm{O} \\
\quad\ \ | \qquad\quad \| \\
\mathrm{H{-}C{-}O{-}C{-}(CH_2)_{10}{-}CH_3} \\
\quad\ \ | \qquad\quad\ \mathrm{O} \\
\quad\ \ | \qquad\quad \| \\
\mathrm{H{-}C{-}O{-}C{-}(CH_2)_{14}{-}CH_3} \\
\quad\ \ | \qquad\quad\ \mathrm{O} \\
\quad\ \ | \qquad\quad \| \\
\mathrm{H{-}C{-}O{-}C{-}(CH_2)_{16}{-}CH_3} \\
\quad\ \ | \\
\quad\ \ \mathrm{H}
\end{array}$$

Glyceryl lauropalmitostearate
(a mixed glyceride)

Exercise 12.1 A small amount of *arachidic* acid is found in butter and in most vegetable oils. The systematic name for arachidic acid is eicosanoic acid. What is its structure?

Oils, on the other hand, are glyceryl esters in which long-chain *unsaturated* acid components predominate. They are liquids and are largely of vegetable origin.

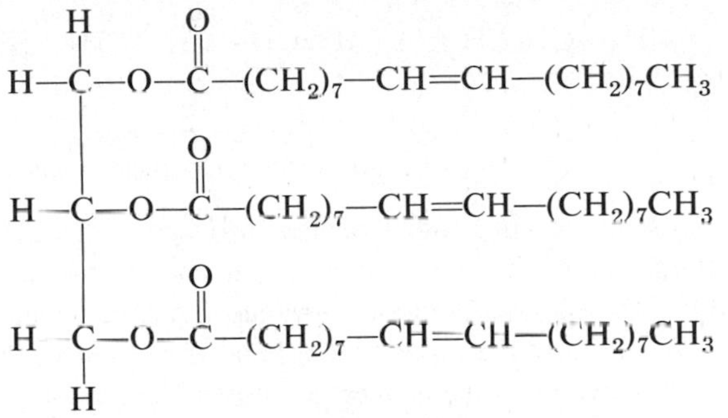

$$
\begin{array}{l}
\mathrm{H{-}\overset{\displaystyle H}{\overset{|}{C}}{-}O{-}\overset{\displaystyle O}{\overset{\|}{C}}{-}(CH_2)_7{-}CH{=}CH{-}(CH_2)_7CH_3} \\
\mathrm{H{-}\overset{|}{C}{-}O{-}\overset{\displaystyle O}{\overset{\|}{C}}{-}(CH_2)_7{-}CH{=}CH{-}(CH_2)_7CH_3} \\
\mathrm{H{-}\underset{\displaystyle H}{\underset{|}{C}}{-}O{-}\overset{\displaystyle O}{\overset{\|}{C}}{-}(CH_2)_7{-}CH{=}CH{-}(CH_2)_7CH_3}
\end{array}
$$

Glyceryl trioleate (Triolein)[2]
(an oil)

The presence of unsaturation in the acid component of a fat tends to lower its melting point. The most important unsaturated acids obtainable by the hydrolysis of oils are the C_{18} acids—**oleic, linoleic,** and **linolenic,** whose structures are given below.

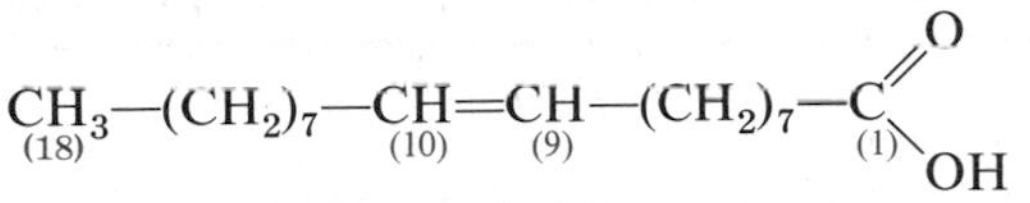

$$
\underset{(18)}{\mathrm{CH_3}}{-}(\mathrm{CH_2})_7{-}\underset{(10)}{\mathrm{CH}}{=}\underset{(9)}{\mathrm{CH}}{-}(\mathrm{CH_2})_7{-}\underset{(1)}{\mathrm{C}}\!\begin{array}{l}\mathrm{{/\!\!/}O}\\ \mathrm{\backslash OH}\end{array}
$$

Oleic acid
((*Z*)-9-Octadecenoic acid)

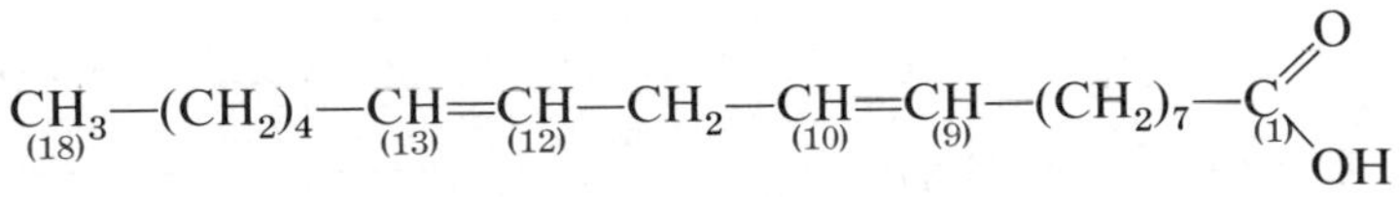

$$
\underset{(18)}{\mathrm{CH_3}}{-}(\mathrm{CH_2})_4{-}\underset{(13)}{\mathrm{CH}}{=}\underset{(12)}{\mathrm{CH}}{-}\mathrm{CH_2}{-}\underset{(10)}{\mathrm{CH}}{=}\underset{(9)}{\mathrm{CH}}{-}(\mathrm{CH_2})_7{-}\underset{(1)}{\mathrm{C}}\!\begin{array}{l}\mathrm{{/\!\!/}O}\\ \mathrm{\backslash OH}\end{array}
$$

Linoleic acid
((*Z,Z*)-9,12-Octadecadienoic acid)

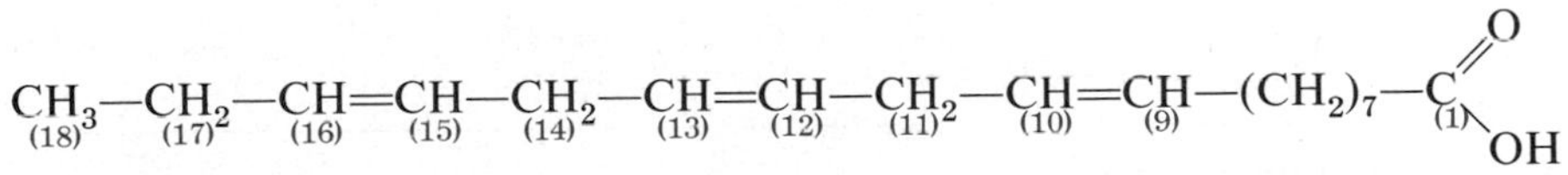

$$
\underset{(18)}{\mathrm{CH_3}}{-}\underset{(17)}{\mathrm{CH_2}}{-}\underset{(16)}{\mathrm{CH}}{=}\underset{(15)}{\mathrm{CH}}{-}\underset{(14)}{\mathrm{CH_2}}{-}\underset{(13)}{\mathrm{CH}}{=}\underset{(12)}{\mathrm{CH}}{-}\underset{(11)}{\mathrm{CH_2}}{-}\underset{(10)}{\mathrm{CH}}{=}\underset{(9)}{\mathrm{CH}}{-}(\mathrm{CH_2})_7{-}\underset{(1)}{\mathrm{C}}\!\begin{array}{l}\mathrm{{/\!\!/}O}\\ \mathrm{\backslash OH}\end{array}
$$

Linolenic acid
((*Z,Z,Z*)-9,12,15-Octadecatrienoic acid)

You will note that the first double bond in each of the above structures is found in the middle of the carbon chain and that other double bonds are farther removed from the carboxyl group. Methylene ($-CH_2-$) units separate one double bond from another, and the unsaturated hydrocarbon portion of these three acids does not represent a conjugated system (Sec. 3.11). Some natural oils do contain acids with systems of alternating single

[2]If all fatty acid residues of a glyceride are the same, the name is shortened by dropping "glyceryl" and changing "tri . . . ate" to "tri . . . in."

and double bonds. One of these is **tung oil,** which on hydrolysis yields **eleostearic** as the principal acid.

$$\overset{}{CH_3}-CH_2CH_2CH_2\overset{(14)}{CH}=\overset{(13)}{CH}-\overset{(12)}{CH}=\overset{(11)}{CH}-\overset{(10)}{CH}=\overset{(9)}{CH}-(CH_2)_7-C\begin{matrix}\nearrow O \\ \searrow OH\end{matrix}$$

Eleostearic acid
((*Z,E,E*)-9,11,13-Octadecatrienoic acid)

Unsaturation in a fatty acid also makes possible *cis,trans*-isomerism depending on the configuration of the hydrogen atoms attached to the doubly bonded carbon atoms. Oleic acid has the *cis* configuration, whereas its isomer, elaidic acid, has the *trans* configuration. When more than one double bond is present in the fatty acid molecule, of course more than two *cis,trans*-isomers are possible. Generally speaking, the *cis*-isomers are the forms found naturally occurring in the unsaturated acid component of food fats and oils. As may be seen from Table 12.1, the most abundant of all saturated acids found in fats is palmitic; the most abundant of the unsaturated acids found in edible oils is oleic.

Exercise 12.2 How many *cis,trans*-isomers are possible for linoleic acid?

TABLE 12.1 Fatty Acid Components of Some Common Fats and Oils

	Component Acids (percent)*						
	Myristic C_{14}	Palmitic C_{16}	Stearic C_{18}	Oleic	Linoleic	Linolenic	Eleostearic
Fats							
Butter	7–10	24–26	10–13	28–31	1.0–2.5	0.2–0.5	
Lard	1–2	28–30	12–18	40–50	7–13	0–1	
Tallow	3–6	24–32	20–25	37–43	2–3		
Edible Oils							
Olive oil		9–10	2–3	73–84	10–12	trace	
Corn oil	1–2	8–12	2–5	19–49	34–62	trace	
Soybean oil		6–10	2–5	20–30	50–60	5–11	
Cottonseed oil	0–2	20–25	1–2	23–35	40–50	trace	
Peanut oil		8–9	2–3	50–65	20–30		
Safflower oil		6–7	2–3	12–14	75–80	0.5–0.15	
Nonedible Oils							
Linseed oil		4–7	2–4	25–40	35–40	25–60	
Tung oil		3–4	0–1	4–15			75–90

*Totals less than 100% indicate the presence of lower or higher acids in small amounts.

Reactions of the Fats and Oils

12.3 Saponification (Soap Preparation)

The glycerides, like other esters, may be hydrolyzed by heating with a solution of sodium or potassium hydroxide. The hydrolysis products are glycerol and the alkali metal salts of long-chain fatty acids (Sec 9.6A). The latter are called soaps, and alkaline hydrolysis is called saponification, whether the term is applied to fats, oils, or simple esters.

12.4 Hydrogenation ("Hardening" of Oils)

Unsaturation in a fat or oil may be diminished by catalytic hydrogenation.

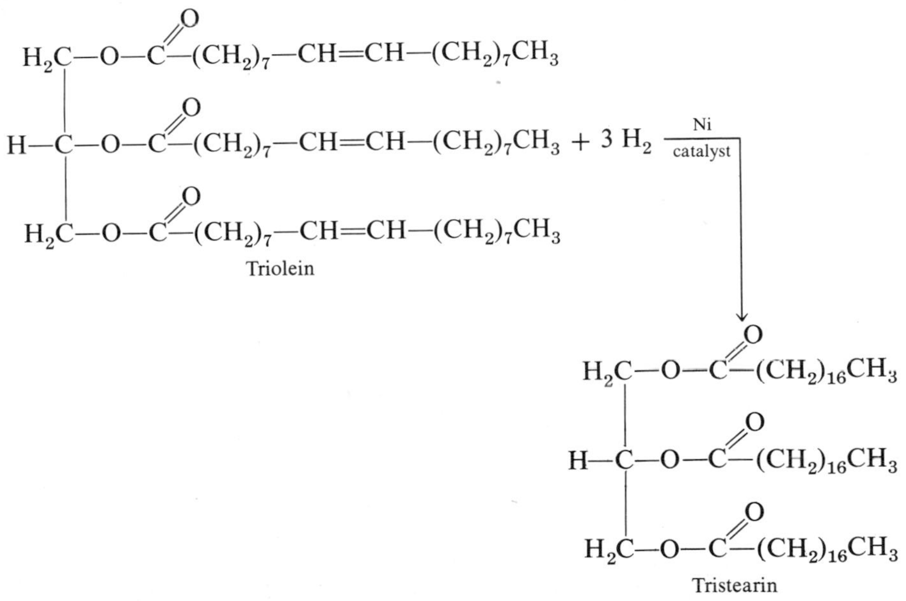

$$\begin{array}{l} H_2C{-}O{-}C({=}O){-}(CH_2)_7{-}CH{=}CH{-}(CH_2)_7CH_3 \\ \quad | \\ H{-}C{-}O{-}C({=}O){-}(CH_2)_7{-}CH{=}CH{-}(CH_2)_7CH_3 \\ \quad | \\ H_2C{-}O{-}C({=}O){-}(CH_2)_7{-}CH{=}CH{-}(CH_2)_7CH_3 \end{array} + 3\ H_2 \xrightarrow[\text{catalyst}]{\text{Ni}}$$

Triolein

$$\begin{array}{l} H_2C{-}O{-}C({=}O){-}(CH_2)_{16}CH_3 \\ \quad | \\ H{-}C{-}O{-}C({=}O){-}(CH_2)_{16}CH_3 \\ \quad | \\ H_2C{-}O{-}C({=}O){-}(CH_2)_{16}CH_3 \end{array}$$

Tristearin

The process is controllable and is used to convert low-melting fats or oils to higher-melting fats of any desired consistency. Because the melting point of a fat increases with saturation, oils may be converted to semisolid fats by hydrogenation. This hydrogenation process is known as **hardening.** Margarines are made by hardening oils to the consistency of butter. When churned with skim milk, fortified with vitamin A, and artificially colored, these butter substitutes have not only the flavor and color of butter, but its nutritional advantages as well. Butter is approximately 80% fat, and margarine, by federal regulation, must also contain not less than 80% fat.

Many of the cooking fats now available have their origin in vegetable oils. When hydrogenation is carried to completion, glycerol and long-chain alcohols are produced. The latter are used in the manufacture of synthetic detergents.

$$\underset{\text{Glyceryl tristearate}}{\begin{array}{l} H_2C{-}O{-}C({=}O){-}(CH_2)_{16}CH_3 \\ H{-}C{-}O{-}C({=}O){-}(CH_2)_{16}CH_3 \\ H_2C{-}O{-}C({=}O){-}(CH_2)_{16}CH_3 \end{array}} + 6\,H_2 \xrightarrow{\text{catalyst}} \underset{\text{Glycerol}}{\begin{array}{c} CH_2OH \\ CHOH \\ CH_2OH \end{array}} + 3\,\underset{\text{1-Octadecanol}}{CH_3(CH_2)_{16}CH_2OH}$$

> ***Exercise 12.3*** Give a series of reactions whereby you might convert 1-octadecanol into a satisfactory detergent.

12.5 Oxidation (Rancidity)

Not only are edible unsaturated fats and oils, when exposed to air and light for long periods of time, subject to slow hydrolysis, but the acid components produced are also subject to oxidative cleavage at the site of unsaturation.

$$\underset{\text{Oleic acid}}{CH_3{-}(CH_2)_7{-}CH{=}CH{-}(CH_2)_7COOH} \xrightarrow{O_2(\text{air})}$$

$$\underset{\text{Pelargonic acid}}{CH_3(CH_2)_7COOH} + \underset{\text{Azelaic acid}}{HO{-}C({=}O){-}(CH_2)_7{-}C({=}O){-}OH}$$

The lower-molecular weight and more volatile acids that are produced by this exposure impart an offensive odor to fats. This condition is known as **rancidity.** If the volatile acids are produced by hydrolysis, the resultant rancidity is known as hydrolytic rancidity. Butter, especially, when left uncovered and out of the refrigerator, easily becomes rancid through hydrolytic rancidity. A substantial portion of the fatty acid components of butterfat is made up of butyric, caproic, caprylic, and capric acids. These are liberated when butter is hydrolyzed and are responsible for the unpleasant odor of rancid butter. Oxidation leading to rancidity (oxidative rancidity) in fats and oils is catalyzed by the presence of certain metallic salts. Proper packaging of foods, therefore, is of the utmost importance. The addition of **antioxidants** will stabilize edible fats for long periods of storage. These inhibitors, or interceptors as they sometimes are called, are themselves easily oxidizable substances. Some are natural products such as δ-tocopherol and the lecithins. The lecithins, as evident from their structures, are themselves glycerides.

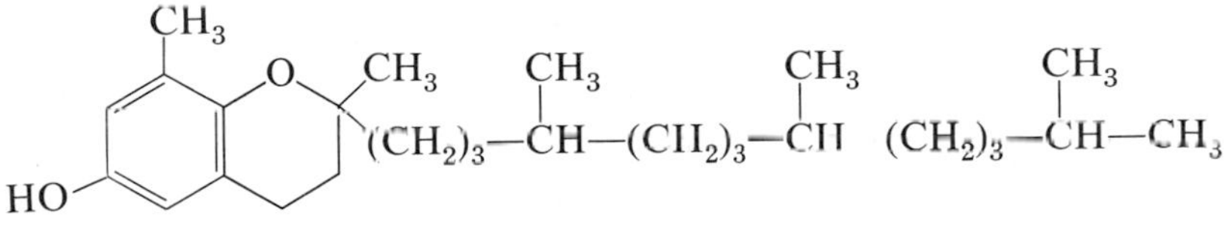

CH_3 ... $(CH_2)_3-CH(CH_3)-(CH_2)_3-CH(CH_3)-(CH_2)_3-CH(CH_3)-CH_3$

δ-Tocopherol

$$H_2C-O-C(=O)-R$$
$$H-C-O-C(=O)-R'$$
$$H_2C-O-P(=O)(O^-)-O-CH_2CH_2\overset{+}{N}(CH_3)_3$$

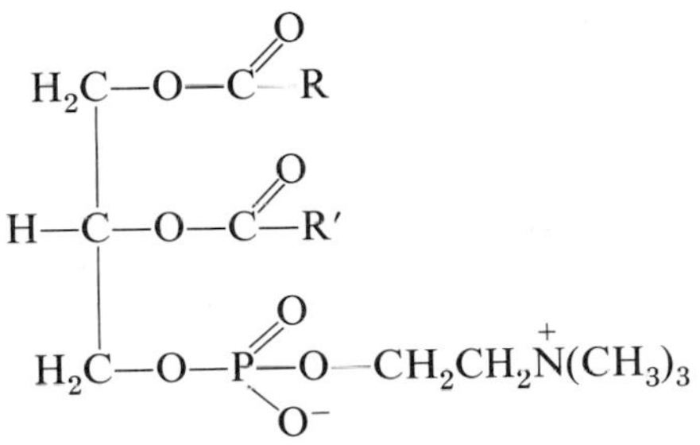

A lecithin

Other antioxidants employed to stabilize fats contain a modified phenolic structure, as does δ-tocopherol (above). Two of these, 3-*tert*-butyl-4-hydroxyanisole (BHA), and 2,6-di-*t*-butyl-4-methylphenol (BHT)[3], are used to stabilize cooking oils at the high temperatures required for the preparation of potato and corn chips and other foods. BHA is also used to stabilize lubricating oils, rubber, and gasoline—all substances containing carbon–carbon unsaturated bonds.

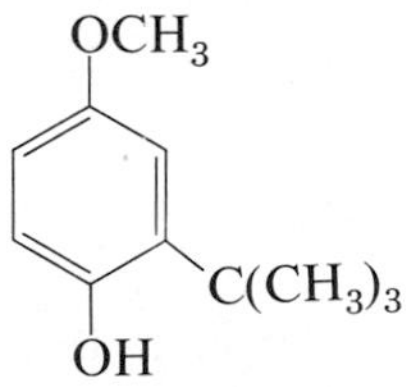

OCH_3, $C(CH_3)_3$, OH

3-*tert*-Butyl-4-hydroxyanisole

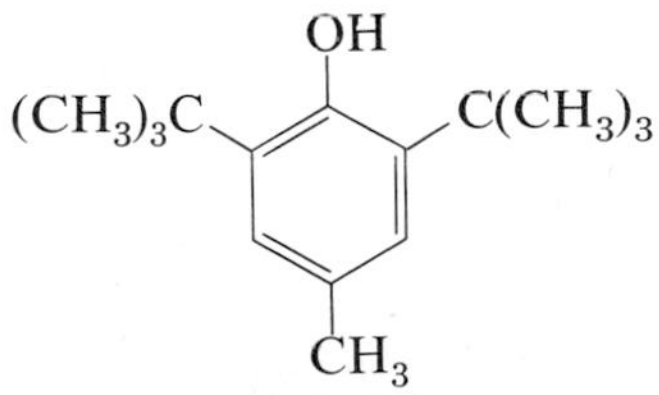

OH, $(CH_3)_3C$, $C(CH_3)_3$, CH_3

2,6-Di-*t*-butyl-4-methylphenol

Antioxidants effectively suppress rancidity when used in only minute amounts (0.01–0.001%).

[3]BHA is **B**utylated **H**ydroxy **A**nisole; BHT is **B**utylated **H**ydroxy **T**oluene.

Exercise 12.4 When a fat is overheated, the strongly lachrymatory vapor of acrolein, $CH_2{=}CH{-}CHO$, makes one's eyes smart. Acrolein is the dehydrated product of glycerol. Write a series of equations (including an enol intermediate) to show how glycerol is converted to acrolein.

12.6 Digestion and Metabolism of Fats

Enzymes called *lipases*[4] are active in the hydrolysis of fats. Gastric lipase, found in the stomach, catalyzes the hydrolysis of fats, but to a small extent. Fats are hydrolyzed mainly in the small intestine, where the environment is slightly alkaline. In this part of the digestive tract, the fat is first emulsified by the bile and then hydrolyzed by the action of pancreatic lipase (steapsin), an enzyme made in the pancreas. The hydrolysis products, glycerol and fatty acids, are absorbed through the wall of the intestinal tract, where recombination to form glycerides (fats) occurs. Fats are transported from the intestinal wall to various parts of the body to be stored as depot fat, used in the formation of protoplasm, or oxidized to supply energy.

The body can make both saturated and monounsaturated fatty acids by modifying dietary fats or synthesizing them from carbohydrates or proteins. However, certain polyunsaturated fatty acids, referred to as ''essential'' fatty acids, cannot be made and must be supplied in the diet.

Fats produce approximately 9.5 kcal of heat per gram when oxidized in the body to carbon dioxide and water. This number of calories is more than twice the energy obtained from the oxidation of an equivalent weight of either protein or carbohydrate.

Fats, once deposited at a storage site, do not remain there for long. The catabolism (degradation) and anabolism (synthesis) of fats represents a dynamic state. Most recent evidence indicates that the essential unit in the synthesis or in the degradation of a fat is the acetyl group,

$$CH_3{-}C({=}O){-}$$

The discovery of coenzyme A (A = acetylation) in 1947 by Fritz Lipmann (1899–1986) of the Rockefeller Institute showed that this substance plays a principal role in metabolism

[4]Enzymes usually are named by appending **ase** to their functions. Thus, an **oxidase** catalyzes chemical combination with oxygen; **dehydrogenase,** the removal of hydrogen. The fats and oils belong to a general classification of natural substances known as *lipids*. Thus, a **lipase** is an enzyme whose function is specific for a lipid, in this case a fat.

via 2-carbon units related to acetic acid. Coenzyme A is a complex molecule composed of four separately identifiable groups.

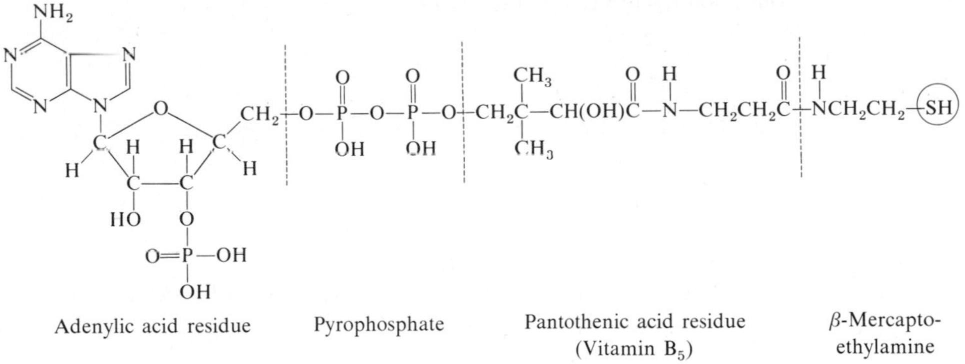

NH_2 (adenine) — ribose ($O=P-OH$, OH on C-3′) — $CH_2-O-\overset{O}{\overset{\|}{P}}(OH)-O-\overset{O}{\overset{\|}{P}}(OH)-O-CH_2C(CH_3)_2-CH(OH)\overset{O}{\overset{\|}{C}}-\overset{H}{N}-CH_2CH_2\overset{O}{\overset{\|}{C}}-\overset{H}{N}CH_2CH_2-SH$

Adenylic acid residue | Pyrophosphate | Pantothenic acid residue (Vitamin B_5) | β-Mercapto-ethylamine

Coenzyme A (CoA)

The terminal sulfhydryl (—SH) group is a very important part of the coenzyme A structure inasmuch as it is involved in transferring (donating or accepting) the acetyl group.

In addition to coenzyme A, two other species are required for the **biosynthesis** of fatty acids—acyl carrier protein (ACP) and an enzyme, β-ketoacyl-ACP-synthase. The structure of ACP resembles that of coenzyme A but differs in the replacement of the adenylic acid residue and one phosphate group by a polypeptide.

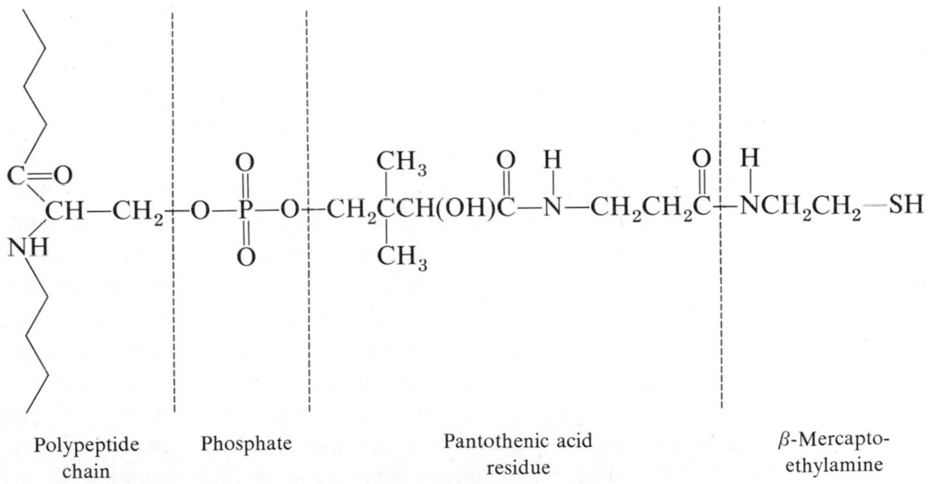

$C=O$, NH (polypeptide chain) $\rangle CH-CH_2-O-\overset{O}{\overset{\|}{\underset{O}{\underset{\|}{P}}}}-O-CH_2C(CH_3)_2CH(OH)\overset{O}{\overset{\|}{C}}-\overset{H}{N}-CH_2CH_2\overset{O}{\overset{\|}{C}}-\overset{H}{N}CH_2CH_2-SH$

Polypeptide chain | Phosphate | Pantothenic acid residue | β-Mercapto-ethylamine

Acyl carrier protein (ACP)

We will not attempt to portray the structure of the enzyme, β-ketoacyl-ACP-synthase, except to point out that it too has a terminal sulfhydryl group; therefore, we will represent its structure by *synthase*—SH. Both the biosynthetic and degradative pathways require acyl derivatives in which the hydrogen attached to the sulfur atom has been replaced by an acyl group in coenzyme A, ACP, and β-ketoacyl-ACP-synthase. These species are interconnected by the following equilibria.

$$2\,CH_3-\overset{\overset{\displaystyle O}{\|}}{C}-SCoA \rightleftharpoons CH_3-\overset{\overset{\displaystyle O}{\|}}{C}-CH_2-\overset{\overset{\displaystyle O}{\|}}{C}-SCoA + CoA-SH$$

Acetyl CoA — Acetoacetyl CoA — CoA

$$CH_3-\overset{\overset{\displaystyle O}{\|}}{C}-SCoA + ACP-SH \rightleftharpoons CH_3-\overset{\overset{\displaystyle O}{\|}}{C}-S-ACP + CoA-SH$$

Acetyl CoA — ACP — Acetyl ACP — CoA

$$CH_3-\overset{\overset{\displaystyle O}{\|}}{C}-S-ACP + \textit{synthase}-SH \rightleftharpoons$$

Acetyl ACP — β-Ketoacyl-ACP-*synthase*

$$CH_3-\overset{\overset{\displaystyle O}{\|}}{C}-S-\textit{synthase} + ACP-SH$$

Acetyl-S-*synthase* — ACP

All the acetyl or acetoacetyl derivatives are simply thioesters (sulfur analogs of esters) with a complex substituent in place of the usual alkyl group. The first equilibrium is the biochemical analog of the Claisen condensation, and the second two equilibria are **transesterification** reactions, that is, reactions in which the ''alkyl'' group of the thioester is exchanged for a different ''alkyl'' group.

In Figure 12.1 the presently accepted sequences of reactions involved in the degradation and biosynthesis of fatty acids are shown for the breakdown and the synthesis of butyric acid. For the *degradation* of a longer-chain fatty acid, we can substitute R for the methyl group in butyric acid and follow the cycle shown by the dotted line. In each pass through the cycle, two carbon atoms are lost from the right-hand end of the molecule and R becomes shorter by two carbon atoms. For the *synthesis* of a longer-chain acid, we follow the cycle shown by the dashed line, substituting the acyl-S-*synthase* at the top for the one at the bottom, and R becomes longer by two carbon atoms. Note the simplicity of the individual steps, reductions and oxidations (dehydrogenations), hydrations and dehydrations. Of course, the actual reactions are much more complex, involving a number of enzymes. If you go on to study biochemistry, you will find that the reactions are beautifully designed to minimize the activational energies for each step, to assure that the overall course of reaction is exothermic, to utilize available sources of energy, and to minimize interaction between synthetic and degradative processes.

FIGURE 12.1 Degradation and Biosynthesis of Fatty Acids

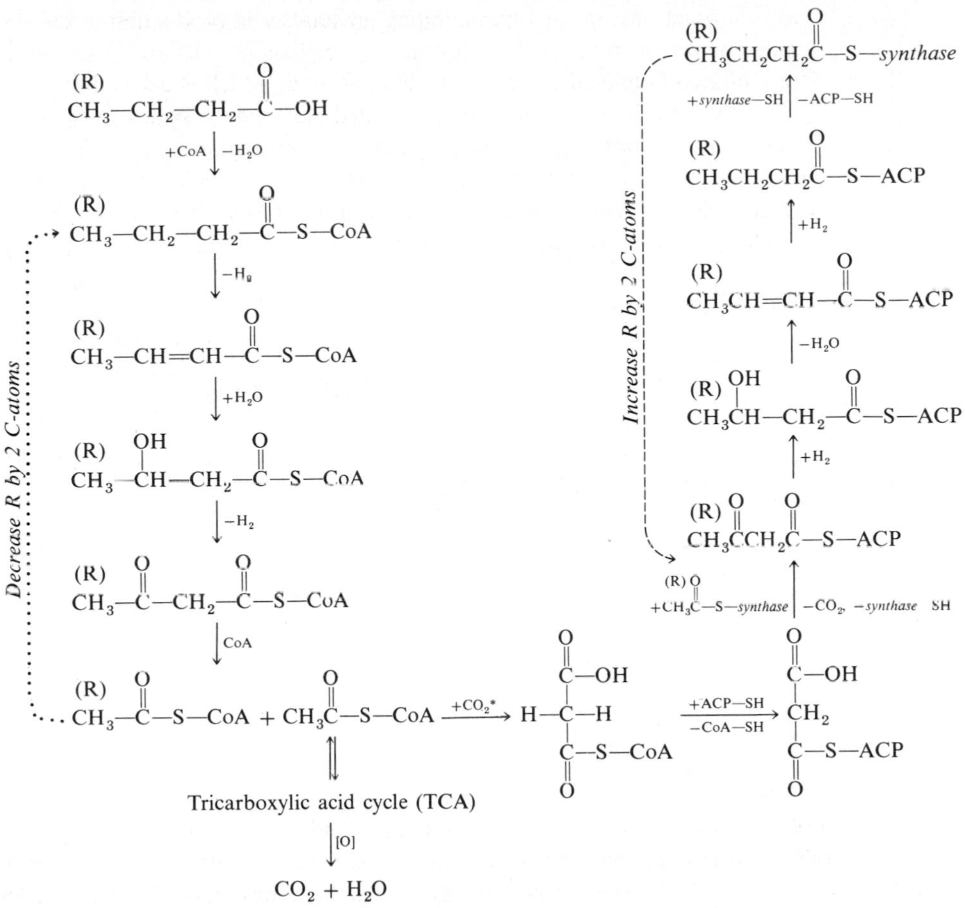

Decrease R by 2 C-atoms (degradation):

(R) $CH_3-CH_2-CH_2-C(=O)-OH$

↓ $+CoA$, $-H_2O$

(R) $CH_3-CH_2-CH_2-C(=O)-S-CoA$

↓ $-H_2$

(R) $CH_3-CH=CH-C(=O)-S-CoA$

↓ $+H_2O$

(R) $CH_3-CH(OH)-CH_2-C(=O)-S-CoA$

↓ $-H_2$

(R) $CH_3-C(=O)-CH_2-C(=O)-S-CoA$

↓ CoA

(R) $CH_3-C(=O)-S-CoA$ + $CH_3C(=O)-S-CoA$ $\xrightarrow{+CO_2^{*}}$ $H-C(H)(COOH)-C(=O)-S-CoA$ $\xrightarrow[-CoA-SH]{+ACP-SH}$ $HOOC-CH_2-C(=O)-S-ACP$

↕

Tricarboxylic acid cycle (TCA)

↓ [O]

$CO_2 + H_2O$

Increase R by 2 C-atoms (biosynthesis):

$HOOC-CH_2-C(=O)-S-ACP$

↑ +(R) $CH_3C(=O)-S-$*synthase*, $-CO_2$, $-$*synthase*$-SH$

(R) $CH_3C(=O)CH_2C(=O)-S-ACP$

↑ $+H_2$

(R) $CH_3CH(OH)-CH_2-C(=O)-S-ACP$

↑ $-H_2O$

(R) $CH_3CH=CH-C(=O)-S-ACP$

↑ $+H_2$

(R) $CH_3CH_2CH_2C(=O)-S-ACP$

↑ +*synthase*$-SH$, $-ACP-SH$

(R) $CH_3CH_2CH_2C(=O)-S-$*synthase*

*As bicarbonate ion, HCO_3^-. The product of this reaction is malonyl CoA, a malonic acid (Secs. 11.2 and 11.3) derivative.

12.7 Prostaglandins

The prostaglandins are a relatively large group of 20-carbon atom acids containing a 5-membered ring that have been isolated from a variety of animal tissues. At least 14 prostaglandins occur in human seminal plasma, and numerous others have been found in many types of cell fluids or synthesized in the laboratory during the past fifteen years. Although present in only trace amounts, the prostaglandins have a number of important biological roles of a hormonal or regulatory nature, such as the regulation of metabolism, stimulation of the contraction of smooth muscles, depression of blood pressure, inhibition of gastric secretion, and dilation of the bronchial vessels. It has even been suggested that aspirin functions, in part, by inhibiting prostaglandin production in the body. The prostaglandins are formed by the biosynthetic cyclization and oxidation of 20-carbon atom unsaturated fatty acids, such as arachidonic acid, which is derived in turn from the essential fatty acid, linoleic acid. An abbreviated biosynthetic pathway is shown in the following equations for the biological preparation of prostaglandin E_2, which along with prostaglandin $F_{2\alpha}$ has seen limited clinical use to induce labor or terminate pregnancy. As shown, oxygen functions as a diradical in this biosynthesis, and a resonance-stabilized allylic radical is formed in the first step by an enzyme-controlled oxidation.

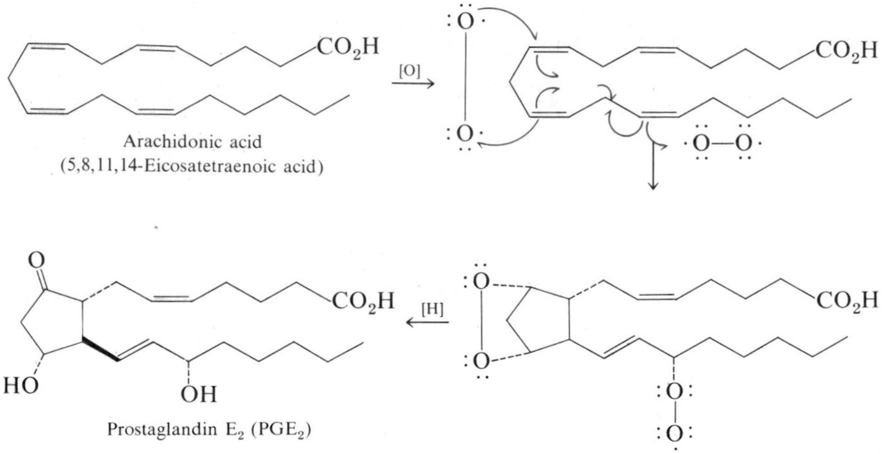

Arachidonic acid
(5,8,11,14-Eicosatetraenoic acid)

Prostaglandin E_2 (PGE_2)

The inhibitory effect of aspirin on prostaglandin production may be due to its antioxidant properties—properties that aspirin probably shares with the other phenolic antioxidants (Sec. 12.5). In some of the other prostaglandins, the carbonyl group is reduced to form the *cis*-diol, or the cyclopentanol hydroxyl group is eliminated by dehydration to form the unsaturated cyclopentenone derivative. The shorthand designations PGE_2 and $PGF_{2\alpha}$ refer to the hydroxy ketone and the cyclopentanediol series, respectively. The subscript number refers to the number of carbon–carbon double bonds present in the side chains, and a subscript α or β indicates the *cis* or *trans* configuration, respectively, of the diol.

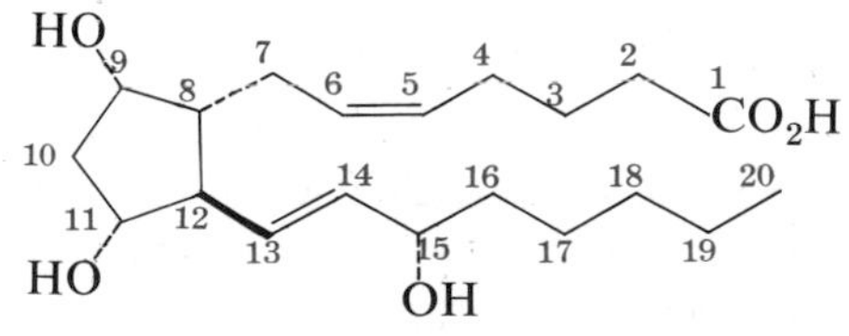

$PGF_{2\alpha}$

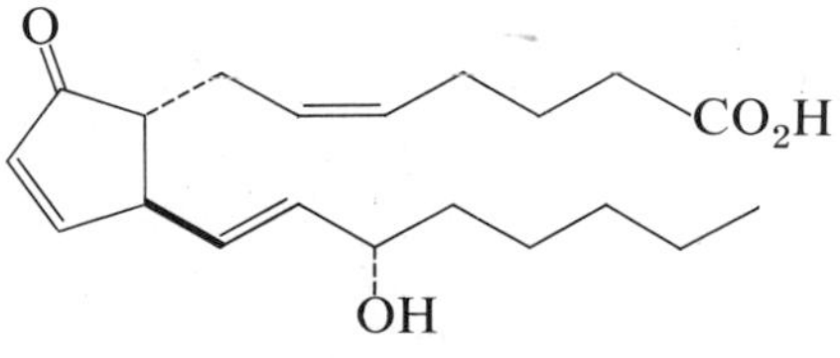

PGA_2

12.8 Phospholipids

The phospholipids are glycerides in which two of the hydroxyl groups are esterified to fatty acid residues and the third to a phosphoric acid moiety. These monophosphate esters are called phosphatidic acids. The fatty acid parts of the ester usually include one saturated and one unsaturated acid residue.

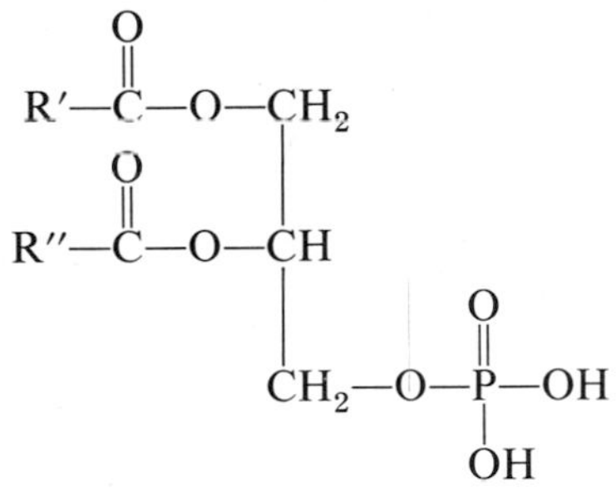

$$\begin{array}{l} R'-\overset{O}{\overset{\|}{C}}-O-CH_2 \\ R''-\overset{O}{\overset{\|}{C}}-O-CH \\ \qquad\qquad\quad CH_2-O-\underset{OH}{\underset{|}{\overset{O}{\overset{\|}{P}}}}-OH \end{array} \qquad R' = C_{17}H_{35} \quad R'' = C_{17}H_{33}$$

Phosphatidic acid

The free phosphatidic acid is seldom found in nature because the phosphoric acid portion still functions as an acid and usually is esterified to another alcohol.

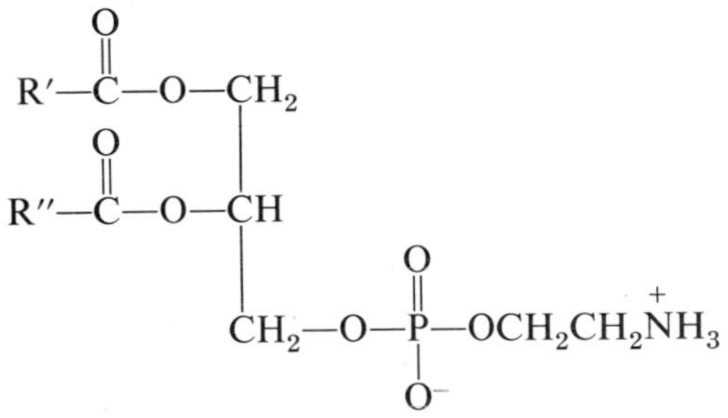

$$\begin{array}{l} R'-\overset{O}{\overset{\|}{C}}-O-CH_2 \\ R''-\overset{O}{\overset{\|}{C}}-O-CH \\ \qquad\qquad\quad CH_2-O-\underset{O^-}{\underset{|}{\overset{O}{\overset{\|}{P}}}}-OCH_2CH_2\overset{+}{N}H_3 \end{array}$$

Phosphatidyl ethanolamine

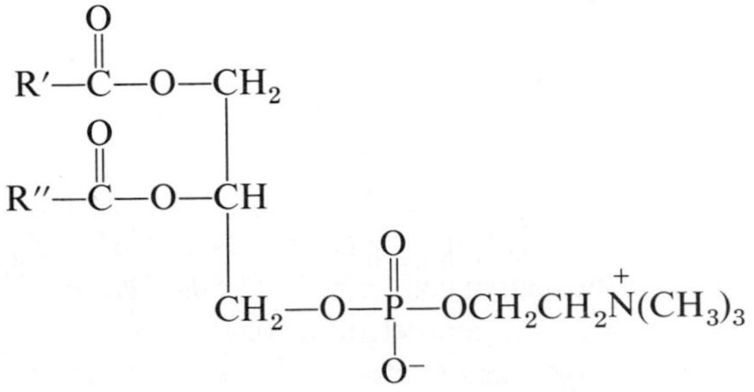

$$\begin{array}{l} R'-\overset{O}{\overset{\|}{C}}-O-CH_2 \\ R''-\overset{O}{\overset{\|}{C}}-O-CH \\ \qquad\qquad\quad CH_2-O-\underset{O^-}{\underset{|}{\overset{O}{\overset{\|}{P}}}}-OCH_2CH_2\overset{+}{N}(CH_3)_3 \end{array}$$

Phosphatidyl choline (a lecithin)

FIGURE 12.2 Cross-section of a phospholipid bilayer in aqueous medium

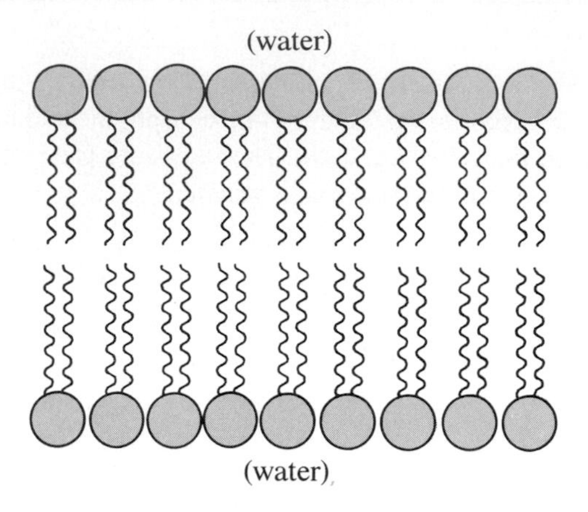

As can be seen from their structures, the ethanolamine and choline esters are salts and, not unlike some of the detergent and soap molecules (Sec. 9.6A), also have a long hydrocarbon nonpolar tail and a polar head. In contact with water the phosphatidic acid esters, like soaps, arrange themselves with the polar head seeking the aqueous environment and the nonpolar tails aligning themselves to form a *bilayer* (see Fig. 12.2). The bilayers of phospholipids make up the framework of cell membranes and play important roles in the chemistry of living systems.

Summary

1. Fats, oils, and waxes belong to a general classification of natural esters known as lipids.
2. Waxes are simple esters in which both the acid and alcohol components are long-chain structures.
3. Fats and oils are long-chain fatty acid esters of glycerol called glycerides. Glycerides may be simple (all fatty acids alike) or mixed (fatty acids different).
4. Fatty acids may be predominantly saturated (as in fats) or unsaturated (as in oils).
5. The reactions of fats and oils may be summarized as follows.
   (a) Alkaline hydrolysis (saponification) of a glyceride yields glycerol and soaps.
   (b) Hydrogenation converts oils to fats. Hydrogenation, if complete, converts a fat to glycerol and long-chain alcohols.
   (c) Hydrolysis and oxidation of *edible* oils and fats results in rancidity. Rancidity in fat and oil food products can be retarded by the addition of antioxidants.

6. The catabolism and anabolism of fats in the animal system appear to involve a sequence of 2-carbon units related to acetic acid.
7. Prostaglandins are cyclic derivatives of 20-carbon atom fatty acids that function as biological regulators.
8. Phospholipids are glycerides containing both fatty acid residues and a phosphoric acid moiety. Phospholipids make up the framework of living tissue.

Supplementary Exercises

12.5 Name the following compounds according to IUPAC rules.

(a) $C_{25}H_{51}{-}C({=}O){-}O{-}C_{12}H_{25}$

(b) $C_{15}H_{31}{-}C({=}O){-}O^{-}Na^{+}$

(c) 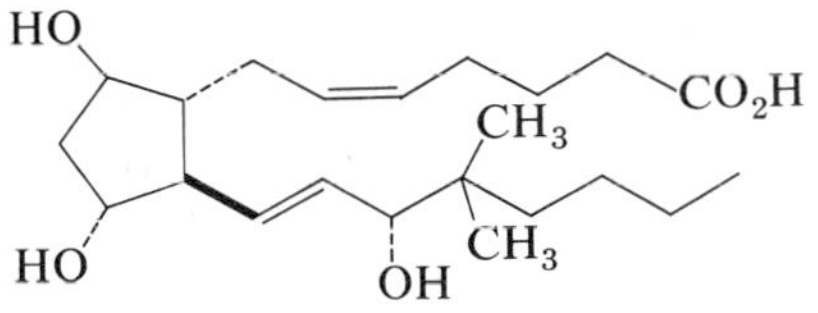

(d) 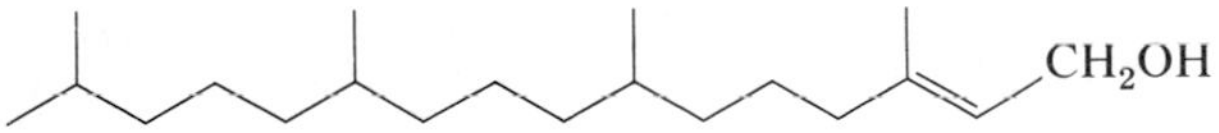

(*Note:* Ignore (R)— and (S)—.)

12.6 Draw the structures for the following:

(a) oleic acid
(b) glyceryl stearate
(c) a soap
(d) a prostaglandin
(e) hexadecyl hexadecanoate
(f) a phosphatidic acid
(g) acetyl CoA

12.7 Draw structures of glycerides that might possibly be those of individual molecules found in

(a) butter
(b) tallow
(c) linseed oil
(d) tung oil
(e) safflower oil
(f) margarine

(*Note:* Consult Table 12.1.)

12.8 By means of equations, illustrate chemical reactions that, when carried out with a fat or an oil, result in:

(a) formation of soap
(b) "hardening" of vegetable oil
(c) rancidity in butter

(*Note:* Use R, R′, R″ to indicate fatty acid residues.)

12.9 A grease spot made by hot butter on a linen napkin may be removed by laundering. If the spot had been made on a woolen flannel skirt or trousers, it should be removed by "dry" cleaning. Explain the mechanics involved in each case.

12.10 Why does soap not lather easily in most deep-well water?

12.11 Before the development of "vinyls," floor coverings were made of ground cork, burlap, and linseed oil. They were called "linoleums." Deduce the origin of the word.

12.12 A molecular diameter measurement of oleic acid gives a value approximately twice that of stearic acid. Explain.

12.13 Explain the difference between each of the following.

(a) hydrolysis and saponification
(b) a fat and an oil
(c) a fat and a wax
(d) oxidative and hydrolytic rancidity
(e) a phospholipid and a fat

12.14 How many pounds of sodium hydroxide would be required to convert 10 pounds of lard into soap? (Average M.W. of lard = 850.) How many pounds of glycerol would result as a by-product? (1 kg = 2.2 pounds.)

12.15 A $C_{20}H_{39}OCOCH_2$—CH_2— ester side chain is attached to a pyrroline nucleus (Table 4.3) in chlorophyll-a. The 20-carbon alcohol, *phytol,* has the structure shown in Exercise 12.5(d). What structural unit is repeated in the phytol molecule and to what general class of natural substances does phytol belong?

12.16 Referring to the previous question, how many optical isomers are possible for phytol? Draw the 7(*R*) and 11(*R*) configurations for phytol using the bold-line, dashed-line convention.

12.17 Why do most of the long-chain fatty acids contain an even number of carbon atoms?

12.18 The β-hydroxybutyric acid formed in the first reductive step of the biosynthesis of fatty acids has the D-configuration, while the β-hydroxybutyric acid formed in the hydration of crotonic acid in the degradation of fatty acids has the L-configuration. Draw the Fischer projections of these enantiomers.

12.19 When the prostaglandin PGE_2 is treated with sodium borohydride, two products result. The ir and nmr spectra of the products show that the carboxyl group and carbon–carbon double bonds have not been reduced. One of the products is $PGF_{2\alpha}$. What is the probable structure of the other product?

13 Amines and Other Nitrogen Compounds

The amines are our principal organic bases. Structurally, they are related to ammonia and may be considered as derivatives of this simple substance. Amines have the general formulas RNH_2, R_2NH, and R_3N in which one, two, or all three of the hydrogen atoms of the ammonia molecule have been replaced by alkyl or aryl groups. Many useful amines are synthetic, but some occur naturally in the decomposition products of nitrogenous substances such as the proteins. Some of our most valuable drugs obtained from plant extracts are characterized by the presence of one or more basic nitrogen atoms in their structures. On the basis of this finding they are assigned the general name ''alkaloids.'' Of course, not all alkaloids are useful drugs.

13.1 Classification and Nomenclature

Amines are classified as primary, secondary, or tertiary according to the number of hydrogen atoms of ammonia that have been replaced.

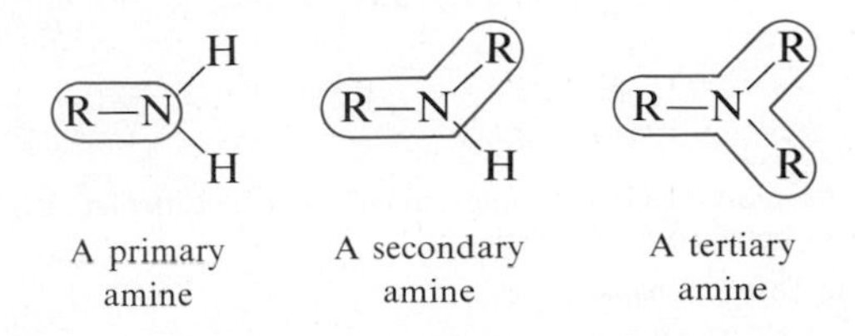

Aliphatic amines of low molecular weight and the aromatic amines are generally known by their common names. These are established simply by prefixing the names of the nitrogen-attached alkyl groups to the word **amine.** If the substituents are identical, the prefixes *di* and *tri* are employed. Examples of each class and their names are given below.

Primary amines:

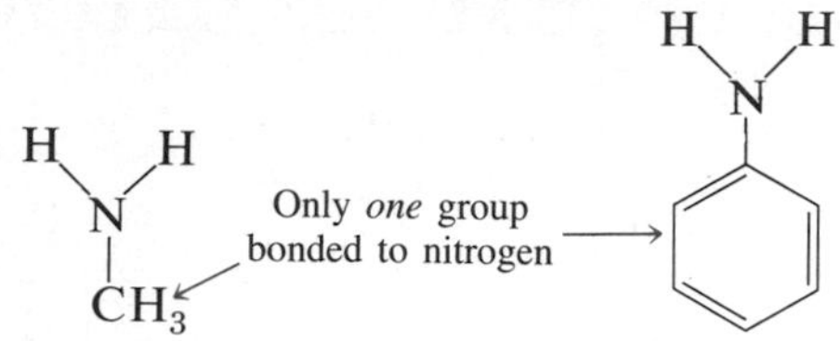

Only *one* group bonded to nitrogen

Methylamine (Methanamine)[2]

Aniline[1] (Benzenamine)[2]

Secondary amines:

Two groups bonded to nitrogen

Dimethylamine

Diphenylamine

Tertiary amines:

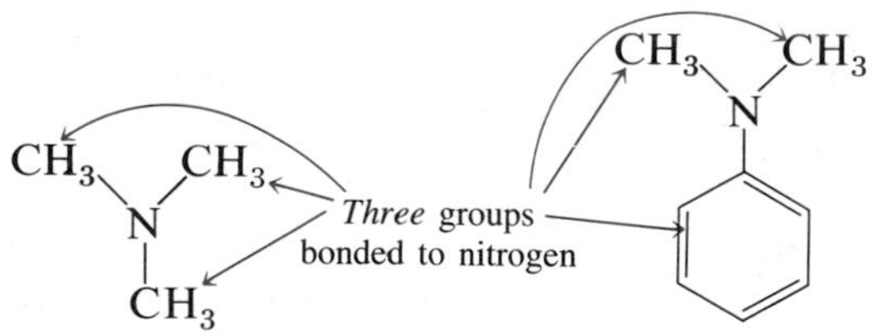

Three groups bonded to nitrogen

Trimethylamine

Dimethylaniline

The prefixes *sec* and *tert,* when part of the name of an amine, refer to the structure of an attached alkyl group, *not* to the class of the amine. For example, *tert*-butylamine and *sec*-butylamine are primary amines because in both amines the group is attached to the *primary* amino group, $—NH_2$.

$$CH_3—C(CH_3)_2—NH_2$$

tert-Butylamine (a primary amine)

$$CH_3—CH_2—CH(CH_3)—NH_2$$

sec-Butylamine (a primary amine)

[1]Aniline, the most important aromatic amine, is named after an early Spanish name for indigo (añil), from which it first was obtained by distillation.

[2]Chemical Abstracts name.

Frequently the amino group ($—NH_2$) is named simply as a substituent within another compound.

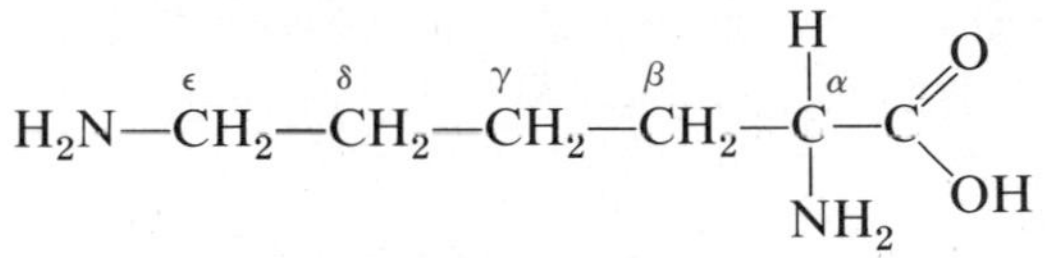

$$\overset{\epsilon}{H_2N}—\overset{\delta}{CH_2}—\overset{\gamma}{CH_2}—\overset{\beta}{CH_2}—CH_2—\overset{\overset{H}{|}\,\alpha}{\underset{\underset{NH_2}{|}}{C}}—C\begin{matrix}\diagup\!\!\!\!\!=O\\ \diagdown OH\end{matrix}$$

α,ϵ-Diaminocaproic acid (Lysine)
(2,6-Diaminohexanoic acid)

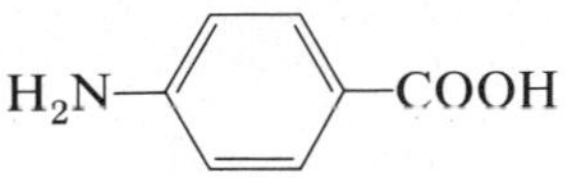

$$H_2N—C_6H_4—COOH$$

p-Aminobenzoic acid
(PABA)

$$H_2N—CH_2—CH_2—CH_2—CH_2—CH_2—CH_2—NH_2$$

1,6-Diaminohexane
(Hexamethylenediamine)

$$H_2N—CH_2—CH_2—OH$$

2-Aminoethanol
(Ethanolamine)

Exercise 13.1 How many different alcohols of each class are obtainable from $C_4H_{10}O$? How many different amines of each class are obtainable from $C_4H_{11}N$?

13.2 Properties of the Amines

The amines are basic compounds because only three of the five electrons in the valence shell of the nitrogen atom are used in covalent bonding. An amino nitrogen with two unshared electrons thus can function as an electron-pair donor (Lewis base). Structures deficient by an electron pair (Lewis acids) may combine with an amine and share these two electrons to produce salts.

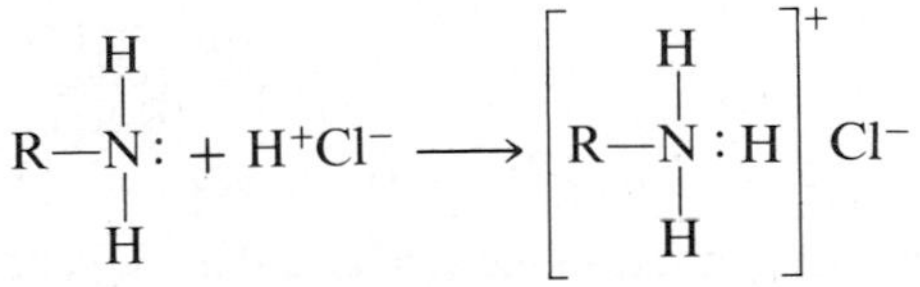

$$R—\overset{\overset{H}{|}}{\underset{\underset{H}{|}}{N}}: + H^+Cl^- \longrightarrow \left[R—\overset{\overset{H}{|}}{\underset{\underset{H}{|}}{N}}:H\right]^+ Cl^-$$

An alkyl
ammonium chloride
(a salt)

Amine salts, for the most part, are water-soluble solids. Their formation from mineral acids thus affords an easy method for the separation of amines from a mixture of water-insoluble compounds. Extraction of a mixture of organic substances with an aqueous acid solution removes the basic amines. Amines, when separated in this manner, may be liberated as free bases from the acid solution by making it strongly alkaline.

$$RNH_3^+Cl^- + Na^+OH^- \longrightarrow RNH_2 + NaCl + H_2O$$

The solution of ammonia in water produces ammonium hydroxide, i.e., ammonium and hydroxide ions. Amines, similarly, combine reversibly with water to produce basic, substituted ammonium hydroxides.

$$NH_3 + H_2O \rightleftharpoons NH_4^+ + OH^-$$

$$K_b = \frac{[NH_4^+][OH^-]}{[NH_3]} = 1.8 \times 10^{-5}$$

$$R{-}\ddot{N}H_2 + H{-}\ddot{\underset{..}{O}}{-}H \rightleftharpoons \left[R{-}NH_3 \right]^+ \ :\ddot{\underset{..}{O}}H^-$$

$$CH_3NH_2 + H_2O \rightleftharpoons \underset{\text{Methylammonium hydroxide}}{CH_3NH_3^+OH^-}$$

$$K_b = \frac{[CH_3NH_3^+][OH^-]}{[CH_3NH_2]} = 5 \times 10^{-4}$$

It is now common practice, particularly in biochemistry, to compare the *acid* strengths of the ammonium ions, which are the conjugate acids of the amines (Sec. 1.13), rather than to compare the *base* strengths of the amines.

$$\underset{\text{Ammonium ion}}{R{-}\overset{+}{N}H_3} + :\ddot{O}H_2 \overset{K_a}{\rightleftharpoons} \underset{\text{Amine}}{R{-}\ddot{N}H_2} + H{-}\overset{..+}{O}{-}H_2$$

$$K_a = \frac{[RNH_2][H_3O^+]}{[RNH_3^+]}$$

The K_b and pK_b of the amine RNH_2 and the K_a and pK_a of its conjugate acid RNH_3^+ are related as shown in the following equations.

$$K_a = \frac{10^{-14}}{K_b} \quad \text{(at 25°C)}$$

$$pK_a = 14 - pK_b$$

These equations lead to the following comparisons: the *stronger* the base, the *larger* the value of K_b, the smaller the value of pK_b, and the *larger* the value of pK_a of the ammonium ion. (See Table 13.1.)

TABLE 13.1 Physical Properties of Some Amines

Name	Formula	bp (mp)°C	pK_a*	pK_b
Ammonia	NH_3	−33.4	9.24	4.76
Methylamine	CH_3NH_2	−6.5	10.62	3.38
Dimethylamine	$(CH_3)_2NH$	7.4	10.77	3.23
Trimethylamine	$(CH_3)_3N$	3.5	9.80	4.20
Ethylamine	$C_2H_5NH_2$	16.6	10.63	3.37
n-Propylamine	$CH_3CH_2CH_2NH_2$	48.7		
Isopropylamine	$(CH_3)_2CHNH_2$	34	10.53	3.47
n-Butylamine	$CH_3CH_2CH_2CH_2NH_2$	77	10.61	3.39
tert-Butylamine	$(CH_3)_3CNH_2$	43.8	10.45	3.55
Aniline	$C_6H_5NH_2$	184	4.60	9.40
Dimethylaniline	$C_6H_5N(CH_3)_2$	193.5	5.21	8.79
p-Toluidine	*p*-$CH_3C_6H_4NH_2$	(43.7)	5.09	8.91
o-Nitroaniline	*o*-$O_2NC_6H_4NH_2$	(71.5)	0.99	13.01

*pK_a for conjugate acid of amine. All values at 25°C.

Exercise 13.2 K_b for *p*-chloroaniline is 1×10^{-10}, for *m*-chloroaniline K_b is 0.3×10^{-10}, and for *o*-chloroaniline K_b is 0.05×10^{-10}.

(a) Which of these three substituted anilines is the strongest base?
(b) Convert each K_b into a pK_b value.

Do not confuse the pK_a of the ammonium ion with the pK_a of the amine itself. For example, ammonia is a very weak acid whose pK_a is about 35 for the dissociation to form amide ion, $NH_2^{:-}$. Amide ion is a powerful base, whose pK_b we can estimate from the equations to be about −21 (corresponding to a K_b of 10^{21}!). Because the amide ions derived from the amines are powerful bases, they are useful reagents in organic chemistry.

A comparison of the pK_b values reveals that methylamine, CH_3NH_2, is a stronger base than ammonia, NH_3. This increased basicity is explained largely in terms of an inductive effect. An alkyl substituent will tend to release or donate electrons to the more electronegative nitrogen atom. This inductive effect can be considered to make the unshared electron pair of the amine nitrogen more readily available for bonding to a proton, or it can be considered to stabilize the ammonium ion by somewhat reducing the net positive charge on the nitrogen atom. However, the inductive effect is not the only factor determining basicity, at least in aqueous solution. Hydrogen bonding to the water molecules may help to stabilize the ammonium ion. Thus, dimethylamine is a stronger base than methylamine, but trimethylamine is a weaker base than methylamine despite three alkyl substituents. This may be the result of the existence of only one hydrogen in the trimethylammonium ion to form hydrogen bonds to water, or the result of the methyl groups inhibiting solvation of the ion by keeping the solvent away from the positively charged nitrogen.

Aromatic amines tend to be weaker bases than the aliphatic amines or ammonia. One explanation of the difference is based on the comparison of the resonance forms for the amine and its ammonium ion, as shown for aniline.

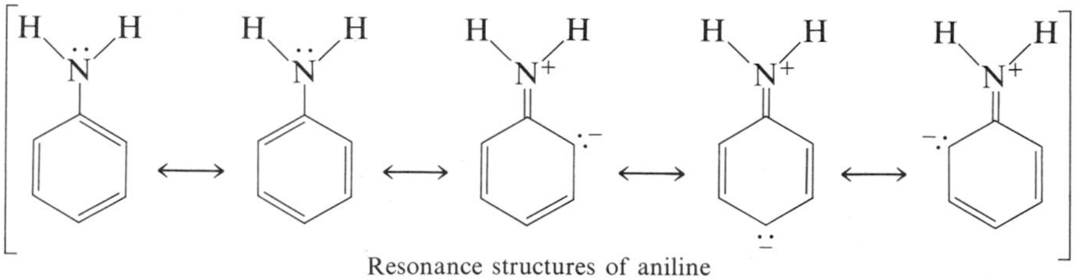

Resonance structures of aniline

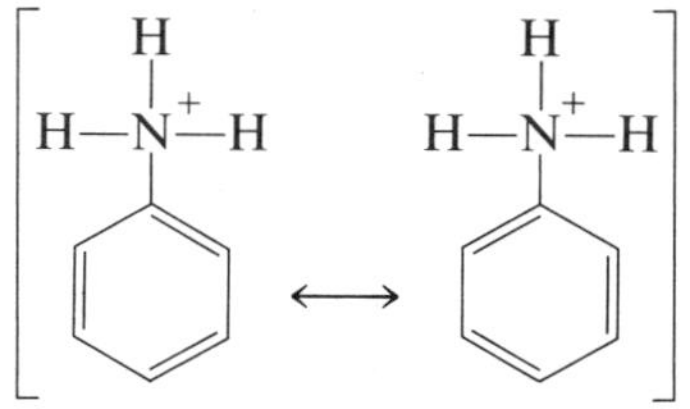

Resonance structures of anilinium ion

Aniline is highly stabilized by resonance interaction of the amino group with the ring; however, when the amino group is protonated, the resultant ammonium ion loses the extra resonance stabilization, making the formation of the ion less favorable. An alternative explanation would be that the resonance interaction shown for the aniline molecule makes the electron pair less available for donation to a base.

Exercise 13.3 Which ring substituents, generally, will increase the basicity of aniline? Which substituents, generally, will make aniline less basic?

Exercise 13.4 The same substituents that increase or diminish the basicity of aniline have the exact opposite effect on the acidity of benzoic acid. Explain.

The lower molecular weight members of the amine family are water-soluble gases with an ammoniacal or fishlike odor. Those containing three to eleven carbons are liquids, and higher homologs are solids. Although the odors of most amines are unpleasant, their ammonium salts are odorless.

13.3 Preparation of Amines

The amines may be prepared by a number of methods. The principal laboratory methods of preparation follow.

A. Alkylation of Ammonia. Alkyl groups may be introduced directly into the ammonia molecule by reaction with alkyl halides. The first product formed is an ammonium salt.

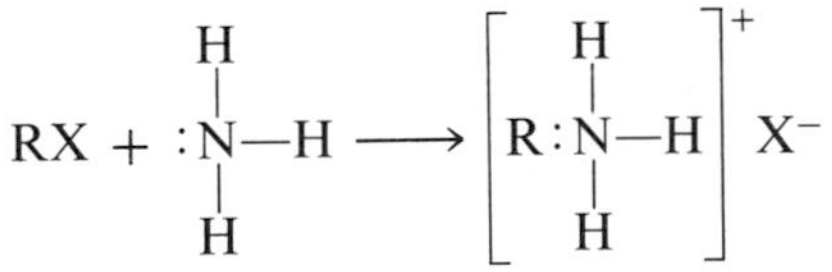

$$RX + :\overset{\displaystyle H}{\underset{\displaystyle H}{N}}-H \longrightarrow \left[R:\overset{\displaystyle H}{\underset{\displaystyle H}{N}}-H\right]^+ X^-$$

Alkylammonium halide

Subsequent treatment of the ammonium salt with a stronger base (NaOH) liberates the free, primary amine.

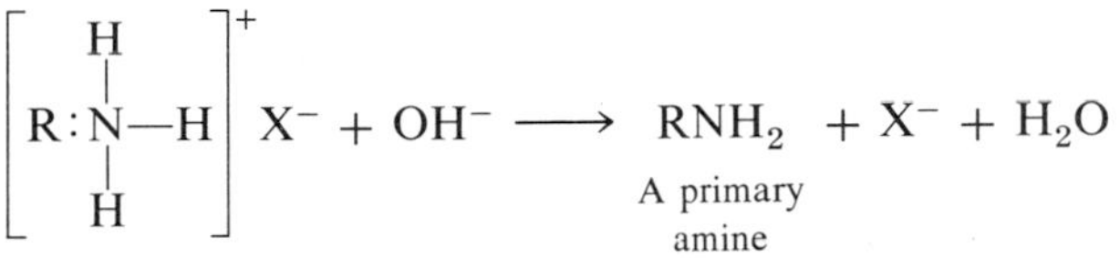

$$\left[R:\overset{\displaystyle H}{\underset{\displaystyle H}{N}}-H\right]^+ X^- + OH^- \longrightarrow \underset{\text{A primary amine}}{RNH_2} + X^- + H_2O$$

The reaction, unfortunately, does not stop at the first stage as illustrated above, but continues until replacement of hydrogen by alkyl groups yields not only the primary amine but the secondary and the tertiary as well. The tertiary amine, with no hydrogen remaining, then may react with a fourth molecule of alkyl halide to produce a quaternary ammonium salt. The following series of reactions illustrates the progressive alkylation of ammonia to produce all of the above products.

$$\begin{aligned}
RNH_2 + RX &\longrightarrow R_2NH_2{}^+X^- \\
R_2NH_2{}^+X^- + NaOH &\longrightarrow R_2NH + NaX + H_2O \\
R_2NH + RX &\longrightarrow R_3NH^+X^- \\
R_3NH^+X^- + NaOH &\longrightarrow R_3N + NaX + H_2O \\
R_3N + RX &\longrightarrow R_4N^+X^-
\end{aligned}$$

A quaternary ammonium salt

B. Reduction of Unsaturated Nitrogen Compounds. Certain organic compounds that already contain nitrogen may be converted to primary amines by reduction methods. Most frequently used starting materials are oximes, nitriles, amides, and nitro compounds. The reduction can be accomplished with lithium aluminum hydride, which will reduce any of the functional groups shown. The reactions below illustrate how the same primary amine may be obtained from four different starting materials.

$$\underset{\text{An aldoxime}}{\text{RCH}{=}\text{N}{-}\text{OH}} \xrightarrow{4[H]} R{-}CH_2NH_2 \xleftarrow{4[H]} \underset{\text{An amide}}{R{-}C({=}O){-}NH_2}$$

$$\underset{\text{A nitrile}}{R{-}C{\equiv}N} \xrightarrow{4[H]} R{-}CH_2NH_2 \xleftarrow{6[H]} \underset{\text{A nitroalkane}}{R{-}CH_2{-}NO_2}$$

Nitroalkanes may be prepared by the reaction of many primary and secondary alkyl bromides or iodides with sodium nitrite in dimethylformamide solution.

Aniline, by far the most important aromatic amine, is prepared by the reduction of nitrobenzene. Industrially, the reaction is carried out with iron and steam. In the laboratory, tin and hydrochloric acid or iron and acetic acid are usually used.

$$\underset{\text{Nitrobenzene}}{C_6H_5NO_2} + 2\,Fe + 4\,H_2O \longrightarrow \underset{\text{Aniline}}{C_6H_5NH_2} + 2\,Fe(OH)_3$$

$$2\,C_6H_5NO_2 + 3\,Sn + 14\,HCl \longrightarrow 2\,\underset{\text{Anilinium chloride}}{C_6H_5NH_3^+Cl^-} + 3\,SnCl_4 + 4\,H_2O$$

The acid salt of aniline, when treated with sodium hydroxide, liberates the free amine.

C. Special Methods for the Preparation of Primary Amines. The German chemist Hofmann[3] discovered that primary amines could be prepared by treating an amide with sodium hypobromite. The reaction involves a rearrangement in which the alkyl or aryl group attached to the carbonyl carbon migrates to the nitrogen atom. Inasmuch as the carbonyl carbon is eliminated as carbon dioxide, the carbon chain of the amide is degraded to produce a primary amine with one carbon atom less than the original amide. The reaction is generally known as the **Hofmann Amide Hypohalite Degradation.** The reaction is believed to proceed through the following steps.

[3]August Wilhelm von Hofmann (1818–1895) was one of the most celebrated chemists of his time. He was noted particularly for his research in the chemistry of the amines. He was professor and the first director of the Royal College of Chemistry at London (1845–1864), professor and director of the laboratory, University of Berlin (1864–1895), and founder of the German Chemical Society (1868).

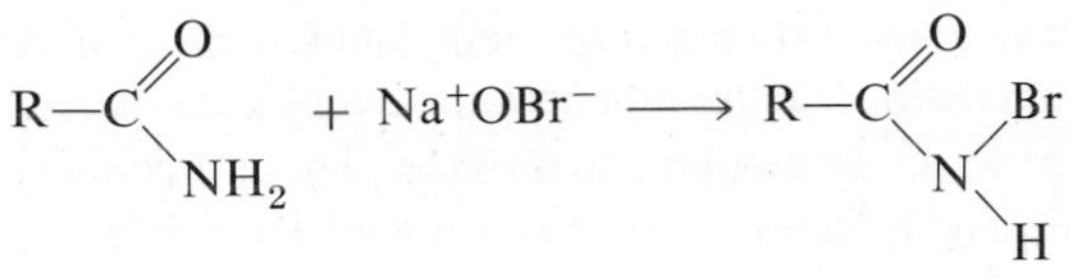

$$R{-}C({=}O){-}NH_2 + Na^+OBr^- \longrightarrow R{-}C({=}O){-}N(H){-}Br$$

An amide — Sodium hypobromite — An N-bromoamide

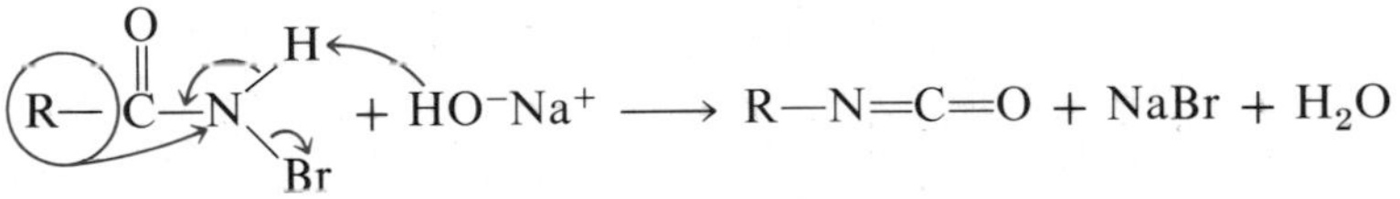

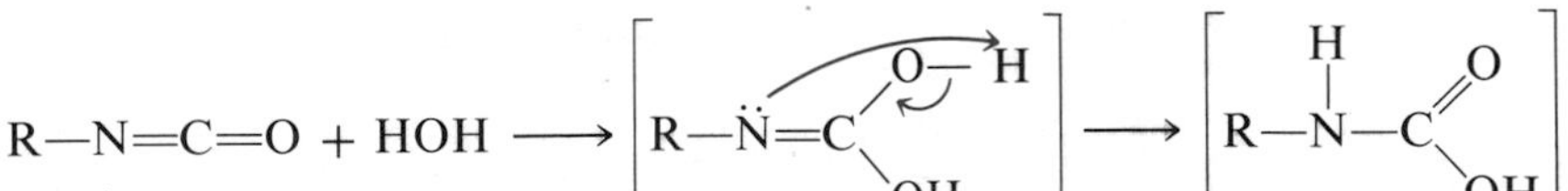

$$R{-}C({=}O){-}N(H)(Br) + HO^-Na^+ \longrightarrow R{-}N{=}C{=}O + NaBr + H_2O$$

$$R{-}N{=}C{=}O + HOH \longrightarrow \left[R{-}\ddot{N}{=}C(OH){-}O{-}H\right] \longrightarrow \left[R{-}N(H){-}C({=}O){-}OH\right]$$

An isocyanate

$$\left[R{-}N(H){-}C({=}O){-}O{-}H\right] \xrightarrow{\text{heat}} R{-}NH_2 + CO_2$$

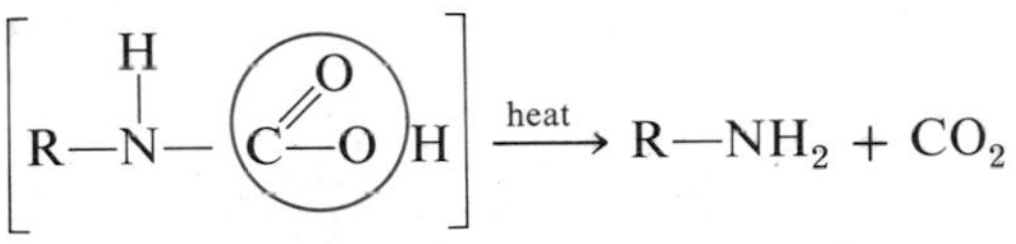

An alkyl-substituted carbamic acid (unstable) — A primary amine

---

Exercise 13.5 Which of the preparative methods outlined in Sec. 13.3 should be employed to accomplish the following conversions and to obtain the products indicated in a high state of purity, i.e., uncontaminated by traces of secondary and tertiary amines.

(a) $CH_2{=}CH{-}CH_2Cl \longrightarrow CH_2{=}CH{-}CH_2CH_2NH_2$
(b) $CH_3CH_2CH_2CH_2COOH \longrightarrow CH_3CH_2CH_2CH_2NH_2$

What reaction mechanism does each of the above reactions illustrate?

13.4 Reactions of Amines

A. Salt Formation. As basic compounds, amines will react with acids to form water-soluble salts. This is a characteristic property of the amines, the usefulness of which has already been pointed out (Sec. 13.2).

B. Alkylation. Primary amines, like ammonia, can be further alkylated with alkyl halides to give secondary and tertiary amines and quaternary ammonium salts. Hofmann discovered that a primary amine, when exhaustively methylated with methyl iodide, produced a quaternary ammonium iodide. On treatment with silver oxide, the substituted ammonium iodide was converted to an ammonium hydroxide. The ammonium base, when heated, decomposed to yield trimethylamine and an olefin with a terminal double bond. The olefin thus had its origin in the alkyl group of the original amine. From the structures of the decomposition products, it was possible to deduce the formula of the original amine. The steps in the reaction are illustrated using ethylamine.

$$\underset{\text{Ethylamine}}{C_2H_5NH_2} + 3\,CH_3I \longrightarrow \underset{\text{Trimethylethylammonium iodide}}{C_2H_5{-}\overset{CH_3}{\underset{CH_3}{\overset{|}{\underset{|}{N^+}}}}{-}CH_3\;I^-} + 2\,HI$$

$$2\,C_2H_5\overset{+}{N}(CH_3)_3\,I^- + Ag_2O + H_2O \longrightarrow \underset{\text{Trimethylethylammonium hydroxide}}{2\,C_2H_5\overset{+}{N}(CH_3)_3\,OH^-} + 2\,AgI$$

$$C_2H_5{-}\overset{CH_3}{\underset{CH_3}{\overset{|}{\underset{|}{N^+}}}}{-}CH_3\;OH^- \xrightarrow{\text{heat}} \underset{\text{Ethylene}}{CH_2{=}CH_2} + \underset{\text{Trimethylamine}}{(CH_3)_3N} + H_2O$$

Exercise 13.6 Suggest a mechanism for the Hofmann elimination. (*Hint:* See Sec. 6.5.) Why does this elimination not take place with simple amines plus sodium hydroxide? (*Hint:* Compare the leaving groups and see Sec. 7.7B.)

C. Acylation.

The acyl group,

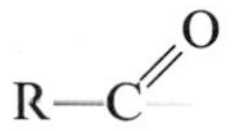

$$R{-}C{\overset{\displaystyle O}{\diagup\!\!\diagup}}$$

may be substituted for hydrogen in both primary and secondary amines by reaction with acid anhydrides or acid chlorides. The use of either reagent produces an N-substituted amide. The reaction is illustrated below with ethylamine.

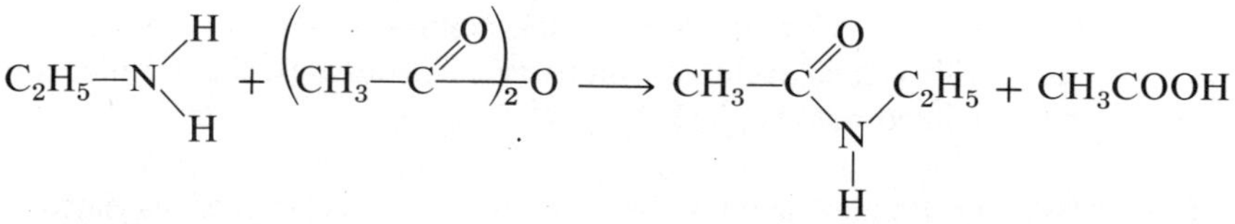

$$\underset{\text{Ethylamine}}{C_2H_5{-}N\begin{matrix}H\\H\end{matrix}} + \underset{\text{Acetic anhydride}}{\left(CH_3{-}C{\overset{O}{\diagup\!\!\diagup}}\right)_2O} \longrightarrow \underset{\text{N-Ethylacetamide}}{CH_3{-}\overset{\overset{O}{\|}}{C}{-}\underset{\underset{H}{|}}{N}{-}C_2H_5} + CH_3COOH$$

The acylation of aniline produces an **anilide.**

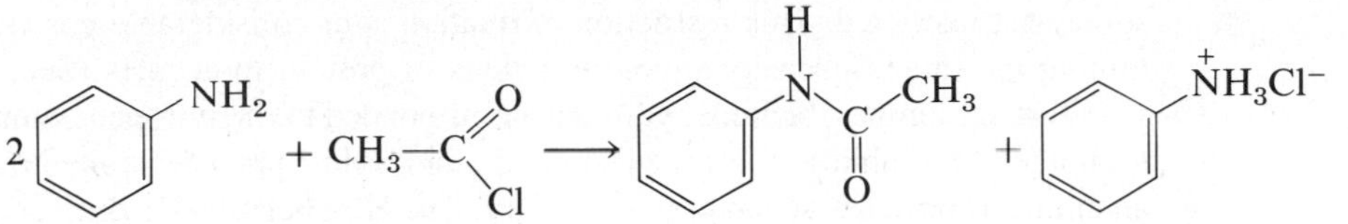

$$2\,C_6H_5NH_2 + CH_3COCl \longrightarrow C_6H_5NHCOCH_3 + C_6H_5\overset{+}{N}H_3Cl^-$$

Aniline Acetyl chloride Acetanilide (mp 114°C) Aniline hydrochloride

The mechanism for the acylation of amines is exactly the same as that illustrated in Section 10.2B.

Although acetanilide is an antipyretic, it is also quite toxic. However, other acetanilide derivatives appear to be both safe and effective. A popular example is acetaminophen ("Tylenol"), which is used as an aspirin substitute.

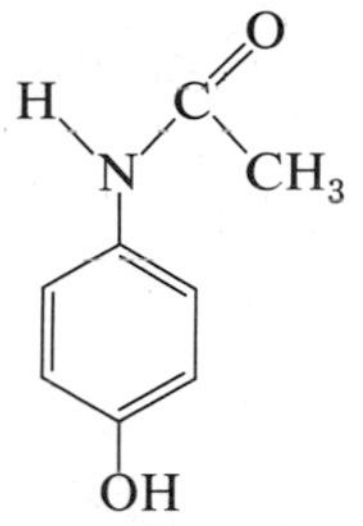

$$HO-C_6H_4-NH-C(=O)-CH_3$$

Acetaminophen

Ammonia and the amines have pyramidal structures, but the amides are planar. The planarity can be attributed to the effects of resonance stabilization of the amide group as shown in the following structures.

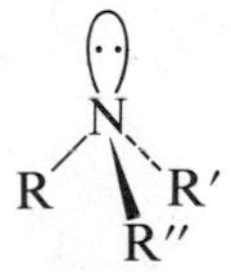

$$R\ddot{N}R'R''$$

Pyramidal nitrogen in amines

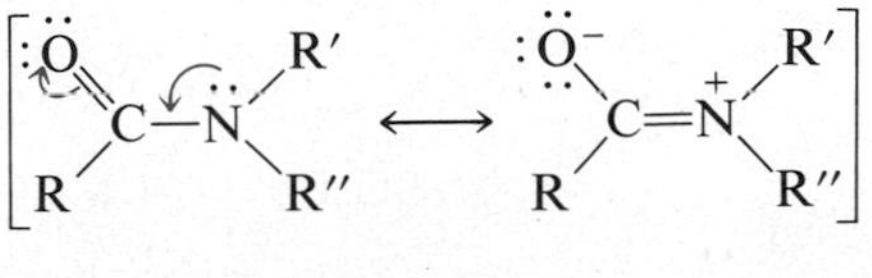

$$\left[R-C(=\ddot{O}:)-\ddot{N}R'R'' \longleftrightarrow R-C(-\ddot{O}:^-)=\overset{+}{N}R'R'' \right]$$

Planar nitrogen in amides

Because the unshared pair of electrons on nitrogen is delocalized by virtue of the resonance interaction, amides are much less basic than amines. Furthermore, there is some double-bond character in the bond between the carbonyl carbon atom and the nitrogen atom. As a result, rotation around the carbon–nitrogen bond of the amide function is somewhat restricted. This restriction of rotation is of considerable consequence in determining the three-dimensional conformations of protein molecules (Sec. 15.7).

Tertiary amines, lacking hydrogen atoms bonded to the nitrogen, cannot be converted to amides. The ability to form substituted amides thus provides a method for classifying an amine as primary, secondary, or tertiary. The **Hinsberg test** is one such method. In the Hinsberg test benzenesulfonyl chloride, $C_6H_5SO_2Cl$, is the acid chloride used. Reaction of a primary amine with benzenesulfonyl chloride yields an N-alkyl sulfonamide that is soluble in an alkaline solution. The hydrogen atom that remains attached to the nitrogen atom of the amide is acidic owing to the electron-withdrawing power of the sulfonyl group. The amide, therefore, is soluble in an alkaline solution.

Replaceable hydrogen

$C_6H_5SO_2Cl + H_2N{-}R \longrightarrow C_6H_5SO_2{-}NH{-}R + HCl$

Benzenesulfonyl chloride

(Soluble in base)

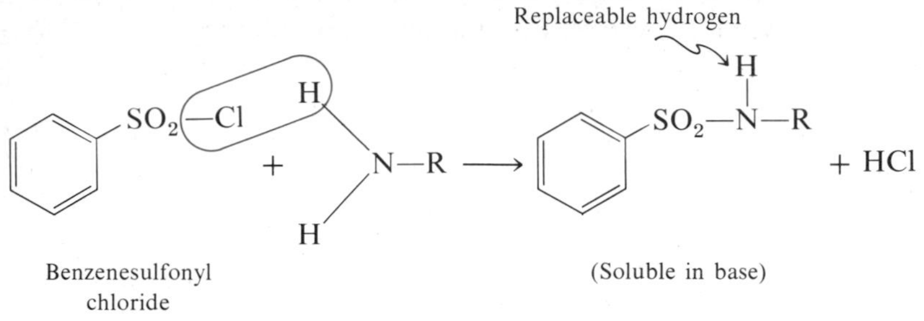

A secondary amine, while reactive with benzenesulfonyl chloride, lacks the acidic hydrogen and forms an *alkali-insoluble* product.

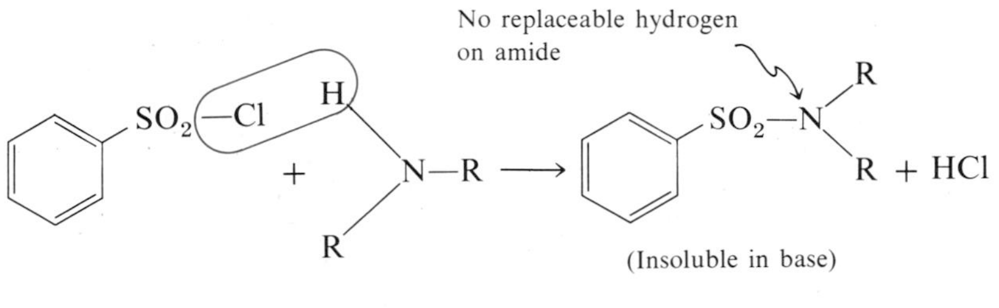

No replaceable hydrogen on amide

$C_6H_5SO_2Cl + HNR_2 \longrightarrow C_6H_5SO_2{-}NR_2 + HCl$

(Insoluble in base)

Tertiary amines fail to react with benzenesulfonyl chloride for the same reason they fail to give acyl derivatives.

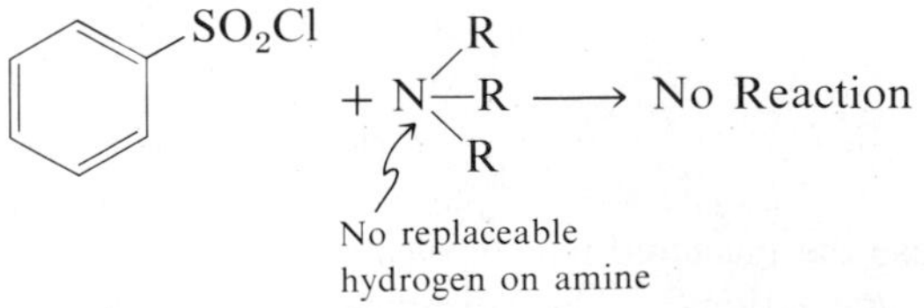

$C_6H_5SO_2Cl + NR_3 \longrightarrow$ No Reaction

No replaceable hydrogen on amine

5-**D**imethyl**a**mino-1-**n**aphthalene**s**ulfon**yl** chloride, called dansyl chloride, reacts with amino acids to give highly fluorescent dansyl derivatives and is used in determining the sequence of amino acids in polypeptides (Sec. 15.8).

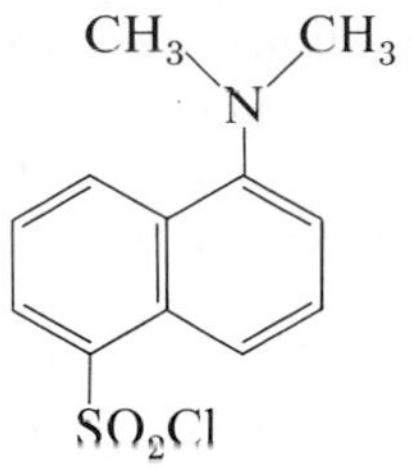

CH_3 CH_3 N SO_2Cl

Dansyl chloride

D. Reaction with Nitrous Acid. Nitrous acid is an unstable substance prepared in solution only when needed, by reaction between a mineral acid and sodium nitrite.

$$Na^+NO_2^- + H^+Cl^- \longrightarrow Na^+Cl^- + HNO_2$$

Nitrous acid

Although the actual mechanism is somewhat more complex, in acid solution nitrous acid can be considered to be in equilibrium with nitrosonium ion, a weakly electrophilic reagent. Nitrosonium ion reacts with the nucleophilic amine to give a series of intermediates formed by successive proton transfers. The final product of the reaction depends upon the class of the amine.

$$H^+ + HO{-}N{=}O \rightleftharpoons H_2O + {}^+N{=}O$$

Nitrosonium ion

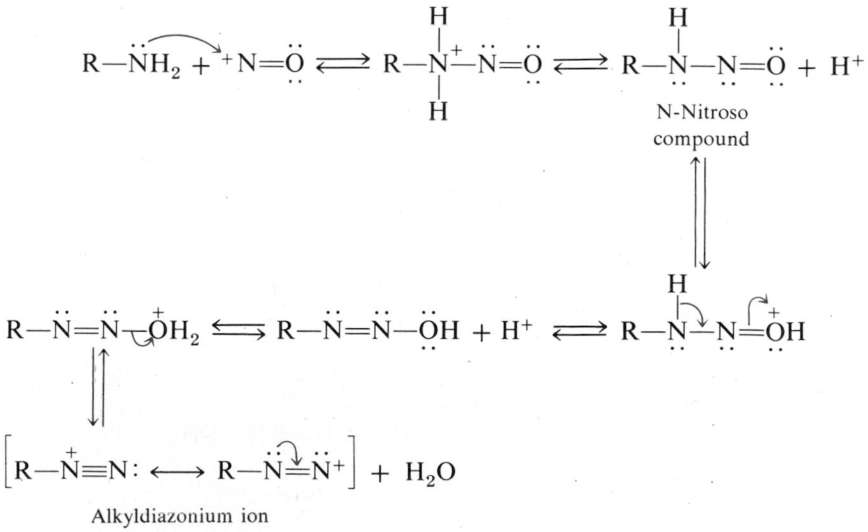

$$R{-}\ddot{N}H_2 + {}^+N{=}\ddot{O} \rightleftharpoons R{-}\overset{H}{\underset{H}{\overset{|+}{N}}}{-}\ddot{N}{=}\ddot{O} \rightleftharpoons R{-}\overset{H}{\overset{|}{\ddot{N}}}{-}\ddot{N}{=}\ddot{O} + H^+$$

N-Nitroso compound

$$\updownarrow$$

$$R{-}\ddot{N}{=}\ddot{N}{-}\overset{+}{O}H_2 \rightleftharpoons R{-}\ddot{N}{=}\ddot{N}{-}\ddot{O}H + H^+ \rightleftharpoons R{-}\overset{H}{\overset{|}{\ddot{N}}}{-}\ddot{N}{=}\overset{+}{O}H$$

$$\updownarrow$$

$$\left[R{-}\overset{+}{N}{\equiv}N: \longleftrightarrow R{-}\ddot{N}{=}\ddot{N}^+\right] + H_2O$$

Alkyldiazonium ion

With primary amines the reaction proceeds to the formation of the diazonium ion. Although such ions are moderately stable in cold, aqueous solution when they are derived from aromatic amines, diazonium ions formed from aliphatic amines are very unstable and decompose to form nitrogen and a carbocation. The carbocation then may react in several different ways. It may combine with a nucleophile, eliminate a proton to form an alkene, or rearrange to a more stable carbocation before following either of the first two pathways (Sec. 7.7B). The routes leading to these different products are indicated by the following series of reactions.

$$\underset{n\text{-Butylamine hydrochloride}}{CH_3CH_2CH_2CH_2{-}NH_2 \cdot HCl} + NaNO_2 \longrightarrow$$

$$\underset{\text{A diazonium ion}}{CH_3CH_2CH_2CH_2{-}\overset{\oplus}{N}{\equiv}N{:}} + H_2O + OH^- + NaCl$$

$$CH_3CH_2CH_2CH_2\overset{\oplus}{N}{\equiv}N \longrightarrow \underset{\text{A carbocation}}{CH_3CH_2CH_2{-}CH_2^{\oplus}} + N_2$$

$$\underset{\substack{n\text{-Butyl chloride}\\(5\%)}}{} \xleftarrow{Cl^-} CH_3CH_2CH_2CH_2^{\oplus} \xrightarrow{H_2O,\ -H^+} \underset{\substack{n\text{-Butyl alcohol}\\(25\%)}}{}$$

$$CH_3CH_2CH_2CH_2^{\oplus} \xrightarrow{-H^+} \underset{\substack{\text{1-Butene}\\(26\%)}}{CH_3{-}CH_2{-}CH{=}CH_2}$$

$$CH_3CH_2CH_2CH_2^{\oplus} \rightleftharpoons CH_3{-}CH_2{-}\overset{\oplus}{C}H{-}CH_3$$

$$\underset{\substack{sec\text{-Butyl chloride}\\(3\%)}}{} \xleftarrow{Cl^-} CH_3{-}CH_2{-}\overset{\oplus}{C}H{-}CH_3 \xrightarrow{H_2O,\ -H^+} \underset{\substack{sec\text{-Butyl alcohol}\\(13\%)}}{}$$

$$CH_3{-}CH_2{-}\overset{\oplus}{C}H{-}CH_3 \xrightarrow{-H^+} \underset{\substack{\text{2-Butene}\\(10\%)}}{CH_3{-}CH{=}CH{-}CH_3}$$

Exercise 13.7 What would be the most probable product if neopentylamine,

$$(CH_3)_3C{-}CH_2{-}NH_2$$

were treated with nitrous acid?

Secondary amines react with nitrous acid to produce the neutral N-nitroso compounds. These are usually yellow oils. Many N-nitrosoamines are known to be carcinogenic.

$$R_2N{-}H + H{-}O{-}N{=}O \longrightarrow R_2N{-}N{=}O + H_2O$$

An N-nitroso compound

The reaction of tertiary amines with nitrous acid is rather complex and of little practical importance. The reaction of aliphatic amines with nitrous acid also has little value except as a diagnostic or as a separatory procedure. In either case the primary amine is destroyed. On the other hand, the reaction of nitrous acid with aromatic primary amines produces intermediates known as diazonium salts.

$$C_6H_5{-}NH_2 + 2\,HCl + NaNO_2 \xrightarrow{0\text{–}5°C} C_6H_5{-}N_2^+Cl^- + NaCl + H_2O$$

Benzenediazonium chloride

A large number of very useful products, difficult if not impossible to arrive at by any other route, are synthesized by way of a diazonium salt. We shall deal with the diazonium salts in a separate section.

Exercise 13.8 A primary aliphatic amine, when treated with nitrous acid, can form an olefin or an alcohol, whereas aniline under the same conditions yields only phenol. Explain.

E. Ring Substitution in Aromatic Amines. The primary amino group of aniline, like the hydroxyl group of a phenol, is very sensitive to oxidation. In order to preserve the amino group in the presence of an oxidizing reagent, it must be "protected" prior to reaction. The amino group usually is protected by acetylation. The following example illustrates this procedure.

EXAMPLE

What sequence of reactions must be followed in order to convert aniline to *p*-nitroaniline?

Solution STEP 1. Acetylate the amino group. This step serves a two-fold purpose: (a) acetylation protects the amino group against oxidation by nitric acid and the formation of various oxidation products. (b) The acetylated amino group will direct our incoming substituent to the *p*-position. On nitration, unacetylated aniline, would give mostly *m*-nitroaniline because aniline would be in the form of the anilinium ion. (See Table 4.2.)

STEP 2. Nitrate acetanilide.

STEP 3. Hydrolyze *p*-nitroacetanilide to *p*-nitroaniline.

The preceding sequence and reagents required are shown in the following condensed form.

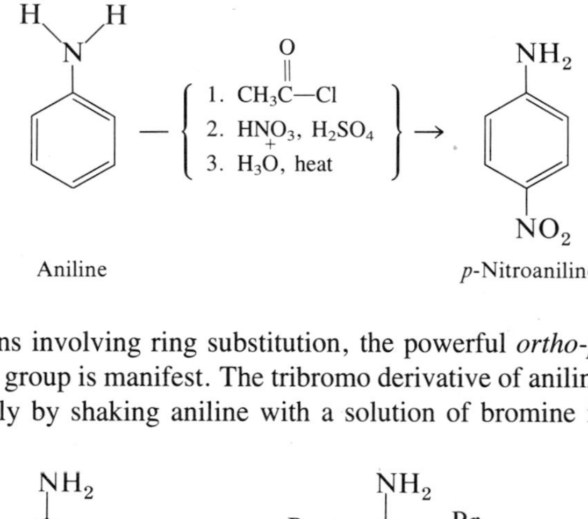

$$\text{C}_6\text{H}_5\text{NH}_2 \xrightarrow{\left\{\begin{array}{l}1.\ \text{CH}_3\overset{\text{O}}{\overset{\|}{\text{C}}}\text{—Cl} \\ 2.\ \text{HNO}_3,\ \text{H}_2\text{SO}_4 \\ 3.\ \text{H}_3\overset{+}{\text{O}},\ \text{heat}\end{array}\right\}} p\text{-NO}_2\text{C}_6\text{H}_4\text{NH}_2$$

Aniline p-Nitroaniline

In all reactions involving ring substitution, the powerful *ortho-para* directive influence of the amino group is manifest. The tribromo derivative of aniline, for example, may be prepared simply by shaking aniline with a solution of bromine in water.

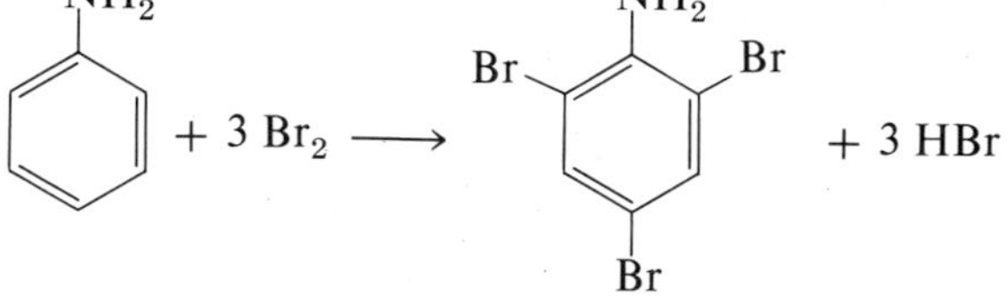

$$\text{C}_6\text{H}_5\text{NH}_2 + 3\ \text{Br}_2 \longrightarrow 2,4,6\text{-Br}_3\text{C}_6\text{H}_2\text{NH}_2 + 3\ \text{HBr}$$

2,4,6-Tribromoaniline

Acetylation of aniline, besides protecting the amino group from oxidation, as explained under nitration, offers the additional advantage of diminishing the reactivity of aniline. For example, monobromoaniline is prepared by brominating acetanilide, followed by removal of the protecting group.

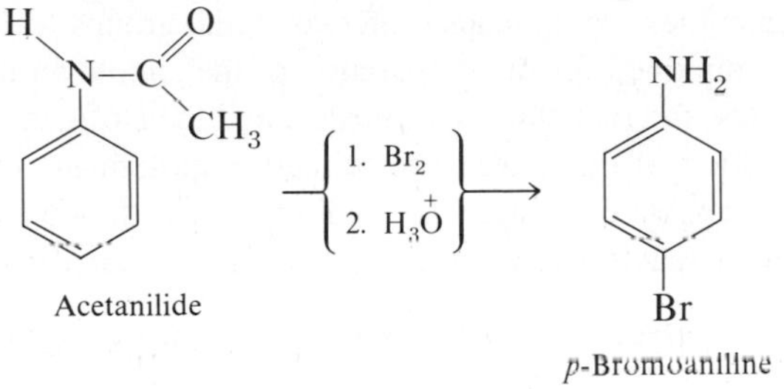

$$\text{Acetanilide} \xrightarrow{\begin{Bmatrix} 1.\ Br_2 \\ 2.\ H_3\overset{+}{O} \end{Bmatrix}} p\text{-Bromoaniline}$$

Acetanilide

p-Bromoaniline

Sulfonation of aniline is a slow reaction in which the aniline hydrogen sulfate salt initially formed appears to undergo rearrangement upon further heating. The product is called sulfanilic acid.

Sulfanilic acid appears to exist largely as an internal salt or a dipolar ion. Inner salts are possible when one group within a molecule is capable of acting as a proton donor (acid) to another within the same molecule capable of acting as a proton acceptor (base). The same phenomenon is shown by the amino acids (Sec. 15.4).

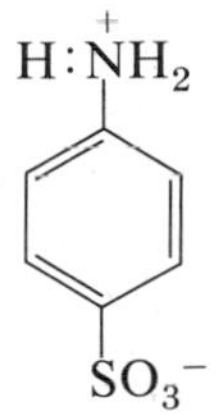

$H{:}\overset{+}{N}H_2$

SO_3^-

Inner salt of sulfanilic acid

Sulfanilic acid is an important intermediate in the synthesis of certain dyes and in the preparation of the *sulfa drugs*. The sulfa drugs were widely used during World War II to prevent infection in wounds and still find some application in medicine. They now have been replaced to a large extent by the antibiotics (Secs. 17.14–17.16). Sulfanilamide is effective internally against streptococci and staphylococci infections. Other sulfa drugs are derivatives of sulfanilamide. In these "sulfas" some other group replaces one hydrogen and appears as an N-substituent in the amide portion of the parent compound.

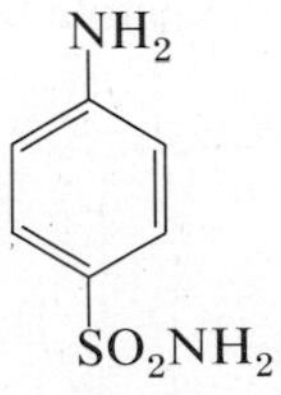

NH_2

SO_2NH_2

Sulfanilamide

13.5 Diamines

The diamines, compounds with two amino groups, can be prepared by methods similar to those employed for the preparation of the simple amines. The reaction of ammonia with dihalides, the reduction of dinitriles, and the Hofmann reaction of diamides all lead to the preparation of diamines. The primary requirement, in each case, is a bifunctional compound capable of being converted into one with two amino groups. Ethylene chloride, for example, reacts with ammonia to yield **ethylenediamine.**

$$ClCH_2CH_2Cl + 4\,NH_3 \xrightarrow{150°} \underset{\text{Ethylenediamine}}{H_2N{-}CH_2CH_2{-}NH_2} + 2\,NH_4Cl$$

The ptomaines, tetramethylenediamine and pentamethylenediamine, are natural, malodorous products formed by bacterial decomposition of protein material such as occurs in decaying flesh. The amino acids ornithine and lysine on breakdown yield these respective diamines, which are also known by the unlovely names *putrescine* and *cadaverine*.

The current industrial procedures for the preparation of hexamethylenediamine, an important intermediate in the synthesis of nylon, illustrate the extremely high level of sophistication in the contemporary chemical industry. Reactions are designed to use the least expensive starting materials, to use a minimum of energy, and to give few, if any, by-products, such as inorganic salts. The two competitive processes begin with syntheses of adiponitrile, one using an electrochemical dimerization of acrylonitrile and the other the addition of hydrogen cyanide to 1,3-butadiene in the presence of a remarkable organometallic nickel catalyst.

$$\underset{\text{Acrylonitrile}}{2\,CH_2{=}CH{-}CN} \xrightarrow[+2\,H^+]{+2\,e^-} \underset{\text{Adiponitrile}}{N{\equiv}C{-}(CH_2)_4{-}C{\equiv}N}$$

$$CH_2{=}CH{-}CH{=}CH_2 + HCN \xrightarrow[\text{1,4-addition}]{} CH_3{-}CH{=}CH{-}CH_2{-}CN$$

HCN ↓ Ni catalyst (*isomerizes double bond; causes anti-Markovnikov addition of HCN*)

$$\underset{\text{Adiponitrile}}{N{\equiv}C{-}(CH_2)_4{-}C{\equiv}N}$$

The reduction of adiponitrile produces hexamethylenediamine.

$$NC{-}(CH_2)_4{-}CN + 4\,H_2 \xrightarrow{Ni} \underset{\text{Hexamethylenediamine}}{H_2N{-}(CH_2)_6{-}NH_2}$$

$$NC{-}(CH_2)_4{-}CN \xrightarrow[\text{heat}]{H_3O^+} \underset{\text{Adipic acid}}{HOOC{-}(CH_2)_4{-}COOH}$$

Adiponitrile on hydrolysis yields adipic acid. The condensation reaction between adipic acid and hexamethylenediamine to produce the polyamide known as Nylon 66 was discussed in Section 11.4.

The aromatic diamines may be prepared by the reduction of the appropriate nitro compounds.

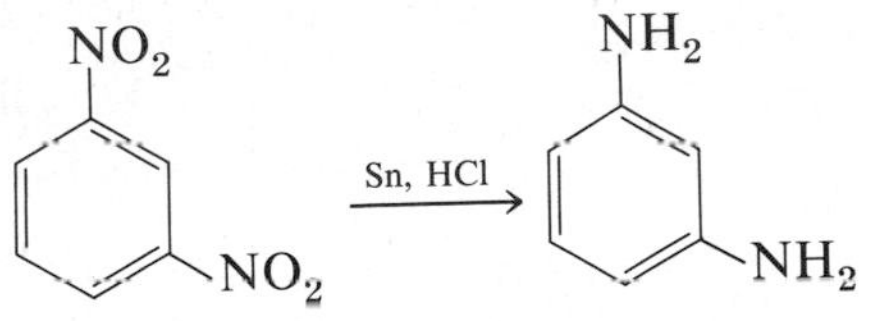

NO_2 / NO_2 —Sn, HCl→ NH_2 / NH_2

13.6 Nitrogen Compounds Used as Dietary Sugar Substitutes

Certain individuals (diabetics) are incapable of producing sufficient insulin for the complete utilization of glucose (Sec. 14.2). Others, through an abnormal hormonal activity, overproduce glucose in the body. In either case, an abnormal amount of this sugar accumulates in the blood and results in a condition known as *hyperglycemia*. For these persons and for those who are overweight and wish to lower their caloric intake, a number of artificial sweeteners are available. These substances are nonnutritive agents with a sweetness many times that of sucrose. *Sucaryl,* the calcium salt of cyclohexylsulfamic acid, and *saccharin,* the imide of *o*-sulfobenzoic acid, are two sweetening agents that have been widely used. However, sucaryl was banned as a food additive in 1969 by the federal government because of the possibility that it might be carcinogenic. The banning of saccharin in 1977 for the same reason led to consumer protests, and this artificial sweetener is now once again widely used.

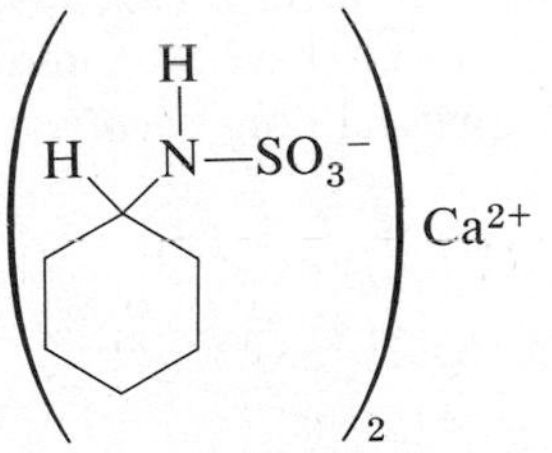

H, H, N—SO_3^-, Ca^{2+}, 2

Sucaryl calcium
(Calcium cyclohexylsulfamate or "Cyclamate")

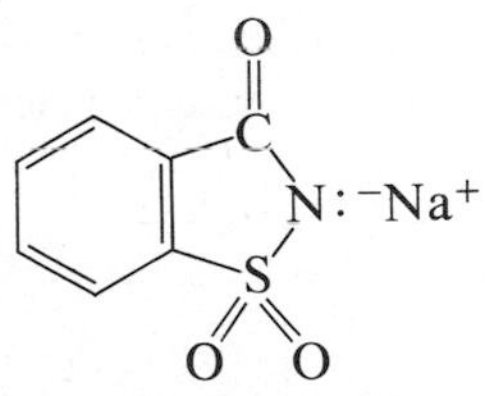

O, C, N:$^-Na^+$, S, O, O

Saccharin sodium
(Sodium benzosulfimide)

Saccharin has a sweetness approximately 500 times that of sucrose, which is assigned a sweetness of 1.0. The sweetness of sucaryl is approximately 30 when measured on the same scale. The most recent addition to the arsenal of sugar substitutes, *aspartame,* is the partially esterified dipeptide, *aspartylphenylalanine methyl ester,* marketed under the names "Equal" or "NutraSweet." Aspartame was approved for sale by the Food and Drug Administration in 1981.

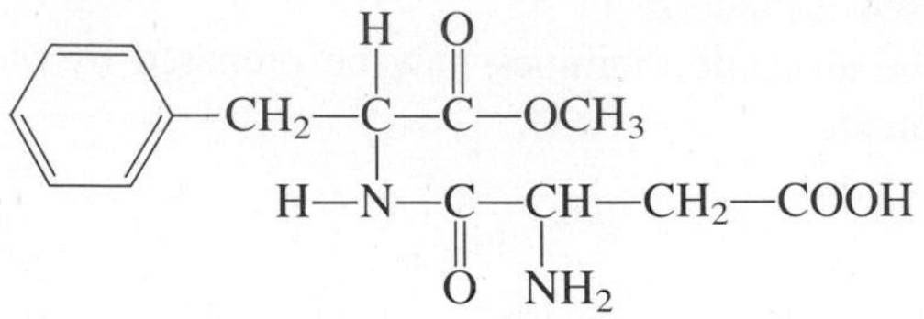

$$C_6H_5{-}CH_2{-}\overset{H}{\underset{|}{C}}{-}\overset{O}{\overset{\|}{C}}{-}OCH_3$$
$$H{-}N{-}\underset{\underset{O}{\|}}{C}{-}\underset{\underset{NH_2}{|}}{CH}{-}CH_2{-}COOH$$

Aspartame

13.7 Amino Alcohols and Some Related Physiologically Active Compounds

The aminoalcohols are found in a number of compounds of potent physiological activity.

Choline, $(CH_3)_3\overset{+}{N}CH_2CH_2OHOH^-$, trimethyl-$\beta$-hydroxyethyl ammonium hydroxide, forms part of the structure of lecithin (Sec. 12.5). Choline appears in lecithin with glycerol as a mixed ester of phosphoric acid and fatty acids. Acetylated choline, or *acetylcholine,* plays a vital role in the generation and conduction of nerve impulses in the body.

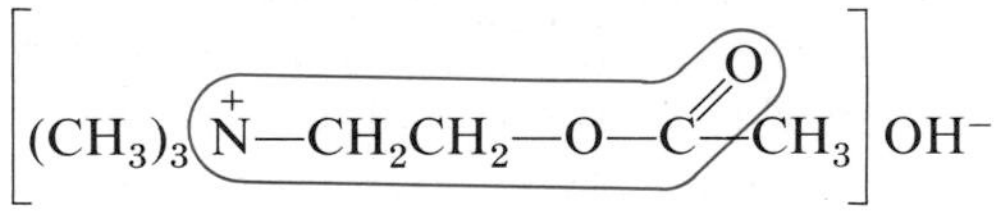

$$\left[(CH_3)_3\overset{+}{N}{-}CH_2CH_2{-}O{-}\overset{O}{\overset{\|}{C}}{-}CH_3\right]OH^-$$

Acetylcholine

The ability of certain local anesthetics to mitigate pain appears to be due, at least in part, to their structural similarity to choline and an ability to replace acetylcholine for a short duration. A number of local anesthetizing agents have in their molecular architecture an aminoalcohol grouping in the form of either an ester or an ether unit. The local anesthetic *procaine,* to which most of us have been introduced while in a dentist's chair, is a synthetic replacement for the natural drug *cocaine* (Sec. 17.3).

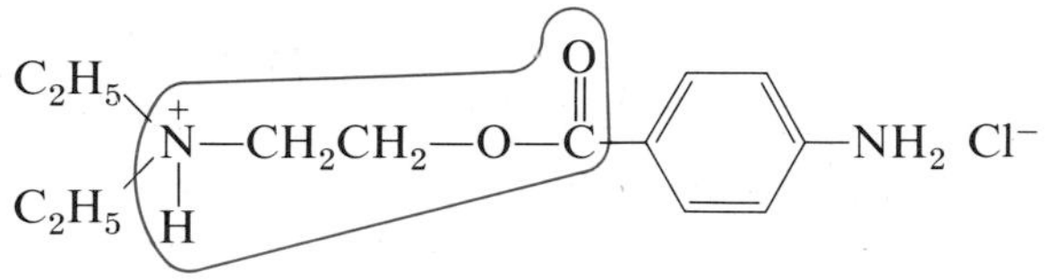

$$(C_2H_5)_2\overset{+}{\underset{\underset{H}{|}}{N}}{-}CH_2CH_2{-}O{-}\overset{O}{\overset{\|}{C}}{-}C_6H_4{-}NH_2\ Cl^-$$

Procaine hydrochloride
(Novocaine)

A large number of amines and amino alcohols, both naturally occurring and synthetic, have an effect upon the sympathetic nervous system and are known as **sympathomimetic agents.** These agents are powerful stimulants and dangerous if used promiscuously. Some may result in habituation when used for prolonged periods. The structures for a few of these are given here.

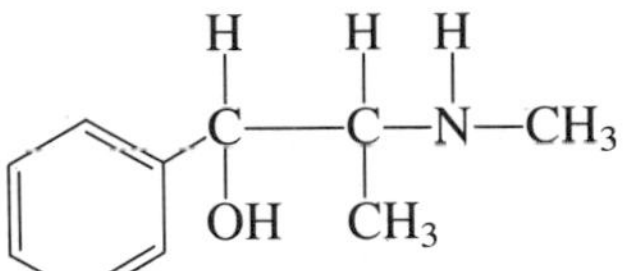

Ephedrine

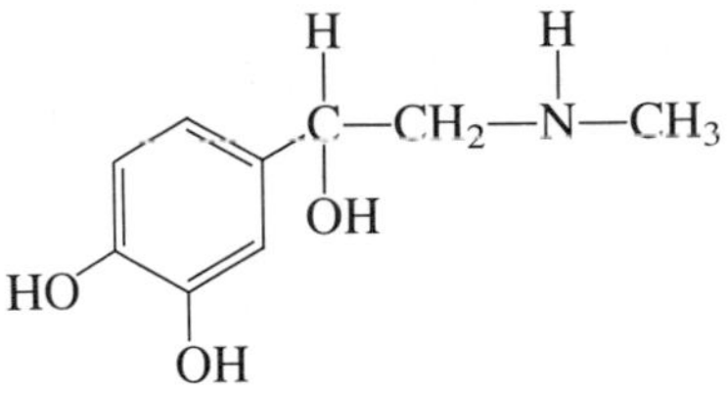

Epinephrine (Adrenalin)
(1-(3,4-Dihydroxyphenyl)-
2-methylaminoethanol)

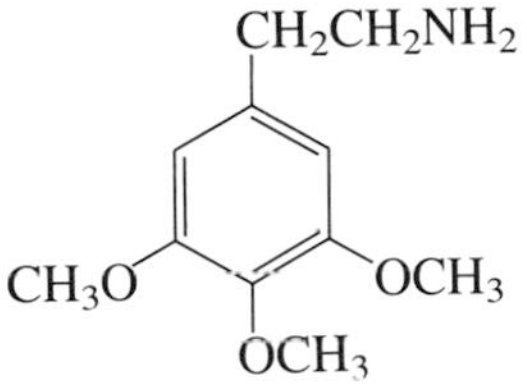

Mescaline

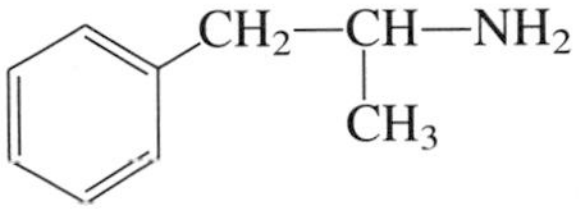

Benzedrine
(β-Aminopropylbenzene)

Exercise 13.9 *Benadryl,* $(C_6H_5)_2CHOCH_2CH_2N(CH_3)_2$, is a synthetic amino ether of benzhydrol (diphenylcarbinol). It is widely used in the treatment of histamine allergies and as a preventive for motion sickness. Suggest a sequence of reactions to show how benadryl might be synthesized using diphenylcarbinol and ethanolamine as starting materials.

13.8 Diazonium Salts

As we have learned, the action of nitrous acid on primary aliphatic amines is of little importance as a preparative reaction. The action of nitrous acid on primary aromatic amines, on the other hand, has a number of applications, which lead to many useful products. The acid salt of an aromatic primary amine, when treated with a cold aqueous solution of sodium nitrite, does not liberate nitrogen as would an aliphatic primary amine. Instead, a water-soluble **diazonium salt** is produced. This reaction is called **diazotization** and is an easy one to carry out.

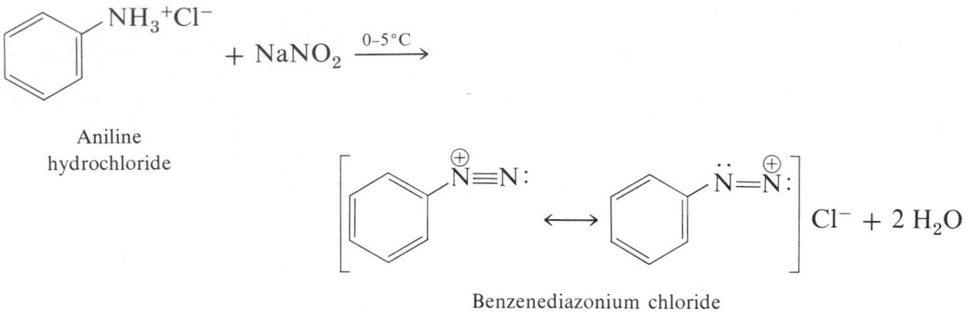

$$C_6H_5NH_3^+Cl^- + NaNO_2 \xrightarrow{0\text{–}5°C} \left[C_6H_5\overset{\oplus}{N}\equiv N: \longleftrightarrow C_6H_5\ddot{N}=\overset{\oplus}{N}: \right] Cl^- + 2\,H_2O$$

Aniline hydrochloride

Benzenediazonium chloride

The benzenediazonium ion usually is written as

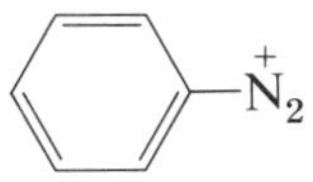

$$C_6H_5\overset{+}{N}_2$$

Although we have illustrated its formation from the hydrochloric acid salt of aniline, the sulfuric acid salt serves equally well but produces the diazonium salt as the sulfate. The diazonium salt, once formed, is not isolated but instead is treated with the reagent required to give the product desired. The reactions of the diazonium salts may be divided into three principal types. In one type of reaction a displacement of the diazonium group by some other group takes place with an accompanying loss of nitrogen. In the second type, a coupling reaction takes place in which the nitrogen atoms are retained as an azo grouping, —N=N—, to become part of a new molecular structure. In the third type of reaction, a reduction, either partial or complete, of the diazonium group occurs. Reactions of each type are illustrated.

A. Displacement Reactions of Diazonium Salts. The following examples illustrate the usefulness of diazonium salts as reaction intermediates, where the diazonium group is replaced by some other group.

(1) $$C_6H_5N_2^+Cl^- + H_2O \xrightarrow{\text{warm}} C_6H_5OH + N_2 + HCl$$

Phenol

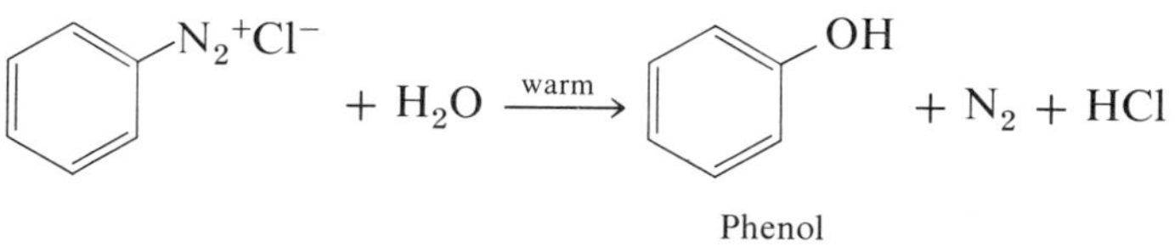

(2)
(3)
(4) $$C_6H_5N_2^+Cl^- \xrightarrow{\begin{pmatrix}CuCl\\ CuBr\\ CuCN\end{pmatrix}} C_6H_5Cl \text{ or } (—Br) \text{ or } (—CN) + N_2$$

Chlorobenzene

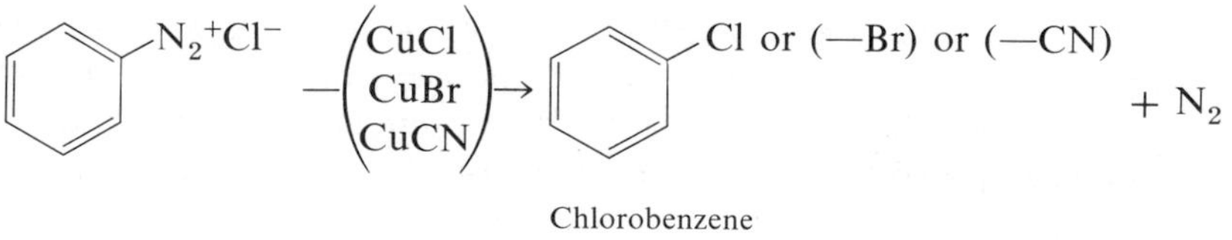

(5) 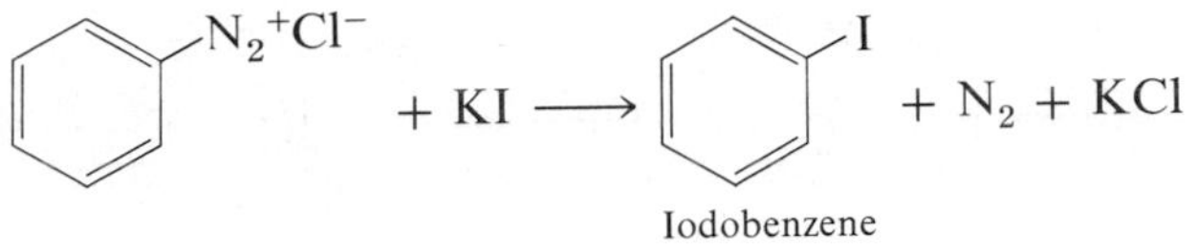 $C_6H_5N_2^+Cl^- + KI \longrightarrow C_6H_5I + N_2 + KCl$

Iodobenzene

(6) 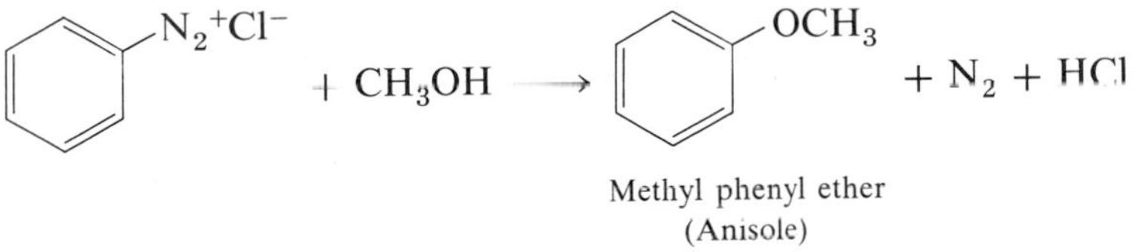 $C_6H_5N_2^+Cl^- + CH_3OH \longrightarrow C_6H_5OCH_3 + N_2 + HCl$

Methyl phenyl ether
(Anisole)

(7) 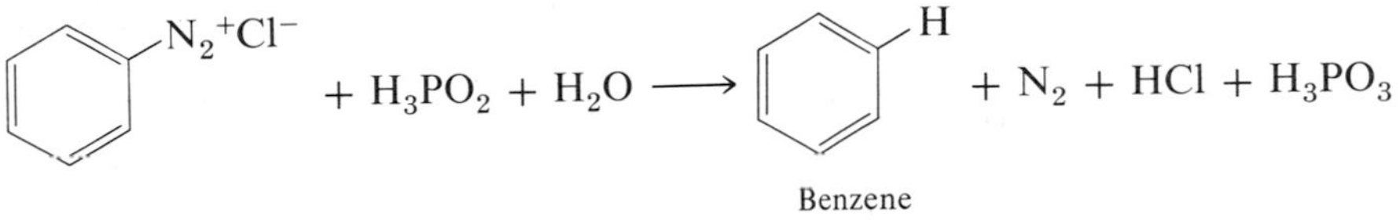 $C_6H_5N_2^+Cl^- + H_3PO_2 + H_2O \longrightarrow C_6H_5H + N_2 + HCl + H_3PO_3$

Benzene

The reactions of a benzenediazonium compound with cuprous salts, reactions (2), (3), and (4) above, usually are referred to as **Sandmeyer reactions.**

Exercise 13.10 Write reactions to show how the product of Equation (4) could be converted into benzylamine and into benzoic acid.

The reduction of nitrobenzene to aniline, followed by diazotization and replacement of the diazonium salt, makes possible an orientation of ring substituents difficult if not impossible to achieve in any other way. For example, it is about the only way to place *ortho-* or *para-*directors *meta-* to each other, or *meta-*directors in *ortho-* or *para* positions to each other.

EXAMPLE

Prepare *m*-bromophenol.

Solution The direct bromination of phenol gives 2,4,6-tribromophenol. Moreover, phenol is a strong *ortho, para-*director. However, we can achieve our result with the following sequence of reactions.

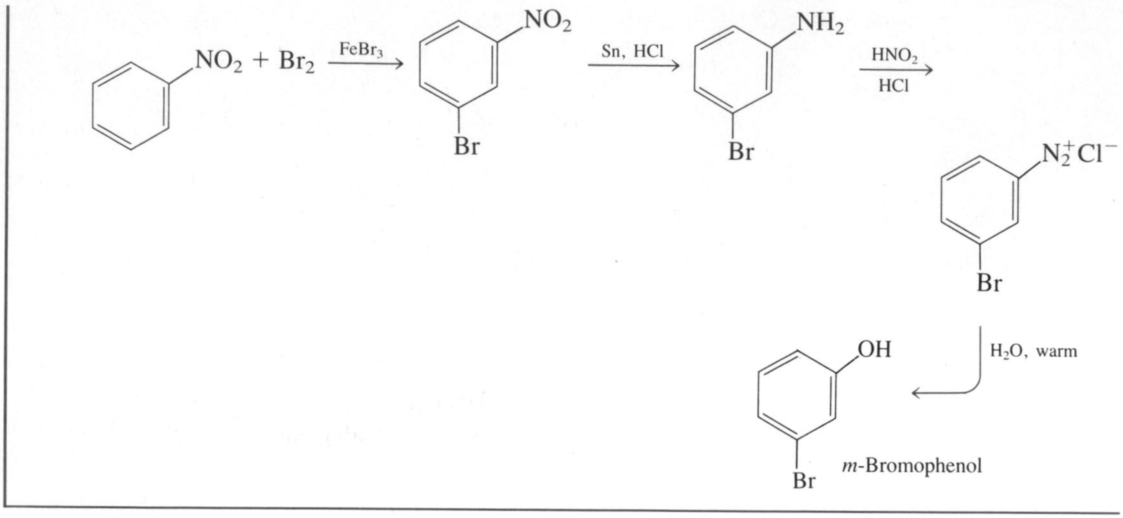

$$C_6H_5NO_2 + Br_2 \xrightarrow{FeBr_3} m\text{-}BrC_6H_4NO_2 \xrightarrow{Sn,\ HCl} m\text{-}BrC_6H_4NH_2 \xrightarrow[HCl]{HNO_2} m\text{-}BrC_6H_4N_2^+Cl^- \xrightarrow{H_2O,\ warm} m\text{-}BrC_6H_4OH$$

m-Bromophenol

B. Coupling Reactions of Diazonium Salts. Diazonium salts readily undergo coupling reactions with phenols and aromatic amines. Since the coupling involves an attack of the positive diazonium ion on a center of high electron density in the benzene ring, the presence of substituents that increase the electron density of the ring enhances the coupling reaction. Conversely, electron-withdrawing substituents on the ring inhibit the coupling reaction or prevent it entirely. Coupling takes place preferably in the position *para* to the activating group. Should the *para* position be occupied, coupling then occurs at one of the *ortho* positions. The coupling reaction is carried out in a neutral, an alkaline, or a weakly acidic solution.

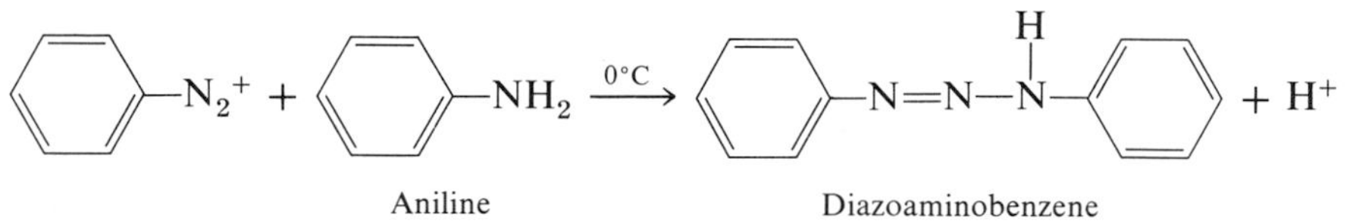

$$C_6H_5N_2^+ + C_6H_5NH_2 \xrightarrow{0°C} C_6H_5N{=}N{-}NH{-}C_6H_5 + H^+$$

Aniline Diazoaminobenzene

The diazoaminobenzene initially formed when aniline couples with the benzenediazonium ion rearranges on heating to *p*-aminoazobenzene.

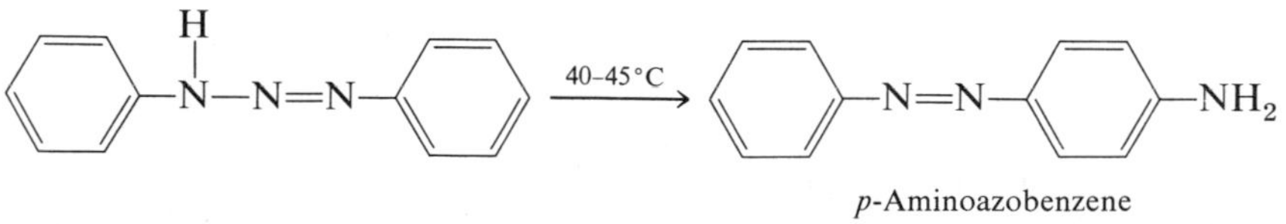

$$C_6H_5NH{-}N{=}N{-}C_6H_5 \xrightarrow{40\text{–}45°C} C_6H_5N{=}N{-}C_6H_4{-}NH_2$$

p-Aminoazobenzene

Coupling will not take place with aniline in a strongly acid solution because, in a strong acid environment, the primary amino group of aniline becomes a positive anilinium ion, which deactivates the ring.

$$C_6H_5NH_2 \underset{OH^-}{\overset{H^+}{\rightleftharpoons}} C_6H_5NH_3^+$$

Ring active (couples readily) — Ring deactivated (does not couple)

Coupling with a phenol is inhibited in a strongly acid solution because the strong activating influence of the phenoxide ion is lacking. Phenols, you will recall, are acidic (Sec. 7.3) and would therefore couple with the benzenediazonium ion most readily in a slightly alkaline solution.

$$C_6H_5OH \underset{H^+}{\overset{OH^-}{\rightleftharpoons}} C_6H_5O^-$$

Weakly active (couples slowly) — Strongly active (couples readily)

When there is a choice of ring positions open to the benzenediazonium ion, the group on the ring with the stronger directive influence prevails.

$$C_6H_5-N_2^+ + \text{4-}CH_3C_6H_4OH \longrightarrow C_6H_5-N{=}N-C_6H_3(OH)(CH_3) + H^+$$

p-Cresol — Benzeneazo-*p*-cresol

The azo compounds are highly colored substances. The azo grouping, for this reason, is the structural feature in a wide variety of dyes called **azo dyes.**

C. Reduction of Diazonium Compounds. The partial reduction of a diazonium salt can be accomplished without the loss of nitrogen by the use of sodium sulfite as the reducing agent. The product of such a reduction is phenylhydrazine, which is a reagent useful in the identification of sugars (Sec. 14.8) and carbonyl compounds (Sec. 8.8A).

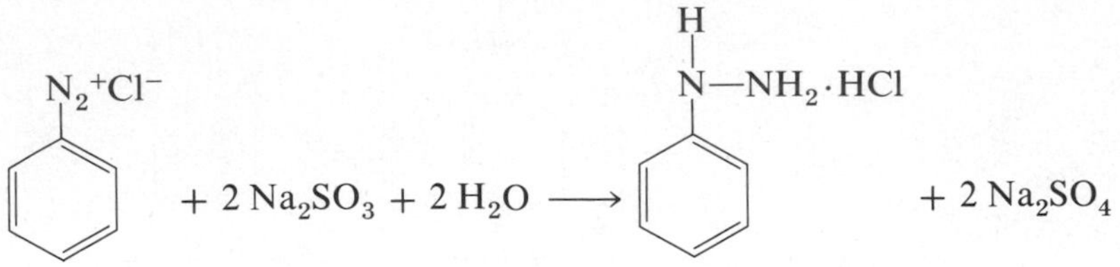

$$C_6H_5N_2^+Cl^- + 2\,Na_2SO_3 + 2\,H_2O \longrightarrow C_6H_5NH{-}NH_2\cdot HCl + 2\,Na_2SO_4$$

Phenylhydrazine hydrochloride

A complete reduction of a diazonium salt that results in its replacement by hydrogen is called **deamination.** Hypophosphorous acid, H_3PO_2, is an excellent reagent for this reaction (Sec. 13.8A, Eq. (7)).

13.9 Azo Dyes

This class represents the largest and most important group of dyes. The characteristic feature of each dye in this class is the chromophoric azo group (—N=N—), which forms part of the conjugated system and joins two or more aromatic rings. Azo dyes are prepared by coupling a diazotized aromatic amine with another aromatic amine or with a phenol or a naphthol (see Sec. 13.8B). This simple reaction can be carried out directly within the fabric, and results in a product complete with chromophore, auxochrome, and conjugated system. One of the first azo dyes to be prepared was *para red,* obtained from diazotized *p*-nitroaniline and *β*-naphthol.

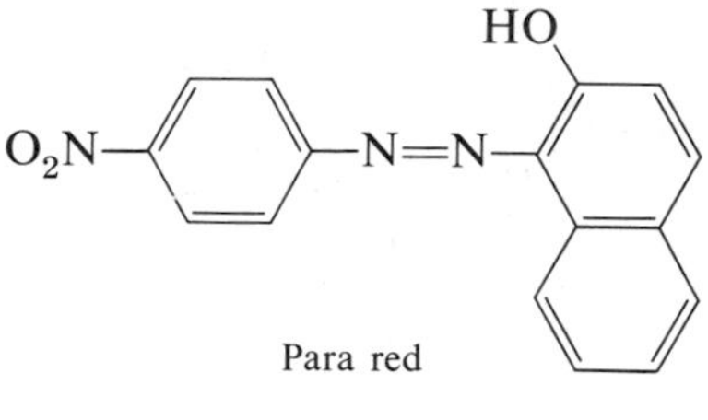

HO

O_2N—N=N—

Para red

More than one aromatic ring, when included in the coupling reaction, extends the conjugated system and deepens the color of a dye. For example, diazotized benzidine (4,4′-diaminobiphenyl) will couple with 1-naphthylamine-4-sulfonic acid to produce a dye called *congo red.* Congo red also is used as a chemical indicator.

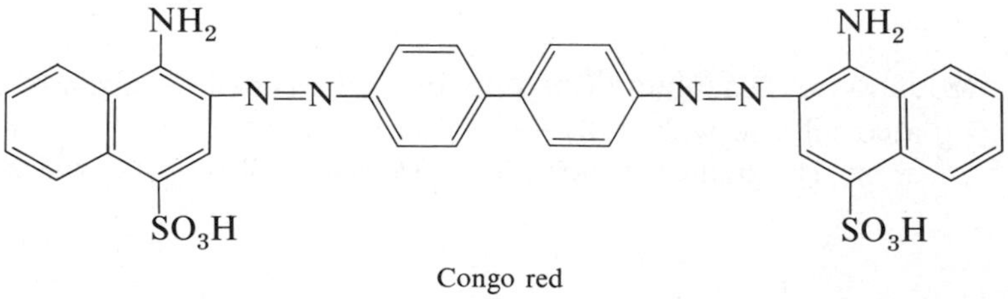

NH_2 —N=N— —N=N— NH_2

SO_3H SO_3H

Congo red

Exercise 13.11 Draw the structure of the dye that results when the diazonium salt of 2-methoxy-5-nitroaniline is coupled to the anilide of 3-hydroxy-2-naphthoic acid.

Summary

1. **Structure**
   Amines may be considered as ammonia derivatives. Amines are classified as primary, secondary, or tertiary; also, as aliphatic or aromatic.
2. **Nomenclature**
   The amines usually are named by naming the alkyl or aryl groups attached to the nitrogen of the amino group, followed by the word "amine."
3. **Physical Properties**
   Amines are weakly basic compounds. The low molecular weight members (C_1–C_2) are gases and low-boiling liquids. They are soluble in acids. Nearly all are foul-smelling.
4. **Preparation**
   The amines may be prepared by
   (a) the direct alkylation of ammonia.
   (b) the reduction of nitriles (—C≡N), amides

   $$\left(-C\overset{\displaystyle O}{\underset{}{\diagup\!\!\!\diagup}}\!\!-NH_2\right)$$

   oximes (—C=NOH), and nitro (—NO_2) compounds.
   (c) Hofmann's hypobromite degradation of amides.
5. Amines react with each of the following reagents to give the product indicated.

	Amine	*Reactant*		*Product*
(a)	RNH_2	+ acids	⟶	salts
(b)	RNH_2	+ acyl halides or (acid anhydrides)	⟶	amides
(c)	RNH_2	+ C_6H_5–SO_2Cl	⟶	alkali-soluble benzenesulfonamide
(d)	R_2NH	+ C_6H_5–SO_2Cl	⟶	alkali-insoluble benzenesulfonamide

(e) $RNH_2 + HNO_2 \longrightarrow N_2$ + alcohols + olefins

(f) $R_2NH + HNO_2 \longrightarrow$ N-nitroso compounds

(g)  $C_6H_5-NH_2 + HNO_2 \xrightarrow[0-5]{HCl}$ diazonium salts

(h)  $C_6H_5-NH_2 + Br_2 \longrightarrow$ *ortho–para* substitution products

(i) $R-CH_2CH_2NH_2 + CH_3I \longrightarrow R-CH_2CH_2\overset{+}{N}(CH_3)_3X^-$

$\xrightarrow[heat]{Ag_2O} R-CH{=}CH_2 + (CH_3)_3N$

6. Diamines may be prepared by the same general methods used to prepare the simple amines.
7. Amino ethanols can be prepared from ethylene oxide and ammonia. Physiologically active compounds frequently contain an aminoalcohol unit.
8. Diazonium salts formed from aromatic primary amines are useful intermediates in organic synthesis. The diazonium group may be
   (a) replaced by: halogen, cyanide, hydroxyl, and alkoxyl.
   (b) coupled to: other aromatic rings. The azo dyes are produced by this reaction.
   (c) reduced to: phenylhydrazine, benzene (deamination).

Supplementary Exercises

13.12 Name the following compounds. If amines, classify each as primary, secondary, or tertiary.

(a) $CH_3-\underset{}{\overset{CH_3}{\overset{|}{N}}}-CH_3$

(b) $C_6H_5-CH_2-NH_2$

(c) $HO-CH_2CH_2N(CH_3)_2$

(d) $H_2N-CH_2CH_2CH_2CH_2CH_2-NH_2$

(e) 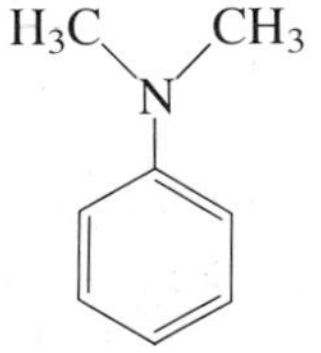 $(H_3C)(CH_3)N-C_6H_5$ (N,N-dimethyl on a benzene ring)

(f)  cyclohexane ring bearing H and NH_2 on one carbon

(g) 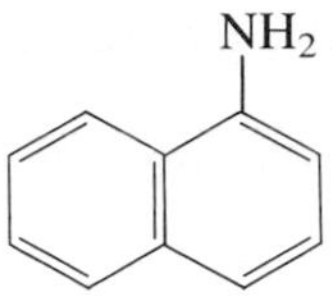 1-naphthylamine structure (NH_2 on naphthalene)

(h) 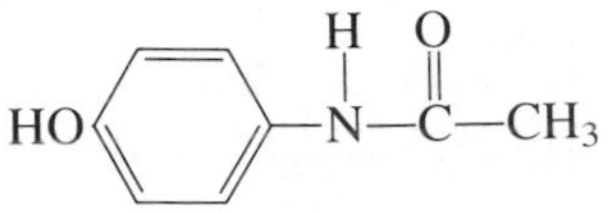 $HO-C_6H_4-N(H)-C(=O)-CH_3$

(i) $H_2N—(CH_2)_4—NH_2$

(j) 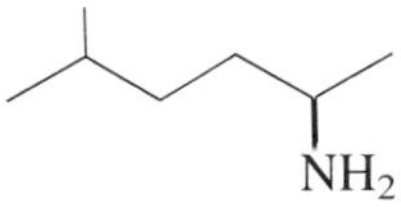 NH_2 (R)— or (S)—?

13.13 Write structures for the following.

(a) aniline
(b) benzenediazonium sulfate
(c) *p*-aminobenzenesulfonic acid (sulfanilic acid)
(d) tetramethylammonium chloride
(e) acetanilide
(f) β-diethylaminoethyl-*p*-aminobenzoate hydrochloride (procaine)
(g) *p*-nitrosodimethylaniline
(h) N-ethylphthalimide
(i) ethyldimethylamine

13.14 Without consulting a table of ionization constants, arrange the following compounds in an order of diminishing basic strength.

(a) aniline, (b) ammonia, (c) dimethylamine, (d) ethylamine, (e) benzamide.

13.15 Taking advantage of acidic, basic, and other properties, outline a procedure for separating a mixture that includes (a) aniline, (b) benzoic acid, (c) acetophenone, and (d) bromobenzene.

13.16 What reagents and conditions are required to accomplish each of the following conversions? (Some require more than one step.)

(a) acetone → 2-aminopropane, (b) ethyl iodide → *n*-propylamine, (c) acetamide → methylamine, (d) *n*-butylamine → trimethylamine + 1-butene, (e) nitrobenzene → aniline, (f) aniline → acetanilide, (g) aniline → benzenediazonium chloride, (h) benzenediazonium chloride → benzonitrile, (i) toluene → 2,4-diaminotoluene, (j) aniline → *p*-aminoazobenzene.

13.17 What simple test tube reactions would serve to distinguish between:

(a) *n*-butylamine and triethylamine
(b) aniline hydrochloride and benzamide
(c) *sec*-butyl alcohol and *sec*-butylamine
(d) aniline and benzylamine
(e) methylaniline and dimethylaniline

13.18 When 1.83 g of an unknown amine was treated with nitrous acid, the evolved nitrogen, corrected to standard temperature and pressure, measured 560 ml. The alcohol isolated from the reaction mixture gave a positive iodoform reaction. What is the structural formula of the unknown amine?

13.19 Give the structures and names of the principal products expected from each of the following reactions.

(a) *p*-chloroaniline + HCl, then $NaNO_2 \xrightarrow{0\text{–}5^\circ}$

(b) benzamide + $LiAlH_4 \longrightarrow$

(c) aniline + acetic anhydride $\longrightarrow$

(d) benzenediazonium chloride + phenol $\xrightarrow{NaOH}$

13.20 Suggest a method for converting *p*-toluic acid (*p*-methylbenzoic acid) into *p*-toluidine (*p*-methylaniline).

13.21 Suggest a reaction mechanism that would account for the following transformation.

$$\text{cyclobutyl}\text{–}CH_2NH_2 \xrightarrow[H^+]{HNO_2} \text{cyclopentyl}\text{–}OH + N_2 + H_2O$$

13.22 Calculate the ratio $[CH_3CH_2NH_2]/[CH_3CH_2NH_3{}^+]$ in water at pH 6, 8, 10, and 12. $\left(\textit{Hint: } pH = pK_a + \log\frac{[A^-]}{[HA]}.\right)$

13.23 *p*-Aminobenzoic acid (PABA) is widely used as a sunscreen in protective creams and tanning lotions. Suggest a sequence of reactions for synthesizing PABA from toluene.

13.24 Draw reasonable structures for the isomeric compounds A, B, and C, $C_{10}H_{13}NO_2$, if:

A:
(a) it is a white solid that is insoluble in cold, dilute acid or base;
(b) refluxing with an alkaline solution produces a gas with a strong ammoniacal odor;
(c) acidification of the hydrolysis mixture liberates an acid. Titration of a 1.5-g sample of the acid required 20 ml of 0.5*N* sodium hydroxide for neutralization.

B:
(a) it is a white solid insoluble in cold, dilute acid or base;
(b) refluxing with sodium hydroxide solution produces an oily layer. The oily layer, when separated and purified, can be diazotized.
(c) acidification of the alkaline hydrolysis mixture, followed by extraction with ether, yields an acid. Titration of a 1.5-g sample of the acid required 40 ml of 0.5*N* sodium hydroxide for neutralization.

C:
(a) it is an oil insoluble in water, acid, or base;
(b) oxidation with sodium dichromate and concentrated sulfuric acid produces a solid compound, $C_8H_5NO_6$, that is soluble in hot water and in cold, dilute base;
(c) strong heating of the compound causes it to melt only.

***13.25** Compound A, $C_8H_{19}N$, gave an nmr spectrum consisting of a nine-proton singlet at δ 1.00, a six-proton singlet at δ 1.17, a two-proton singlet at δ 1.28, and a two-proton singlet at δ 1.42. When compound A was treated with D_2O, the peak at δ 1.42 was no longer present. Suggest a structure for compound A.

***13.26** Compound B, $C_6H_{15}N$, gave an nmr spectrum consisting of a one-proton, rather broad singlet at δ 0.67, a twelve-proton doublet at δ 1.00, and a two-proton septet at δ 2.88. Suggest a structure for compound B.

***13.27** According to Table 18.3, aniline has uv bands at 230 and 280 nm in water solution, which are found at 203 and 254 nm in *acidic* water solution. Explain.

***13.28** A compound C, C_7H_9N, was soluble in dilute hydrochloric acid. The acid solution was cooled in an ice bath and a second solution of sodium nitrite was added. No noticeable change was observed, but when the mixture was allowed to come to room temperature, bubbles evolved and an oil appeared on the surface. A sample of the oil was found to be soluble in sodium hydroxide solution.

Propose a structure for C that corresponds to the nmr spectrum below.

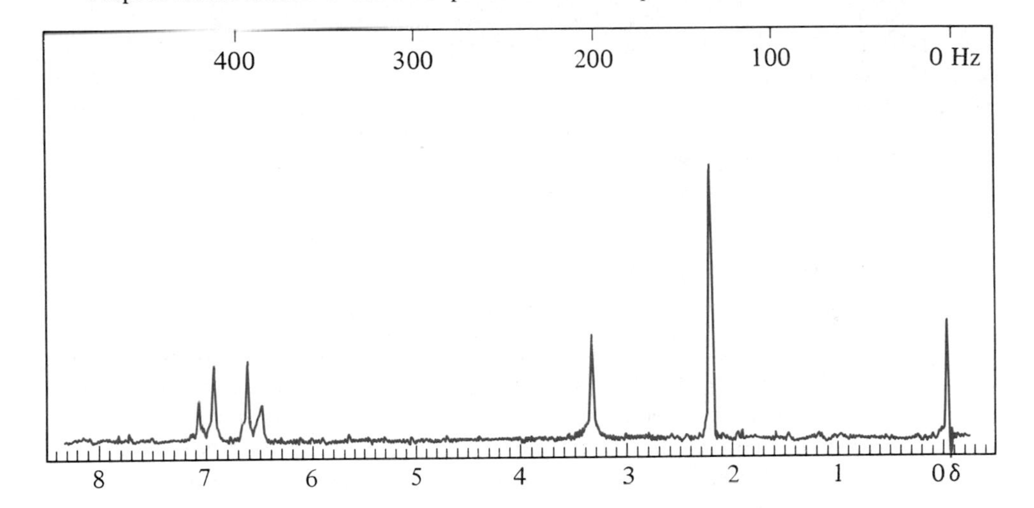

14 Carbohydrates

The name **carbohydrate** originated from the French ''hydrate de carbon.'' Early analyses of a number of naturally occurring compounds of this class gave empirical formulas in which the ratios of carbon, hydrogen, and oxygen appeared to indicate hydrates of carbon of the type $C_xH_{2y}O_y$ or $C_x(H_2O)_y$ where x and y may be the same or different. For example, the simple sugar **glucose** (grape sugar) has a molecular formula of $C_6H_{12}O_6$, and **sucrose** (cane sugar) the formula $C_{12}H_{22}O_{11}$, but such definite ratios of water to carbon are not to be found in every carbohydrate. Therefore, the name carbohydrate, as is often the case in chemical nomenclature, is not descriptive but has persisted and probably will be with us for a long time.

The carbohydrates comprise a great class of natural substances that includes the sugars, starches, cellulose, and related products. Carbohydrates may be described as polyhydroxy aldehydes or polyhydroxy ketones, or as substances that, when hydrolyzed, give polyhydroxy aldehydes or polyhydroxy ketones.

14.1 Classification and Nomenclature

The carbohydrates may be subdivided conveniently into three principal classes: monosaccharides, oligosaccharides (Gr. *oligos,* a few; L. *saccharum,* sugar), and polysaccharides.

Monosaccharides include all sugars that contain a single carbohydrate unit—that is, one incapable of producing a simpler carbohydrate on further hydrolysis. Most of the monosaccharides are five- and six-carbon structures. The five-carbon monosaccharides are called **pentoses,** and those of six-carbons are called **hexoses.** The suffix *-ose* is a generic designation of any sugar. The monosaccharides are frequently referred to as simple sugars and are either polyhydroxy aldehydes or polyhydroxy ketones. If an aldehyde, the name **aldose** is applicable; if a ketone, the name **ketose** is often used to describe the sugar.

Oligosaccharides consist of two or a relatively small number of monosaccharide units joined by acetal linkages between the aldehyde or ketone group of one simple sugar and a hydroxy group of another. This kind of coupling in sugar chemistry gives rise to what is called a **glycosidic** linkage. Hydrolysis of an oligosaccharide yields the simple sugar components. **Disaccharides** are composed of two simple sugar units, **trisaccharides** of three, etc.

Polysaccharides consist of hundreds or even thousands of monosaccharide units joined together through glycosidic linkages to form macro molecules, or polymers.

14.2 Glucose: A Typical Monosaccharide

Glucose (frequently called dextrose because of its dextrorotation) is the most important of the monosaccharides. Not only is it the most widely occurring sugar, but in free or combined form it is perhaps the most abundant of organic compounds. Glucose is the end product of the hydrolysis of starch and cellulose and is closely associated with the metabolic processes. Glucose is a main source of energy for all living organisms. It comprises 0.08–0.1% of the blood content of all normal mammals and is one of the few organic compounds that may be injected as a food directly into the bloodstream. Our best introduction to carbohydrate chemistry is offered by a review of the chemical and physical properties of glucose.

14.3 Structure of Glucose

The experimental evidence that led to the elucidation of the glucose structure provides one of the most fascinating chapters in organic chemistry. Combustion analysis and a molecular weight determination established the molecular formula of glucose as $C_6H_{12}O_6$. Structural evidence was supplied by the following: (a) reaction with acetic anhydride produced a crystalline pentacetate and suggested the presence of five hydroxyl groups; (b) reaction with hydroxylamine (Sec. 8.8) produced an oxime and suggested the presence of a carbonyl group; (c) mild oxidation with bromine in an aqueous solution yielded gluconic acid ($C_6H_{11}O_5COOH$), an **aldonic acid,** and indicated an aldehyde function at one end of the molecule. The latter observation was reenforced when the addition of HCN yielded a cyanohydrin, the hydrolysis of which, followed by reduction, produced heptanoic acid.

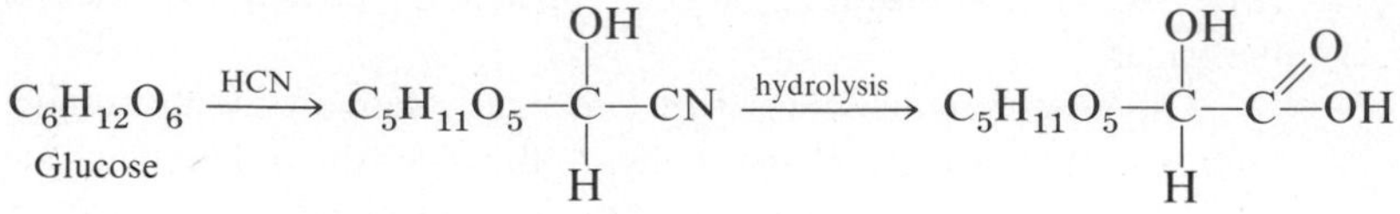

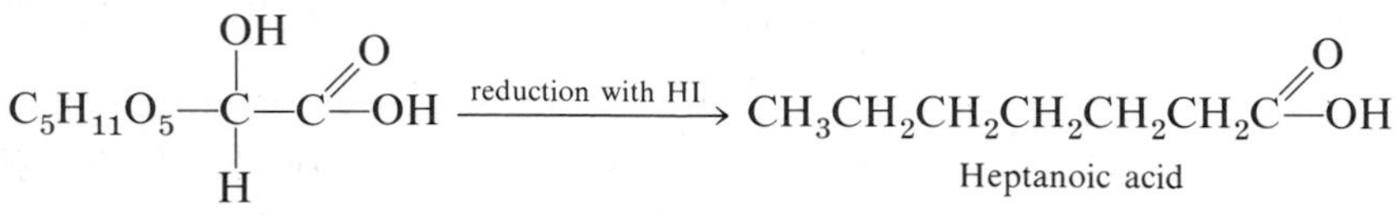

$$\underset{\text{Glucose}}{C_6H_{12}O_6} \xrightarrow{\text{HCN}} C_5H_{11}O_5-\underset{\displaystyle H}{\overset{\displaystyle OH}{C}}-CN \xrightarrow{\text{hydrolysis}} C_5H_{11}O_5-\underset{\displaystyle H}{\overset{\displaystyle OH}{C}}-\overset{\displaystyle O}{\overset{\|}{C}}-OH$$

$$C_5H_{11}O_5-\underset{\displaystyle H}{\overset{\displaystyle OH}{C}}-\overset{\displaystyle O}{\overset{\|}{C}}-OH \xrightarrow{\text{reduction with HI}} \underset{\text{Heptanoic acid}}{CH_3CH_2CH_2CH_2CH_2CH_2\overset{\displaystyle O}{\overset{\|}{C}}-OH}$$

The above reactions indicate a six-carbon chain with a terminal aldehyde group. A six-carbon aldehyde with five hydroxyl groups can have a stable structure only if each hydroxyl group is attached to a different carbon atom. The structure of glucose thus was proposed as the following:

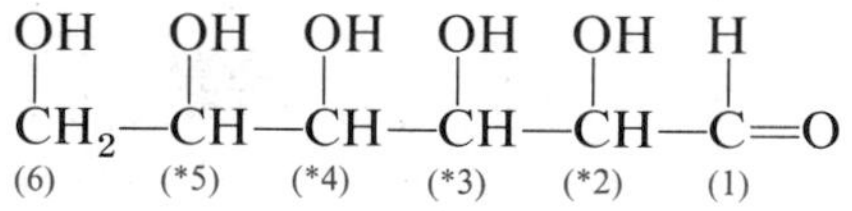

$$\underset{(6)}{\overset{\displaystyle OH}{CH_2}}-\underset{(*5)}{\overset{\displaystyle OH}{CH}}-\underset{(*4)}{\overset{\displaystyle OH}{CH}}-\underset{(*3)}{\overset{\displaystyle OH}{CH}}-\underset{(*2)}{\overset{\displaystyle OH}{CH}}-\underset{(1)}{\overset{\displaystyle H}{C}}=O$$

Glucose

14.4 Configuration of Glucose

Inspection of the above formula reveals the presence of four different asymmetric carbon atoms. No less than sixteen ($2^4 = 16$) optical isomers with the above structure are possible. Only two other aldohexoses (mannose and galactose) occur in nature. Which of the sixteen possible structures is glucose? Which is mannose? Which is galactose? The answers to these perplexing questions were obtained through a brilliant series of syntheses and degradation studies performed by a research group headed by Emil Fischer between 1891 and 1896. Most of the remaining thirteen aldohexoses were synthesized, and the configurations of all known isomers elucidated. To review all the chemistry that finally led to this great accomplishment is beyond the scope of this text, but a simplified scheme may be written to show how all eight D-forms of the aldohexoses were obtained.

The reference standard chosen for relating configurations of optically active compounds is D-glyceraldehyde (Sec. 5.4). The carbon chain of D-glyceraldehyde can be lengthened into a sugar molecule through repeated cyanohydrin formation. A lengthening of the carbon chain in this manner is known as the **Kiliani synthesis.**

```
     H                        CN              CN
     |                        |               |
     C=O                  H—C—OH        HO—C—H
     |             HCN        |               |
 H—C—OH          ———→    H—C—OH   +    H—C—OH
     |                        |               |
     CH2OH                    CH2OH           CH2OH
```

D (+) Glyceraldehyde

(A mixture of optically active diastereoisomers. Both retain the D configuration.)

```
     CN                      COOH                      H
     |                       |                         |
 H—C—OH                  H—C—OH                    C=O
     |                       |                         |
 H—C—OH                  H—C—OH                H—C—OH
     |                       |                         |
     CH2OH                   CH2OH                 H—C—OH
                                                       |
                hydrolysis              reduction[1]   CH2OH
                                                   D-Erythrose

     CN                      COOH                      H
     |                       |                         |
 HO—C—H                  HO—C—H                    C=O
     |                       |                         |
 H—C—OH                  H—C—OH                HO—C—H
     |                       |                         |
     CH2OH                   CH2OH                 H—C—OH
                                                       |
                                                       CH2OH
                                                   D-Threose
```

The schematic structures on the following page show how the configurations of the pentoses are related to those of their parent structures, the tetroses. Similarly, the configurations of the hexoses may be related to those of their antecedents, the pentoses. The hydroxyl groups in each structure are denoted by a short horizontal spur, and the carbon chain as a vertical line. The dotted line encloses only that portion of each structure that has a configuration identical with that of its parent. Mirror images of all structures shown would have resulted had we used L-glyceraldehyde as our starting material.

[1]Conversion to lactone and reduction with sodium borohydride at pH 3–5.

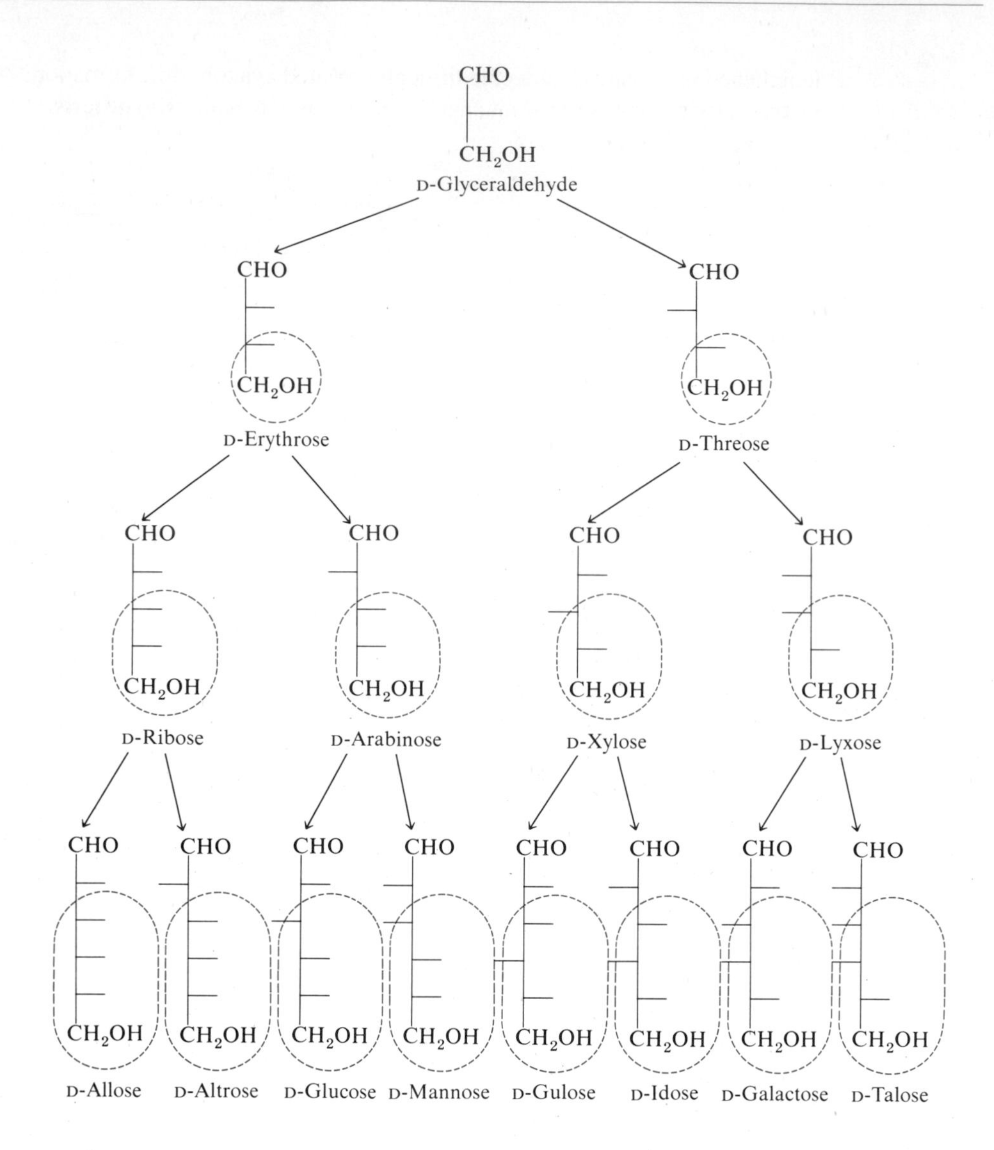

Exercise 14.1 The oxidation of glucose by strong nitric acid converts both the aldehyde and the primary alcohol group into carboxyl groups. The resulting dicarboxylic acid, called **glucaric acid,** was found to be optically active. Why did this result eliminate the structure shown for galactose as being that of glucose?

The Fischer projection formula for glucose is usually written as on the following page and is assigned a D-configuration (Sec. 5.4). The ''tail'' or lower extremity of the formula is $HOCH_2$—, that is, the hydroxymethyl group:

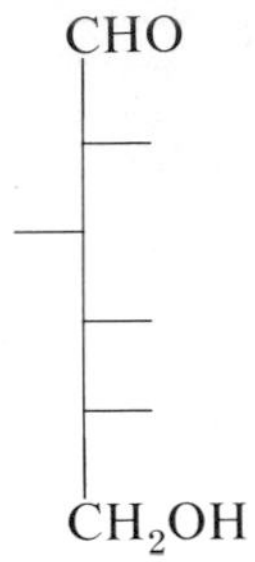

CHO

CH_2OH

Exercise 14.2 Designate the configuration about carbon 2 of erythrose as (*R*) or (*S*).

14.5 Cyclic Structure of Sugars. Mutarotation

Several properties of glucose are not explainable by the open-chain formula. For example, when D-glucose is heated with methanol in the presence of hydrogen chloride, the expected dimethyl acetal is not obtained. Instead, two optically active isomeric compounds result, each containing but *one* methoxyl group. Acetals such as those shown in structures II and III are named **glycosides.** If glycosides are formed from a sugar and a nonsugar hydroxy compound, the nonsugar component is called an **aglycone. Glucoside** is a name specifically assigned to a glycoside produced from glucose. The easily oxidized aldehyde group no longer exists in a glucoside. The methyl glucosides (structures II and III), therefore, are incapable of reducing Fehling's solution or Tollens' reagent. A second observation, unexplainable by the straight-chain formula, was that two crystalline forms of D-glucose could be isolated. One form, designated the α-form, crystallized from a concentrated aqueous solution at 30°C, decomposed at its melting point (147°C), and, in a freshly prepared solution, showed a specific rotation of +113°. The other form, designated the β-form, crystallized from a hot, glacial acetic acid solution, melted at 148–150°C, and, in a freshly prepared solution, showed a specific rotation of +19°. The rotation of an aqueous solution of either the α- or the β-form of glucose, when allowed to stand, was found to change. The rotation of each solution reached the equilibrium value of +52°. Such a change in rotation on standing is called **mutarotation.** Mutarotation is caused by a change in configuration of one asymmetric center in each isomer. The explanation for this behavior for glucose rests in the fact that the glucose molecule exists largely in one of two cyclic hemiacetal forms. Ring closure can result easily through intramolecular acetal formation when the hydroxyl group of carbon 5 is brought into close proximity to the carbonyl group. Zigzag carbon chains, as illustrated in structures IV and V, provide this condition. You will note that two different modes of addition are possible.

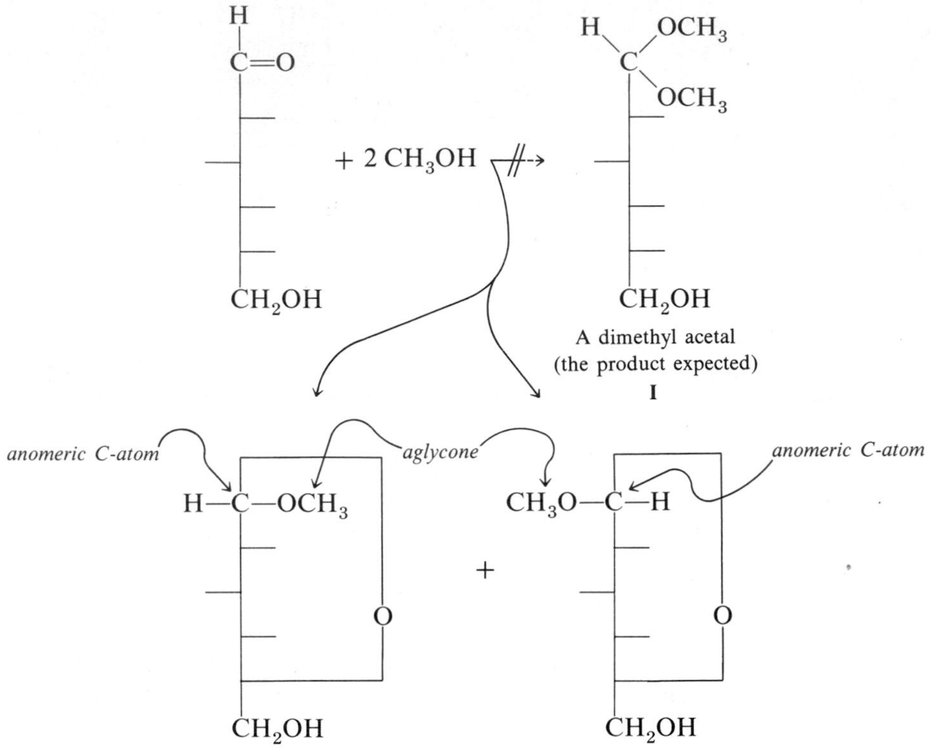

H

C=O

CH_2OH

$+ 2\,CH_3OH$

H OCH_3

C

OCH_3

CH_2OH

A dimethyl acetal
(the product expected)
I

anomeric C-atom

aglycone

anomeric C-atom

$H—C—OCH_3$

O

CH_2OH

+

$CH_3O—C—H$

O

CH_2OH

α-Methylglucoside
$[\alpha]_D^{20} = +159°$
II

β-Methylglucoside
$[\alpha]_D^{20} = -34°$
III

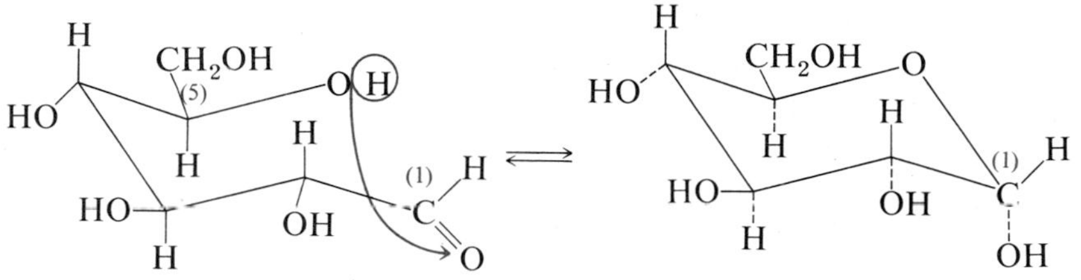

H, HO, CH_2OH, (5), O, H, H, H, HO, OH, (1), H, C, O, H

D-Glucose IV

H, HO, CH_2OH, O, H, H, (1), H, HO, OH, C, H, OH

α-D-Glucose

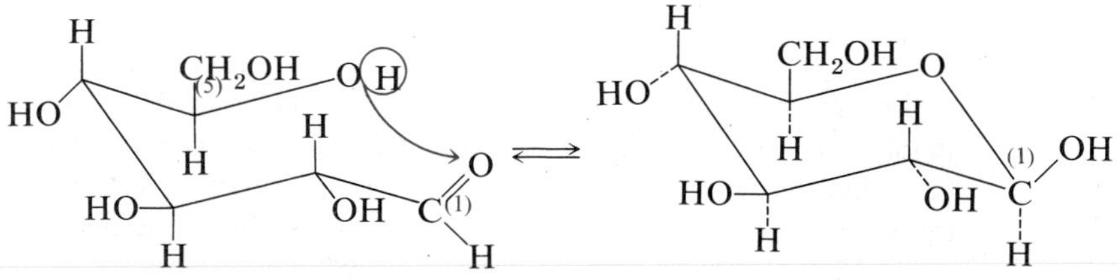

H, HO, CH_2OH, (5), O, H, H, H, O, HO, OH, C, (1), H, H

D-Glucose V

H, HO, CH_2OH, O, H, H, (1), OH, HO, OH, C, H, H

β-D-Glucose

Carbon 1 in each of the cyclic hemiacetal structures shown represents a new asymmetric center, and two diastereoisomers thus are possible: α- and β-D-glucose. Such stereoisomers are called **anomers,** and the hemiacetal carbon (or acetal carbon in the glycosides) is called the **anomeric** carbon atom. Enough of the open-chain (about 1%) form of glucose is present in the equilibrium mixture to give some of the reactions typical of the aldehyde group, but glucose exists preferentially in the cyclic form. The structures below represent typical ways of formulating the glucose molecule.

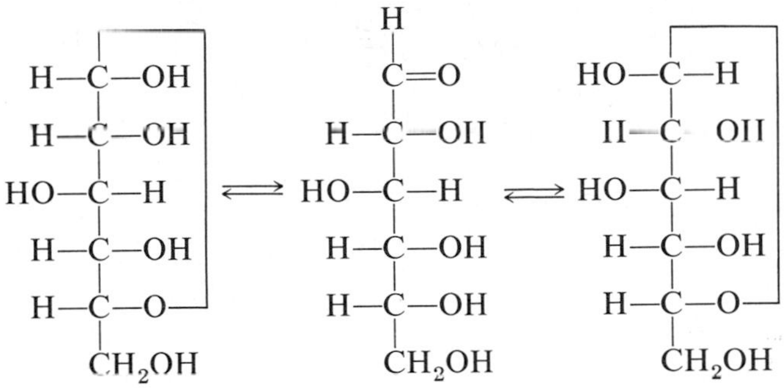

α-D-Glucose ⇌ D-Glucose ⇌ β-D-Glucose

The Haworth[2] formulas for the cyclic forms of glucose are drawn as planar, hexagonal slabs with darkened edges toward the viewer. Hydroxyl groups and hydrogen atoms are shown either as above (solid bonds) or below (dotted bonds) the plane of the hexagon.

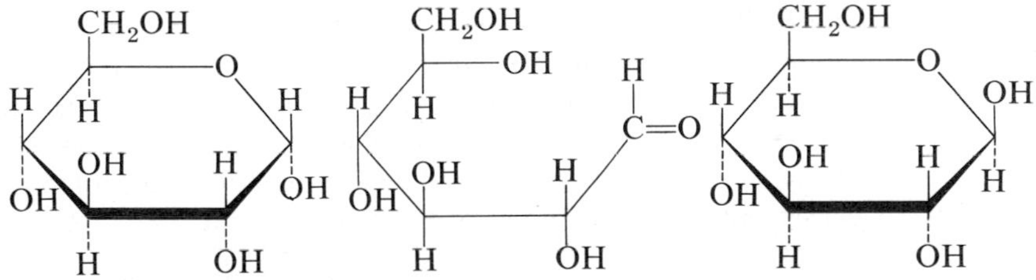

α-D-Glucose (α-D-Glucopyranose) ⇌ D-Glucose ⇌ β-D-Glucose (β-D-Glucopyranose)

The *pyranose* designation for the cyclic forms of glucose indicates a structural similarity to the 6-membered, heterocyclic **pyran** ring,

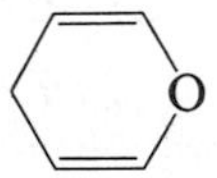

A *furanose* designation for a sugar indicates a 5-membered ring with a structural relationship to the heterocycle, **furan** (Sec. 4.9),

[2]Walter Haworth (1883–1950), Professor of Chemistry, University of Birmingham. Winner of the Nobel Prize in Chemistry in 1937.

The Haworth structures are easy to draw, but one must not forget that the 6-membered ring is very much like that of cyclohexane and can pucker (Sec. 2.9). Correct cyclic structures for glucose are indicated by the chair representations that follow. Of the two structures illustrated below for α-D-glucopyranose, the conformation in which most of the bulkier groups are bonded equatorially (structure VII) appears to be the more stable one. Similar conformations for β-D-glucopyranose can be drawn by interchanging the positions of the hydrogen atom and the hydroxyl group on carbon 1.

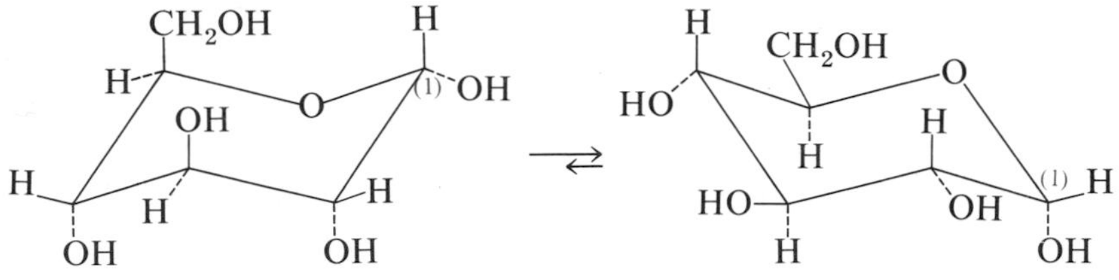

VI $\rightleftarrows$ VII

α-D-Glucopyranose

14.6 Fructose

Fructose, $C_6H_{12}O_6$, also called *levulose* because it is levorotatory, is the most widely distributed ketose. Fructose is found along with glucose in the juices of ripe fruits and in honey. It also occurs with glucose as a component of the disaccharide *sucrose*.

A comparison of the structures of glucose and fructose shows that fructose has only three asymmetric carbon atoms, and the configurations about these are identical to those of corresponding carbon atoms in the glucose chain.

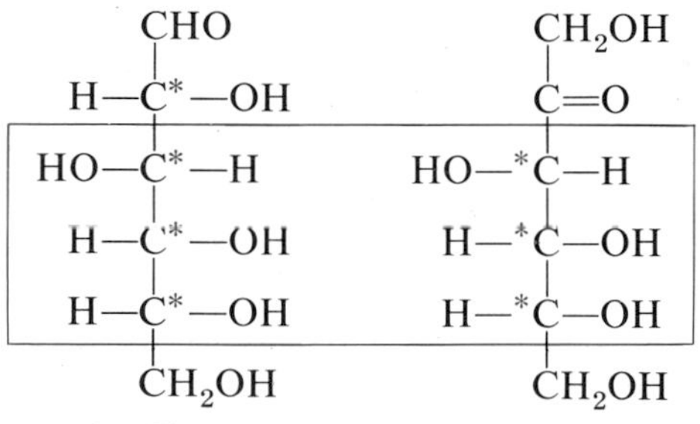

D(+)Glucose	D(−)Fructose
CHO	CH_2OH
H—C*—OH	C=O
HO—C*—H	HO—*C—H
H—C*—OH	H—*C—OH
H—C*—OH	H—*C—OH
CH_2OH	CH_2OH

Fructose, unlike glucose, cannot be oxidized by aqueous bromine and, as a keto sugar, one would not expect fructose to reduce Fehling's solution. However, the alkalinity of Fehling's solution is sufficient to cause a rearrangement of fructose to glucose or mannose, both easily oxidizable sugars.

The cyclic form of fructose may be that of either a pyranose or a furanose. In more complicated structures of which fructose is a part, it is found in the furanose form.

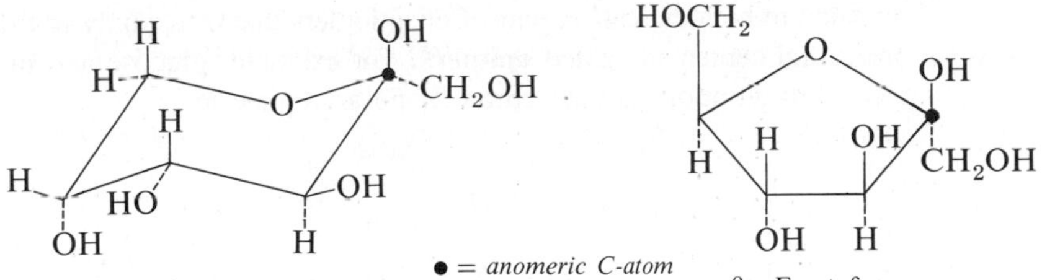

β-D-Fructopyranose

● = *anomeric C-atom*

β-D-Fructofuranose

14.7 Reactions of the Hexoses

A number of reactions involving both the carbonyl and the hydroxylic groups of the monosaccharides were reviewed in previous sections. Oxidation to aldonic acids by bromine, Tollens' reagent, and Fehling's solution, the addition of hydrogen cyanide, the formation of acetals and acetates, all were reactions helpful in elucidating the structure of glucose. In addition to the preceding reactions, the following also are important in sugar chemistry.

14.8 Reaction with Phenylhydrazine. Osazone Formation

The carbonyl function in a sugar molecule will, if free to react, condense with phenylhydrazine to produce a phenylhydrazone. The phenylhydrazones of sugars, unlike the phenylhydrazones of simple compounds containing a carbonyl function, usually are difficult to isolate. On the other hand, when an excess of phenylhydrazine is used, it forms with sugars a diphenylhydrazone derivative called an **osazone.** The osazones of sugars are easily crystallized and can be used for identification purposes. When excess phenylhydrazine is used, the carbinol adjacent to the carbonyl, that is, carbon No. 2 in aldoses but carbon No. 1 in ketoses, becomes oxidized to a second carbonyl group. The carbonyl formed by this oxidation, along with the one originally present, condenses with phenylhydrazine to produce the osazone. The mechanism of the oxidation step is not entirely clear, but it appears that for osazone formation to occur, the ratio of phenylhydrazine to sugar in the reaction mixture must be at least 3:1 because one molecule of phenylhydrazine is changed to aniline and ammonia. Inasmuch as only carbons 1 and 2 are involved in osazone formation, D-glucose, D-mannose, and D-fructose all form identical osazones.

The configurations of carbon atoms 3, 4, and 5 in each of these three sugars are thus revealed to be identical. A pair of enantiomers that differ only in the configuration about one chiral center are called **epimers.** For example, glucose and mannose differ only in their configurations about carbon 2 and are epimeric.

$$
\begin{array}{ccccc}
\begin{array}{c} H \\ | \\ {}^{(1)}C{=}O \\ | \\ H{-}\underset{(2)}{C}{-}OH \\ | \\ HO{-}\underset{(3)}{C}{-}H \\ | \\ H{-}\underset{(4)}{C}{-}OH \\ | \\ H{-}\underset{(5)}{C}{-}OH \\ | \\ {}^{(6)}CH_2OH \end{array}
& \text{or} &
\begin{array}{c} H \\ | \\ C{=}O \\ | \\ HO{-}C{-}H \\ | \\ HO{-}C{-}H \\ | \\ H{-}C{-}OH \\ | \\ H{-}C{-}OH \\ | \\ CH_2OH \end{array}
& \text{or} &
\begin{array}{c} CH_2OH \\ | \\ C{=}O \\ | \\ HO{-}C{-}H \\ | \\ H{-}C{-}OH \\ | \\ H{-}C{-}OH \\ | \\ CH_2OH \end{array}
\quad + 3\,C_6H_5\overset{H}{\overset{|}{N}}{-}NH_2
\\
\text{D-Glucose} & & \text{D-Mannose} & & \text{D-Fructose}
\end{array}
$$

$$
\longrightarrow
\begin{array}{l}
\quad H \qquad\; H \\
\quad | \qquad\quad | \\
\quad C{=}N{-}N{-}C_6H_5 \\
\quad | \qquad\quad H \\
\quad | \qquad\quad | \\
\quad C{=}N{-}N{-}C_6H_5 + C_6H_5NH_2 + NH_3 + 2\,H_2O \\
\quad | \\
HO{-}C{-}H \\
\quad | \\
H{-}C{-}OH \\
\quad | \\
H{-}C{-}OH \\
\quad | \\
\quad CH_2OH
\end{array}
$$

Osazone of D-Glucose,
D-Mannose, and D-Fructose

Exercise 14.3 Criticize the following statement: "Two hexoses that react with phenylhydrazine and yield identical osazones are epimers."

Exercise 14.4 Each pair of sugars produced from their antecedents in the scheme on page 388 became epimeric at carbon 2. Are any of the eight hexose D-forms epimeric with glucose at carbon atoms other than number 2?

14.9 O-Acylation and O-Alkylation

The free hydroxyl groups of monosaccharides and polysaccharides show many of the normal reactions of alcohols. Thus, treatment of α-D-glucose with excess acetic anhydride gives penta-O-acetyl-α-D-glucose.

$(CH_3CO)_2O$

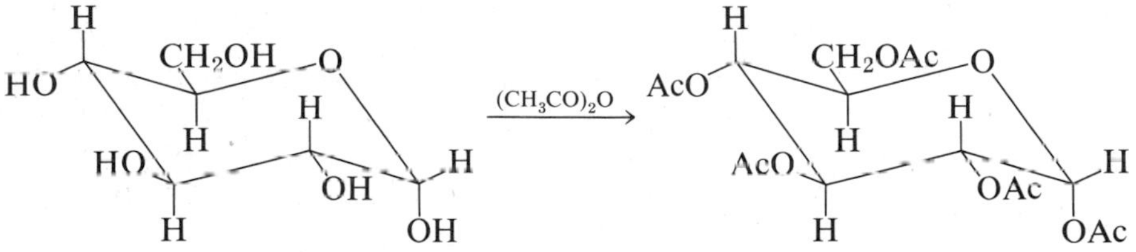

α-D-Glucose

Penta-O-acetyl-α-D-glucose

Ac = $CH_3C(=O)-$

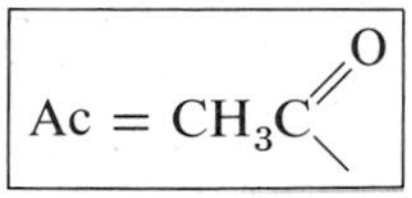

Methylation of the hydroxyl group on the anomeric carbon atom of a monosaccharide is accomplished with the use of methanol and hydrogen chloride (Sec. 14.5). Methylation of the remaining free hydroxyl groups requires the use of a modified Williamson synthesis (Sec. 7.11), that is, by treatment with methyl iodide and silver oxide or with dimethyl sulfate and 30% aqueous sodium hydroxide. Thus, methylation of methyl α-D-glucopyranoside yields methyl 2,3,4,6-tetra-O-methyl-D-glucopyranoside. Mild acid hydrolysis may be used to cleave the glycosidic linkage without affecting the other O-alkyl groups, which are ethers. Methylation is commonly used to establish the position of substituents on the carbohydrate molecule or to determine whether a glycoside has the furanose and pyranose ring structure.

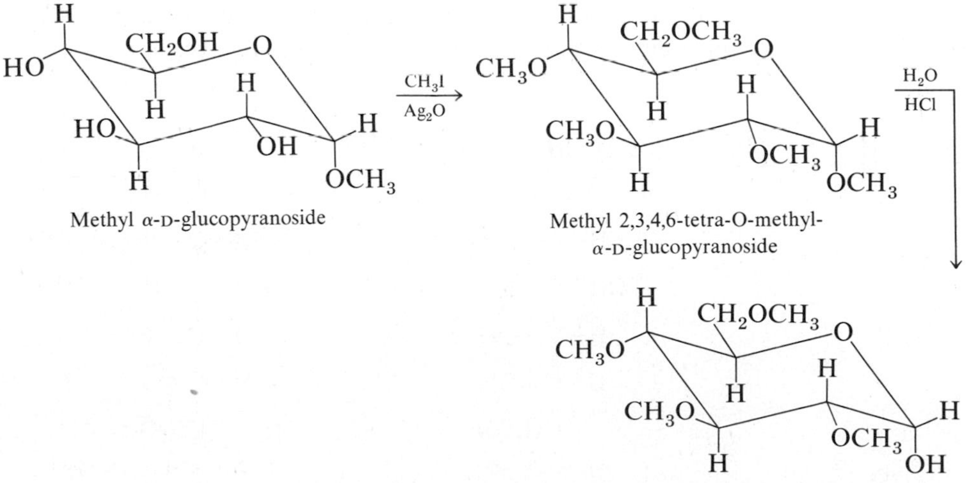

CH_3I, Ag_2O

Methyl α-D-glucopyranoside

Methyl 2,3,4,6-tetra-O-methyl-α-D-glucopyranoside

H_2O, HCl

2,3,4,6-Tetra-O-methyl-D-glucopyranose

Exercise 14.5 Open the ring of 2,3,4,6-tetra-O-methyl-α-D-glucopyranoside (the hydrolysis product in the previous reaction) and draw the methylated glucose as an open-chain structure. How did this series of reactions help establish the pyranose structure of D-glucose?

14.10 Fermentation of Sugars

Emil Fischer found that of the sixteen aldohexoses, only those found in nature (D-glucose, D-mannose, and D-galactose) could be fermented by yeast. The three simple sugars just named, along with D-fructose, are acted upon by the enzyme **zymase** to yield the same decomposition products—ethyl alcohol and carbon dioxide. The reactions by which alcohol is produced from a sugar are complex, but the net result may be indicated by the equation below.

$$C_6H_{12}O_6 \xrightarrow{\text{zymase}} 2\ C_2H_5OH + 2\ CO_2$$

The fermentation of sugar, the oldest chemical reaction known, provided the ancients with wine and the leavening action for making bread.

It is interesting to note that the same three aldoses that are fermentable also form 2-amino derivatives called "osamines." Where found in nature, these amino sugars are in the N-acetylated form. N-Acetyl-D-glucosamine is a component of *chitin,* the horny polysaccharide that forms part of the outer shell of crustaceans.

$$\begin{array}{c} \text{H} \\ | \\ \text{C}{=}\text{O} \\ | \\ \text{H—C—NH}_2 \\ | \\ \text{HO—C—H} \\ | \\ \text{H—C—OH} \\ | \\ \text{H—C—OH} \\ | \\ \text{CH}_2\text{OH} \end{array}$$

D-Glucosamine

$$\begin{array}{c} \text{H} \\ | \\ \text{C}{=}\text{O} \quad\quad \text{O} \\ | \quad\quad\quad \| \\ \text{H—C—NH—C—CH}_3 \\ | \\ \text{HO—C—H} \\ | \\ \text{H—C—OH} \\ | \\ \text{H—C—OH} \\ | \\ \text{CH}_2\text{OH} \end{array}$$

N-Acetyl-D-glucosamine

14.11 Oligosaccharides

Disaccharides, the most important of the oligosaccharides, may be regarded as glycosides. However, unlike the simple methylglucosides (Sec. 14.5), the hydroxylic compound that is coupled through the glycosidic link is a second monosaccharide unit—not an aglycone. A molecule of water is split out from two monosaccharide units to form a disaccharide. Restoration of the water by hydrolysis of the disaccharide reforms the two monosaccharide components. Hydrolysis may be effected by dilute acids or by enzymes. The most important of the disaccharides and the only ones we shall consider are *sucrose, maltose, cellobiose,* and *lactose.*

14.12 Sucrose

Sucrose, ordinary table sugar, is obtained from the juices extracted from sugar cane and sugar beets. The juice from either source contains about 14–25% of sucrose. To a small extent, sucrose is obtained from the sap of certain species of maple trees. Maple sugar is marketed largely as a syrup. Sucrose is dextrorotatory but on hydrolysis produces equimolar quantities of D-glucose (dextrose) and D-fructose (levulose). During the hydrolysis of sucrose the sign of the specific rotation changes from positive to negative, or is said to **invert.** The hydrolysis mixture is called *invert sugar*. The enzyme *invertase,* carried by the bee, is able to accomplish the same result in the production of honey.

$$\underset{\substack{\text{Sucrose} \\ [\alpha]_D^{20} = +66.5^\circ}}{C_{12}H_{22}O_{11}} \xrightarrow[\text{(or } invertase\text{)}]{\text{acid hydrolysis}} \underbrace{\underset{[\alpha]_D^{20} = +52^\circ}{\text{D-Glucose}} + \underset{[\alpha]_D^{20} = -92^\circ}{\text{D-Fructose}}}_{\text{Invert sugar}}$$

$$\frac{+52 - 92}{2} = [\alpha]_D^{20} = -20^\circ$$

The sweetness of honey is largely due to the presence of fructose, which is approximately three times as sweet as glucose. Invert sugar shows less tendency to crystallize than does sucrose and for this reason is used to a large extent in the manufacture of candy. Sucrose will not reduce Fehling's solution and therefore is classified as a *nonreducing* disaccharide. The α-D-glucopyranose ring, as shown in the sucrose structure on page 398, is joined to the β-D-fructofuranose ring through a glycosidic link. The anomeric carbon atoms of glucose and of fructose are involved in this union. A "head-to-head" arrangement of this type leaves no potential aldehyde function to be oxidized and explains why sucrose is nonreducing. A junction at these ring positions further explains why sucrose fails to form an osazone with phenylhydrazine and why it fails to show mutarotation.

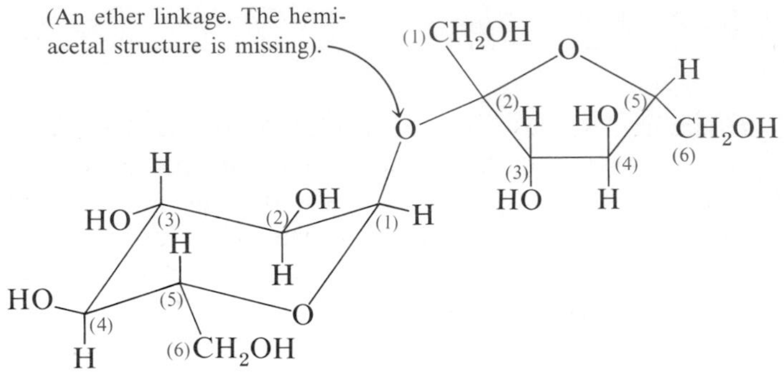

(An ether linkage. The hemiacetal structure is missing).

α-D-Glucose unit β-D-Fructose unit

Sucrose
(α-D-Glucopyranosyl β-D-fructofuranoside)

Sucralfate is a basic aluminum salt of sucrose heptasulfate in which the hydrogen of each OH group of sucrose has been replaced by $SO_3^{-}Al_2(OH)_5^{+}$. This promising new medicinal product was approved for use in 1981 and is said to be effective in the prevention of the recurrence of gastric ulcers.

14.13 Maltose

Maltose, or malt sugar, is a disaccharide produced when starch is hydrolyzed by malt *diastase,* an enzyme found in sprouting barley. Another enzyme, *maltase,* selectively splits the alpha-glycosidic link and completely hydrolyzes maltose to yield two D-glucose units. Maltose is a *reducing sugar*. This chemical evidence suggests the presence of an aldehyde group either uncombined or in equilibrium with the hemiacetal form. In agreement with this, maltose forms an osazone and also exhibits mutarotation. Maltose is dextrorotatory and gives at equilibrium a specific rotation value of +136°. The structure of maltose is that of two glucose units in a "head-to-tail" arrangement joined through an α-linkage from carbon 1 of one glucose unit to carbon 4 of a second glucose unit.

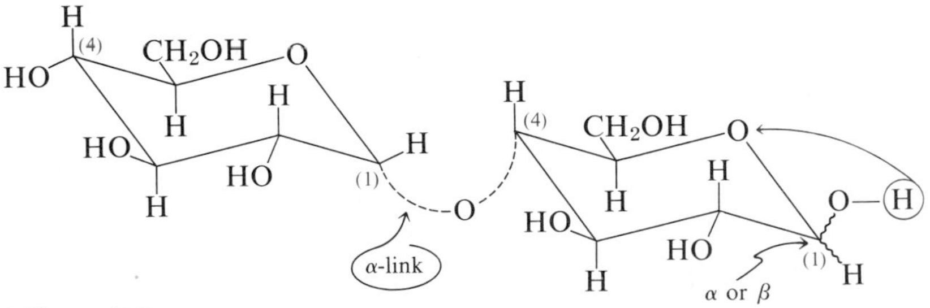

α-link

α or β

$[\alpha]_D^{20} = +136°$

Maltose

(Potential aldehyde group in hemiacetal structure)

14.14 Cellobiose

Cellobiose is a disaccharide that is obtained by the hydrolysis of cellulose. It is a reducing sugar consisting of two glucose units joined as in maltose, *but through a β-linkage*. The enzyme maltase is incapable of hydrolyzing cellobiose. In all other respects the behavior of cellobiose is identical to that described for maltose.

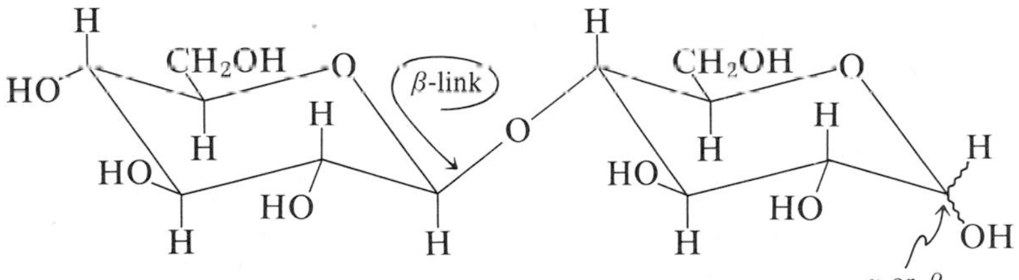

Cellobiose

14.15 Lactose

Lactose is known as milk sugar because it is present in the milk of mammals. It is present in cows' milk to the extent of about 5% and in human milk to about 7%. It is produced commercially as a by-product in the manufacture of cheese. Lactose is a reducing sugar, forms an osazone, and exhibits mutarotation. Lactose is dextrorotatory and, at equilibrium, gives a specific rotation of +55°. Lactose, when hydrolyzed by mineral acids or by the action of the enzyme *lactase*, produces equimolar quantities of D-glucose and D-galactose. Lactose is a β-galactoside in which the anomeric 1 carbon of galactose is joined through a β-linkage to the number 4 carbon of glucose in a "head-to-tail" arrangement. This is shown below.

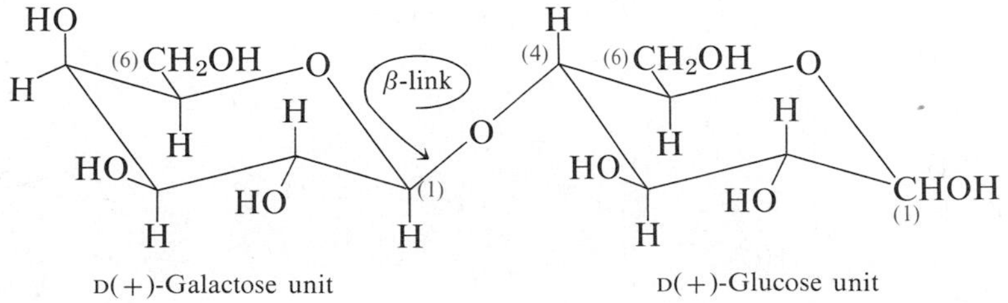

D(+)-Galactose unit D(+)-Glucose unit

Lactose

14.16 Polysaccharides

The polysaccharides are high molecular weight (25,000–15,000,000) natural polymers in which hundreds or even thousands of pentose or hexose units have been joined through glycosidic linkages. The most important polysaccharides are starch, glycogen, inulin, and cellulose.

14.17 Starch, Glycogen, and Inulin

Starch is the reserve carbohydrate of most plants. It comprises the major part of all cereal grains and most plant tubers, where it is stored. Starch is used as a principal food source throughout the world. Glycogen is the reserve carbohydrate of animals, and a relatively small amount is stored in the liver and muscles. Structurally, starch and glycogen are similar. Both have the empirical formula $(C_6H_{10}O_5)_n$, and both, when completely hydrolyzed, yield glucose. Starch and glycogen are only partially hydrolyzed to maltose by the enzyme *amylase*. When a paste made of starch and water is heated, two fractions may be separated. One, called the **amylose fraction,** is water-soluble and has a molecular weight range of 20,000–225,000. A second fraction, called **amylopectin,** is water-insoluble and has a molecular weight range of 200,000–1,000,000.

Amylose can be hydrolyzed almost completely to maltose by the enzyme β-amylase. This hydrolysis indicates that the maltose units arise from glucose units joined through alpha 1–4 linkages. Amylopectin, on the other hand, is hydrolyzed to maltose to a much lesser degree by the same enzyme. Chemical evidence indicates that considerable branching occurs in the amylopectin fraction of the starch molecule. Such branches are formed from glucose units linked through carbon atoms 1 and 6. The glycosidic link at these positions resists hydrolysis by β-amylase. A segment of a branched starch molecule is shown.

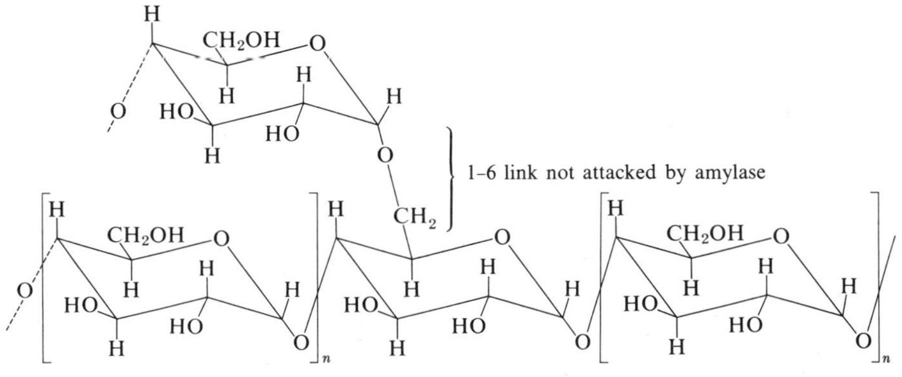

1-6 link not attacked by amylase

A segment of the starch molecule

Inulin is a polysaccharide comparable to starch. It is found in the underground stems (tubers) of the Jerusalem artichoke and in dahlia roots. Inulin, on hydrolysis, yields fructose as its end product.

14.18 Cellulose

Cellulose comprises the skeletal material of plants and is the most abundant organic substance found in nature. It is the chief constituent of wood and cotton. Cotton, almost pure cellulose, is the principal source of cellulose used as fiber for fabrics. The cellulose content of wood is approximately 50 per cent. The separation of cellulose from other plant components is an important commercial process upon which the textile, paper, and plastics industries are largely dependent.

The general formula for cellulose, like that for starch, may be written $(C_6H_{10}O_5)_n$, but the numerical value of n in the formula for cellulose is much larger than that for starch. Methylation studies indicate that the structure of cellulose, unlike that of starch, is largely unbranched.

Complete hydrolysis of cellulose produces D-glucose. Partial hydrolysis produces cellobiose (Sec. 14.14), cellotriose, and higher oligosaccharides. The presence of β-glycosidic linkages establishes the structure of cellulose as

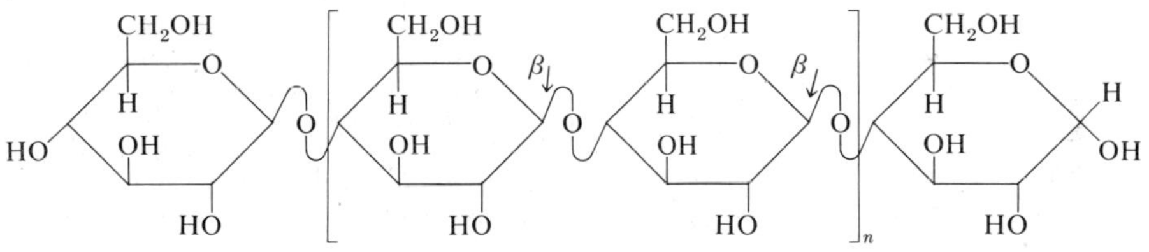

Structure of cellulose unit

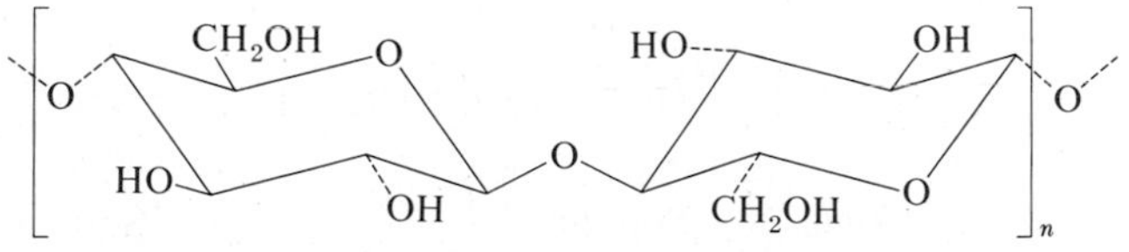

Conformation of cellulose

Humans are incapable of utilizing cellulose for food because their digestive juices lack enzymes capable of hydrolyzing the β-glycosidic linkage. Ruminants, or cud-chewing animals, are able to digest cellulose because certain microorganisms present in their compartmented stomachs cause a preliminary hydrolysis of cellulose before it reaches the intestine. Certain lower orders of animals such as snails and termites, having similar assistance, are also able to feed upon cellulose.

14.19 Derivatives of Cellulose

The derivatives of cellulose, like those of any polyhydric alcohol, are principally esters and ethers. Each $C_6H_{10}O_5$ unit of cellulose has three hydroxyl groups available. The conversion of some or all of these groups to ethers or to ester functions alters the properties of cellulose remarkably. Table 14.1 lists some of the currently used derivatives and their applications. The first three entries in this listing are water-soluble derivatives of cellulose, and the last three are water-insoluble. Usually two or three of the OH groups in each glucose unit are replaced by the substituent listed.

Cellulose nitrates are formed when cellulose is nitrated in the presence of sulfuric acid. Nitration can be controlled to esterify only some or nearly all hydroxyl groups, as shown.

$$\underset{\text{Cellulose}}{[C_6H_7O_2(OH)_3]_n} \xrightarrow{2n\text{HNO}_3,\ H_2SO_4} \underset{\text{Pyroxylin}}{[C_6H_7O_2(OH)(ONO_2)_2]_n}$$

Guncotton is a product obtained when cellulose is almost completely nitrated under conditions carefully controlled to prevent degradation of the cellulose molecule. Guncotton contains about 12–13% of nitrogen, is explosive, and is used in the manufacture of smokeless powder.

Cellulose acetate is obtained by the reaction of cellulose with acetic anhydride. Like the nitrates, cellulose acetate also is an ester but is not explosive. A viscous solution of cellulose acetate, when extruded through the fine openings of a die, called a ''spinneret,'' produces thin fibers. From such fibers the ''acetate'' fabrics are woven. Photographic film also is prepared from cellulose acetate. Cellulose from cotton linters or wood pulp, when treated with sodium hydroxide and carbon disulfide, can be converted into a product

TABLE 14.1 Common Cellulose Derivatives

Substituent	Name	Applications
$—O—CH_3$	Methylcellulose	Water-base paints, paper coatings, ice cream, adhesives, inks
$—OCH_2CH_2OH$	Hydroxyethylcellulose	Water-base paints, binder in fabrics, adhesives, ice cream, inks
$—OCH_2CO_2Na$	Sodium carboxymethylcellulose	Soil-suspending agent in detergents, textile sizing, oil-drilling mud, paper coatings
$—OCOCH_3$	Cellulose acetate	Films, lacquers, fibers
$—ONO_2$	Cellulose nitrate	Lacquers, fabric coatings, film, explosives
$—OCH_2CH_3$	Ethyl cellulose	Coatings, plastics

known as a **xanthate.** Xanthates with dilute alkali form a heavy, viscous solution called "viscose." Cellulose in the form of *rayon* is regenerated when viscose is extruded through a spinneret into a bath of dilute mineral acid. Extrusion of viscose through a long, thin slit produces *cellophane,* still a major packaging material. The chemical steps involved in the formation of viscose rayon may be illustrated by applying the same reactions to a simple alcohol.

$$ROH + NaOH + CS_2 \longrightarrow \underset{\text{A xanthate}}{RO{-}\overset{\overset{\displaystyle S}{/\!/}}{C}{-}S^-Na^+} + H_2O$$

$$RO\overset{\overset{\displaystyle S}{/\!/}}{C}{-}S^-Na^+ + H_2SO_4 \longrightarrow ROH + CS_2 + NaHSO_4$$

Excellent cellulose sponges may be made from xanthate by reconstituting the cellulose in the presence of crystals of sodium sulfate of various sizes. The sodium sulfate is washed out, leaving a spongelike material.

Ethyl cellulose is prepared by a modified Williamson synthesis. Alkali-cellulose, produced as in the viscose process, is treated with ethyl chloride to produce the ether.

14.20 Digestion and Metabolism of Carbohydrates

The digestion of carbohydrates involves the enzymatic hydrolysis of carbohydrates to produce the simple sugars: glucose, fructose, and galactose. These simple sugars are absorbed into the bloodstream and are transported first to the liver and eventually to the muscles. At these sites, glycogen can be synthesized and temporarily stored. These carbohydrate stores, when called upon to supply energy, are hydrolyzed to glucose. A series of enzymatic reactions cleaves the six-carbon glucose chain into two three-carbon pyruvic acid molecules,

$$CH_3{-}\overset{\overset{\displaystyle O}{\|}}{C}{-}\overset{\overset{\displaystyle O}{\|}}{C}OH$$

Oxidative decarboxylation of pyruvic acid, in association with CoA (Sec. 12.6), forms acetyl CoA, which then enters the tricarboxylic acid cycle as shown in Fig. 14.1. All carbohydrates, in meeting the energy requirements of the body, ultimately are oxidized to carbon dioxide and water.

FIGURE 14.1 The Tricarboxylic Acid Cycle

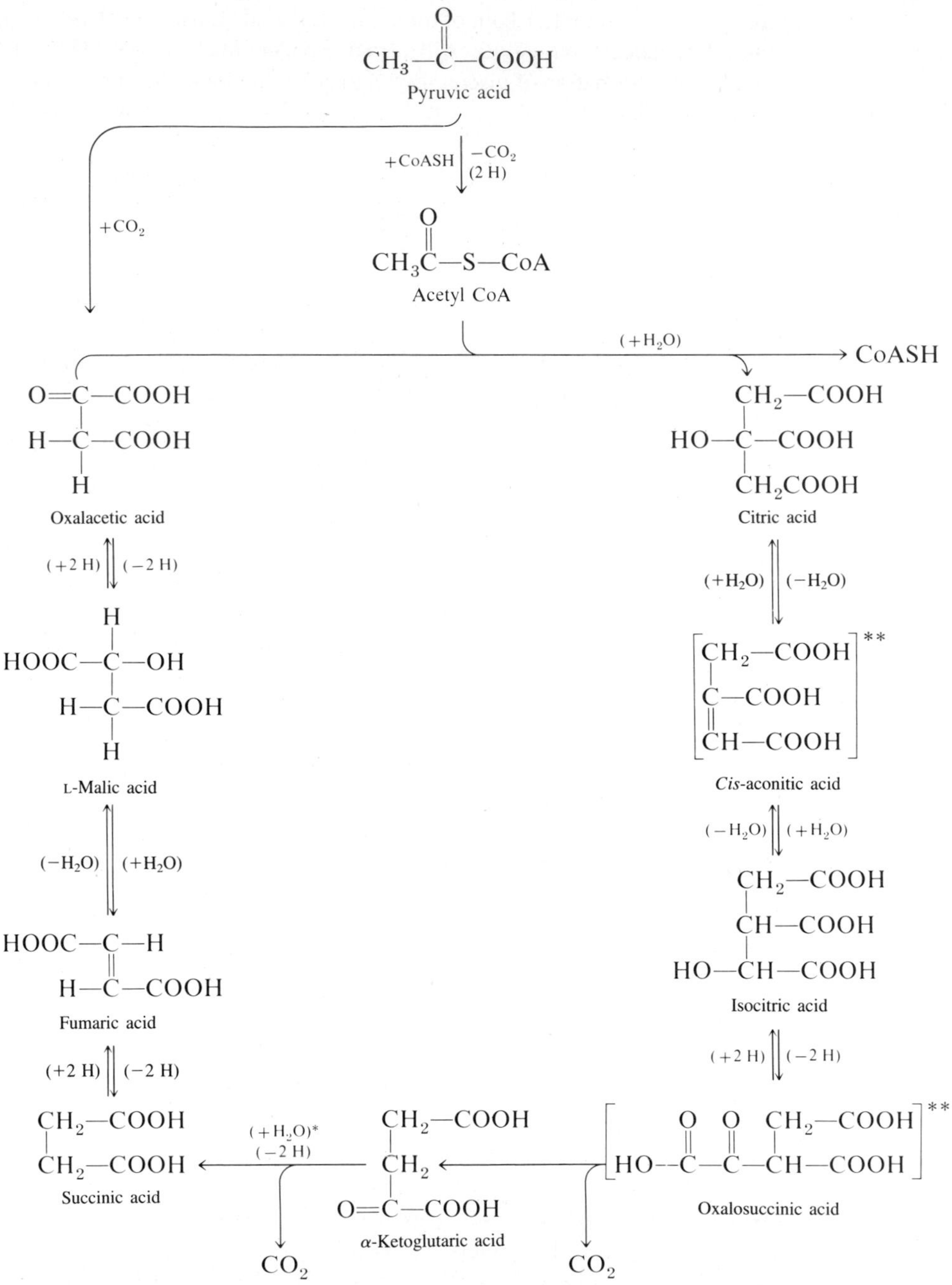

$CH_3-C(=O)-COOH$
Pyruvic acid

$+CO_2$

$+CoASH$ ↓ $-CO_2$ (2 H)

$CH_3C(=O)-S-CoA$
Acetyl CoA

$(+H_2O)$ → CoASH

$O=C-COOH$
$H-C-COOH$
H
Oxalacetic acid

CH_2-COOH
$HO-C-COOH$
CH_2COOH
Citric acid

$(+2\ H)$ ⇅ $(-2\ H)$

$(+H_2O)$ ⇅ $(-H_2O)$

H
$HOOC-C-OH$
$H-C-COOH$
H
L-Malic acid

[CH_2-COOH
$C-COOH$
$CH-COOH$]**
Cis-aconitic acid

$(-H_2O)$ ⇅ $(+H_2O)$

$(-H_2O)$ ⇅ $(+H_2O)$

$HOOC-C-H$
$H-C-COOH$
Fumaric acid

CH_2-COOH
$CH-COOH$
$HO-CH-COOH$
Isocitric acid

$(+2\ H)$ ⇅ $(-2\ H)$

$(+2\ H)$ ⇅ $(-2\ H)$

CH_2-COOH
CH_2-COOH
Succinic acid

← $(+H_2O)$* $(-2\ H)$ → CO_2

CH_2-COOH
CH_2
$O=C-COOH$
α-Ketoglutaric acid

← → CO_2

[$HO-C(=O)-C(=O)-CH(CH_2-COOH)-COOH$]**
Oxalosuccinic acid

* Succinoyl-CoA is an intermediate in this step. ** Enzyme-bound intermediates.

Summary

1. Carbohydrates are polyhydroxy aldehydes or polyhydroxy ketones.
2. Carbohydrates, in general, may be classified as mono-, oligo-, and polysaccharides. Monosaccharides may be classified functionally as aldoses or ketoses, according to chain length as pentoses or hexoses.
3. Glucose is the most important monosaccharide. It is an aldohexose but exists largely in a cyclic hemiacetal structure. It exhibits *mutarotation* and has an equilibrium rotation value of +52°.
4. Important reactions of glucose include
   (a) reduction of Fehling's solution.
   (b) formation of a hemiacetal (glucoside) with an alcohol.
   (c) formation of an osazone with phenylhydrazine.
   (d) formation of a pentacetate with acetic anhydride.
   (e) addition of HCN to form a cyanohydrin.
   (f) fermentation to ethyl alcohol and carbon dioxide.
   (g) formation of ethers with dimethyl sulfate and a base.
5. The disaccharides sucrose, maltose, cellobiose, and lactose are the most important oligosaccharides. Sucrose and maltose are α-glucosides. Lactose is a β-galactoside. Maltose and lactose reduce Fehling's solution, form osazones, and exhibit mutarotation; sucrose does none of these. The hydrolysis of sucrose ''inverts,'' or changes, its rotation value from (+) to (−). Invert sugar is hydrolyzed sucrose.
6. Polysaccharides are high molecular weight polymers in which thousands of basic pentose or hexose units are combined. The most important polysaccharides are starch, glycogen, and cellulose.
7. Starch is composed of thousands of glucose units joined predominantly through α-linkages at carbons 1 and 4. It is the reserve carbohydrate of plants.
8. Glycogen is the reserve carbohydrate of animals. It is stored in the liver and muscle.
9. As with starch, cellulose is composed of glucose units, but the connections between glucose molecules in cellulose are β-linkages. Moreover, cellulose appears to be an unbranched structure.
10. Cellulose derivatives (esters and ethers) are useful plastics. Among these are
    (a) cellulose nitrates, which are used in smokeless powder, lacquers, and other surface coatings.
    (b) cellulose acetate, which is used in textiles and photographic film.
    (c) ethyl cellulose, which is used in a variety of plastic articles.
    (d) water-soluble ethers, such as methyl cellulose, hydroxymethyl cellulose, and carboxymethyl cellulose, used in water-base paints, foods, inks, and adhesives, and as soil suspension agents.
11. Cellulose that has been regenerated from viscose (a cellulose xanthate) may be formed into *rayon* and *cellophane*.
12. The digestion of carbohydrates results in the production of simple sugars.
13. The metabolism of simple sugars involves a series of degradation reactions via the tricarboxylic acid cycle.

Supplementary Exercises

14.6 Define each of the following terms: (a) furanose, (b) invert sugar, (c) glycosidic linkage, (d) mutarotation, (e) pyranose, (f) epimer, (g) anomeric carbon, (h) lactase, (i) glucaric acid, (j) aglycone.

14.7 Write Fischer open-chain projection formulas for D-glucose and D-fructose. Draw cyclic structures for each and show the manner in which they are joined to form sucrose, maltose, and lactose.

14.8 Write balanced equations to show the products formed when glucose is caused to react with each of the following:

(a) phenylhydrazine
(b) Fehling's solution
(c) CH_3OH, anhydrous HCl
(d) HNO_3
(e) acetic anhydride
(f) zymase in yeast
(g) Br_2 (aqueous)
(h) hydroxylamine

14.9 Explain why both maltose and lactose are reducing sugars but sucrose is not.

14.10 Explain the apparent contradiction in the following statement: "Fructose is not a reducing sugar but may be oxidized by Fehling's solution."

14.11 Draw structures for each of the following.

(a) methyl α-D-glucoside
(b) sucrose
(c) a structural unit of starch
(d) a structural unit of cellulose

14.12 Using R—OH to represent a cellulose unit, write equations illustrating the preparation of

(a) cellulose acetate
(b) smokeless powder
(c) viscose rayon
(d) ethyl cellulose

14.13 Which of the following tests would serve to distinguish between the sugars in each of the pairs listed below: (a) bromine (aqueous), (b) Fehling's solution (c) phenylhydrazine, (d) a polarimetric measurement.

(A) glucose and fructose
(B) glucose and mannose
(C) maltose and lactose
(D) sucrose and maltose

14.14 An optically active hexose, (A), $C_6H_{12}O_6$, was degraded to an optically active pentose (B), $C_5H_{10}O_5$, by the following series of reactions:

$$\text{H(O=)C–CHOH} \xrightarrow{Br_2,\ H_2O} \text{HO(O=)C–CHOH} \xrightarrow{H_2O_2,\ Fe^{3+}} \text{H(O=)C–} + CO_2 + H_2O$$

$$(B) + HNO_3 \longrightarrow C_5H_8O_7 \text{ (optically active)}$$

Compound (B) was degraded to an optically active tetrose (C), $C_4H_8O_4$, through the same series of reactions as shown above.

$$\text{(C)} + HNO_3 \longrightarrow \textit{meso} \text{ tartaric acid}$$

Give possible structures for (A) and (B).

14.15 To which carbohydrates or their derivatives do the following phrases apply: (a) highly explosive, (b) stored in the liver, (c) a polyfructofuranoside, (d) on hydrolysis yields equimolar amounts of galactose and glucose, (e) regenerated cellulose, (f) a polyglucoglucoside.

14.16 When naturally occurring aldotetrose A is reduced with sodium borohydride, *meso*-1,2,3,4-butanetetrol is formed. What are the name and structure of A?

14.17 The α- and β-anomers of D-mannose have the optical rotations $[\alpha]_D = +29.3°$ and $[\alpha]_D = -16.3°$, respectively, in aqueous solution. On standing, solutions of either anomer mutarotate to an equilibrium value of $[\alpha]_D = +14.5°$. Calculate the relative amounts of each anomer present at equilibrium.

14.18 For the aldohexose, D-gulose, draw the perspective (Haworth) formulas for the following:

(a) α-D-Gulose
(b) α-L-Gulose
(c) Anomer of (a)
(d) Epimer of (a)
(e) A reducing disaccharide formed from D-glucose and D-gulose in which the C-4 hydroxyl of D-glucose is one point of attachment.

14.19 A pentose of the D-family gave an optically active glycaric acid (a dicarboxylic acid (Exercise 14.1)) on oxidation with nitric acid. The pentose was converted by the Kiliani reaction into a pair of diastereoisomeric hexoses, one of which gave an optically active glycaric acid when oxidized with nitric acid and the other of which gave an optically inactive glycaric acid. What are the names and configurations of the pentose and the two hexoses?

14.20 Two aldoses A and B give the same osazone. Oxidation of A gives an optically inactive glycaric acid, and oxidation of B gives an optically active glycaric acid. Treatment of D-glucose with hydrogen cyanide followed by hydrolysis gives a mixture of the glycaric acids obtained from A and B. From these observations deduce the structures of A and B.

14.21 Oxidative degradation of a glycoside with periodic acid (Exercise 8.32) may be used to determine whether the glycoside is a furanoside or pyranoside. Thus, treatment of methyl α-D-glucopyranoside with periodic acid gives a dialdehyde and formic acid, while similar treatment of methyl α-D-arabinofuranoside gives the same dialdehyde but no formic acid. Write equations for these oxidative degradations and explain the formation of the products.

14.22 On treatment with periodic acid, a 0.243-g sample of cellulose gave 0.00100 mmole (1.00×10^{-6} mole) of formic acid. What is the approximate number of glucose units in each molecule of the cellulose? (*Hint:* There will be $(2n + 2)$ glucose units based on the structure given in Sec. 14.18.)

14.23 A disaccharide, $C_{11}H_{20}O_{10}$, may be hydrolyzed by α-glucosidase (an enzyme that is a specific catalyst for the hydrolysis of α-glucosides) to yield a hexose and a pentose. The disaccharide does not reduce Fehling's solution. Methylation of the disaccharide with methyl bromide and silver oxide, followed by acid hydrolysis, yields 2,3,4,6-tetra-O-methyl-D-glucose and a tri-O-methyl pentose. Oxidation of the latter with bromine water gives 2,3,4-tri-O-methyl-D-xylonic acid. From these data deduce the structure of the disaccharide and write equations for the transformations described.

14.24 Glucose is the precursor of ascorbic acid (Vitamin C). In what respects is ascorbic acid like glucose? Like fructose? In what respects is it different? To what class of hydroxy acid derivatives does this kind of structure belong?

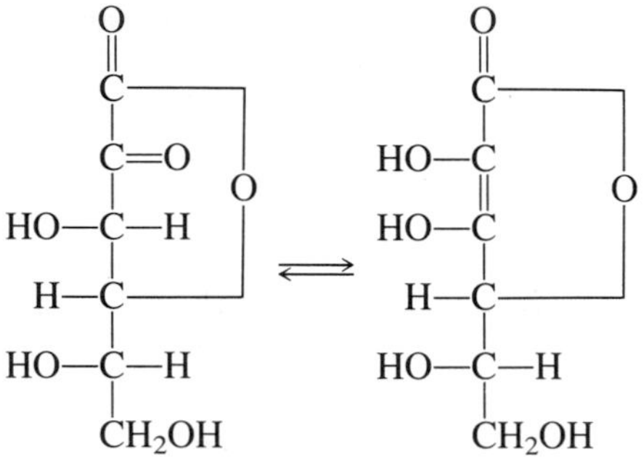

```
          O                              O
          ‖                              ‖
          C ────────┐                    C ────────┐
          |         |                    |         |
          C=O       O               HO—C           O
          |         |                    ‖         |
     HO—C—H         |               HO—C           |
          |         |                    |         |
       H—C ────────┘        ⇌         H—C ────────┘
          |                              |
     HO—C—H                         HO—C—H
          |                              |
         CH2OH                          CH2OH
```

Ascorbic acid
(Vitamin C)

15 Amino Acids, Peptides, and Proteins

The name protein has its origin in the Greek word *proteios,* meaning "of first importance." The name is well chosen, for proteins are the basis of protoplasm and comprise the underlying structure of all living organisms. Proteins, in the form of muscle, skin, hair, and other tissue, make up the bulk of the body's nonbony structure. In addition to providing such structural material, proteins have many other functions. Proteins, as enzymes, catalyze biochemical reactions; as hormones, they regulate metabolic processes; as antibodies, they resist and nullify the effects of toxic substances. Such specialized functions illustrate the great importance of proteins.

Proteins are high molecular weight, long-chain polymers made up largely of various amino acids linked together. The constituent amino acids are obtained when a protein is hydrolyzed by dilute acids, by dilute alkalis, or by protein-digesting enzymes. It would be difficult to consider the properties of molecules as complex as the proteins without first examining the properties of the α-amino acids from which they are constructed.

Amino Acids

15.1 Structure of Amino Acids

Nearly all amino acids obtained from plant and animal proteins have an amino group on the carbon atom alpha (α) to the carboxyl function. An α-amino acid has the following general formula.

$$R-\underset{NH_2}{\overset{H}{\underset{|}{\overset{|}{C}}}}-\overset{O}{\overset{\|}{C}}-OH$$

The R in the general formula for an α-amino acid may be a hydrogen, a straight- or branched-chain aliphatic group, an aromatic ring, or a heterocyclic nitrogen-containing ring. Most amino acids have one amino group and one carboxyl group and are usually classified as neutral amino acids. A few have a second amino group joined to other carbon atoms in the molecule and show basic properties. Others contain a second carboxyl group and behave as acids. With the exception of glycine, all amino acids contain at least one center of asymmetry and are optically active. Amino acids of protein origin all possess the L-configuration. With few exceptions, the body is able to utilize completely only the L-isomers of those amino acids that it is not able to synthesize itself.

15.2 Nomenclature and Classification by Structure

Of the known α-amino acids, about twenty have been found to be constituents of the more common plant and animal proteins. The structures for these, together with their names and abbreviations, are given in Table 15.1.

TABLE 15.1 Amino Acids Derived from Proteins

A.1. Neutral Amino Acids with Nonpolar R Group

Name (Abbreviation)	Formula
1. **Glycine (Gly)*** (Aminoacetic acid)	$H-CH(NH_2)-COOH$
2. **Alanine (Ala)** (α-Aminopropionic acid)	$CH_3-CH(NH_2)-COOH$
3. **Valine (Val)** (α-Aminoisovaleric acid)	$CH_3-CH(CH_3)-CH(NH_2)-COOH$
4. **Leucine (Leu)** (α-Aminoisocaproic acid)	$CH_3-CH(CH_3)-CH_2-CH(NH_2)-COOH$

*Because the R group on glycine is hydrogen, it has little effect on the rest of the molecule, and glycine is often classified with the neutral amino acids having a polar R group.

TABLE 15.1 *Continued*

A.1. Neutral Amino Acids with Nonpolar R Group

5. **Isoleucine (Ile)** (α-Amino-β-methylvaleric acid)	$CH_3—CH_2—CH(CH_3)—CH(NH_2)—COOH$
6. **Phenylalanine (Phe)** (α-Amino-β-phenylpropionic acid)	$C_6H_5—CH_2—CH(NH_2)—COOH$
7. **Tryptophan (Trp)** (α-Amino-β-(3-indolyl) propionic acid)	(3-indolyl, N—H)$—CH_2—CH(NH_2)—COOH$
8. **Proline (Pro)**† (2-Pyrrolidine carboxylic acid)	$H_2C—CH_2$, H_2C, $CHCOOH$, N—H (ring)
9. **Methionine (Met)** (α-Amino-γ-methylthiobutyric acid)	$CH_3—S—CH_2—CH_2—CH(NH_2)—COOH$

A.2. Neutral Amino Acids with Polar R Group

10. **Serine (Ser)** (α-Amino-β-hydroxypropionic acid)	$HOCH_2—CH(NH_2)—COOH$
11. **Threonine (Thr)** (α-Amino-β-hydroxybutyric acid)	$CH_3—CH(OH)—CH(NH_2)—COOH$
12. **Cysteine (Cys)** (α-Amino-β-mercaptopropionic acid)	$HS—CH_2—CH(NH_2)—COOH$

†Proline and hydroxyproline are *imino acids*. The nitrogen atom, although joined to the α-carbon, is part of a ring. An imino nitrogen bears only one hydrogen atom but can still take part in the formation of proteins.

TABLE 15.1 *Continued*

A.2. Neutral Amino Acids with Polar R Group	
13. **Tyrosine (Tyr)** (α-Amino-*p*-hydroxyhydrocinnamic acid)	$HO-C_6H_4-CH_2-CH(NH_2)-COOH$
14. **Asparagine (Asn)** (β-Carbamoyl-α-aminopropionic acid)	$H_2N-C(=O)-CH_2-CH(NH_2)-COOH$
15. **Glutamine (Gln)** (γ-Carbamoyl-α-aminobutyric acid)	$H_2N-C(=O)-CH_2-CH_2-CH(NH_2)-COOH$
B. Basic Amino Acids	
16. **Histidine (His)** (α-Amino-β-5-imidazolylpropionic acid)	$H-C{=}C(-CH_2-CH(NH_2)-COOH)$, ring: $N{=}C(H)-NH$
17. **Lysine (Lys)** (α,ε-Diaminocaproic acid)	$CH_2(NH_2)-CH_2-CH_2-CH_2-CH(NH_2)-COOH$
18. **Arginine (Arg)** (α-Amino-γ-guanidinovaleric acid)	$H_2N-C(=NH)-NH-CH_2CH_2CH_2-CH(NH_2)-COOH$
C. Acidic Amino Acids	
19. **Aspartic acid (Asp)** (Aminosuccinic acid)	$HOC(=O)-CH_2-CH(NH_2)-COOH$
20. **Glutamic acid (Glu)** (α-Aminoglutaric acid)	$HOC(=O)-CH_2-CH_2-CH(NH_2)-COOH$

15.3 Nutritive Classification of Amino Acids

A nutritive classification of the α-amino acids has resulted from nutritional experiments carried out on laboratory animals. It is not a strict classification but varies with the requirements of the different species of animals tested. The omission of certain amino acids in the diet of some animals prevents their normal growth and development. The animal organism either is incapable of synthesizing certain amino acids or is incapable of synthesizing them in sufficient quantity to maintain a normal state of good health. The effects of such malnutrition disappear when the missing amino acids are supplied in the diet. On the basis of such studies, α-amino acids are divided into two categories as **essential** or **nonessential.** Table 15.2 lists the α-amino acids in these two general classes.

TABLE 15.2 The Amino Acids According to Nutritional Requirements

Essential (indispensable)	Nonessential (dispensable)
Arginine*	Alanine
Glycine†	Asparagine
Histidine‡	Aspartic acid
Isoleucine§	Cysteine
Leucine§	Glutamic acid
Lysine§	Glutamine
Methionine§	Proline
Phenylalanine§	Serine
Threonine§	Tyrosine
Tryptophan§	
Valine§	

*Required for optimum growth in the rat and chick.
†Required for optimum growth in the chick.
‡Required by all subhuman species tested and by infants, but not by adult humans.
§Required by all species tested, including adult humans.

15.4 Properties of the Amino Acids. Isoelectric Point

The amino acids are colorless crystalline solids and have melting points (with decomposition) in excess of 200°C. Most amino acids are soluble in water but sparingly soluble in organic solvents. These properties are not characteristic of most simple organic acids or simple amines but are more like those of salts.

A neutral amino acid possesses an amino group and a carboxyl group, and thus can behave as either a base or an acid. An amino acid in an alkaline solution reacts like an acid to produce a *metal salt.*

$$\mathrm{R{-}\overset{H}{\underset{NH_2}{C}}{-}C(=O)OH} + \mathrm{Na^+OH^-} \longrightarrow \mathrm{R{-}\overset{H}{\underset{NH_2}{C}}{-}C(=O)O^-} + \mathrm{Na^+} + \mathrm{H_2O}$$

An amino acid anion

If an alkaline solution of an amino acid is electrolyzed, the anion of the amino acid salt migrates toward the anode or positive electrode.

An amino acid in an acidic solution behaves like a base to form an *amine salt.*

$$\mathrm{R{-}\overset{H}{\underset{\ddot{N}H_2}{C}}{-}C(=O)OH} + \mathrm{H^+Cl^-} \longrightarrow \mathrm{R{-}\overset{H}{\underset{N^+{:}H(H)_2}{C}}{-}C(=O)OH} + \mathrm{Cl^-}$$

An amino acid cation

If an acidic solution of an amino acid is electrolyzed, the cation of the amine salt migrates to the cathode or negative electrode.

The hydrogen ion concentration at which the amino acid shows no net migration to either electrode is called the **isoelectric point.** Isoelectric points are given in pH values and vary from low values for acidic amino acids (pH 3 for aspartic acid) to high values for basic amino acids (pH 10.8 for arginine). Neutral amino acids do not have isoelectric points at the neutral figure (pH 7.0), as might be expected, but slightly on the acid side. At its isoelectric point, a ''neutral'' amino acid is completely ionized, the proton shifting from the carboxyl group to the amino group to produce an inner salt or **dipolar ion.**

$$\underset{\text{I}}{\mathrm{R{-}\overset{H}{\underset{\ddot{N}H_2}{C}}{-}C(=O)O^-}} + \mathrm{H_2O} \underset{\mathrm{OH^-}}{\overset{\mathrm{H^+}}{\rightleftharpoons}} \underset{\text{II}}{\mathrm{R{-}\underset{N^+{:}H(H)_2}{C}{-}C(=O)O^-}} \underset{\mathrm{OH^-}}{\overset{\mathrm{H^+}}{\rightleftharpoons}} \underset{\text{III}}{\mathrm{R{-}\overset{H}{\underset{\overset{+}{N}H_3}{C}}{-}C(=O)OH}}$$

Dipolar ion

Exercise 15.1 The amino acid cation (second equation in Sec. 15.4) has two acidic groups. What are they? Which group is more acidic—that is, will give up a proton more readily when base is added to a solution of the amino acid hydrochloride? (*Hint:* Observe the changes in the dipolar ion (same section) with a change in pH.)

Dipolar ions, as the preceding equation illustrates, are *amphoteric* and can act as either acids or bases.

The pK_a of the carboxylic acid group can be determined by adding one-half equivalent of aqueous sodium hydroxide, or enough to half-neutralize the carboxyl group, to a solution of the amino acid cation (III) and measuring the pH. At this point the solution will contain approximately equal amounts of III and the dipolar ion (II). Addition of a full equivalent of base will convert most of the III into II and bring the pH of the solution to approximately the isoelectric point. Addition of a further half-equivalent of base will half-neutralize the ammonium ion (conjugate acid of the amino group), at which point the pH will equal (approximately) the pK_a of the ammonium ion. The solution will contain roughly equal amounts of II and the carboxylate salt (I). In general, the isoelectric pH will be the average of the two pK_a values—that is, one-half their sum. Some pK_a data are given in Table 15.3.

TABLE 15.3 Ionization Constants of the Amino Acids in Water at 25°C*

Amino Acid	pK_1	pK_2	pK_3
Alanine	2.29	9.74	
Arginine	2.01	9.04	12.48
Asparagine	2.02	8.80	
Aspartic acid	2.10	3.86	9.82
Cysteine	2.05	8.00	10.25
Glutamic acid	2.10	4.07	9.47
Glutamine	2.19	9.13	
Glycine	2.35	9.78	
Histidine	1.77	6.10	9.18
Isoleucine	2.32	9.76	
Leucine	2.33	9.74	
Lysine	2.18	8.95	10.53
Methionine	2.28	9.21	
Phenylalanine	2.58	9.24	
Proline	2.00	10.60	
Serine	2.21	9.15	
Threonine	2.09	9.10	
Tryptophan	2.38	9.39	
Tyrosine	2.20	9.11	10.07
Valine	2.29	9.72	

*pK_1 is for the COOH. pK_2 is for the second COOH, if present; otherwise, it is for the ammonium ion. pK_3 is for an ammonium ion or other weak-acid function.

15.5 Synthesis of Amino Acids

Interest in obtaining the individual amino acids in pure form for use in nutrition experiments has led to a number of synthetic procedures for the preparation of amino acids. Some of these are adaptations of methods previously reviewed for the preparation of simple primary amines (Sec. 13.3). Of the many methods that have been employed for the preparation of amino acids, we shall consider only two. These two methods are general methods and are satisfactory for the preparation of *some* amino acids. No single method has been developed for the synthesis of *all* amino acids.

A. Direct Amination[1] of α-Halogen Acids. In this method the halogen atom of an α-halo acid is replaced by an amino group directly by treating the acid with ammonia. In practice, a large excess of ammonia is used to prevent the formation of a disubstituted ammonia derivative.

$$CH_3-\underset{\underset{Br}{|}}{\overset{\overset{H}{|}}{C}}-C\begin{matrix}\nearrow O \\ \searrow OH\end{matrix} + 2\,NH_3 \longrightarrow CH_3-\underset{\underset{NH_2}{|}}{\overset{\overset{H}{|}}{C}}-C\begin{matrix}\nearrow O \\ \searrow OH\end{matrix} + NH_4Br$$

α-Bromopropionic acid → D,L-Alanine

B. The Hydrolysis of α-Amino Nitriles (Strecker Synthesis). In the **Strecker synthesis** of an amino acid, an aldehyde is treated with ammonium cyanide (ammonia and hydrogen cyanide). The aminonitrile that results from this reaction is then hydrolyzed to the amino acid. The method lends itself very well to the preparation of simple neutral amino acids. The preparation of D,L-alanine by the Strecker synthesis begins with acetaldehyde.

$$CH_3-\overset{\overset{H}{|}}{C}=O + HCN + NH_3 \longrightarrow CH_3-\underset{\underset{NH_2}{|}}{\overset{\overset{H}{|}}{C}}-CN + H_2O$$

Acetaldehyde → Aminonitrile

$$CH_3-\underset{\underset{NH_2}{|}}{\overset{\overset{H}{|}}{C}}-CN + 2\,H_2O \longrightarrow CH_3-\underset{\underset{NH_2}{|}}{\overset{\overset{H}{|}}{C}}-C\begin{matrix}\nearrow O \\ \searrow OH\end{matrix} + NH_3$$

D,L-Alanine

[1] A reaction that introduces the amino group into the molecule is called *amination*.

Any synthesis of amino acids (except glycine) leads to a racemic mixture and must be followed by resolution if the natural form of the amino acid is required.

Exercise 15.2 Using acrylic acid, $CH_2{=}CHCOOH$, and any other reagents you might require, outline a synthesis leading to aspartic acid.

15.6 Reactions of the Amino Acids

The reactions of the amino acids are, in general, reactions characteristic of both carboxylic acids and primary amines.

A. Esterification. All amino acids can be esterified. Emil Fischer utilized this reaction as early as 1901 as a technique for the separation of the constituent amino acids obtained from a protein hydrolysate. The protein was hydrolyzed to its constituent amino acids, the mixture was esterified, and the liquid amino acid esters were separated by fractional distillation. Fischer's method is illustrated by the following equations.

$$CH_3{-}CH(NH_2){-}C({=}O)OH + HCl + C_2H_5OH \longrightarrow CH_3{-}CH(\overset{+}{N}H_3){-}C({=}O)OC_2H_5 + H_2O + Cl^-$$

$$CH_3{-}CH(\overset{+}{N}H_3){-}C({=}O)OC_2H_5 + OH^- \longrightarrow \underset{\text{Ethyl ester of alanine}}{CH_3{-}CH(NH_2){-}C({=}O)OC_2H_5} + H_2O$$

B. Reaction with Nitrous Acid. The amino acids, with the exception of proline, react with nitrous acid to liberate nitrogen gas. This reaction is the basis for the **Van Slyke method** for determining "free" amino groups (uncombined α-amino groups) in protein material. The reaction gives an index to the number of uncombined $—NH_2$ groups such as would be provided by certain basic amino acids.

$$\underset{\text{Lysine unit}}{\left({-}NH{-}CH[(CH_2)_3{-}CH_2{-}NH_2]{-}C({=}O){-} \right)} + HNO_2 \longrightarrow \left({-}NH{-}CH[(CH_2)_3{-}CH_2OH]{-}C({=}O){-} \right) + N_2 + H_2O$$

(a "free" amino group)

C. Reaction with Acid Halides or Acid Anhydrides. The amino group of amino acids is readily converted to an amide by reaction with an acid halide or anhydride. Thus, glycine reacts with benzoyl chloride to produce hippuric acid. Benzoates, when ingested, are rendered water-soluble and voided from the body in the urine as hippuric acid.

$$C_6H_5-\overset{O}{\overset{\|}{C}}-Cl + H-\underset{H}{N}-\underset{H}{\overset{H}{C}}-COOH \longrightarrow C_6H_5-\overset{O}{\overset{\|}{C}}-\overset{H}{N}-\underset{H}{\overset{H}{C}}-COOH + HCl$$

Benzoyl chloride Glycine Hippuric acid

D. Reaction with Ninhydrin. Amino acids react with a ninhydrin solution (triketohydrindene hydrate) to produce purple compounds. The reaction is of value in the assay of protein material and can be used for the quantitative determination of amino acids. The following sequence of reactions illustrates how ninhydrin converts an amino acid into an aldehyde and carbon dioxide.

$$2\ \text{Ninhydrin} + H_2N-\underset{H}{\overset{R}{C}}-C\overset{O}{\underset{OH}{}} + OH^- \longrightarrow$$

Ninhydrin

$$\text{(Colored product)} + R-C\overset{O}{\underset{H}{}} + CO_2 + 4\,H_2O$$

(Colored product)

E. Peptide Formation. Amino acids can be condensed with each other to form peptides (Sec. 15.7). However, the sequence of the individual amino acids in a peptide cannot be controlled unless steps are taken to block reaction at an amino group or at a carboxyl function. Without such protective measures, random condensation can occur. In practice, when amino acids are to be joined into peptides, the amino acid to appear first in the peptide chain (counting from left to right) must have its amino group protected. The protecting group is removed after condensation is completed. The method is illustrated in the preparation of the dipeptide, alanylglycine (Ala-Gly), with both amino and carboxyl group protection.

STEP 1. Amino group of alanine protected.

$$C_6H_5-CH_2-O-C(=O)-Cl + H_2N-CH(CH_3)-COOH \longrightarrow HCl + C_6H_5-CH_2-O-C(=O)-NH-CH(CH_3)-COOH$$

Benzyl chloroformate (Benzyl chlorocarbonate or carbobenzoxy chloride) — Alanine — Benzyloxycarbonylalanine (Carbobenzoxyalanine) Cbz · Ala

STEP 2. Carboxyl group of glycine protected.

$$H_2N-CH_2-C(=O)-OH + C_6H_5-CH_2OH \xrightarrow[HCl]{} H_2N-CH_2-C(=O)-OCH_2-C_6H_5$$

Glycine — Benzyl alcohol — Glycine benzyl ester

Next the two protected amino acids may be condensed using the generally useful coupling reagent, dicyclohexylcarbodiimide (DCC), which is a specific catalyst for the condensation of carboxylic acids with alcohols and amines.

STEP 3. Condensation.

$$R-C(=O)-OH + R'-NH_2 + C_6H_{11}-N{=}C{=}N-C_6H_{11} \longrightarrow R-C(=O)-NH-R' + C_6H_{11}-NH-C(=O)-NH-C_6H_{11}$$

Carboxylic acid — Amine — Dicyclohexylcarbodiimide — Amide — Dicyclohexylurea (*very insoluble*)

$$C_6H_5-CH_2-O-C(=O)-NH-CH(CH_3)-C(=O)OH + H-NH-CH_2-C(=O)-O-CH_2-C_6H_5 \xrightarrow{DCC} C_6H_5-CH_2-O-C(=O)-NH-CH(CH_3)-C(=O)-NH-CH_2-C(=O)-O-CH_2-C_6H_5$$

Finally, both protecting groups are removed by catalytic **hydrogenolysis,** a procedure that is specific for the reduction of benzyl esters to the acid plus toluene.

STEP 4. Hydrogenolysis.

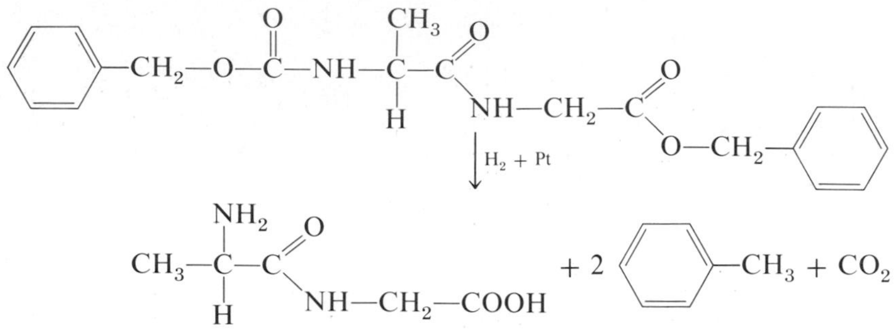

$$C_6H_5{-}CH_2{-}O{-}\overset{O}{\overset{\|}{C}}{-}NH{-}\underset{H}{\overset{CH_3}{C}}{-}\overset{O}{\overset{\|}{C}}{-}NH{-}CH_2{-}\overset{O}{\overset{\|}{C}}{-}O{-}CH_2{-}C_6H_5 \xrightarrow{H_2\ +\ Pt} CH_3{-}\underset{H}{\overset{NH_2}{C}}{-}\overset{O}{\overset{\|}{C}}{-}NH{-}CH_2{-}COOH + 2\ C_6H_5{-}CH_3 + CO_2$$

Alanylglycine (a dipeptide) Toluene

Peptides and Proteins

15.7 Structure and Nomenclature

In a protein molecule, the α-amino acids are joined together through amide linkages formed between the amino group of one acid molecule and the carboxyl group of another. Such amide linkages are called peptide links and serve to unite hundreds of amino acid residues in a protein molecule. When only two α-amino acids are joined, as in alanylglycine (Sec. 15.6E), the product is a dipeptide. A dipeptide, with free amino and carboxyl groups on opposite ends of the chain, can unite at either end with a third α-amino acid to form a tripeptide. A tripeptide can form a tetrapeptide, and so on, until finally a long-chain polypeptide results. If n is the number of different amino acids in a polypeptide, the number of possible sequences is *n factorial* ($n!$). Thus, three different amino acids could combine in one of six different sequences ($3 \times 2 \times 1 = 6$). One such sequence is illustrated below.

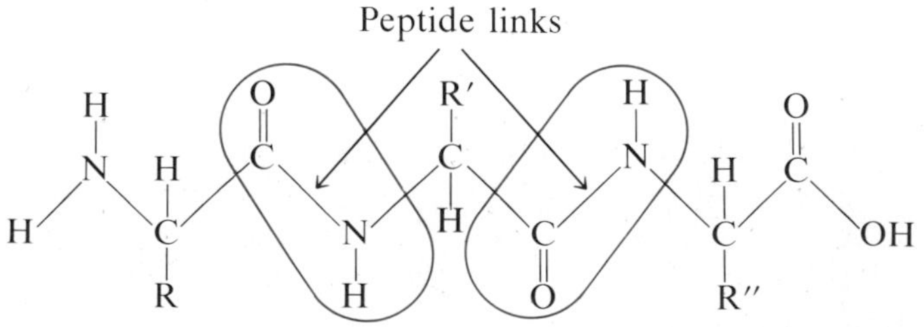

Peptide links

$$H_2N{-}\underset{R}{\overset{H}{C}}{-}\overset{O}{\overset{\|}{C}}{-}\underset{H}{N}{-}\underset{H}{\overset{R'}{C}}{-}\underset{O}{\underset{\|}{C}}{-}\overset{H}{N}{-}\underset{R''}{\overset{H}{C}}{-}\overset{O}{\overset{\|}{C}}{-}OH$$

A variation in sequence, as well as a variation in the number and kind of amino acids joined, makes possible an almost infinite number of arrangements. It must not be thought, however, that the manner in which amino acids are joined in nature to form the polypeptide links in proteins is a haphazard one. On the contrary, the amino acids that comprise a protein are joined uniformly to give specificity to certain types of tissue within each organism.

Peptides are named as derivatives of the C-terminal amino acid, which still has a free carboxyl group. The C-terminal amino acid is the last one, counting from left to right. The method of naming is illustrated by the following tetrapeptide.

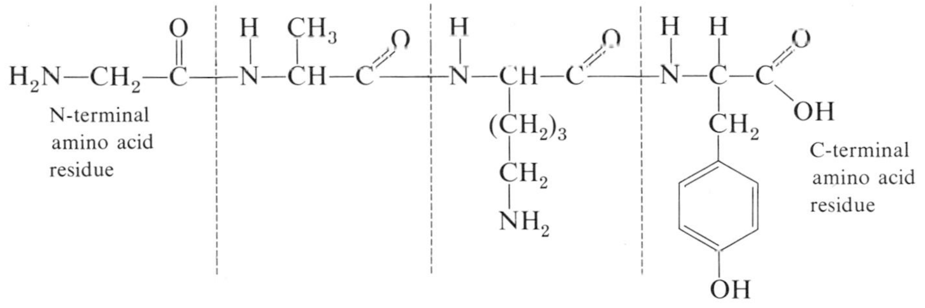

$$H_2N-CH_2-C(=O)-NH-CH(CH_3)-C(=O)-NH-CH((CH_2)_3CH_2NH_2)-C(=O)-NH-CH(CH_2C_6H_4OH)-COOH$$

N-terminal amino acid residue

C-terminal amino acid residue

Glycyl-alanyl-lysyl-tyrosine

15.8 Sequence of Amino Acids in Peptides

Determination of the amino acid sequence of a natural peptide is a tedious and laborious process but has been accomplished in a few cases. In practice, the N-terminal group of a peptide is ''tagged'' by reaction with Sanger's reagent, 2,4-dinitrofluorobenzene, or with dansyl chloride (Sec. 13.4C) before the peptide is hydrolyzed.

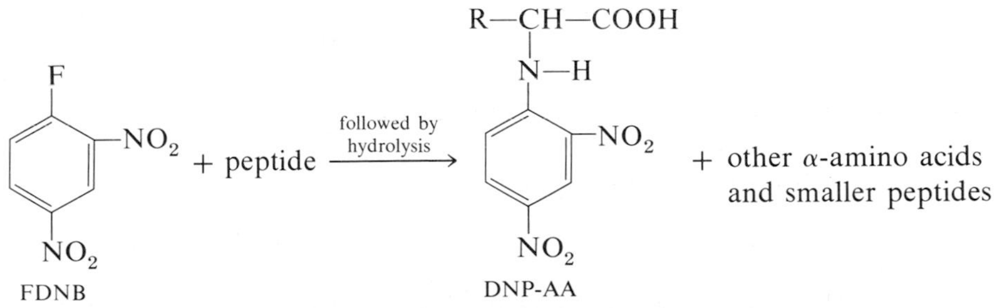

$$\text{FDNB (2,4-}(NO_2)_2C_6H_3F) + \text{peptide} \xrightarrow[\text{hydrolysis}]{\text{followed by}} \text{DNP-AA } (R-CH(COOH)-NH-C_6H_3(NO_2)_2) + \text{other } \alpha\text{-amino acids and smaller peptides}$$

The tagged amino acid is liberated by hydrolysis. Since it is colored, it is easily distinguished from the other amino acids that comprise the peptide structure. The advantage of the dansyl chloride procedure, which is chemically similar to the Sanger method, is that it yields easily detected fluorescent products and is thus about 100 times as sensitive as the Sanger method.

Exercise 15.3 Write equations similar to those given for the Sanger method, substituting dansyl chloride for Sanger's reagent.

A third procedure, called the **Edman degradation,** may be used, not only for the identification of N-terminal groups but also for the establishment of the sequence of amino acids in the polypeptide chain by a *stepwise* cleavage of the amino acid from the N-terminus. The reagent used is phenylisothiocyanate, which gives a product that can be split by treatment with anhydrous acid to form the phenylthiohydantoin derivative of the N-terminal amino acid.

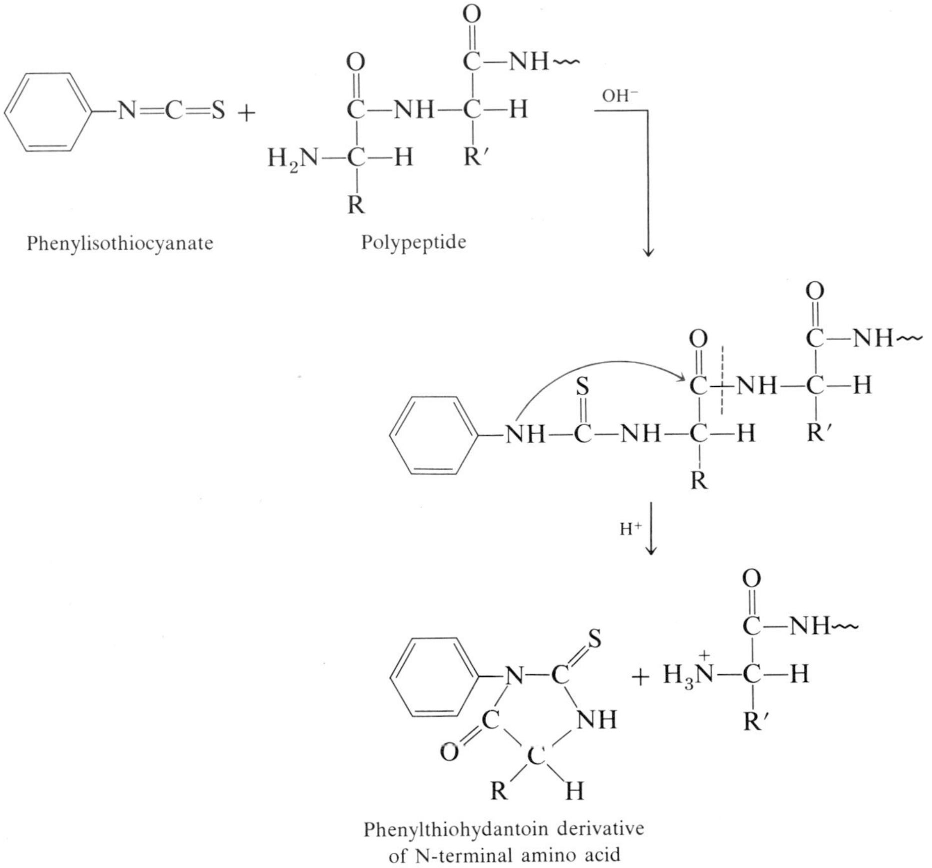

Phenylisothiocyanate

Polypeptide

OH^-

H^+

Phenylthiohydantoin derivative
of N-terminal amino acid

The process can be repeated to remove the new N-terminal amino acid and has been automated to permit the rapid determination of sequences of up to 20 amino acids by the sequential removal and identification of amino acids. Other methods are available for the sequential degradation of polypeptides from the C-terminal end. To illustrate the general procedure, let us review the work that led to the elucidation of the *glutathione* structure.

Glutathione is a tripeptide isolated from yeast. A complete enzymatic or acid hydrolysis of the peptide revealed that it was composed of only three α-amino acids: L-glutamic acid, L-cysteine, and glycine. If each of these α-amino acids had but one amino group and one carboxyl group, the number of possible sequences would be six (Sec. 15.7). However, glutamic acid (see Table 15.1) has two carboxyl groups. This increases the total number of possible sequences to twelve, for we do not know whether the α-carboxyl or the γ-carboxyl group of glutamic acid is involved in a peptide link. Mild hydrolysis of glutathione gave two dipeptides. One of these, on further treatment, yielded cysteine and glutamic acid; the other gave cysteine and glycine. This much is now known about glutathione: the cysteine structure is linked to both glycine and glutamic acid. But how? A little pencil work now will show that only the three sequences I, II, and III could possibly fit that of glutathione.

Cys-Glu

I $$H_2N{-}CH_2{-}\overset{O}{\overset{\|}{C}}{-}\overset{H}{\overset{|}{N}}{-}\underset{CH_2SH}{\underset{|}{CH}}{-}\overset{O}{\overset{\|}{C}}{-}\overset{H}{\overset{|}{N}}{-}\underset{CH_2CH_2COOH}{\underset{|}{CH}}{-}COOH$$

Gly-Cys

Cys-Gly

II $$H_2N{-}\underset{CH_2CH_2COOH}{\underset{|}{CH}}{-}\overset{O}{\overset{\|}{C}}{-}\overset{H}{\overset{|}{N}}{-}\overset{CH_2SH}{\overset{|}{CH}}{-}\overset{O}{\overset{\|}{C}}{-}\overset{H}{\overset{|}{N}}{-}CH_2COOH$$

α-(COOH) Glu-Cys

Cys-Gly

III $$H_2N{-}\underset{COOH}{\underset{|}{CH}}{-}CH_2CH_2\overset{O}{\overset{\|}{C}}{-}\overset{H}{\overset{|}{N}}{-}\overset{CH_2SH}{\overset{|}{CH}}{-}\overset{O}{\overset{\|}{C}}{-}\overset{H}{\overset{|}{N}}{-}CH_2COOH$$

γ-(COOH) Glu-Cys

The tagging technique identified glutamic acid as the N-terminal amino acid. This finding eliminates structure I as a possibility. The synthesis of the remaining two dipeptides is now in order. In one of these the α-carboxyl of glutamic acid must be linked to cysteine, and in the other the γ-carboxyl must be joined. The latter structure proved to be identical with that of the dipeptide obtained by hydrolysis. The correct structure for glutathione is thus established as that shown by the tripeptide III.

Once the proper sequence of amino acids in a peptide is known, a synthesis for it usually follows. The historic accomplishment of duVigneaud[2] and his co-workers in determining the oxytocin structure provides such an outstanding example.

Oxytocin is a peptide hormone produced by the posterior lobe of the pituitary gland. It has the function of causing a contraction of smooth muscle, particularly uterine muscle, and finds application in obstetrics. In addition to this function, the hormone also promotes the flow of milk from the mammary glands.

Oxytocin, on hydrolysis, produced one molecule each of leucine, isoleucine, proline, glutamic acid, tyrosine, aspartic acid, cystine, glycine, and three equivalents of ammonia. The sequence of the α-amino acids in oxytocin was determined, and the structure shown in Fig. 15.1 was assigned to it. The research team subsequently synthesized an octapeptide of the same sequence. This outstanding achievement represents the first synthesis of a peptide hormone.

Another brilliant example of determining the sequence of amino acids in a natural polymer was that performed by Sanger[3] of England. He and his associates, after years of diligent investigation and by the use of ingenious techniques, were able to elucidate the amino acid sequence of the hormone insulin. Until about 1955 the amino acid sequence in any polypeptide within the protein range was unknown. The molecular weight of beef insulin was determined to be 5,734, and its composition that of forty-eight amino acid residues of sixteen different kinds!

Exercise 15.4 Why could not bromobenzene be used for tagging the N-terminal α-amino acid of a peptide? (*Hint:* See Sec. 6.5.)

15.9 Composition and Structure of Proteins

Proteins differ from carbohydrates and fats in elementary chemical composition. All proteins contain, in addition to carbon, hydrogen, and oxygen, other elements in the approximate percentages as follows: nitrogen (15%), sulfur (1.0%), and phosphorus (0.5%). The molecular weights of proteins are unbelievably high, ranging from 10,000 to 10,000,000.

Acceptance of the peptide hypothesis of protein structure, first proposed in 1902 by Fischer and Hofmeister, had led to many other questions regarding protein structures.

[2]Vincent duVigneaud (1901–1979), Cornell University College of Medicine. Winner of the Nobel Prize in Chemistry (1955).

[3]Frederick Sanger (1918–   ), Cambridge University. Winner of the Nobel Prize in Chemistry (1958).

FIGURE 15.1 Oxytocin*

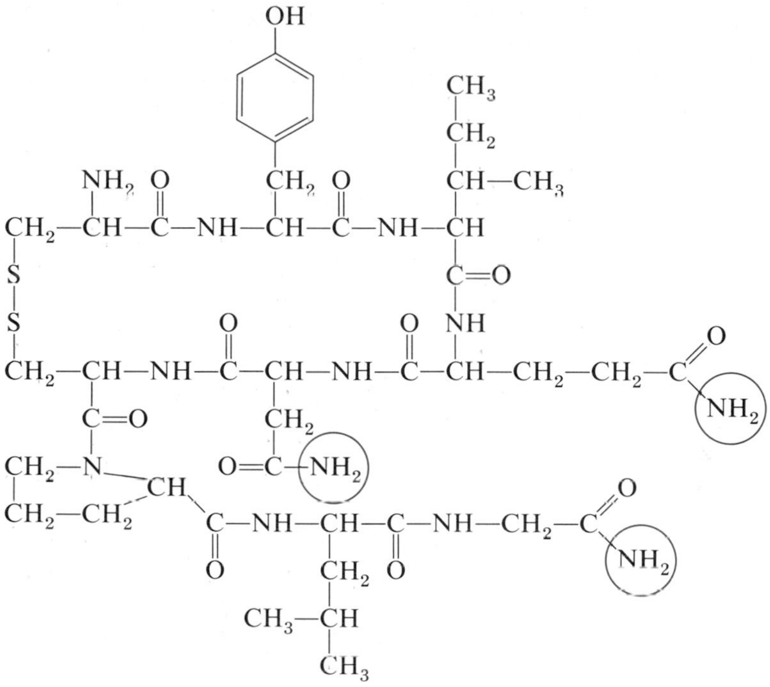

*Areas circled in color indicate the free amino groups of the amides of glycine, glutamic acid, and aspartic acid.

How are peptide chains held together in the protein structure? How are they arranged in space, or what shapes do they assume? X-ray analysis and the persistent efforts of numerous investigators have revealed the answers to some of these questions.

In order to help simplify the study of molecules as large and complex as the proteins, biochemists have adopted a classification of protein structure as *primary, secondary, tertiary,* and *quaternary*. This classification does not always allow a clear-cut distinction between some structures, but it is a helpful one nonetheless.

A primary structure is simply one in which the α-amino acids are covalently bonded in a polypeptide chain, with their sequence in the chain having been genetically determined. A primary structure is thus an integral part of any other structure.

The long polypeptide chains that comprise the fibrous proteins of hair and wool (the α-keratins) are arranged in right-handed helical coils with 3.6 amino acid units per turn. The coils are held together by hydrogen bonding between the amide hydrogen atom in one peptide link and the carbonyl oxygen atom of another peptide link three amino acid units beyond the first. Such a coil is called an α-helix, and represents a secondary structure.

Figure 15.2 shows the helical arrangement of α-amino acids in a polypeptide chain where *intrachain* H-bonds pull the polypeptide into a coil-like configuration.

FIGURE 15.2 The Right-handed α-Helix

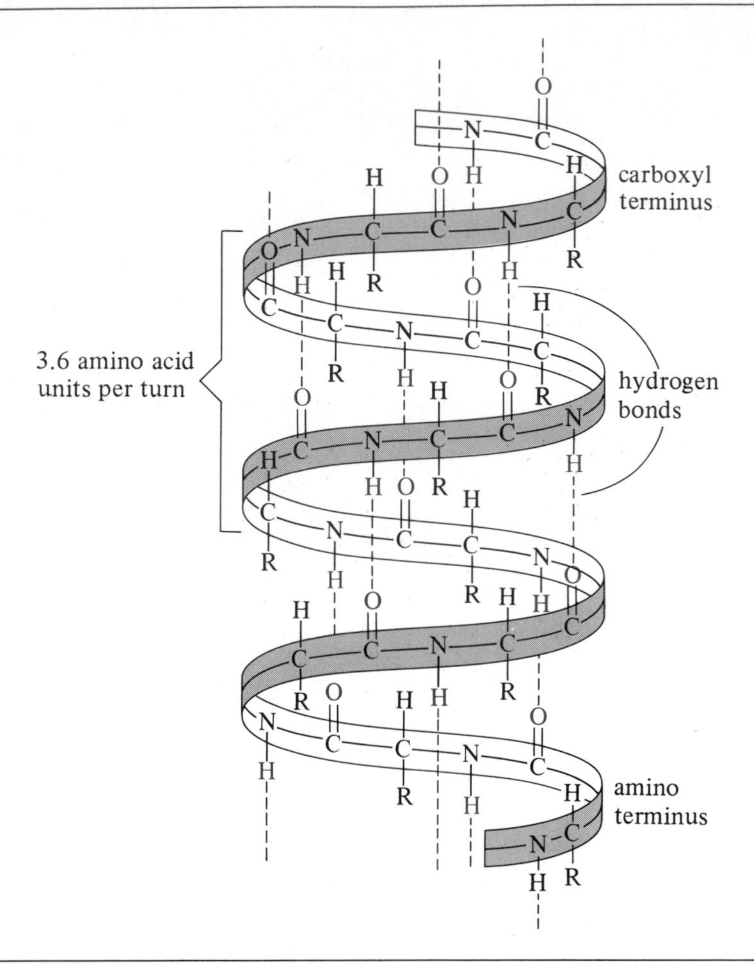

In the case of hair and wool, all the polypeptide chains run in the same direction, from N-terminal to C-terminal. The chains are twisted into a ropelike strand and are held together by the formation of disulfide, —S—S—, cross-links between cysteine units in adjacent polypeptide chains. A structure of this type, which is held together by forces other than or in addition to hydrogen bonds, is called a tertiary structure.

In silk fibroin (a β-keratin) the polypeptide chains are grouped together in side-by-side chains joined by hydrogen bonds *between* chains. Adjacent polypeptides run in opposite directions, or antiparallel. The overall appearance of the β-conformation is that of a *pleated sheet,* which you may visualize by studying the schematic structure below in which the *interchain* hydrogen bonds are shown by dotted lines. Note that the R groups of the amino acids project out from the pleated surface, while the smaller hydrogen atoms on the α-carbon atoms of the amino acid units are located in the folds of pleats. No disulfide cross-links are found in β-keratins.

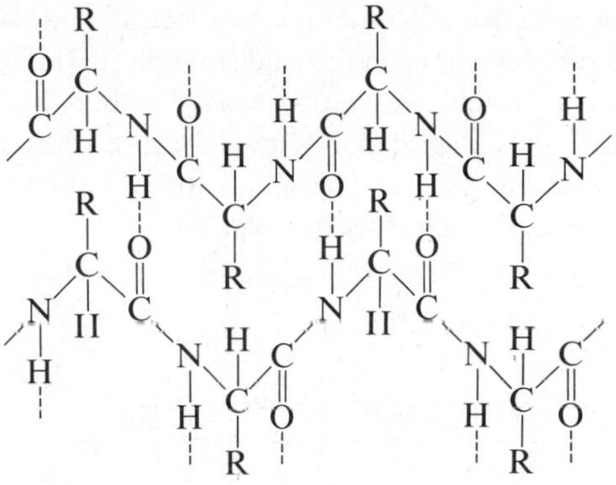

The collagen of connective tissue consists of three kinked, left-handed helices held together by hydrogen bonds. The globular proteins such as the catalytic proteins (enzymes) and transport proteins—myoglobin, hemoglobin, and cytochrome C—are more complex structures in which the chains are folded into three-dimensional globular shapes, as implied by the name. In the oligomeric globular proteins, two or more chains (four in hemoglobin) are intertwined or bunched together in an aggregate. The globular proteins are classified as quaternary structures.

Segments of these folded chains may have the α-helix structure or the β-conformation. Generally, the folds are so arranged as to have polar groups on the exterior to promote water solubility, and nonpolar hydrophobic groups inside. The catalytic or transport function of a protein requires the presence of a unique structural feature called the **active site.** Thus, for example, in an enzyme the substrate[4] is bound to the active site while being acted upon. To discover how enzymes and the transport proteins function, and to determine the nature of the active sites, are among the major objectives of contemporary research in biochemistry.

15.10 Separation of Protein Mixtures

You will note that in both the α-helix and the pleated-sheet structures of polypeptides, every third atom in the peptide chain is attached to an R group that extends to the outside of these structures. Certain of these side chains will contain carboxyl groups if an amino acid in the peptide link is a diacid such as aspartic acid. Likewise, a side chain will contain a basic amino group if an amino acid in the peptide link is a dibasic one such as lysine. The presence of acidic and basic groups in the side chains makes possible the formation of positive and negative charges along the peptide chain, as illustrated below.

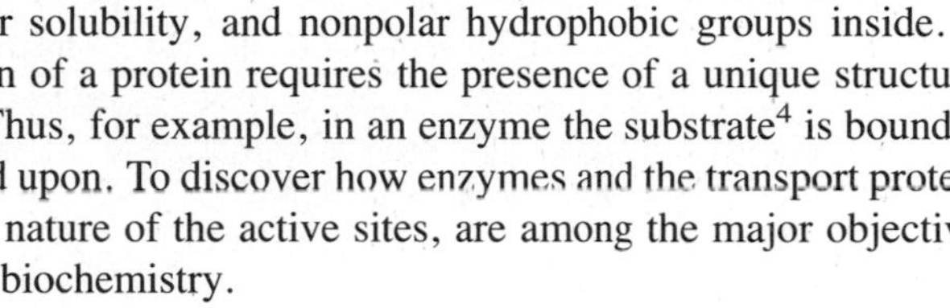

$$\sim \mathrm{NH{-}CH[(CH_2)_4NH_3^+]{-}C(=O)} \sim \left(\mathrm{NH{-}CHR{-}C(=O)}\right)_n \sim \mathrm{NH{-}CH(CH_2COO^-){-}C(=O)} \sim$$

[4]The term *substrate* refers to the organic compound undergoing a reaction with a given reagent.

The behavior of a protein in an electric field will depend upon the number and kind of charges present and upon the acid strength (pH) of the solution. By adjusting the pH of the solution, a protein will move toward either the anode or the cathode. How fast this migration takes place depends upon the isoelectric point of the protein and upon the nature of the other side chains. The separation of a mixture of proteins (or amino acids) electrolytically is called **electrophoresis.**

15.11 Simple and Conjugated Proteins

Proteins may be classified as simple proteins or as conjugated proteins. Simple proteins are those which yield, on hydrolysis, only α-amino acids. Albumin in eggs, gluten in wheat, keratin in hair, and collagen in connective tissue are examples of simple proteins. Conjugated proteins are those which, on hydrolysis, yield other compounds in addition to α-amino acids. Such nonprotein materials are called **prosthetic** groups. Hemoglobin is an example of a conjugated protein. The prosthetic group, in this case, is the iron-containing porphyrin structure called heme.

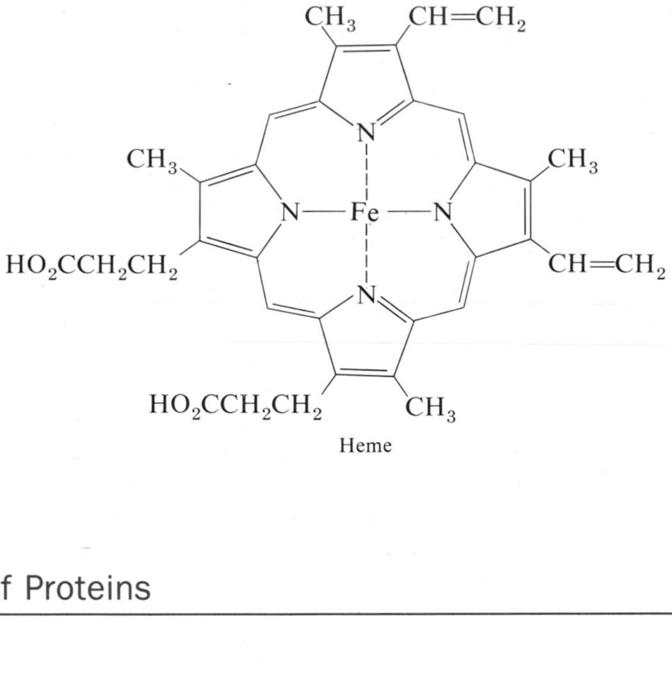

Heme

15.12 Properties of Proteins

When heated, proteins coagulate, or precipitate. The protein albumin, found in egg white, is a common example of a protein easily coagulated by heat. The salts of certain heavy metals (silver, mercury, lead) also cause proteins to precipitate. The immediate ingestion of protein material such as egg white as an antidote for accidental heavy-metal poisonings is based on this behavior. The cauterizing action of silver nitrate is another application. Proteins also are precipitated by certain acids. Precipitation of protein by any agent is sometimes an irreversible change, and the precipitated protein is then said to be **denatured.**

A number of chemical tests on proteins produce color reactions. One of these, the **biuret test,** produces a pink or purple color when an alkaline solution of protein material is treated with a very dilute cupric sulfate solution. The test is specific for multiple peptide links and is not given by α-amino acids. It is a convenient test and is often used on a protein hydrolysate to determine the completeness of hydrolysis.

The **xanthoproteic test** is a yellow color reaction produced when a protein is treated with concentrated nitric acid. The test is simply a nitration of the aromatic ring of certain amino acids (tyrosine, phenylalanine, and tryptophan) and is one recognized by every student as the familiar nitric acid stain. Another color test for proteins is the **Millon test.** The reagent used for this test is a mixture of mercuric and mercurous nitrates. A protein, when treated with Millon's reagent and heated, produces a red color. Any phenolic compound will give the reaction, and the Millon test, like others that give colored reactions, is dependent upon the presence of certain individual amino acids in the protein molecule.

Exercise 15.5 Suggest a reason why it used to be a legal requirement that the eyes of newborn infants be treated with a silver nitrate solution.

15.13 Nutritional Importance of Proteins

Proteins provide one of the major nutrients for the body, but their utilization differs from that of the fats and carbohydrates. Whereas fats and carbohydrates are used primarily to supply heat and energy, the proteins are used mainly to repair and replace worn-out tissue. Such repairs and replacements are made of protein material that the animal organism has synthesized from other ingested proteins. Unlike certain of the lower plants, animals are not capable of "fixing"[5] atmospheric nitrogen or of converting ammonium or nitrate salts to proteins. Animals obtain their protein by eating plants that have synthesized protein material or by eating other animals that have eaten such plants. The digestion of proteins to α-amino acids by the body supplies the required building material from which the animal's own protein can be formed. So far as is known, and in contrast to carbohydrate and fat depots, there are no body depots of proteins that serve as stores and have no other function.

[5]The conversion of atmospheric nitrogen to nitrates or other nitrogenous compounds. Certain free-living soil bacteria and others that live in nodules on the roots of leguminous plants (peas, beans, clover) are able to "fix" and store nitrogen in the form of nitrates.

15.14 Metabolism of Proteins

The digestion of proteins leads to mixtures of simple amino acids and polypeptides of varying lengths. Digestion destroys the specificity of a protein and frees the constituent amino acids for the synthesis of new proteins that suit the requirements of the individual. Amino acids not required for such syntheses are converted to other necessary foods such as carbohydrates and fat, and ultimately are oxidized to yield energy. The metabolic changes involved in these conversions are very complex.

The amino acids produced by the digestion of a protein are absorbed through the intestinal wall into the blood and are transported to the liver. Certain of the amino acids then proceed from the liver to other tissues. Amino acids are required by the cells for the synthesis of proteins, enzymes, certain hormones, and other nitrogen-containing substances. Body tissues are capable of synthesizing the nonessential amino acids by removing the required amino groups from other amino acids. The amino group taken from one acid is transferred to an α-keto analog of the amino acid to be synthesized. This transfer is called **transamination** and is illustrated below.

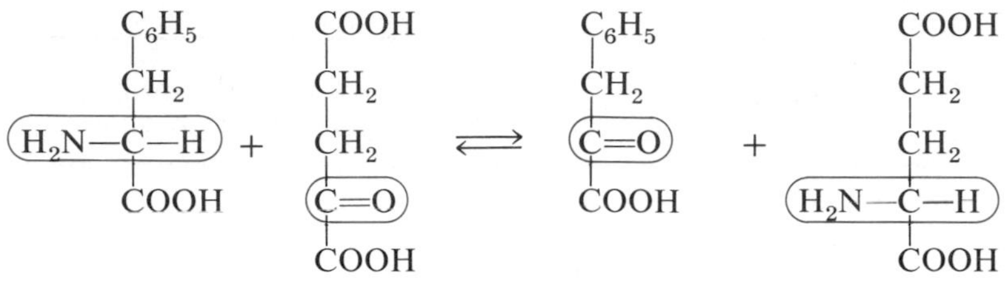

$$\begin{array}{c} C_6H_5 \\ | \\ CH_2 \\ | \\ \boxed{H_2N{-}C{-}H} \\ | \\ COOH \end{array} + \begin{array}{c} COOH \\ | \\ CH_2 \\ | \\ CH_2 \\ | \\ \boxed{C{=}O} \\ | \\ COOH \end{array} \rightleftharpoons \begin{array}{c} C_6H_5 \\ | \\ CH_2 \\ | \\ \boxed{C{=}O} \\ | \\ COOH \end{array} + \begin{array}{c} COOH \\ | \\ CH_2 \\ | \\ CH_2 \\ | \\ \boxed{H_2N{-}C{-}H} \\ | \\ COOH \end{array}$$

Phenylalanine — α-Ketoglutaric acid — Phenylpyruvic acid — Glutamic acid

Transamination of phenylalanine with α-ketoglutaric acid is normally a minor pathway in the metabolism of phenylalanine, the major pathway being oxidation of phenylalanine to tyrosine catalyzed by phenylalanine 4-monooxygenase. About 1 in every 10,000 human beings lacks this enzyme, owing to a genetic defect. In children the excess phenylpyruvic acid impairs normal brain development, and serious mental retardation results. This condition, called phenylketonuria (PKU), was responsible, in part, for the 10-year delay in the approval of the new artificial sweetener *aspartame,* Sec. 13.6, which is about 180 times as sweet as sugar and is said to have no bitter aftertaste. When used as a sugar substitute aspartame helps to provide a part of the daily requirements of aspartic acid and phenylalanine.

The amino group of α-amino acids, if not required for the synthesis of new amino acids, can be oxidatively removed. This process is called **deamination.**

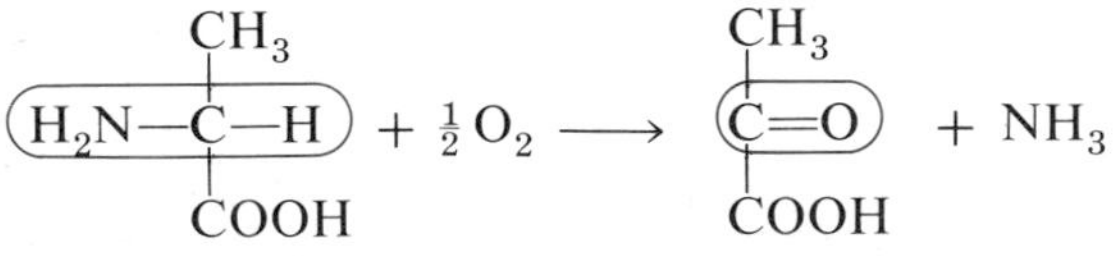

$$\begin{array}{c} CH_3 \\ | \\ \boxed{H_2N{-}C{-}H} \\ | \\ COOH \end{array} + \tfrac{1}{2}O_2 \longrightarrow \begin{array}{c} CH_3 \\ | \\ \boxed{C{=}O} \\ | \\ COOH \end{array} + NH_3$$

Alanine — Pyruvic acid

The ammonia formed by deamination combines with carbon dioxide through a series of enzymatic reactions to produce urea.

$$2\,NH_3 + CO_2 \longrightarrow \underset{\text{Urea}}{H_2N{-}\overset{\overset{\displaystyle O}{\|}}{C}{-}NH_2} + H_2O$$

Urea is eliminated from the body by way of the urine and is a major end product of protein metabolism.

The carbon skeleton of an amino acid, after the amino group is removed by either of the processes described, may be used in the synthesis of other amino acids, or it may enter the pathways of carbohydrate and fatty acid metabolism. The latter course proceeds by way of the tricarboxylic acid cycle (Sec. 14.22). Thus, the deaminated amino acid, like a fat or a carbohydrate, may ultimately be converted to carbon dioxide, water, and energy.

15.15 Improper Metabolism of Proteins. Allergies

The inability of some persons to accomplish complete hydrolysis of certain protein material may result in the absorption of minute amounts of unchanged protein from the intestinal tract. Such metabolic failures cause the individual to become extremely sensitive or allergic to certain foods. When present even in minute amounts and eaten unknowingly as ingredients in other foods, such allergens can have very distressing effects. Sneezing, hives, eczema, and general discomfort result. The old phrase ''one man's food is another's poison'' has some basis in fact. Proteins injected as serums or as antibiotics sometimes, when incompatible with the individual, produce even more serious results. Incompatibilities of the kind described appear to be genetically related and part of the heredity of the individual. The complex chemistry that controls heredity rests in the area of nucleo proteins (Chapter 16) and is currently the area of greatest excitement to biochemists.

Summary

1. Proteins are high molecular weight natural polymers made up largely by combination of various α-amino acids.
2. Approximately twenty amino acids comprise the bulk of plant and animal protein. Names and structures for the common amino acids are listed in Table 15.1.
3. Amino acids may be classified according to properties as:
   (a) neutral (one amino group and one carboxyl group)
   (b) acidic (more than one carboxyl group per amino group)
   (c) basic (more than one amino group per carboxyl function)

4. The α-amino acids, with the exception of glycine, are optically active.
5. The α-amino acids obtained from plant or animal protein have the L-configuration.
6. Amino acids may be grouped under a nutritive classification into two categories:
   (a) **essential** (required in the diet because the animal organism is incapable of synthesizing it)
   (b) **nonessential** (not required in the diet)
7. Amino acids have the properties of both carboxylic acids and primary amines. Amino acids form inner salts or dipolar ions.
8. Amino acids have an isoelectric point.
9. Individual amino acids can be separated from protein hydrolysates. Amino acids can be synthesized by:
   (a) amination of α-halogen acids with ammonia
   (b) the Strecker synthesis
10. Amino acids generally show reactions characteristic of both the primary amines and the carboxylic acids. The principal reactions of amino acids are:

Starting compound: $R-CH(NH_2)-COOH$

$$R-CH(NH_2)-COOH \xrightarrow{R'OH,\ H^+} \text{(esters)} \quad R-CH(NH_2)-COOR'$$

$$R-CH(NH_2)-COOH \xrightarrow{Na^+OH^-} \text{(sodium salt of a carboxylic acid)} \quad R-CH(NH_2)-COO^-Na^+$$

$$R-CH(NH_2)-COOH \xrightarrow{H^+Cl^-} \text{(amine hydrochloride salt)} \quad R-CH(\overset{+}{N}H_3Cl^-)-COOH$$

$$R-CH(NH_2)-COOH \xrightarrow{HNO_2\ \text{(Van Slyke)}} \text{(free } N_2 \text{ and hydroxy acids)} \quad R-CH(OH)-COOH + N_2\uparrow$$

$$R-CH(NH_2)-COOH \xrightarrow{R'-C(=O)-Cl} \text{(acyl or aroyl derivatives)} \quad R'-C(=O)-NH-CH(R)-COOH$$

$$R-CH(NH_2)-COOH \xrightarrow{\text{Other amino acids}} \text{(peptides)}$$

$$R-CH(NH_2)-COOH \xrightarrow{\text{Ninhydrin}} \text{(colored products)}$$

11. Two or more amino acids joined through amide linkages form peptides. A peptide is named as a derivative of the amino acid with a C-terminal carboxyl function.
12. Proteins, when hydrolyzed, yield amino acids and polypeptides.

13. Proteins are classified as simple proteins or conjugated proteins. Simple proteins on complete hydrolysis give only amino acids. Conjugated proteins on hydrolysis give, in addition to amino acids, nonprotein *prosthetic* groups.
14. Proteins are easily precipitated, or coagulated, by heat, by acids, and by certain heavy metals. Such coagulation is sometimes irreversible and is called *denaturation*.
15. Proteins are used in the body mainly to repair and replace worn-out tissue.
16. The digestion and metabolism of protein material begins with its hydrolysis to α-amino acids. The α-amino acids, by **transamination** and/or **deamination,** are converted into other α-amino acids or become oxidized to carbon dioxide and water.

Supplementary Exercises

15.6 Draw the structures and name two sulfur-containing α-amino acids. Which of these can form disulfide bonds?

15.7 Beginning with a three-carbon alcohol, outline all necessary steps in the preparation of D,L-alanine.

15.8 Complete the following equations.

(a) $CH_3-\overset{\displaystyle H}{\underset{\displaystyle NH_2}{\overset{|}{\underset{|}{C}}}}-COOH + NaNO_2 + HCl \longrightarrow$

(b) $C_6H_5-C(=O)Cl + H_2NCH_2COOH \longrightarrow$

(c) $HS-CH_2-\overset{\displaystyle H}{\underset{\displaystyle NH_2}{\overset{|}{\underset{|}{C}}}}-COOH + I_2 \longrightarrow$

(d) $CH_3-\overset{\displaystyle H}{\underset{\displaystyle NH_2}{\overset{|}{\underset{|}{C}}}}-COOH + NaOH \longrightarrow$

(e) $C_6H_5-CH_2-O-C(=O)Cl + CH_3-\overset{\displaystyle H}{\underset{\displaystyle NH_2}{\overset{|}{\underset{|}{C}}}}-C(=O)OH \longrightarrow$

(f) Product of (e) + $SOCl_2 \longrightarrow$

(g) Product of (f) + alanine $\longrightarrow$

15.9 Draw all tripeptides that could possibly be synthesized from glycine, alanine, and phenylalanine.

15.10 A kilogram of human hemoglobin contains approximately 3.33 grams of iron. If each molecule of hemoglobin contains four heme structures (the iron-containing prosthetic group), what is the minimum molecular weight of this protein?

15.11 A Van Slyke nitrogen determination made on a solution of 8.74 mg of an unknown amino acid liberated 2.50 mL of N_2 at 740 mm and 25°. What is the minimum molecular weight of the amino acid? If the isoelectric point of this unknown amino acid occurs at a pH value of 10.8, what amino acid could it be?

15.12 Write the structures of the following peptides.

(a) cysteylleucylproline
(b) histidylalanyllysine
(c) aspartame (aspartylphenylalanine methyl ester)

15.13 A peptide on complete hydrolysis yielded five different amino acids in the following amounts: one unit each of serine and glycine, two units each of arginine and phenylalanine, and three units of proline. The N-terminal amino acid was identified as arginine. Partial hydrolysis liberated di- and tripeptides of the following compositions:

Ser-Pro-Phe	Gly-Phe-Ser	Pro-Gly
Phe-Ser-Pro	Pro-Phe-Arg	Arg-Pro

Deduce and write out the structure of this nonapeptide.

15.14 Calculate the isoelectric pH values for the following amino acids.

(a) glycine
(b) alanine
(c) glutamine
(d) serine

15.15 A solution turns red litmus blue and the indicator phenolphthalein slightly pink. The solution also contains the amino acid, glycine. If the solution were electrolyzed, would the glycine migrate to the cathode or the anode?

15.16 What is the probable net charge (such as $+\frac{1}{2}$) on the following amino acids in solutions of (1) pH 1.0, (2) pH 2.2, (c) pH 4.0, and (4) pH 9.8.

(a) alanine
(b) aspartic acid
(c) asparagine

15.17 Draw the structures of the following amino acids as they would appear in solutions of (1) pH 3.0 and (2) pH 10.

(a) lysine
(b) glutamic acid
(c) alanine
(d) glycine

15.18 The dipolar (zwitterionic) form of arginine has the following structure. Suggest an explanation for the protonation of the δ-guanadino group (which is responsible for the high pK_3) rather than the α-amino group.

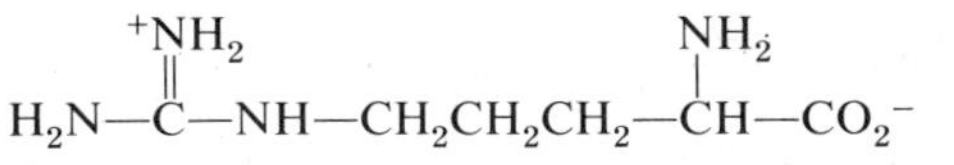

$$H_2N-\overset{\overset{+}{N}H_2}{\overset{\|}{C}}-NH-CH_2CH_2CH_2-\overset{NH_2}{\overset{|}{C}H}-CO_2^-$$

15.19 The use of amino acids labeled with radioactive isotopes is a valuable technique in biochemistry. Show how the following substances could be prepared from commercially available sources of ^{14}C: $Ba^{14}CO_3$ and $Na^{14}CN$.

(a) alanine labeled at the C-1 position (carboxyl group)
(b) alanine labeled at the C-2 position (α-carbon atom)
(c) aspartic acid labeled at the C-4 position (second carboxyl group)
(d) phenylalanine labeled at the C-3 position (β-carbon atom)

15.20 Draw the structures of the products obtained in the following reactions.

(a) Ala-Leu-Cy(SH)-Gly-Ser-Leu-His-Cy(SH)-Phe $\xrightarrow{I_2}$

(b) Ala-His-Gly-Arg $\xrightarrow[\text{2. HCl}]{\text{1. } C_6H_5N{=}C{=}S}$

(c) Cbz-Phe + Ala—$OCH_2C_6H_5$ $\xrightarrow{\text{DCC}}$

15.21 At a pH value of 3, would the peptide of Exercise 15.13 migrate towards the anode or towards the cathode in an electrophoretic separation?

15.22 The small packets of low-calorie sweetener of the type currently found in many restaurants list the contents but also a cautionary note to phenylketonurics: "Contains phenylalanine." Explain.

15.23 The enzyme *ribonuclease* contains 124 amino acid residues. What would be the *minimum* molecular weight of this enzyme?

15.24 An intravenous anesthetic of short duration called *etomidate* is similar in pharmacologic effect to sodium pentothal, but bears a structural relationship to histidine. It has the chemical name (*R*)-(+)-ethyl 1-(1-phenylethyl)imidazole-5-carboxylate. Draw a pseudo-three-dimensional structure for this anesthetic to show the (*R*)-configuration. (*Hint:* See Tables 4.3 & 15.1.)

16 Nucleic Acids

The nucleic acids are high molecular weight natural polymers found in the nuclei and cytoplasm of all living cells, where they play a highly specialized role in life chemistry. The nucleic acids not only direct the synthesis of specialized cellular material in each individual according to a predetermined code, but also control the genetic continuity of the species.

Your background knowledge of sugars, amino acids, and peptides has prepared you to understand and, to a degree, participate in the excitement that the study of nucleic acids has brought to the field of biochemistry. In no other part of your semester's study will you find organic chemistry more closely related to biology than in the area of nucleic acids, for it is here that the creative capabilities of the living laboratory are so wondrously manifest.

16.1 Structure of Nucleic Acids

The structure of a nucleic acid has been determined, as have the structures of many other natural substances, through a series of degradative studies in which the whole structure is broken down into its component parts. Once the parts of an unknown have been identified, a determination of the complete structure usually follows. There are two families of nucleic acids, but the hydrolysis of either type results in the production of three different chemical substances—orthophosphoric acid, a five-carbon sugar, and four different heterocyclic basic compounds. The sugar D-ribose (Sec. 14.4) is obtained from **ribonucleic acid** (RNA), the nucleic acid found largely in the cytoplasm. The sugar 2-deoxy-D-ribose is obtained from **deoxyribonucleic acid** (DNA), the constituent of the genetic material in the cell nucleus. The heterocyclic bases found as parts of the nucleic acid structure are derivatives of either pyrimidine or purine (Sec. 16.3).

The nucleic acid polymer actually is a polyester structure in which phosphoric acid–sugar units are repeated over and over. This polyphosphate chain in the nucleic acid polymer is sometimes described as the ''backbone'' structure. One of five different heterocyclic bases is bonded to each sugar molecule in the ester unit via an N-glycosidic link. The following schematic drawing shows the fundamental nucleic acid structure.

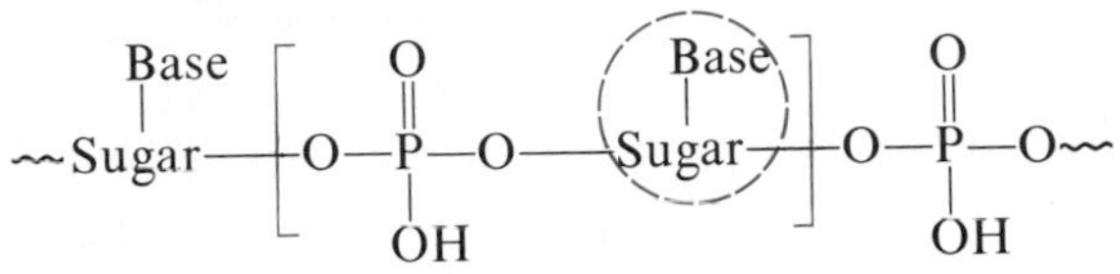

A section of the polynucleotide chain

The portion of the polynucleotide enclosed by brackets in the structure above shows all three components of a nucleic acid—base, sugar, and phosphoric acid unit. This portion is called a *nucleotide*. A nucleotide must include the phosphoric acid structure; otherwise it would not be a nucleic acid. The N-glycosidic portion shown enclosed in a broken circle—the sugar with attached base but lacking the phosphoric unit—is called a *nucleoside*. Examples of both nucleotides and nucleosides will be given in following sections.

Let us now examine the component parts of a nucleic acid, and then we will be able to understand better how the parts combine to form DNA and RNA, and also how the nucleic acids can perform their vital roles.

16.2 The Sugar-Phosphate Ester

The sugar components of DNA and RNA differ only at the number 2 carbon atom of the furanose ring. In ribose the number 2 carbon atom has bonded to it both hydrogen and a hydroxyl group. The hydroxyl group is lacking at this position in deoxyribose as is indicated by the prefix *2-deoxy*.

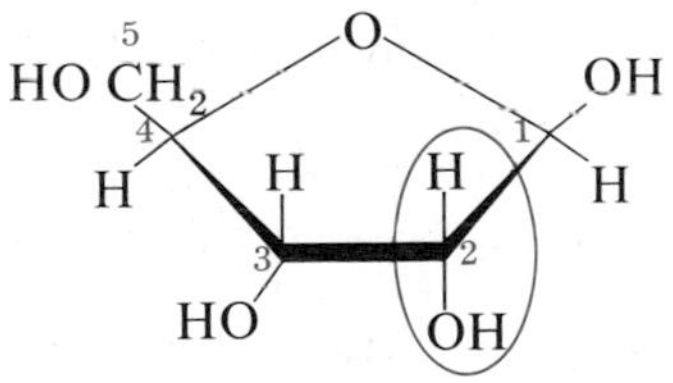

Ribose
(β-D-Ribofuranose)

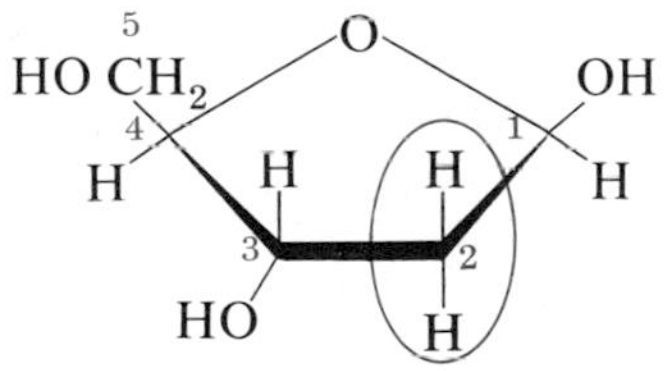

Deoxyribose
(2-Deoxy-β-D-ribofuranose)

Both ribose and deoxyribose have the β-configuration (Sec. 14.5) at the number 1 carbon atom, and in both RNA and DNA the nitrogen bases are attached to the sugar molecule at this position. Also, in both nucleic acids each phosphoric acid molecule forms

FIGURE 16.1 A Segment of the DNA Polynucleotide "Backbone"

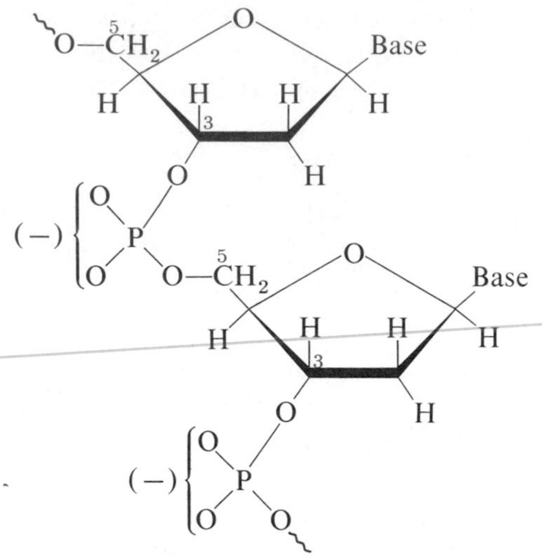

a diester by joining carbon 5 of one sugar molecule to carbon 3 of the next. The mode of esterification in DNA is illustrated in Figure 16.1.

16.3 Purines and Pyrimidines

The basic portions of nucleic acids are derivatives of two heterocyclic structures, purine and pyrimidine, and are the parts of nucleic acids responsible for the storage and transmission of genetic information. Although there are a number of minor, but important, bases, five constitute the major bases found in nucleotides. Two of these, **adenine** and **guanine,** are purines and are found in both RNA and DNA. The other three, **thymine, cytosine,** and **uracil,** are pyrimidines. DNA contains thymine, and RNA contains uracil. Cytosine is common to both nucleotides.

The structures and numbering systems of the two parent heterocycles are shown, along with the derivatives of each that occur in RNA and DNA.

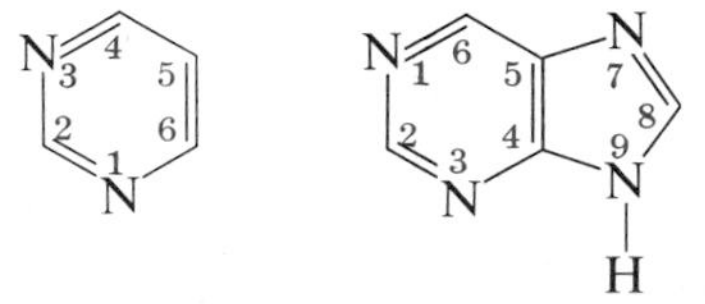

Pyrimidine Purine

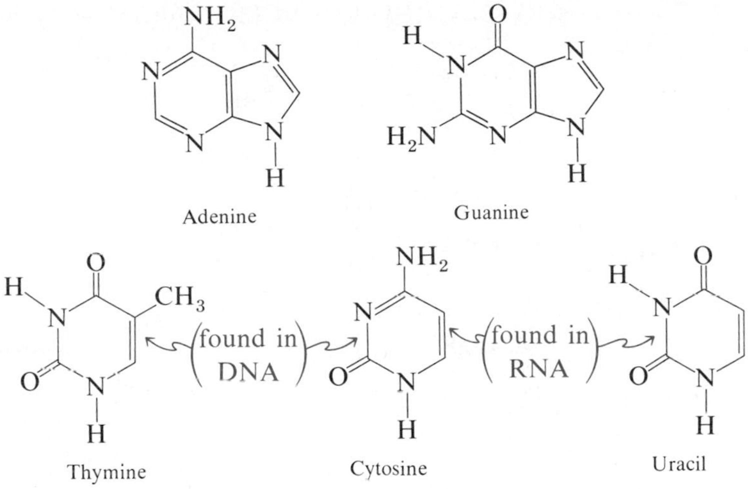

Adenine Guanine

Thymine (found in DNA) Cytosine (found in RNA) Uracil

> ***Exercise 16.1*** Xanthine (2,6-dihydroxypurine) may be made from guanine. With what reagents must guanine be reacted to effect this change?
>
> ***Exercise 16.2*** Draw the structure for caffeine, 1,3,7-trimethylxanthine.

16.4 Nucleotides and Nucleosides

The structural units in both RNA and DNA are called nucleotides and, as has been pointed out earlier, are the phosphate esters of a substituted pentose, the substituent being either a purine or a pyrimidine derivative attached at carbon atom number 1′ of the sugar ring.[1] The point of attachment in the heterocyclic base is at position 1 in the pyrimidine and at position 9 in the purine. A nucleoside is only the substituted sugar portion of a nucleotide and structurally is analogous to a glycoside (Sec. 14.5), except that here we have a nitrogen atom instead of an oxygen atom bonded at carbon 1′. Examples of both nucleoside and nucleotide formation are illustrated on the following page.

The name of a nucleotide incorporates that of its nitrogen base, using the suffix *ylic* and followed by *acid* as a separate word. The name of a nucleoside is similar to that given a nucleotide with the same sugar and base except that the name ends with the suffix *idine*

[1]In order to distinguish between two different ring systems present within a nucleic acid, primed numbers will henceforth be used to designate positions in the sugar, unprimed numbers to designate ring positions in the heterocyclic base.

SUGAR + HETEROCYCLIC BASE ⟶ NUCLEOSIDE

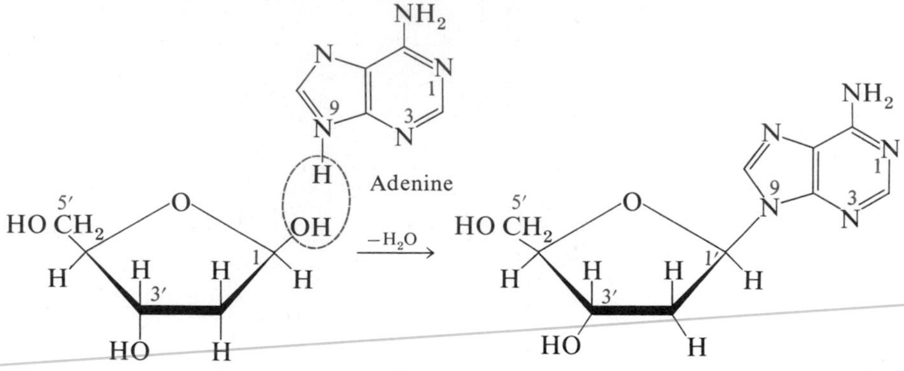

NH$_2$ N N 1 9 3 N N H Adenine

5′ HO CH$_2$ O OH H H H 1 3′ H HO H $\xrightarrow{-H_2O}$ 5′ HO CH$_2$ O 1′ H H H 3′ H HO H NH$_2$ N N 1 9 3 N N

Deoxyribose

Deoxyadenosine
(A nucleoside found in DNA)

PHOSPHORIC ACID + NUCLEOSIDE ⟶ NUCLEOTIDE

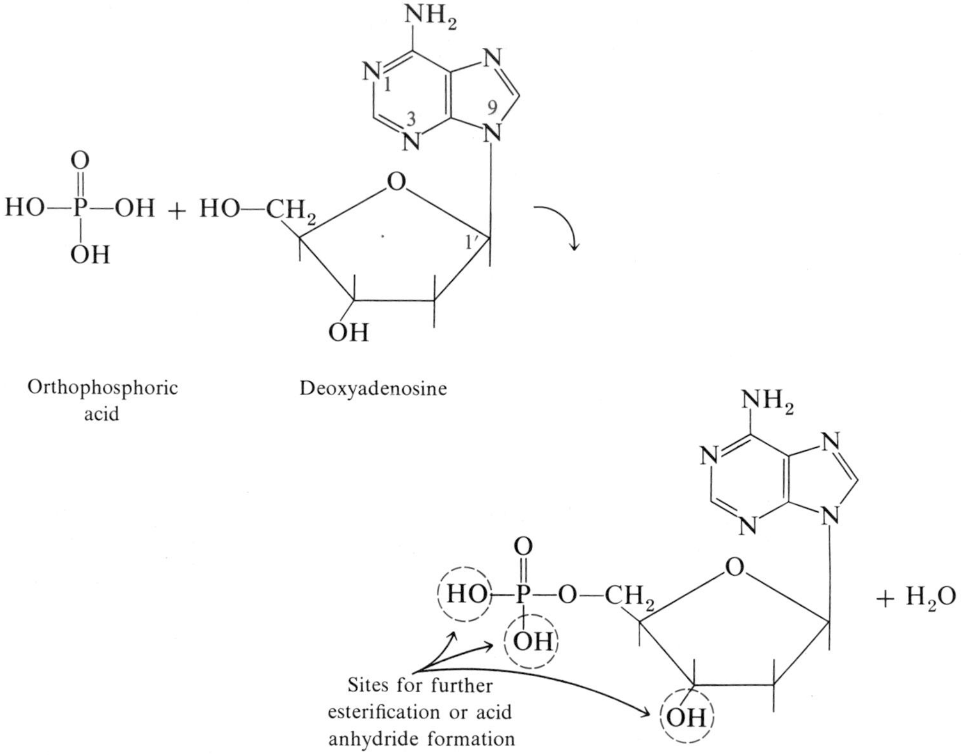

NH$_2$ N 1 N 9 3 N N

O HO—P—OH + HO—CH$_2$ O 1′ OH OH

Orthophosphoric acid

Deoxyadenosine

NH$_2$ N N N N

O HO—P—O—CH$_2$ O OH OH + H_2O

Sites for further esterification or acid anhydride formation

Deoxyadenylic acid
(A nucleotide found in DNA)

TABLE 16.1 Some Nucleosides and Nucleotides

	Base	Pentose	Nucleoside	Nucleotide
Purines	Adenine	Ribose	Adenosine	Adenylic acid
	Guanine	Deoxyribose	Deoxyguanosine	Deoxyguanylic acid
Pyrimidines	Uracil	Ribose	Uridine	Uridylic acid*
	Thymine	Deoxyribose	Deoxythymidine	Deoxythymidylic acid†
	Cytosine	Deoxyribose	Deoxycytidine	Deoxycytidylic acid

*Found only in RNA.
†Found only in DNA.

if the base portion is a pyrim*idine*, with *osine* if the base portion is a purine. Of course the word acid does not apply. Names of structural units found in RNA and DNA are given in Table 16.1.

Exercise 16.3 Which nucleotide unit is contained in Coenzyme A?

16.5 The DNA Polymer

One of the most significant observations that led to a deduction of the DNA structure was the finding that the ratio of adenine to thymine and of guanine to cytosine in the polynucleotide was almost 1:1. Such equimolar quantities of pyrimidines and purines suggested that the bases must be present in the DNA structure in pairs, but how could such equivalence be accomplished? The answer to this question was supplied by Watson and Crick,[2] who built a model that appeared to embody all facts known about DNA. The Watson–Crick model was based upon measurements and calculations made from X-ray diffraction data obtained by M. Wilkins from crystallographic studies on a prepared sample of DNA polymer. To account for the molecular ratio of unity between pyrimidine and purine derivatives, Watson and Crick proposed that the DNA structure is not that of a simple helix as are many other protein structures, but has instead the form of a *double* helix. A

[2]James D. Watson (1928–), Professor of Biology, Harvard University, and Francis H. Crick (1916–), British Research Council, shared with M. H. F. Wilkins (1916–), Kings College, University of London, the Nobel Prize for Medicine and Physiology (1962).

double-helix structure would allow the nitrogen bases in each DNA strand to be projected toward the center. Uniform intermolecular hydrogen bonding would then be possible between adenine in one strand of the helix and thymine in the other strand, and also between cytosine in one strand and guanine in the other. Such hydrogen bonding between complementary strands would result in connecting both strands somewhat as the treads connect the two rising parts of a spiral staircase. A structure of this kind not only would require base pairing, but also would provide the stability that holds the two strands together. The hydrogen bonding capabilities of complementary base pairs are shown in Figure 16.2.

FIGURE 16.2 Hydrogen Bonding Between Base Pairs in DNA

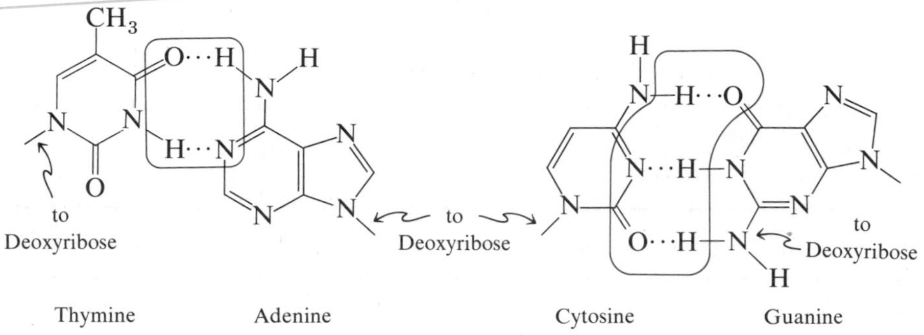

Watson and Crick also postulated that the two polynucleotide strands entwine about a common axis and are right-handed spirals. The 3′, 5′-internucleotide phosphodiester bridge (Fig. 16.1) runs in one direction for one strand and in the opposite direction for the other. The approximate diameter of the double helix was revealed to be about 20 Å and consisted of ten nucleic acid residues per turn—each turn measuring a distance of 34 Å. A diagrammatic segment of the double helix with base pairings between strands is shown in Figure 16.3.

16.6 Replication of DNA

The concept of guanine–cytosine and adenine–thymine pairing between two coiled polynucleotide chains not only offered a plausible structure for DNA, but also suggested the mechanism whereby genetic information is transmitted from one generation to the next.

FIGURE 16.3 A Segment of the DNA Double Helix

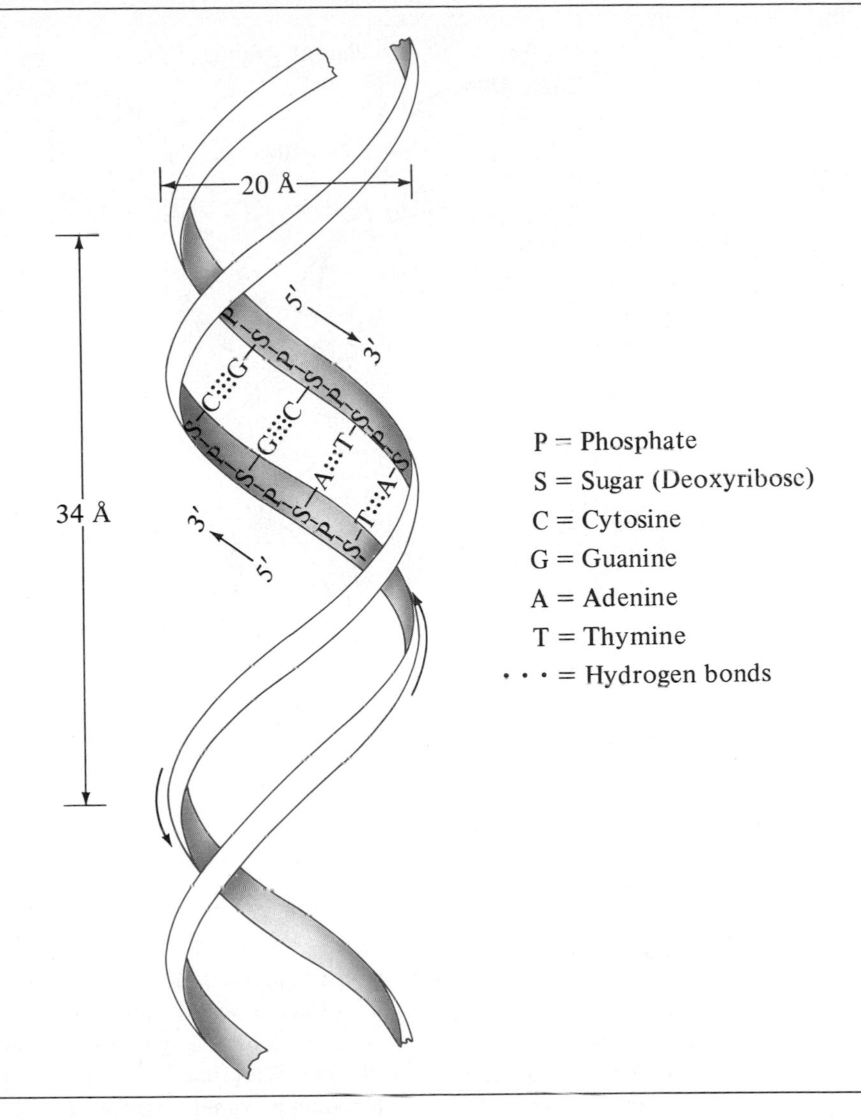

The biosynthesis of new strands of DNA containing the same sequence of base pairs as the parent can result if the parent DNA double helix uncoils and each single strand functions as a template on which a new complementary strand of DNA forms. In this way two new daughter strands result with the same base-pair sequence and hereditary information as were present in the parent. As each new strand is formed, hydrogen bonding between complementary bases leads to an identical replica of the parent. A schematic mechanism of DNA replication is shown in Figure 16.4.

FIGURE 16.4 The Schematic Representation of DNA Replication

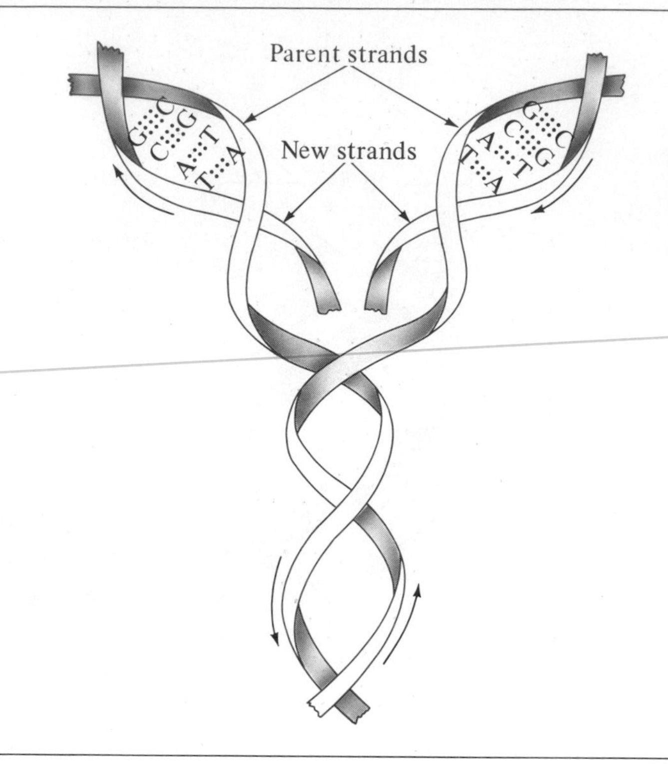

16.7 The Genetic Code

Protein material designed to perform a particular function—hair, enzymes, muscle tissue, etc.—must be formed from amino acids selected and combined in a certain sequence. The genetic information that must be programmed for the synthesis of such specialized material is embodied in the *sequence* of the four bases found in the DNA strand and is referred to as the *genetic code*. How is such coding accomplished for the proper selection of the twenty different amino acids from which proteins are synthesized? Obviously, one nucleotide could not code for only one particular amino acid since, in that case, we could use only four different amino acids. The coding information, therefore, must lie in *combinations* of nucleotides. Let us abbreviate with capital letters the names of the four bases found in DNA and consider combinations of bases in groups of two, e.g., AT, TG, CG, AA. By so doing we could possibly form 4^2, or sixteen, combinations—not quite enough. However, if we use combinations of three nucleotides in different sequences, we could form 4^3, or sixty-four, coding combinations—more than enough to accommodate all twenty amino acids. In fact, we would have enough combinations to allow more than one coding triplet for some amino acids and additional combinations that do not appear to code

TABLE 16.2 m-*RNA* Codons Proposed for Amino Acids Used in Protein Synthesis

Amino acid	Codons	Amino acid	Codons
Alanine	GCA, GCC, GCG, GCU	Lysine	AAA, AAG
Arginine	AGA, AGG, CGA, CGC, CGG, CGU	Methionine	AUG
Asparagine	AAC, AAU	Phenylalanine	UUC, UUU
Aspartic acid	GAC, GAU	Proline	CCA, CCC, CCG CCU
Cysteine	UGC, UGU	Serine	UCA, UCC, UCG, UCU, AGC, AGU
Glutamic acid	GAA, GAG	Threonine	ACA, ACC, ACG, ACU
Glutamine	CAA, CAG	Tryptophan	UGA, UGG
Glycine	GGA, GGC, GGG, GGU	Tyrosine	UAC, UAU
Histidine	CAC, CAU	Valine	GUA, GUC, GUG, GUU
Isoleucine	AUA, AUC, AUU		
Leucine	CUA, CUC, CUG, CUU, UUA, UUG		

for any. Coding units made up of such triplet combinations are called *codons* and may be thought of as three-lettered words from which an entire sentence may be constructed—the sentence, in this case, being the polypeptide chain. Codons for the twenty amino acids are given in Table 16.2. The code is usually expressed in terms of codons for messenger RNA.

16.8 RNA

The primary structure of RNA differs from that of DNA mainly in the nature of the nucleosides esterified. RNA incorporates the base uracil instead of thymine. Except for lacking the 5-methyl substituent, uracil is structurally similar to thymine, and its hydrogen-bonding characteristics allow it also to pair with adenine. Thus uracil appears in the RNA strand instead of thymine.

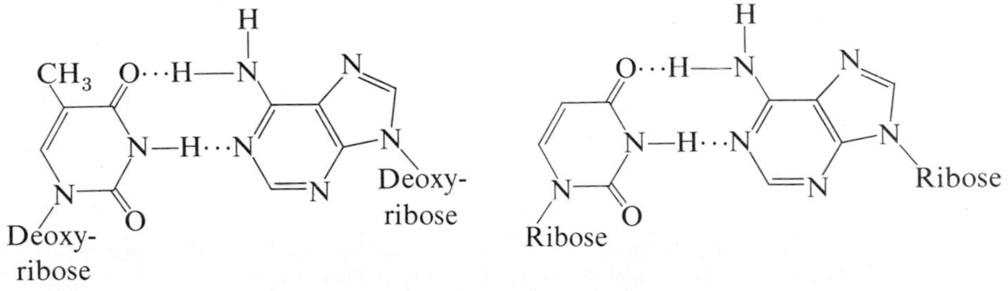

Thymine-Adenine pairing in DNA

Uracil-Adenine pairing in RNA

Another difference between DNA and RNA lies in their secondary structures. RNA molecules exist mainly as single strands, rather than in the form of double helices, and therefore, unlike DNA, may have a pyrimidine–purine ratio other than 1:1.

Three different types of RNA molecules appear to be involved in the vital business of protein synthesis. One of these, **ribosomal RNA** (*r*-RNA), is of the greatest molecular weight and is associated with the ribosomes, or nucleoprotein particles, in the cytoplasm. The ribosomes provide the site for protein synthesis. A second type of RNA, called **messenger RNA** (*m*-RNA), is a polynucleotide of lower molecular weight than *r*-RNA and carries the transcribed genetic code from the DNA in the nucleus of the cell to the ribosomes at the protein-building site. A third type of RNA is called **transfer RNA** (*t*-RNA). Transfer RNA has a molecular weight of approximately 30,000 and is the smallest of the polynucleotides. Because it is soluble, it also is referred to as **soluble RNA** (*s*-RNA). The function of *t*-RNA is to select and to carry the amino acids required for a specific protein to the correct *m*-RNA-ribosome site. The amino acid selected by a *t*-RNA molecule is the one called for by the codon in the *m*-RNA. The triplet of complementary bases in *t*-RNA that corresponds to this ''order from above'' is called an anticodon. All *t*-RNA molecules, though possibly of a different base composition individually, have as a common feature the terminal base triplet C-C-A at the 3′-end. In Figure 16.5 the unfolded two-dimensional structure of yeast alanine *t*-RNA, as deduced by Holley,[3] is shown. All *t*-RNAs have similar cloverleaf structures, sometimes with an extra arm. In its active, three-dimensional conformation, the cloverleaf is twisted into a more compact and specific form. The role that each type of RNA molecule plays in the synthesis of a protein is discussed in Sec. 16.9.

The code is read from the 5′ to the 3′ direction. The coding triplets for that strand of DNA responsible for transcribing the genetic message to *m*-RNA are complementary to those in *m*-RNA, and T replaces U in DNA. In the anticodons of *t*-RNA, the coding triplets are also complementary to those in *m*-RNA; however, the first base (called the ''wobble'' base) of the anticodon may not be the exact complement of the third base in the codon and may be one of the minor bases (see Fig. 16.5.). The various interrelationships are shown in the following diagram and in Figures 16.4 and 16.5.

DNA

5′-end *3′-end*

AAC—AGA—AGT—ACG—TCC—GGC—AAA—TCC—GTA

m-*RNA*

3′-end *5′-end*

UUG—UCU—UCA—UGC—AGG—CCG—UUU—AGG—CAU

[3]Robert W. Holley (1922–), Cornell University, Ithaca, New York (now at the Salk Institute, La Jolla, Ca.), Nobel Prize in Medicine and Physiology (1968).

t-*RNA*

5′-end *3′-end*

AAC IGA AGU ICG UCC IGC AAA UCC GΨA

*I and Ψ are minor bases.

Amino acid

Carboxyl end *Amino end*

Val—Ser—Thr—Arg—Gly—Ala—Phe—Gly—Tyr

FIGURE 16.5 Cloverleaf Structure of Yeast Alanine t-*RNA*

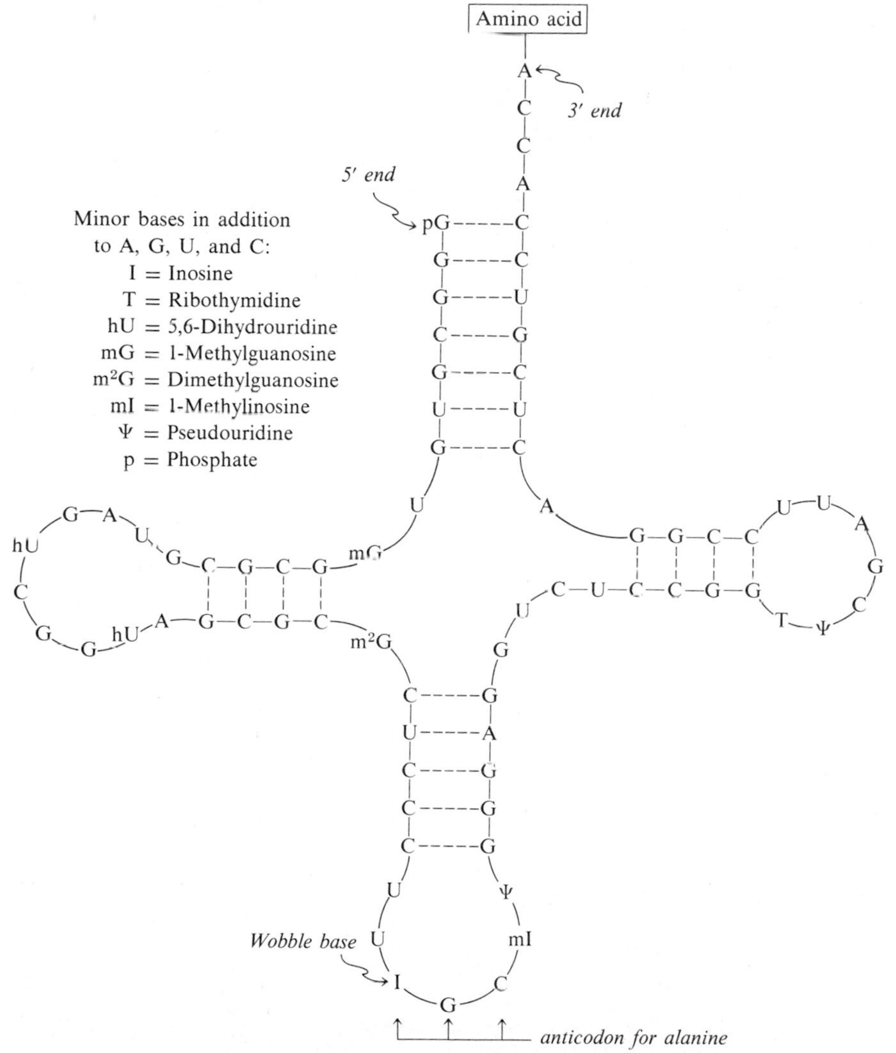

Amino acid

3′ end

5′ end

Minor bases in addition to A, G, U, and C:

- I = Inosine
- T = Ribothymidine
- hU = 5,6-Dihydrouridine
- mG = 1-Methylguanosine
- m^2G = Dimethylguanosine
- mI = 1-Methylinosine
- Ψ = Pseudouridine
- p = Phosphate

Wobble base

anticodon for alanine

16.9 Protein Synthesis

The first step in protein synthesis (see below) is the enzyme-initiated reaction of an amino acid with adenosine triphosphate (ATP) to form a phosphoric-amino acid anhydride com-

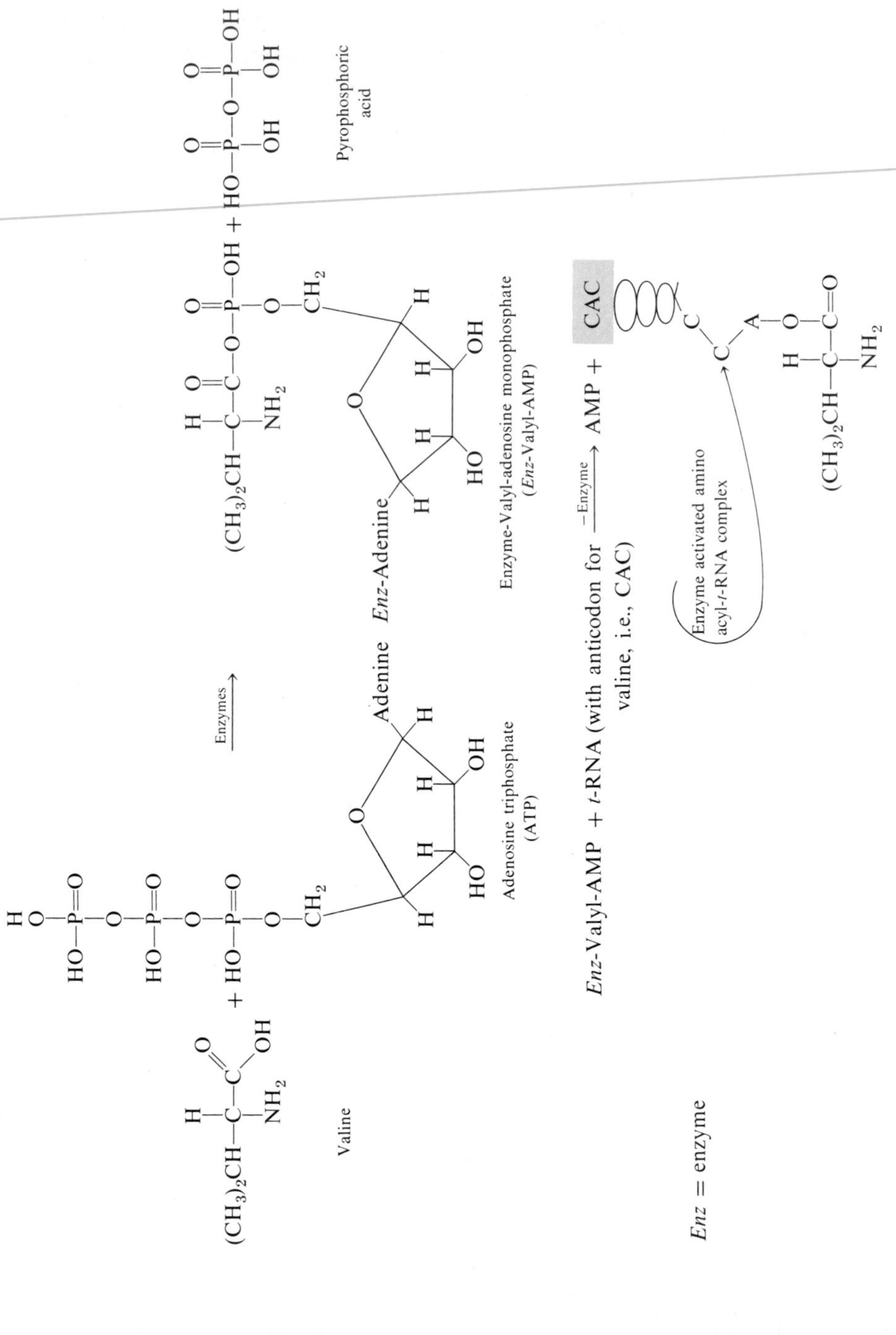

$(CH_3)_2CH{-}CH(NH_2){-}COOH$ + HO–P(=O)(OH)–O–P(=O)(OH)–O–P(=O)(OH)–O–CH_2–(ribose)–Adenine $\xrightarrow{\text{Enzymes}}$ $(CH_3)_2CH{-}CH(NH_2){-}C(=O){-}O{-}P(=O)(OH){-}O{-}CH_2$–(ribose)–*Enz*-Adenine + HO–P(=O)(OH)–O–P(=O)(OH)–OH

Valine

Adenosine triphosphate (ATP)

Enzyme-Valyl-adenosine monophosphate (*Enz*-Valyl-AMP)

Pyrophosphoric acid

Enz-Valyl-AMP + *t*-RNA (with anticodon for valine, i.e., CAC) $\xrightarrow{-\text{Enzyme}}$ AMP + CAC–C–C–A–O–C(=O)–$CH(NH_2){-}CH(CH_3)_2$

Enzyme activated amino acyl-*t*-RNA complex

Enz = enzyme

plex bound to the active site of the enzyme. This enzyme-bound amino acid complex then reacts with a *t*-RNA molecule to produce an activated amino acid ester at the 3′ position of the terminal adenylic acid in the C-C-A "tail." The series of chemical changes on page 448 shows the attachment of valine to *t*-RNA.

The attached valine ester is then transferred by the *t*-RNA to the ribosomes, where it is combined according to a predetermined code with another amino acid previously added to the polypeptide chain (see diagram below). After the *t*-RNA has delivered and deposited an amino acid, it returns for others. The exact details of this transfer of amino acids

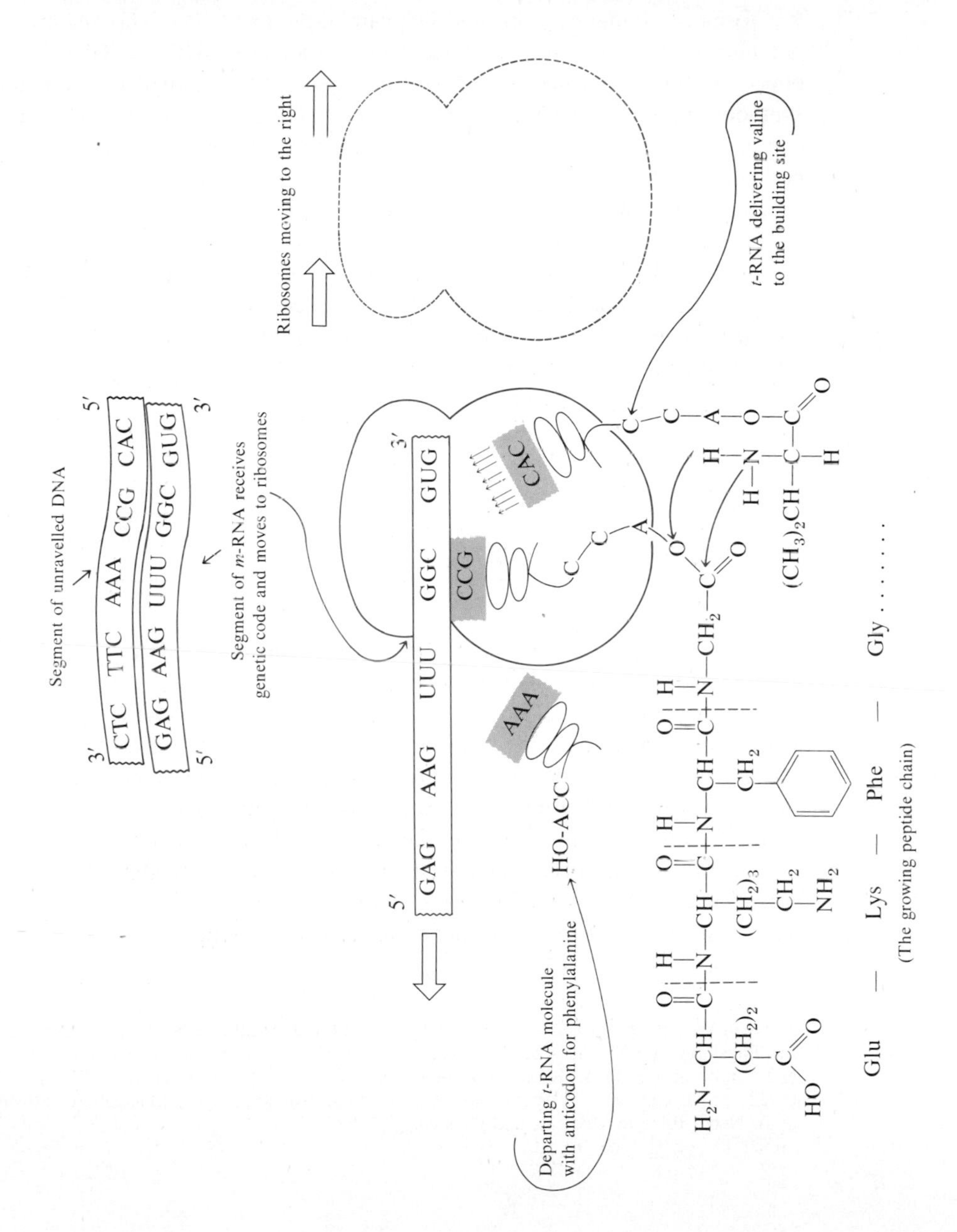

and their incorporation into an ever-growing polypeptide structure are yet unknown. It appears, however, that *m*-RNA is synthesized at the site of a portion of unravelled DNA and has had transcribed to it the genetic information necessary for the proper sequential alignment of certain amino acids at the protein-building site.

Experiments carried out by Nirenberg, Khorana, and Ochoa[4] provided some insight into the role played by *m*-RNA in protein synthesis. Nirenberg combined a synthetic RNA molecule composed entirely of polyuridylic acid (UUUUUUUUU . . .) with the twenty amino acids used in protein synthesis and found that only the peptide, polyphenylalanine, was formed. Similar experiments were carried out by Ochoa. Khorana prepared synthetic polynucleotides with repeating dinucleotides, such as UGUGUGUGUG . . . . Polynucleotides of this type contain two alternating triplets, UGU and GUG, along the chain. This synthetic messenger gave no homopolypeptide but led only to the formation of the alternating polypeptide containing valine and cysteine, Val-Cys-Val-Cys-Val-Cys . . . . Experiments of the latter type definitely established the existence of the triplet code, and the deciphering of the code followed quickly thereafter.

The mechanism of protein synthesis and the roles played by DNA and the different RNA molecules as shown in the illustrations in this chapter certainly are an oversimplification of the extremely complex and highly specialized function of the nucleic acids in protein synthesis. Much research remains to be done before more of the details of this intricate chemistry can be discovered. Some of the finest scientific minds of our time are dedicated to the attainment of this goal.

16.10 Miscoding and Mutations

One would imagine that any biosynthesis involving hundreds of amino acid residues of twenty different kinds, all combined in a great variety of different sequences, would also involve the possibility of errors. Such indeed is the case, and errors do occur. Correct genetic information rests in the proper sequence of bases in the DNA molecule; and when one base in the sequence appears in an altered form, then the genetic message also is altered by a triplet that miscodes. Such miscoding in DNA results in an erroneous transcription to *m*-RNA and finally in the wrong amino acid sequence for the protein under construction. Sickle-cell anemia is an example of such miscoding. A valine unit appears instead of glutamic acid, which normally occurs as a side chain in the hemoglobin molecule. If such genetic errors or mutations should prove beneficial, a better species could

[4]Marshall W. Nirenberg (1927–), National Institutes of Health, Bethesda, Md., Nobel Prize in Medicine and Physiology (1968); Servero Ochoa (1905–), New York University School of Medicine, New York, N.Y. (now at La Roche Research Institute, Nutley, N.J.), Nobel Prize in 1959; Har Gobind Khorana (1922–), University of Wisconsin, Madison, Wisconsin (now at the Massachusetts Institute of Technology), Nobel Prize in Medicine and Physiology (1968).

result. On the other hand, if such genetic defects leave the organism handicapped or weakened to the extent that it will not reproduce, the species will die.

It is believed that the bases in the DNA sequence might be altered structurally by exposure to high-energy radiation. Certain of the bases present in DNA can exist in tautomeric forms (pairs of equilibrium isomers, Sec. 8.5) in which one isomeric structure is much more stable than the other. Conceivably, the absorption of high-energy radiation such as that from X rays, γ-rays, or cosmic rays could transform the structure of a base into that of its higher-energy tautomer. Should this occur, then the altered structure of the normal base in DNA no longer would be able to hydrogen-bond with the proper complementary base in *m*-RNA, and miscoding would result. An example of such a tautomeric shift and resultant miscoding is illustrated for adenine below.

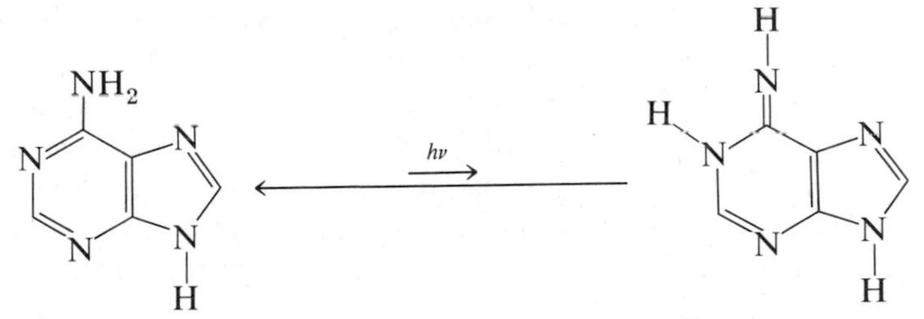

(Stable amino form) $\underset{}{\overset{h\nu}{\rightleftharpoons}}$ (Unstable imino form)

Adenine

Adenine $\begin{cases} \text{(amino form) + Thymine} \rightarrow \text{normal pairing} \\ \text{(imino form) + Cytosine} \rightarrow \text{abnormal pairing} \end{cases}$

Mutations are also thought to be the result of chemical changes. You will recall from our study of amines that nitrous acid may be used to convert a primary amine to an alcohol according to the following reaction:

$$R{-}NH_2 + \{NaNO_2 + HCl\} \longrightarrow R{-}OH + H_2O + NaCl + N_2$$

In the same manner nitrous acid could also possibly convert adenine to hypoxanthine and cytosine to uracil, since both contain primary amino groups.

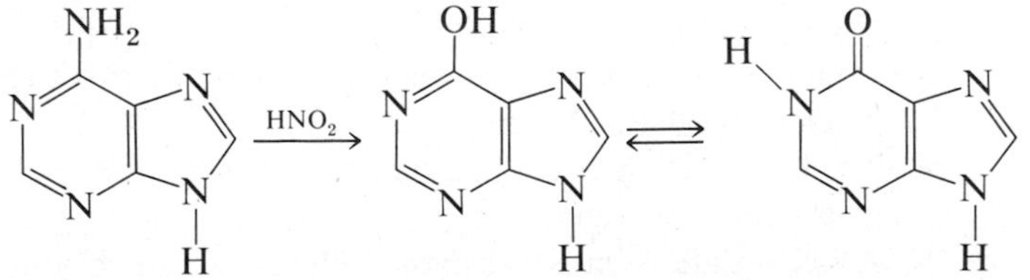

Adenine $\xrightarrow{HNO_2}$ Enol form $\rightleftharpoons$ Keto form

Hypoxanthine

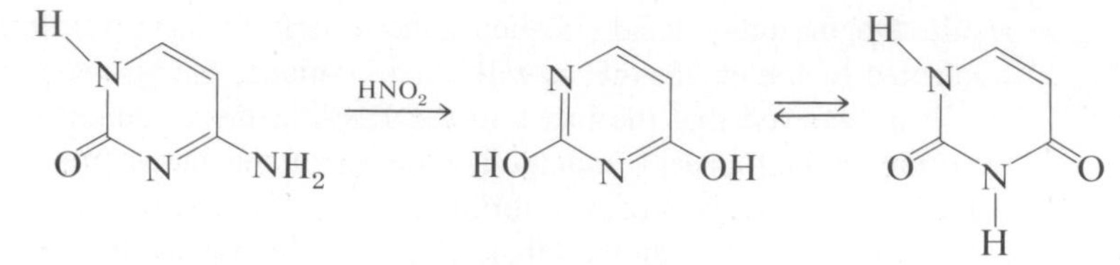

Cytosine $\xrightarrow{HNO_2}$ Enol form $\rightleftarrows$ Keto form

Uracil

Although mutations have been effected in viral proteins through the use of nitrous acid, it is doubtful that the reactions shown here play much part in human or animal carcinogenesis. However, nitrous acid can react with amino compounds in the stomach to form N-nitrosoamines (Sec. 13.4D), substances known to be powerful carcinogens. Presumably, the N-nitrosoamines or their further decomposition products can modify the normal bases in DNA by alkylation reactions involving OH and NH groups. Whether or not the N-nitrosoamines are important factors in human carcinogenesis is the subject of considerable current research. Other carcinogens may also function by the alkylation of the DNA bases, for example, by reaction with the lower molecular weight alkyl halides. Many otherwise innocuous substances are thought to be converted by the body's oxidation system (principally the cytochrome P450 system found in the liver) to substances capable of altering DNA; thus, many substances labeled carcinogens are not the true cancer-causing agents, but are their precursors.

Exercise 16.4 The thio analog of hypoxanthine, 6-mercaptopurine, is used in the treatment of leukemia. Draw its structure.

Summary

1. The nucleic acids are base-substituted sugar esters of phosphoric acid.
2. The two principal types of nucleic acids are *ribose nucleic acid* (RNA), and *deoxyribose nucleic acid* (DNA).
3. Both ribose and deoxyribose appear in nucleic acids as N-glycosides of five different heterocyclic bases.
4. The base portion of the nucleic acid molecule may be either a pyrimidine derivative or a purine derivative.
5. A nucleoside is an N-substituted glycoside; a nucleotide is an N-substituted glycosidic ester of phosphoric acid.
6. The DNA polymer has the form of a double helix with the pyrimidine and purine base structures oriented toward the axis of the helix.

7. Hydrogen bonding between complementary bases makes possible the double-helix DNA structure, as well as replication of the DNA molecule.
8. The genetic code is embodied in DNA but transcribed to *m*-RNA.
9. A sequence of three bases in the DNA polymer may code for only one of the amino acids used in protein synthesis.
10. Amino acids are carried to the protein building site by *t*-RNA.
11. Alteration of the base structures in DNA alters their hydrogen-bonding characteristics. Such changes can result in an erroneous sequence that leads to miscoding.

Supplementary Exercises

16.5 Distinguish between the following.

(a) pyrimidine ring system and a purine ring system
(b) a nucleoside and a nucleotide
(c) ribose and deoxyribose
(d) a codon and an anticodon

16.6 Draw structures for the following.

(a) deoxyadenosine
(b) adenylic acid
(c) cytidine
(d) cytidylic acid
(e) adenosine triphosphate
(f) a nucleotide found only in RNA
(g) a nucleotide found only in DNA

16.7 If ATP is *completely* hydrolyzed, what chemical compounds result?

16.8 The ratio of cytosine to guanine in the DNA double helix is 1:1, but the ratio of cytosine to adenine need not be unity. Explain.

16.9 Why do neither adenine and cytosine nor guanine and thymine form complementary base pairs in DNA?

16.10 How does *m*-RNA differ from *t*-RNA

(a) in base composition?
(b) in function?

16.11 If only cytosine and guanine are used to form 3-base sequences, and if each sequence could code for a different amino acid, theoretically for how many amino acids could codes be formed? How many amino acids are actually coded for with only these bases in sequence?

16.12 Draw the imine form of adenine and show how it could then pair with cytosine rather than with thymine.

16.13 If the unstable tautomer of adenine (identified by an asterisk) were to appear in DNA in the sequence C-$\overset{*}{\text{A}}$-G, which amino acid would be erroneously built into a peptide chain?

16.14 Which codons correspond to a nucleotide sequence in a segment of the DNA chain and when misread result in sickle-cell anemia? Which codons would normally constitute a proper sequence?

16.15 Which amino acid sequences will be found in peptides formed by ribosomes in response to the synthetic *m*-RNA's given below? Assume that the polypeptide chain will begin with the codon on the left

(a) GGUGCAAAGUCCUGACACAUA
(b) UGCGUAAUACUUUUCCCUAAU
(c) AAACCCUUUUUCCCUAGGGAA

16.16 One segment of a strand of DNA consists of the following sequence going in the 5′ to 3′ direction (the code is read in this direction):

AAACACTTCTCGACGATCGGCGGCTAC

(a) Write the sequence of bases in the other strand of DNA.
(b) Write the sequence of bases in the *m*-RNA transcribed from the first strand of DNA.
(c) Write the sequence of amino acids called for by the message in the *m*-RNA.

16.17 Draw the anticodon sequence of the *t*-RNA and the codon sequence of the *m*-RNA that would be required for the synthesis of oxytocin (Fig. 15.1).

17 Natural Products

The following sections that are grouped together under the general heading of natural products deal with compounds that occur in nature and possess properties intimately related to life processes. Some of these substances have profound effects upon the nervous system. Others are necessary for the normal growth and good health of the individual. Still others are agents counteractive in the treatment of disease.

Although the laboratory synthesis of many of the substances described has been accomplished, their structures are rather complex, and the chemistry of most is difficult. For these reasons our treatment of these natural substances will necessarily be brief and only descriptive.

The Alkaloids

17.1 Introduction to the Alkaloids

The alkaloids are naturally occurring nitrogenous substances of plant origin that possess marked physiological properties. The term alkaloid means ''alkalilike'' and originates in the fact that nearly all alkaloids are nitrogen heterocycles with basic properties. Alkaloid chemistry represents an extremely complex study and one to which volumes have been devoted. In this book we can examine only a few members of each general class.

17.2 Alkaloids with Isolated Five- and Six-Membered Heterocyclic Systems

Coniine, 2-*n*-propylpiperidine, has a relatively simple structure when compared to those of other alkaloids.

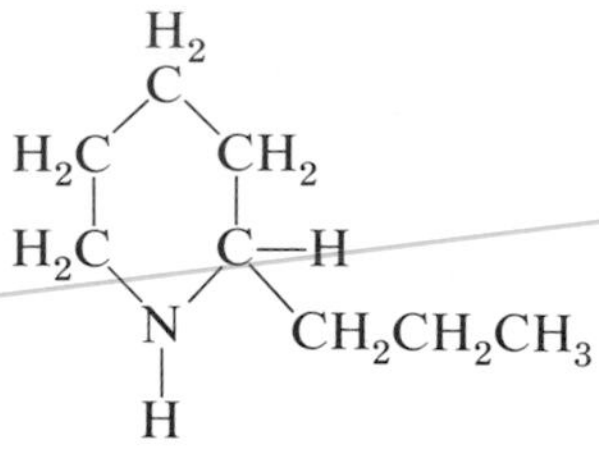

Coniine

Coniine is the toxic substance in *Conium maculatum* and certain other closely related herbs. Its common name is "hemlock," but, of course, this is not the hemlock tree of our forests. Students of history will remember reading of hemlock as the death potion Socrates was forced to drink.

Meperidine, a synthetic phenylpiperidine sometimes called "Demerol," has a physiological action similar to, but less marked than, that of morphine (Sec. 17.4). The analgesic property of meperidine was accidently discovered when an injection into a rat caused the animal to manifest the same symptoms as those produced by morphine.

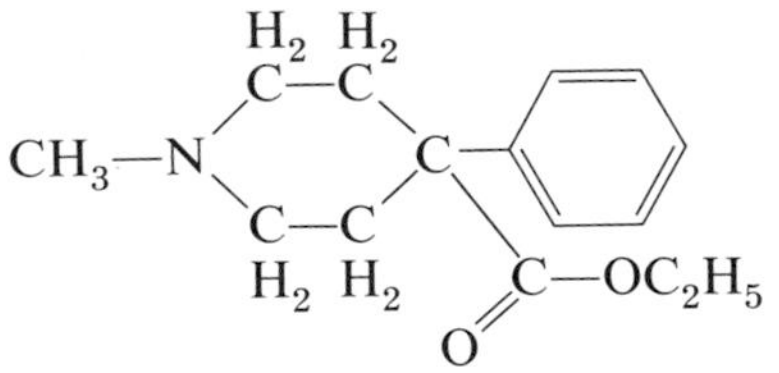

Meperidine

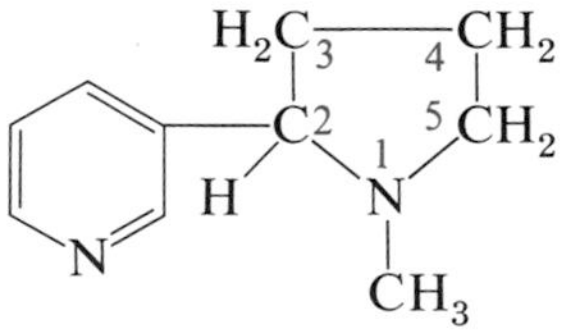

Nicotine

Nicotine, one of the principal alkaloids found in the tobacco plant, is a β-pyridine derivative in which N-methylpyrrolidine is joined at its 2-position to the pyridine ring.

Nicotine is extremely toxic to animals when ingested. It kills a number of insects on contact and is used in sprays against a number of leaf-sucking pests.

17.3 Alkaloids Containing Bridged Heterocyclic Systems

Cocaine and *atropine* belong to a class of alkaloids known as the **tropane alkaloids.** Cocaine is obtained from coca leaves and is a powerful anesthetic but has the disadvantages of being very toxic and habit-forming. A synthetic substitute, *procaine* (sometimes called novocaine), incorporates the beneficial properties of cocaine without the undesirable side effects. The part of the cocaine structure enclosed in the formula (solid line) appeared to be responsible for its anesthetic property and served as a pattern in early attempts to synthesize local anesthetics.

However, in the evolution of hundreds of compounds sought for their anesthetic properties, it was discovered that a benzoic acid ester is not necessarily a requisite for anesthetic action. *Xylocaine,* and more recently *mepivacaine,* are amides that retain certain structural similarities to cocaine.

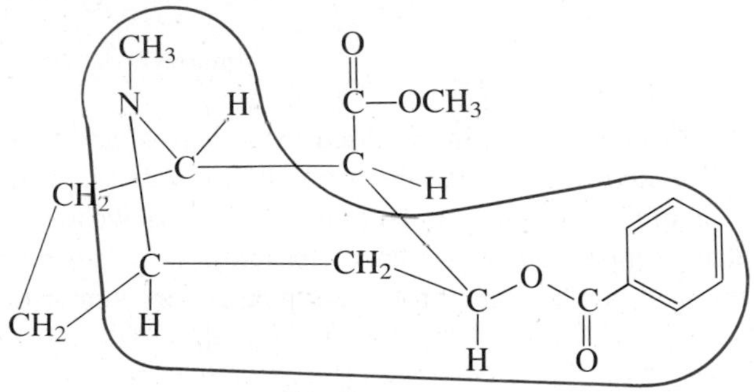

Cocaine

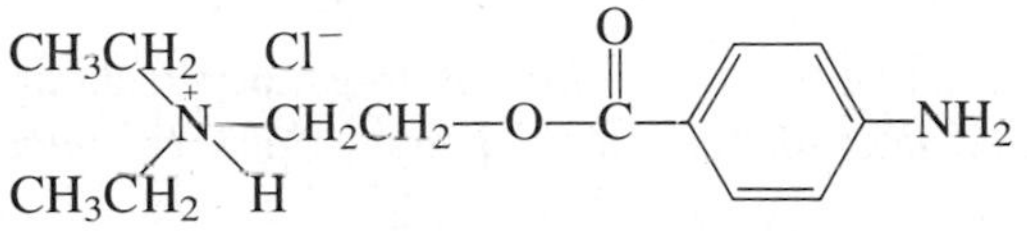

β-Diethylaminoethyl-*p*-aminobenzoate hydrochloride
(Procaine)

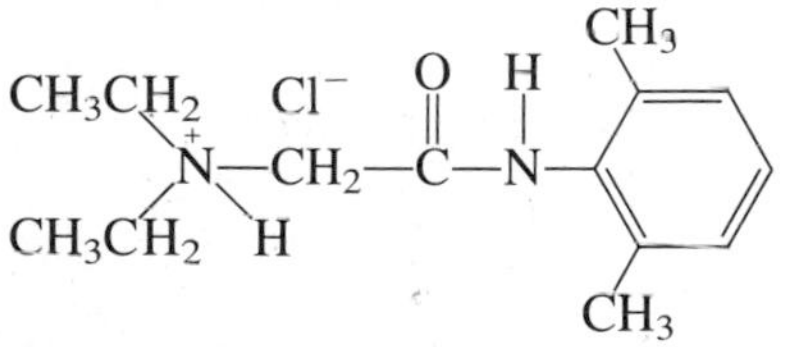

Diethylaminoacet-(2,6-dimethylanilide) hydrochloride
(Xylocaine)

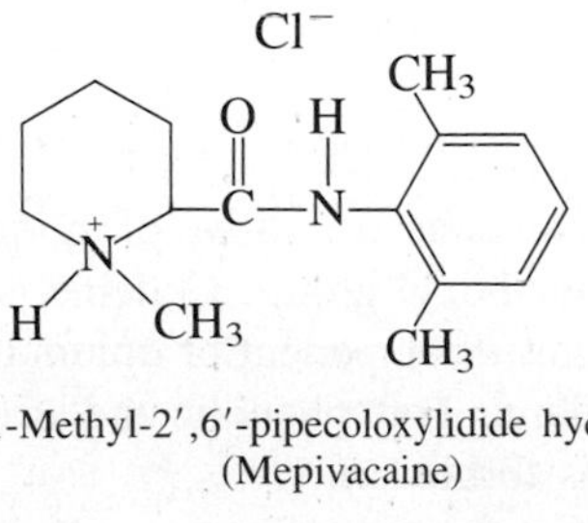

($\pm$)-1-Methyl-2′,6′-pipecoloxylidide hydrochloride
(Mepivacaine)

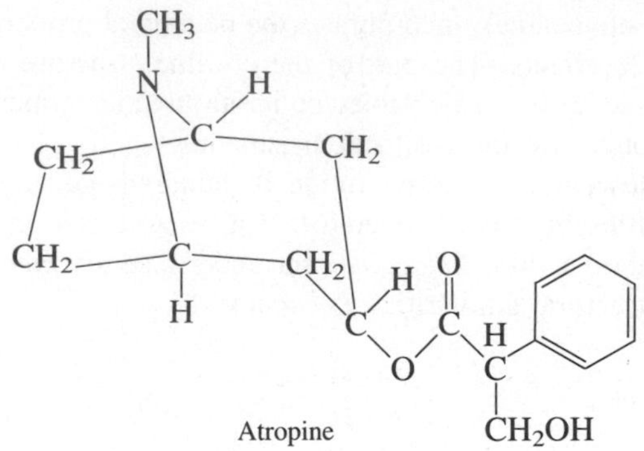

Atropine

Atropine occurs in the dried roots, leaves, and tops of the belladonna plant, *Atropa belladonna*. It is used in the form of its sulfate salt as a mydriatic for the dilation of the pupil of the eye. The action of the drug was known to early Europeans, and the belladonna plant, supposedly, was named "beautiful lady" in reference to the cosmetic effects of enlarged pupils. Structurally, atropine is very similar to cocaine.

17.4 Alkaloids with Fused Heterocyclic Systems

Morphine, one of the most useful of drugs, occurs along with twenty-four other alkaloids in the dried latex of the opium poppy, *Papaver somniferum*. Morphine is a powerful analgesic, but its use, like that of other narcotics, leads to addiction. A synthetic compound called *methadone* was developed by the Germans during World War II as a substitute for morphine, and was found to be an even more effective analgesic than the natural product. While the use of methadone also leads to addiction, it does not produce the same mental and physical deterioration suffered by heroin addicts. By treatment with daily oral doses of methadone, heroin addicts have been socially rehabilitated but not always returned to a drugfree status.

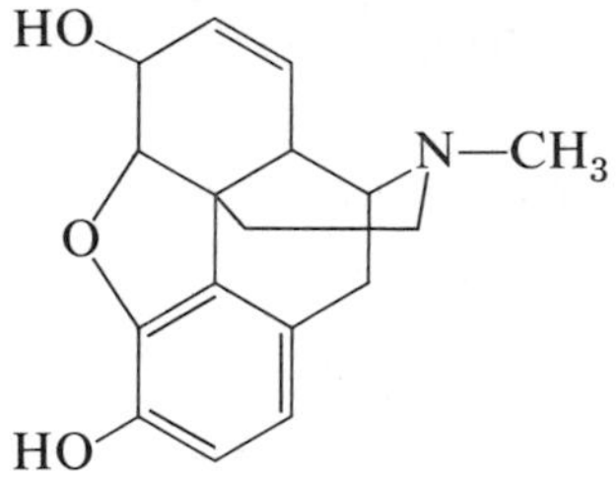

Morphine

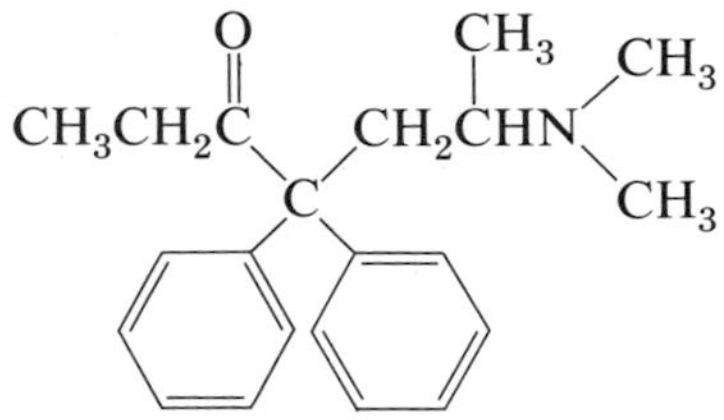

Methadone

Codeine is an ether derivative of morphine in which the phenolic hydroxyl has been replaced by a methoxyl group. Codeine is widely used as a cough supressant.

Heroin is not a component of opium but is the diacetyl derivative of morphine produced by acetylating both phenolic and alcoholic hydroxyl groups. Its manufacture in the United States is forbidden.

Ergot alkaloids are alkaloids produced by the fungus *ergot,* a parasitic growth on rye and other cereals. They are amides of lysergic acid, a structure that includes the indole nucleus. Perhaps the most notorious of the ergot alkaloids is the synthetic diethyl amide of lysergic acid known as LSD. The oral administration of as little as 50 μg (0.000,050 g) of this amide produced psychotic symptoms in humans resembling schizophrenia. It is believed by psychopharmacologists that the drug is antagonistic to serotonin, a substance naturally present in brain tissue.

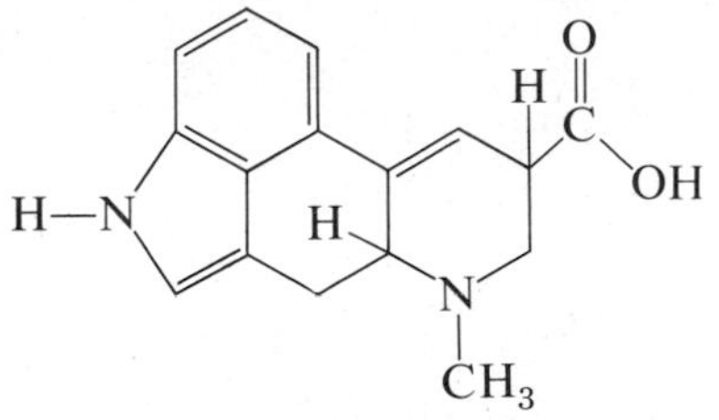

Lysergic acid

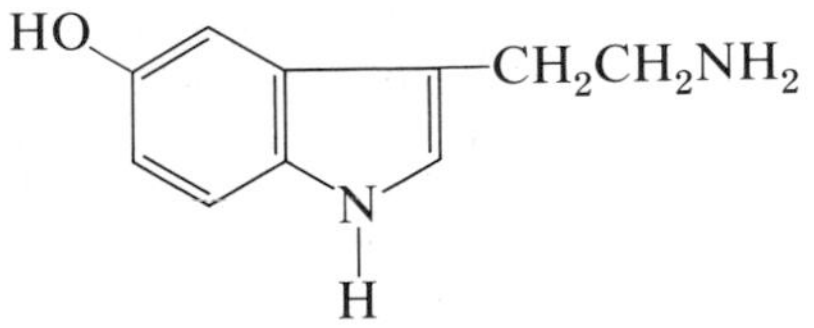

Serotonin

Reserpine is a complex polycyclic system which, like the ergot alkaloids, also includes the indole nucleus. Reserpine is one of the alkaloids obtained from the extracts of a species of Indian snake root called *Rauwolfia serpentina.* Such extracts have been used for centuries by the natives of southern Asia to treat a variety of disorders including snakebite, dysentery, insanity, and epilepsy. Reserpine has remarkable tranquilizing powers and is called a behavioral drug. Its synthesis was accomplished by Woodward[1] in 1956.

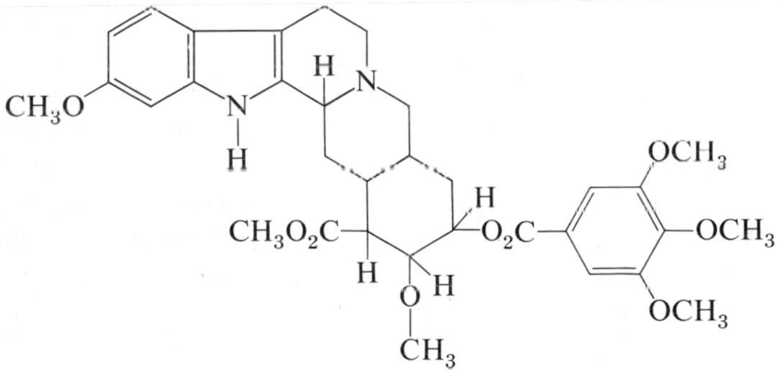

Reserpine

Tubocurarine is an alkaloid that contains two isoquinoline residues. It is the active principle of curare, a paralyzing arrow and dart poison used by South American aborigines. Tubocurarine and other curariform drugs act as skeletal muscle relaxants and have been used as valuable aids during surgery. A lethal dose of the drug results in respiratory failure. The tubocurarine structure includes two quaternary ammonium ions and in this

[1]Robert B. Woodward (1917–1979), Professor of Chemistry, Harvard University. Winner of the Nobel Prize in Chemistry (1965).

respect bears a functional similarity to choline (Sec. 13.7).

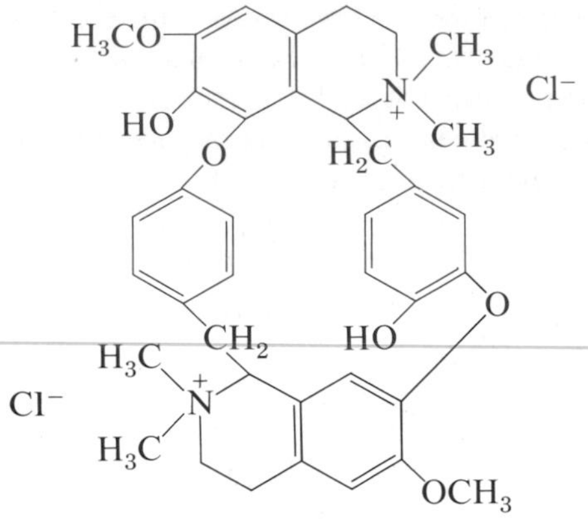

Tubocurarine chloride

Vitamins

17.5 Introduction to the Vitamins

Fats, carbohydrates, and proteins comprise the bulk of the animal diet, but certain other nutrients in minute amounts are also necessary constituents of the diet. The lack of these trace nutrients in the animal diet results in deficiency diseases manifested by improper growth, metabolism, and behavior. Among these essential dietetic constituents are the vitamins. Many enzymes require the assistance of relatively small, nonprotein molecules called coenzymes in order to carry out their catalytic functions. Some of the water-soluble vitamins serve as components of a larger coenzyme molecule. The fat-soluble vitamins do not appear to serve in coenzyme roles but have other important functions. At present the full scope of the functions of some vitamins is not known with any certainty. Vitamins in this category include vitamins A, C, D, and E; thus for our purposes the vitamins may be defined simply as abundant, potent, naturally produced substances that the animal organism requires but usually is incapable of synthesizing. The need for a certain vitamin varies with the species, and a vitamin required by one animal may not always be required by another. Vitamin research is one of several areas in which organic chemistry and nutrition are interrelated. It is a vast and complex field of study, and our discussion will be limited to only a brief review of the principal vitamins.

17.6 Historical

One of the major hardships encountered by the early explorers—Columbus, Vasco da Gama, Jacques Cartier, Henry Hudson, and others—on their expeditions into the new world was the disease *scurvy*. Scurvy, caused by a lack of vitamin C, was especially common among sailors on long voyages without fresh food. It was recognized as early as 1750 to be a deficiency disease, and British naval surgeons, in their attempts to treat it, observed the beneficial effects of fruit juices—especially citrus fruits. They recommended fruit as a regular part of a sailor's ration, but the recommendation went unheeded by the British Admiralty for almost fifty years. Finally, the inclusion of limes in ships' provisions eliminated scurvy as one of a British seaman's hazards. The name "limey" in reference to a British seaman is still heard today.

The first serious inquiry into the cause and effects of a vitamin deficiency began with the observations of Dr. Eijkman, a Dutch physician working in Java. In 1897 he observed that a diet of polished rice caused beriberi among the natives. This disease is characterized by a polyneuritis, muscular atrophy, and general debilitation. The protective principle against this disease, now known as thiamine, appeared to have been removed in the rice polishings. The symptoms of beriberi disappeared when the rice polishings were restored to the diet. Funk, in 1911, succeeded in isolating from rice polishings the vital substance. On analysis, it was found to contain nitrogen, and Funk concluded that the chemical structures of these essential agents were those of amino compounds. He therefore named them **vitamines** (vital amines). The name since has been shortened by dropping the final "e." As used today the term designates a number of substances, some of which lack nitrogen entirely.

17.7 Fat-Soluble Vitamins

The fat-soluble vitamins are those soluble in fats and in fat solvents. Included in this classification are vitamins A, D, E, and K. The role of vitamins as food accessories usually is considered in relation to nutritional deficiencies. However, doses of the fat-soluble vitamins, when given far in excess of normal requirements, can have toxic effects. Vitamin poisoning occurred in a number of arctic explorers, who became seriously ill after eating polar bear liver. There also have been numerous cases of vitamin poisoning in infants. Young mothers, eager to fulfill all vitamin requirements of their first offspring, sometimes give overdoses of fat-soluble vitamins. Poisoning by water-soluble vitamins is not possible because any amounts not required are voided from the body in urine.

Each of the fat-soluble vitamins is discussed briefly in the following sections.

(a) **Vitamin A** may be obtained from the coloring matter of many green and yellow vegetables. Vitamin A, as such, is not found in plants, only β-carotene, its precursor or *provitamin*. The β-carotene molecule (Sec. 3.13), when cleaved in the center of the linear chain and converted at each end to an alcohol function, yields two molecules of vitamin A. Other sources of vitamin A are fish-liver oil, the livers of other animals, eggs, butter, and cheese.

Vitamin A_1, or *retinol,*[2] is oxidized in the body to the aldehyde 11-*cis-retinal,* which combines with a protein called *opsin* to produce a light-sensitive substance *rhodopsin.*

[2]There are two vitamins A. One is known as vitamin A_1, or retinol, and the other as vitamin A_2, or 3-dehydroretinol. Vitamin A_2 is found in the liver oils of freshwater fish and differs structurally from vitamin A_1, found in the livers of cod and other saltwater fish, by having a second double bond between carbons 3 and 4. Physiologically, the two vitamins have the same activity, and both are called vitamin A.

Rhodopsin is located in the retina and, upon absorption of radiation within the visible range, causes an isomerization of *cis*-retinal to the *trans* form.

A deficiency of vitamin A causes night blindness—an inability to see in dim light. Another disease of the eye known as *xerophthalmia,* in which the tear glands cease to function, results from a lack of vitamin A.

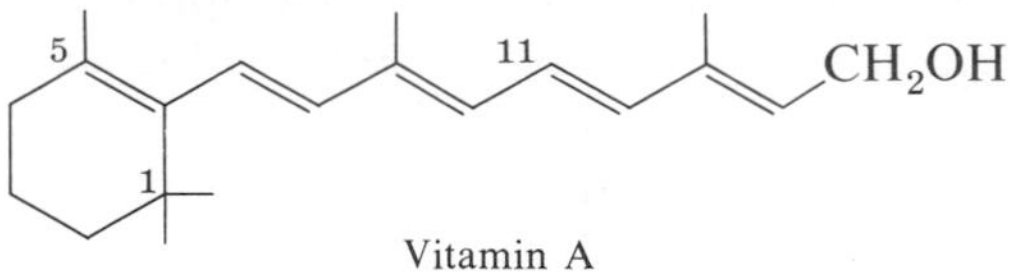

CH_2OH

Vitamin A
(*trans*-retinol)

(b) **Vitamin D** is sometimes referred to as the ''antirachitic'' vitamin. It is related to the proper deposition of calcium phosphate and controls the normal development of the teeth and bones. There are ten or more compounds that have antirachitic properties and are designated D_1, D_2, D_3, etc. Vitamin D from fish oils is D_3, while that produced by irradiation of the skin with ultraviolet or sunlight is D_2. Vitamin D_2 is known as *calciferol* and is derived from ergosterol, a plant sterol (Sec. 17.11). Milk is fortified with vitamin D by irradiation. Vitamin D is produced in or on the skin by irradiation of the sterols present there. Sunlight supplies *not* the vitamin, only the necessary radiation. Vitamin D_2 has the following structure.

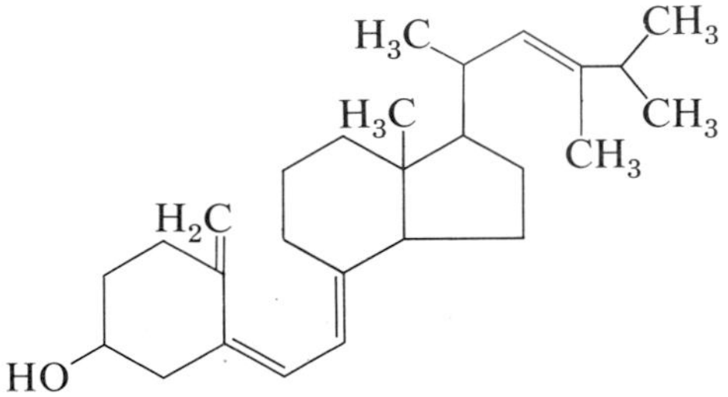

Calciferol (vitamin D_2)

(c) **Vitamin E,** sometimes called the fertility factor, is related to the proper functioning of the reproductive system. Vitamin E is found in the nonsaponifiable fraction of vegetable oils such as corn-germ oil, cottonseed oil, wheat-germ oil, and peanut oil. It also occurs in green leafy vegetables. As in the case of the A and D vitamins, there is more than one form of vitamin E. Four different structures called *tocopherols* have vitamin E activity. These are designated α-, β-, γ-, and δ-tocopherols. The structure of α-tocopherol, the most potent, is shown.

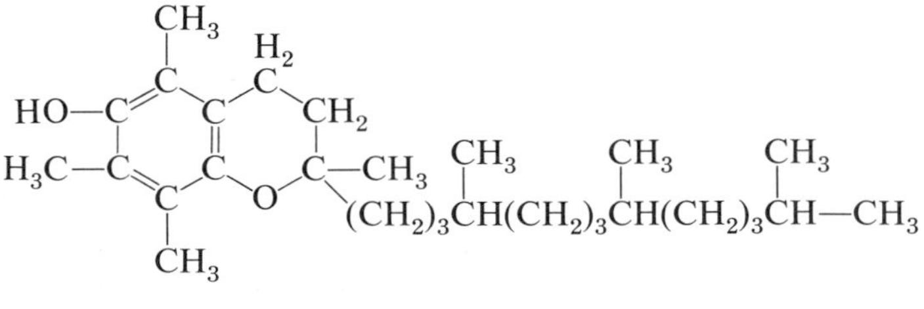

α-Tocopherol

Vitamin E, as was pointed out earlier (Sec. 12.5), is also used as an antioxidant for the prevention of oxidative rancidity in vegetable oils.

(d) **Vitamin K** is the antihemorrhagic factor related to the blood-clotting mechanism. This vitamin is especially important from a surgical standpoint. There are at least two K vitamins. Vitamin K_1 is obtained from the alfalfa leaf; vitamin K_2 is produced by bacterial action in the intestinal canal. The structure of vitamin K_1 is shown.

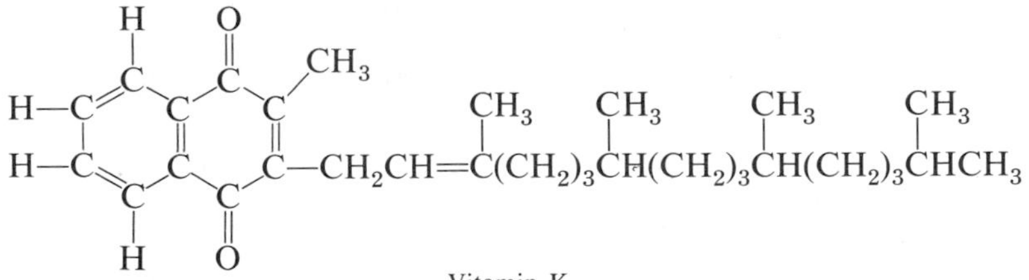

```
        H   O
        |   ||
        C   C   CH3
H—C//   \C/  \C/            CH3       CH3          CH3          CH3
   |     ||   ||             |         |            |            |
H—C\\   /C\  /C—CH2CH=C(CH2)3CH(CH2)3CH(CH2)3CHCH3
        C   C
        |   ||
        H   O
```

Vitamin K_1

17.8 Water-Soluble Vitamins

The water-soluble vitamins include vitamin C and all other vitamins designated B. The latter are collectively referred to as the vitamin B complex. The water-soluble vitamins bear less resemblance to each other than do the fat-soluble vitamins. Whereas the fat-soluble vitamins are isoprenoid structures and largely of hydrocarbon composition, the water-soluble vitamins all posses polar groupings to render them water-soluble.

(a) **Vitamin C,** now called **ascorbic acid,** is the vitamin that prevents scurvy. A person with scurvy suffers pain in the joints and hemorrhages from the mucous membranes of the mouth. The gums especially are affected and become red, ulcerated, and even gangrenous. Ascorbic acid is derived from glucose through a series of enzyme-catalyzed oxidations and reductions. Humans and primates cannot provide the necessary enzymes required for this conversion; therefore, vitamin C must be provided in the diet. Ascorbic acid is abundantly found in citrus fruits, tomatoes, green peppers, and parsley. Because of its acidic properties and ease of oxidation, it should be obtained from fresh sources, since it loses its potency when heated or exposed to air for any length of time. Ascorbic acid is believed to exist in the form of an enediol.

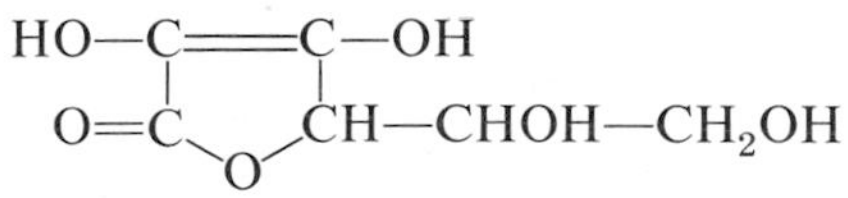

```
HO—C════C—OH
   |    |
 O=C    CH—CHOH—CH2OH
    \  /
     O
```

Ascorbic acid

(b) **The B Vitamin Complex.** Vitamins designated as B vitamins were at one time all thought to be the same. The isolation of each new factor from vitamin B preparations has led to a designation of B_1, B_2, B_3, etc., for each new factor. The B vitamins appear to play an important role in energy metabolism. A brief discussion of the principal B vitamins is given in the following sections.

Vitamin B_1 (thiamine), a deficiency of which causes beriberi in humans, is present in whole cereal grains, legumes, lean meat, nuts, and yeast. The vitamin has been found to contain both the pyrimidine and thiazole heterocyclic systems. The structure of thiamine has been determined, and the vitamin has been prepared. Thiamine is usually prepared in the form of an acid salt.

N S
CH_3—C C—$NH_3^+Cl^-$ CH C—CH_2CH_2OH
N C—CH_2—N—C—CH_3
CH +
Cl^-

Thiamine hydrochloride
(Vitamin B_1)

Thiamine occurs in nature either as the pyrophosphoric acid ester or as the free vitamin.

OH OH
N S
CH_3—C C—NH_2 CH C—CH_2CH_2—O—P—O—P—OH
N C—CH_2—N—C—CH_3 O O
CH +
Cl^-

Thiamine hydrochloride pyrophosphate

Vitamin B_2 (riboflavin) is an orange-yellow, crystalline compound widely distributed in nature. It is found in milk, lean meats, liver, fish, eggs, and leafy vegetables. A lack of riboflavin in the diet causes an inflammation of the lips, dermatitis, and a dryness and burning of the eyes, accompanied by a sensitivity of light.

The structure of riboflavin is shown.

H CH_2—CHOH—CHOH—CHOH—CH_2OH
C N N
H_3C—C C C C=O
H_3C—C C C NH
C N C
H O

Riboflavin

Niacin (nicotinic acid), another of the B vitamins, is the antipellagra factor. Pellagra is a disease characterized by dermatitis, a pigmentation and thickening of the skin, and soreness and inflammation of the tongue and mouth. Niacin may be found in most of the same foods that supply riboflavin. Especially rich sources of niacin are lean meats and liver. Although present in whole cereal grains, niacin is lost in the milling process. The niacin structure, as the amide, is part of the enzyme *nicotinamide-adenine-dinucleotide* (NAD). The structural formula of niacin, shown below, is relatively simple when compared to those of other vitamins. The oxidation of nicotine (Sec. 17.2) or β-picoline (3-methylpyridine) produces nicotinic acid, or niacin.

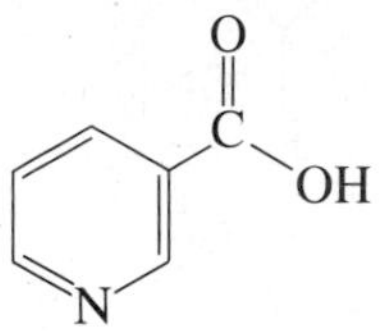

Niacin
(Nicotinic acid)

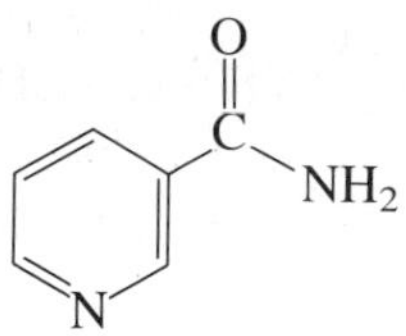

Nicotinamide

Vitamin B_6 (pyridoxine) is a vitamin whose function appears to be intimately related to the proper metabolism of fats and amino acids. Meat, fish, egg yolk, and whole cereal grains are rich sources of vitamin B_6. The exact requirement of vitamin B_6 for adult humans has not been established, but a lack of pyridoxine is the diet of experimental animals leads to dermatitis, anemia, and epileptic seizures. The structures of pyridoxine and two of its derivatives that also show vitamin B_6 activity are given below.

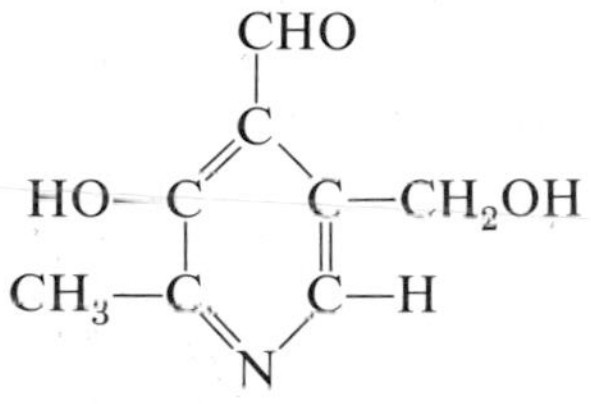

Pyridoxal

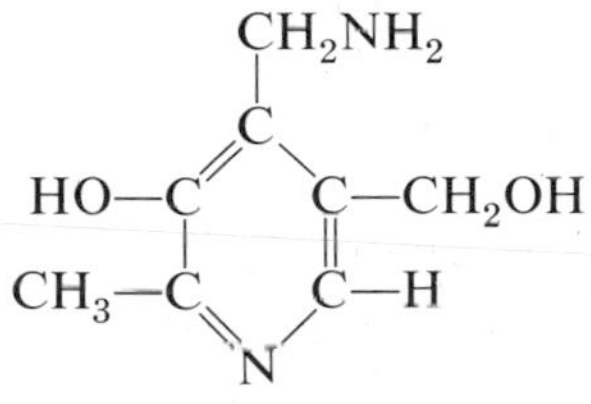

Pyridoxamine

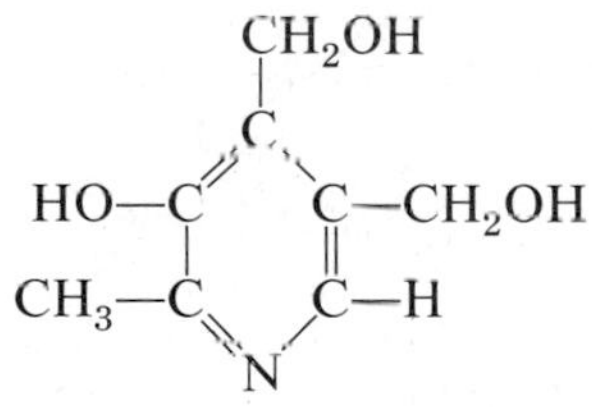

Pyridoxine

Vitamin B_{12} is the vitamin that prevents pernicious anemia. It is a dark red crystalline compound with a complex structure. Like heme and chlorophyll-a, it also contains a porphyrin nucleus, but a unique feature of its structure is the presence of trivalent cobalt. Vitamin B_{12} was the first cobalt-containing organic compound to be found. It is sometimes referred to as *cobalamin*. It is found most abundantly in liver but also occurs in meat, eggs, and seafoods. The structure of vitamin B_{12} is shown on the following page.

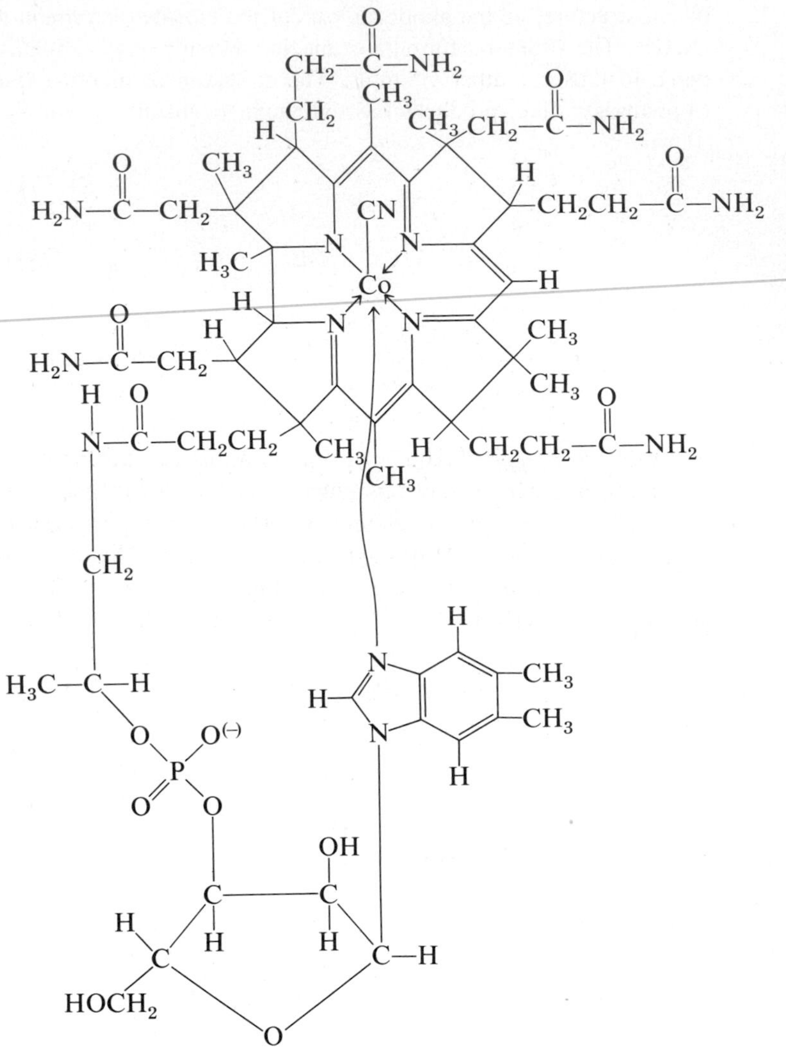

Vitamin B_{12} (Cyanocobalamin)

The Steroids

17.9 Introduction to the Steroids

The steroids are a family of compounds widely distributed in plants and animals. Common to the structure of all compounds of this class is a tetracyclic framework composed of the phenanthrene nucleus (Sec. 4.4) to which is fused at the 1,2-positions a cyclopentene ring.

The rings in the steroid molecule usually are not aromatic but often contain one or more isolated double bonds. The total structure of one steroid differs from that of another, usually by a variation in the side chain or by a variation in the number and type of functional groups. To the family of steroids with this common ring system belong the sterols, the sex hormones, the bile acids, and other biologically important materials.

For purposes of nomenclature, the steroid ring skeleton is numbered as shown.

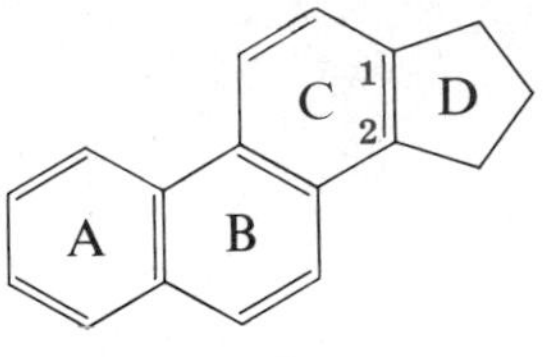

1,2-Cyclopentenophenanthrene

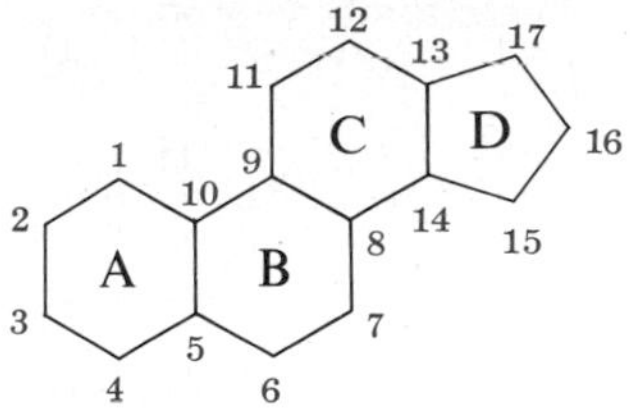

The steroid ring system

17.10 Cholesterol

The **sterols** are solid alcohols that possess a hydroxyl group at position 3, a double bond between carbons 5 and 6, a side chain on carbon 17, and methyl groups joined to ring carbons numbered 10 and 13. *Cholesterol,* $C_{27}H_{46}O$, one of the most widely distributed sterols, is found in almost all animal tissue but is particularly abundant in the brain, the spinal cord, and gallstones. Deposition of cholesterol or its derivatives in the arteries (hardening of the arteries) restricts the flow of blood, causes high blood pressure, and leads to some forms of cardiovascular disease. The structure of cholesterol is shown, with asymmetric carbon atoms indicated by asterisks.

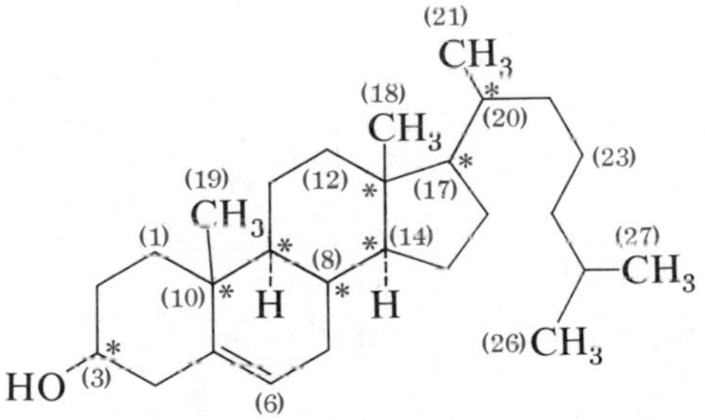

Structure and numbering system of cholesterol

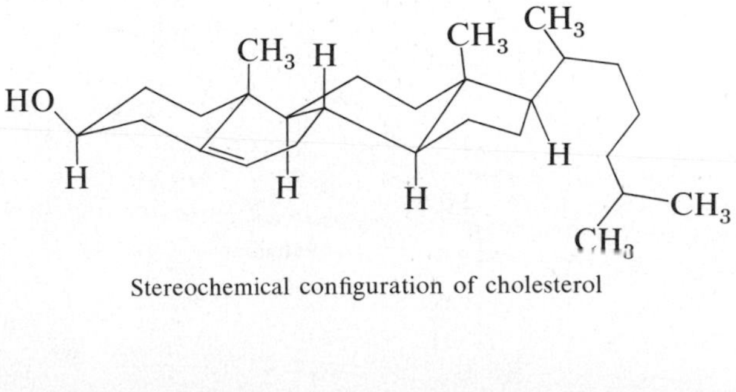

Stereochemical configuration of cholesterol

17.11 Ergosterol

Although cholesterol is found only in animals, a large number of closely related compounds known as **phytosterols** are found in plants. One of these, *ergosterol,* $C_{28}H_{44}O$, is produced by yeast. Ergosterol is of particular interest because, when irradiated, it yields calciferol, vitamin D_2.

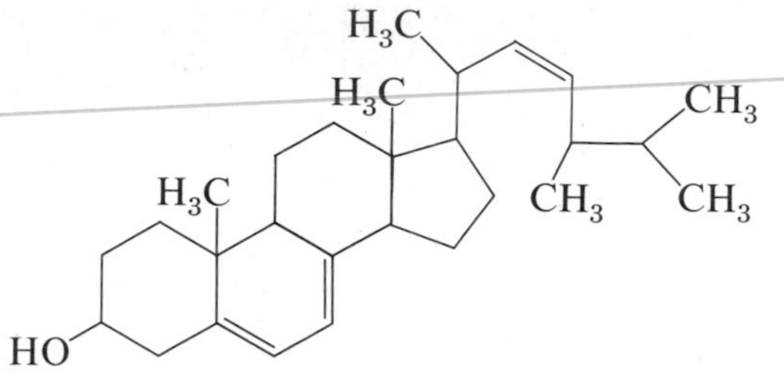

Ergosterol

17.12 Sex Hormones

The male and female sex hormones are structurally related steroids responsible for the development of sex characteristics and sexual processes in animals. Sex hormones are produced in the gonads (ovaries and testes) when the latter are stimulated by other gonadotropic hormones. The female sex hormones are involved in the menstrual cycle, the changes in the uterus, and the preparation for and maintenance of pregnancy.

The structures and names of the principal female sex hormones are shown below.

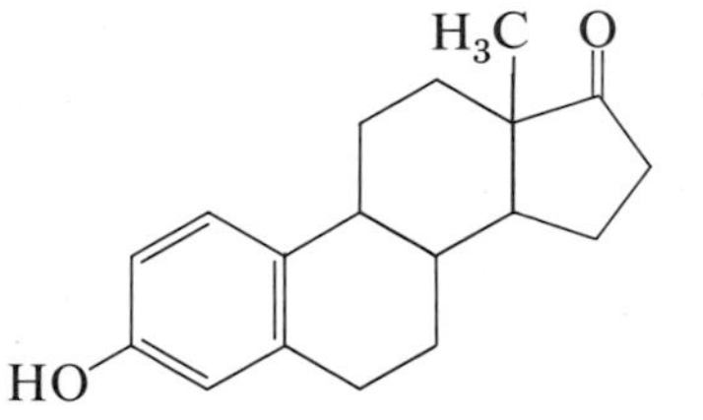

Estrone

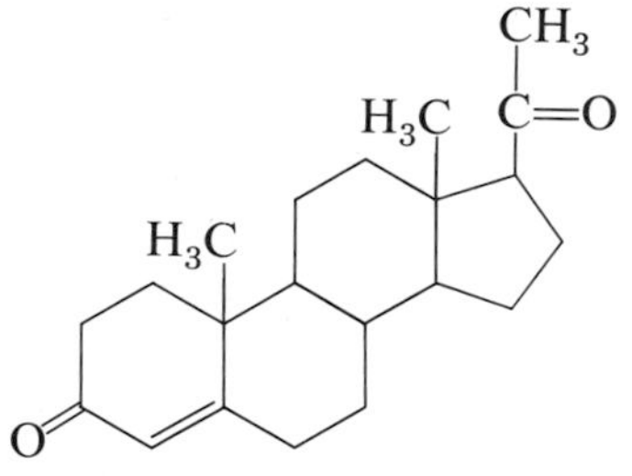

Progesterone

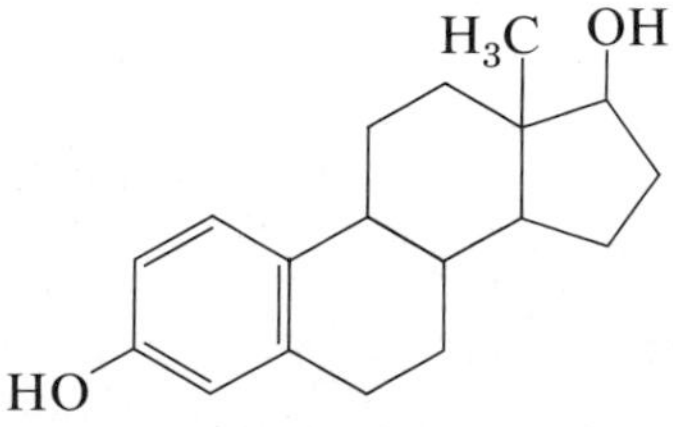

Estradiol

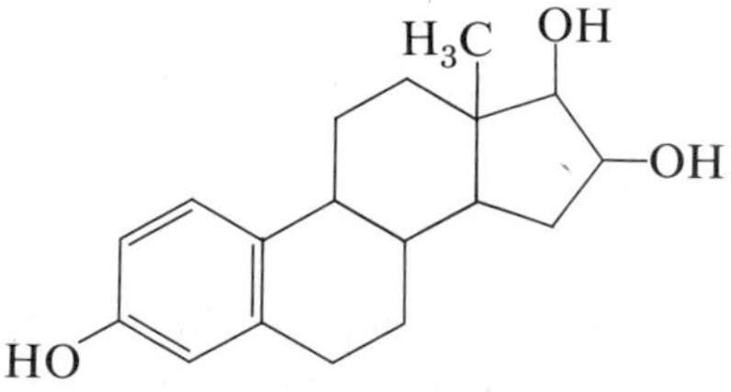

Estriol

Oral contraceptives have been developed that contain synthetic compounds structurally similar to progesterone and estriol but modified chemically to permit easier assimilation into the bloodstream. These synthetic agents, when taken orally, suppress ovulation and mimic pregnancy.

The structures of two synthetic progesterones and a synthetic estrogen[3] are shown. Both types of synthetic hormones are used either in combined form or sequentially.

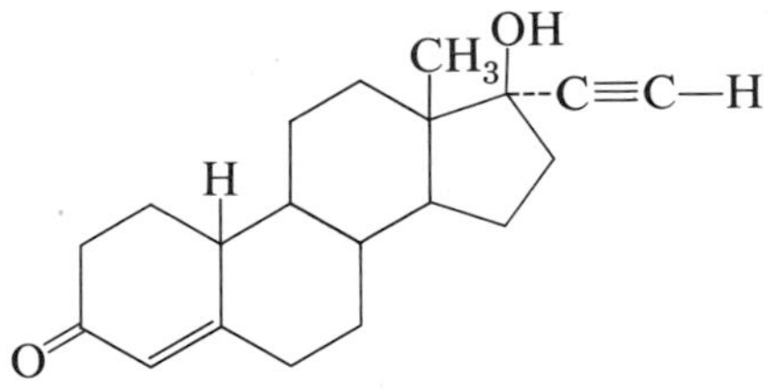

(a)
Norethindrone

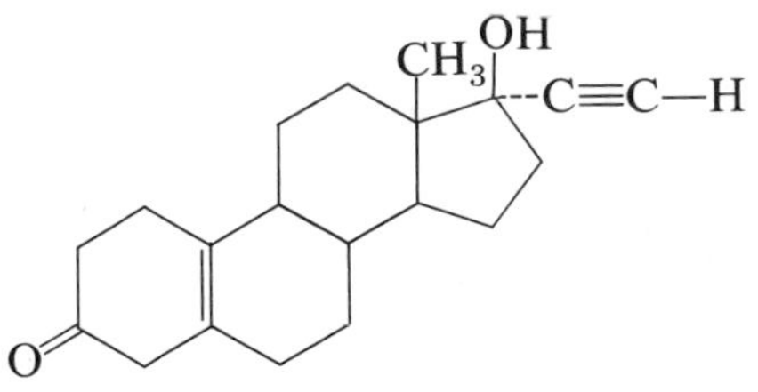

(b)
Norethynodrel

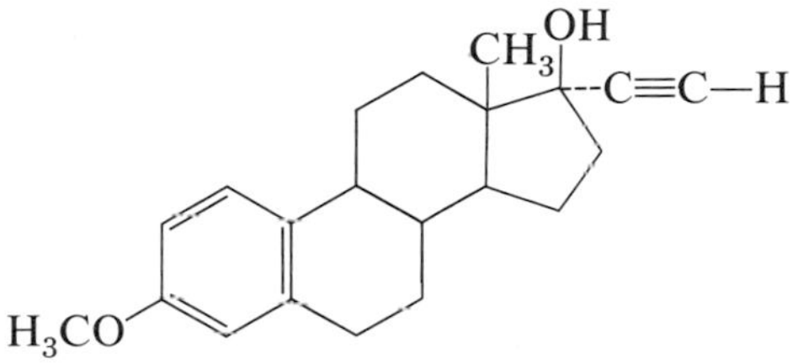

(c)
Mestranol

Orthonovum = (a) + (c)
Enovid = (b) + (c)

The male sex hormones, called **androgens,** except for the absence of an aromatic ring, are very similar in structure to the female hormones. The male sex hormones control the development of the male genital tract and the secondary male characteristics such as beard and voice. The structures of the principal male sex hormones are shown below.

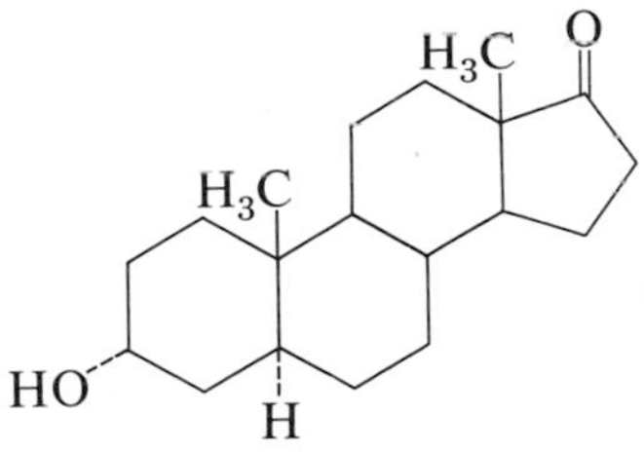

Androsterone

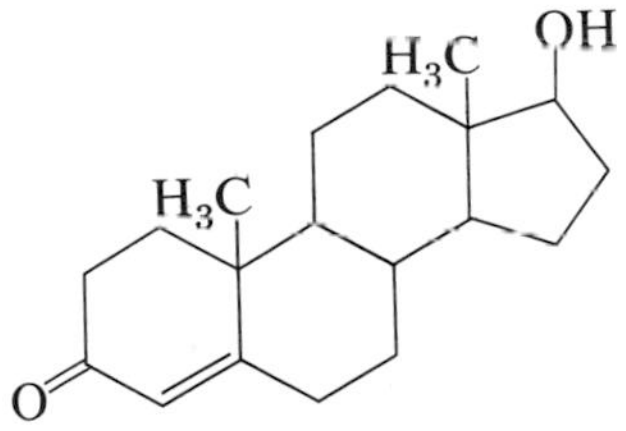

Testosterone

[3]Estrogen is a generic term for a substance that induces estrus—the cyclic phenomenon of the female reproductive system.

17.13 Adrenal Steroids

Cortisone, and its derivative, *17-hydroxycorticosterone,* are two steroids produced by the adrenal cortex. These compounds have been used with beneficial results in the treatment of inflammatory and allergic diseases. The structures of cortisone and its reduced form, 17-hydroxycorticosterone, are shown.

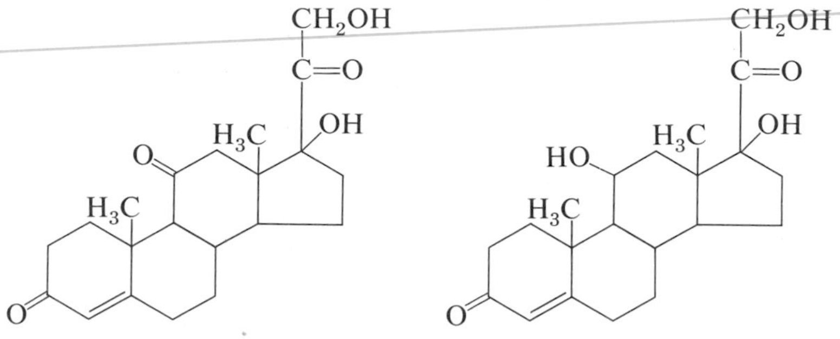

CH_2OH, C=O, OH, H_3C, O, H_3C, O

Cortisone

CH_2OH, C=O, OH, H_3C, HO, H_3C, O

17-Hydroxycorticosterone

Antibiotics

Our discussion of natural products would not be complete without a brief treatment of the antibiotics. These chemotherapeutic agents are potent antibacterials that possess the power to inhibit the growth of, or destroy, microorganisms.

17.14 Penicillins

Penicillin, the first antibiotic to be isolated from a mold, was introduced into clinical practice in 1941 with remarkable results. It was found to be active against a number of *gram-positive*[4] microorganisms of the cocci type and against spirochetes. A number of penicillins have been prepared. Their skeletal structures are the same, but they differ from one another in the character of the side chain R (see formula for penicillin). Commercial

[4]*Gram-positive* refers to organisms that stain ''positive'' when treated with gram stain; *gram-negative* refers to those that stain ''negative'' to the same reagent. A *broad-spectrum antibiotic* is one that may be effective against both gram-negative and gram-positive organisms.

preparations of penicillin are mainly penicillin G (R = $C_6H_5CH_2$—, benzyl).

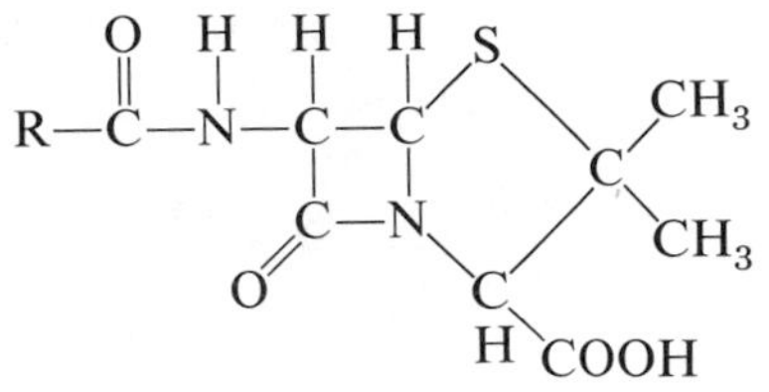

Penicillin

17.15 Streptomycin

A more recent antibiotic is **streptomycin,** whose structure is that of a trisaccharide. Streptomycin is active against *gram-negative* bacteria and is used in tuberculosis therapy. The structure of streptomycin is indicated below.

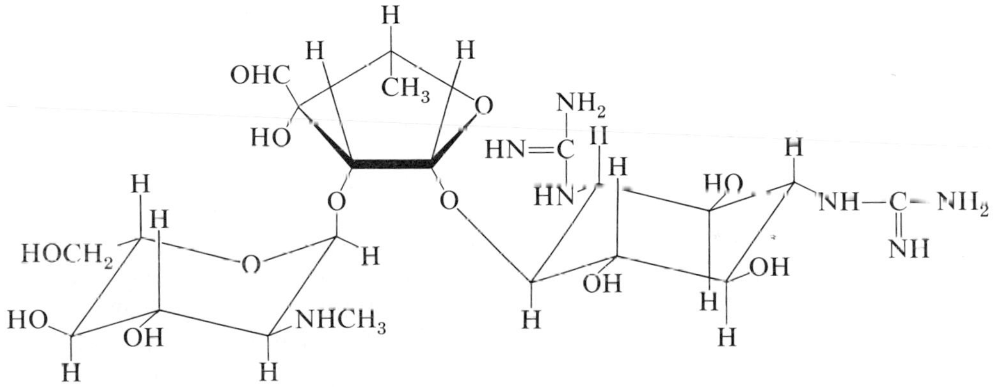

Streptomycin

17.16 Tetracyclines

The **tetracyclines** comprise a family of compounds, each member of which has a four-fused-ring structure. Each member differs from the other only in minor detail. Tetracycline, its 7-chloro derivative, *aureomycin,* and its 5-hydroxy derivative, *terramycin,* are *broad-spectrum* antibiotics widely used against a number of bacterial and viral diseases.

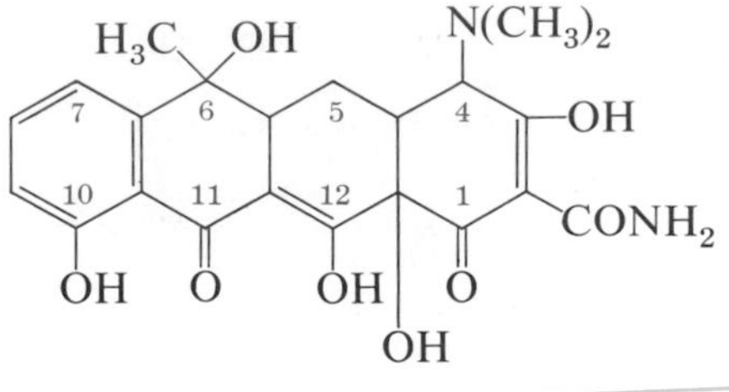

Tetracycline

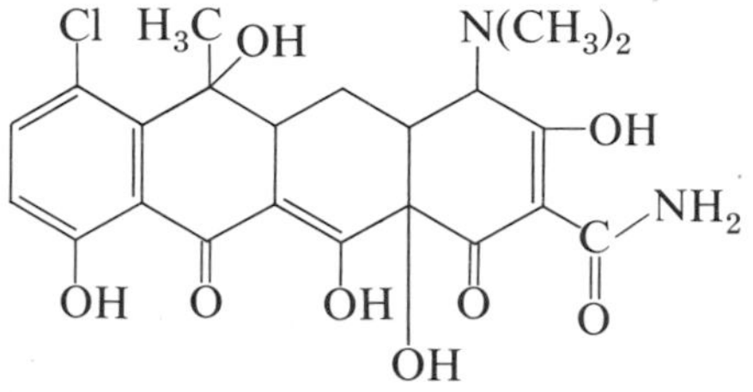

Aureomycin

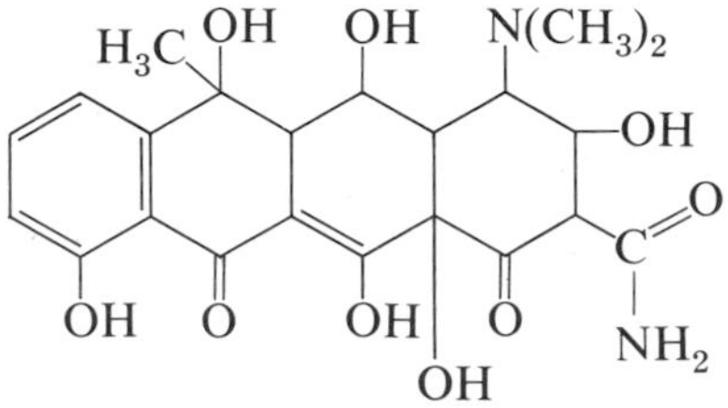

Terramycin

Summary

Alkaloids

1. The alkaloids are nitrogen-containing plant products that possess marked physiological properties.
2. Nearly all the alkaloids are made up of more than one heterocyclic ring. Such rings usually are fused rings—that is, two rings have one or more bonds in common.
3. Many of our most valuable drugs are obtained from alkaloids.
4. The promiscuous use of alkaloids is dangerous and usually leads to addiction.

Vitamins

5. The vitamins are trace nutrients required in the diet of animals for proper growth, development, and good health.
6. The absence of sufficient amounts of vitamins in the diet results in deficiency diseases.
7. Vitamins are generally classified as fat-soluble or water-soluble.
8. The fat-soluble vitamins are vitamins A, D, E, and K; the water-soluble vitamins are vitamin C and those usually designated B vitamins.

Steroids

9. The steroids are natural products that have in common the 4-cycle, skeletal structure of 1,2-pentenophenanthrene.
10. Steroids with one or more hydroxyl groups may be called *sterols*.
11. The most abundant steroids of animal origin are cholesterol, ergosterol, the sex hormones, and cortisone.

Antibiotics

12. The antibiotics are antibacterials used in chemotherapeutic agents to inhibit the growth of, or to destroy, microorganisms. A number are of plant (mold) origin.
13. Commonly used antibiotics are the penicillins, streptomycin, and those of the tetracycline family—aureomycin and terramycin.

Supplementary Exercises

17.1 Would caffeine (Exercise 16.2) fit our definition of an alkaloid? Explain.

17.2 Consider the structure of procaine (Sec. 17.3) and show the different steps that Einhorn must have taken when he synthesized the compound in 1912.

17.3 What property common to the water-soluble vitamins permits water solubility?

17.4 To which general class of compounds does ascorbic acid (Vitamin C) belong? In which general class of compounds could we classify all other water-soluble vitamins?

17.5 Offer an explanation for one's inability to see the aisle, much less find a seat, upon first entering a darkened theater from a brightly lighted street.

17.6 Cholesterol (Sec. 17.10) is shown both as a planar structure and in its natural conformation. We have been told why cyclohexane rings prefer a chair conformation, but why is the D ring not a planar pentagonal structure?

17.7 In the biosynthesis of cholesterol from squalene, the triterpene lanosterol is an important intermediate. Apparently the immediate precursor of lanosterol is squalene oxide. Suggest a mechanism for the formation of lanosterol from squalene oxide. The reaction is acid-catalyzed (H^+) and involves rearrangements of the type discussed in Sec. 7.7B.

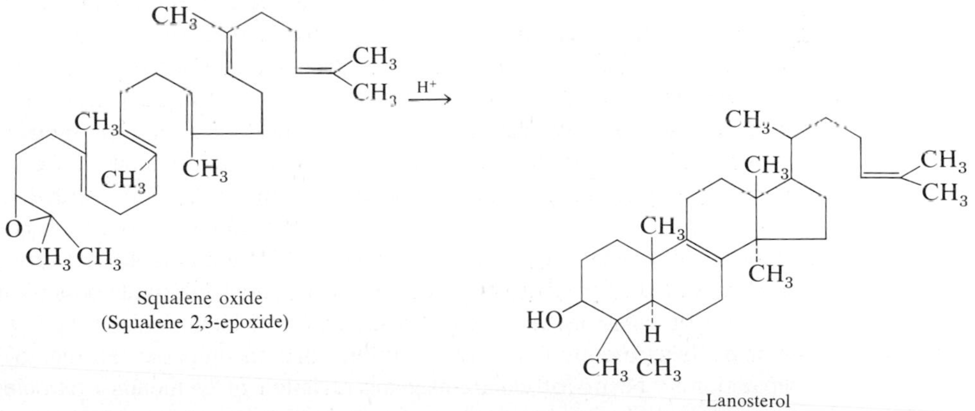

$\xrightarrow{H^+}$

Squalene oxide
(Squalene 2,3-epoxide)

Lanosterol

17.8 Nicotinamide adenine dinucleotide (NAD) is a very descriptive name for the coenzyme. The two nucleotides are connected via a phosphoric acid anhydride bridge. Draw the structure of NAD. Show that it is a salt by placing plus and minus signs on the appropriate atoms.

18 Determination of Molecular Structure; Spectroscopy

Let us assume that we have in hand a small amount of a substance that we believe to be a pure compound. It may be the product of a reaction we have carried out in the laboratory, or it may be a naturally occurring material we have isolated from a plant or animal source after the expenditure of considerable time, labor, and ingenuity. It may be one of the several million organic compounds that have already been prepared or isolated and whose structures are known, or it may be a new substance with exciting chemical or biological properties whose structure has yet to be determined. How do we go about the task of identifying our compound if it is one of those already known, or of establishing its structure if the compound is an unknown substance? During the first century of the development of modern organic chemistry, the identification of known compounds or the determination of the structure of new substances was based on the use of a few physical and biological properties (mp, bp, refractive index, color, odor, taste, etc.) and a host of chemical tests and reactions. It was not uncommon for a chemist to spend months or years on the structural characterization of a single important compound of complex structure. However, during the past thirty years the determination of the structure of an organic compound has been greatly simplified by the development of spectroscopic methods of analysis and structure determination. The combination of **spectroscopy,** other physical properties, and, if need be, chemical tests or reactions has made possible the rapid determination of most organic compounds, often with no more than a few milligrams of sample. Spectroscopy is a very broad term that encompasses all methods based on the separation of beams of electromagnetic radiation or elementary particles or ions. Our discussion will be concentrated on **absorption spectroscopy,** which may be defined as a procedure based on the selective absorption of electromagnetic radiation by a system and the experimental determination of the frequencies and amounts of radiation absorbed.

18.1 General Characteristics

Of the many types of spectroscopy known, those most frequently used at present in the analysis and determination of structure of organic molecules are infrared (ir), ultraviolet or ultraviolet-visible (uv-vis), and nuclear magnetic resonance (nmr) spectroscopy. Although these three spectroscopic techniques differ in the details of their theory, instrumentation, and application, they are based on the same fundamental principle: the *selective* absorption of energy from an electromagnetic wave by the molecule being studied, accompanied by a change in the energy state of the molecule.

A. Frequency and Wavelength. Although the nature of electromagnetic waves is complex, as they have both electric and magnetic properties, we will treat electromagnetic radiation as a simple wave, represented as in Fig. 18.1(a). Let us assume that the diagram is an instantaneous photograph of a wave as it travels through space at the velocity of light ($c = 3 \times 10^{10}$ cm/sec = 186,000 mi/sec), and that the length along the abscissa (x-axis)

FIGURE 18.1 Spectroscopy: (a) Relationship between Wavelength and Frequency, (b) Promotion from Ground to Excited State upon Absorption of Electromagnetic Radiation

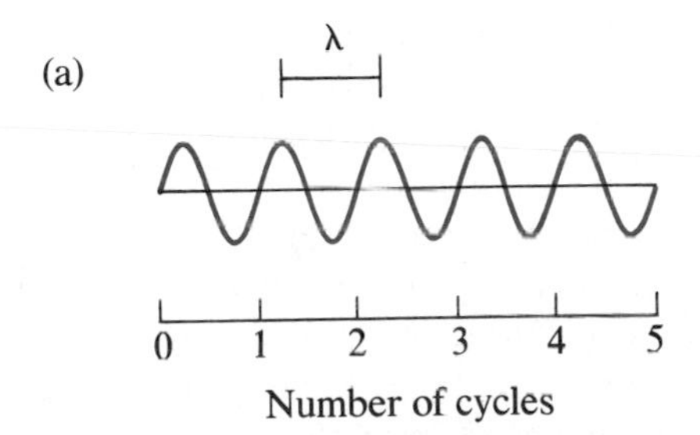

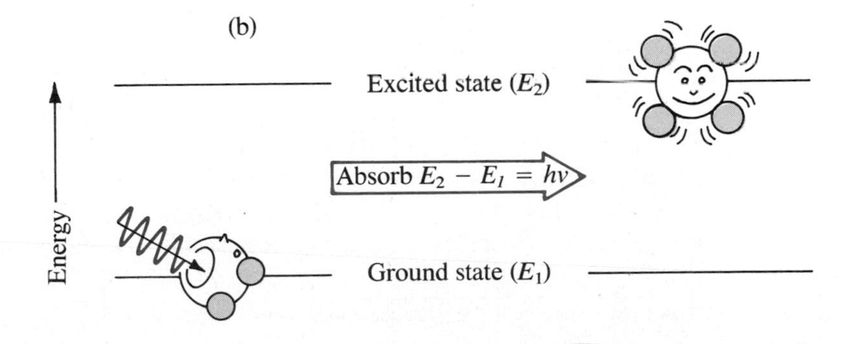

is the distance light travels in one second. Since the amplitude of the wave increases to a maximum and decreases to a minimum five times during the one second, we can say that it has gone through five cycles or that the frequency (ν) of the wave is 5 cycles per second (5 cps) or 5 hertz (5 Hz). A common alternative unit of frequency is the wave number ($\tilde{\nu}$), the number of waves per centimeter, which is defined by the equation.

$$\tilde{\nu} = \frac{\nu}{c}$$

Next, we define the wavelength (λ) of the electromagnetic wave to be the distance between any two identical points on two successive cycles of the wave. For simplicity we have chosen two successive crests of the wave. The three terms, frequency, wave number, and wavelength, are interrelated, as shown by the following equations.

$$c = \lambda\nu \quad \text{and} \quad \tilde{\nu} = \frac{1}{\lambda} \qquad \text{where } c = \text{velocity of light}$$

The names of some spectroscopic units have been changed in recent years as physicists have sought to attach names of famous physicists to the units. Also, because of the broad span of the electromagnetic spectrum, the preferred units for a given type of spectroscopy may be different from those used in another. The most common units are tabulated in Table 18.1.

B. Absorption of Electromagnetic Energy. In the earlier chapters we found that electrons in atoms and in molecules were restricted to discrete atomic or molecular orbitals; that is, they were found only in certain discrete electronic energy states. The same type of restriction applies to the other energy states of molecules; therefore, molecules can occupy only certain discrete electronic, magnetic, vibrational, or rotational energy levels and cannot be found at intermediate energy levels. We can represent these restrictions in a general way by a simple energy diagram, as shown in Fig. 18.1(b). Let us assume that the laws of physics dictate that the energy of the molecule will be such that it can be found only in two energy states of a given type, E_1 or E_2. The energy of the molecule cannot take on values intermediate between E_1 and E_2. Normally, the molecule will be found in its lowest energy level (**ground state**). However, if the molecule is allowed to absorb electromagnetic energy *of the proper frequency*, it may be ''promoted'' to the higher energy level

TABLE 18.1 Common Spectroscopic Units

Frequency	Wavelength units
1 Hz = 1 hertz = 1 cps*	1 cm = 10^{-2} m
1 kHz = 1000 Hz	1 mm = 10^{-3} m
1 MHz = 10^6 Hz	1 micrometer (μm) = 1 micron* (μ) = 10^{-6} m
1 wave number = 1 cm^{-1}	1 nanometer (nm) = 1 millimicron* (mμ) = 10^{-9} m
	1 Angstrom (Å)* = 10^{-10} m

*Indicates older unit; we will not use these older units.

(**excited state**). Very probably it will not stay there for long. Energy will be lost in the form of heat or in some other way, and the molecule will return to its ground state, ready to absorb more electromagnetic energy. Planck's Law tells us that the "proper frequency" of electromagnetic radiation to bring about the promotion must be:

$$E = h\nu$$

where E is the amount of energy absorbed (the difference in energy between the two levels, $E_2 - E_1$) and h is Planck's constant (6.6262×10^{-34} joules/Hz for ν in Hz).

In a typical spectroscopic analysis using an **absorption spectrometer,** a sample is placed in the path of a beam of electromagnetic radiation (such as light). The frequency of the radiation is varied, and the amount of radiation that is absorbed by the sample, if any, at each frequency is determined by the spectrometer, which plots the results in the form of a **spectrum.** Of course, radiation (energy) is absorbed only when the frequency and the energy required for a transition (change in energy level) obey Planck's Law. The spectrum usually consists of one or more **peaks** or **bands** (broad or overlapping peaks), whose frequency and size (amplitude, intensity) can be correlated with some structural feature in the molecule.

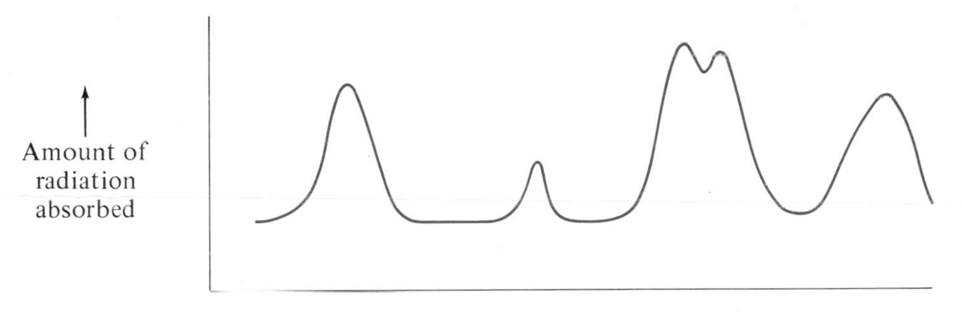

For a given organic molecule there are many energy states; thus, electromagnetic energy of many frequencies may be absorbed by the molecule. Our problem is to relate the frequencies and energy absorption in a qualitative and quantitative fashion to the structure of the molecule. Fortunately, the absorption of light in the uv-vis portion of the electromagnetic spectrum (Fig. 18.2) can be related to the different **electronic energy states** associated with the promotion of an electron from one orbital to another, and the absorption of light in the ir region of the electromagnetic spectrum can be correlated with the different **vibrational** and **rotational energy states** associated with the stretching and bending of bonds. Absorption of electromagnetic energy in the high-frequency radio portion of the spectrum in the presence of a strong magnetic field is dependent on the **magnetic environment** of certain atoms in the molecule and makes possible one of the most versatile of all analytical techniques, nmr spectroscopy.

FIGURE 18.2 Electromagnetic Spectrum

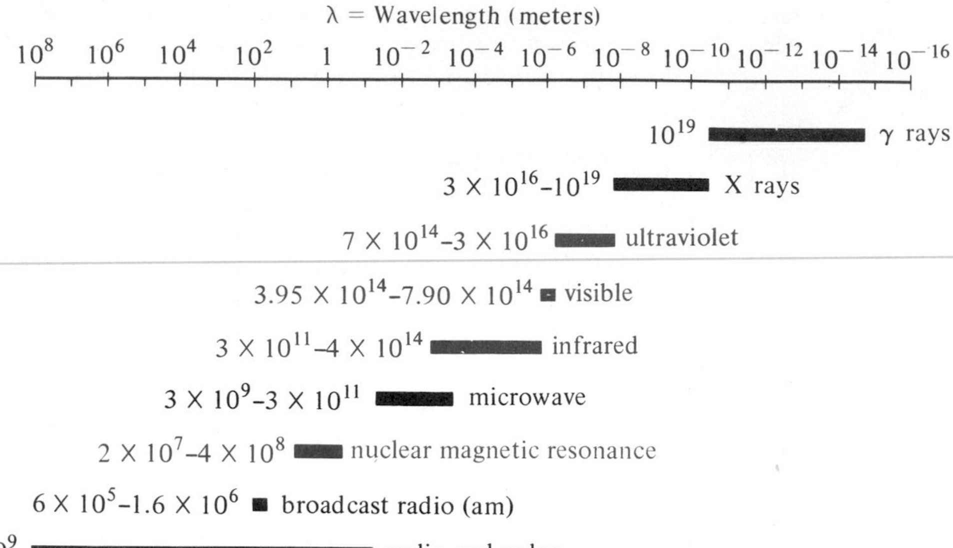

Exercise 18.1 For chemists a more useful form of Planck's Law is:

$$E \text{ (kcal/mole)} = 9.545 \times 10^{-14} \times \text{frequency (Hz)}.$$

Using this form of the law, select a frequency from each of the uv-vis, ir, and nmr regions of Fig. 18.2 and calculate the energy required for the transition corresponding to that frequency.

18.2 Nuclear Magnetic Resonance (nmr) Spectroscopy

The nuclei of isotopes with either odd atomic numbers or odd atomic masses, or both, show a nuclear property called "spin," somewhat analogous to the spin of an electron. One result of this spin is that these nuclei tend to behave as tiny magnets when placed in a very strong magnetic field. Some of the isotopes that behave in this fashion and are useful in organic chemistry, biochemistry, or biology are the common isotopes of hydrogen (^{1}H) and of phosphorus (^{31}P), and the low-abundance isotopes of hydrogen (deuterium, ^{2}D), carbon (^{13}C), oxygen (^{17}O), and nitrogen (^{15}N). Unforunately, some of the common isotopes of the elements of interest to organic chemists or biochemists do not show this behavior, for example, the common isotopes of carbon (^{12}C) and of oxygen (^{16}O). These isotopes do not give rise to nmr spectra. In this text we are most interested in the hydrogen nucleus (the proton) and in proton nmr spectroscopy. By analogy with the familiar compass needle, we will use a small arrow to represent the tiny nuclear magnet corresponding to the hydrogen nucleus, the arrow head indicating the north pole. Nuclear magnets are very weak; therefore, in a weak magnetic field, such as the earth's magnetic field of about 0.5 gauss, they will be oriented at random (Fig. 18.3(a)); but when placed in a *very strong* magnetic field (of the order of 14,000–80,000 gauss), a majority of the nuclear magnets will become aligned with the applied field (Fig. 18.3(b)) as this is the more stable state, the state of lowest energy. If this assembly of nuclear magnets is irradiated with electromagnetic radiation of *exactly the right frequency*, some of the nuclei will absorb sufficient energy from the electromagnetic radiation to reverse their orientation with respect to the magnetic field. Their energy will increase, and they will be

FIGURE 18.3 Nuclear Magnetic Resonance: (a) Orientation of Nuclei in the Absence of an External Magnetic Field, (b) Orientation of Nuclei in a Strong External Magnetic Field, (c) Absorption of Electromagnetic Energy Changes Orientation of a Nucleus

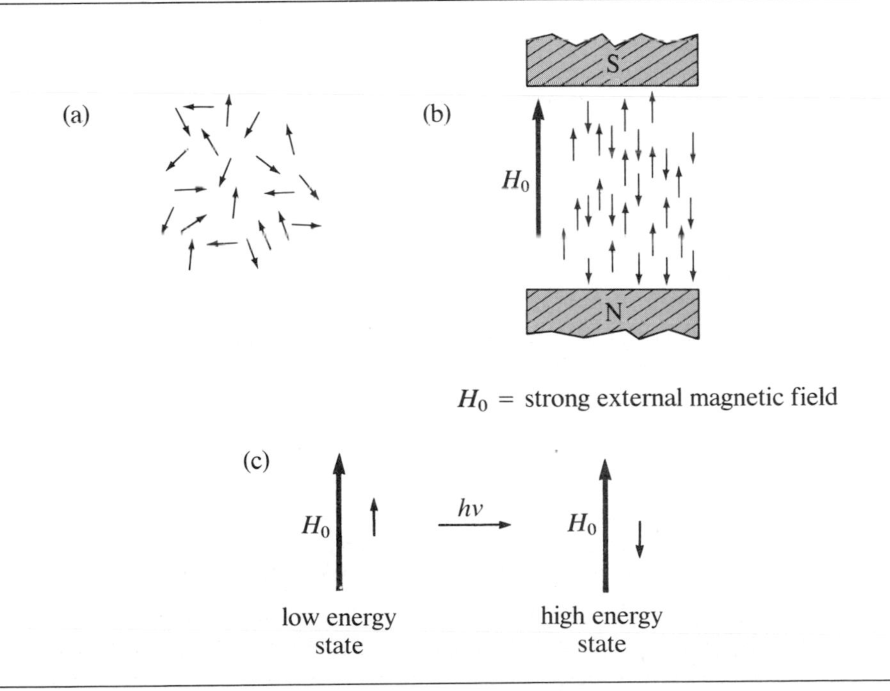

promoted to the less stable, higher energy state (Fig. 18.3(c)). When *both* the strength of the magnetic field *and* the frequency of the electromagnetic radiation are such that energy can be absorbed by the nuclei,[1] the system is said to be in resonance, hence the name, nuclear magnetic resonance.

What we have said thus far really applies only to isolated protons, bare hydrogen nuclei, all of which are in resonance at the same value of the magnetic field strength (H_0) for a given value of frequency (ν), or vice versa. Fortunately, in molecules the nuclei of the various hydrogen atoms present tend to absorb energy from the electromagnetic radiation at slightly different frequencies, which depend on the molecular *magnetic* environments in which the atoms are located. This occurs because, in a strong magnetic field, the electrons in bonds and in unshared pairs tend to **shield** the hydrogen nuclei from the full effects of the magnetic field. This shielding varies with the location of the individual nuclei in the molecule. Thus, protons that are in the same magnetic environment due to the structure of the molecule will absorb energy at the same frequency, whereas those in different magnetic environments will absorb energy at different frequencies. For our purposes protons that are not distinguishable chemically will be considered to be in the same magnetic environment, will be called **chemically equivalent protons,** and will absorb energy at the same frequency. For example, in 2-chloro-2,4,4-trimethylpentane there are three sets of equivalent protons, as shown by the circle, rectangle, and triangle surrounding them.

$$CH_3-\underset{Cl}{\overset{CH_3}{C}}-CH_2-\underset{CH_3}{\overset{CH_3}{C}}-CH_3$$

2-Chloro-2,4,4-trimethylpentane

Within each set the protons are equivalent, but no proton in one set is chemically or magnetically equivalent to any of the protons in the other sets. Therefore, this molecule should absorb electromagnetic radiation at three slightly different frequencies, if the magnetic field is held constant, or at three slightly different magnetic field strengths, if the frequency is held constant.

Exercise 18.2 Label the sets of chemically nonequivalent protons in (1) isopentane, (2) methylcyclohexane, (3) toluene, and (4) 1-butene.

[1]The requirement for resonance in *proton* nmr is that the magnetic field strength at the nucleus (in kilogauss) must equal 0.235 times the frequency (in MHz).

A highly simplified schematic of an nmr spectrometer is shown in Fig. 18.4. The sample, as a liquid or solution (in CCl_4, $CDCl_3$, or D_2O, or some other *nonprotonic* solvent), is contained in a slender glass tube inserted into a wire coil situated in a strong, highly uniform (homogeneous) magnetic field. A radio frequency oscillator sends a high-frequency current to the coil, which functions as a tiny transmitting antenna bathing the sample in the electromagnetic radiation required for the experiment. To determine an nmr spectrum, the magnetic field is held constant, and the radio frequency is varied. On some instruments the frequency is held constant, and the magnetic field is varied. As each chemically, and hence magnetically, nonequivalent proton in the sample comes into resonance, electromagnetic energy is absorbed, causing a change in the flow of current through the coil. The change in current is monitored by a sensitive detector and is recorded as a plot of energy absorbed (ordinate) versus the frequency (abscissa). The difference in the two values of frequency required to bring two protons in different molecular environments into resonance is called the **chemical shift.** Thus the abscissa of an nmr spectrum is usually marked in chemical shift units. Generally, the spectrum will consist of a relatively small number of "peaks" of varying size (area). Since each proton in the molecule is absorbing about the same amount of electromagnetic radiation, the area of

FIGURE 18.4 Schematic Diagram of an nmr Spectrometer

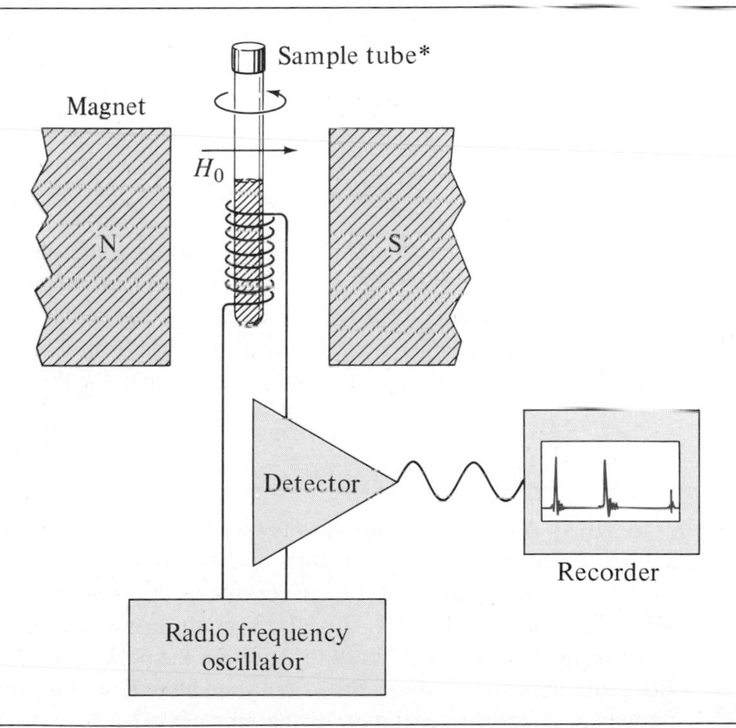

*The sample tube is spun by a small air turbine to average out inhomogeneities in the magnetic field.

each peak will be roughly proportional to the number of protons responsible for the peak. Therefore, the nmr instrument is equipped with an ''integrator'' to measure the area of each peak. The **integral** is usually shown as a separate tracing on the same sheet as the spectrum and looks like a series of steps. The height of the steps in the integral is roughly proportional to the areas under the corresponding peaks, hence to the number of protons responsible for the peaks.

It is the practice in nmr spectroscopy to make chemical shift measurements relative to a standard, which is added to each sample for this purpose. The most widely used standard is tetramethylsilane, $(CH_3)_4Si$, which is called TMS and was chosen because its twelve protons are equivalent and are more highly shielded than the protons in most organic compounds. Therefore, TMS gives a strong, 12-proton, single peak, which is found at a lower frequency (higher field) than, and is well separated from, the peaks due to the protons in the sample.

```
       CH3
       |
  CH3—Si—CH3
       |
       CH3
```

Generally, chemical shifts are reported in frequency units ($\Delta\nu$) relative to TMS *for a specific spectrometer frequency* (ν_0), which *must* be specified, or in δ units that are independent of the magnitude of the spectrometer's nominal operating frequency. Thus

$$\Delta\nu = \text{distance from TMS in Hz (at a specified } \nu_0\text{)}$$

$$\delta = \frac{\text{distance from TMS in Hz}}{\text{spectrometer frequency in MHz}} = \frac{\Delta\nu}{\nu_0}$$

The unit of δ is parts per million (ppm); however, the unit is usually not specified in present practice. Positive values of δ correspond to chemical shifts **downfield** (lower field, higher frequency) from TMS, and seldom-seen negative values correspond to chemical shifts **upfield** (higher field, lower frequency) from TMS. Since δ is independent of the spectrometer frequency, it is the preferred unit for the chemical shift.

Exercise 18.3 An nmr spectrum shows two peaks (in addition to that of TMS) at δ 1.45 and δ 1.55. Calculate the chemical shifts of these two peaks relative to TMS and relative to each other in frequency units ($\Delta\nu$ in Hz) for spectrometer frequencies of 60 MHz and 220 MHz. Plot the 60-MHz and 220-MHz spectra, one above the other, on a sheet of graph paper using 1-inch vertical lines to represent the peaks and plotting frequency units (Hz) along the abscissa (x-axis). Based on your plot, how does increasing the spectrometer frequency affect the separation of the nmr peaks?

Figure 18.5 shows the nmr spectrum of methyl *tert*-butyl ketone run at a frequency of 60 MHz with TMS as the internal standard.

$$CH_3-\overset{\displaystyle CH_3}{\underset{\displaystyle CH_3}{\overset{|}{\underset{|}{C}}}}-\overset{\displaystyle O}{\overset{\|}{C}}-CH_3$$

Methyl *tert*-butyl ketone

Note that there are two traces: the lower trace is the nmr spectrum, and the small stair-step trace at the top is the integral. For accuracy, the integral usually is made as large as possible, overlapping the spectrum; but, for clarity, in Fig. 18.5 the integral was reduced below its normal size. There are three peaks in the spectrum, two for the sets of nonequivalent protons in the sample and one for the TMS protons, the latter always appearing at the extreme right, the so-called **upfield** region. On the original spectrum the steps in the integral were measured and found to be 6.5 and 20.0 mm in height. Therefore, we can calculate the *ratio* of the protons "in the two peaks" to be 20.0:6.5 or about 3:1. Unless

FIGURE 18.5 The nmr Spectrum of Methyl *tert*-Butyl Ketone

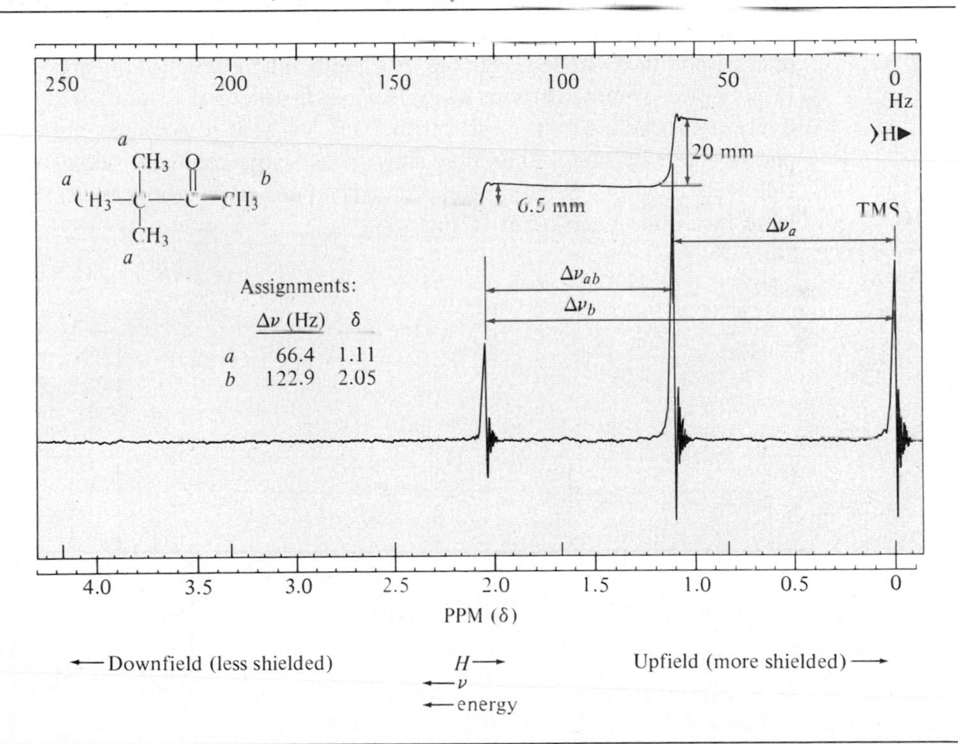

we have information from some other source, we cannot distinguish between ratios of 3:1, 6:2, 9:3, etc. In this instance we know what our sample is and that the ratio must be 9:3. The integral also tells us which peak is due to the methyl group (3 protons) and which to the *tert*-butyl group (9 protons).

From Fig. 18.5 it should be clear that the term chemical shift ($\Delta\nu$) is sometimes used in two different ways: (1) to indicate the difference in resonance frequencies between the peaks of the sample and that of TMS and (2) to indicate the difference in resonance frequencies between the peaks of the sample molecule itself. Unless it is clearly specified otherwise, as in Exercise 18.3, we will begin using the first of these to describe our nmr spectra. The chemical shift assignments for the two peaks, relative to TMS, are given in the figure. Thus, in the jargon of nmr spectroscopy, we describe our spectrum as consisting of a 9-proton **singlet** at δ 1.11 and a 3-proton singlet at δ 2.05, where singlet means a single peak. We use the term "singlet" rather than "peak" because the nmr spectra of many organic compounds, even relatively simple ones, are much more complex than that of Fig. 18.5. To illustrate some of this complexity, we will examine the nmr spectrum of ethanol (ethyl alcohol) after making a few observations on the relationship of chemical shift to structure.

Usually, the chemical shifts of protons in an organic molecule are closely correlated with the structure of the molecule. For example, the chemical shift of a CH_2 group in any molecule, R—CH_2—O—R′ where R and R′ are simple, saturated alkyl groups, will usually fall within a narrow range of delta values near δ 3.4. The chemical shift of a specific proton will depend largely on its location relative to hetero atoms (atoms other than carbon or hydrogen), double or triple bonds, aromatic rings, or functional groups. Protons two or more carbons away from such structural features will usually show relatively small chemical shifts, less than δ 1.7, whereas protons that are attached to or are a part of these structural units may show rather large chemical shifts, as illustrated in the following diagram. Some typical proton chemical shifts are listed for ready reference in a table inside the back cover of this text.

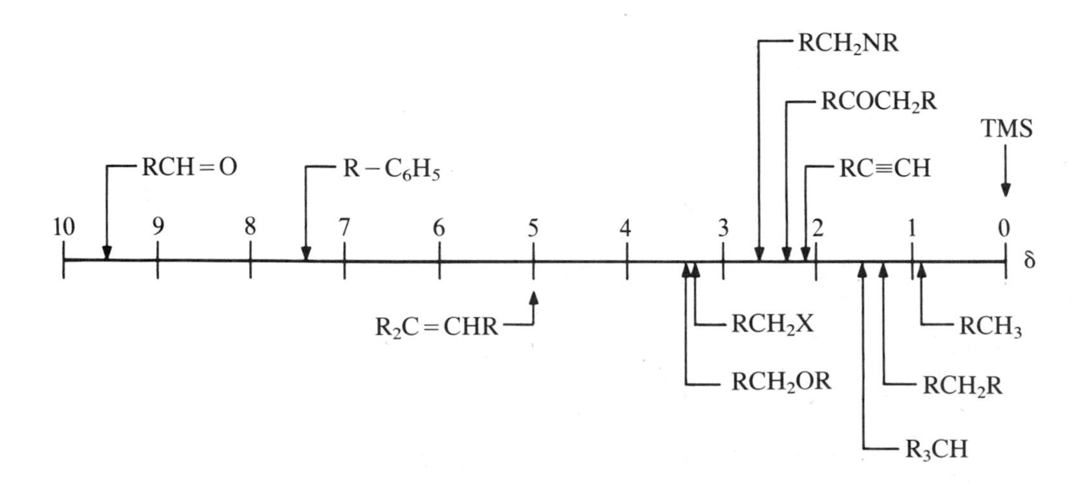

Exercise 18.4 Based on the table inside the back cover, predict the approximate chemical shifts of the protons in the following compounds:

(a) toluene, (b) propene (propylene), (c) neopentane,

(d) $CH_3CH_2—\overset{\overset{\large O}{\|}}{C}—H$, (e) propyne, (f) $CH_3—CO_2H$.

The nmr spectrum of ethanol is given in Fig. 18.6. The integral is not shown, but we have used it to identify the three sets of equivalent protons. This spectrum is described as a three-proton **triplet** at δ 1.22, a two-proton **quartet** at δ 3.70, and a one-proton singlet at δ 2.9. Thus, where we might have expected three singlets, we actually observe two **multiplets** and a singlet. The **splitting** of singlets into multiplets is a result of the interaction of the proton being observed with the protons on the *next adjacent* atom. Recall that the latter protons are also behaving like tiny magnets. If they are close enough to the proton being observed, they can affect its absorption of electromagnetic radiation in such a way as to cause it to come into resonance at two or more frequencies. Because the ultimate cause of splitting is nuclear spin, the phenomenon is often called **spin-spin coupling,** and the nuclei (protons) involved are said to be **coupled.**

FIGURE 18.6 The nmr Spectrum of Ethyl Alcohol in $CDCl_3$

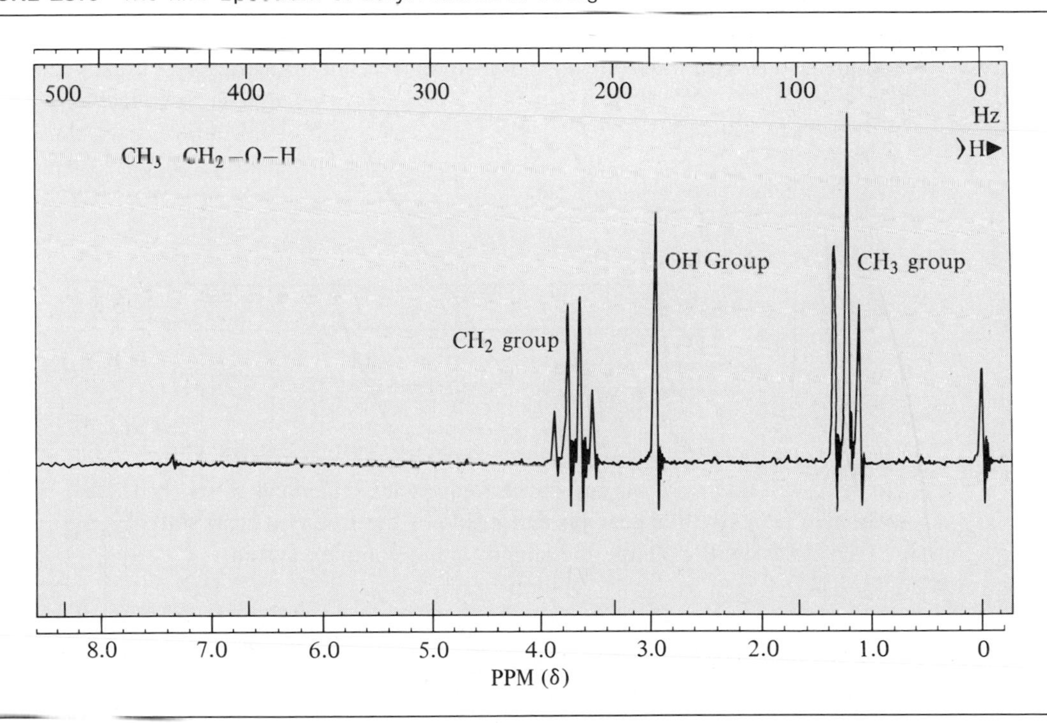

To explain splitting, we will first consider the following simple system of two adjacent, magnetically nonequivalent protons, H_a and H_b, in which the chemical shift of H_a is downfield of that of H_b.

—C—C—
H_a H_b

We begin at low field and increase the main magnetic field, H_0, moving from left to right as shown in the following simplified spectrum. In the diagram we represent H_0 with a large arrow and the magnetic fields of protons H_a and H_b with small arrows, h_a and h_b. When we are observing the chemical shift of H_a, H_b will *not* be in resonance; therefore, in about half of the molecules present the H_b protons will be aligned with H_0, and in about half the H_b protons will be oriented in the opposite direction. In those molecules in which the H_b protons are aligned with H_0, the field at H_a will always be slightly greater than H_0. Therefore, H_a will come into resonance at a somewhat lower value of H_0. In other words, if proton H_a was expected to come into resonance when $H_0 = (H_0)_a$ in the absence of other protons, now H_a will be expected to come into resonance when $H_0 = (H_0)_a - h_a$. By a similar reasoning, it can be seen that in those molecules in which H_b is oriented against H_0, the field at H_a will be less than H_0. Therefore, H_a will come into resonance at a field value of $H_0 = (H_0)_a + h_b$. Since both orientations of H_b are almost equally probable, the H_a peak will be split into two peaks, a doublet, of almost equal intensity. Since proton H_a will affect the chemical shift of H_b in a similar fashion, the H_b peak will also be split into a doublet. The spectrum will look something like the following.

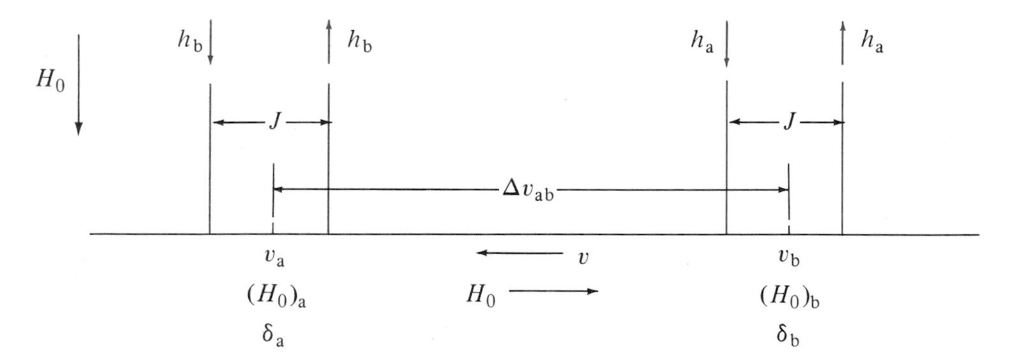

No peak will appear at the resonance frequencies (chemical shifts) of H_a and H_b. However, two peaks will appear on either side of the true chemical shifts.

Next, we will examine the slightly more complex system

—C—C—H_b
H_a H_b

in which H_a is again downfield of the two protons, H_b. As the magnetic field, H_0, is increased, the total field at H_a is now affected by the orientation of *two* H_b protons. These protons can be oriented in four ways, represented schematically as: (1) ↑↑, (2) ↓↓, (3) ↑↓, and (4) ↓↑. Orientation (1) increases and orientation (2) decreases the magnetic field at H_a. In orientations (3) and (4) the fields of the two H_b protons cancel each other, so the combinations have no effect on the field at H_a. Therefore, we expect *three* peaks for proton H_a, the first resulting from orientation (1), the second from orientations (3) and (4), and the third from orientation (2). Since the four orientations are almost equally probable, the second, or middle, peak should be twice as intense as the first and third because the second peak is the result of *two* different, but equivalent, orientations. We also expect the second peak to come at H_a's resonance frequency (chemical shift), since the effects of the H_b fields have been cancelled for this peak. As we continue to increase H_0 and approach the resonance condition for the two H_b protons, these protons sense H_a in its two possible orientations. Therefore, as in the first system we examined, the H_b protons will appear as a doublet symmetrically disposed about the H_b chemical shift. Our complete nmr spectrum will consist of a 1:2:1 triplet and a 1:1 doublet in which the ratio of *total* areas of the triplet:doublet will be 1:2. The spectrum will be approximately as follows:

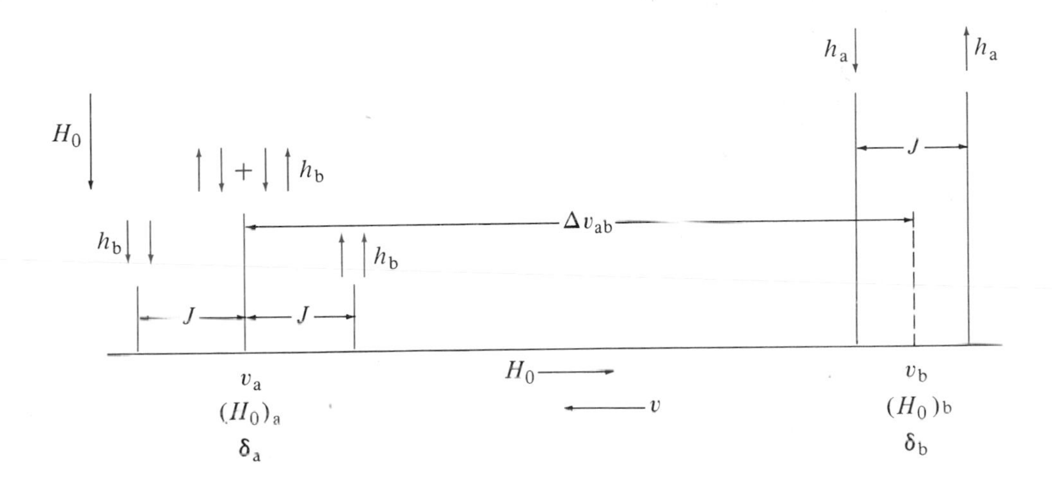

Although we could continue to examine more complex systems as we have done, fortunately it is usually possible to predict the number of peaks in a multiplet (the **multiplicity**) by counting the protons on the next adjacent carbon atoms, the so-called neighboring protons. The multiplicity is one greater than the number of neighboring protons:

$$\text{multiplicity} = n + 1, \text{ where } n = \text{number of neighboring protons}$$

For example, on the carbon atom next to the methyl group of ethanol there are two protons; therefore,

$$\text{multiplicity of } CH_3 = (2 + 1) = 3 \text{ (triplet)}$$

On the carbon atom next to the methylene (CH_2) group there are three protons; therefore,

$$\text{multiplicity of } CH_2 = (3 + 1) = 4 \text{ (quartet)}$$

In calculating the multiplicity of the methylene group, we did not count the proton on the oxygen atom, although it is indeed a neighboring proton. The reason for this is that, in order to observe a proton in a given location relative to the rest of the molecule, the proton must remain in that location for a rather long time (0.1–1 sec or longer). If the proton should change its position or location rather rapidly, the nmr spectrometer will report only its average location. Very often the protons of OH and NH groups are **exchanging,** moving rapidly from one molecule to another (especially in the presence of the traces of acids or bases that are almost always present). There are two effects of this chemical exchange. First, the chemical shifts of OH and NH protons tend to vary more widely than those of protons on carbon atoms and are more sensitive to experimental conditions. Second, the rapid exchange of protons between OH groups or NH groups in the sample tends to eliminate the effects of coupling of the OH or NH protons with the protons of adjacent CH groups. Thus, in the spectrum of ethanol, as run under ordinary conditions, the CH_3 and CH_2 groups split each other according to the $(n + 1)$ rule, but the OH group does not split the CH_2 group, nor is it split by the CH_2 group.

Another exception to the multiplicity rule is the instance of two protons with the same chemical shift. Nuclei with the same chemical shift do not split each other. They are *coupled* but not *split;* thus these two terms are not synonyms. In the spectrum of 1,2-dimethoxyethane (Fig. 18.7), the four protons in the two CH_2 groups have the same

FIGURE 18.7 The nmr Spectrum of 1,2-Dimethoxyethane in $CDCl_3$

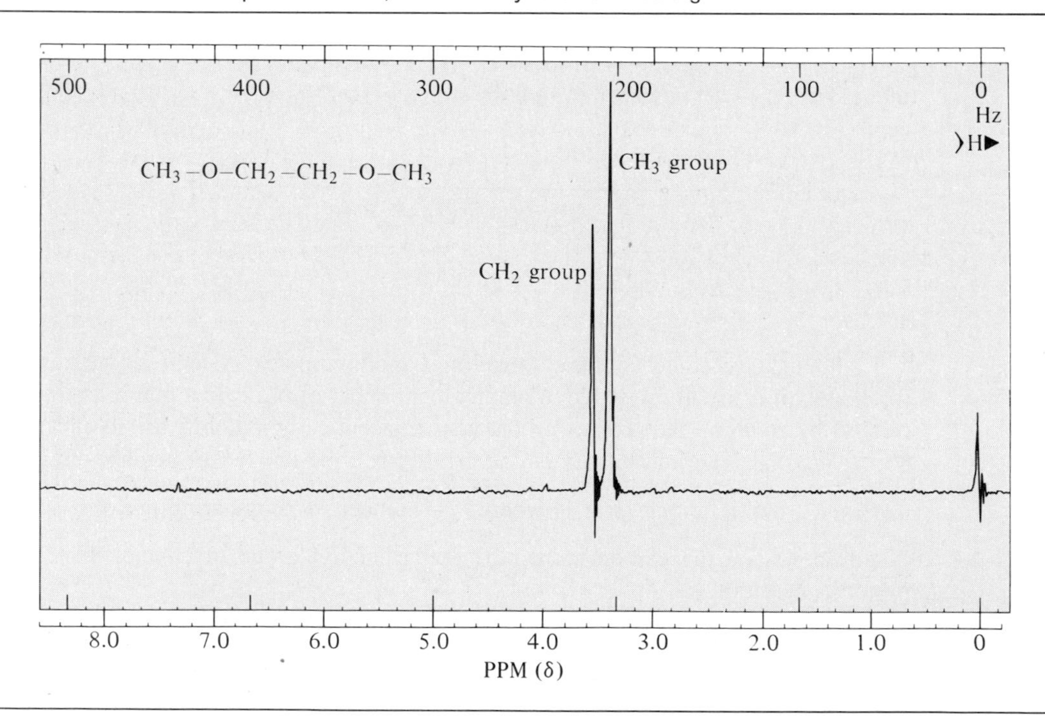

chemical shift; therefore, they do not split each other and appear as one peak, as do the six protons in the two CH_3 groups.

When splitting is observed, the separation of the peaks in each multiplet will be identical and equal to J Hz where J is called the **coupling constant.** As its name implies, J is a constant for a given coupling and, *unlike the chemical shift,* is independent of the spectrometer frequency. Typically, J is of the order of 0–15 Hz for protons. Some selected typical values of coupling constants for protons in common structural units are given in a table inside the back cover of this text.

Exercise 18.5 Using vertical lines for peaks, draw simple diagrams that will show the expected splitting patterns in the proton nmr spectra of the following compounds. Try to get the shifts in the correct order (greatest δ-value to the left), but don't be concerned with the exact chemical shift values.

(a) 1,1-dibromo-2,2-dichloroethane, (b) 1,1,2-trichloroethane, (c) 1,1-dichloroethane, (d) chloroethane, (e) 2-chloropropane.

When run on older routine nmr instruments (60–90 MHz), the spectra of some organic molecules, even simple molecules, are often too complex to analyze by the $(n + 1)$ rule. However, with *current* routine instruments (180–200 MHz) or with research instruments (360–500 MHz), most organic molecules do give relatively simple spectra. The difference between nmr spectra run at 60 MHz and 200 MHz for the simple molecule 1-bromo-2-chloroethane can be seen from the partial spectra in Figure 18.8. For this molecule the chemical shift is about 12.6 Hz at 60 MHz and about 42 Hz at 200 MHz. Whenever the chemical shift is not much larger than the coupling constant, here about 7 Hz, the spectrum may be much more complex than predicted by the $(n + 1)$ rule. However, with high-frequency ("high-field") instruments, many complex spectra can be simplified. For 1-bromo-2-chloroethane at 200 MHz, the spectrum is beginning to resemble the pair of triplets predicted by the $(n + 1)$ rule.

Although the most abundant isotope of carbon, ^{12}C, does not give rise to nmr signals, all naturally derived organic compounds contain about 1.1% of their carbon content in the form of ^{13}C. In spite of the low abundance of ^{13}C and an inherent low sensitivity to detection, methods and instruments are now available for routine determination of ^{13}C nmr spectra, often called cmr spectra. The theory and instrumentation for carbon nuclear magnetic resonance (cmr) spectroscopy is essentially the same as that for proton nuclear magnetic resonance (pmr) spectroscopy as currently practiced in most major laboratories. The instruments depend heavily on computers to gather and process spectral data. Although the interpretation of cmr spectra is not difficult, it is somewhat complex and beyond the scope of this text. However, a typical cmr spectrum run on a sample of

FIGURE 18.8 The nmr Spectra of 1-Bromo-2-chloroethane in $CDCl_3$ at 60 and 200 MHz

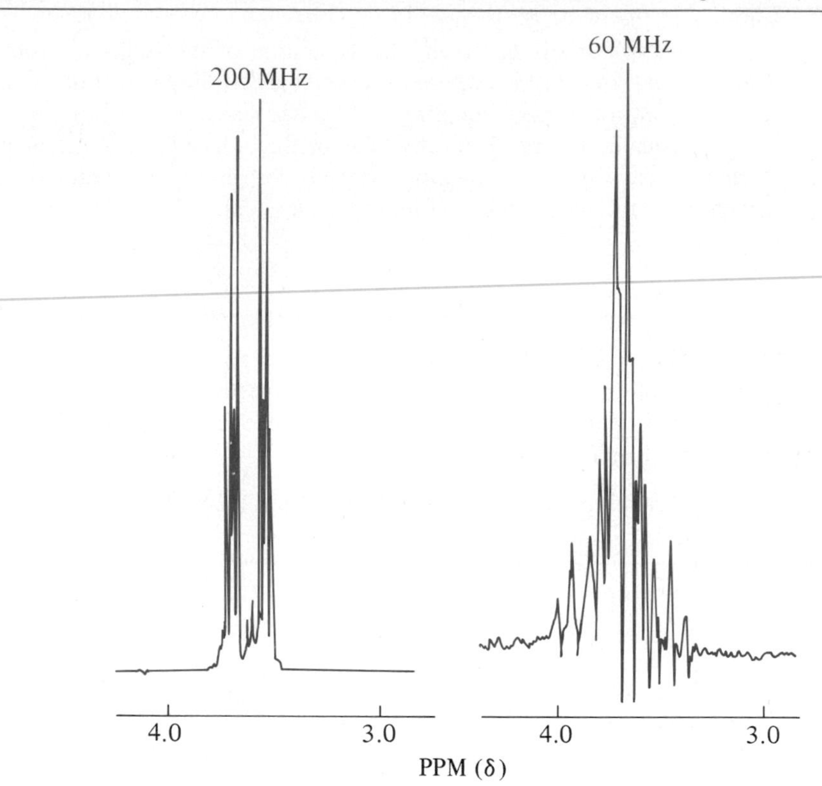

ordinary table sugar, sucrose ($C_{12}H_{22}O_{12}$) (Sec. 14.12), is given in Figure 18.9. In this spectrum the ^{13}C atoms are **decoupled** from the protons in the compound; thus, each peak corresponds to one carbon atom, although peak areas do not tell us the number of carbon atoms per peak in this type of nmr spectroscopy. Two of the peaks are poorly separated, but each of the twelve carbon atoms does give rise to a separate peak.

Nuclear magnetic resonance is not only a useful spectroscopic technique, it is also a powerful new diagnostic tool used by radiologists to obtain cross-sectional pictures of soft tissues in a safe, noninvasive manner in live human beings. This new tool is called **nuclear magnetic resonance imaging,** or NMRI, by the scientists studying it. However, to avoid unduly alarming patients, radiologists usually drop the word ''nuclear,'' and refer to the technique as MRI. The objective of this procedure is to obtain a picture, for example, on a computer or TV screen, of a cross-section of some part of the body, brain, heart, chest, etc. As you may know, the surface of a computer screen is a grid that resembles a piece of graph paper. Each tiny square is a **pixel** that may be turned on or off or made to appear as a different color. By analogy, a slice of human anatomy may be regarded as resembling an ice-cube tray. Each compartment or volume element is a **voxel.** In the MRI experiment, we make an nmr measurement for each voxel in the slice, obtain-

FIGURE 18.9 The ^{13}C nmr Spectrum of Sucrose in D_2O (25.2 MHz)

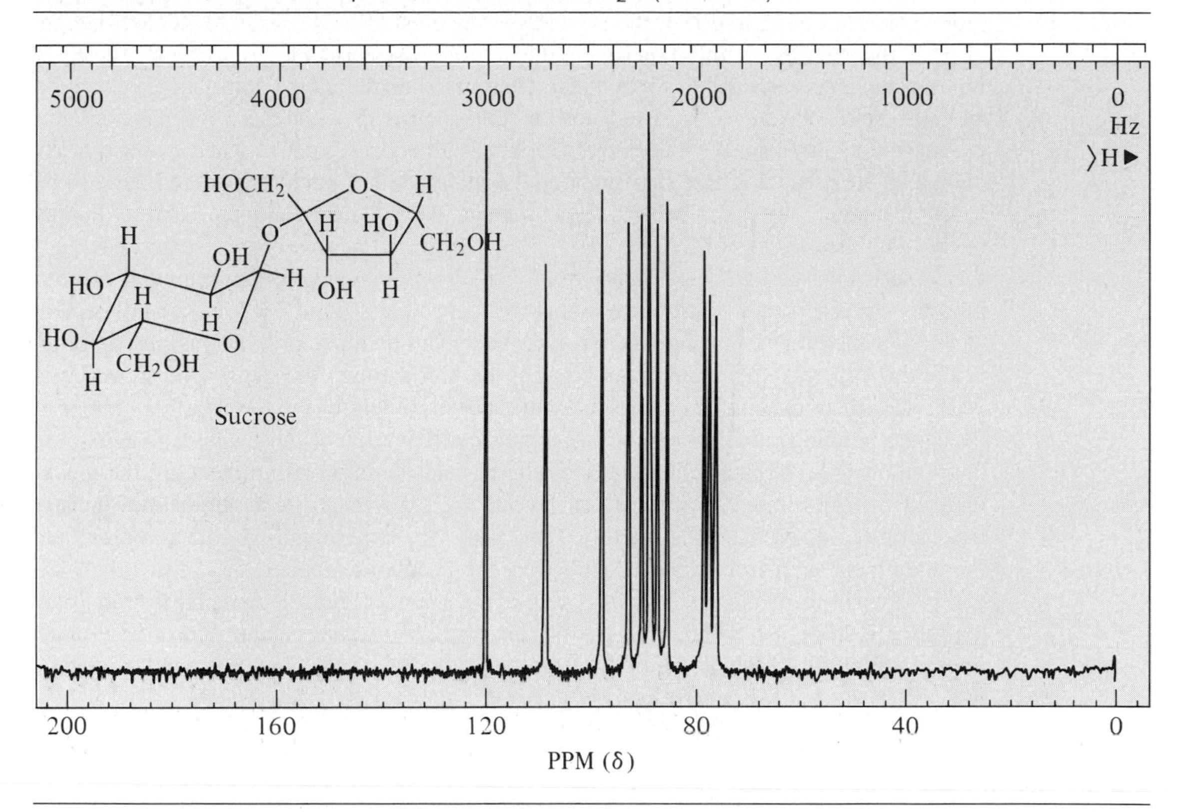

ing a number for each. This number will determine how dark or light, or red or blue, etc., the corresponding pixel on the computer screen will be (Fig. 18.10). The number we obtain will depend, in general, on three factors: the number of protons in the voxel and two **relaxation times,** T_1 and T_2, of the excited protons. As we learned earlier, when a

FIGURE 18.10 Magnetic Resonance Image Representing the nmr Properties of a Group of Voxels of a Cross-Sectional Slice of the Brain

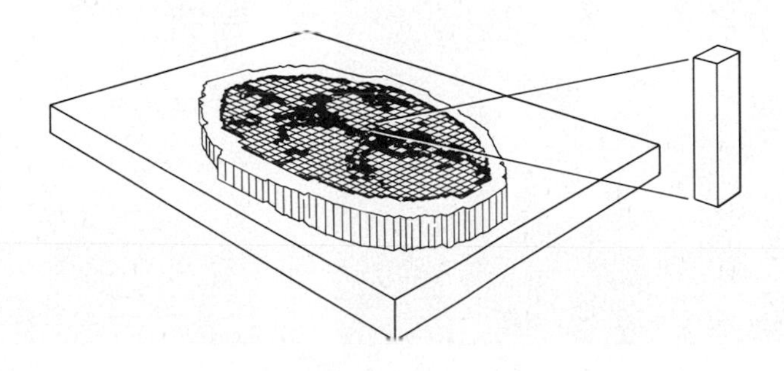

species is promoted to an excited state, it does not remain there long but returns to the ground state. In order to do so, the excited species must lose its energy to its environment in some manner. In the nmr experiment this return process is called relaxation. There are two mechanisms responsible for relaxing the excited nuclei called spin-lattice relaxation and spin-spin relaxation. We need not be concerned with what these mechanisms are, except to note that they are associated with times, T_1 and T_2, which measure how quickly the nuclei are relaxed. These depend not only on the kind of nucleus involved but also on its environment, the latter being very important in biological samples, such as human tissue. The measurement of T_1 and T_2 is a relatively routine matter in research laboratories. The intensity of the MRI signal from a voxel will depend on the number of protons present, whereas the duration of the signal will depend on T_1 and T_2. Through adjustment of the nmr instrument, the signal from each voxel can be made proportional to any or all of the three factors. When a human being is the nmr sample, the density of protons in a voxel and the relaxation times depend on the water content of the tissues; the number of hydrogen-bonding sites per voxel; the geometric distribution of these sites; the geometric distribution of hydrophobic, or water-repelling, patches on protein molecules; the movement of the large biomolecules present during the experiment; relaxation times in fatty tissues in the voxel; and other factors. Obviously, the experiment is quite complex, and the equipment is quite expensive. However, the results are spectacular. Thus, the MRI can be used in a noninvasive manner to diagnose cancer, brain edema and infarctions, abnormal distribution of fat, iron storage diseases, hemorrhages, and blood flow problems. Figures 18.10 and 18.11 are illustrations of the use of MRI as a powerful diagnostic tool.

FIGURE 18.11 Magnetic Resonance Images of the Human Brain *(Photos courtesy of General Electric Company.)*

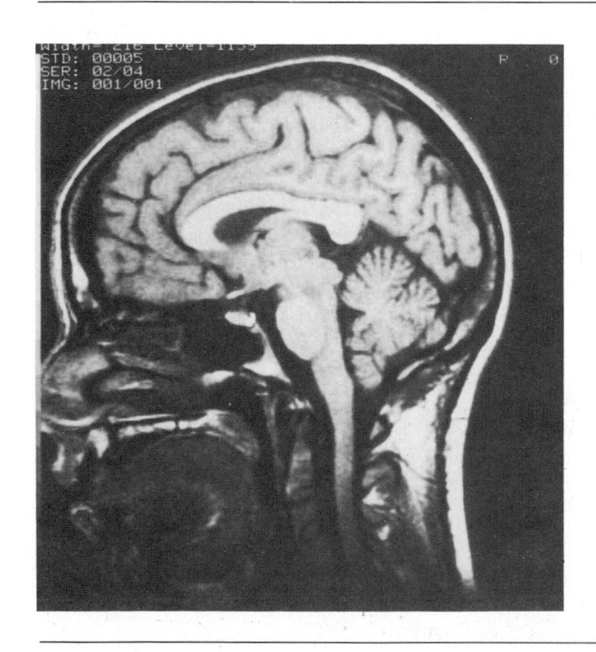

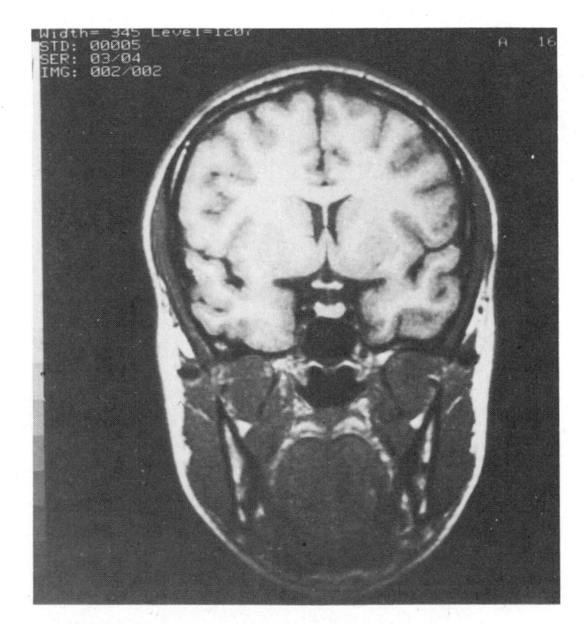

18.3 Infrared (ir) Spectroscopy

Although we tend to describe organic molecules as rather rigid structures with fixed bond lengths and precise bond angles, the bond lengths and bond angles in drawings and models are only average values of interatomic distances and angles, because the atoms within the actual molecules are in a constant state of motion. Bonds are stretching and contracting in a periodic fashion. Bond angles are opening and closing, and groups of atoms are rotating relative to each other about their common bond axes. The types of intramolecular motion can be separated into two classes: (1) vibration, which includes stretching and bending motions of bonds, and (2) rotation. We will be concerned principally with molecular vibrations, because it is from such motions and their dependence on the absorption of infrared light that the organic chemist can obtain the most useful structural information.

For each organic molecule there are discrete vibrational energy levels, and the frequencies (and amplitudes) of molecular vibrations are related to these energy levels and to the absorption of infrared light by Planck's Law, $E = h\nu$, where the frequency (ν) now refers both to the frequency of the infrared radiation absorbed and to the frequency of the vibrational motion. At room temperature most of the molecules of an organic compound will be in the lowest vibrational energy level, and the amplitude of the vibration will be at a minimum. When the molecule is irradiated with electromagnetic radiation of exactly the right frequency near 10^{14} Hz, that is, with light in the infrared region, the molecule can absorb some of the radiation and become promoted to the next higher energy level, in which the frequency of the molecule's vibration will be equal to that of the infrared light and the amplitude of the vibration will be increased.

A graphical plot of the infrared energy absorbed (as the ordinate) versus the frequency or wavelength of the infrared light (as the abscissa) is called an infrared (ir) spectrum. In part because the use of the fundamental frequency unit (Hz) gives such large numbers, infrared frequencies are usually reported in wave numbers. For example, if we are observing the stretching of an O—H bond in an alcohol (R—OH), the frequency (ν) of the vibration will be near 10^{14} Hz. Conversion of this frequency to its equivalent wave number gives

$$\tilde{\nu} = \frac{\nu}{c} = \frac{10^{14}\ \text{sec}^{-1}}{3 \times 10^{10}\ \text{cm sec}^{-1}} = 3333\ \text{cm}^{-1}$$

It is current and acceptable practice to describe the OH-stretching vibration of our alcohol at $\tilde{\nu} = 3333\ \text{cm}^{-1}$ as occurring at a *frequency* of 3333 cm^{-1} (wave numbers, centimeters to the minus one, or reciprocal centimeters). Although technically incorrect, this practice is based on the assumption that division by c is understood. Although best current practice requires that ir band positions be reported in frequency units, at present most ir spectra will display both frequency and wavelength scales; however, when a contemporary instrument is used, only the frequency will vary linearly across the chart. The common unit of wavelength in ir spectroscopy is the micrometer (μm) (or micron, μ). The intensity of absorption is most commonly reported as percent transmission (that is, 100% = *no* light absorbed; 0% = *all* light absorbed).

In organic chemistry most instruments operate over that part of the infrared region from about 4000–650 cm^{-1}. These numbers run backwards, from 4000 down to 650, because the design of early instruments favored the use of wavelength over frequency. Other special features of ir spectra can be seen by examination of Fig. 18.12. First, it is apparent that the peaks or bands have been plotted upside down relative to nmr (or uv-vis) spectra. This also is a matter of custom and one with which you will rapidly become familiar. Second, many of the absorption bands are relatively broad compared with nmr peaks. This broadening is, in part, the result of molecular rotations occurring along with molecular vibrations.

Block diagrams of typical infrared and ultraviolet-visible spectrometers are given in Fig. 18.13. Both types of instrument consist of three basic units: (1) a **source** of continuous or polychromatic radiation that is a mixture of light of all the frequencies required; (2) a **monochromator,** which selects from the continuous radiation monochromatic light of the desired frequency, or, more realistically, of a very narrow range of frequencies centered on the desired frequency; and (3) a detector, usually called a **photometer.** Polychromatic light from the source is passed through both a sample and a reference compartment, although the reference compartment will contain a reference cell filled with pure solvent only if the sample is being studied in solution. As the two polychromatic light beams pass through the monochromator, the monochromator selects one by one the monochromatic frequencies to be transmitted to the photometer. The photometer then determines the intensity of the transmitted light in the sample beam and compares it with that of the reference beam, thereby compensating for variations in source intensity and the

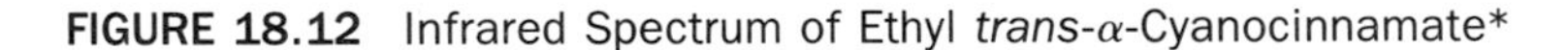

FIGURE 18.12 Infrared Spectrum of Ethyl *trans*-α-Cyanocinnamate*

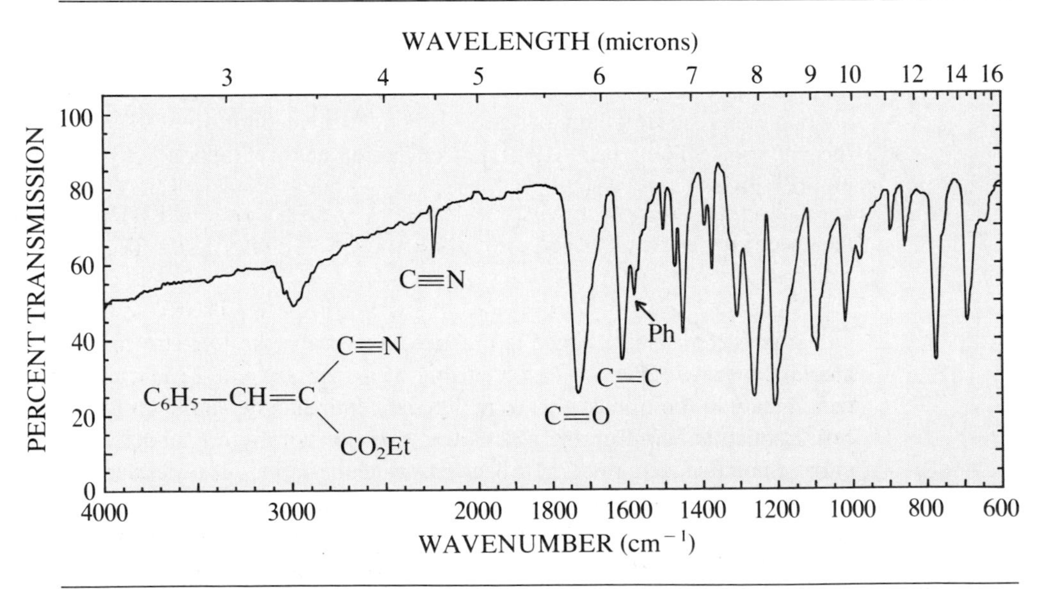

*Run on a solid sample ground with potassium bromide powder and pressed into a clear wafer.

FIGURE 18.13 Block Diagrams of (a) Infrared and (b) Ultraviolet-Visible Spectrometers

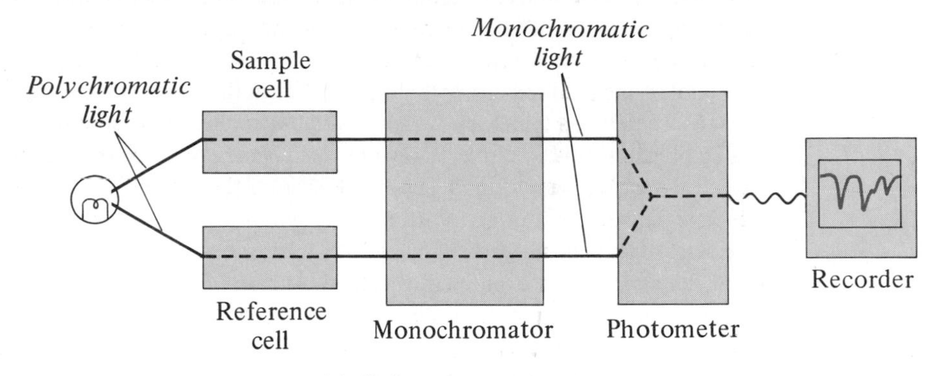

(*a*) Infrared spectrometer

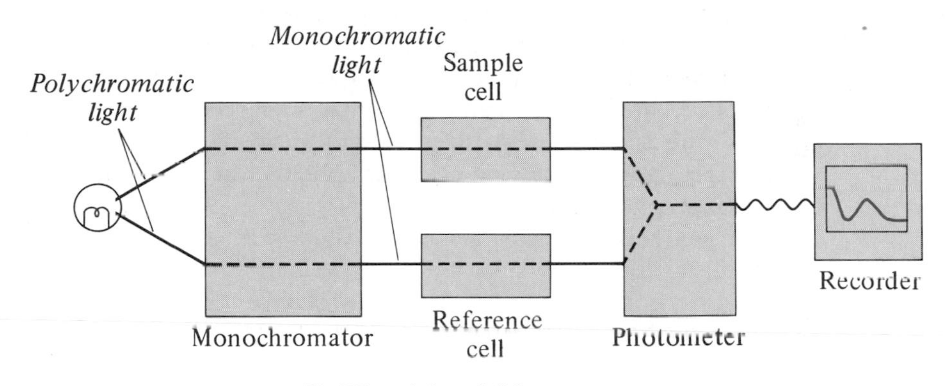

(*b*) Ultraviolet-visible spectrometer

absorption of ir light by the air and by the solvent, if any. Electronic signals from the monochromator (frequency) and photometer (intensity) are fed to a recorder, which plots the spectrum. Infrared spectra may be obtained with samples in solid, liquid, gaseous, or solution form.

Since there are many bonds and bond angles in all but the simplest organic molecules, there are many possible vibrational motions. Thus ir spectra tend to be rather complex. Usually, we do not try to interpret an entire ir spectrum as we would an nmr spectrum, but rather take advantage of the empirical observation that many molecules containing a certain functional group or structural unit will absorb ir light in a rather narrow frequency range characteristic of that group or structural unit. Such absorptions are called **characteristic group frequencies** or vibrations. For example, any compound that contains a carbonyl group, C=O, will have a strong ir band in the region between 1870 and 1630 cm^{-1}. In effect, we may treat *some* molecular vibrations as though they were caused by the stretching or bending of an individual bond.

Although ir spectroscopy can be used as a powerful probe of structure, most organic chemists tend to use ir spectroscopy as an adjunct to nmr spectroscopy for the detection of specific functional groups and structural units rather than for a complete structural analysis. Table 18.2 gives a brief overview of characteristic group frequencies. A somewhat larger table, intended for use with the exercises in this text, may be found inside the back cover. More detailed tables are widely available in chemical handbooks, in special collections in libraries, and in numerous books on spectroscopy. Thus, a typical approach to an ir spectral problem would involve a quick scan of the regions above 1600 cm^{-1} to determine whether or not any of the stretching vibrations of the functional groups listed are present. If there is no absorption due to OH, NH, C≡C, C≡N, C═O, or C═C, then the rest of the spectrum is examined for other structural clues, such as the strong absorption characteristic of ethers in the 1300–1000 cm^{-1} region, or of nitro groups (NO_2) at 1550–1330 cm^{-1}, or of C—Br or C—Cl at 850–650 cm^{-1}. Analysis of the region below 1500 cm^{-1} is more difficult because of the large number of the absorptions of various types that are found in this region, sometimes called the ''fingerprint'' region; that is, each molecule not only tends to have its own ''fingerprint'' in this region but also can often be conclusively identified by this ''fingerprint'' if the spectrum is on record.

Figure 18.12 gives the ir spectrum of a compound often used as an ''unknown'' in the authors' laboratory. This compound, ethyl α-cyanocinnamate, has four functional groups whose stretching vibrations show up fairly clearly in its ir spectrum: the cyano (nitrile) group at 2240, the ester carbonyl group at 1730, the carbon–carbon double bond at 1615, and the phenyl ring (less obviously) at 1580 cm^{-1}. A skilled spectroscopist could identify other stretching or bending vibrations, but those listed have sufficed for the compound's identification by many students.

Apart from its use in structural determination, ir spectroscopy has been used to study electronic effects in molecules by examination of changes in the frequencies of stretching vibrations when electron-withdrawing or electron-releasing groups are introduced on or near a functional group, to study hydrogen bonding (Sec. 7.3), and to explore the effects of the various types of strain in organic molecules (Sec. 2.9).

TABLE 18.2 Correlation of Infrared Absorption Frequencies and Functional Groups

$\tilde{\nu}$ (cm^{-1})	Compounds	Characteristic Group
3700–3100	Alcohols, amines, phenols, amides	OH, NH
3100–2800	Alkanes, alkenes, arenes	CH
3000–2500	Carboxylic acids	OH
2400–2000	Alkynes,* nitriles	C≡C, C≡N
1870–1630	Aldehydes, ketones, carboxylic acids and their derivatives	C═O
1690–1560	Alkenes*	C═C
1615–1515	Arenes	aromatic ring
1300–1000	Ethers, alcohols, phenols, esters	C—O

*Ir peak may be very weak or absent if molecule is symmetrical about double or triple bond.

Exercise 18.6 If an organic compound has the following properties, which functional group(s) may be present in the molecule? (a) a hydrocarbon absorbs strongly at 1660 cm^{-1}; (b) a hydrocarbon shows weak absorption near 2150 cm^{-1} and reacts with water and sulfuric acid in the presence of mercury salts to give a compound absorbing strongly at 1730 cm^{-1}; (c) a hydrocarbon shows absorption at 1650, 990, and 910 cm^{-1} and is cleaved by ozone to a product absorbing at 1730 cm^{-1}; (d) a compound shows no absorption above 1600 cm^{-1} but reacts with alcoholic potassium hydroxide to give a product showing very weak absorption at 1665 cm^{-1} and showing only one peak in its nmr spectrum.

18.4 Electronic Spectroscopy [Ultraviolet-Visible (uv-vis) Spectroscopy]

The absorption of electromagnetic radiation in the ultraviolet (uv) and visible (vis) regions of the electromagnetic spectrum is a property of all organic molecules, and is the result of an electronic excitation in which an electron is promoted from its lowest energy or ground state to a higher electronic state. A common approximation of such excitations is that they correspond to a transition of an electron from a bonding (σ or π) or nonbonding (n) molecular orbital to an antibonding (σ^* or π^*) orbital. The relationships between the energies of the various types of orbitals and the possible transitions between energy levels are shown schematically in Fig. 18.14. As in the other forms of spectroscopy, the exact frequency of "uv-vis" light required to cause an electron to be promoted is given by Planck's Law.

FIGURE 18.14 (a) Electronic Energy Levels and Transitions, (b) $\pi \rightarrow \pi^*$ Transition in 1,3-Butadiene

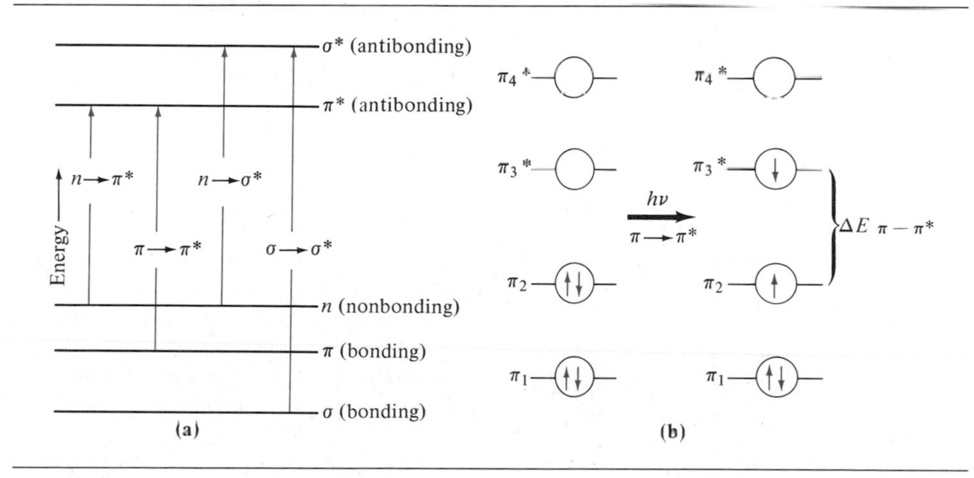

In electronic or uv-vis spectroscopy, the present practice in this country is to use wavelength (λ) rather than frequency (ν) or wave number ($\tilde{\nu}$) units. In other countries wave numbers may be used. The common units of wavelength in uv-vis spectroscopy are either nanometers (nm, formerly millimicrons (mμ)) or angstroms (Å), although the use of the latter is being discouraged.

The uv-vis region includes wavelengths of about 50–800 nm. Because the human eye can see light only in the 400 to 750-nm region, and because below about 190 nm we must employ vacuum (or other) techniques to avoid the strong absorption by atmospheric oxygen, electronic spectroscopy is divided for convenience into three regions: vacuum or far uv (50–200 nm), near uv (200–400 nm), and visible (400–800 nm). We will be concerned with only the latter two regions.

The promotion energies involved in uv-vis spectroscopy range from about 40 to over 250 kcal/mole. Thus, the energy absorbed may be greater than that required for the dissociation of some bonds (Table 2.4). It is for this reason that we sometimes use uv light to initiate free-radical reactions.[2] Indeed, a whole branch of organic chemistry called photochemistry is based on the reactions of organic compounds that take place when these molecules are irradiated with uv-vis light. Similarly, the concern about the possible depletion of the ozone layer in the upper atmosphere due to excessive use of aerosol sprays (Sec. 6.3) or nitrogen-supplying fertilizers is based on the supposition that the strong absorption of uv radiation from the sun by ozone helps protect us from certain types of skin cancer caused by uv-vis light–induced chemical changes (Sec. 16.10).

Electronic excited states have short lifetimes (10^{-7} sec or less). As the excited molecule returns to its ground state, the absorbed energy may be (1) reemitted as light of a longer wavelength than the original exciting radiation (fluorescence or phosphorescence), (2) converted to thermal energy (heat), or (3) used to initiate a photochemical reaction. In part because of these various possibilities, uv-vis spectrometers are generally constructed so that the light from the source is separated into monochromatic light before passage through the sample (Fig. 18.13) rather than after as in ir spectroscopy.

In uv-vis spectroscopy, as in ir spectroscopy, sharp absorption peaks or lines are rarely observed. This is because transitions between electronic energy levels are affected to some extent by the vibrational and rotational motions of the atoms in molecules. Also the number of peaks present may be relatively small, and the variation in their intensity very large. The intensity of absorption in the uv-vis region is proportional to the number of absorbing molecules in the light path (Beer-Lambert law), and is generally described in terms of the following equation:

$$\log \frac{I_0}{I} = \varepsilon c l = A$$

I_0 = intensity of incident light
I = intensity of transmitted light
c = concentration in moles/liter
l = path length of cell in centimeters
A = absorbance reading from the spectrometer
ε = molar absorption coefficient (molar extinction coefficient)

[2] As an example we can rewrite the first equation of Sec. 2.8-B as a $\sigma \rightarrow \sigma^*$ transition followed by bond cleavage: $\mathrm{Cl{-}Cl} \xrightarrow{uv} \mathrm{Cl\overset{*}{\cdot}Cl} \longrightarrow 2\ \mathrm{Cl}\cdot$

Usually uv spectra are reported in terms of the values of the wavelength (λ_{max}) at which ε reaches a maximum (ε_{max}). A spectrum may contain several λ_{max} and their corresponding ε_{max}. The intensity of uv-vis bands (as measured by ε_{max}) varies widely (over a range of 1–10^5); therefore, more than one tracing may be required to display the spectrum.

Of the electronic transitions represented in Figure 18.14, only the $n \rightarrow \pi^*$ and $\pi \rightarrow \pi^*$ transitions are generally useful, as the other transitions fall below the wavelength limits of common instrumentation or have very low ε_{max} values. The $n \rightarrow \pi^*$ transition involves the excitation of an electron from a nonbonding (*n*) orbital (unshared pair orbital) to an antibonding π^* orbital. These low-intensity (ε about 10–100) bands occur at relatively long wavelengths and are important factors in determining the colors of certain dyes and pigments. A compound showing an $n \rightarrow \pi^*$ transition band usually contains one or more of the following structural units:

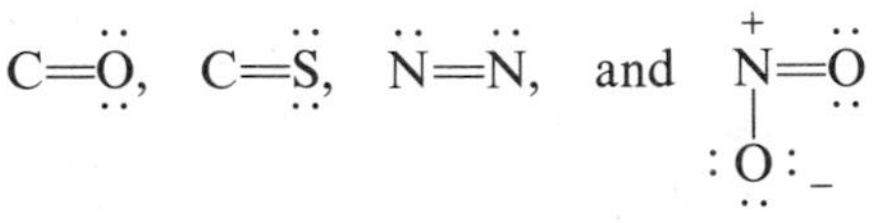

$$\mathrm{C{=}\ddot{\underset{..}{O}}},\quad \mathrm{C{=}\ddot{\underset{..}{S}}},\quad \mathrm{\ddot{N}{=}\ddot{N}},\quad \text{and}\quad \overset{+}{\mathrm{N}}{=}\ddot{\underset{..}{\mathrm{O}}}\ \ (\mathrm{N{-}:\ddot{\underset{..}{O}}:^{-}})$$

The most important transitions from the standpoint of structure determination are the $\pi \rightarrow \pi^*$ transitions, which involve the excitation of an electron from a π orbital to a π^* orbital. Compounds having double or triple bonds or aromatic rings undergo such transitions; however, the absorption bands may fall outside the limits of our instrument unless the π system is a conjugated system. As the length of the conjugated system increases, the wavelength and usually the intensity of absorption increase. The yellow-to-orange color of carotene (Sec. 3.13) (λ_{max}, 452 nm) from carrots and the red color of lycopene (λ_{max}, 469) from tomatoes and paprika are due to the extended conjugation of the system. The explanation for this phenomenon is that, as the π system becomes more extended, the energy difference between the highest occupied orbital (π) and the lowest unoccupied (π^*) orbital decreases. The uv-vis spectra of retinol (Vitamin A) (five conjugated double bonds) and *trans*-β-carotene (eleven conjugated double bonds) shown in Figure 18.15 on the following page and the listing of some typical absorption bands in Table 18.3 (page 501) illustrate these observations.

A structural unit in a molecule that can undergo an electronic transition is called a **chromophore** (Gr. *chroma,* light; *phorein,* to bear). A nonchromophoric group, usually containing unshared electron pairs, that causes a shift in the wavelength of an absorption peak and usually an increase in intensity is called an **auxochrome** (Gr. *auxanein,* to increase). The absorption bands of the chromophores in a number of simple molecules are given in Table 18.3. Chromophores and auxochromes are also important in determining the colors of dyes. Of course, when a dye absorbs visible light, what the eye sees is the complement of the color absorbed. Thus, absorption of the longer wavelengths (red, orange, yellow) results in a blue or blue-green color; absorption of the shorter wavelengths (violet, blue) results in an orange or yellow color. Furthermore, if the absorption band is narrow and intense, the color will be brilliant and clear; however, if the band is broad or if there are many overlapping bands, the color will be muddy or dull. The

FIGURE 18.15 Ultraviolet Spectra of Retinol (Vitamin A, 1.32×10^{-5} M in Methanol) and *trans*-β-Carotene (5.96×10^{-6} M in Hexane)

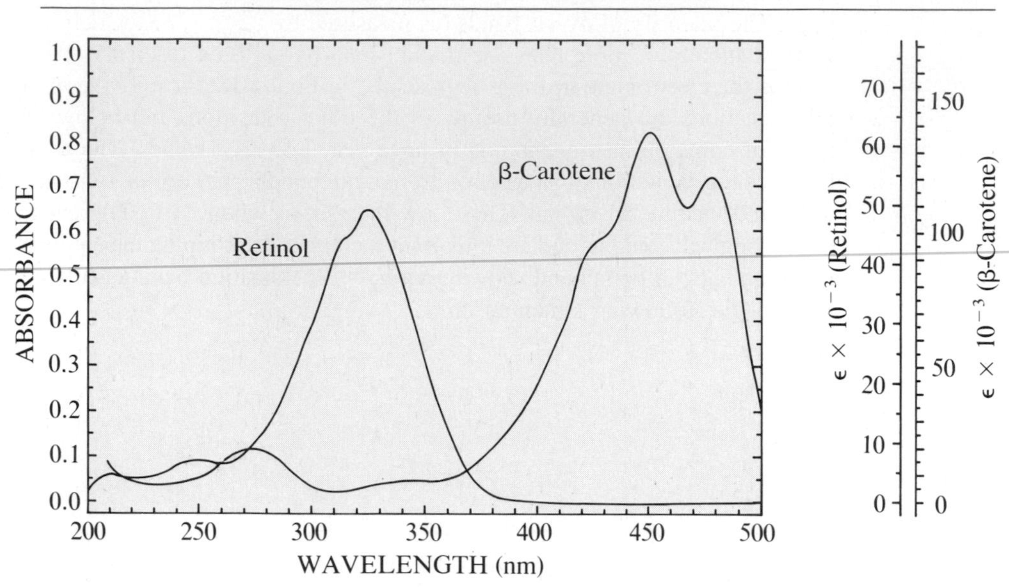

following are examples of common chromophores.

—N=O (Nitroso) —NO_2 (Nitro) —N=N— (Azo) —C(=O)—(CH=CH)$_n$—C(=O)— (Dicarbonyl, (n = 0 or some integer)) *p*-Quinoid *o*-Quinoid

Common auxochromes are the —OH, —OR, —NH_2, and —NR_2 groups. The auxochromes have at least one unshared pair of electrons that may interact with and extend any conjugation already present.

The most common structural application of uv-vis spectroscopy is the detection of conjugation and the determination of its nature and extent. However, at present the analytical applications far outnumber structural applications, and uv-vis spectrometry is widely used in the assay of vitamins, antibiotics, enzyme activity, etc., and in the determination of rates of organic reactions. Measurement of the absorption at 280 nm is a rapid and convenient method of estimating the protein content of a solution, and absorption at 260 nm has been used to assay nucleotide (Sec. 16.4) content and to monitor the denaturation of DNA (Sec. 16.5).

TABLE 18.3 Electronic Absorption Bands

Compound	Transition	λ_{max} (nm)	ε_{max}	Solvent
$CH_3—CH_3$	$\sigma \longrightarrow \sigma^*$	135	7000	vapor
CH_3CH_2OH	$n \longrightarrow \sigma^*$	181	325	vapor
CH_3I	$n \longrightarrow \sigma^*$	259	400	hexane
$(CH_3CH_2)_2NH$	$n \longrightarrow \sigma^*$	193	193	hexane
$CH_3CH{=}CHCH_3$ (*trans*)	$\pi \longrightarrow \pi^*$	178	13,000	vapor
$CH_3(CH{=}CH)_3CH_3$	$\pi \longrightarrow \pi^*$	275	30,000	hexane
$CH_3(CH{=}CH)_4CH_3$	$\pi \longrightarrow \pi^*$	310	76,500	hexane
$CH_3(CH{=}CH)_5CH_3$	$\pi \longrightarrow \pi^*$	342	122,000	hexane
$CH_3(CH{=}CH)_6CH_3$	$\pi \longrightarrow \pi^*$	380	146,500	chloroform
$HC{\equiv}CH$	$\pi \longrightarrow \pi^*$	173	600	vapor
$CH_3—C(=O)—CH_3$	$\pi \longrightarrow \pi^*$	190	1000	cyclohexane
	$n \longrightarrow \pi^*$	275	22	
$CH_3—C(=O)—C(=O)—CH_3$	$\pi \longrightarrow \pi^*$	282	19	ethanol
	$n \longrightarrow \pi^*$	420	10	
$CH_3—C(=O)OH$	$n \longrightarrow \pi^*$	204	41	ethanol

Compound	λ_{max} (nm)	ε_{max}	λ_{max} (nm)	ε_{max}	λ_{max} (nm)	ε_{max}	Solvent
Benzene	184	68,000	204	8,800	254	250	hexane
Chlorobenzene			210	7,500	257	170	ethanol
Toluene	189	55,000	208	7,900	262	260	hexane
Aniline			230	8,600	280	1,400	water
			203	7,500	254	2160	water + HCl
Phenol			211	6,200	270	1,450	water
			236	9,400	287	2,600	water + NaOH
Nitrobenzene*			252	10,000	280	1,000	hexane
4-Nitrophenol			226	6,900	318	10,000	water
Acetophenone†			243	13,000	279	1,200	ethanol

*Also an $n \rightarrow \pi^*$ band at 330 nm (ε 140).
†Also an $n \rightarrow \pi^*$ band at 315 nm (ε 55).

Exercise 18.7 Phenol is colorless, and nitrobenzene is pale yellow, but 4-nitrophenol is bright yellow. Draw the structure of 4-nitrophenol and label the chromophores and auxochromes in this molecule. Write resonance forms for 4-nitrophenol that show how the auxochrome interacts with the chromophore(s) to extend the conjugated system.

18.5 Mass Spectroscopy; X-Ray Crystallography

Two other methods for structural investigation are of such importance to the organic chemist that they require at least brief mention: mass spectroscopy and X-ray crystallography. In mass spectroscopy a very small sample of an organic compound is bombarded by a high-energy beam of electrons in a positively charged, evacuated chamber. The electron beam removes an electron from the molecule, forming a radical cation $M^{+\cdot}$, called the **molecular ion.** This ion generally has so much excess energy that one or more of its bonds break, yielding a large number of **fragment ions** (and neutral radicals or smaller molecules).

$$\underset{\text{Molecule}}{R''\!-\!\overset{\overset{\textstyle R'}{|}}{\underset{\underset{\textstyle R'''}{|}}{C}}\!:\!R} - e^- \longrightarrow \underset{\text{Molecular ion}}{\left[R''\!-\!\overset{\overset{\textstyle R'}{|}}{\underset{\underset{\textstyle R'''}{|}}{C}}\!\cdot\!R\right]^{+\cdot}} \longrightarrow \underset{\text{Radical}}{R\cdot} + \underset{\text{Fragment ion}}{R''\!-\!\overset{\overset{\textstyle R'}{|}}{\underset{\underset{\textstyle R'''}{|}}{C}}\!+}$$

The positively charged ions are repelled by the high positive charge on the chamber and are ejected out of the chamber and propelled at high velocity down a curved tube into the field of an electromagnet. In a magnetic field a charged particle follows a curved path dependent on the *intensity* of the field and the *mass* of the particle. By varying the field intensity in a regular fashion, each positive ion's path can be controlled to cause that ion to strike an ion collector, which counts the ions electronically. A recorder in the system plots the intensity (ion count) of each ion against its mass number (that is, the magnetic field intensity required to cause the particle to strike the collector). The result is a mass spectrum, which consists of a series of many peaks at the mass numbers of the fragment ions. Molecules tend to fragment in predictable ways; thus, the fragmentation pattern shown by the mass spectrum can often be used to deduce the structure of the molecule.

Furthermore, we can often determine not only the molecular weight of the molecule but also its molecular formula. This is done by measuring the **exact mass** of the molecular ion to about four decimal places. Since the actual masses of the common atoms are

H = 1.007825, C = 12.000000, O = 15.994915, and N = 14.003074, the exact masses can be used to distinguish between molecular ions having the same nominal mass. For example, cyclohexanone and 1,3-cyclopentanedione have the same nominal molecular weight, 98. However, their exact molecular weights are sufficiently different, as shown below, that we can distinguish between the two compounds by mass spectroscopy.

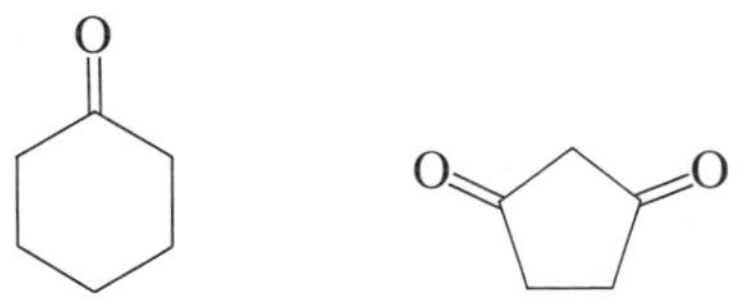

	Cyclohexanone	1,3-Cyclopentanedione
Molecular formula:	$C_6H_{10}O$	$C_5H_6O_2$
Molecular weight:	98.0732	98.0368

Because of the large number of fragment ions and the many possible fragmentation pathways, analysis of mass spectra can be complex. However, it is often relatively easy to explain the presence of the more intense peaks in a mass spectrum using the ideas developed earlier in this text. Thus, we would expect that radical cations would cleave so as to give principally the more stable cationic fragments. Many fragmentations involve the cleavage of a single bond, as shown above. In these, we expect simple cations to form most easily in the order: tertiary > secondary > primary > methyl. We expect resonance-stabilized cations, such as allyl and benzyl, to form more readily than those lacking this stabilizing factor. For example, the following fragmentations are often observed:

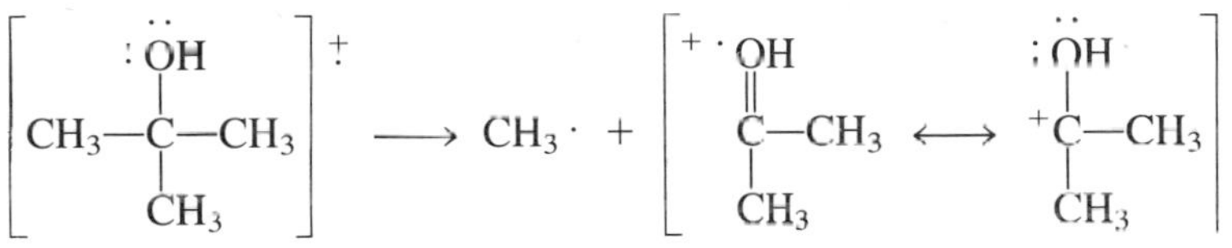

$$[(CH_3)_3C\text{–}\ddot{O}H]^{+\cdot} \longrightarrow CH_3\cdot + [(CH_3)_2C{=}\overset{+}{O}H \longleftrightarrow (CH_3)_2\overset{+}{C}\text{–}\ddot{O}H]$$

protonated acetone cation

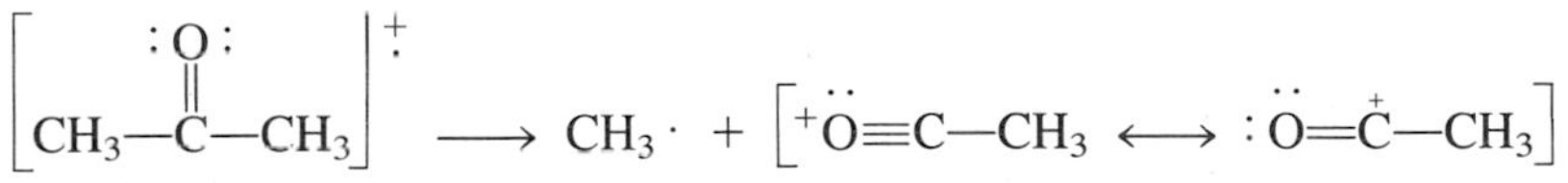

$$[CH_3\text{–}C(=\ddot{O})\text{–}CH_3]^{+\cdot} \longrightarrow CH_3\cdot + [\overset{+}{O}{\equiv}C\text{–}CH_3 \longleftrightarrow \ddot{O}{=}\overset{+}{C}\text{–}CH_3]$$

acylium cation

$$[C_6H_5\text{–}CH_2CH_2CH_3]^{+\cdot} \longrightarrow C_6H_5\text{–}CH_2^+ + \cdot CH_2CH_3$$

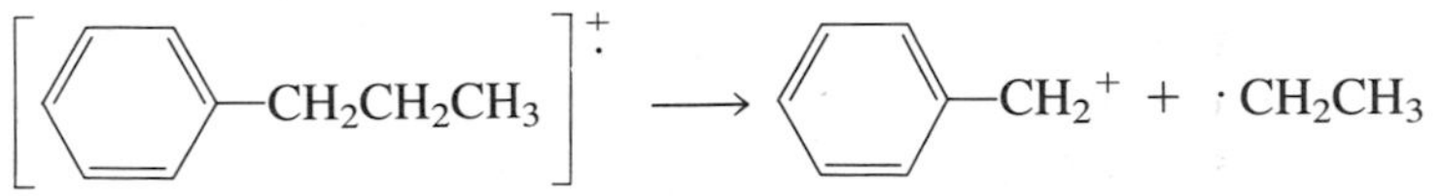

benzyl cation

When two bonds are broken in a fragmentation process, elimination of a neutral molecule is a common pathway, as shown in the following examples:

Reverse Diels–Alder

Limonene radical cation → Isoprene + Isoprene radical cation

$[CH_2CH_2CH_2C(=O)CH_2CH_2CH_3]^{+\cdot} \longrightarrow CH_2{=}CH_2 + [CH_2{=}C(OH)CH_2CH_2CH_3]^{+\cdot}$

Ethene an enol radical cation

The last of the above examples illustrates a common fragmentation pathway of ketones and esters called the McLafferty rearrangement.

Exercise 18.8 Using the examples given above as models, account for the appearance of the peaks at m/e values (mass-to-charge ratios)[3] of 45 and 59 in the mass spectrum of 2-butanol [Fig. 18.16(a)]. (The molecular ion may be a very weak peak or even absent in the mass spectra of alcohols.) Also account for the appearance of the peaks at m/e values of 43, 58, and 85 in the mass spectrum of 2-hexanone [Fig. 18.16(b)].

When a beam of X rays is directed at a *single crystal* of an organic compound at the proper angle, the X rays are scattered by the atoms of the molecule, which are arranged in a regular pattern throughout the crystal. The scattered X rays tend to be arranged in a pattern of alternating regions of high and low intensity, depending on the arrangement of the atoms in the crystal. If the scattered X rays are allowed to fall on a piece of photographic film, an array of spots of varying darkness (density) in a definite pattern, called a diffraction pattern, appears on the developed film. If the angle of the X-ray beam is varied in a regular way, the nature of the pattern changes. These patterns may also be collected

[3]The position of a peak in a mass spectrum depends on both the mass and the charge of the cation; that is, the mass spectrometer cannot distinguish between a cation with a mass of 100 and a charge of +1 and a cation with a mass of 200 and a charge of +2. Although cations with charges greater than +1 are rare, masses are reported as mass/charge values or m/e values.

FIGURE 18.16 Mass Spectra of (a) 2-Butanol and (b) 2-Hexanone

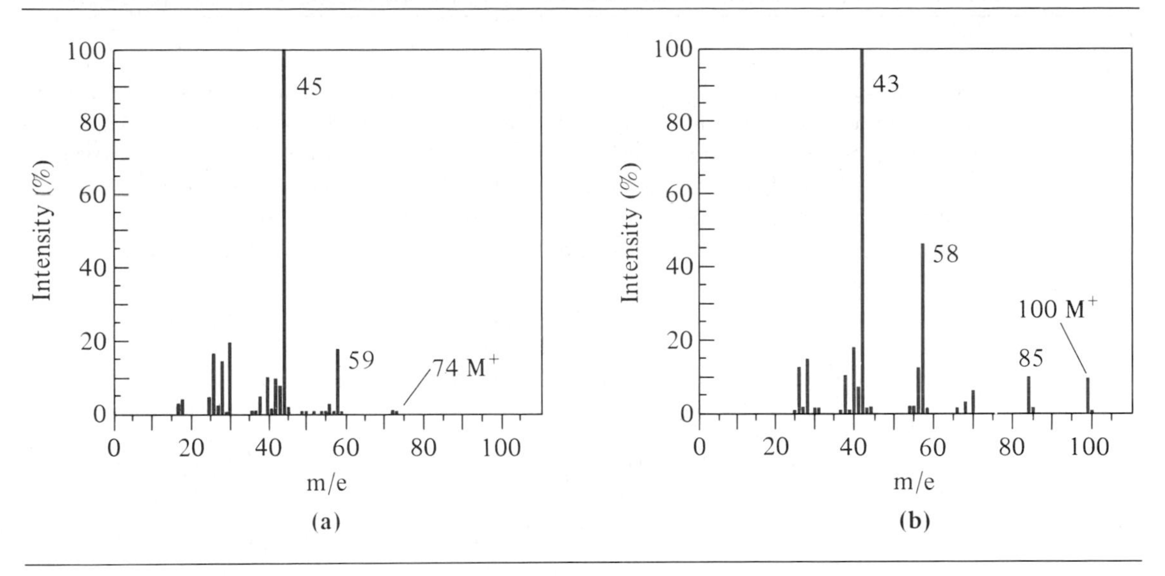

electronically and fed directly to a small computer. Analysis of thousands of the spots by the computer gives an unequivocal structure for the molecule in many instances. The analysis takes from several days for a small molecule to many months for a large molecule such as a protein. However, for small molecules, not only does the computer give accurate bond lengths and bond angles, but also from a recorder connected to the computer we can obtain a perspective drawing of the molecule. An example of such a drawing is given below (left). X-ray crystallography is the ultimate technique for the determination of the precise geometry of molecules in the crystalline state below (right).

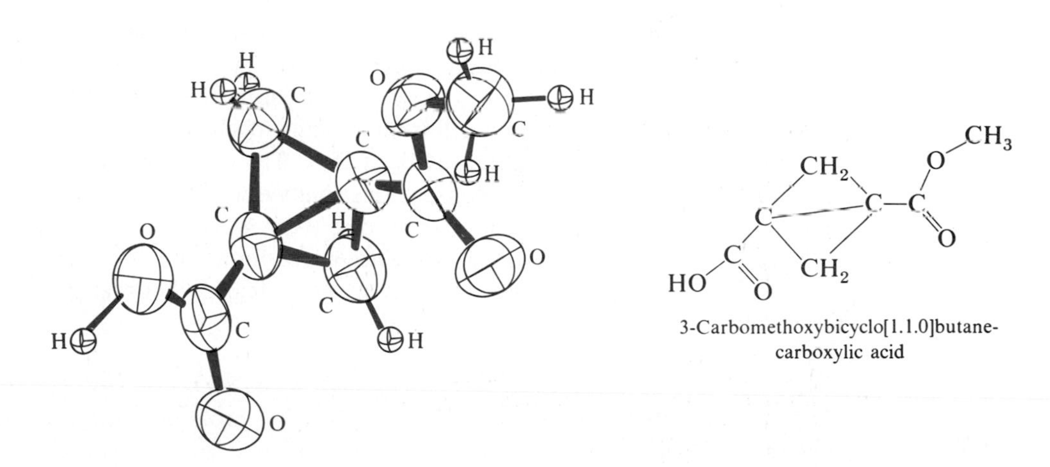

18.6 Solving a Structural Problem

Let us now apply some of the techniques described in the previous sections to the characterization of an ''unknown'' compound, that is, to the determination of its structure. This compound is a colorless liquid (bp, 221°), containing only carbon, hydrogen, and oxygen. The mass spectroscopy laboratory reports that the exact mass of the molecular ion is 148.0884 ± 0.0015 and that, based on this exact mass, the molecular formula is probably $C_{10}H_{12}O$. The ir, nmr, and uv spectra of the compound are given in Figures 18.17(a), (b), and (c) respectively.

Before analyzing the spectra, it would be wise for us to use the molecular formula to calculate the number of rings and/or double bonds in the molecule, which we will represent by the symbol $\ominus$ (a ring with a double bond through it). For a molecule of the formula, $C_cH_hN_nO_oX_x$ (where X = halogen)

$$\ominus = \text{number of rings and/or double bonds} = c - \frac{h}{2} + \frac{n}{2} - \frac{x}{2} + 1$$

The formula gives the total number of rings and/or double bonds (C=C, C=O, N=N, etc.), where a triple bond (C≡C or C≡N) is equivalent to two double bonds and a benzene ring is equivalent to a ring plus three double bonds. For our ''unknown''

$$\ominus = 10 - \frac{12}{2} + 1 = 5$$

Generally, a molecule with such a high value of $\ominus$ is likely to be aromatic; however, the possibility of an unsaturated acyclic compound cannot be ignored.

Next we turn to the ir spectrum to determine, if we can, what functional groups may be present. The strong band at 1690 cm^{-1} is characteristic of the C=O group, according to Table 18.2. From the table inside the back cover we find that the aromatic ketones (ArCOR, 1700–1650 cm^{-1}) have absorption bands in the same region as our unknown. The bands at 3100, 1600, and 1480 cm^{-1} are suggestive of an aromatic compound.

Rather than try to analyze the ir spectrum in greater detail, which could be done in this instance, we next examine the nmr spectrum. The one-proton septet at δ 3.47 requires six neighboring protons (two CH_3 groups?), and the six-proton doublet at δ 1.17 requires a single neighboring proton. Putting this information together we conclude that our compound contains an isopropyl group, $(CH_3)_2CH$—. The complex multiplet between 440 and 480 Hz is probably that of a benzene ring, but we will not try to analyze this higher order multiplet in detail. With a benzene ring ($\ominus$ = 4) and an isopropyl group and a carbonyl (C=O) group ($\ominus$ = 1), we have accounted for all of our carbon atoms, hydrogen atoms, and the oxygen atom as well as our $\ominus$ value of 5. Thus, at this stage we have the following structual units

>C=O possibly Ar—C(=O)—R

[benzene ring] probably monosubstituted

$(CH_3)_2CH$—

FIGURE 18.17 (a) Infrared Spectrum of an Unknown (Neat Liquid)

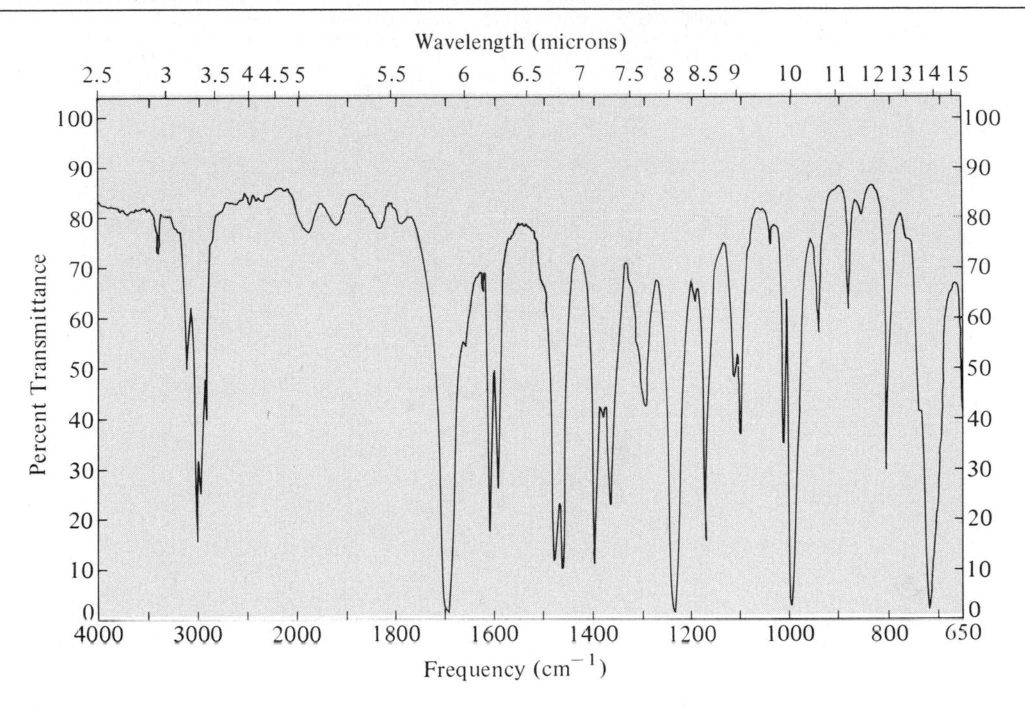

(b) nmr Spectrum of an Unknown ($CDCl_3$ Solution)

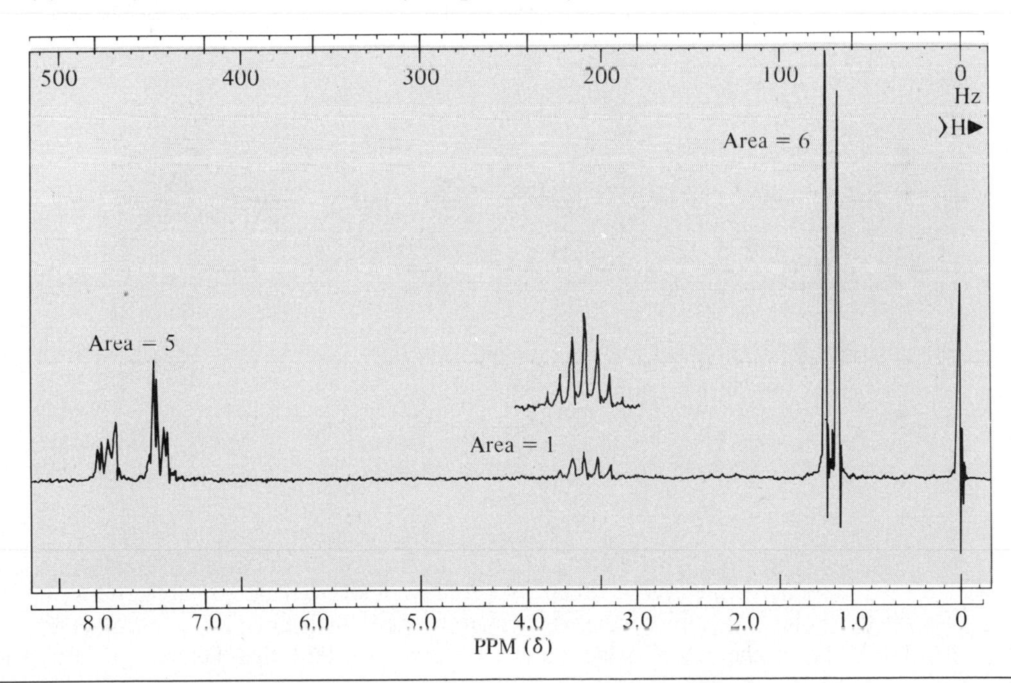

FIGURE 18.17 *continued* (c) Ultraviolet Spectrum of an Unknown (Ethyl Alcohol Solution)

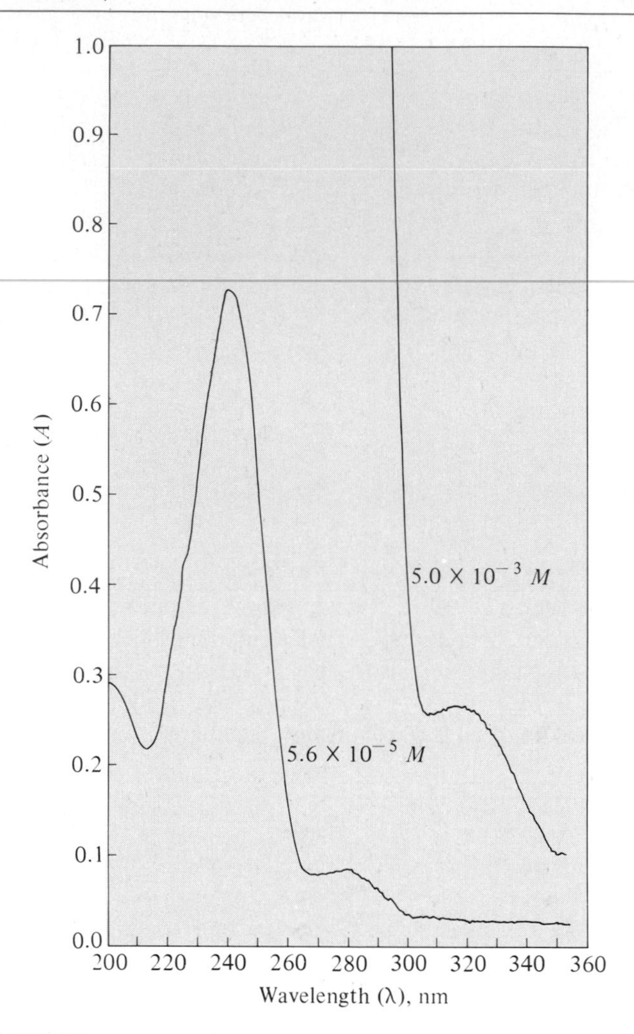

Putting the units together in a logical fashion we arrive at the structure

$$C_6H_5-\overset{\overset{\large O}{\|}}{C}-\underset{CH_3}{\overset{CH_3}{C}}-H$$

Isopropyl phenyl ketone

If our structure is to be acceptable, it must fit all the data, not just those we have used to this point. Therefore, we next check to be sure that the chemical shift data are reasonable. From the table inside the back cover we find that ketones of the structure

$R_2CH—COAr$ have chemical shifts near δ 3.5, which is quite close to the observed value, δ 3.47. The doublet at δ 1.17 has no exact counterpart in the table but is within the range of δ 0.9–1.5 for alkyl groups attached to carbon. Thus the chemical shift values are reasonable ones for isopropyl phenyl ketone.

The uv spectrum serves to confirm our assignment of structure. The bands at 240 and 280 nm are characteristic of substituted benzenes, and the very weak band at 318–320 nm could be the $n \rightarrow \pi^*$ band of a ketone. Indeed, the spectrum appears to resemble that of acetophenone (methyl phenyl ketone) (Table 18.3).

All of the data seem to be consistent with the structural assignment that we have made. Naturally, not all structural problems are so readily solved, but many are no more difficult than this one, and all structural problems look simpler as we learn through practice and experience.

Summary

1. Organic absorption spectroscopy is based on the correlation of structure with the selective absorption of electromagnetic radiation by organic molecules.
2. The principal types of absorption spectroscopy used in organic chemistry are
   (a) Nuclear magnetic resonance (nmr) spectroscopy, in which absorption is related to the magnetic environment of protons or other "magnetic" nuclei in the molecule.
   (b) Infrared (ir) spectroscopy, in which absorption is related to the vibrations of bonds in the molecule.
   (c) Electronic (ultraviolet-visible, (uv-vis)) spectroscopy, in which absorption is related to the promotion of electrons from the lowest (ground) electronic energy state to a higher electronic state.
3. Nuclear magnetic resonance spectroscopy:
   (a) The **chemical shift** is the difference in the **resonance frequencies** of two nuclei at a constant value of the magnetic field.
   (b) Proton chemical shifts are usually measured relative to a standard, tetramethylsilane (TMS).
   (c) Proton chemical shifts are characteristic of the chemical environment of the protons. Protons in the same chemical environment are said to be chemically equivalent (chemically indistinguishable).
   (d) The area under an nmr peak is roughly proportional to the number of protons responsible for the peak.
   (e) Protons on adjacent carbon atoms may **couple** (or exchange spin information) with each other, resulting in a **splitting** of singlets into doublets, triplets, etc.
   (f) In simple systems the number of peaks (multiplicity) in a multiplet will be $n + 1$, where n is the number of protons on the adjacent carbon atom(s).
   (g) If a proton changes location or position in a molecule rapidly, the nmr spectrometer reports an average position; if it changes slowly, the spectrometer may report both (or all) positions as separate peaks.

4. Infrared spectroscopy:
   (a) The frequency of absorbed infrared light can be correlated with the **stretching** and **bending** motions (vibrations) of bonds in functional groups and small, specific structural units.
   (b) Many molecules containing a given functional group or structural unit will absorb infrared light of a **characteristic group frequency;** therefore, characteristic group frequencies can be used to establish the presence or absence of specific functional groups.
5. Electronic spectroscopy:
   (a) The absorption of uv-vis light depends on the presence in the molecule of **chromophores,** structural units that can undergo an electronic transition.
   (b) Transitions occur from occupied bonding or nonbonding orbitals to unoccupied antibonding orbitals.
   (c) The most useful transitions are the $n \rightarrow \pi^*$ and $\pi \rightarrow \pi^*$ transitions.
   (d) The $\pi \rightarrow \pi^*$ transitions may be used in the detection of conjugation and the determination of its nature and extent.
   (e) **Auxochromes** are nonchromophoric groups that cause a shift in the wavelength of an absorption peak.
6. Mass spectroscopy may be used to determine molecular weight and molecular formula and to provide structural information based on the fragmentation patterns of organic molecules.
7. X-ray crystallography is used to determine the precise geometry of molecules in the crystalline state.

Supplementary Exercises

18.9 The nmr pattern of an ethyl group that is not coupled to any other protons is one of the most common patterns in nmr spectroscopy. A typical example is the ethyl group in ethyl iodide (CH_3CH_2I). The nmr spectrum of ethyl iodide consists of two groups of peaks at δ 1.83 and δ 3.20, respectively. Answer the following questions about this spectrum:

(a) What is the multiplicity of the group of peaks at δ 1.83?
(b) What is the multiplicity of the group of peaks at δ 3.20?
(c) Which group of peaks corresponds to the CH_3 group?
(d) Which group of peaks corresponds to the CH_2 group?
(e) What is the chemical shift of the CH_2 group in Hz at 60 MHz?
(f) What is the chemical shift of the CH_2 group in Hz at 100 MHz?
(g) The coupling constant (J) in the 60-MHz spectrum is 7 Hz. What is the coupling constant in the 100-MHz spectrum?

18.10 The *approximate* relative areas of peaks in a multiplet can be predicted by using Pascal's triangle (a memory device used in algebra to remember the binomial coefficients). The triangle may be extended to any size using only addition. Examine the first four rows and add three more rows.

n	Pascal's Triangle	Example
0	1	n = number of neighboring protons
1	1 1	For $n = 2$, triplet is expected
2	1 2 1	From triangle, triplet areas = 1:2:1
3	1 3 3 1	

18.11 Using lines of appropriate height and spacing (δ and J), sketch the nmr spectra expected for the following compounds.

(a) $CH_3—CH(Br)—CH_3$

(b) $CH_3—O—CH_2CH_3$

(c) $CH_3—CH(Br)—CH(Br)—CH_3$

(d) $H—C{\equiv}C—CH_2—Br$

(e) $Cl—CH_2—CH_2—CH_2—Cl$

(f) $CH_3—CH(Br)—CH_2—CH(Br)—CH_3$

18.12 Estimate the chemical shifts (in Hz and in δ values) and coupling constants (in Hz) for the peaks in the 60-MHz spectra in:

(a) Figure 18.6
(b) Figure 18.7
(c) Figure 18.17(b)

18.13 Calculate ε_{max} for *trans*-β-carotene (Fig. 18.15) (cell path length = 1.0 cm). Check your results against the ε scale provided at the right side of the spectrum.

18.14 Calculate ε_{max} for each band in the uv spectrum of isopropyl phenyl ketone [Figure 18.17(c)] (cell path length = 1.0 cm).

18.15 Hooke's law for relating the frequency of vibration of two masses, m_1 and m_2, connected by a spring of force constant (strength) k, is sometimes used to illustrate the validity of the ball and spring model for molecular stretching vibrations. Reduced to molecular dimensions and units, Hooke's law tells us that

$$\tilde{\nu} = \frac{1}{2\pi c}\sqrt{\frac{k(m_1 + m_2)}{m_1 \times m_2}} = 1303\sqrt{\frac{k(m_1 + m_2)}{m_1 \times m_2}}\,cm^{-1}$$

where k is the bond force constant (strength of the bond). For single, double, and triple bonds k = 5, 10, and 15 mdyne/Å (the proper units for the equation), respectively. Using this approximation, calculate the stretching frequencies of the following bonds. Compare your results with the values in Table 18.2.

(a) C=C
(b) C≡C
(c) C≡N
(d) C—H
(e) O—H
(f) C=O
(g) C—O
(h) N—H

18.16 Compound A contains nitrogen and chlorine and has a molecular weight of 89.5. The compound gave the ir and nmr spectra shown in Figure 18.18. From these data deduce the structure of compound A.

FIGURE 18.18 (a) Infrared Spectrum of Compound A (Neat Liquid)

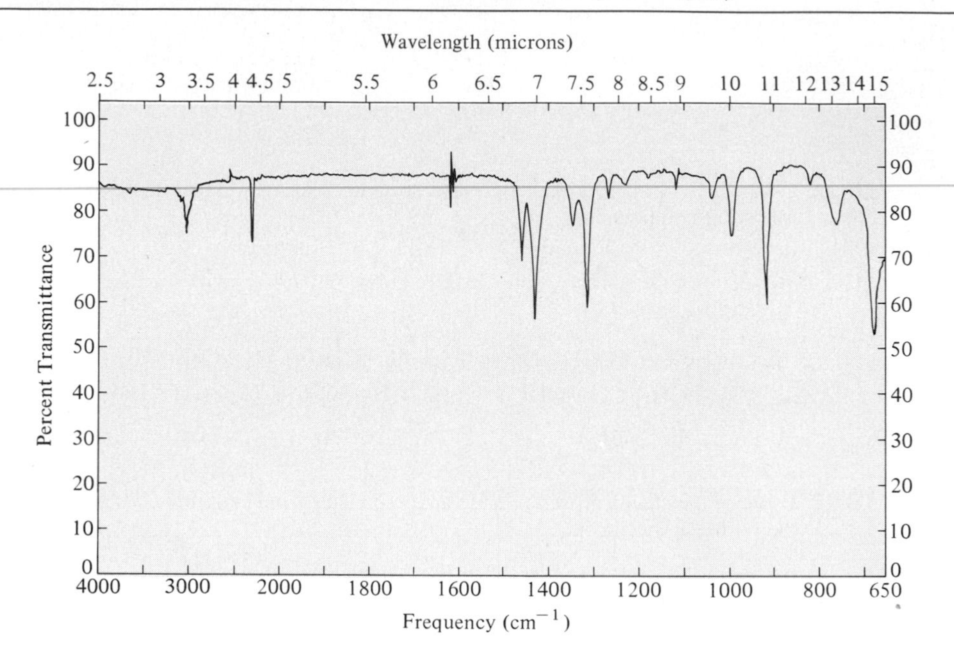

(b) nmr Spectrum of Compound A (CCl_4 Solution)

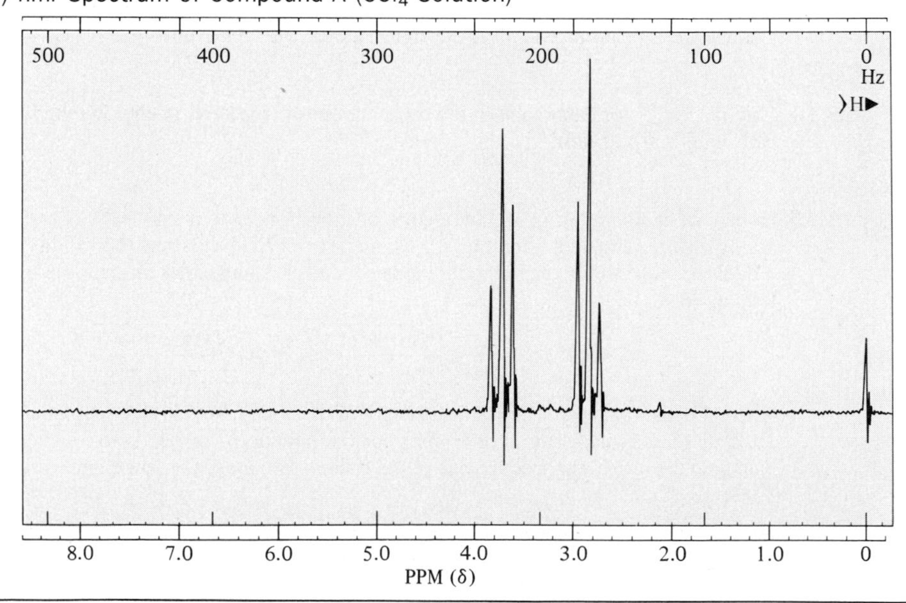

18.17 Compound B, $C_9H_8O_2$, gave the ir and nmr spectra shown in Figure 18.19. Suggest a reasonable structure for compound B, including the proper stereochemistry, if necessary.

FIGURE 18.19 (a) Infrared Spectrum of Compound B ($CHCl_3$ Solution, 1.0-mm Cell)

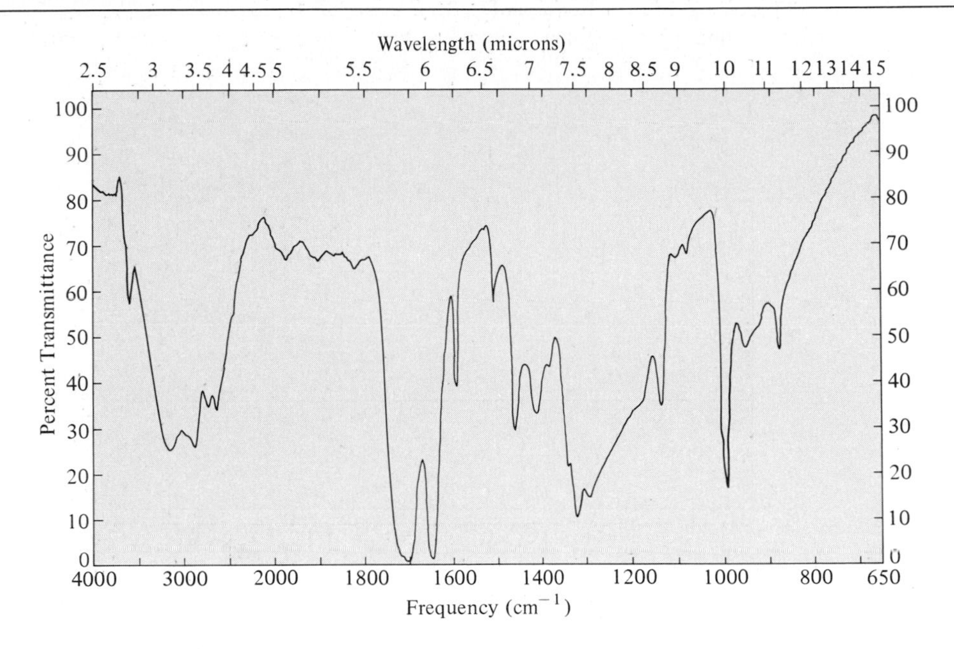

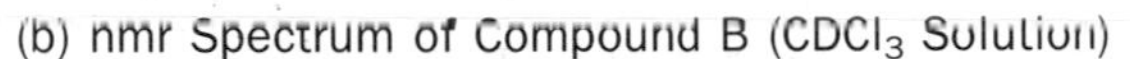
(b) nmr Spectrum of Compound B ($CDCl_3$ Solution)

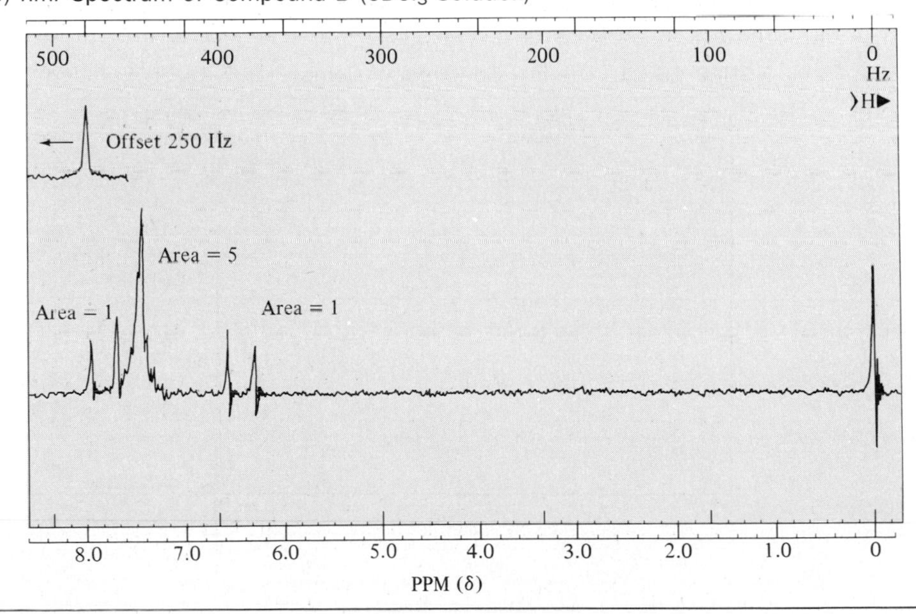

18.18 Compound C, $C_5H_{11}Cl$, gave the nmr spectrum shown in Figure 18.20. Suggest a reasonable structure for compound C.

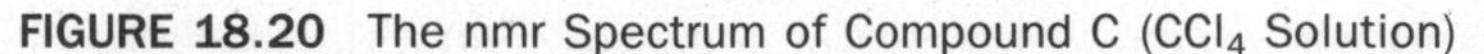

FIGURE 18.20 The nmr Spectrum of Compound C (CCl_4 Solution)

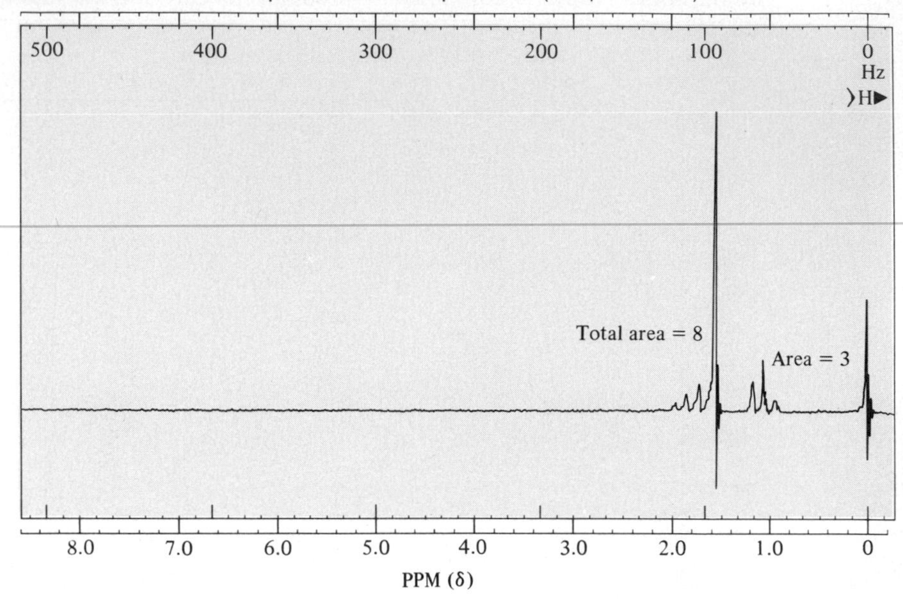

18.19 Compound D, $C_4H_9O_2Cl$, gave the nmr spectrum shown in Figure 18.21. Suggest a reasonable structure for compound D.

FIGURE 18.21 The nmr Spectrum of Compound D ($CDCl_3$ Solution)

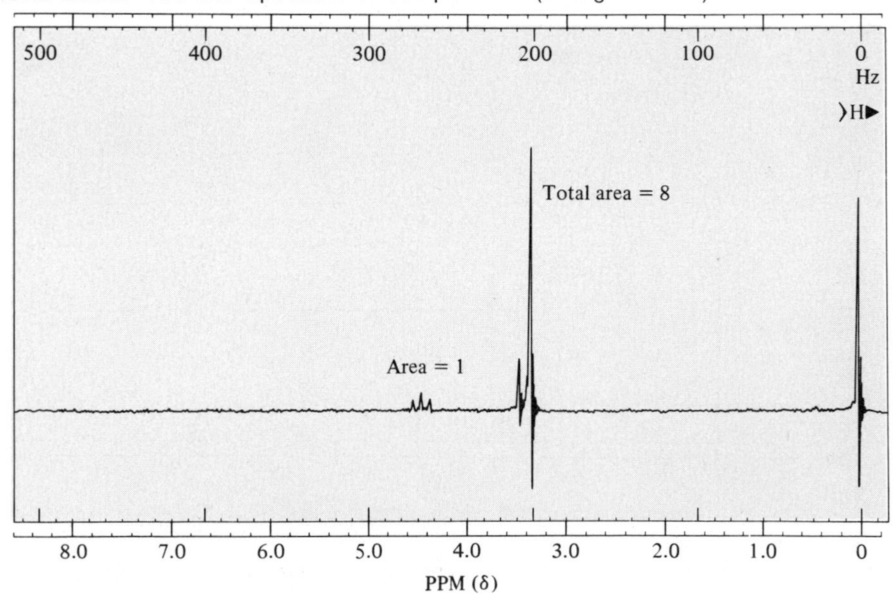

18.20 Compound E, $C_5H_{11}Cl$, gave the nmr spectrum shown in Figure 18.22. Suggest a reasonable structure for compound E.

FIGURE 18.22 The nmr Spectrum of Compound E ($CDCl_3$ Solution)

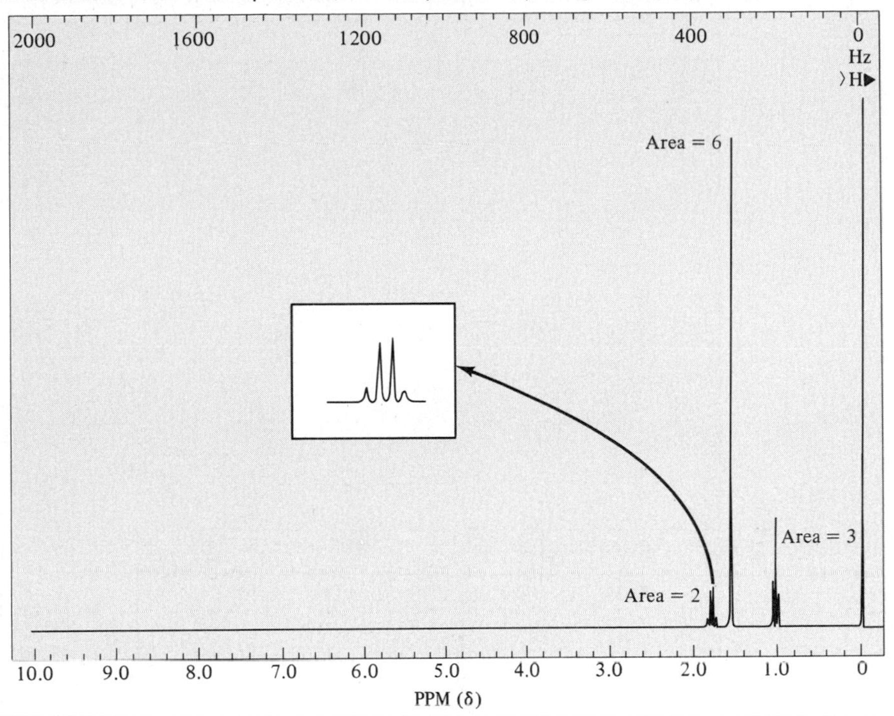

18.21 Compound F, $C_5H_9O_2Br$, gave the nmr spectrum shown in Figure 18.23. Suggest a reasonable structure for compound F.

FIGURE 18.23 The nmr Spectrum of Compound F ($CDCl_3$ Solution)

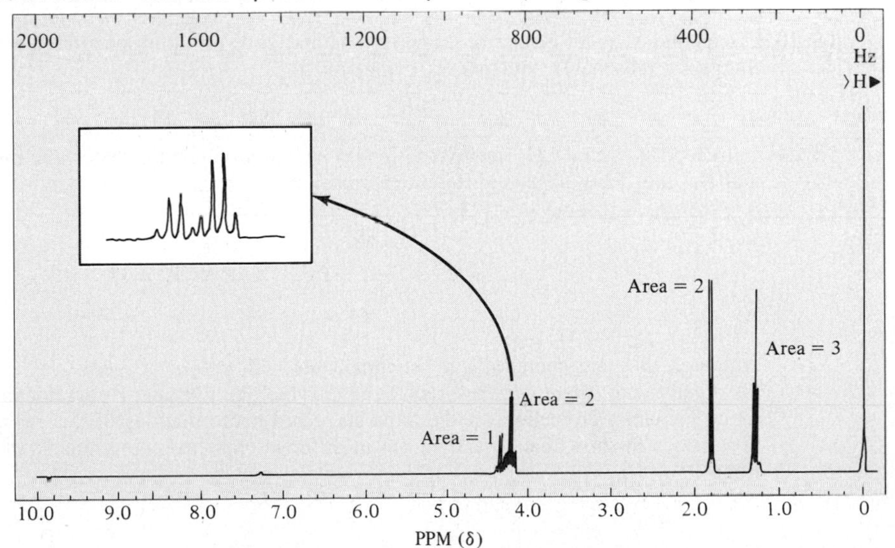

18.22 Compound G, $C_{10}H_{12}$, gave the nmr spectrum shown in Figure 18.24. Suggest a reasonable structure for compound G.

FIGURE 18.24 The nmr Spectrum of Compound G ($CDCl_3$ Solution)

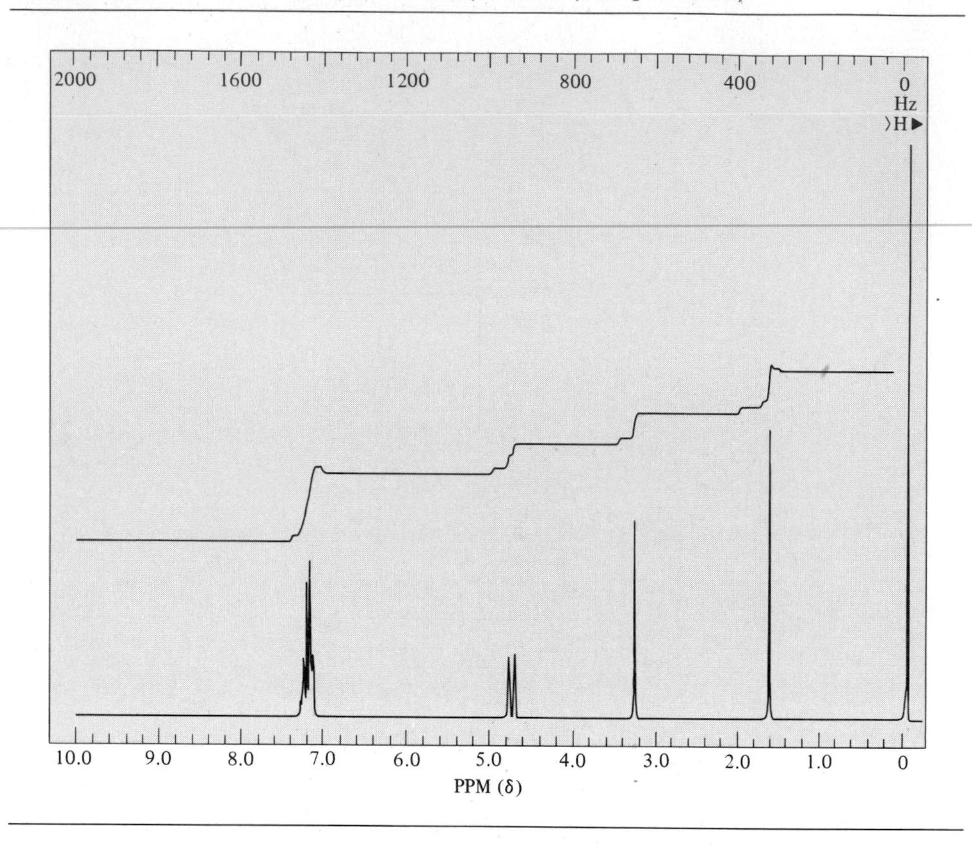

18.23 Compound H is an isomer of compound G and gave the nmr spectrum shown in Figure 18.25. Suggest a reasonable structure for compound H.

18.24 In molecules with a CH_2 (methylene) group near an asymmetric center, the methylene protons, H_a and H_b, are likely to have different chemical shifts.

$$\begin{array}{ccccccc} & & H_a & & H_c & & \\ & & | & & | & & \\ Z & — & C & — & C & — & Y \\ & & | & & | & & \\ & & H_b & & X & & \end{array} \qquad X \neq Y;\ Z \neq H$$

Although they are chemically indistinguishable with *achiral* reagents, H_a and H_b may not react identically with *chiral* reagents (such as an enzyme); thus H_a and H_b are not chemically equivalent. Draw Newman projections of the three staggered conformations of this molecule, and use these projections to show that H_a and H_b are in different chemical environments in the three conformations.

FIGURE 18.25 The nmr Spectrum of Compound H ($CDCl_3$ Solution)

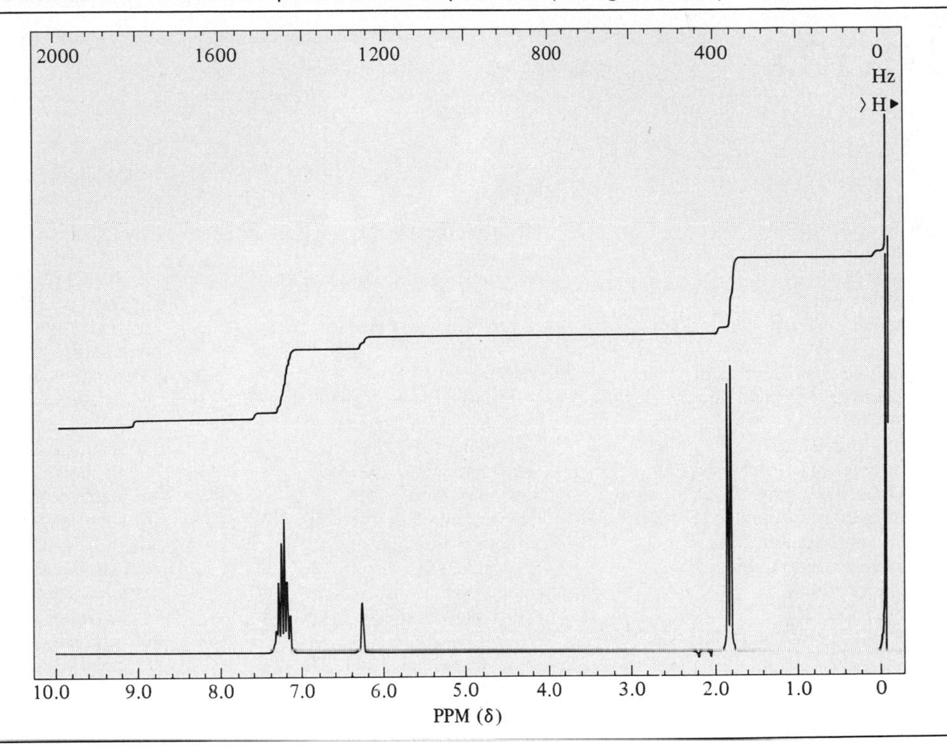

Index

0 1 2 3 4 5 6 7 8 9

Characteristic Group Infrared Frequencies (Stretching Vibrations)

$\tilde{\nu}$ (cm^{-1})	*Compounds*			*Characteristic Vibrations*
3700–3100	Alcohols, amines, phenols, amides			—O—H, —N—H
3100–3000	Aromatic hydrocarbons, unsaturated hydrocarbons			—C=C—H
3000–2800	Saturated hydrocarbons			—C—H
3000–2500	Carboxylic acids (broad)			—C(=O)—O—H
3000–2200	Amine salts			$\geq\overset{+}{N}$—H
2400–2000	Acetylenes, nitriles			—C≡C—, —C≡N
1870–1630	Carbonyl compounds			>C=O
	RCHO	1740–1720		
	ArCHO	1720–1680		
	R_2C=O	1740–1700		
	ArCOR	1700–1650		
	RCOOH	1800–1740	(monomer)	
		1720–1680	(dimer)	
	ArCOOH	1700–1680	(dimer)	
	RCO_2^-	1650–1550	(two bands)	
		1440–1350		
	RCO_2R	1750–1730		
	$ArCO_2R$	1800–1760		
	$RCONH_2$	1700–1670	(solution)	
		1680–1630	(solid)	
1690–1560	Alkenes			>C=C<
1615–1515	Aromatic rings			
1550–1200	Nitro compounds; methylene groups (—CH_2—), methyl groups (—CH_3 bending)			—$\overset{+}{N}$(=O)—O^-, —CH_2—, —CH_3
1300–1000	Ethers, alcohols, phenols, esters			—C—O—
1000–650	Alkenes (C—H bending)			—C—H
	R—CH=CH_2	1000–980; 920–900		
	R_2C=CH_2	900–880		
	trans-R—CH=CH—R	980–950		
	cis-R—CH=CH—R	750–650		
	R_2C=CHR	830–780		